Hydroboration and Organic Synthesis

9-Borabicyclo[3.3.1]nonane (9-BBN)

Ranjit S. Dhillon

Hydroboration and Organic Synthesis

9-Borabicyclo[3.3.1]nonane (9-BBN)

With 196 Figures and 260 Tables

Springer

Prof. Dr. Ranjit S. Dhillon
Department of Biochemistry and Chemistry
Punjab Agricultural University
Ludhiana-141 004
India

e-mail: dhillonr@rediffmail.com

Library of Congress Control Number: 2006935689

ISBN 978-3-540-49075-3 Springer Berlin Heidelberg New York
DOI 10.1007/978-3-540-49076-0

Springer is a part of Springer Science + Business Media
springeronline.com

Printed in Germany

Coverdesign: WMXDesign GmbH, Heidelberg
Typesetting & production: LE-TEX Jelonek, Schmidt & Vöckler GbR, Leipzig, Germany
Printed on acid-freepaper 2/3141/YL - 543210

This book is dedicated to

Prof. Akira Suzuki, Professor Emeritus
Hokkaido University, Sapporo Japan

Preface

Organoborane chemistry expanded very rapidly after the hydroboration process developed by Nobel Laureate, Late Prof. Herbert C. Brown. Organoborane intermediates can be converted to almost every class of functionality present in organic molecules. This continent has contributed immensely for the welfare of mankind and is continuing to do so.

Among the various hydroborating reagents, 9-borabicyclo[3.3.1]nonane dimer $(9\text{-BBN})_2$ has found the most extensive use because of its unique properties, convenient preparation, commercial availability and enormous synthetic applications.

Many research papers and some review articles have been published on the synthetic importance of 9-BBN and its derivatives, *B*-R-9-BBN. However, no single manuscript is available on the burgeoning literature of *B*-R-9-BBN for the convenient access to working chemists, teachers and students.

The aim of the book is to provide organoborane chemistry of *B*-R-9-BBN in a fast, organized and illustrated, easily readable, and introductory fashion.

The second aim to prepare this book is to classify a variety of organic reactions of *B*-R-9-BBN and describe these reactions in a clear manner so that organic chemists can use for the synthesis of intricate molecules.

I hope the book will stimulate the chemists to explore, further, the potential of *B*-R-9-BBN intermediates for developing new synthetic methodologies and in designing organic syntheses. It is my opinion that organization of topics of the book will attract advanced organic chemistry students, industrial and academic chemists.

I am grateful to Prof. Akira Suzuki, Professor Emeritus, Hokkaido University, Sapporo, Japan who introduced me to this subject and without his help it would not have been possible to complete this book. I thank Prof. S. Hara of Hokkaido University, Dr. (Ms.) V.K. Gautam, Dr. (Ms.) Shikha, Ms. Urvashi and my wife Sukhjinder for their help. I also thank Editorial Board of Springer-Verlag for their help and suggestions.

Ludhiana, January 2007 Ranjit S. Dhillon

Contents

1	**Introduction**	1
2	**General Remarks**	3
3	**Preparation and Properties**	5
4	**Kinetic Studies**	17
4.1	Hydroboration Kinetics of Alkenes	18
4.2	Hydroboration Kinetics of Alkynes	27
4.2.1	Relative Rates between Monohydroboration and Dihydroboration of Alkynes	30
4.3	Hydroboration Kinetics of Haloalkenes	31
4.4	Hydroboration Kinetics of Haloalkynes	38
4.5	Reduction Kinetics	41
4.6	Kinetics of Complex Formation	48
4.7	Kinetics of Protonolysis	51
5	**Hydroboration**	59
5.1	Hydroboration of Alkenes	59
5.1.1	Hydroboration of Acyclic Alkenes	59
5.1.1.1	Terminal Olefins	61
5.1.1.2	Internal Olefins	61
5.1.1.3	*cis* and *trans* Isomers	62
5.1.1.4	Effect of α-Methyl Substituents	64
5.1.1.5	Effect of α-Conjugated Substituents	65
5.1.2	Hydroboration of Cyclic Alkenes	68
5.1.3	Hydroboration of Chloro-, Acetate-, and Acetal-Functionalized Alkenes	73
5.1.4	Hydroboration of Allylsilanes	76
5.1.5	Hydroboration of Chiral Allyl Amines and Chiral Allyl Alcohols	78
5.1.6	Hydroboration of Alkenylheterocycles	84
5.1.7	Hydroboration of Enamines	96
5.1.8	Hydroboration of Heterocyclic Olefins	99
5.2	Hydroboration of Alkynes	111
5.2.1	Hydroboration of 1-Alkynes	112

5.2.2 Hydroboration of 1-Silyl-2-Alkyl-1-Alkynes ... 116
5.2.3 Hydroboration of 1-Halo-1-Alkynes ... 120
5.3 Hydroboration of Dienes ... 126
5.4 Hydroboration of Allenes ... 130
5.5 Hydroboration of Enynes and Diynes ... 136

6 Synthesis of Alcohols ... 143
6.1 Synthesis of Saturated Alcohols ... 143
6.2 Synthesis of Allylic Alcohols ... 153
6.3 Synthesis of Homoallylic Alcohols ... 155
6.4 Synthesis of Propargylic Alcohols ... 163
6.5 Synthesis of Homopropargylic Alcohols ... 166
6.6 Synthesis of Silyl Alcohols ... 172
6.6.1 Saturated Alcohols ... 173
6.6.2 Allylic Alcohols ... 175
6.6.3 Homoallylic Alcohols ... 177
6.6.4 Homopropargylic Alcohols ... 179
6.6.5 Diyne Alcohols ... 180
6.6.6 Endiyne Alcohols ... 181
6.6.7 Diendiyne Alcohols ... 181
6.6.8 Allenic Alcohols ... 183
6.6.9 Allenene Alcohols ... 186
6.6.10 Allenyne Alcohols ... 187
6.6.11 Allendienyne Alcohols ... 188
6.7 Synthesis of Heterocyclic Alcohols ... 188
6.8 Synthesis of Amino Alcohols ... 193
6.9 Asymmetric Synthesis of Alcohols ... 194

7 Synthesis of Aldehydes and Ketones ... 213
7.1 Synthesis of Aldehydes ... 213
7.2 Synthesis of Ketones ... 219
7.2.1 Synthesis of Saturated and Aromatic Ketones ... 219
7.2.2 Synthesis of Enones ... 227
7.2.2.1 Synthesis of (*E*)-β-γ-Enones ... 227
7.2.2.2 Synthesis of γ,δ-Enones ... 227
7.2.3 Synthesis of γ,δ-Ynones ... 229
7.2.4 Synthesis of Dienones ... 231
7.2.5 Synthesis of Conjugated Enynones ... 231
7.2.6 Synthesis of β-Ketosilanes ... 233
7.2.7 Synthesis of 1-Metallacyclohexan-4-ones ... 233
7.2.8 Synthesis of Chiral Ketones ... 234

8 Synthesis of Carboxylic Acids ... 237
8.1 Synthesis of Unsaturated Acids ... 237
8.2 Synthesis of β-Amino Acids ... 237

9 Synthesis of Esters ... 241
9.1 Synthesis of Achiral Esters ... 241
9.2 Synthesis of Chiral Esters ... 242
9.3 Synthesis of (*E*)-β,γ-Unsaturated Esters ... 246
9.4 Synthesis of α-Amino Acid Esters ... 248

10 Synthesis of Nitriles ... 253
10.1 Synthesis of Achiral Nitriles ... 253
10.2 Synthesis of Chiral Nitriles ... 255
10.3 Synthesis of (*E*)-β,γ-Unsaturated Nitriles ... 257

11 Synthesis of (*E*)-β,γ-Unsaturated Amides ... 259

12 Synthesis of Amines ... 261
12.1 Synthesis of Chiral Amines ... 261
12.2 Synthesis of Homopropargylic Amines ... 266
12.3 Synthesis of Secondary Amines ... 267

13 Synthesis of Halides ... 271
13.1 Synthesis of Halides *via* Hydroboration ... 271
13.2 Synthesis of Halides *via* Haloboration ... 273
13.2.1 Synthesis of 2-Halo-1-Alkenes ... 273
13.2.2 Synthesis of (*Z*)-1-Alkynyl-2-Halo-1-Alkenes ... 276
13.2.3 Synthesis of (*Z*)-δ-Halo-γ,δ-Unsaturated Ketones ... 277

14 Synthesis of Dialkylsulfides ... 283

15 Synthesis of Thiophene Oligomers ... 285

16 Synthesis of Cyclopropanes and Cyclobutanes ... 287

17 Synthesis of Borinanes ... 291

18 Synthesis and Transformations of Butterflyboranes: *cis*-Bicyclo[3.3.0]oct-1-yldialkylboranes ... 297

19 Synthesis of α-Bromoboranes ... 303

20 Synthesis of Borinates ... 307
20.1 Enol Borinates ... 307
20.1.1 Synthesis of (*E*)- and (*Z*)-Enolborinates from Saturated Ketones 307
20.1.2 Stereoselective Synthesis of (*Z*)-Enol Borinates from α,β-Unsaturated Ketones ... 312
20.1.3 Synthesis of Stable *cis*-*B*-Vinyl-9-Oxa-10-Borabicyclo[3.3.2]decane (*cis*-*B*-Vinyl-OBBD) Derivatives ... 313

20.1.4 Synthesis of Stable *trans-B*-Vinyl-9-Oxa-10-Borabicyclo[3.3.2]decane (*trans-B*-Vinyl-OBBD) ... 314
20.1.5 Markovnikov Vinylborinates: Synthesis of *B*-Vinyl-9-Oxa-10-Borabicyclo[3.3.2]decane (Vinyl-10-OBBD) ... 314
20.2 Synthesis of Alkynylborinates ... 316

21 Synthesis and Transformation of Polymers ... 321

22 Synthesis of Alkali Metal 9-Boratabicyclo[3.3.1]nonane (Li, K, and Na 9-BBNH) ... 323

23 Synthesis of *B*-R-9-BBN Not Available *via* Hydroboration ... 327

24 Synthesis of Unsaturated Compounds ... 337
24.1 Synthesis of Alkenes ... 337
24.1.1 Synthesis of Acyclic, Cyclic, and Heterocyclic Alkenes ... 337
24.1.2 Synthesis of (*Z*)-Alkenylsilane and (*Z*)-Alkenyltin ... 347
24.1.3 Synthesis of (*Z*)-1-Halo-1-Alkenes ... 351
24.2 Synthesis of Dienes ... 352
24.2.1 Synthesis of 1,ω-Dienes ... 352
24.2.2 Synthesis of Internal 1,3-Dienes ... 361
24.2.3 Synthesis of Nitrogen-Containing Heterocyclic (*Z*)-1,3-Dienes ... 364
24.2.4 Synthesis of Sulfur-Substituted 1,3-Dienes ... 366
24.2.5 Synthesis of Stannyl Dienes ... 368
24.2.6 Synthesis of *B*-Isoprenyl Derivatives ... 369
24.3 Synthesis of Enynes ... 370
24.3.1 Stereoselective Synthesis of Conjugated (*E*)- and (*Z*)-Enynes ... 370
24.3.2 Stereoselective Synthesis of Silylated (*E*)- and (*Z*)-Enynes ... 371
24.4 Synthesis of Endiynes ... 373
24.4.1 Stereoselective Synthesis of (*E*)- and (*Z*)-Endiynes ... 373
24.4.2 Stereoselective Synthesis of (*E*)- and (*Z*)-Silylated Endiynes ... 373
24.5 Synthesis of 5-Methylene-1,3-Cyclohexadienes (*o*-Isotoluenes) and 1,2,4,6-Heptatetraenes: (Diene-Allenes) ... 375
24.6 Synthesis of Enyne-Allenes, Dienyne-Allenes, and Trienyne-Allenes, and Their Cycloisomerization ... 376
24.7 Synthesis of Silylated *Z,Z*-Diendiynes and *Z,Z,Z,Z*-Tetraentetraynes, and Their Cycloisomerization ... 381

25 Reduction ... 397
25.1 With 9-BBN·THF ... 397
25.2 With 9-BBN·Py ... 408
25.3 With Li 9-BBNH ... 410
25.4 With *B*-Siamyl-9-BBN ... 414
25.5 With Lithium 9,9-Di-*n*-Butyl-9-Borabicyclo[3.3.1]nonanate ... 415
25.6 With 9-BBN Derivatives (as Catalysts) ... 422

26 Asymmetric Reduction ... 427
26.1 Alpine-Borane ... 428
26.1.1 Reduction of Aldehydes ... 428
26.1.2 Reduction of Ketones ... 433
26.1.2.1 Reduction with Alpine-Borane, under Pressure ... 433
26.1.2.2 Reduction with Alpine-Borane as Neat or in Excess ... 437
26.1.2.3 Reduction of Prochiral Ketones ... 437
26.1.2.4 Reduction of α-Haloketones ... 441
26.1.2.5 Reduction of Ketoesters ... 442
26.1.2.6 Reduction of α,β-Unsaturated Ketones ... 444
26.1.2.7 Reduction of Propargyl Ketones ... 445
26.1.2.8 Reduction of Propargylic Ketones with α-Chiral Centers ... 449
26.1.2.9 Reduction of Acylcyanide ... 451
26.2 NB-Enantrane ... 452
26.3 Eapine-Borane and Prapine-Borane ... 453
26.4 NB-Enantride ... 459
26.5 Eapine-Hydride ... 461
26.6 K-Glucoride ... 462
26.7 K-Xylide ... 468
26.8 K9-OThx-9-BBNH ... 472
26.9 Lithium Di-*n*-Butyl Ate Complex of 9-BBN ... 473
26.10 Li 9-BBNH ... 475
26.11 Comparative Data of Asymmetric Reducing Agents ... 476

27 Cleavage of Ethers ... 487

28 *trans*-Metalation ... 491

29 Separation of Isomers ... 499

30 Diels-Alder Reaction ... 501

31 Suzuki Reaction ... 523
31.1 Mechanism of the Suzuki Catalytic Cycle ... 554

32 Miscellaneous Reactions ... 559

Subject Index ... 573

1 Introduction

Hydroboration constitutes one of the most important and facile methods for the synthesis of organoboranes from unsaturated compounds [1]. The organoboranes serve as valuable intermediates for the synthesis of a wide variety of organic compounds. Thus, there is a huge interest in exploring their chemistry, and the hydroboration reactions of major significance in synthetic organic chemistry have been reviewed [1–15]. Selective hydroboration with various hydroborating agents and their application in organic synthesis has also been reviewed [16].

References

1. (a) Brown HC (1969) Hydroboration. Benjamin, New York; (b) Brown HC (1972) Boranes in organic chemistry. Cornell University Press, Ithaca, New York; (c) Cragg GML (1973) Organoboranes in organic synthesis. Dekker, New York; (d) Pelter A, Smith K, Brown HC (1988) Borane reagents. Academic, London; (e) Onak TK (1975) Organoborane chemistry. Academic, London
2. (a) Brown HC (1975) Organic syntheses via boranes. Wiley, New York [reprinted as vol. 1 by Aldrich (1999); Catalog no. Z 40,094-7]; (b) Brown HC Zaidlewicz M (2001) Organic synthesis via boranes, vol. 2, recent developments. Aldrich, Milwaukee; (c) Suzuki A, Brown HC (2002) Organic synthesis via boranes, vol. 3. Suzuki coupling. Aldrich, Milwaukee; (d) Mikhailov BM, Bubnov YN (1984) Organoboron compounds in organic synthesis (English translation). Harwood, Utrecht
3. Matteson DS (1995) Stereodefined synthesis with organoboranes. Springer, Berlin Heidelberg New York
4. Ramachandran PV, Brown HC (2001) ACS Symp Ser 783
5. Brown HC (1974) Aldrichim Acta 7:43
6. Pelter A, Smith K (1979) In: Barton DHR, Ollis WD (eds) Comprehensive organic chemistry, vol. 3. Pergamon, Oxford, p 689
7. Brown HC, Zaidlewicz M, Negishi E (1982) In: Wilkinson G, Stone FGA, Abel EW (eds) Comprehensive organometallic chemistry. Pergamon, Oxford
8. Negishi EJ (1976) Organomet Chem 108:281
9. Weill-Raynal J (1976) Synthesis 633
10. Brown HC, Campbell JB Jr (1981) Aldrichim Acta 14:3
11. Pelter A (1982) Chem Soc Rev 11:191

12. Suzuki A (1982) Acc Chem Res 15:178
13. Suzuki A (1983) Topics Curr Chem 112:69
14. Smith K, Pelter A (1991) In: Trost BM, Fleming I (eds) Comprehensive organic synthesis, vol.8. Pergamon, Oxford, p 703
15. Soderquist JA (1995) In: Paquette LA (ed) Encyclopedia of reagents for organic synthesis, vol. 1. Wiley, New York, p 622
16. Suzuki A, Dhillon RS (1986) Topics Curr Chem 130:23

2 General Remarks

The hydroboration of unhindered carbon–carbon double bonds with diborane leads conveniently to trialkylboranes (Eq. 2.1).

$$RCH{=}CH_2 \quad + \quad BH_3 \longrightarrow (RCH_2CH_2)_3B \tag{2.1}$$

The presence of another unhindered carbon–carbon double or triple bond or functional group often affords a complex mixture of products resulting from competing hydroboration. As a result, the hydroboration of dienes, enynes, or functionally substituted alkenes produces a considerable amount of other products, which is highly undesirable for the subsequent utilization of the resulting organoborane. In addition, the stereoselective addition of borane is very poor in the absence of steric bulk, as shown:

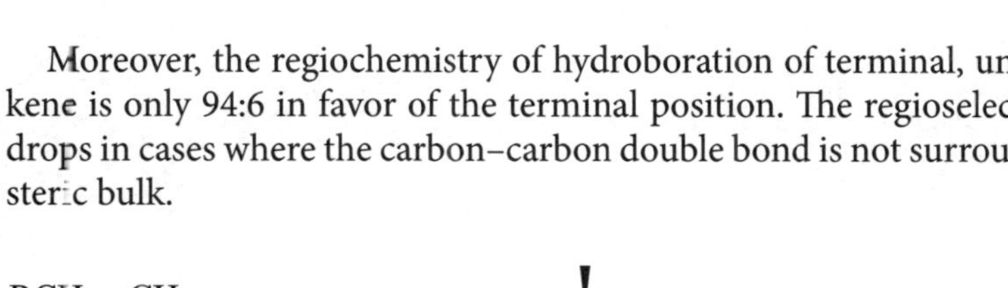

Moreover, the regiochemistry of hydroboration of terminal, unhindered, alkene is only 94:6 in favor of the terminal position. The regioselectivity further drops in cases where the carbon–carbon double bond is not surrounded by large steric bulk.

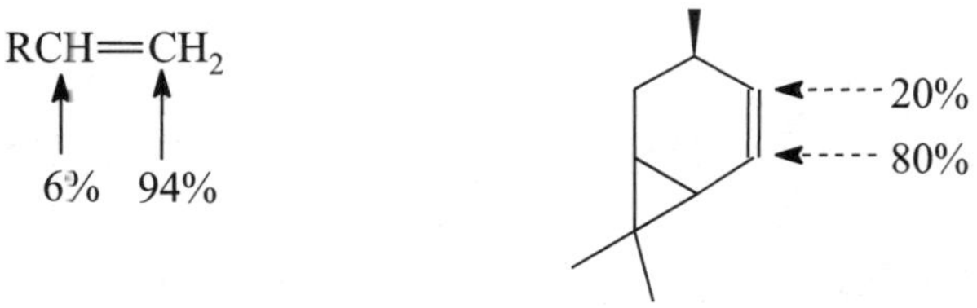

In addition, in many synthetic reactions that involve trialkylboron, only one or two alkyl groups are utilized, and this results in the loss of valuable alkyl groups.

The competitive reduction of functional groups and the formation of minor hydroboration products can be minimized by the use of disiamylborane (Sia_2BH) or other dialkylboranes as hydroborating agents instead of diborane [1–3]. However, these reagents are relatively unstable and must be freshly prepared prior to use, a serious handicap for their utilization. The 9-borabicyclo[3.3.1]nonane dimer $(9\text{-BBN})_2$, which was first identified by Köster [4], has been found to be an unusual dialkylborane with some valuable properties [5–8]. The utility of this unusual heterocyclic dialkylborane, and the rearrangement of ate complexes of it have been reviewed [6, 7]. Since then, 9-BBN has been extensively studied and utilized by synthetic organic chemists. It has also been converted into various derivatives for asymmetric reduction and other reactions. Among the various hydroborating agents, 9-BBN has found the most extensive use in various reactions due to its unique properties, thermal stability, and convenient preparation.

References

1. Brown HC, Unni MK (1968) J Am Chem Soc 90:2902
2. Brown HC, Gallivan RM Jr (1968)J Am Chem Soc 90:2906
3. Brown HC, Sharp RL (1968) J Am Chem Soc 90:2915
4. Köster R (1960) Angew Chem 72:626
5. Brown HC, Knights EF, Scouten CG (1974) J Am Chem Soc 96:7765
6. (a) Brown HC, Lane CF (1977) Heterocycles 7:453; (b) Kramer GW, Brown HC (1977) ibid. 7:487
7. Köster R, Yalpani M (1991) Pure Appl Chem 63:387
8. Soderquist JA (1995) In: Paquette LA (ed) Encyclopedia of reagents for organic synthesis, vol. 1. Wiley, New York, p 622

3 Preparation and Properties

Köster [1] was the first to report the preparation, isolation, and characterization of 9-borabicyclo[3.3.1]nonane as a dimer; $(9\text{-BBN})_2$ has been obtained from the thermal disproportionation of tetra-*n*-propyldiborane and *B*-*n*-propyl-9-BBN (Eq. 3.1). *B*-alkyl-9-BBN preparation is itself a two-step process, thus restricting the utility of the 9-BBN for synthetic purposes.

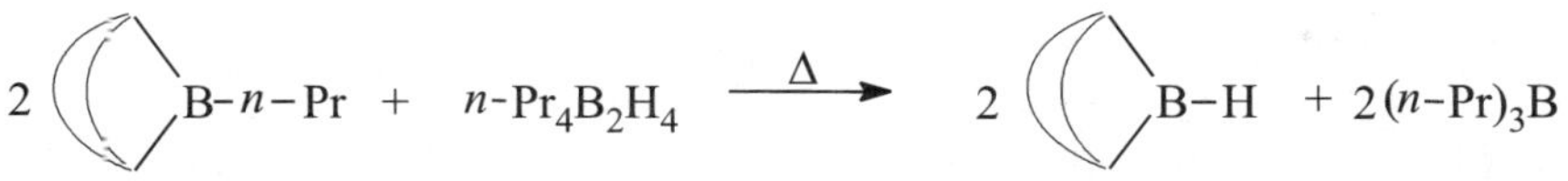

(3.1)

However, the direct and convenient synthesis of $(9\text{-BBN})_2$ has been reported by Knights and Brown [2], and this development opened the door for its application in hydroboration [2–5]. The synthesis involves the cyclic hydroboration of 1,5-cyclooctadiene with a borane–tetrahydrofuran (THF) complex [2, 3] in a 1:1 ratio, followed by refluxing the mixture at 65 °C, thus producing a solution containing $(9\text{-BBN})_2$ in ca. 90% yield.

In fact, the borane adds to 1,5-cyclooctadiene to afford 1,4- and 1,5-isomers in a 30:70 mixture. With simple thermodynamic considerations, it is apparent that the 1,4-isomer, which has a seven-membered ring fused to a five-membered ring, is less stable. Consequently, the 1,4-addition product is easily isomerized to the 1,5-isomer at 65 °C (Eq. 3.2). This process affords a microcrystalline product with a melting point (m.p.) of 142 °C. This material is further purified by vacuum sublimation, with an increase in m.p. to 152–155 °C [5].

9-BBN exists as the dimer [A] both in the vapor state and in a crystalline solid state. The chair–chair conformation of the dimer and the B–H bridge has been confirmed by spectral studies [5] and crystal structure determination [6].

IR (THF)[5]: 1,490 (w), 1,567 (s) cm^{-1}. The IR of $(9\text{-BBN})_2$ exhibits a strong absorption at 1,567 cm^{-1} either as a mineral oil mull of the solid or in solution (THF, benzene, hexane), indicating a B–H bridge. This confirms that 9-BBN must exist as the dimer [A]. ^1H NMR (benzene–TMS) exhibits a broad singlet at δ 1.8. ^{11}B NMR (THF)[5] shows absorption at δ –28 relative to external $BF_3{\cdot}OEt_2$; (benzene) at δ –28 relative to external $BF_3{\cdot}OEt_2$ (broad singlet). The mass spec-

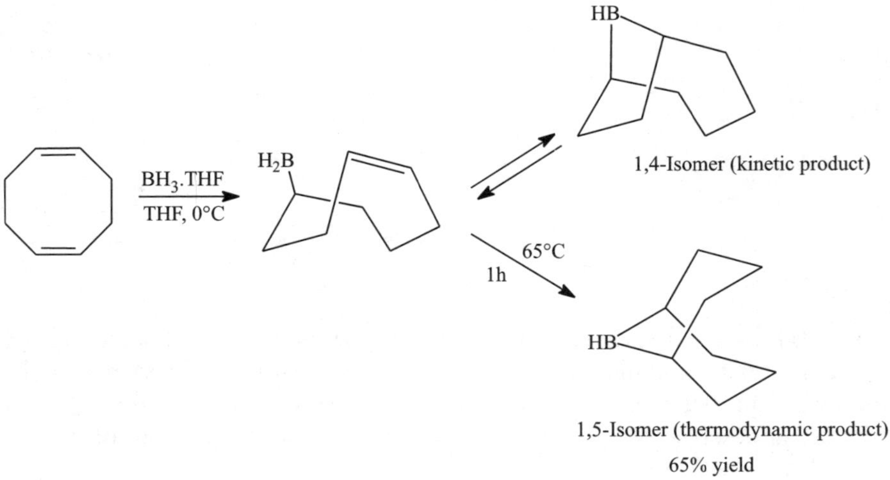

(3.2)

trum [5] (70 eV) shows a prominent cluster of peaks (*m*/*e* 242, 243, 244) in the approximate ratio 1:8:16. This is the expected ratio of molecules containing two boron atoms, since the natural abundance of ^{10}B is 20% and that of ^{11}B is 80%.

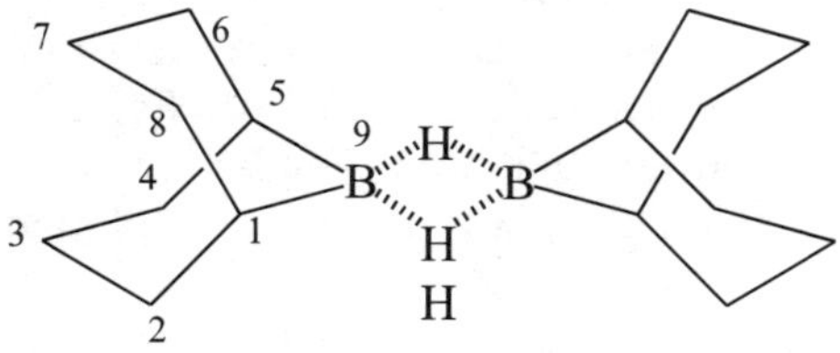

[A]

In solvents like carbon tetrachloride, hexane, benzene, and diethylether, 9-BBN exists exclusively as the dimer. However, in THF and Me_2S an equilibrium between the $(9\text{-BBN})_2$ dimer and solvent-complexed 9-BBN monomer (9-BBN-solvent) is observed [7]. From the kinetic studies (*vide infra*), it has been proved that the 9-BBN monomer is actually the hydroborating agent. For convenience, 9-BBN is represented in the shorthand notation as shown in [B].

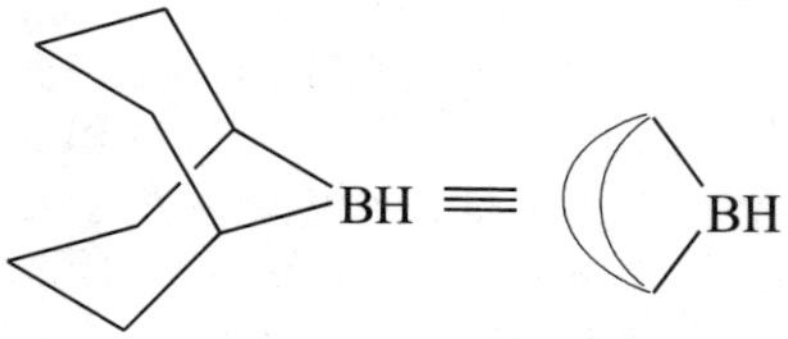

[B]

The preparation of $(9\text{-BBN})_2$ from BH_3-THF suffers from some practical difficulties:

1. The choice of BH_3-THF as the reagent used for hydroboration necessarily requires that THF be used as the reaction solvent.
2. $(9\text{-BBN})_2$ is significantly soluble in THF and exists in an equilibrium with a 9-BBN-THF complex (Eq. 3.3):

$$(9\text{-BBN})_2 \xrightleftharpoons{\text{THF}} 2\ 9\text{-BBN-THF} \tag{3.3}$$

3. The microcrystalline product $(9\text{-BBN})_2$ obtained from THF solvent occasionally contains minor amounts of impurities which render it pyrophoric. These factors, thus, diminish the yield and purity of 9-BBN.

   The complexation of 9-BBN with basic solvents is summarized in Table 3.1 [8].

Table 3.1 Complexation of 9-BBN with basic solvents [8]

Base	Complex (%)	^{11}B chemical shift (J^{11}_{B-H})
THF	14	13.9 (~90 Hz)
SMe_2	46	3.9 (107 Hz)
NC_5H_5	100	−0.7 (88 Hz)

The molar solubility of 9-BBN dimer is summarized in Table 3.2 [8].

Table 3.2 Molar solubility of dimeric 9-BBN in representative solvents [8]

Solvent	Temperature	
	0 °C	25 °C
Monoglyme	0.01	0.07
Diglyme	<0.01	0.04
1,4-Dioxane	0.03	0.07
1,3-Dioxolane	<0.01	0.04
Diethyl ether	0.09	0.18
THF	0.12	0.29
CH_2Cl_2	0.11	0.28
$CHCl_3$	0.21	0.50
CCl_4	0.15	0.36
Pentane	0.13	0.23

Table 3.2 (*Continued*) Molar solubility of dimeric 9-BBN in representative solvents [8]

Solvent	Temperature	
	0 °C	25 °C
Hexane	0.11	0.25
Benzene	0.19	0.36
Cyclohexane	0.03	0.08
Toluene	0.14	0.33
Dimethylsulfide	–	0.6

These studies also reveal that the hydroboration of 1,5-cyclooctadiene using borane–dimethylsulfide to prepare a solution of 9-BBN can be carried out in solvents other than THF [9].

Consequently, an efficient route for the synthesis of a high purity crystalline $(9\text{-BBN})_2$ dimer has been reported [8], and it has been found that 1,2-dimethoxyethane (monoglyme) possesses a major advantage as the reaction solvent over THF. Accordingly, the cyclic hydroboration of 1,5-cyclooctadiene is carried out in this solvent using a borane–methylsulfide complex as a hydroborating agent. Removal of dimethylsulfide yields 88% of 9-BBN dimer in large crystal form, with more than 99% purity (m.p. 153–155 °C).

Brown and Mandal [10] have reported another convenient procedure for the preparation of $(9\text{-BBN})_2$ using borane-1,4-thioxane (BOT) as the hydroborating reagent. BOT is readily synthesized [11] by adding diborane to 1,4-thioxane (Eq. 3.4). It is a stable liquid at 25 °C and crystallizes out at 0 °C with a m.p. of 11–15 °C. Neat BOT is 8 M in borane. ^{11}B NMR of BOT exhibits only one absorption at δ –23 (relative to $BF_3{\cdot}OEt_2$), supporting the boron–sulfur coordination ($BH_3{\cdot}SMe_2$, δ –20.3; $BH_3{\cdot}$THF, δ +1) [10] BOT as a neat reagent is indefinitely stable at 0 °C.

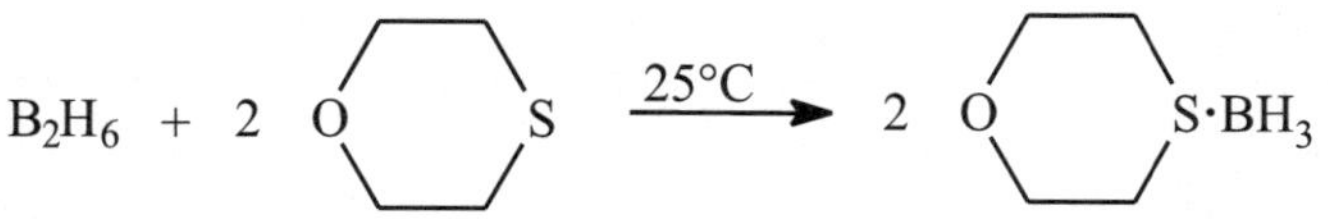

BOT

(3.4)

The hydroboration with BOT can be carried out in a wide variety of solvents, such as THF, diethylether, methylene chloride, and pentane, or with neat reagents at 0 or 25 °C.

1,5-Cyclooctadiene undergoes hydroboration in a 2-M THF solution of BOT. The hydroboration is complete after 0.5 h at 25 °C, and thermal isomerization in refluxing THF leads to complete formation of 9-borabicyclo[3.3.1]nonane after 1 h (Eq. 3.5). After completion of the reaction, the supernatant liquid is removed, crystals are washed with pentane, and recrystallization from THF gives pure $(9\text{-BBN})_2$ dimer (m.p. 153 °C).

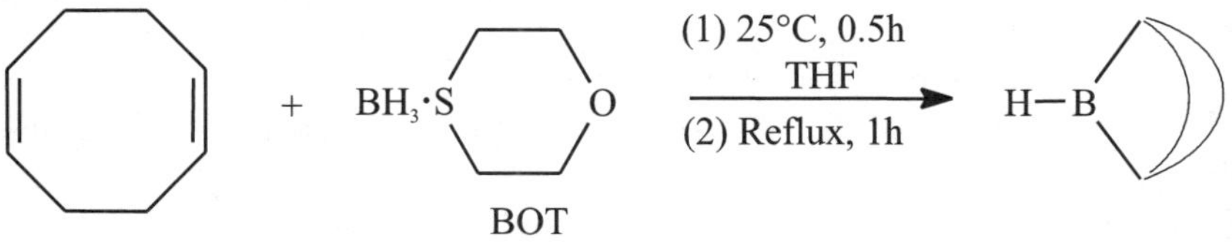

(3.5)

The neat reaction is performed by the addition of 1,5-cyclooctadiene to BOT at 0 °C. Hydroboration is complete after 0.25 h at 0 °C and after 0.25 h at 25 °C. After thermal isomerization at 165–170 °C for 1 h, 1,4-thioxane is distilled completely from the reaction mixture, leaving $(9\text{-BBN})_2$ as a pure, stable crystalline solid. Both procedures afford $(9\text{-BBN})_2$ almost as a pure solid, which on recrystallization from THF, affords $(9\text{-BBN})_2$ crystals, m.p. of 153 °C. Moreover, it is now commercially available [12] in crystalline and in solution forms.

In an indirect method for the synthesis of 9-BBN, Brown and Kulkarni [13] have revealed that exchange between borane-methylsulfide (BMS) and *B*-MeO-9-BBN is effected to get 9-BBN and $B(OMe)_3$ as the reaction products (Eq. 3.6).

$$3[B\text{-MeO-9-BBN}] + BH_3\text{–}SMe_2 \longrightarrow \frac{3}{2}[\text{9-BBN}]_2 + B(OMe)_3 \tag{3.6}$$

1,2-Dimethoxyethane (MG) is a superior solvent for crystallization [8] of 9-BBN; Soderquist and Negron [14] reported the reaction of *B*-MeO-9-BBN with $H_3B\text{-}SMe_2$ in a 2:1 stoichiometry, employing MG as the reaction solvent to afford 9-BBN in 93% yield (m.p. of 154–156 °C; Eq. 3.7).

$$2[B\text{-MeO-9-BBN}] + H_3B\text{–}SMe_2 \xrightarrow{\text{MG}} [\text{9-BBN}]_2 + (MeO)_2BH + (MeO)_3B \tag{3.7}$$

Both crystalline 9-BBN and its solutions are indefinitely stable when stored under an inert atmosphere, and no noticeable change in activity has been recorded even after more than 4 years. The stability of colorless, crystalline $(9\text{-BBN})_2$ toward air oxidation is unique among dialkyl boranes. As a consequence, a fresh, unopened bottle of commercial crystalline $(9\text{-BBN})_2$ can be opened in the air and the entire contents can be transferred rapidly to a nitrogen-flushed flask with minor loss of activity. However, small amounts should not be removed in air at frequent intervals, but must be weighed out and transferred with approximately the same precautions and relative convenience accompanying the utilization of sodium borohydride and lithium aluminum hydride. On the other hand, 9-BBN in the solution form is very reactive to both water and oxygen. Such a solution must be rigorously protected from the atmosphere for both preparative and quantitative studies in the manner utilized in handling air-sensitive reagents [15].

The inertness of crystalline $(9\text{-BBN})_2$ toward oxygen is due to the unusual stability of the B–H–B bridge in the dimer and is supported by the observation that neat *B*-methoxy-9-BBN, solid *B*-chloro-9-BBN, and solid *B*-hydroxy-9-BBN are all pyrophoric. Moreover, it is found that *B*-alkyl-9-BBN derivatives are very reactive toward oxygen, more so than the corresponding trialkylboranes [16]. Hence, with the B–H–B bridge no longer present, the exposed boron atom makes these derivatives of 9-BBN unusually reactive toward oxygen.

The boiling point of 9-BBN, 195 °C (12 mm), is unusually high in contrast to dialkylboranes. Simple dialkylboranes dissociate and distill as the monomer, whereas 9-BBN must distill as the dimer. Dissociation of the dimer is accompanied by an increase in C–B–C angle from 111.8 to 120°. Such an increase is readily accommodated in acyclic dialkylboranes but would be resisted in the rigid bicyclic structure. The unusual stability of the $(9\text{-BBN})_2$ is reflected in the chemistry of 9-BBN.

Another remarkable property of 9-BBN where it distinguishes itself from other dialkylboranes is its thermal stability. 9-BBN can be distilled at 195 °C (12 mm) or heated for 24 h at 200 °C under nitrogen, without loss of hydride activity [2, 5], in sharp contrast to other dialkylboranes. For instance, dicyclohexylborane decomposes at 180–200 °C, yielding cyclohexene and polymeric boranes, and disiamylborane undergoes isomerization [5, 17] at 75 °C.

The 9-BBN and its 10 B-substituted derivatives have been examined and had their structures confirmed by their ^{13}C NMR data [18]. 9-BBN, *B*-Cl-9-BBN, *B*-OMe-9-BBN, *B*-*t*-Bu-9-BBN, and *B*-CH_2-CH_2SiMe_3-9-BBN, exhibit ^{13}C NMR signals that confirm that C-2, C-4, C-6, and C-8 ring carbon of the bicyclic system are identical, and a separate second signal for C-3 and C-7 of the ring reveals the symmetric structure of these derivatives. The bulkier *B*-alkyl groups containing a chiral center reveal different signals: one for C-2 and C-6, and the second for C-4 and C-8, corresponding to the asymmetric environment of the bicyclic structure (Table 3.3) [18].

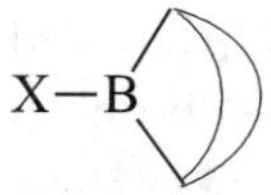

Table 3.3 ^{11}B and ^{13}C NMR spectral data for B-substituted derivatives of 9-BBN [18]. Spectra are recorded in $CDCl_3$ solvent. No attempt is made to determine the chemical shifts of the very broad signals attributed to B-bonded carbons

	^{11}B	^{13}C		
X (Compound)		C-2,4,6,8	C-3,7	Other carbons[a]
H (1)	28	33.4	23.9	
Cl (2)	81.5	34.2	23	
OCH_3 (3)	53.2	33.1	23.2	53.3
CH_2CH_3 (4)	88	33.2	23.4	8
$CH_2CHCH_2CH_3$ (5)	87.3	33.4	23.2	25.3, 13.9
$CH_3CHCH(CH_3)_2$ (6)	88.4	33.8 33.6	23.3	30, 23.7 21.2, 10.9
$C(CH_3)_3$ (7)	85.8	33.6	23.2	25.9
$C_6H_5CHCH_2CH_3$ (8)	83.9	33.9 33.7	23.1	131.3, 129.1 128.4, 125.7 14.3, 1.1
$CH_2CH_2Si(CH_3)_3$ (9)	86.4	33.4	23.3	8.6, –2
$CH(CH_3)Si(CH_3)_3$ (10)	85.9	33.2 32.7	23.2	10.3, –0.9
$CH_2CH(CH_3)Si(CH_3)_3$(11)	88.9	33.6 33.3	23.4	17.9, 15.2 –3.6

[a]Assignments for the "other" carbons are made based on decoupling experiments. In the order given in the table they are the following: *3*, OCH_3; *4*, CH_3; *5*, CH_2CH_3 (coincident signals); *6*, CH, 3-CH_2, 3-CH_3, 1-CH_3; *7* CH_3; *8*, aromatic (1, 3, 2, 4) CH_2, CH_3; *9*, 2-CH_2, $Si(CH_3)_3$; *10*, CH_3, $Si(CH_3)_3$; *11*, CH_3, CH, $Si(CH_3)_3$.

9-BBN forms a stable isolable 1:1 complex with pyridine and exhibits in its ^{13}C spectrum the nonequivalence of two halves of the bicyclic ring system.

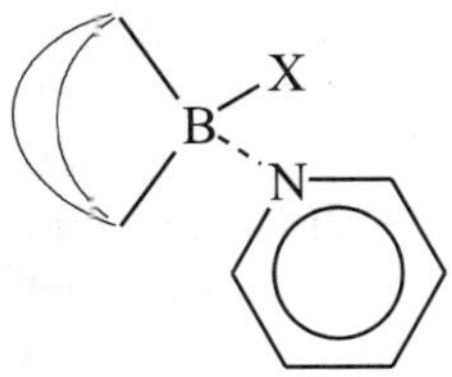

One of the two halves is *syn* to the pyridine ring, and the other is *anti* to the pyridine ring. The C-2 and C-4 positions occupy a *gauche* relationship to the pyridine, and these carbons correspond to the upfield signal at 29 ppm [19] (Table 3.4).

Similarly, the *B*-Cl-9-BBN derivative also forms a 1:1 complex with pyridine and gives separate signals for *syn* and *anti* halves (relative to pyridine [Py]) of the 9-BBN ring system. Quaternary nitrogen exerts a slightly greater upfield effect than does chlorine; thus the C-2 and C-4 positions are assigned to 30.8-ppm absorbance [19] (Table 3.4) [18]. The ^{13}C NMR spectrum of *B*-OCH_3-9-BBN and pyridine reveals incomplete formation of a 1:1 complex but full equivalence of the two halves of the 9-BBN ring. The spectrum of *B*-Et-9-BBN shows the complete formation of a 1:1 complex, and the ^{13}C spectrum reveals full equivalence of the two halves, which is attributed to rapid dissociation and recombination of the complex.

Table 3.4 ^{11}B and ^{13}C NMR spectral data for pyridine complexes of B-substituted derivatives of 9-BBN [18]. Spectra are recorded in $CDCl_3$ solvent. No attempt is made to determine the chemical shifts of the very broad signals attributed to boron-bonded carbons

	^{11}B $^{13}C^a$			
X (Compound)		C-2,4,6,8	C-3,7	Py[b]
H (12)	−0.7[c]	35.1	25.4	145.9, 138.8, 125.3
		29	24.8	
Cl (13)	9.5	31.9	23.9	145, 141.7, 126.3
		30.8	23.7	
CH_2CH_3 (14)	1.3	31.6	25	146.3, 138.3, 124.8
$CH_3CHCH_2CH_3$ (15)	1.4	31.2	24.7	147, 137.9, 124.1
$CH_3CHCH(CH_3)_2$ (16)	−0.1	31.6	24.6	147.1, 138.2, 124.4
		31.2		
$CH_2CH_2Si(CH_3)_3$ (17)	2.1	31.6	25.1	146.1, 138.3, 124.8

[a]Alkyl carbons are assigned as in Table 3.3. The corresponding signals (parts per million) are observed at 8.5 (*14*); 24.7, 14.2 (1-CH_3), 12.9 (4-CH_3) (*15*); 26.3, 26.1, 18.1, 7.8 (*16*); 9.4, −1.7 (*17*).
[b]Order follows α, β, γ.
[c]Doublet, $J = 88$ Hz.

However, with bulky groups such as *t*-Bu and 1-trimethylsilylethyl, the complex formation with pyridine is incomplete, even with excess pyridine.

The ^{13}C NMR spectra of the interaction of 9-BBN with two series of amines with regular increasing steric requirements are studied (1) for the role of strain as a factor, (2) for the stability of addition compounds formed, and (3) for their exchange with an amine. One set of a Py bases includes Py and 2-methyl-, 2-ethyl, 2-isopropyl-, and 2-*tert* butylpyridine; the second set is aliphatic amines such as *n*-propylamine, isopropylamine, diethylamine, diisopropylamine, and triethylamine, which are of increasing steric requirements and are examined for their interaction with 9-BBN. Quinuclidine (QN), a base with relatively low steric requirements, is compared with triethylamine, a base with very large steric requirements.

These two families of amines reveal four types of behavior: (1) formation of stable complexes with no observable exchange, even with excess amine (Py, *n*-$PrNH_2$, *i*-$PrNH_2$, Et_2NH, QN); (2) formation of stable complexes with rapid exchange of amine (2-MePy); (3) formation of partially dissociated complexes with rapid exchange (2-EtPy, 2-*i*-PrPy, *i*-Pr_2NH); and (4) no detectable interaction of 9-BBN with amine (2-*t*-BuPy, Et_3N). These results are summarized in Table 3.5 [20].

Table 3.5 Summary of ^{13}C NMR observations on the interaction of 9-BBN with amines [20]

Associated 100%, no exchange	Associated 100%, rapid exchange	Partially associated, rapid exchange	Not associated
Py	2-MePy	2-EtPy 2-*i*-PrPy	2-*t*-BuPy
n-$PrNH_2$ *i*-$PrNH_2$ Et_2NH QN		*i*-Pr_2NH	Et_3N

In general, there is a regular progression from (1) to (2) to (3) to (4) along these four types of behavior, with increasing steric requirements in both series of amines. Consequently, triethylamine fails to show any interaction with 9-BBN, whereas QN forms a stable that does not exchange with excess amine. These studies confirm the earlier results of the stabilities of these addition compounds conducted with IR spectroscopic methods [21]. However, the ^{13}C NMR provides considerable additional information about the exchange or lack of exchange in addition compounds that are completely associated under the experimental conditions. Consequently, the ^{13}C NMR method separates these compounds into two separate classes, (1) and (2), providing a more sensitive probe into the effects of steric strains over in systems where association is essentially complete.

Soderquist and Najafi [22] have reported the selective monoxidation of B-substituted derivatives of 9-BBN to afford in good yield the exclusive formation of 9-oxa-10-borabicyclo[3.3.2]decane products (Eq. 3.8; Table 3.6). The reaction proceeds smoothly using 1 equiv of anhydrous trimethylamine *N*-oxide (TMANO) in $CHCl_3$ at 0 °C.

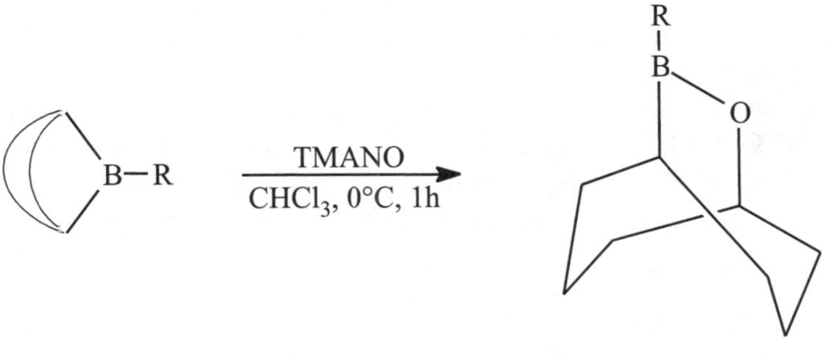

(3.8)

Table 3.6 Oxidation of B-substituted 9-BBN derivatives with 1 equiv of TMANO [22]

R of *B*-R-9-BBN	Yield of isolated 9-oxa-10-borabicyclo[3.3.2] series[a]	Reaction temp (°C)	Time (h)
Me	81	0	1
n-Hx	84	0	1
c-Hx	84	0	1
t-Bu	80	25	20
CH_2SiMe_3	93	0	1
$C(SiMe_3)$=CHMe (Z)	87	0	1
OMe	85	0	1
O-(*n*-Hx)	86	0	1

[a]Reactions are carried out by the dropwise addition of a solution of TMANO in $CHCl_3$ to the organoborane.

The reaction involves the exclusive ring B–C bond oxidation, and this specific process has a conformational dependence. The selectivities are rationalized in terms of the required antiperiplanar relationship of the B–C ring bond, which undergoes oxidation (case 1).

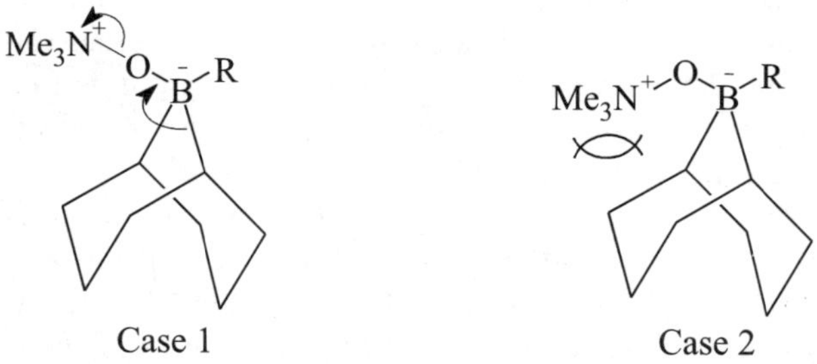

It is significant to mention that 9-oxa-10-borabicyclo[3.3.2]decane derivatives, selectively, transfer only the alkyl group of *B*-alkyl and consequently, are valuable intermediates in the syntheses of variety of compounds (*vide infra*).

Matos and Soderquist [23] reported the synthesis of $(9\text{-BBN-D})_2$ through the reduction of MeO-9-BBN with $LiAl(OEt)D_3$, and the latter facilitates the separation of the 9-borata complex from the insoluble $[AlD(OMe)(OEt)]_x$ (Eq. 3.9) [24]. The treatment of the borohydride with TMSCl affords crystalline $(9\text{-BBN-D})_2$ in a 31% yield.

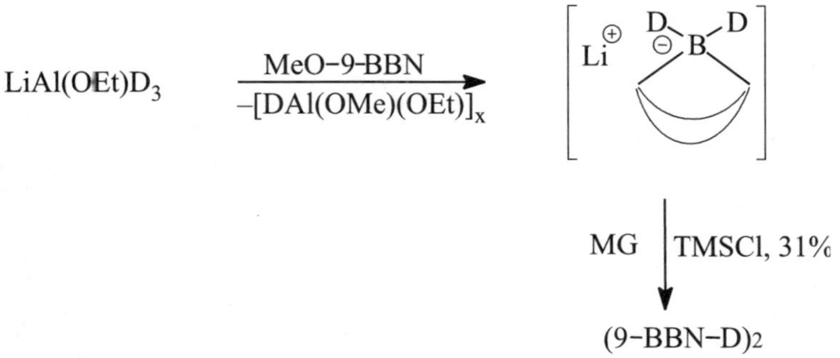

(3.9)

References

1. (a) Köster R (1960) Angew Chem 72:626; (b) Köster R, Griaznov G (1961) Angew Chem 73 171
2. Knights EF, Brown HC (1968) J Am Chem Soc 90:5280
3. Soderquist JA, Negron A (1991) Org Synth 70:169
4. In sharp contrast, 1,5-cyclooctadiene on hydroboration with equimolar amount of thexylborane followed by oxidation gives an 80:20 mixture of *cis*-1,4- and *cis*-1,5-cyclooctanediols in 93% yield. Brown HC, Neigishi E J (1972) Am Chem Soc 94:3567
5. Brown HC, Knights EF, Scouten CG (1974) J Am Chem Soc 96:7765
6. Brauer DJ, Kruger C (1973) Acta Crystallogr B 29:1684
7. Brown HC, Wang KK, Scouten CG (1980) Proc Natl Acad Sci USA 77:698
8. Soderquist JA, Brown HC (1981) J Org Chem 46:4599
9. Brown HC, Mandal AK, Kulkarni SU (1977) J Org Chem 42:1392
10. Brown HC, Mandal AK (1992) J Org Chem 57:4970
11. Brown HC, Mandal AK (1980) Synthesis 153
12. Aldrich Chemical, Milwaukee, Wisconsin
13. Brown HC, Kulkarni SU (1979) J Organomet Chem 168:281
14. Soderquist JA, Negron A (1987) J Org Chem 52:3441
15. Lane CF, Kramer GW (1977) Aldrichim Acta 10:11
16. Brown HC, Midland MM (1971) Chem Commun 699
17. Brown HC, Zweifel G (1966) J Am Chem Soc 88:1433
18. Brown HC, Soderquist JA (1980) J Org Chem 45:846
19. Wehrli FW, Wirthlin T (1976) Interpretation of carbon-13 NMR spectra. Heyden, London
20. Brown HC, Wang KK (1980) J Org Chem 45:1748
21. Brown HC, Wang KK (1979) Recl Trav Chim Pays Bas 98:117
22. Soderquist JA, Najafi MR (1986) J Org Chem 51:1330
23. Matos K, Soderquist JA (1998) J Org Chem 63:461
24. Singaram B, Cole TE, Brown HC (1984) Organometallics 3:1520

4 Kinetic Studies

Kinetic studies provide valuable information in the areas of both mechanistic and synthetic chemistry concerning the effects of substituents in alkenes and alkynes. The effects of substituents that donate or withdraw or polarize electrons of C=C or C≡C provide information regarding the mechanism of hydroboration. On the other hand, relative rates of hydroboration of substituted or unsubstituted C=C or C≡C give synthetic chemists improved means of predicting the selective hydroboration of C=C or C≡C or their functionalized derivatives. 9-BBN has proven to be the best candidate for the investigation of mechanism and kinetics of hydroboration because:

1. It has high thermal stability and purity.
2. It is convenient to handle compared to other boranes as it has lower sensitivity to oxygen and water vapors.
3. With only one center per boron, its overall reaction with an alkene involves only one dissociation step and one hydroboration step; in contrast, borane (BH_3) has three consecutive addition reactions, three redistribution equilibria, and five monomer–dimer equilibria [1].
4. 9-BBN reactions can be studied in solvents such as carbon tetrachloride, cyclohexane, and benzene whereas 9-BBN exists exclusively as dimer, thus eliminating complexation [2] with solvents and simplifying the kinetics.
5. 9-BBN is highly regio- and stereoselective, which assures the study of only one reaction.
6. In addition, the progress of the reaction can easily be monitored *via* IR where disappearance of the 1,570 cm^{-1} absorption of bridged B–H bonds occurs.

Consequently, kinetic studies of 9-BBN have yielded very useful data for its relative reactivity toward various types of unsaturation and the effect of solvent on the reaction.

It is significant to mention that in compounds that display first-order kinetics, the values differ in THF and CCl_4 solvents. The first-order rate constant in THF is about 10 times that in CCl_4 solvent. This is due to the catalytic effect of THF [3] on the $(9\text{-BBN})_2$ dimer that breaks the B–H bridge bond [4].

4.1 Hydroboration Kinetics of Alkenes

The hydroboration reactions of $(9\text{-BBN})_2$ with more reactive olefins have been found to be the first-order kinetics (Eq. 4.1) [1].

$$-\frac{d[(9\text{-BBN})_2]}{dt} = k_1[(9\text{-BBN})_2] \tag{4.1}$$

and three-halves kinetics, with less reactive olefins (Eq. 4.2) [2].

$$-\frac{d[(9\text{-BBN})_2]}{dt} = k_{\frac{3}{2}}[(9\text{-BBN})_2]^{\frac{1}{2}}[\text{olefin}] \tag{4.2}$$

The kinetics for hydroboration of alkenes are conducted in various solvents such as carbon tetrachloride, hexane, cyclohexane, benzene, and THF. 9-BBN exists predominantly as the dimer $(9\text{-BBN})_2$ [2]. After the addition of olefins, at 25 °C, the aliquots from the reaction mixture are removed after appropriate intervals of time, quenched with an excess methanol, and analyzed by GLC for residual olefin. All operations are performed under nitrogen until identical rates are observed for more reactive olefins such as 1-hexene, 2-methyl-1-pentene, 3,3-dimethyl-1-butene, and cyclopentene, and variation of olefin concentration does not alter the rate. These results establish that the reaction is first order (Eq. 4.1). Typical data for cyclopentene and cyclohexene are presented in Table 4.1 [1].

Table 4.1 Rate data and rate constants for the hydroboration of cyclopentene (0.400 M) and cyclohexene (0.400 M) with $(9\text{-BBN})_2$ (0.200 M) in carbon tetrachloride at 25 °C [1]

Time (s)	Cyclopentene[a] (M)	$10^4 k_1$[b] s^{-1}	Time (s)	Cyclohexene[a] (M)	$10^4 k_{3/2}$[b] $\frac{l^{1/2}}{mol^{-1/2}\,s^{-1}}$
0	0.400		0	0.400	
298	0.382	1.50	6,001	0.339	0.321
1,205	0.332	1.54	15,380	0.262	0.343
2,713	0.263	1.55	21,605	0.225	0.344
4,540	0.202	1.51	42,494	0.148	0.338
6,297	0.153	1.52	61,769	0.108	0.336
9,001	0.102	1.52	72,007	0.096	0.324

[a] Concentration of $(9\text{-BBN})_2$ is one half that of olefin.
[b] Calculated from the equations

$$k_1 t = \ln[\tfrac{b}{(b-2x)}] \text{ and } k_{3/2} t = \sqrt{2}[(b-2x)^{-1/2} - b^{-1/2}] \quad (4.3)$$

where b is the initial concentration of olefin and $b-2x$ is the concentration at time (t).

The first-order rate constants are quite similar in solvents other than THF, which are 1.52 in CCl_4, 1.97 in hexane, 1.52 in cyclohexane, and 1.95 in benzene, and 2 in diethylether. However, in THF, k_1 is considerably larger, at 10.8.

The less reactive olefins such as cyclohexene, 1-methylcyclohexene, and 2,3-dimethyl-2-butene exhibit rates that are slower and vary with concentration and the structure of individual olefins. The kinetics establishes these reactions to be first order in olefins and one-half order in the 9-BBN dimer (Eq. 4.2). The calculated three-halves-order rate constants also do not change as the reaction proceeds. The rate constants observed both for the first-order and three-halves-order kinetics are summarized in Table 4.2 [1].

These studies unequivocally establish that the hydroboration of alkenes with $(9\text{-BBN})_2$ proceeds through the prior dissociation of the dimer to the monomer (Eq. 4.4), followed by the reaction of monomer with the unsaturated substrate (Eq. 4.5).

$$(9\text{-BBN})_2 \underset{k_{-1}}{\overset{k_1}{\rightleftharpoons}} 2\ 9\text{-BBN} \quad (4.4)$$

$$9\text{-BBN} + \text{alkene} \xrightarrow{k_2} B\text{-alkyl-9-BBN} \quad (4.5)$$

Table 4.2 First-order- and three-halves-order rate constants for the hydroboration of representative olefins with $(9\text{-BBN})_2$ in carbon tetrachloride at 25 °C [1]

Olefin[a]	$10^4 k_1$, s^{-1}	$10^4 k_{3/2}$, $l^{1/2}\ M^{-1/2}\ s^{-1}$
1-Hexene	1.54	
2-Methyl-1-pentene	1.53	
3,3-Dimethyl-1-butene	1.45	
Cyclopentene[b]	1.52	
Cyclohexene[b]		0.323
1-Methylcyclohexene		0.051
2,3-Dimethyl-2-butene		0.02

[a] Rate constants in table are for initial concentration of olefin (0.4 M), and $(9\text{-BBN})_2$ (0.2 M).

[b] Variation of the initial concentrations of the olefin and $(9\text{-BBN})_2$ do not change the observed rate constants, significantly: cyclopentene (0.4 M), $(9\text{-BBN})_2$ (0.1 M), $10^4 k_1$ 1.58; cyclopentene (0.2 M), $(9\text{-BBN})_2$ (0.1 M), $10^4 k_1$ 1.58; cyclohexene (0.4 M), $(9\text{-BBN})_2$ (0.1 M), $10^4 k_{3/2}$ 0.324; cyclohexene (0.2 M), $(9\text{-BBN})_2$ (0.1 M), $10^4 k_{3/2}$ 0.345.

This mechanism leads to the following kinetic expression (Eq. 4.6) utilizing the usual steady state approximation.

$$-\frac{d[(9\text{-BBN})_2]}{dt} = k_1\,[(9\text{-BBN})_2]\left(\frac{\frac{1}{2}k_2\,[\text{olefin}]}{k_{-1}\,[9-\text{BBN}] + \frac{1}{2}k_2\,[\text{olefin}]}\right) \tag{4.6}$$

If $1/2\,k_2$ [olefin] >> k_{-1} [9-BBN] Eq. 4.6 reduces to Eq. 4.1, the reaction behaves like a unimolecular reaction and exhibits first-order kinetics. However, if $1/2\,k_2$ [olefin] << k_{-1} [9-BBN], Eq. 4.6 reduces to Eq. 4.2. The reaction exhibits three-halves-order kinetics. Consequently, kinetics reveals that the hydroboration with dimeric 9-BBN of representative alkenes proceeds through prior dissociation of the dimer to the monomer, leading to simplified kinetic expression, as expressed in Eqs. 4.1 and 4.2 for reactive alkenes and for less reactive alkenes, respectively. However, for olefins like 2-methyl-2-butene and *cis*-3-hexene $1/2\,k_2$ [olefin]$\approx k_{-1}$ $[9\text{-BBN}]_2$ and the kinetics fail to follow the simplified rate expression Eqs. 4.1 and 4.2.

To test the proposed dissociation mechanism with $(9\text{-BBN})_2$, a detailed study is conducted using quantitative IR spectrometry [3]. $(9\text{-BBN})_2$ exhibits a very strong IR absorption at 1,570 cm^{-1} due to B–H bridges [4]. This method is convenient and more reliable and thus has an advantage over tedious GLC analyses.

The hydroboration kinetics are studied by addition of alkenes to the solution of $(9\text{-BBN})_2$ in the solvent maintained at 25 °C. The reaction mixtures are pumped through a sodium chloride IR cell. The rates of the disappearance of B–H bridges of $(9\text{-BBN})_2$ at 1,570 cm^{-1} are monitored by quantitative IR spectrometry. The absorbance is recorded on chart paper.

In IR studies, an exponentially decaying curve of the absorbance of B–H bridges is noticed which also reveals that the reaction follows first-order kinetics. Six representative points on the exponentially decaying curve are calculated (Table 4.3) [3].

$$[(9\text{-BBN})_2]_t = [(9\text{-BBN})_2]_o\left[\frac{(c-b)}{(a-b)}\right] \tag{4.7}$$

where $[(9\text{-BBN})_2]_o$ is the initial concentration of $(9\text{-BBN})_2$ and $[(9\text{-BBN})_2]_t$ is its concentration at time *t*; *a*, is the initial absorbance; *c*, is the absorbance at time *t*; and *b* is the background absorbance

$$k_{3/2}t = 2^{1/2}\,[(b-2x)^{-1/2} - b^{-1/2}] \tag{4.8}$$

where *b* is the initial concentration of cyclopentene and *b*–2*x*, is the concentration at time *t*.

Table 4.3 Rate data and rate constants for the hydroboration of cyclopentene (0.4 M) with $(9\text{-BBN})_2$ (0.2 M) in CCl_4 at 25 °C [3]

t (s)	$[(9\text{-BBN})_2]$[a] (M)	[Cyclopentene] (M)	$10^4 k_1$[b] (s^{-1})
0	0.2	0.4	–
500	0.185	0.370	1.55
1,000	0.173	0.346	1.46
3,000	0.131	0.262	1.42
5,000	0.096	0.191	1.48
7,000	0.07	0.139	1.5
9,000	0.05	0.1	1.54

[a] Calculated from Eq. 4.7
[b] Calculated by Eq. 4.8.
[c] The three-halves-order rate constant obtained by the quenching method, is 0.323×10^{-4} $M^{-1/2}$ s^{-1}

The first-order constants remain essentially the same with varying initial concentrations of $(9\text{-BBN})_2$ and cyclopentene (Table 4.4) [3].

The IR exponentially decaying curve for various more reactive alkenes in different solvents (Table 4.5) [3] also shows a good agreement with the corresponding first-order rate constants realized by the quenching method.

Table 4.4 Effect of concentration on the rate constants for the hydroboration of cyclopentene, cyclohexene, and *cis*-3-hexene with $(9\text{-BBN})_2$ in carbon tetrachloride at 25 °C [3]

	Initial concentration (M)		$10^4 k_1$ (s^{-1})	$10^4 k_{3/2}$ ($M^{-1/2}$ s^{-1})	$10^4 k_2$[b] (M^{-1} s^{-1})
	Alkene	$(9\text{-BBN})_2$			
Cyclopentene	0.4	0.2	1.54	2.41	7.76
	0.4	0.1	1.58	1.21	4.81
	0.2	0.1	1.58	3.54	16.4
Cyclohexene	0.4	0.2	0.194	0.314	1.05
	0.4	0.1	0.456	0.324	1.58
	0.2	0.1	0.176	0.345	1.37
cis-3-Hexene[a]	0.4	0.2	1.07	1.69	5.52
	0.4	0.1	1.45	1.08	4.75
	0.2	0.1	0.925	2.07	9.49

[a] The kinetic data do not fit well to any of the integrated kinetic expressions.
[b] The second-order rate constants.

Table 4.5 First-order rate constants for the hydroboration of representative alkenes with $(9\text{-BBN})_2$ in various solvents at 25 °C [3]

Alkene	$10^4 k_1$ (s^{-1})				
	CCl_4	THF	Cyclohexane	Benzene	Et_2O
1-Hexene	1.54 (1.53)[a]	13.9 (14.2)[a]	1.51 (1.45)[a]	2.05	2.83
2-Methyl-1-pentene	1.46 (1.54)	13.7 (14.3)	1.51 (1.51)	1.99	2.81
3,3-Dimethyl-1-butene	1.45 (1.45)	14 (13.2)	1.48 (1.43)	2.02	2.80
Cyclopentene	1.54 (1.52)	11.8 (10.8)[b]	1.46 (1.52)	2.06	2.77

[a] Numbers in *parentheses* are from quenching method.
[b] The reaction is not a clean first-order reaction.

The kinetic data obtained by quantitative IR spectrometry also reveal that for more reactive alkenes (cyclopentene, 2-methyl-1-pentene, 1-hexene, and 3,3-dimethyl-1-butene), the rate-determining step is the dissociation of the dimer, thereby showing first-order kinetics (Eq. 4.1). For the less reactive alkenes such as cyclohexene, the reaction of monomer also becomes the rate-determining step, and thus exhibits three-halves-order kinetics (Eq. 4.2; Table 4.6) [3], first order in cyclohexene, and half order in $(9\text{-BBN})_2$. For certain alkenes such as 2-methyl-2-butene and *cis*-3-hexene, neither of these two steps are a decisive rate-determining step. Therefore, the reaction exhibits intermediate kinetic behavior between that of first-order and three-halves-order kinetics.

Table 4.6 Rate data and rate constants for the hydroboration of cyclohexene (0.400 M) with $(9\text{-BBN})_2$ (0.200 M) in carbon tetrachloride at 25 °C [3]

t (s)	$(9\text{-BBN})_2$[a] (M)	Cyclohexene (M)	$10^4 k_{3/2}$,[b, c] ($M^{-1/2} s^{-1}$)
0	0.200	0.400	–
12,000	0.148	0.295	0.307
18,000	0.129	0.258	0.304
24,000	0.114	0.228	0.303
30,000	0.101	0.202	0.303
42,000	0.081	0.161	0.308
72,000	0.049	0.098	0.316

[a] Calculated from Eq. 4.7.
[b] Calculated from Eq. 4.8.
[c] The three-halves-order rate constant obtained by the quenching method is $0.323 \times 10^{-4} M^{-1/2} s^{-1}$.

The reported 9-BBN preference for *trans* isomer [5], on reinvestigation via the kinetics of regiospecificity hydroboration of isomeric *cis*- and *trans*-alkenes by 9-BBN in THF, reveal that 9-BBN predictably hydroborates neither the *cis* nor the *trans* isomer selectively (Table 4.7) [6]. These isomers obey three-halves-order kinetic, first order in alkene, and half order in $(9\text{-BBN})_2$.

Table 4.7 Relative reactivities and rate constants for the hydroboration via 9-BBN of several *cis*–*trans* pairs at 25 °C in THF solvent [6]

Alkene	Structure	Relative reactivity	k_{cis}/k_{trans}	$k_{3/2}(cis)/k_{3/2}(trans)$	$10^4 k_{3/2}$ ($M^{-1/2} s^{-1}$)
1-Hexene		100			14.3 (first order)
cis-3-Hexene		0.68	2.1	1.7	4.53
trans-3-Hexene		0.32			2.65
cis-4-Methyl-2-pentene	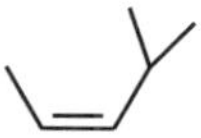	0.53			4.20
trans-4-Methyl-2-pentene	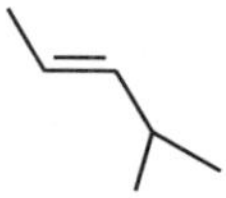	0.16	3.3	3.3	1.28
cis-1-Phenylpropene		0.024	0.39	0.39	0.227
trans-1-Phenylpropene	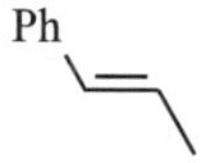	0.062			0.579
cis-2,5-Dimethyl-3-hexene	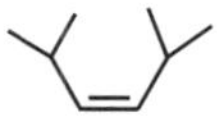	0.0017	0.076	0.6	0.011
trans-2,5-Dimethyl-3-hexene	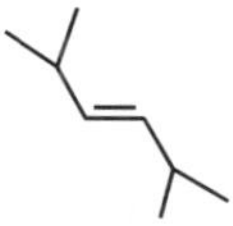	0.022			0.183
cis-1,4-Dichloro-2-butene		0.0144	4.4	2.3	0.218
trans-1,4-Dichloro-2-butene		0.0033			0.096
cis-1-Chloro-1-butene		0.0093	3	2	0.115
trans-1-Chloro-1-butene	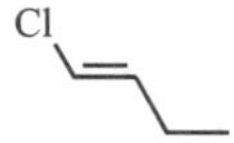	0.0031			0.058

The bulkier groups, because of greater steric requirements, preferably direct hydroboration at the less hindered carbon (Chart 4.1) [6] except in the case of 1-phenylpropene. Also the directive influence is less in *trans* isomers of alkenes.

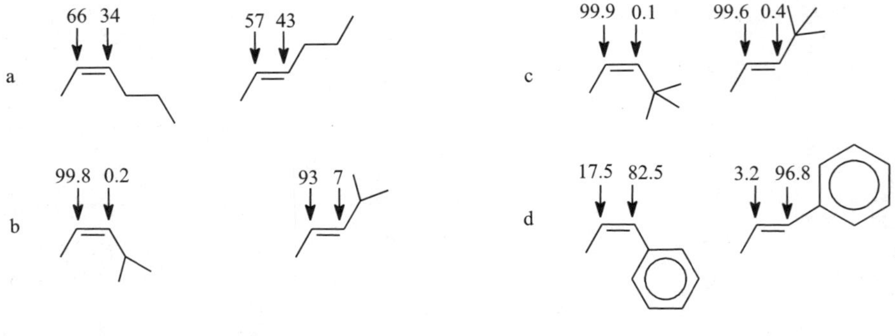

Chart 4.1

The C-4 Me in an alkene (Chart 4.1) increases the placement of boron at the C-2 position and also decreases the rate of hydroboration (Chart 4.2). Apparently, the steric effect and hyperconjugation by the methyl group, which would tend to direct hydroboration to the C-3 position, are outweighed due to steric effects of the alkyl group.

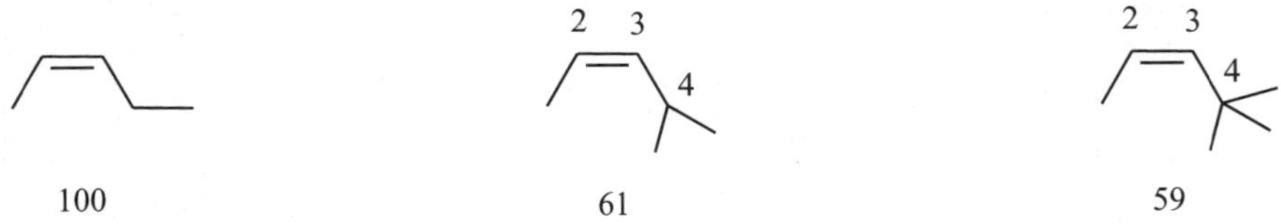

Chart 4.2

However, 1-phenylpropene undergoes hydroboration at the C-1 position. This is not what one would expect based on the steric effect of a phenyl group. This is because of the combined effects of phenyl conjugation (–K) and methyl hyperconjugation (+K), which act to decrease the amount of electron density at the C-2 position and to increase it at the C-1 position (Fig. 4.1) [6]. Evidently, these mesomeric effects are strong enough to override the steric effects of the phenyl group.

The directing effect of bulkier alkyl groups is pronounced in *cis*-alkene due to their nonplanar geometry as compared to *trans*-alkene. This allyl strain in *cis*-alkene is termed the $A^{(1,3)}$ strain [7]. Consequently, the *trans* isomer of 1-phenylpropene and *trans*-2,4-dimethyl-3-hexene isomers react at a faster rate than do the corresponding *cis* isomers (Chart 4.3) [6].

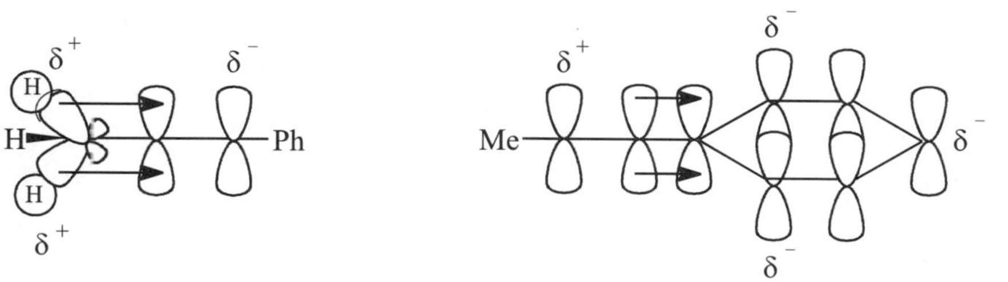

Fig. 4.1 Diagram to show combined effects of phenyl conjugation (–K) and methyl hyperconjugation (+K) in 1-phenylpropene [6]

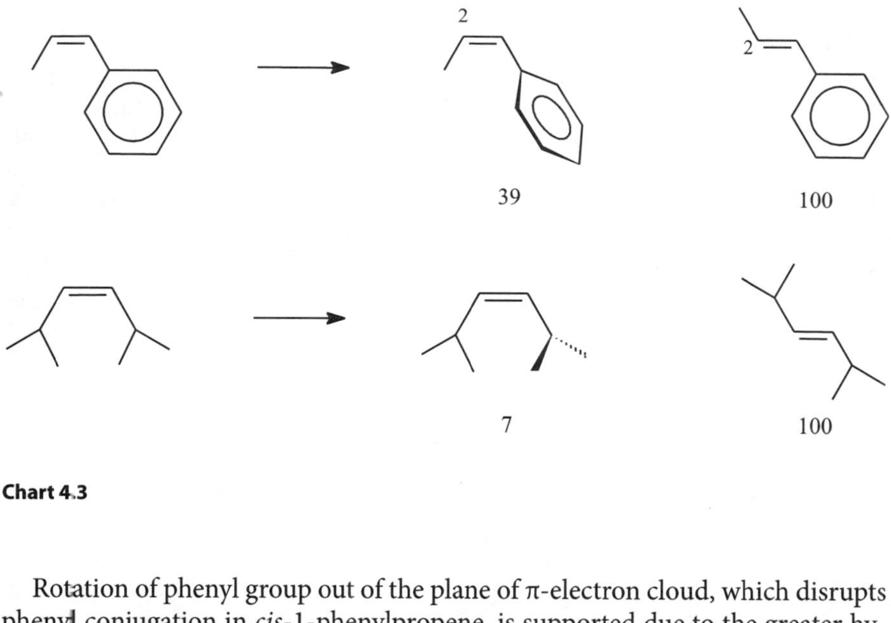

39

100

7

100

Chart 4.3

Rotation of phenyl group out of the plane of π-electron cloud, which disrupts phenyl conjugation in *cis*-1-phenylpropene, is supported due to the greater hydroboration at the C-2-postion (17.5%), than in the *trans* isomer (3.2%). However, the rate-retarding steric effects of phenyl rotation prevail.

In the styrene system, the rate-increasing effects of C-2-Me are larger. Here the $A^{(1,2)}$ strain [7] causes rotation of the phenyl ring out of the plane of the π system, disrupting the rate-retarding phenyl conjugation. Consequently, 2-methyl substitution in this system causes a larger rate increase than it does in 1-hexene. In this system, steric and hyperconjugative effects work together and direct hydroboration to C-1. Evidently, methyl hyperconjugation (+K) and removal of phenyl conjugation (–K) act together to override the steric effects due to phenyl rotation.

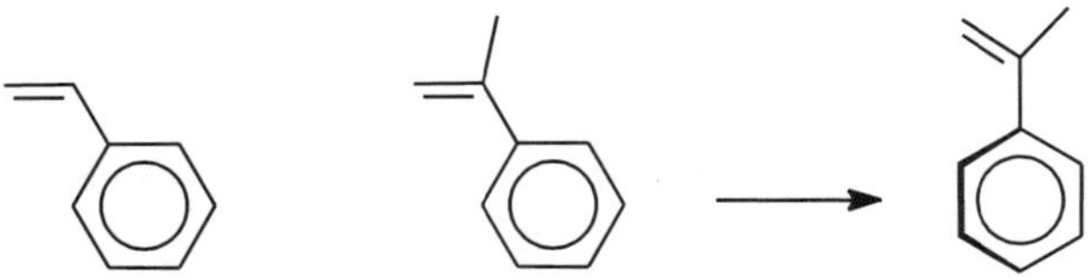

1.00 5.62

It is seen that C-2-methyl substitution in open-chain olefins increases the relative reactivity by a factor of about 2 due to hyperconjugative effects [8].

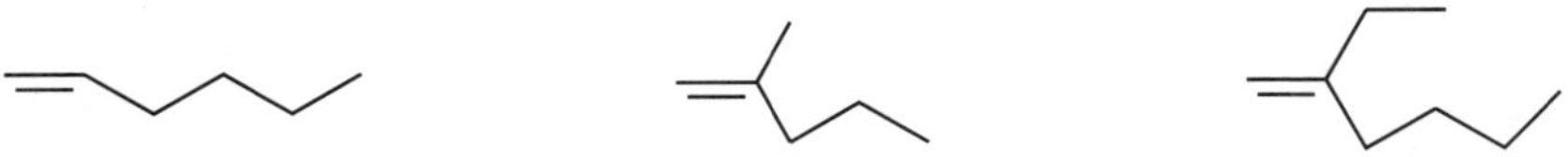

1.00 1.96 1.68

On the other hand [6], in cyclic systems such as cyclopentene and cyclohexene, identical methyl substitution retards the rate of hydroboration. In a cyclic system, the steric retardation by methyl far outweighs the hyperconjugative contribution. Here, the interaction between the methyl group and α-methylene unit causes rotation of the methyl group, disrupting the hyperconjugation, which otherwise contributes to rate enhancement. In the case of an ethyl group, the steric interaction with α-methylene causes the rotation of the ethyl group so that the terminal methyl group is moved away. By doing so, some hyperconjugation is gained. The following representations indicate the kind of interactions that are responsible for these phenomena [6].

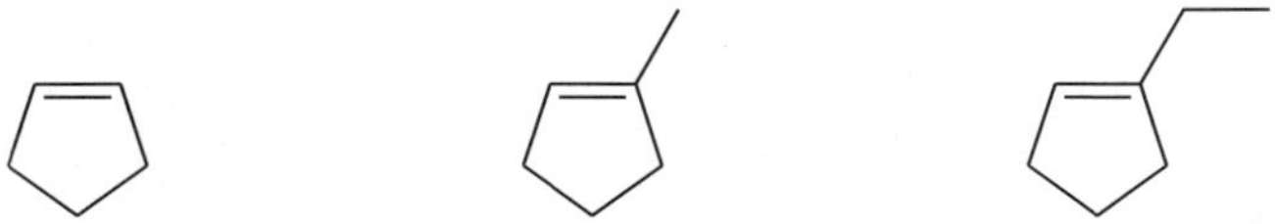

1.00 0.261 0.421

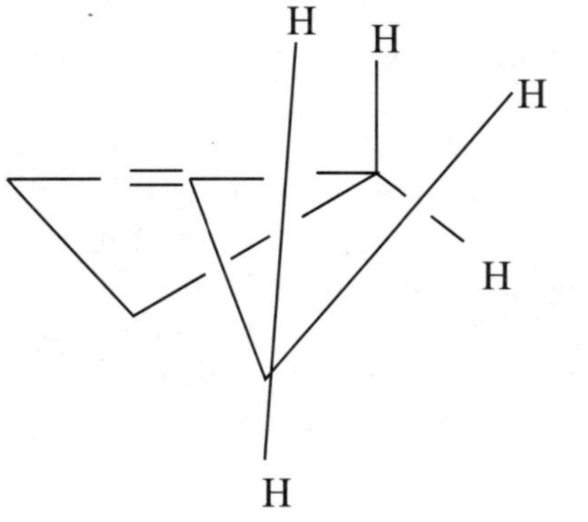

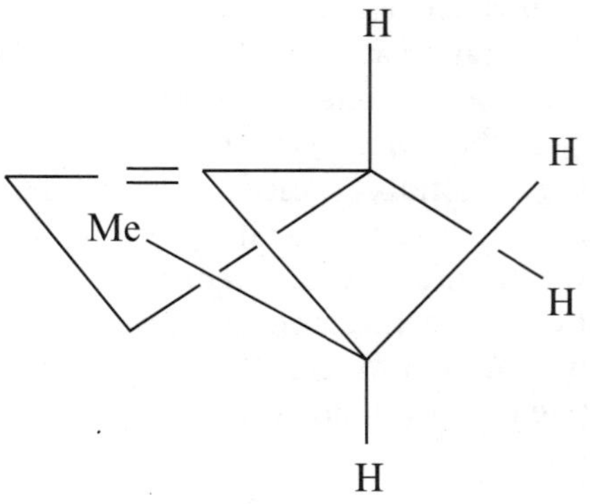

The reaction rate is governed by two electronic modes [8, 9], inductive (I) and conjugative (K), and by steric effect. In most cases increasing steric effect decreases the reaction rate (Chart 4.4).

100 0.68

Chart 4.4

However, in β-substituted olefins, the steric effect is slightly overridden by the inductive and hyperconjugation of the methyl group to place more electron density [10] at C-1 than at C-2 (Chart 4.5).

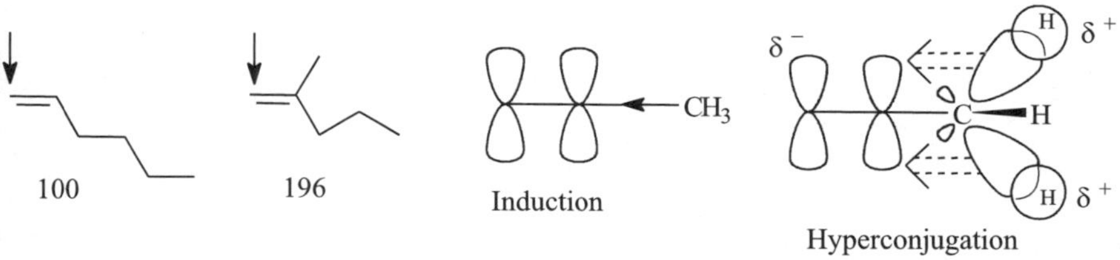

Chart 4.5

The results are summarized in Table 4.8 [8].

The effect of an alkyl group, α, on the site of hydroboration is different than if it is β. β-Substituent (type A) increases the reaction rate by factor of ~1.87, whereas alkyl substituent α (type B) lowers the rate of reaction by an average of ~170. The calculation, e.g., for three alkyl substituents, two α and one β, to the site of hydroboration are done as $\frac{(\alpha\ \text{effect})^2}{(\beta\ \text{effect})} = \frac{\downarrow 171^2}{\uparrow 1.87} = 1.56\times10^4$ (rate reduction). The estimate is in good agreement with the experimentally observed effect of $\downarrow 1.64\times10^4$ (rate reduction).

4.2 Hydroboration Kinetics of Alkynes

The hydroboration of alkynes with $(9\text{-BBN})_2$ exhibits kinetics similar to those for the hydroboration of alkenes. The kinetics are studied (1) by pumping the reaction mixture through a sodium chloride IR cell and monitoring the disappearance of B–H bridges of $(9\text{-BBN})_2$ at 1,570 cm^{-1} by quantitative IR spectrometry [1]; and (2) by quenching the aliquots of the reaction mixture periodically with excess methanol and analyzing by GLC for residual alkyne. For the more

Table 4.8 A breakdown of the effects of alkyl groups that influence the change in reaction rate in alkenes [8]

Alkyl-substituted compound	Parent	Factor by which relative reactivity changes	Type	Hyperconjugative effects of additional alkyl group(s) at position relative to site of hydroboration α	β	Inductive effects on π bond	Steric effects of additional alkyl group(s) at position relative to site of hydroboration α	β
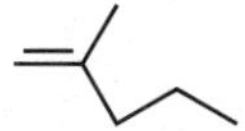		↑1.96	A	0	1	1	0	1
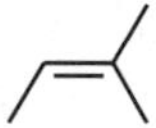		↑1.77						
		↓156	B	1	0	1	1	0
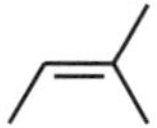	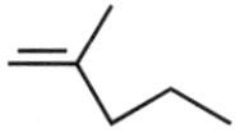	↓173						
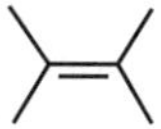	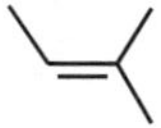	↓185						
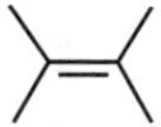	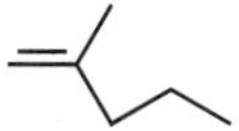	↓3.21×10^4	2 of B	2	0	2	2	0
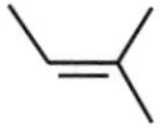		↓88.5	1 of A and 1 of B	1	1	2	1	1
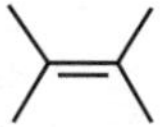		↓105						
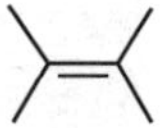		↓1.64×10^4	2 of B and 1 of A	2	1	3	2	1

Table 4.9 First-order rate constants for the hydroboration of alkynes with $(9\text{-BBN})_2$ in various solvents at 25 °C [2]

	$10^4 k_1$ (s^{-1})			
Alkyne	THF	CCl_4	Cyclohexane	Benzene
1-Hexyne	14.1	1.52	1.48	1.98
3-Methyl-1-butyne	14	-	-	-
Cyclohexylethyne	14.3	-	-	-
3,3-Dimethyl-1-butyne	13.9	-	-	-

reactive alkynes such as 1-hexyne, 3-methyl-1-butyne, cyclohexylethyne, and 3,3-dimethyl-1-butyne, each reaction exhibits first-order kinetics, first order in $(9\text{-BBN})_2$ only, and the data are summarized in Table 4.9 [2].

Apparently, in these solvents (Table 4.9) [2], the rate of dissociation of 9-BBN dimer is the rate-determining step. The much higher first-order rate constant in THF is of immense interest.

Although the hydroborations of 1-hexyne, 3-methyl-1-butyne, 3,3-dimethyl-1-butyne, and cyclohexylethyne all exhibit the same first-order rate of reaction, they have very different relative reactivities toward $(9\text{-BBN})_2$ (Table 4.10) [2].

The relative reactivities obtained by the competitive studies are the relative values of k_2 of the alkynes. The fact that these alkynes exhibit different k_2 values, and yet have almost identical k_1 values, further supports the dissociation mechanism.

Table 4.10 Relative reactivities of representative alkynes and alkenes toward $(9\text{-BBN})_2$, Sia_2BH, and $HBBr_2{\cdot}SMe_2$ at 25 °C [2]

Alkyne or alkene	Relative reactivity		
	THF, $(9\text{-BBN})_2$	THF, Sia_2BH	CH_2Cl_2, $HBBr_2{\cdot}SMe_2$
1-Hexene	100	100	100 (of 1-Octene)
cis-3-Hexene	0.68	1.85	20 (of *cis*-4-Octene)
1-Hexyne	15.3	373	290
1-Decyne	18	-	-
3-Methyl-1-butyne	48.5	-	-
3,3-Dimethyl-1-butyne	86	-	-
Cyclohexylethyne	41	-	-
Phenylethyne	1.41	-	-
3-Hexyne	0.64	225	5,900
4,4-Dimethyl-2-pentyne	0.63	-	-
Diphenylethyne	1.77×10^{-3}	-	-

It is clear from Table 4.10 that unlike the corresponding alkenes, the branching in terminal alkynes at the 3 position increases the relative reactivity in the following order: 1-hexyne < 3-methyl-1-butyne < 3,3-dimethyl-1-butyne. On the other hand, the relative reactivities of corresponding alkenes decrease with branching in the following order: 1-hexene > 3-methyl-1-butene > 3,3-dimethyl-1-butene [3].

The rate of hydroboration of diphenylethyne, with $(9\text{-BBN})_2$ is slower than that of 1-hexyne in CCl_4 at 25 °C. The reaction exhibits three-halves-order kinetics, first order in diphenylethyne and one-half order in $(9\text{-BBN})_2$. The kinetic behavior; intermediate between first order and three-halves order is observed with phenylethyne in CCl_4 at 25 °C, which is between that of 1-hexyne and diphenylethyne.

It is important to note that unlike the disiamylborane dimer [4] and the dibromoborane–dimethylsulfide complex [5], $(9\text{-BBN})_2$ does not exhibit a large increase of relative reactivities toward alkynes. The relative reactivity of 1-hexyne is only one seventh that of 1-hexene, and 3-hexyne shows about the same relative reactivity as does *cis*-3-hexene. Consequently, one can selectively hydroborate the terminal alkene in the presence of internal alkyne [6].

4.2.1 Relative Rates between Monohydroboration and Dihydroboration of Alkynes

Unlike the disiamylborane dimer [4] and the dicyclohexylborane dimer [7], the hydroboration of a terminal alkyne, such as 1-hexyne with equimolar ratio of 9-BBN produces, also, a considerable percentage of *gem* dibora derivatives (Scheme 4.1, Eq. 4.10) [8].

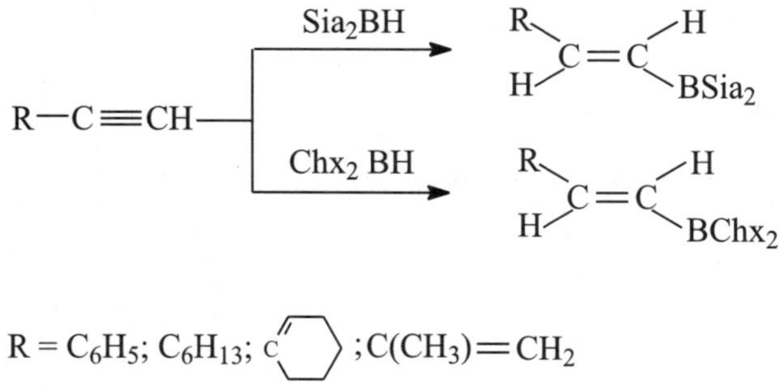

Scheme 4.1

The studies reveal that the ratio between k_2 (Eq. 4.9) and k_2' (Eq. 4.10) for terminal alkynes is relatively small, and this ratio for internal alkynes is much larger (Table 4.11) [2] and thus explains the much larger percentage of dihydroboration for terminal alkynes. In fact, the percentage of mono- and dihydroboration are in direct correlation with k_2/k'_2. A computer program using the Runga-Kutta numerical method [9] is written to establish the k_2/k'_2 relationship [2]. The correlation between $\frac{k_2}{k'_2}$ and the percentage of monohydroboration is obtained by this calculation method.

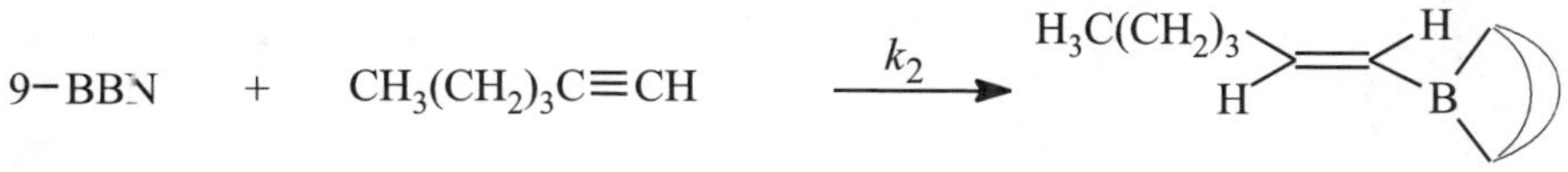

(4.9)

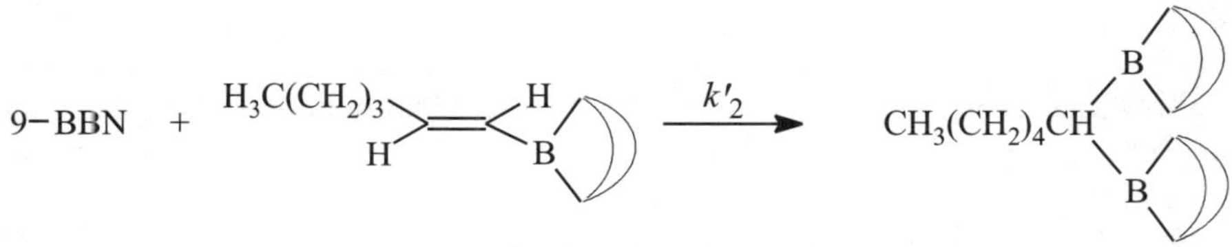

(4.10)

Table 4.11 Relative rates between mono- and dihydroboration of alkynes with $(9\text{-BBN})_2$ in THF at 25 °C [2]

Alkyne	Relative rate k_2/k'_2
1-Hexyne	1.92
3-Hexyne	193

The calculated and experimental data [2] for mono- and dihydroboration during the entire reaction of 1-hexyne are in good agreement.

The data reveal [2] that 6 equiv at 25 °C (2 equivat 0 °C) of 1-hexyne is required to achieve 90% monohydroboration. These calculations are extended to other alkynes. The above information is very useful for synthetic application and one can calculate the amount of excess alkyne required to achieve a relatively clean monohydroboration [2].

4.3 Hydroboration Kinetics of Haloalkenes

Allylic haloalkenes with terminal C=C undergo hydroboration primarily at C-1 position, whereas all other haloalkenes place boron on the carbon β to the halogen. However, the product undergoes rapid elimination of *B*-halo-9-BBN, followed by rapid rehydroboration with another 9-BBN as shown in Scheme 4.2. It is important to mention that in the reaction of 1 equiv of each of haloalkene and 9-BBN, no haloalkyl-9-BBN is detected by ^{11}B NMR spectroscopy.

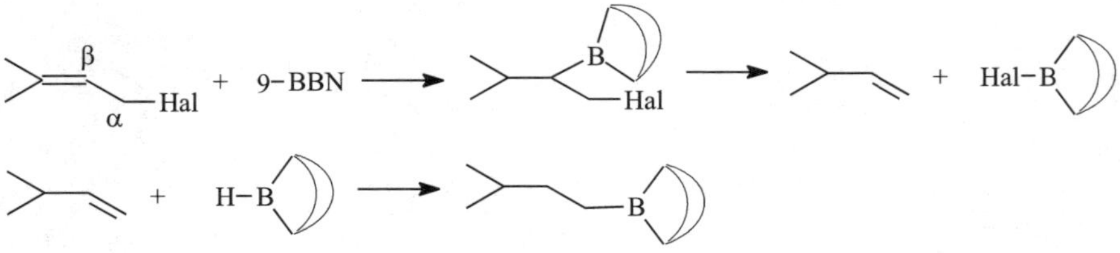

Scheme 4.2

The kinetic results of haloalkenes and their unsubstituted parent compounds are summarized in Table 4.12 [1].

Table 4.12 Relative reactivities and rate constants for the hydroboration of haloalkenes and their parent compounds in CCl_4 and THF at 25 °C [1]

Alkene	Relative reactivities	$10^4 k_1$, s^{-1} or $10^4 k_{3/2}$, $M^{-1/2} s^{-1}$		
		CCl_4	THF	
2-Methyl-1-pentene	196	1.53	14.3	First-order kinetics
1-Hexene	100	1.54	14.2	
Methallyl chloride	13.6	1.65	13.5	
Allyl iodide	7.1	1.63	13.6[a]	
Allyl bromide	4.6	1.57	12[a]	
Allyl chloride	4	1.55	12.3[a]	
2-Methyl-2-butene	1.13	–[a]	6.91[a]	Intermediate kinetics
cis-3-Hexene	0.68	–[a]	4.53	
1-Chloro-3-methyl-2-butene	0.64	–	3.34	
1-Bromo-3-methyl-2-butene	0.61	–	2.90	
trans-3-Hexene	0.32	–	2.65	
cis-1,4-Dichloro-2-butene	0.0144	0.161	0.218	
cis-1-Chloro-1-butene	0.0093	–	0.115	
2,3-Dimethyl-2-butene	0.0061	0.02	0.049	Three-halves-order kinetics
trans-1,4-Dichloro-2-butene	0.0033	0.024	0.048	
trans-1-Chloro-1-butene	0.0031	–	0.029	

[a] The reaction displays at least some intermediate behavior in the indicated solvent.

The results of Table 4.12 [1] reveal that terminal haloalkenes with an unhindered end of double bond are faster reacting, and these display first-order kinetics. Most of the haloalkenes with internal C=C follow kinetics of three-halves order, first order in haloalkenes, and one-half order in $(9\text{-BBN})_2$. Several of other haloalkenes display intermediate kinetic behavior.

Like the other slower reacting haloalkenes, the first hydroboration of the compounds is followed by rapid elimination and rehydroboration and displays three-halves-order kinetics. Apparently, the first hydroboration is rate determining, with elimination and rehydroboration taking place at a much faster rate (Eqs. 4.11–13).

$$(9\text{-BBN})_2 \underset{k_{-1}}{\overset{k_1}{\rightleftharpoons}} 2\ 9\text{-BBN}$$
$$9\text{-BBN} + \text{haloalkene} \xrightarrow{k_2} B\text{-haloalkyl-9-BBN} \tag{4.11}$$

$$B\text{-haloalkyl-9-BBN} \xrightarrow{\text{fast}} \text{alkene} + B\text{-halo-9-BBN} \tag{4.12}$$

$$9\text{-BBN} + \text{alkene} \xrightarrow{\text{fast}} B\text{-alkyl-9-BBN} \tag{4.13}$$

To compare the rate of hydroboration for compounds that undergo elimination–rehydroboration with that of the other haloalkenes, the rate of first hydroboration step (Eq. 4.11) must be determined. This is obtained by the following kinetic expression (Eq. 4.14).

$$\frac{-d[(9\text{-BBN})_2]}{dt} = k_1[(9\text{-BBN})_2]\left(\frac{k_2[\text{haloalkene}]}{k_{-1}[9\text{-BBN}] + k_2[\text{haloalkene}]}\right) \tag{4.14}$$

If k_2 [haloalkene] >> k_1 [9-BBN], Eq. 4.14 reduces to Eq. 4.1. If k_2 [haloalkene]<<k_{-1} [9-BBN], three-halves-order rate expression is represented by Eq. 4.15.

$$\frac{-d[(9\text{-BBN})_2]}{dt} = 2k_{3/2}[(9\text{-BBN})_2]^{1/2}\ [\text{haloalkene}] \tag{4.15}$$

where $k_{3/2}$ is the three-halves-order rate constants for hydroboration of unsubstituted alkenes. From the above equations and approximations, the three-halves-order rate expressions for the substituted and unsubstituted cases differ only by a factor of 2. Consequently, rate constants of the first hydroboration step can be calculated for compounds that exhibit rapid elimination–rehydroboration.

Table 4.13 Effect of halogen substitution on the relative reactivity of alkenes in hydroboration with $(9\text{-BBN})_2$ [1]

Haloalkene	Parent	Type of halogen	Position of hydroboration relative to halogen	Factor by which relative reactivity of haloalkene is lowered		
				Cl	Br	I
		Allylic	γ	25	21.7	14.1
	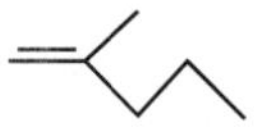	Allylic	γ	13.6		
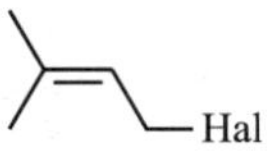	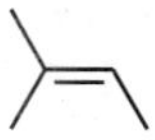	Allylic	β	1.77	1.85	
		Allylic	β, γ	48		
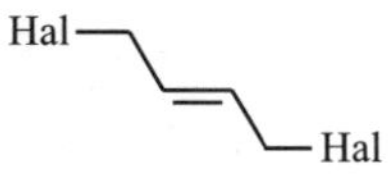	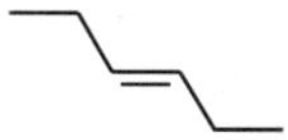	Allylic	β, γ	100		
		Vinylic	β	73		
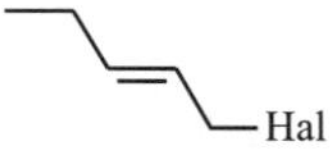		Vinylic	β	103		

On comparing the relative reactivity of a haloalkene with that of a parent compound, the effect of halogen substituent has been determined (Table 4.13) [1].

The electronic effects become significant in allyl halides. The Cl, Br, or I substituent at the γ position (allyl halides) has a rate reducing effect of 25, 21.7, or 14.1, respectively, as compared with 1-hexene.

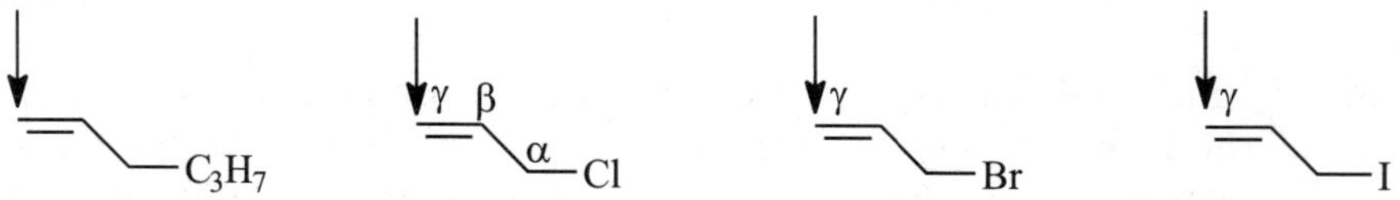

100 4.0 4.6 7.1

The order of rate reduction Cl > Br > I is the same as the increasing electron-withdrawing effect, (cumulative inductive and mesomeric) of the halogen. The effect is slightly reduced in methallyl chloride.

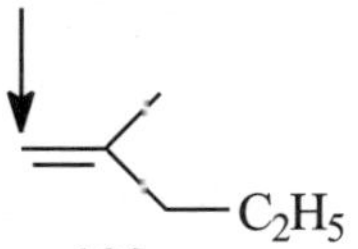
100

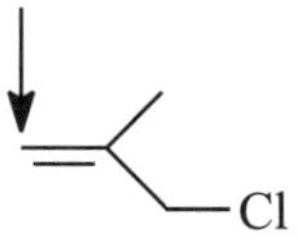
6.9

In the 1-halo-3-methyl-2-butene, hydroboration is directed to the position β to the halogen substituent, with a rate-retarding effect of about 2, much less as compared with the above examples where the hydroboration occurs at the γ position.

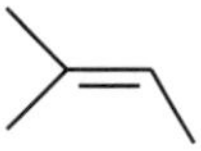
100

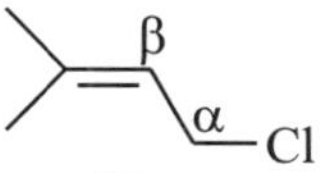
57

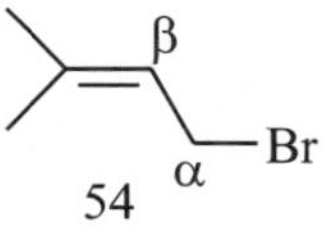
54

Table 4.14 Breakdown of the effects of chlorine substituents that influence the electron availability in allylic chlorides [1]

Allyl chloride	Parent	Increases in indicated interactions in going from parent to substituted compound: Factor by which relative reactivity is lowered	Inductive interaction of Cl with π bond (–I)	Hyperconjugative effect of Cl at position relative to chlorine: $(-K_\beta)$	$(-K_\gamma)$
		25	1	0	1
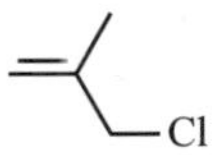	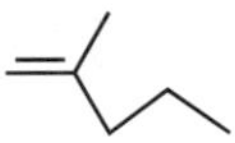	13.6	1	0	1
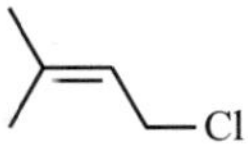	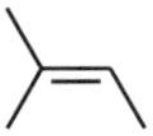	1.77	1	1	0
		48	2	1	1
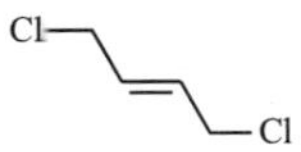	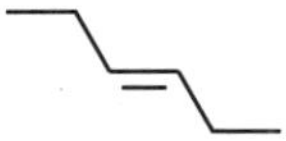	100	2	1	1

In 1,4-dichloro-2-butene, two equivalent sites are available for hydroboration. Both are β to one chlorine and γ to others and exhibit cumulative effect of β- and γ-chlorine substitution. Employing the data for allyl chloride and 1-chloro-3-methyl-2-butene (Table 4.14) [1], the rate change is calculated as (↓25) (↓1.77) = ↓44, which is in good agreement with experimentally observed values of ↓48.

The relative data in another form is as:

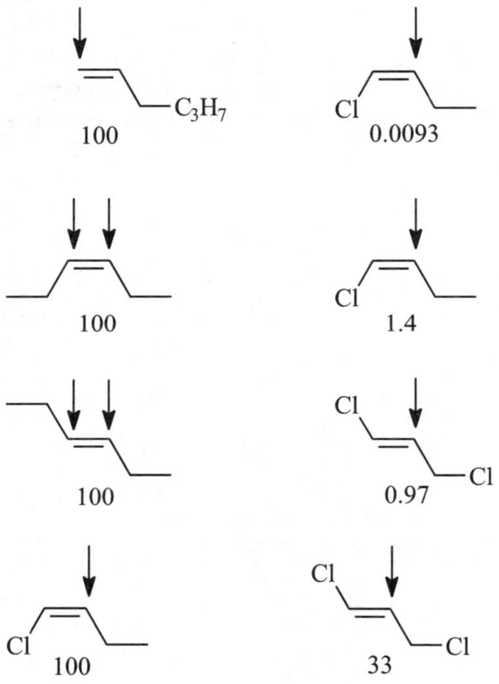

Chart 4.6

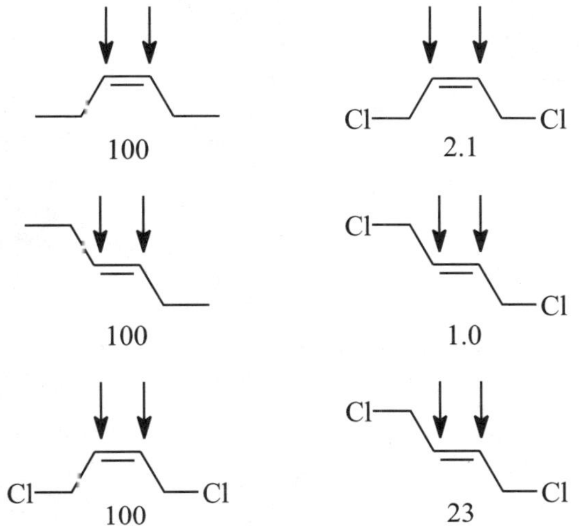

Chlorine substitution at the terminal position (vinyl chlorides) has a dramatic rate-retarding influence on rate of reaction. Moreover, the hydroboration is forced to take place at C-2 position (Chart 4.6) [1]. In each pair (for comparison) the higher rate is chosen as 100.

In allyl halide, both inductive and hyperconjugative are electron withdrawing and thus rate reducing at either the β or the γ position.

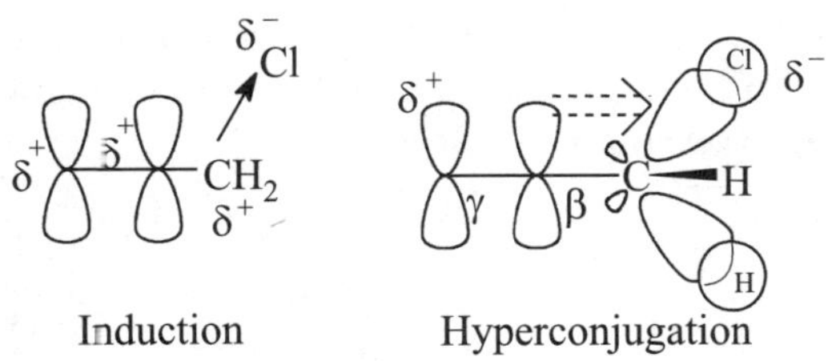

Relative to propene

In vinyl chloride, inductive effect of chlorine reduces the electron availability of the π bond to retard severely the rate of reaction. On the other hand, conjugation returns some electron density to the site β to chlorine, which becomes the preferred site of hydroboration.

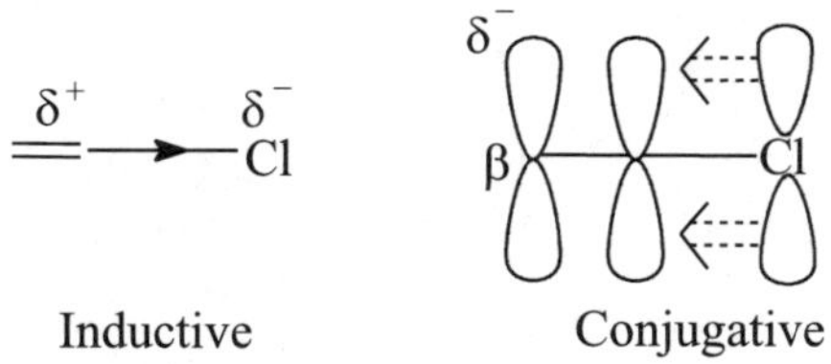

Relative to ethylene

In all the above cases, the hydroboration proceeds *via* dissociation of the dimer (Eq. 4.4) and subsequent reaction of the monomer with the unsaturated compound (Eq. 4.16).

$$\text{9-BBN} + \begin{pmatrix}\text{alkene}\\ \text{haloalkene}\\ \text{alkyne}\end{pmatrix} \longrightarrow B-\begin{pmatrix}\text{alkyl}\\ \text{haloalkyl}\\ \text{alkenyl}\end{pmatrix}-\text{9-BBN} \qquad (4.16)$$

4.4
Hydroboration Kinetics of Haloalkynes

Unlike 1-alkynes, which give considerable amount of dihydroboration at the 1 position [1], the 1-halo-1-alkynes afford only monohydroborating products and hydroboration occurring at the 1 position (Eq. 4.17) [2].

Table 4.15 Relative reactivities and rate constants for the hydroboration with $(9\text{-BBN})_2$ of haloalkynes and their parent compounds in CCl_4 at 25 °C [2]

Haloalkyne	Relative reactivities		Rate constants	
	 = 100	 =100	$10^4 k_1$, s^{-1} or $10^4 k_{3/2}$, $M^{-1/2} s^{-1}$	Type of kinetics
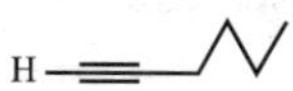 [a]	15.3	100	1.52	First order
	1.88	12.3	Intermediate	
H—≡—Ph [a]	1.41	9.22	Intermediate	
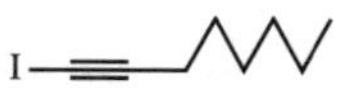	1.31	8.56		

Table 4.15 *(Continued)* Relative reactivities and rate constants for the hydroboration with (9-BBN)$_2$ of haloalkynes and their parent compounds in CCl_4 at 25 °C [2]

Haloalkyne	Relative reactivities		Rate constants	
	 = 100	 =100	$10^4 k_1$, s^{-1} or $10^4 k_{3/2}$, $M^{-1/2} s^{-1}$	Type of kinetics
I—≡—Ph	0.515	3.36	Intermediate	Intermediate
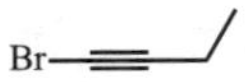	0.304	1.99		
	0.303	1.98		
	0.250	1.63		
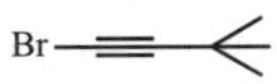	0.196	1.28		
	0.155	1.01	Intermediate	
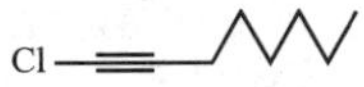	0.110	0.721	0.651	
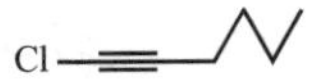	0.0645	0.422	0.372	Three-halves order
Br—≡—Ph	0.0189	0.124	0.149	
Cl—≡—Ph	0.00838	0.0548	0.057	

[a] Data from Wang KK, Scouten CG, Brown HC (1982) J Am Chem Soc 104:531.

Table 4.16 The effect of halogen substitution on the reactivity of various alkynes toward 9-BBN [2]

Haloalkyne	Parent[a]	Factor by which reactivity is lowered		
		Cl	Br	I
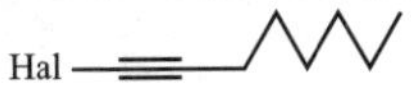		164	59.6	13.7
Hal—≡—Ph	H—≡—Ph	168	74.4	2.74
	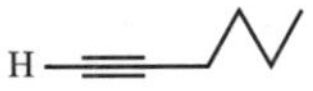	237	99	8.13
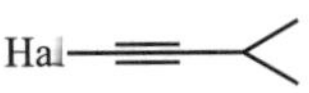	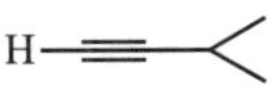		194	
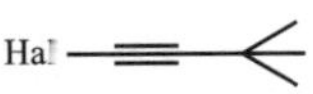	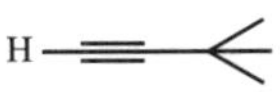		439	

[a] Data from, Wang KK, Scouten CG, Brown HC (1982) J Am Chem Soc 104:531.

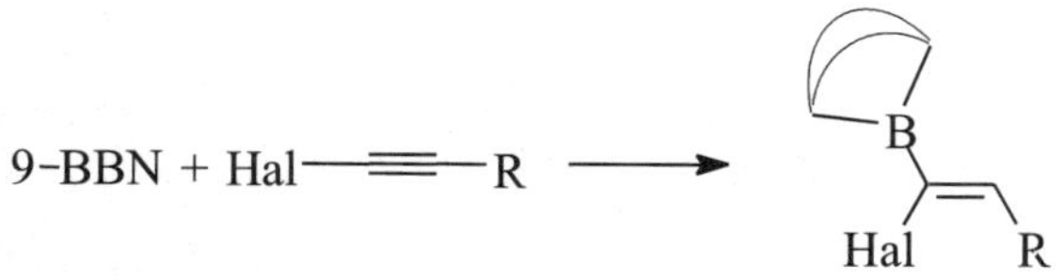

(4.17)

The kinetic results [2] for hydroboration of haloalkynes and their unsubstituted parent alkynes are shown in Table 4.15.

Even internal alkynes like 3-hexyne undergoes dihydroboration to yield a *gem* derivative, whereas 1-haloakynes undergo only monohydroboration at the 1 position. This reveals that the electronic rather than the steric factor is more important in limiting the reaction of 9-BBN with 1-haloalkyne to the monohydroboration product only.

The relative effects of halogens in 1-halo-1-alkynes are summarized in Table 4.16 [2].

Thus, in general, replacing a hydrogen or an alkyl group with a group, which reduces electron availability; such as halogen or phenyl results in decrease the rate of reaction. The time required for these reactions (0.4 M) to 95% completion are as follows: 1-chloro-1-hexyne, 2.5 days; 1-bromo-1-hexyne, 11 h; 1-iodo-1-hexyne, 8 h. The relative time, thus, indicates that the reaction is useful for synthetic chemist.

It is noteworthy that none of the 1-halo-1-alkynes including 1-iodo-1-alkynes give enough hydroboration at the C-2 position to be detected by GC or NMR. However, both terminal and internal alkynes give the *gem* dibora products, suggesting it must be electronic rather than steric effects causing the *gem* products to be preferred. This is explained in terms of change the alternation effect, which would be the result of the combined inductive and mesomeric effects. The effect of a boron substituent in the direction of hydroboration of an alkene is opposite to that of halogens. A vinyl halide favors placement of boron at the 2 position, whereas a vinylborane favors placement of boron at the 1 position. This is because the inductive and conjugative effects of halogen and boron act in an opposite manner (as depicted below). Consequently, hydroboration occurs at the site of greatest electron availability, which results primarily due to mesomeric effect.

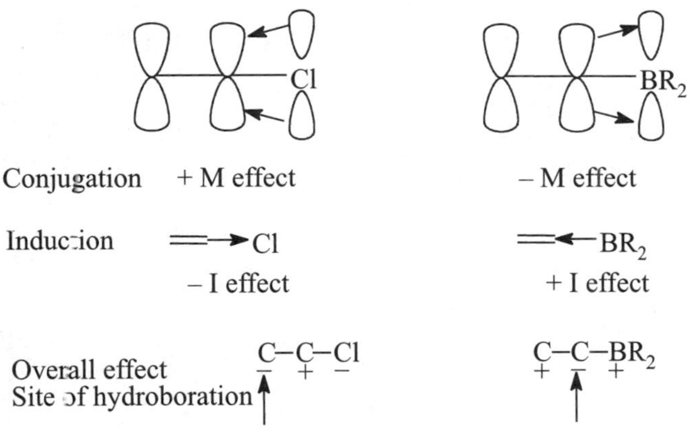

The contrasting regiochemistry of hydroboration for 1-chloroalkynes and 1-chloroalkenes is explained that the +M effect [3], in the chloroalkyne is a small one relative to the stronger charge pattern of the alkyne moiety, which places more electron density at the 1 position.

The steric and electronic effects of halogen substituents bring about dramatic rate decrease in the order Cl > Br > I, the trend one expects if the induction by

Table 4.17 Rate constants for the reduction of representative aldehydes and ketones with $(9\text{-BBN})_2$ at 25 °C [1]

Compound	$10^4 k_1$, s^{-1}			$10^4 k_{3/2}$, $M^{-1/2}\,s^{-1}$
	THF	CCl_4	C_6H_{12}	THF
Hexanal	14.2	1.42	1.41	
Propanal	12.8			
2-Methylpropanal	13.2			
2,2-Dimethylpropanal	11.7			
2-Phenylpropanal	12.4			
Benzaldehyde	13.7			
p-Methoxybenzaldehyde	13.1			
p-Chlorobenzaldehyde	15.2			
Cyclohexanone	14.4	1.41	1.49	
2-Methylcyclohexanone	12.8			
Cyclopentanone	11.7			
Cycloheptanone	12.7			
Acetone	13.7			
Norbornanone	12.7			
Acetophenone	12.8			
p-Methylacetophenone	11.9			
p-Methoxyacetophenone	11.9			
2,4-Dimethyl-3-pentanone				0.98

halogens are important. Since the iodine has the smallest effect, the electronic effect must outweigh the steric effects in cases of bromine and chlorine.

Table 4.18 Relative reactivities for the reduction of representative aldehydes and ketones with $(9\text{-BBN})_2$ in THF at 25 °C [1]

	Relative rate	
Compound	1-Hexene = 100	Hexanal = 100
p-Methoxybenzaldehyde	1,154	142
p-Tolualdehyde	911	112
Hexanal	813	100
Benzaldehyde	672	87
p-Chlorobenzaldehyde	494	64
2,2-Dimethylpropanal	479	62
Cyclohexanone	197	25.5
Cyclopentanone	51	6.61
p-Methoxyacetophenone	32.1	4.15
Cycloheptanone	22	2.85
Norbornanone	19.6	2.54
p-Methylacetophenone	16.8	2.18
Acetone	13.8	1.78
2-Butanone	11.3	1.46
Acetophenone	9.8	1.27
2-Heptanone	9	1.17
3-Methyl-2-butanone	8.9	1.15
Benzophenone	7.2	0.93
p-Chloroacetophenone	7	0.9
3,3-Dimethyl-2-butanone	6.9	0.89
Cyclooctanone	3.9	0.5
Camphor	3.6	0.46
3-Pentanone	3	0.39
2,4-Dimethyl-3-pentanone	0.94	0.122

4.5 Reduction Kinetics

The kinetic reduction of a number of aldehydes and ketones with $(9\text{-BBN})_2$ is studied at 25 °C. The reduction of aldehydes and reactive ketones follows first-order kinetics (Table 4.17) [1], thus supporting the dissociation mechanism.

$$\text{9-BBN} + \text{Aldehyde/Ketone} \xrightarrow{k_2} \text{Product}$$

In THF solvent, a much larger first-order rate constant, similar to hydroboration [2], is observed. This solvent effect is attributed [2] to the direct attack of the THF molecule on $(9\text{-BBN})_2$, which breaks the B–H bridge bond [3] (Table 4.17) [1].

As in the case of hydroboration, the reduction rate of sterically hindered ketones such as 2,4-dimethyl-3-pentanone is considerably slower and exhibits three-halves-order kinetics, first order in ketone, and one-half order in $(9\text{-BBN})_2$.

Many ketones such as 2-methylcyclopentanone, 2-heptanone, 3-methyl-2-butanone, camphor, cyclooctanone, benzophenone, *p*-chloroacetophenone, 3,3-dimethyl-2-butanone, and 2-methyl-3-pentanone do not show either first-order or three-halves-order kinetics cleanly. Apparently, in these cases, the rates of dissociation of $(9\text{-BBN})_2$ and reduction by 9-BBN monomer are comparable, which leads to complex kinetic behavior.

Although many aldehydes and ketones exhibit essentially the same first-order constant, they by no means have the same reactivity toward $(9\text{-BBN})_2$. The relative rates of reduction of representative aldehydes and ketones are determined by competitive method (Table 4.18) [1], since the kinetic study cannot reveal the effect of structure on reactivity.

A comparison behavior of the acidic reducing agent 9-BBN with that of extensively studied [4] nonacidic $NaBH_4$ has been conducted [6]. The data reveal that reduction of carbonyl compounds with 9-BBN is less susceptible to steric effects than is $NaBH_4$, in spite of the bulky nature of the former reagent. The lower susceptibility of 9-BBN to steric effect indicates that the boron atom of 9-BBN monomer is coordinated with carbonyl oxygen of the substrate during reduction; such a coordination will keep the bulky cyclooctyl moiety of 9-BBN away from the alkyl groups of the substrate including the acyclic ketones. The relative reduction rate factor of $(9\text{-BBN})_2$ and $NaBH_4$ is shown in Charts 4.7 and 4.8.

	$(9\text{-BBN})_2$	$NaBH_4$
Cyclopentanone	26	4.35
Cyclohexanone	100	100
Cycloheptanone	11	0.64
Cyclooctanone	2	0.049

Chart 4.7

	$(9\text{-BBN})_2$	$NaBH_4$
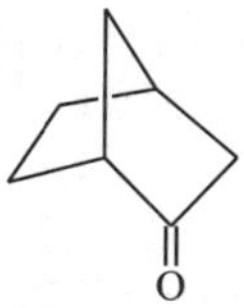	1.0	1.0
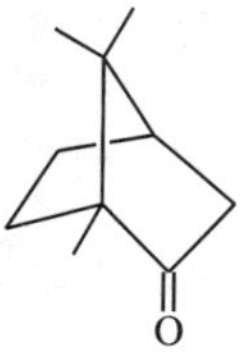	0.18	0.011

Chart 4.8

Increasing the steric hindrance on one side of the carbonyl function in ketones brings a modest decrease, while on both sides leads to a considerable rate decrease. For example, for two methyl groups at the α position in acetone, i.e., 3-methyl-2-butanone, there is only modest decrease as compared with 3-pentanone, having methyl groups on both the α positions of acetone. The relative rates are listed in Chart 4.9.

	Relative rate
Acetone	1
3-Methyl-2-butanone	0.67
3-Pentanone	0.2
2,4-Dimethyl-3-pentanone	0.069

Chart 4.9

Electron-withdrawing substituents decrease and electron-releasing ones increase the rate of reduction of aldehydes and ketones with 9-BBN (Chart 4.10) [1]. These facts strictly suggest that the boron atom of 9-BBN monomer is complexed with the carbonyl oxygen during the reduction process.

X	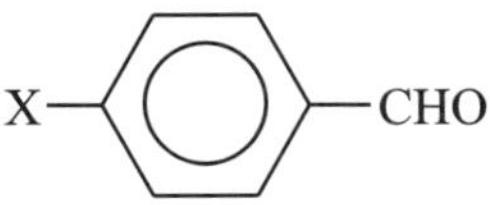	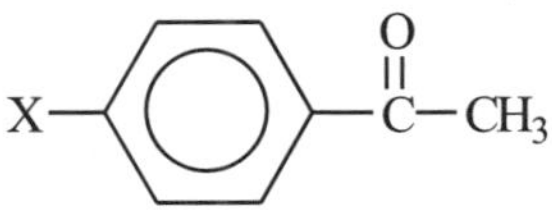
OMe	1.63	3.27
Me	1.29	1.72
H	1.00	1.00
Cl	0.74	0.71

Chart 4.10

The opposite electronic effects are exhibited in $\bar{B}H_4$ reduction [5].

B-3-Pinanyl-9-BBN is a highly effective asymmetric reducing agent for aldehydes [6] and for acetylenic ketones [7], with enantiomeric excess approaching 100%. Similarly, *B*-3-methyl-2-butyl-9-BBN shows high chemoselective abilities to reduce aldehydes in the presence of unhindered ketones [8]. The reaction is proposed to proceed *via* a cyclic mechanism reminiscent of the Meerwein-Ponndorf-Verley (MPV) path [9, 10]. Unlike most other organometallics or metal alkoxides that undergo similar type of reaction, the organoborane reduction is not complicated by oligomerization of the reagent, reversibility of the reaction, addition of the organometallic to the carbonyl, or other problems.

Table 4.19 Second-order rate constants for the reduction of *p*-substituted benzaldehydes with *B*-*n*-octyl-9-BBN [12]

Substituent	Temp (°C)	*k*, $M^{-1} s^{-1} \times 10^4$	Substituent	Temp (°C)	*k*, $M^{-1} s^{-1} \times 10^4$
NO_2	0	66.4±4.1	H	55	227±15
NO_2	15	173±18	CH_3	25	21.7±1.1
NO_2	25	311±41	OCH_3	25	8.19±0.64
CN	25	233±37	OCH_3	40	19.8±0.6
Cl	25	55.5±0.7	OCH_3	55	38.9±2.3
H	25	45.8±1.1	$N(CH_3)_2$	25	0.825[a]
H	40	105±3			

Table 4.20 Relative rate constants for the reduction of *p*-substituted benzaldehydes with *B*-3-pinanyl-9-BBN and *B*-*n*-octyl-9-BBN at 25 °C [12]

	k_X/k_H	
Substituent	*B*-3-Pinayl-9-BBN[a]	*B*-*n*-Octyl-9-BBN[b]
NO_2	2.3	6.8
CN	–	5.1
Cl	1.8	1.2
H	1	1
CH_3	0.84	0.47
OCH_3	0.35	0.18
$N(CH_3)_2$	0.17	0.018

[a] From competition experiments.
[b] From data in Table 4.19.

However, with slower-reacting and sterically hindered carbonyl compounds, the organoborane reduction may take an alternate pathway involving a prior dehydroboration process (Eqs. 4.18, 4.19) [11].

$$R_3B \longrightarrow R_2BH + \text{Olefin} \tag{4.18}$$

$$R_2BH + O{=}CHR \longrightarrow R_2BOCH_2R \tag{4.19}$$

The rate of reduction of aldehydes with *B*-alkyl-9-BBN compounds is greatly dependent on the structure of the alkyl group on boron [9]. In general, rate of reduction is very fast when alkyl group has tertiary β-hydride, while with a primary β-hydride provides a very slow rate. For example, *B*-3-pinanyl-9-BBN reacts rapidly and exothermally at room temperature and the VPC and NMR techniques have proven futile, since the reaction is so fast. Consequently, the

Table 4.21 Activation parameter for the reduction of *p*-substituted benzaldehydes with *B*-*n*-octyl-9-BBN [12]

Substituent	E_{act} kcal mol^{-1}	$\Delta G_{298}^{\ddagger}$ kcal mol^{-1}	$\Delta H_{298}^{\ddagger}$ kcal mol^{-1}	$\Delta S^{\ddagger}$ cal deg^{-1} mol^{-1}
NO_2	10.1±1.7	22.3±3.5	9.5±1.7	−43±6
H	10.4±0.8	23.2±1.7	9.8±0.8	−45±3
OCH_3	10.5±1	23.7±2.2	9.1±1	−49±4

slower reacting *B*-*n*-octyl-9-BBN is examined [12], where the formation of 1-octene (Eq. 4.20) is conveniently monitored by VPC.

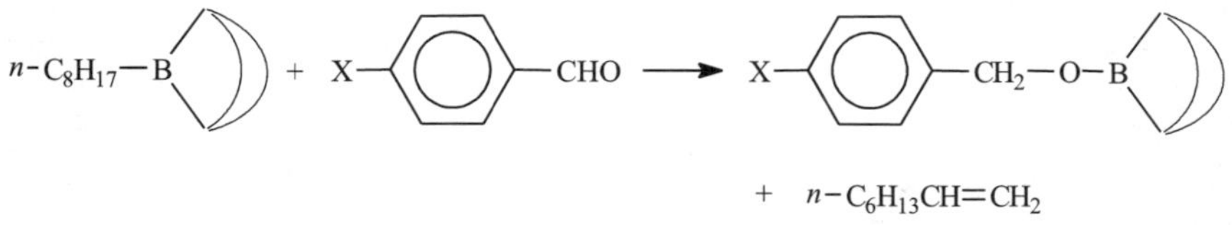

(4.20)

The reaction of *B*-*n*-octyl-9-BBN shows second-order rate constants with *para*-substituted benzaldehydes (Table 4.19) [12].

The relative rate constants for the reduction at 25 °C are given in Table 4.20 [12]. As the reductions with *B*-3-pinanyl-9-BBN are too fast, the relative rates are obtained by competition experiments.

A Hammett plot of the log of the relative rate constants versus σ+ [13] gives a linear correlation with σ of +1.03 (correlation coefficient of 0.9995).

The activation parameters are listed in Table 4.21 [12].

The reaction of tri-*n*-butylborane with benzaldehydes is reported to require forcing conditions of 100–150 °C [10], though thermodynamically, reaction should be very favorable as the conversion of B–C bond to B–O bond is energetically favorable. However, activation parameters (Table 4.21) suggest that the major barrier to this reaction is entropy. The larger negative entropies of activation (–43 to –49 eV) indicate that reduction proceeds through a highly ordered transition state.

It is postulated that the reduction of benzaldehydes proceeds predominately through a cyclic process [9]. This conclusion is based on the second-order kinetic data, the change in rate with structural and electronic changes in the aldehydes, and the results of asymmetric reduction. However, greater steric effects, both in the organoboranes and substrate, lead predominantly to the dehydroboration process.

To determine the rate of reduction, it is important that the B–C–C–H group forms a *syn* coplanar arrangement [9]. For example, *B*-cyclopentyl and *B*-*trans*-2-methyl-cyclopentyl-9-BBN react very fast with benzaldehyde, whereas *B*-cyclohexyl and *B*-*trans*-2-methyl-cyclohexyl-9-BBN are usually slow reducing agents. To achieve the *syn* coplanar B–C–C–H arrangement, the boat cyclohexane conformation is favored; as in the case of *B*-4-isocaranyl-9-BBN, the reduction is rapid (Scheme 4.3) [12].

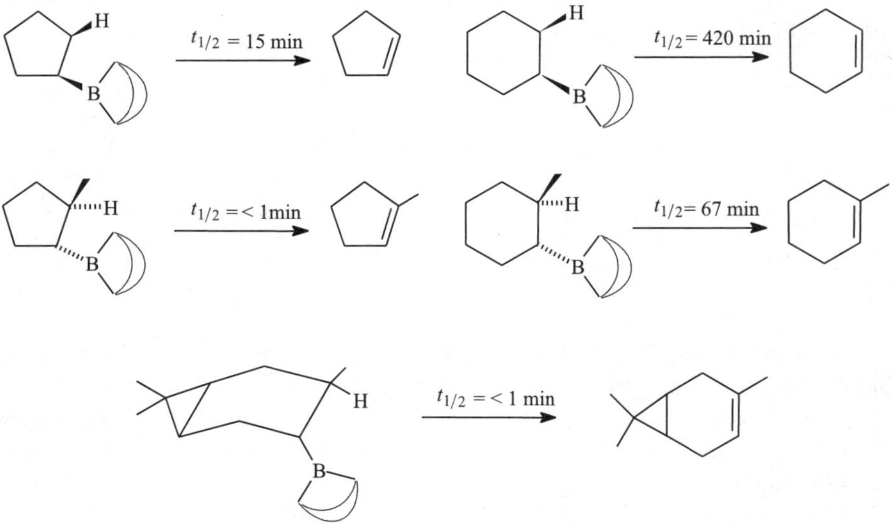

Scheme 4.3. Reaction of *B*-alkyl-9-BBN compounds with benzaldehyde to form olefin [5] ($t_{1/2}$ is the time for 50% completion of reaction at reflux THF (65 °C)).

The *syn* coplanar arrangement probably allows a maximum overlap of the developing π system of the displaced olefin. This need for coplanar B–C–C–H arrangement is also probably the cause for the large negative entropies.

The rate of reduction with *B*-*n*-octyl-9-BBN of *p*-substituted benzaldehydes correlates with σ^+ (ρ + 1.03) [12]. Relative rates of *p*-substituted benzaldehydes with *B*-3-pinanyl-9-BBN gave σ of + 0.49. Electron-withdrawing groups increase the rate of reduction of substituted benzaldehydes. This effect is presumably also important for alkynyl ketones [7], where the small steric size of the acetylene and its negative inductive effect can combine to make the reduction possible.

It is postulated that the reaction involves an organoborane–carbonyl oxygen complex as an intermediate in the reaction. The hydride transfer, k_2, is presumably the rate-determining step (Eq. 4.21) [12].

Initial concentration		Rate constants	
$(9\text{-BBN})_2$	Py	10^2k_1, s^{-1}	10^2k_2, $M^{-1}\,s^{-1}$
0.065	0.26	0.209	1.13
0.065	0.13	0.069	1.2

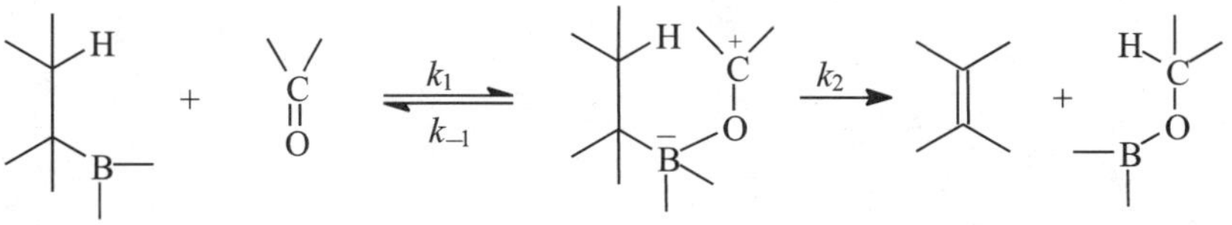

(4.21)

4.6 Kinetics of Complex Formation

The complex formation of $(9\text{-BBN})_2$ with representative amines, which are more powerful nucleophiles than alcohols, are studied [1]. The symmetrical cleavage of the bridged B–H bonds is reported to yield 9-BBN-amine complexes [2]. The ^{11}B NMR spectrum of 9-BBN-Py [3] and ^{13}C NMR spectra of representative 9-BBN-amine reveal the formation of these complexes [4].

Table 4.22 Kinetic data for the complex formation of amines with $(9\text{-BBN})_2$ in cyclohexane at 25 °C [1]

Amine	$10^2 k_2$, $M^{-1}\,s^{-1}$	$10^4 k_1$, s^{-1}
Pyrrolidine	27.5	
Piperidine	5.63	
n-Butylamine	1.73	
Neopentylamine	1.4	
Pyridine	1.2	
Benzylamine	0.95	
Cyclohexylamine	0.51	
Cycloheptylamine	0.45	
Cyclopentylamine	0.44	
sec-Butylamine[a]	–	–
2-Methylpyridine	–	1.46
N-Methylpyrrolidine	–	1.4
di-*n*-Butylamine	–	1.61
Quinuclidine	–	1.37
t-Butylamine	–	1.42
Aniline	–	1.37

[a] The data do not fit the integrator rate expression of either second or first order.

Table 4.23 Effect of solvent on the complexation of amines with $(9\text{-BBN})_2$ at 25 °C [1]

Amine	Cyclohexane	Benzene	THF
Second-order kinetics, $k\times10^2$ M^{-1} s^{-1}			
Pyrrolidine	27.5	86.3	101.5
Piperidine	5.63	10.2	15.7
Pyridine	1.2	1.8	4.03
First-order kinetics, $k\times10^4$ s^{-1}			
2-Methylpyridine	1.46	2.16	14.1
N-Methylpyrrolidine	1.4	1.92	13.3
Di-*n*-butylamine	1.61	2.37	14.2

$$(9\text{-BBN})_2 + \text{amine} \underset{k'_{-2}}{\overset{k'_2}{\rightleftharpoons}} 9\text{-BBN-amine} + 9\text{-BBN} \quad (4.22)$$

$$9\text{-BBN} + \text{amine} \xrightleftharpoons{\text{fast}} 9\text{-BBN-amine} \quad (4.23)$$

Unlike hydroboration or reduction [5], the complex formation of pyridine (0.130 M) with $(9\text{-BBN})_2$ (0.065 M) in cyclohexane at 25 °C displays second-order kinetics. Moreover, the rate of reaction is faster than the rate of dissociation of $(9\text{-BBN})_2$. The first half-life of the reaction is only 10.7 min ($k_2 = 1.20\times10^{-2}$ M^{-1} s^{-1}, Table 4.21) which is about eight times shorter than is the half-life of the dissociation of $(9\text{-BBN})_2$ ($t_{1/2} = 78$ min, $k_1 = 1.50\times10^{-4}$ s^{-1}).This suggests that the complex formation with Py cannot proceed through dissociation mechanism. Thus, bimolecular direct mechanism can explain the observed kinetic behavior (Eqs. 4.22, 4.23). As 9-BBN-Py complex formation is more than 99% complete, reverse reaction of Eqs. 4.22 and 4.23 are ignored [6]. Although dissociation and the direct attack mechanisms operate simultaneously, the latter one is the predominant pathway, as is evident by kinetic observations [1]. The second-order rate constants remain also unaffected by doubling the initial concentration of pyridine, whereas the first-order rate constants change very much.

The results of complexation of $(9\text{-BBN})_2$ by several other amines are given in the Table 4.22.

Sterically unhindered amines such as pyrrolidine, piperidine, *n*-butylamine, and *neo*-pentylamine, etc., all react faster than Py, and the reaction exhibits second-order kinetics, indicating that the rate-determining step involves the direct reaction between the dimer and the amine. Sterically more-hindered amines, such as *tert*-butylamine, di-*n*-butylamine, and quinuclidine, exhibit first-order kinetics, first order in $(9\text{-BBN})_2$. Apparently, in these amines the reaction proceeds by dissociation of the dimer, followed by the reaction of the amine with the monomer. *sec*-Butylamine displays intermediate kinetic behavior. Thus, the mechanism of this reaction is strongly affected by moderate changes in the steric requirement and also on the nucleophilicity of the amines.

To understand the role of solvents in the reaction, complexation reactions of some amines are studied [1] in cyclohexane, benzene, and THF (Table 4.23). The reaction proceeds faster in THF and benzene, irrespective of the mechanism. The rate of complexation is ~10 times more in THF than in noncomplexing cyclohexane solvent. THF assists in breaking up the bridged B–H bonds of the $(9\text{-BBN})_2$ [7].

The increase in the rates of complexations proceeding by the direct attack mechanism is about four times for THF, compared with cyclohexane, as that observed in reaction proceeding through the dissociation mechanism.

Consequently, the systematic kinetic studies [1] of the reaction of alkenes, alkynes, aldehydes and ketones, alcohols, and amines with $(9\text{-BBN})_2$ reveal the following spectrum of mechanistic pathways:

1. Second-order kinetics:

$$(9\text{-BBN})_2 + \text{N} \xrightarrow{k_2} 9\text{-BBN}\cdot\text{N} + 9\text{-BBN}$$

$$9\text{-BBN} + \text{N} \xrightarrow{\text{fast}} 9\text{-BBN}\cdot\text{N}$$

2. First-order kinetics:

$$(9\text{-BBN})_2 \xrightarrow{k_1} 2\ 9\text{-BBN}$$

$$9\text{-BBN} + \text{N} \xrightarrow{\text{fast}} \text{product}$$

3. Three-halves-order kinetics:

$$(9\text{-BBN})_2 \underset{k_{-1}}{\overset{k_1}{\rightleftharpoons}} 2\ 9\text{-BBN}$$

$$9\text{-BBN} + \text{N} \xrightarrow{\text{slow}} \text{product}$$

Less hindered amines react by mechanism (1). More hindered amines follow mechanism (2). Reactive alkenes react by mechanism (2). Less reactive alkenes follow mechanism (3). And, in certain cases, intermediate kinetic behavior is observed. Clearly, the nucleophilicity of the substrate plays a very important role in determining the reaction pathway.

4.7 Kinetics of Protonolysis

The kinetic of protonolysis of $(9\text{-BBN})_2$ with representative alcohols and phenols are followed [1] in THF and CCl_4 at 25 °C. The protonolysis of $(9\text{-BBN})_2$ in CCl_4 with *tert*-butyl alcohol exhibits first-order kinetics, first order in $(9\text{-BBN})_2$, and is in good agreement with those obtained in alkenes [1], alkynes [1], and reduction of aldehydes and ketones [2].

$$(9\text{-BBN})_2 \underset{k_{-1}}{\overset{k_1}{\rightleftharpoons}} 2\ 9\text{-BBN}$$

$$9\text{-BBN} + t\text{-BuOH} \xrightarrow{\text{fast}} t\text{-BuO-9-BBN} + H_2$$

$$(9\text{-BBN})_2 + CH_3OH \xrightarrow{k_2} B\text{-}CH_3O\text{-9-BBN} + 9\text{-BBN} + H_2 \quad (4.24)$$

$$9\text{-BBN} + CH_3OH \longrightarrow B\text{-}CH_3O\text{-9-BBN} + H_2 \quad (4.25)$$

However, with unhindered alcohols (in carbon tetrachloride) such as methanol and ethanol, there is a competition between the dissociation pathway and a pathway involving direct attack of the alcohol on the dimer (Eqs. 4.24, 4.25). The reaction when carried out under pseudounimolecular conditions (with excess ROH) at several different concentrations, and k_{obsd} plotted versus [ROH], linear plots are realized. The rate constants for the dissociation pathways (k_1) and the direct attack pathway (k_2) can be evaluated from these plots.

The rate equation for the overall process is Eq. 4.26.

$$-\frac{d[(9\text{-BBN})_2]}{dt} = k_1[(9\text{-BBN})_2] + k_2[CH_3OH][(9\text{-BBN})_2 \quad (4.26)$$

As mentioned (Table 4.24), the k_1 values are essentially the same for all three alcohols and agree well with rate constants for the dissociation of $(9\text{-BBN})_2$ in noncomplexing solvents such as hexane, CCl_4, and cyclohexane. As expected, k_2 decreases in the order methanol > ethanol > isopropyl alcohol (Table 4.24) [1].

Table 4.24 Kinetic data for the protonolysis of $(9\text{-BBN})_2$ by unhindered alcohols in CCl_4 at 25 °C [1]

Alcohol	k_1, 10^{-4} s^{-1}	k_2, 10^{-5} M^{-1} s^{-1}
Methyl	1.45	7
Ethyl	1.44	3.3
Isopropyl	1.42	1

The rate constants were obtained under pseudo–first-order conditions, and the k_{obsd} were plotted versus the initial concentration of the alcohol. The linear plots had correlation coefficient better than 0.995.

The protonolysis of $(9\text{-BBN})_2$ in THF with hindered and unhindered alcohols and phenols follow first-order kinetics (Table 4.25) [1], indicating that the dominant pathway involves prior dissociation of $(9\text{-BBN})_2$.

Table 4.26 Relative rates for the protonolysis of $(9\text{-BBN})_2$ by representative alcohols and phenols in THF at 25 °C [1]

	Relative rate	
Compound	1-Decene = 100	Methanol = 100
p-Nitrobenzyl alcohol	1,360	1,470
p-Chlorobenzyl alcohol	465	502
Benzyl alcohol	400	432
1-Decene	100	108
Methanol	92.6	100
p-Methoxybenzyl alcohol	84	90.7
Ethanol	62.9	67.9
Isopropyl alcohol	43.7	47.2
Cyclopentanol	42.9	46.3
Cyclohexanol	41.7	45
Cycloheptanol	35.3	38.1
tert-Butyl alcohol	24.7	26.7
2,3-Dimethyl-2-butanol	15.7	17
p-Methoxyphenol	13.4	14.5
Triethylcarbinol	12.5	13.5
o-Cresol	12.5	13.5
Phenol	12.1	13.1
2,6-Dimethylphenol	10.2	11
p-Nitrophenol	2.6	2.8
o-*tert*-Butylphenol	1.1	1.2
2,6-Diisopropylphenol	1.1	1.2
2,6-Di-*tert*-Butylphenol	0	0

Table 4.25 Kinetic data for the protonolysis of $(9\text{-BBN})_2$ with representative alcohols and phenol in THF at 25 °C [1]

Compound	k_1, $10^{-4}\ s^{-1}$	Compound	k_1, $10^{-4}\ s^{-1}$
Methanol[a]	15.4	*tri-n*-Octylcarbinol	14
1-Hexanol[a]	14.1	2,2,4-Trimethyl-3-pentanol	15
tert-Butyl alcohol	14.2	Phenol[b]	11.4
2,3-Dimethyl-2-butanol	15.2		

[a] Data from Brown HC, Krishnamurthy S, Yooon NM (1976) J Org Chem 41:1778.
[b] Shows slight intermediate behavior.

In order to understand the effects of structure on the protonolysis reaction, the relative rates of protonolysis of number of representative alcohols and phenols are determined in THF at 25 °C, utilizing competitive experiments. The major difficulty encountered here is rapid *B*-alkoxy-9-BBN product exchange with alcohols (Eq. 4.27).

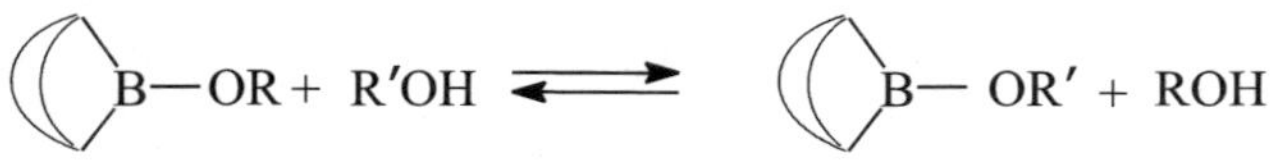

(4.27)

However, it is possible to compare the relative rate of protonolysis of $(9\text{-BBN})_2$ by an alcohol with that of hydroboration of an alkene of suitable reactivity, such as 1-decene, employing the competition method. Consequently, equimolar quantities of an alcohol and an alkene are allowed to react with an insufficient amount of 9-BBN, and the amount of H_2 evolved is measured. The amount of alcohol reacted is calculated from the hydrogen evolved. From this quantity; the amount of alkene reacted is deduced. The relative rates are then obtained by employing the following Ingold-Shaw expression [3].

$$\frac{k_{\text{alcohol}}}{k_{\text{alkene}}} = \frac{\ln[\text{alcohol}]_{\text{initial}} - \ln[\text{alcohol}]_{\text{final}}}{\ln[\text{alkene}]_{\text{initial}} - \ln[\text{alkene}]_{\text{final}}}$$

The data are summarized in Table 4.26 [1].

The data indicate the increasing steric hindrance in the alcohol decreases the protonolysis rate. The same trend is observed for *ortho*-substituted phenols.

CH_3OH	$(CH_3)_3COH$	$(C_2H_5)_3COH$
1.0	0.27	0.14

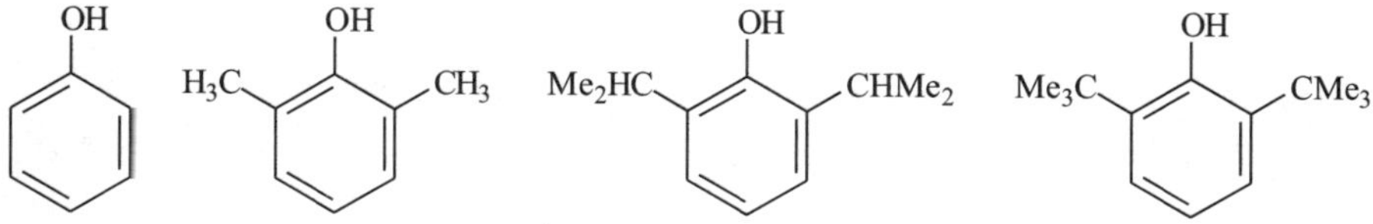

1.0 0.85 0.09 0

As far as electronic factors are concerned benzyl alcohol is considerably more reactive than methanol. This reflects that –I effect of phenyl substituent increases the acidity of the O–H bond.

CH_3OH	$PhCH_2OH$	PhOH
1.00	4.32	0.13

However, this cannot be sole factor for phenol, which is far more acidic, yet its rate of protonolysis is far slower. One of the probable reasons is that protonolysis proceeds through a prior coordination of oxygen atom of the alcohol with the boron.

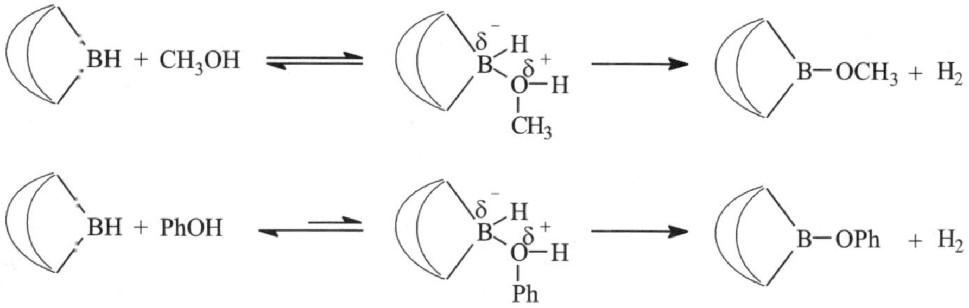

The observed rates thus seem to be a complex function of the relative ability of the oxygen atom of the alcohol or the phenol to coordinate with the boron and relative acidity of the contained proton. The much lower donor properties of the oxygen atom in phenol thus result in a markedly decreased formation the complex, resulting in a decreased rate of formation of H_2.

The effect of substitution both in benzyl alcohol and phenol are surprising and are found to have opposite effect.

In benzylic alcohols, electron-withdrawing substituents enhance and electron-releasing ones decrease, the rate of protonolysis, suggesting that the dissociated 9-BBN monomer forms a complex with alcohol which loses H_2.

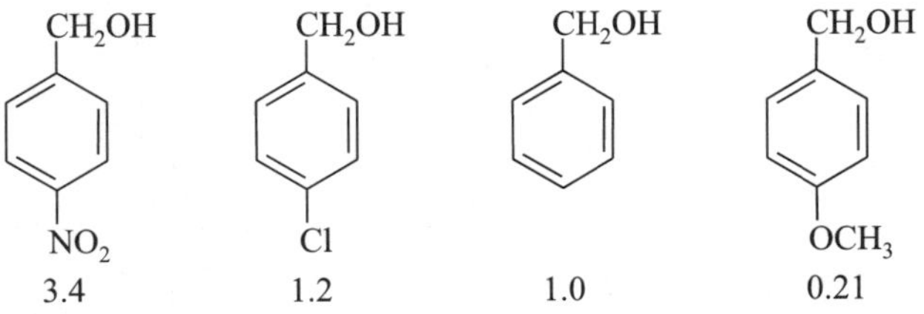

Phenols protonolyze $(9\text{-BBN})_2$ considerably slower than do alcohols. Moreover, opposite electronic effects are observed. In case of phenols, probably, the complex formation is difficult due to their poor nucleophilicity, leading to decreased rates of protonolysis.

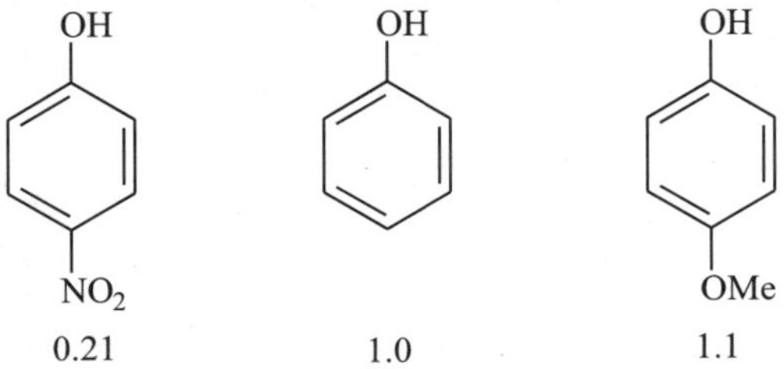

References

1. Pasto DJ, Lepeska B, Balasubramaniyan V (1972) J Am Chem Soc 94:6090
2. Brown HC, Liotta R, Scouten CG (1976) J Am Chem Soc 98:5297
3. Brown HC, Wang KK, Scouten CG (1980) Proc Natl Acad Sci USA 77:698
4. Wang KK, Brown HC (1982) J Am Chem Soc 104:7148

Section 4.1

1. Sia_2BH also exists predominantly as the dimer in both Et_2O and THF, see Brown HC, Klender G (1962) J Inorg Chem 1:204
2. Brown HC, Scouten CG, Wang KK (1979) J Org Chem 44:2589
3. Wang KK, Brown HC (1980) J Org Chem 45:5303
4. Brown HC, Kinghts EF, Scouten CG (1974) J Am Chem Soc 96:7765
5. Brown HC, Liotta R, Scouten CG (1976) J Am Chem Soc 98:5297
6. Brown HC, Nelson DJ, Scouten CG (1983) J Org Chem 48:641
7. (a) Johnson F (1965) J Am Chem Soc 87:5492; (b) Johnson F (1968) Chem Rev 68:375
8. Nelson DJ, Brown HC (1982) J Am Chem Soc 104:4907
9. Ingold CK (1969) Structure and mechanism in organic chemistry, 2nd edn. Bell, London
10. Graham GD, Freilich SC, Lipscomb WN (1981) J Am Chem Soc 103:2546

Section 4.2

1. Wang KK, Brown HC (1980) J Org Chem 45:5303

2. Wang KK, Scouten CG, Brown HC (1982) J Am Chem Soc 104:531
3. Brown HC, Liotta R, Scouten CG (1976) J Am Chem Soc 98:5297
4. Brown HC, Moerikofer AW (1963) J Am Chem Soc 85:2063; (b) Brown HC, Zwefel G (1961) J Am Chem Soc 83:3834
5. Brown HC, Campbell JB Jr (1980) J Org Chem 45:389
6. Brown CA, Coleman RA (1979) J Org Chem 44:2328
7. Zweifel G, Clark GM, Polston NL (1971) J Am Chem Soc 93:3395
8. Brown HC, Scouten CG, Liotta R (1979) J Am Chem Soc 101:96
9. Kuo SS (1972) Computer applications of numerical methods. Addison-Wesley, Reading, Massachusettes

Section 4.3

1. Nelson DJ, Brown HC (1982) J Am Chem Soc 104:4907
2. Ingold CK (1969) Structure and mechanism in organic chemistry, 2nd edn. Bell, London
3. Graham GD, Freilich SC, Lipscomb WN (1981) J Am Chem Soc 103:2546

Section 4.4

1. Wang KK, Scouten CG, Brown HC (1982) J Am Chem Soc 104:531
2. Nelson DJ, Blue CD, Brown HC (1982) J Am Chem Soc 104:4913
3. Delavarenne SU, Viehe HG (1969) Chemistry of acetylenes. Dekker, New York

Section 4.5

1. Brown HC, Wang KK, Chandrasekharan J (1983) J Am Chem Soc 105:2340
2. Brown HC, Wang KK, Scouten CG (1980) Proc Natl Acad Sci USA 77:698
3. Wang KK, Brown HC (1982) J Am Chem Soc 104:7148
4. (a) Brown HC, Wheeler OH, Ichikawa K (1957) Tetrahedron 1:214; (b) Brown HC, Ichikawa K (1957) ibid. 1:221; (c) Brown HC, Ichikawa K (1962) J Am Chem Soc 84:373; (d) Brown HC, Muzzio J (1996) ibid. 88:2811
5. Bowden K, Hardy M (1966) Tetrahedron 22:1169
6. (a) Midland MM, Tramontano A, Zderic SA (1977) J Am Chem Soc 99:5211; (b) Midland MM, Greer S, Tramontano A, Zderic SA (1979) ibid. 101:2352
7. Midland MM, McDowell DC, Hatch RL, Tramontano A (1980) J Am Chem Soc 102:867
8. Midland MM, Tramontano A (1978) J Org Chem 43:1470
9. Midland MM, Tramontano A, Zderic SA (1978) J Organomet Chem 156:203
10. Mikhailov BM, Bubnov Yu N, Kiselev VG (1966) J Gen Chem USSR (English translation) 36:65
11. (a) Midland MM, Petre JE, Zderic SA (1979) J Organomet Chem 182 C53; (b) Midland MM, Petre JE, Zderic SA, Kazubski A (1982) J Am Chem Soc 104:528
12. Midland MM, Zderic SA (1982) J Am Chem Soc 104:525
13. Brown HC, Kamoto Y (1958) J Am Chem Soc 80:4979

Section 4.6

1. Brown HC, Chandrasekharan J, Wang KK (1983) J Org Chem 48:3689
2. Brown HC, Kulkarni SU (1977) Inorg Chem 16:3090
3. Brown HC, Soderquist JA (1980) J Org Chem 45:846
4. Brown HC, Wang KK (1980) J Org Chem 45:1748

5. Brown HC, Chandrasekharan J, Wang KK (1983) J Org Chem 48:2901
6. Brown HC, Wang KK (1979) Recl Trav Chem Pays Bas 98:117
7. Wang KK, Brown HC (1982) J Am Chem Soc 104:7148

Section 4.7

1. Brown HC, Chandrasekharan J, Wang KK (1983) J Org Chem 48:2901
2. Brown HC, Wang KK, Chandrasekharan J (1983) J Am Chem Soc 105:2340
3. Ingold CK, Shaw FR (1927) J Chem Soc 130:2918

5 Hydroboration

5.1 Hydroboration of Alkenes

5.1.1 Hydroboration of Acyclic Alkenes

Simple alkenes, such as 1-hexene, undergo hydroboration with BH_3 to place 6% of the boron at the 2 position and 94% on the terminal position and this distribution is not significantly influenced by branching [1].

Moreover, the polyfunctional nature of BH_3, its relative low selectivity and its low steric requirements cause various difficulties. To overcome these problems, various types of borane derivatives have been developed [2]. The regioselectivity [2] of addition of these boranes to an alkene is dependent on both steric and electronic effects exerted by the substituents present on C=C and also on the bulkiness of the hydroborating agent [3].

However, the instability of most of the hydroborating agents and their derivatives has limited their use in hydroboration and their subsequent synthetic applications. On the other hand, both 9-BBN and *B*-R-9-BBN compounds exhibit remarkable thermal stabilities. The hydroboration with 9-BBN is carried out at 25 °C or at a higher temperature to achieve essentially complete hydroboration of alkenes of widely different structural types [4] as compared with disiamylborane, where the reaction is normally carried out at 0 °C. In general, the hydroboration with 9-BBN is carried out in THF as the solvent, but carbon tetrachloride, benzene, and hexane are also employed for the reaction.

The reactions of 9-BBN with terminal alkenes in THF are complete at room temperature in 2 h, whereas the internal alkenes undergo hydroboration at reflux (65 °C). Under these conditions almost all the alkenes are quantitatively converted to *B*-R-9-BBN. The hydroboration of tetrasubstituted 2,3-dimethyl-2-butene, alkene has been achieved in 8 h at 65 °C (Eq. 5.1) [5].

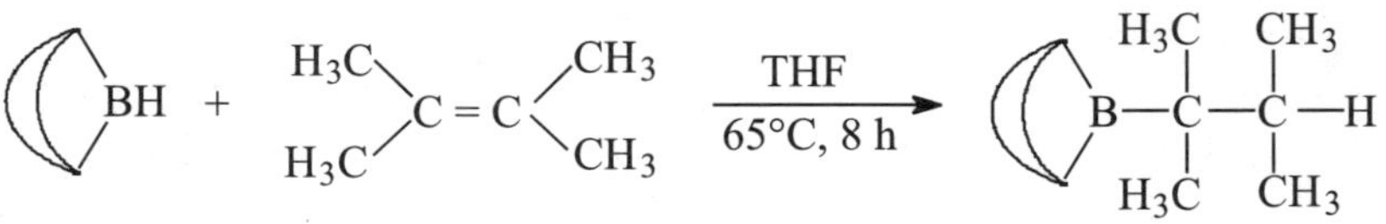

(5.1)

9-BBN exhibits a remarkable regiospecificity in the hydroboration of alkenes [6] (Chart 5.1).

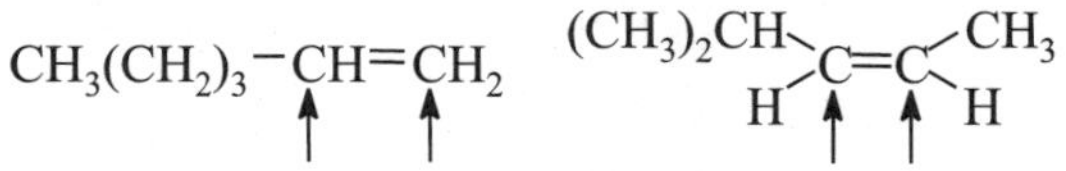

BH_3	6%	94%	43%	57%
Sia_2BH	1%	99%	3%	97%
9-BBN	–	>99.9%	0.2%	99.8%

Chart 5.1

The facile reaction of 9-BBN with highly substituted alkenes indicates that the reagent is less steric than disiamylborane. However, the heterocyclic structure of 9-BBN is rigid where the steric crowding in the transition state cannot be relieved by internal rotation. Thus, the rigid bicyclic 9-BBN reagent has proven to be more sensitive to subtle differences in steric environment than the more flexible, acyclic disiamylborane.

Trialkylboranes (*B*-R-9-BBN) derived from hydroboration of alkenes with 9-BBN are remarkably stable to thermal isomerization [6]. The alkylborane, derived from (*Z*)-hex-3-ene by hydroboration with 9-BBN, requires the heating at 150 °C for 163 h to attain the equilibrium distribution of boron along the hexyl chain. At the same temperature the isomerization of the alkylborane derived from (*Z*)-hex-3-ene using borane–THF is complete in 1 h.

The relative reactivities of various olefins toward hydroboration by 9-BBN in THF solution have been determined [7], and the results are illustrated below:

5.1.1.1
Terminal Olefins

The rate of hydroboration of 1-alkene ($RCH=CH_2$) is nearly independent of the length of the straight chain. However, branching in the R group, decreases the rate of hydroboration:

	$CH_3(CH_2)_3-CH=CH_2$	$CH_3CH(CH_3)CH=CH_2$	$CH_3C(CH_3)_2CH=CH_2$
9-BBN	1.00	0.50	0.23
Sia_2BH	1.00	0.57	0.047

Branching at position more remote from the double bond, such as at position C-4, has little or no effect on the rate.

5.1.1.2
Internal Olefins

The rate of 9-BBN hydroboration of *cis*-2-pentene is 100 times slower than that of 1-hexene.

	$CH_3(CH_2)_3CH=CH_2$	$CH_3CH_2C(H)=C(H)CH_3$
9-BBN	100	1.00
Sia_2BH	50	1.00

One-side branching of internal olefins only moderately reduces the rate. The behavior follows from the fact that 9-BBN reacts almost exclusively at the less hindered site of the double bond.

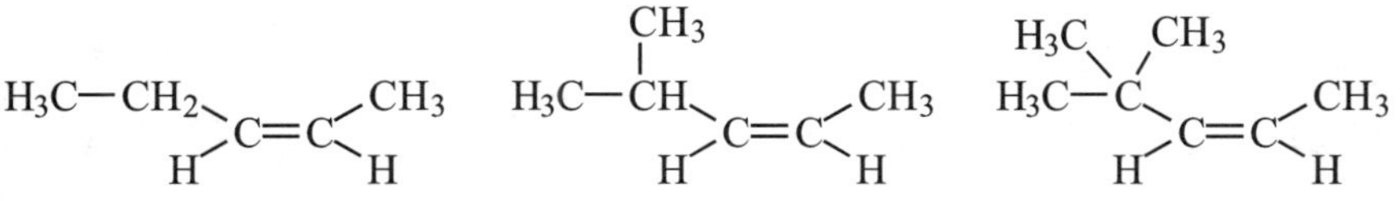

9-BBN	1.00	0.61	0.59

The branching on both sides of the double bond drops the rate of hydroboration sharply.

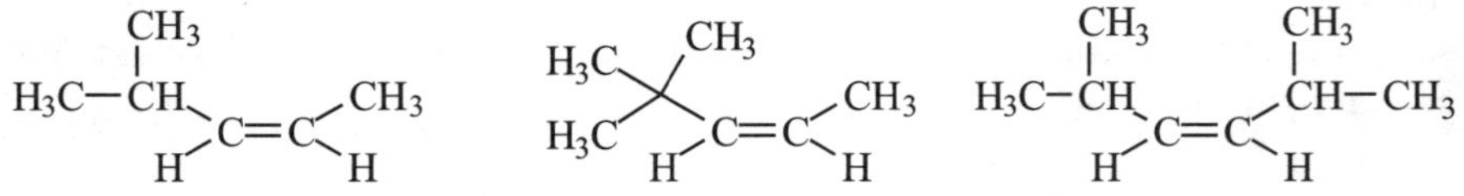

9-BBN	160	100	1.00
Sia_2BH	6.25	1.00	1.25

5.1.1.3
cis and *trans* Isomers

The *cis*-alkenes undergo hydroboration with disiamylborane at a rate considerably faster than that of *trans*-alkenes. However, the reverse is true for 9-BBN, although, the selectivity is low—only a factor of 2 or 3.

9-BBN	1.00	2.16
Sia_2BH	10.00	1.00

This phenomenon is general and holds for other isomers with highly branched alkyl groups.

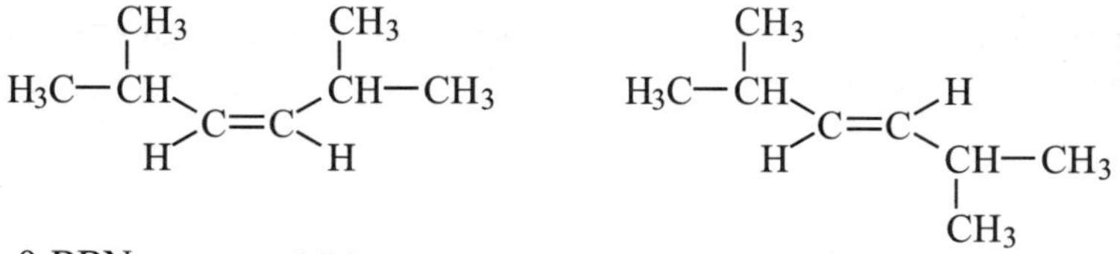

9-BBN	1.00	2.90
Sia_2BH	2.50	1.00

The relative rates of some *cis–trans* pairs are summarized in Table 5.1 [8].

Table 5.1 Relative reactivities and rate constants for the hydroboration *via* 9-BBN of *cis–trans* pairs at 25 °C in THF [8]

Alkene	Structure	Relative reaction (1-hexene = 100)	k_{cis}/k_{trans}
1-Hexene		100	
cis-3-Hexene		0.68	
trans-3-Hexene	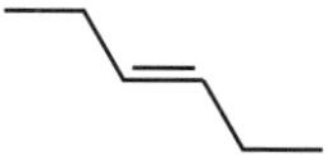	0.32	2.1
cis-4-Methyl-2-pentene	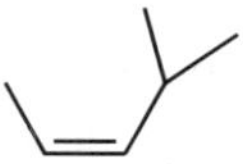	0.53	3.3
trans-4-Methyl-2-pentene	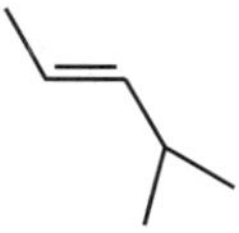	0.16	
cis-1-Phenylpropene		0.024	0.39
trans-1-Phenylpropene	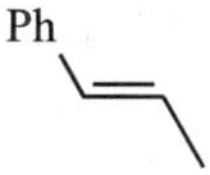	0.062	
cis-2,5-Dimethyl-3-hexene	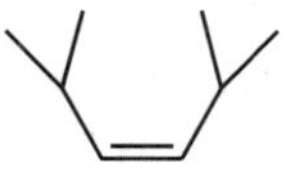	0.0017	0.076
trans-2,5-Dimethyl-3-hexene	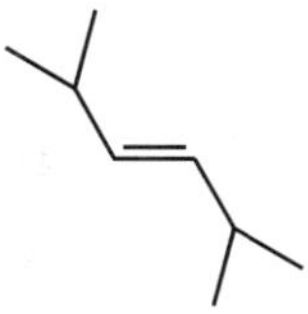	0.022	

5.1.1.4
Effect of α-Methyl Substituents

An α-methyl substituent increases the rate toward 9-BBN over that of the parent 1-alkene due to hyperconjugative affect. However, opposite is true for disiamylborane.

The ethyl substituent, on the other hand, increases the steric interactions with the 9-BBN moiety.

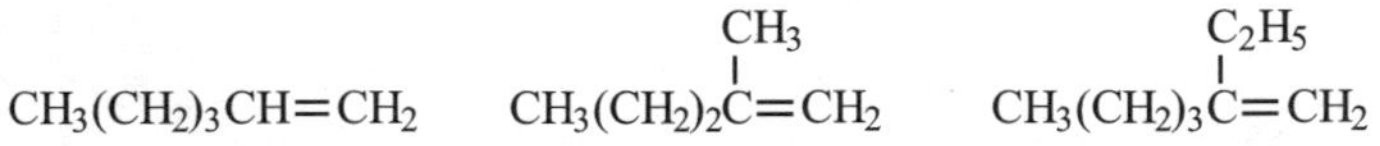

9-BBN	1.00	1.96	1.68
Sia_2BH	1.00	0.049	—

The electronic factor thus controls in the case of 9-BBN, which has the relative openness of the boron atom. The inductive effect of the methyl group increases the availability of electron of the double bond. The transition state involves the development of electron deficiency at C-2, with partial hydridic character at the >B–H moiety.

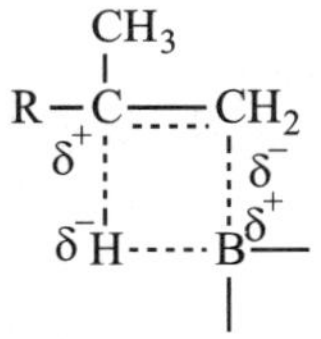

Transition state

The much greater steric congestions in disiamylborane must swamp these small electronic contributions, and the steric factor prevails.

Addition of an α-methyl group in butene with internal C=C has little effect on the rate with 9-BBN, whereas the rate with disiamylborane is very low. In 2,3-dimethyl-2-butene there is drastic drop in rate with 9-BBN, and disiamylborane fails to react.

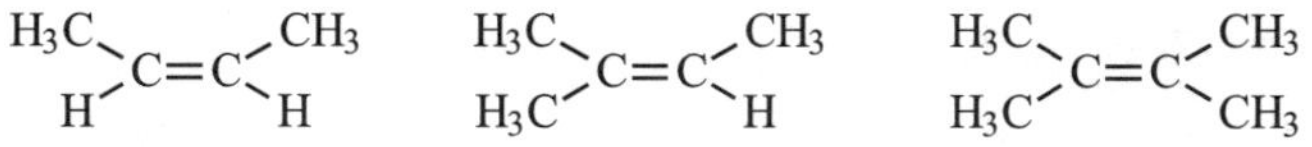

9-BBN	1.00	0.90	0.0064

5.1.1.5
Effect of α-Conjugated Substituents

Both aromatic and cyclopropyl groups conjugate strongly with an electron-deficient center. Styrene, the simplest olefin with phenyl substituent, is much less reactive than the simple 1-alkene.

9-BBN	1.00	40
Sia_2BH	1.00	36

On the other hand, cyclopropyl enhances the reactivity of the double bond.

9-BBN	2.3	1.0

The electron-donating substituent on aromatic ring enhances the rate, whereas the effect is opposite for electron-withdrawing substituent.

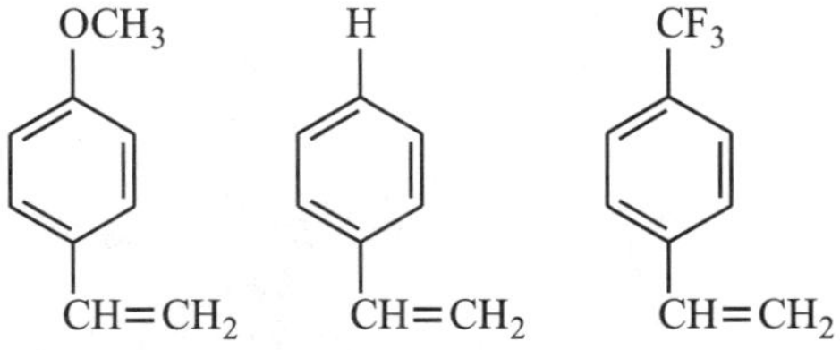

9-BBN	2.95	1.00	0.2
Sia_2BH	1.4	1.00	0.9

The relative rates of reaction of *p*-CH_3O, *p*-CH_3, unsubstituted, *p*-F, *p*-Cl, *m*-Cl, and *m*-NO_2 systems with 9-BBN are determined as shown in Chart 5.2 [9].

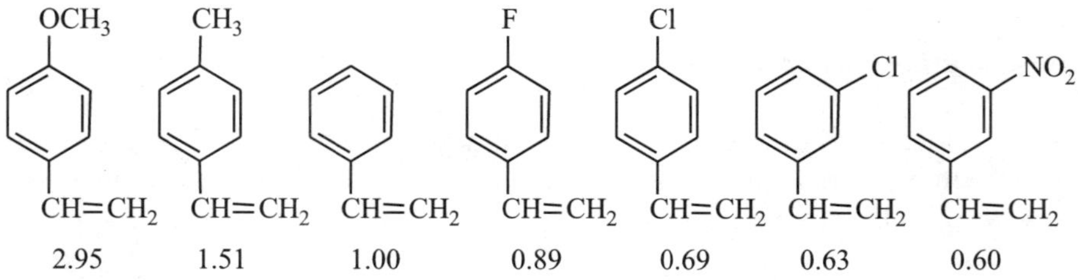

Chart 5.2

A Hammett plot, using σ^+, is reasonably linear with $\rho = -0.49$, which corresponds to a small buildup of positive charge at the α-carbon of the styrene and reaction with $B^{\delta+}-H^{\delta-}$ bond. In all these cases the β-phenylethyl alcohol is formed in more than 97.4% yield.

The electron-withdrawing substituents on cyclopropyl decrease the rate:

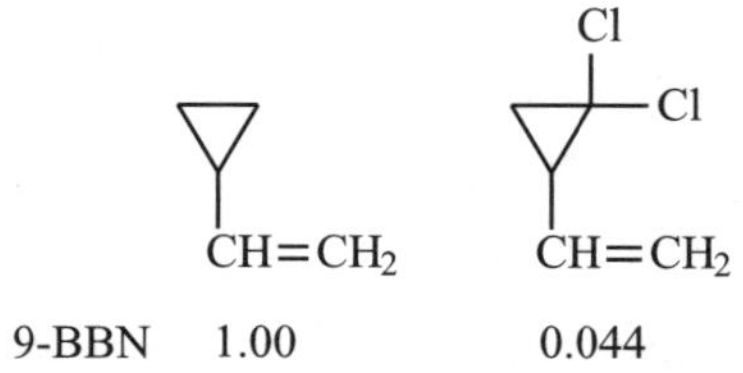

An α-methyl substituent enhances the rate of both styrene and vinylcyclopropane derivatives.

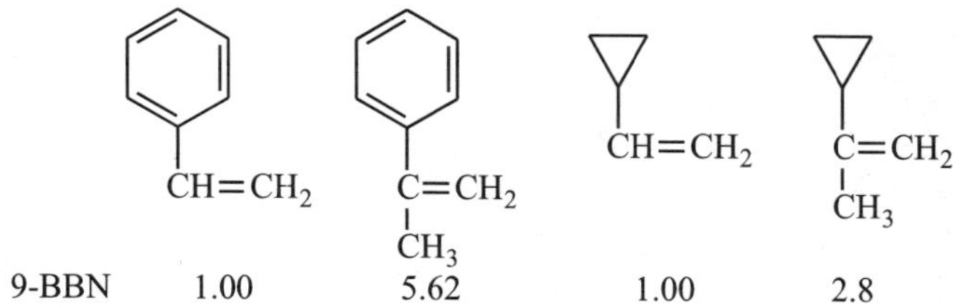

It has been found that cyclopropyl group largely overcomes the deactivating effect of phenyl ring.

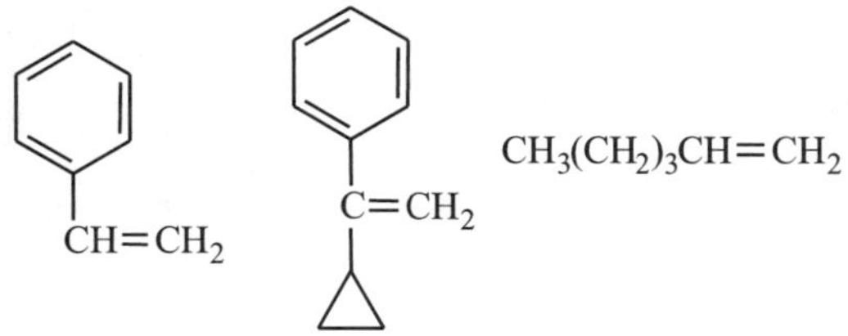

9-BBN 0.025 0.28 1.0

The hydroboration at the C-1 position in phenylpropene with 9-BBN is contrary to factors of steric effect. This is reasoned due to the combined effects of phenyl conjugation (–K) and methyl hyperconjugation (+K), which act to decrease the amount of electron density at the 2 position and to increase it at the 1 position (Fig. 5.1) [8].

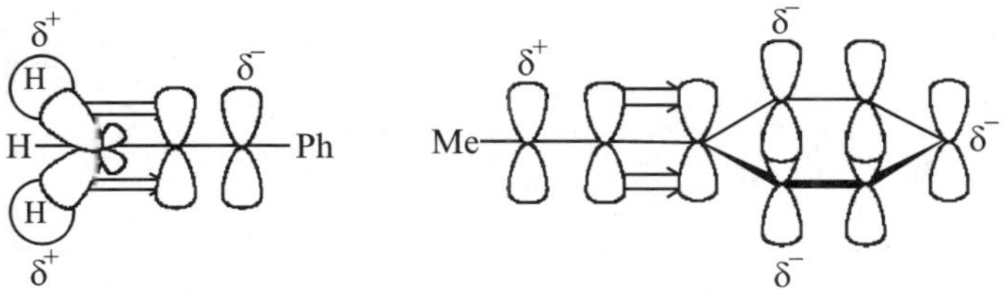

Fig. 5.1 Diagram to show combined effects of phenyl conjugation (–K) and methyl hyperconjugation (–K) in 1-phenylpropene

Because of nonpolar geometry in *cis*-1-phenylpropene, it displays enhanced steric effects as compared with *trans*-1-phenylpropene. Rotation of the phenyl group out of the plane of the π system in *cis*-1-phenylpropene also leads to a greater amount of hydroboration at the C-2 position (17.5%) than in the *trans* isomer (3.2%).

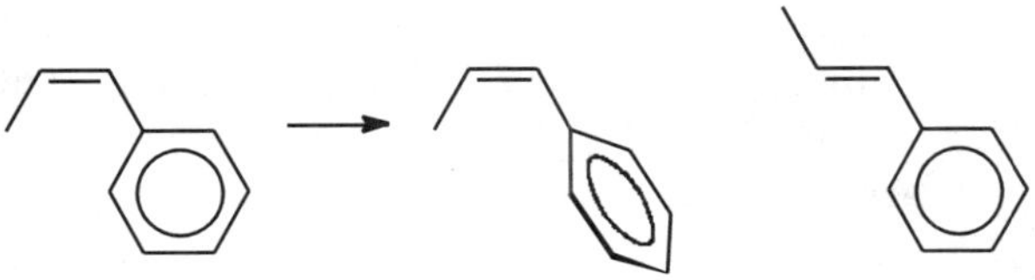

39 100

The relative reactivities of olefins with 9-BBN have been summarized in decreasing order (Table 5.2) [7].

Table 5.2 Relative reactivities of representative olefins toward 9-BBN in THF at 25 °C (1-hexene = 1) [7]

2-Cyclopropylpropene	6.37	Styrene	0.025
Vinylcyclopropane	2.3	1-Methylcyclopentene	0.015
2-Methyl-1-pentene	1.94	*trans*-3-Hexene	0.012
Norbornene	1.45	*cis*-2-Pentene	0.01
1-Dodecene	1.12	*cis*-2-Butene	9.5×10^{-3}
1-Octene	1.1	2-Methyl-2-butene	8.6×10^{-3}
1-Decene	1.08	*trans*-4-Methyl-2-pentene	7.6×10^{-3}
1-Hexene	1	*cis*-4-Methyl-2-pentene	6.1×10^{-3}
3-Methyl-1-butene	0.50	*cis*-3-Hexene	5.6×10^{-3}
p-Methoxystyrene	0.349	*p*-Trifluoromethylstyrene	5.2×10^{-3}
α-Cyclopropylstyrene	0.28	*cis*-4,4-Dimethyl-2-pentene	3.8×10^{-3}
3,3-Dimethyl-1-butene	0.236	Cyclohexene	6.7×10^{-4}
2-*p*-Tolylpropene	0.21	*trans*-Propenylbenzene	6.3×10^{-4}
α-Methylstyrene	0.14	*cis*-Propenylbenzene	2.5×10^{-4}
2,2-Dichlorocyclopropylethene	0.1	*trans*-2,5-Dimethyl-3-hexene	1.1×10^{-4}
Cycloheptene	0.076	1-Methylcyclohexene	1.1×10^{-4}
Cyclopentene	0.072	4-*tert*-Butylcyclohexene	7.7×10^{-5}
Cyclooctene	0.069	2,3-Dimethyl-2-butene	6.1×10^{-5}
		cis-2,5-Dimethyl-3-hexene	3.8×10^{-5}

The relative reactivities of alkenes are influenced by the steric, inductive, and hyperconjugative factors of the alkyl group. In Table 5.3, changes in relative reactivity toward hydroboration with 9-BBN has been summarized [10] for acyclic alkyl-substituted alkenes. The effect of alkyl group is obviously different if it is to the site of hydroboration than if it is β to the site. A few empirical rules have been drawn from the available data. An alkyl group α to the site of hydroboration (type B, Table 5.3) has an effect of decreasing the rate of reaction by an average of about 170, whereas the β-substituent (type A, Table 5.3) has an effect of increasing the rate of reaction by a factor of ≈1.87.

These empirical rules can be employed to demonstrate the regularity of the effects, e.g., two α- and one β-alkyl group at the site of hydroboration would figure this effect as (α effect)^2/(β effect) = $\downarrow 171^2/\uparrow 1.87 = \downarrow 1.56\times10^4$ and is rate reduction. The calculated effect is in good agreement with that experimentally observed, $\downarrow 1.64\times10^4$ (rate reduction), in comparing the relative reactivity of tetramethylethylene with that of 1-hexene.

5.1.2 Hydroboration of Cyclic Alkenes

The relative rates of hydroboration [7] of cyclic alkenes depend on the presence of strain [11] on the carbon–carbon double bond. The double bond of cyclopentene is considerably more strained than the double bond of cyclohexene, and this strain is responsible for higher reactivity of cyclopentene.

Table 5.3 A breakdown of the effects of alkyl groups that influence the change in reaction rate in alkenes [10]

Alkyl-substituted alkene	Parent	Factor by which relative reactivity changes	Type	Hyperconjugative effects of additional alkyl substituent(s) at position relative to site of hydroboration			Steric effect of additional alkyl groups at position relative to site of hydroboration	
				α	β	Inductive effects on π bond	α	β
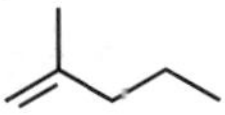	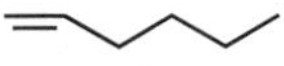	↑1.96	A	0	1	1	0	1
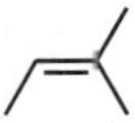		↓1.77						
		↓156	B	1	0	1	1	0
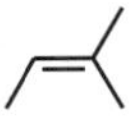	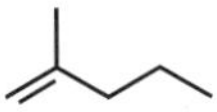	↓173						
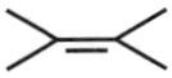	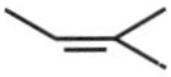	↓185						
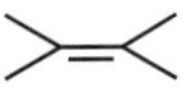	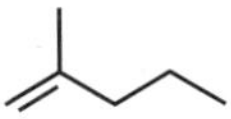	↓3.21×10^4	2 of B	2	0	2	2	0
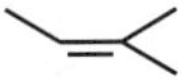		↓88.5	1 of A & 1 of B	1	1	2	1	1
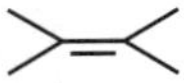		↓105						
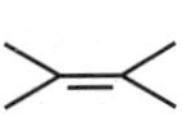	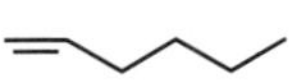	↓1.64×10^4	2 of B & 1 of A	2	1	3	2	1

9-BBN	1.00	107
Sia_2BH	1.00	140

Similarly, high-strained carbon–carbon double bonds [11] of cycloheptene and cyclooctene also exhibit [7] reactivities remarkably higher than that of cyclohexene toward both 9-BBN and Sia_2BH. However, Sia_2BH distinguishes considerably among the three strained olefins, whereas 9-BBN does not. Norbornene containing a highly strained double bond is even more reactive toward 9-BBN.

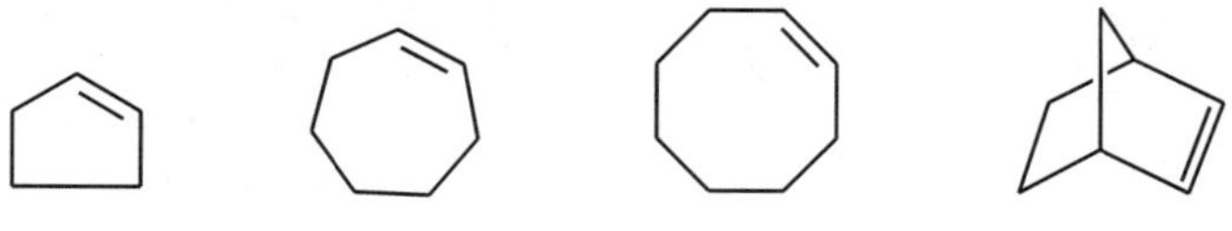

9-BBN	1.0	1.10	0.96	20.0
Sia_2BH	1.0	18.60	68.60	–

Unlike α-methyl–substituted acyclic terminal olefins, in the cyclic systems such as cyclopentene and cyclohexene, similar methyl substitution retards the rate of reaction [8]. Here, the steric interactions between the methyl group and the α-methylene unit causes rotation of the methyl group, resulting in the disruption of the hyperconjugation, which otherwise contributes to the rate enhancement. The disruption of conjugation is indicated by the representations as shown in Chart 5.3 [12].

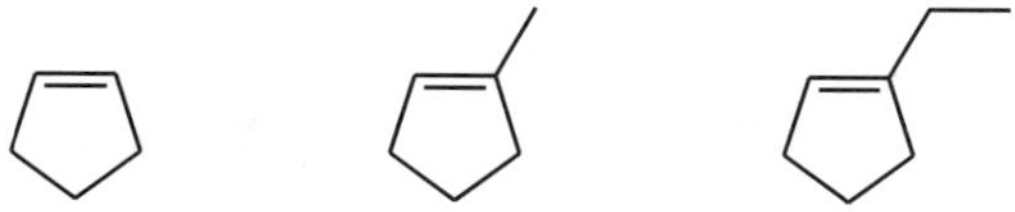

1.0	0.261	0.421

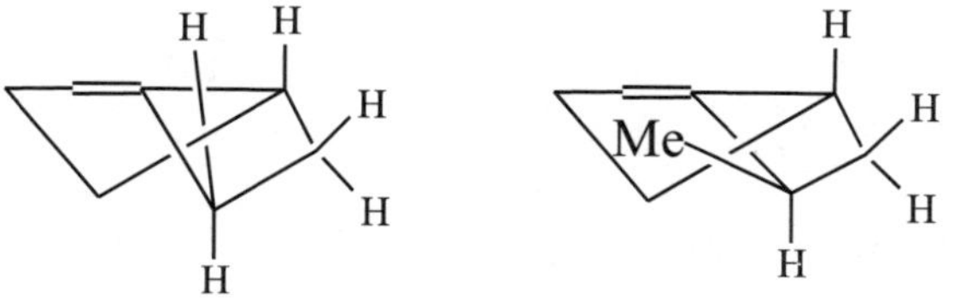

Chart 5.3

Ethyl substituent reduces the rate of retardation. In this case, the steric interaction with α-methylene causes rotation of the ethyl group so that terminal methyl group is moved away. Consequently, some conjugation is gained, and the rate is higher than that of methylcyclopentene. The diagram (Chart 5.3) indicates the interaction, which is responsible for the increased hyperconjugation.

9-BBN exhibits remarkable stereoselectivity toward cyclic olefins, e.g., norbornene reacts to afford *exo* isomer almost exclusively (Eq. 5.2) [13], whereas 7,7-dimethylnorbornene leads to the formation of *endo* isomer (Eq. 5.3) [14].

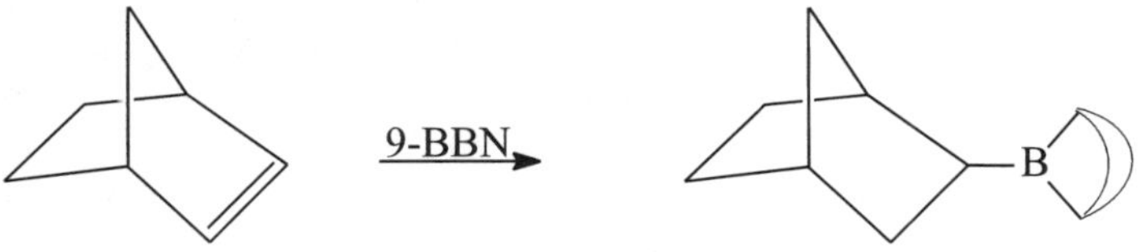

99.5% (*exo*)

(5.2)

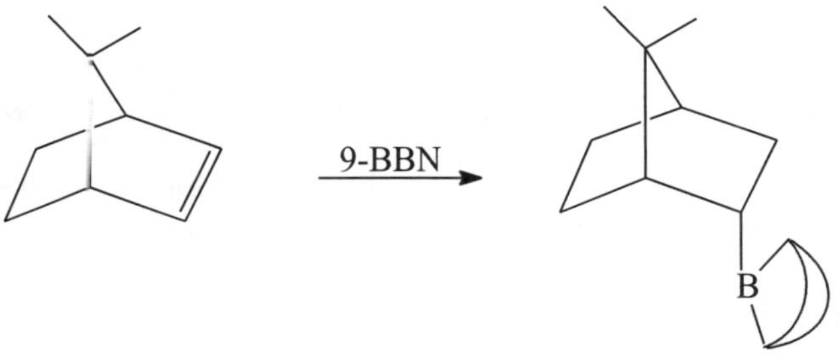

97% (*endo*)

(5.3)

It has been observed that in general, hydroboration of 1-alkylcycloalkenes with 9-BBN affords the corresponding *trans*-2-alkylcycloalkyl-9-BBN in essentially quantitative yields [15]. The reactions of 1-methycyclohexene [10] and 1-methylcyclooctene [7] are exemplified in Eqs. 5.4 and 5.5, respectively. However, the organoborane derived from 1-methylcyclooctene with borane or disiamylborane undergoes rapid isomerization [16].

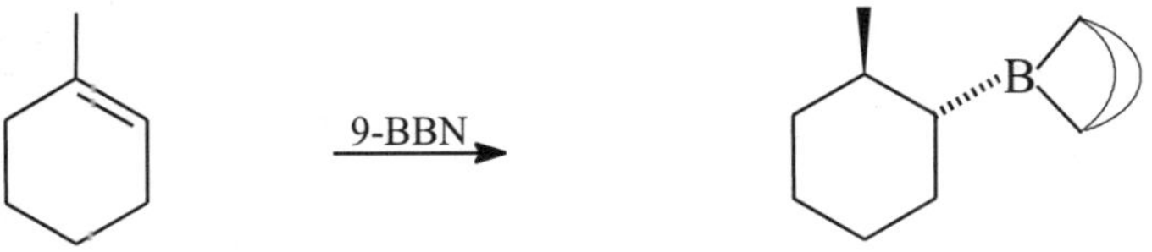

(5.4)

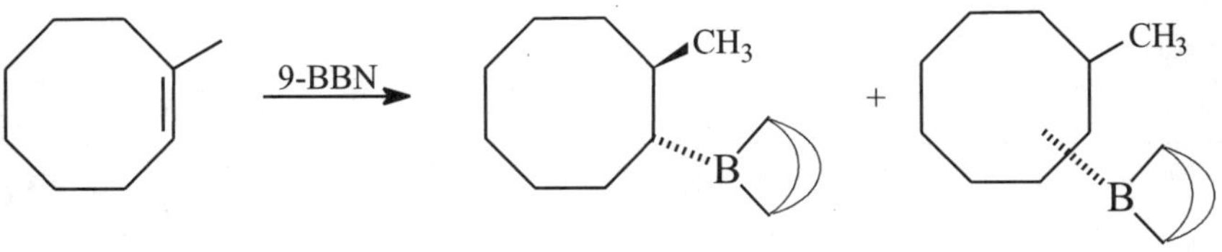

(5.5)

The hydroboration of 3-alkyl cycloalkenes with diborane or disiamylborane form significant amounts of the *cis*-1,2-isomer. However, it is important to note that none of this isomer is produced when 3-alkylcycloalkenes are hydroborated [17] with 9-BBN. This amazing selectivity shown by 9-BBN with 1-methyl and 3-methylcycloalkenes gives, exclusively, the *trans*-dimethylcycloalkyl-9-BBN derivative [18]. It is evident from 1,3-dimethylcyclopentene, which is exclusively converted into *t*-2,*t*-5-dimethylcyclopentanol (Eq. 5.6).

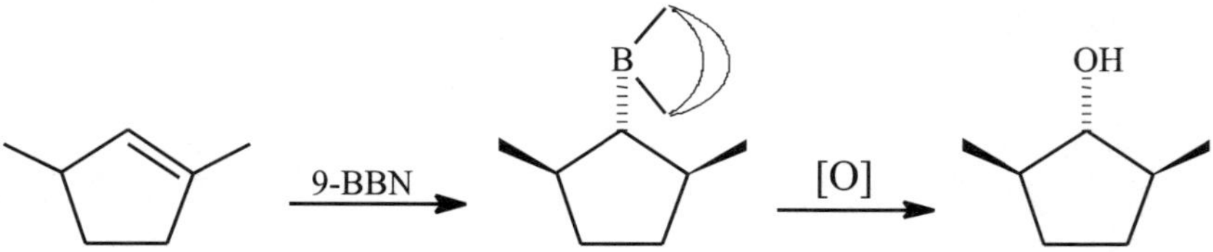

(5.6)

The time required to achieve essentially the complete hydroboration of olefins of different structural types with 9-BBN is summarized in Table 5.4 [4].

Table 5.4 Reaction of various olefins with 9-BBN in THF at 25 °C [4]

	Olefin reacted (%)		
Olefins[a]	2 h	4 h	24 h
1-Hexene	100		
3,3-Dimethyl-1-butene	100		
Styrene	94	100	
2-Methyl-1-pentene	100		
cis-3-Hexene	81	92	100
cis-4,4-Dimethyl-2-pentene	77	90	99
2-Methyl-2-butene	91		
2,3-Dimethyl-2-butene	11	15	33
Cyclopentene	100		
Cyclohexene	30	46	88
Cycloheptene	100		
Cyclooctene	100		
Norbornene	100		

[a] Olefins concentration is 0.4 M. An equivalent amount of 9-BBN is used.

5.1.3
Hydroboration of Chloro-, Acetate-, and Acetal-Functionalized Alkenes

9-BBN reduces many of the functional groups at a relatively slower rate [18] than that of hydroboration.

However, electronegative substituents present in an alkene influence significantly the direction of the addition of the hydrogen–boron across the carbon–carbon double bond [3c]. Terminal alkenes containing remote functional groups are hydroborated with remarkable regioselectivity (≥98% terminal) [19], producing the corresponding stable organoboranes. 9-BBN hydroborates the functionalized allylic alkenes containing a terminal double bond so as to place the boron essentially at the 1 position (≥97%). The directive effect is further enhanced (≥ 99%) in the case of β-methylallyl derivatives. The data are summarized in Chart 5.4.

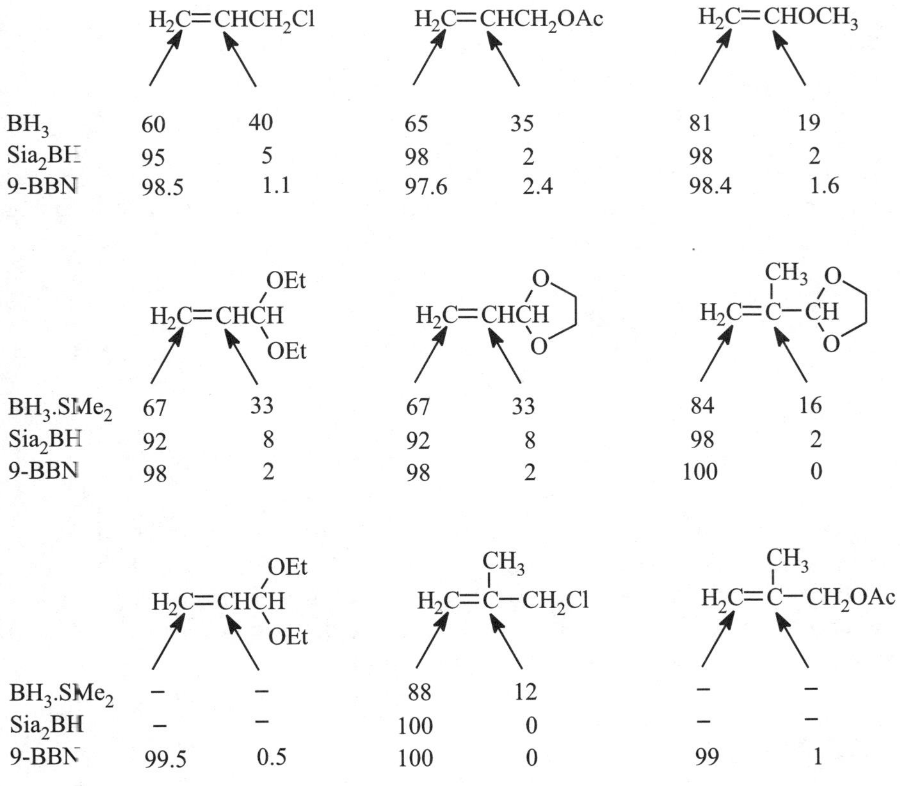

Chart 5.4

However, all other allylic haloalkenes with an internal double bond undergo hydroboration by placing the boron β to the halogen substituent. These products immediately undergo elimination, followed by rapid rehydroboration of the new terminal double bond (Eqs. 5.7, 5.8) [13].

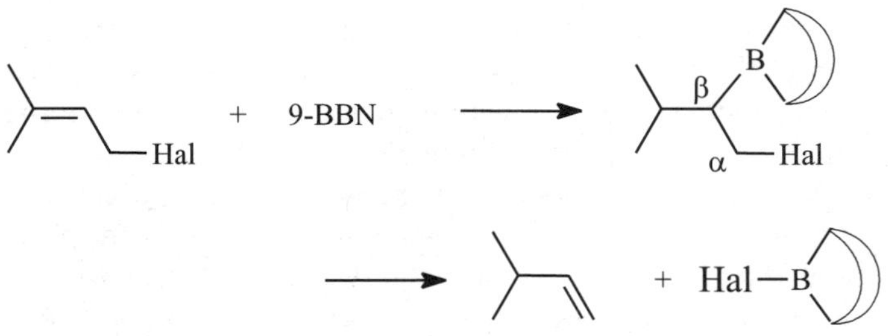

(5.7)

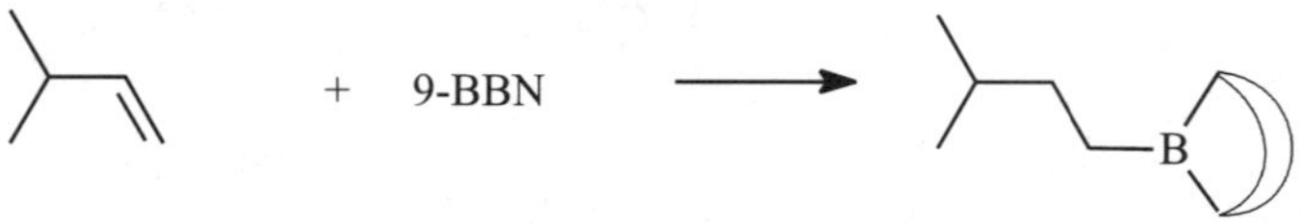

(5.8)

The electronic effects of the hydroboration of allyl halides have been studied and it is found that a Cl, Br, or I substituent at the γ position has a rate-reducing effect of 25, 21.7, or 14.1, respectively [11].

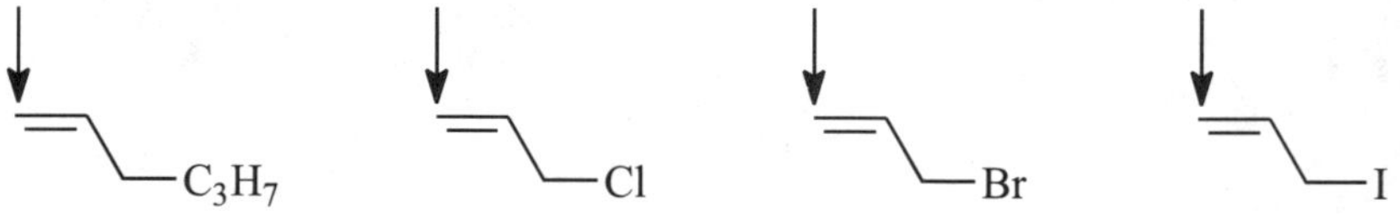

100 4.0 4.6 7.1

The order of rate reduction Cl > Br > I is in the order of increasing electron-withdrawing effect (inductive and mesomeric) of the halogen. However, in a 1-halo-3-methyl-2-butene system, the rate-retarding effect of halogen substituent on hydroboration at the β position is a factor of ≈2, much less as explained above, where hydroboration occurs, at the γ position.

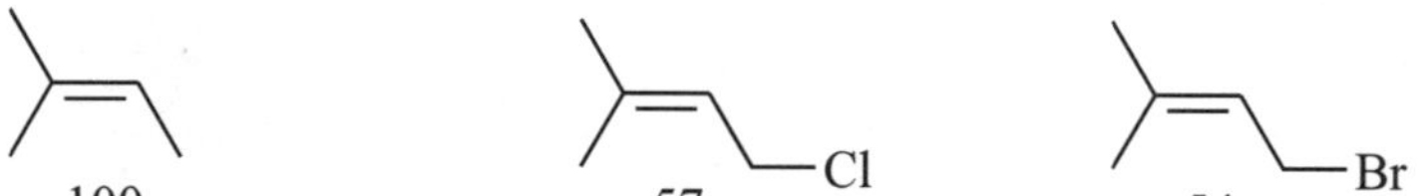

100 57 54

In 1,4-dichloro-2-butene, there are two sites for hydroboration, β to one chlorine and γ to the other, and involves rapid elimination and rehydroboration. The rate reduction of hydroboration is rather large when compared with the corresponding hydrocarbon; it is 48 with respect to *cis*-alkene and is 100 with respect to *trans*-alkene [5].

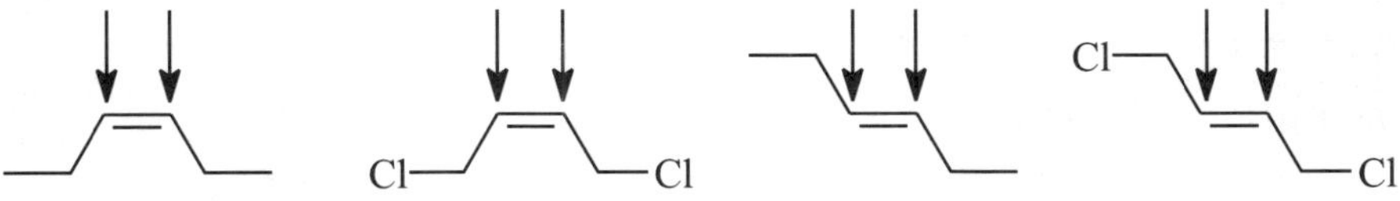

100 2.1 100 1.0

When chlorine is directly attached to double bond, there is dramatic rate reduction by a factor more than 10^4, and the hydroboration with 9-BBN occurs at the 2 position.

100 0.0093

The relative rates of other 1-halo-1-alkenes are given below in comparison with the parent alkene.

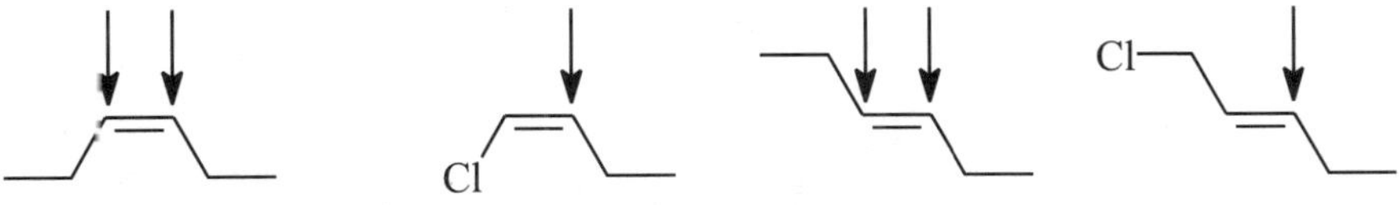

100 1.4 100 0.97

The relative reactivities of dichloro- and monochloro-*cis*-, *trans*-butenes are summarized in Table 5.5 [8].

The rate-retarding effect of the allylic halogen substituent is accounted both for inductive and hyperconjugation. Both of these are electron withdrawing and therefore, rate reducing at either the β position or γ position [10].

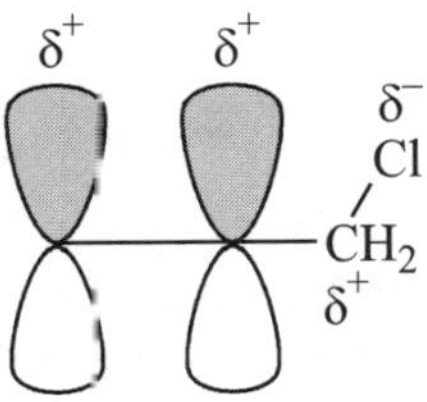

Induction

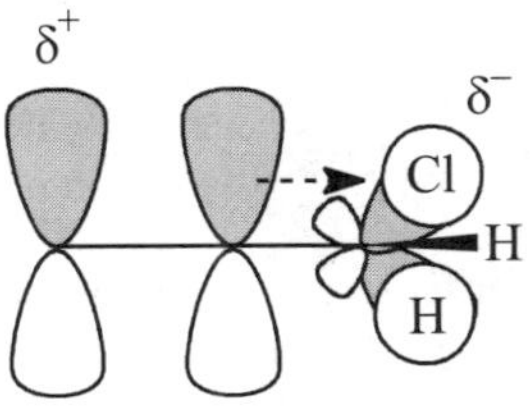

Hyperconjugation

Relative to propene

Table 5.5 The relative reactivities of rate constants for the hydroboration *via* 9-BBN of *cis–trans* pairs at 25 °C in THF [8]

Alkene	Structure	Relative reaction 1-hexene = 100	k_{cis}/k_{trans}
cis-1,4-Dichloro-2-butene		0.0144	4.4
trans-1,4-Dichloro-2-butene	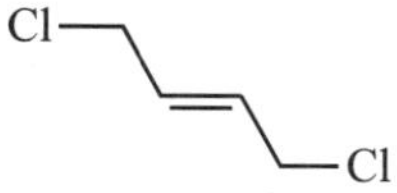	0.0033	
cis-1-Chloro-1-butene		0.0093	3
trans-1-Chloro-1-butene	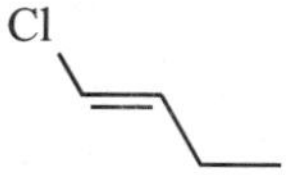	0.0031	

In a vinylic system, the chlorine not only retards the rate of reaction but direct the hydroboration to the 2 position. The inductive effect of chlorine here reduces the electron availability of the π bond, thus severely retarding the rate of the reaction, while some conjugation returns some electron density to the site β to the chlorine, which then becomes the preferred site of hydroboration.

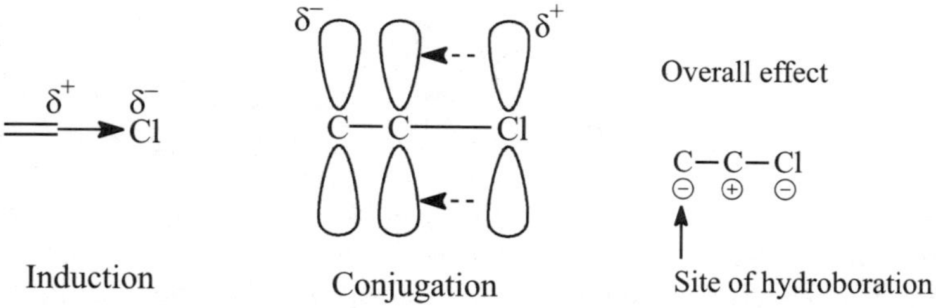

Relative to ethylene

5.1.4 Hydroboration of Allylsilanes

The silyl group is both larger and more electropositive than carbon, and these factors direct the hydroboration of allylsilanes to place exclusively the boron on C-3 of the allyl unit [20], in contrast to 1-butene, which gives 6% of hydroboration with diborane on C-2 [1]. Fleming and Lawrence [20] have compared the placement of boron of diborane and 9-BBN on a variety of allylsilanes as shown in Scheme 5.1.

The highly selective hydroboration of allylsilanes with 9-BBN affords the products having the silicon and boron in a 1,3-relationship (Scheme 5.1) [20]. As already evident [4] the addition of 9-BBN is affected significantly by steric factors.

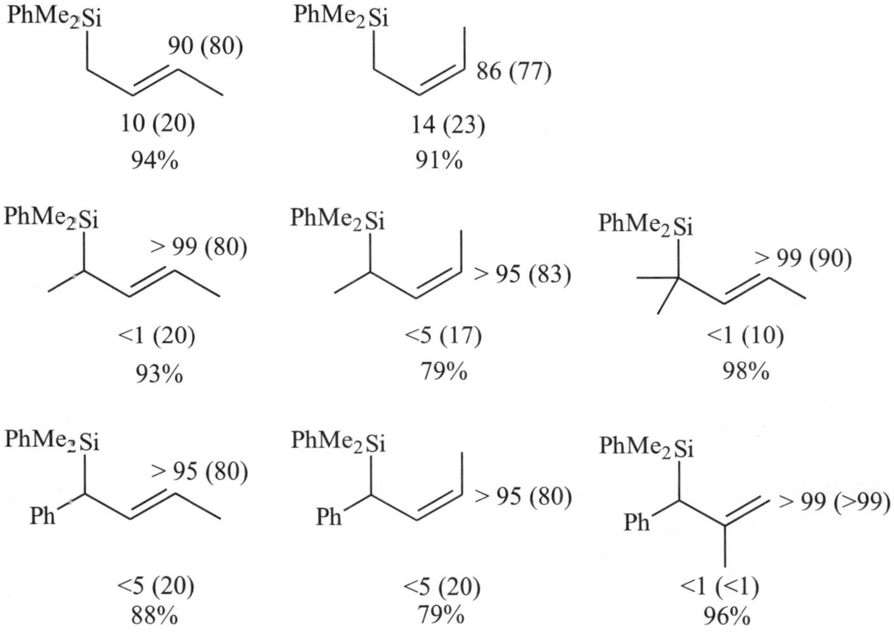

Scheme 5.1 Regiochemistry of hydroboration of allylsilanes using 9-BBN (in *parentheses* of BH_3 [21])

Consequently, the synthetic utility of both different metals, silicon and boron, which are in a 1,3-relationship, can be exploited as both can be replaced, independently of each other by a hydroxy group. The boron in the usual way, using alkaline hydrogen peroxide and the phenyldimethylsilyl group by electrophilic removal of the phenyl group, followed by treatment with peracid [22]. This pair of reactions can also be carried out in one pot [23], with the overall sequence to construct 1,3-diols from allylsilanes.

Fleming and Lawrence [24] have further demonstrated that hydroboration of allylsilane is highly stereoselective with 9-BBN and the selective conversion to 1,3-diols is shown in Scheme 5.2.

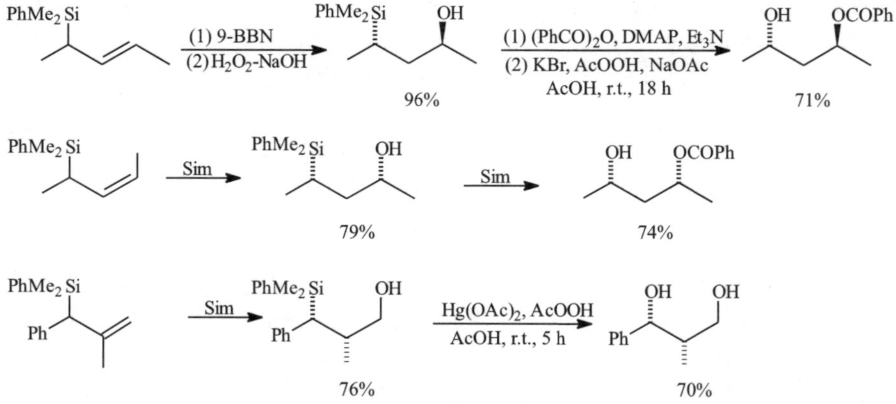

Scheme 5.2

5.1.5
Hydroboration of Chiral Allyl Amines and Chiral Allyl Alcohols

Substrate-controlled diastereoselective hydroboration of protected chiral allylic alcohols [25–27] or amines [28, 29] with 9-BBN gives almost always *anti* selective products. On the other hand, catalyzed hydroboration in most of the cases using catecholborane as hydroborating agent tends to be *syn* selective [28–30] (Eq. 5.9).

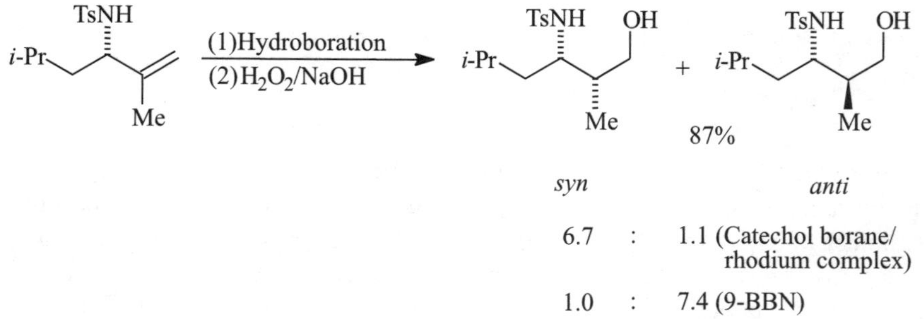

(5.9)

The diastereoselectivity is, however, lowered in cases of less hindered amines. The amines are prepared from (*S*)-amino acids *via* methylenation [31] of *N*-tosylaminoketones (Eq. 5.10).

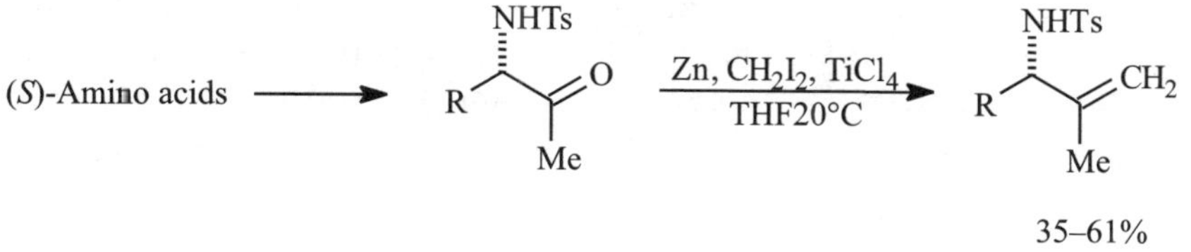

(5.10)

The catalyzed hydroboration of allylic amines conforms to the model wherein predominant *syn* product arises from approach of the rhodium complex *anti* to the group that is the largest and has the best σ-electron–withdrawing capacity, i.e., the NHTs entity. Uncatalyzed hydroboration with 9-BBN of the same substrate (except when R = *i*-Pr) shows poor *anti* selectivity [28], as is seen for protected allylic alcohols. The observations for allylic amines reflect two opposing factors: (1) the electronic influences favoring a reaction conformation in which the R functionality orients *anti* to the approaching borane, and (2) steric demands that guide the NHTs group to this same position [31, 32]. With larger alkyl (*i*-Pr) substrate, the steric differences between isopropyl and NHTs groups are outweighed by electronic factors, and *anti* selectivity results as shown (Fig. 5.2) in the reactive conformation. Though steric demands of *i*-Pr and Ph groups are similar, but latter is less σ-donating; hence, no *anti* selectivity is observed when R = Ph. The other alkyl groups are smaller, so the opposing steric and electronic effects balance, and very little selectivity is observed.

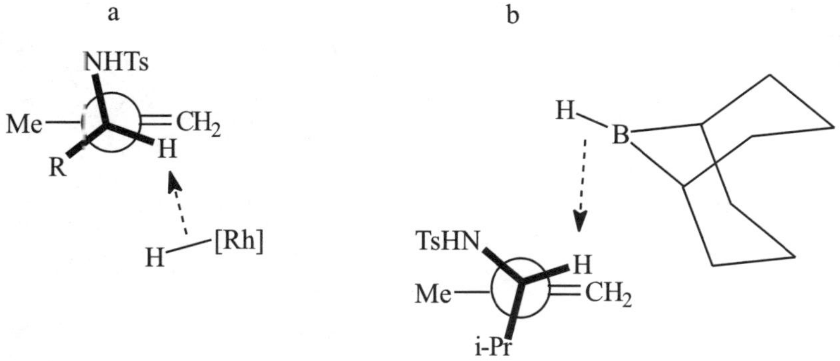

Fig. 5.2 Reactive conformations in catalyzed and 9-BBN by hydroboration of *N*-tosyl–protected allylic amine derivatives. **a** Preferential orientation in catalyzed hydroboration of *n*-tosyl–protected allylic amine derivatives. **b** Preferential orientation in 9-BBN hydroboration of *N*-tosyl–protected allylic amine substrate (R = *i*-Pr)

To increase the steric demands and to remove the acidic NH group, *N*-tosylated allylamine derivatives are converted to NTsBn, using benzylbromide–potassium carbonate–catalytic tetra-*n*-butylammonium iodide in acetone (Eq. 5.11), a reaction that is more convenient and affords the products in excellent yields.

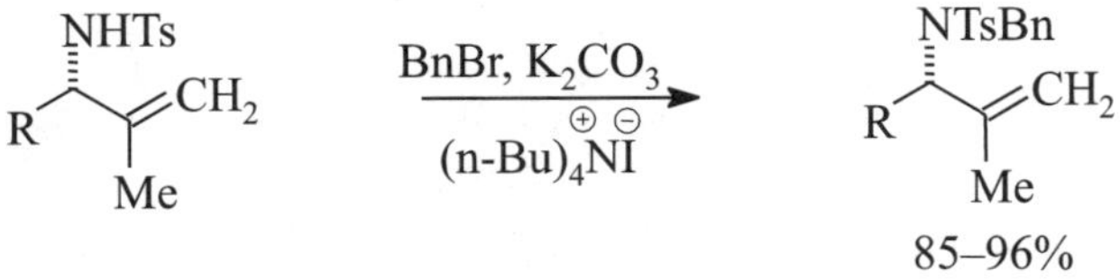

(5.11)

This enhanced bulk around nitrogen accentuates steric effects in hydroboration of these alkenes. Electronic parameters reinforce these steric effects in the catalyzed process for *syn* selectivity. Hindrance in 9-BBN hydroborations excludes NTsBn from the same face of the alkene as the borane, even from the outside position; hence, these conversions are also *syn* selective. The borane, on the other hand, is smaller, so it can tolerate the NBnTs group from the outside position; consequently *anti* selectivity is observed from this reaction conformation.

The above results are different from those of Still's pioneering study [33], and which are also beyond the realm of Houk's calculations [31, 32] for hydroborations of chiral allylic compounds.

The excellent *syn* selectivity of chiral allylamine derivatives, using catalyzed hydroboration with 9-BBN and borane–THF complex, can be explained by the following reactive conformations (Fig. 5.3).

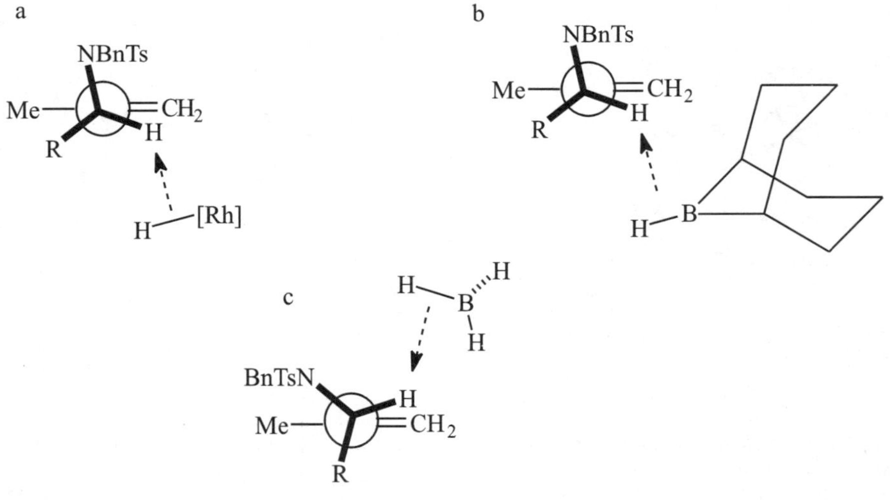

Fig. 5.3 Relative conformations with catalyzed reagent, 9-BBN, and borane reveal (Table 5.6) [28] that: **a** preferential orientation in catalyzed hydroboration is governed by steric and electronic effects, **b** preferential orientation in 9-BBN hydroboration is governed predominantly by steric effects, and **c** preferential orientation in BH_3 hydroboration is governed predominantly by electronic effects

Table 5.6 Hydroboration of *N*-(benzyltosyl) substrate [28]

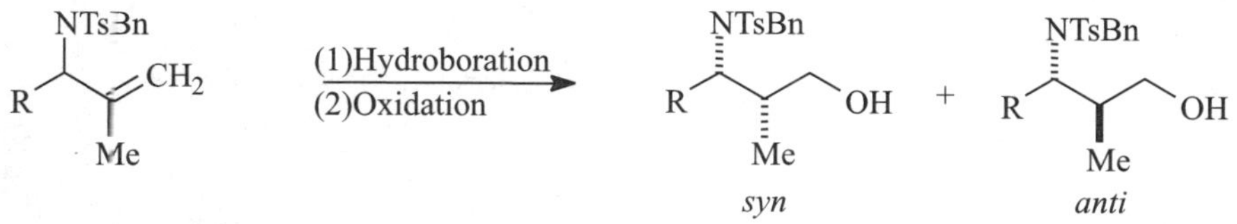

Product R	Method	*syn*:*anti*
$PhCH_2$	$C_6H_4O_2BH$/cat [Rh]	18:1
$PhCH_2$	9-BBN	13:1
$PhCH_2$	BH_3	1:17
Bu	$C_6H_4O_2BH$/cat [Rh]	10:1
Bu	9-BBN	7:1
Bu	BH_3	1:21
$^iPrCH_2$	$C_6H_4O_2BH$/cat [Rh]	6:1
$^iPrCH_2$	9-BBN	25:1
$^iPrCH_2$	BH_3	1:20
$^tBuO_2CCH_2$	9-BBN	5.3:1.0
$^tBuO_2CCH_2$	BH_3	1:52

Burgess and Ohlmeyer [30] have reported that electronic effects are important in catalyzed hydroboration, e.g., allylic acetates are hydroborated with less *syn* selectivity than allylic trifluoroacetate is [25], and proposed the general model [A] (Fig 5.4) for catalyzed hydroboration of chiral allylic alcohols. The model predicts that the $OCOCF_3$ substituent (good σ acceptors) will preferentially orientate *anti* to the approaching rhodium complex. The largest of the other two substituents on the chiral center will occupy the outside position, and the smallest will reside in the inside (crowded) site and thus, *syn* selectivity will result.

[A]

Fig. 5.4 Electronic factors orient catalyzed hydroboration

However, according to Houk [34] uncatalyzed hydroboration (Eq. 5.12) with 9-BBN tends to be *anti* selective due to the following most reactive conformation.

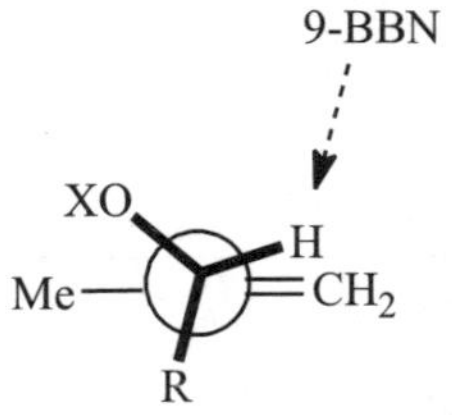

[B]

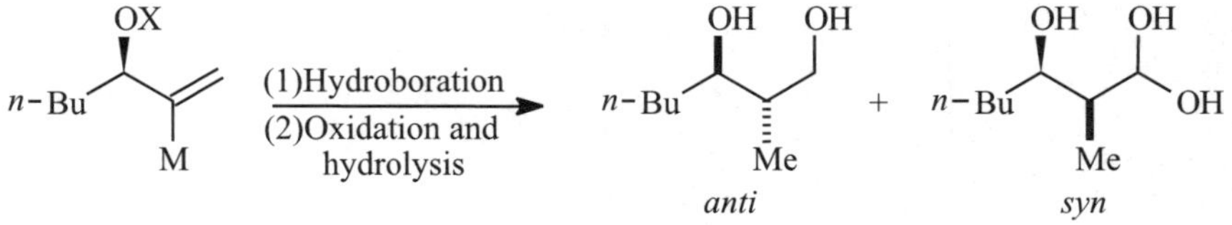

(5.12)

The results are summarized in Table 5.7 [35].

Table 5.7 Catalyzed and uncatalyzed hydroboration of chiral allylic alcohols [35]

X	$HBO_2C_6H_4$/Rh-cat (*syn*:*anti*)	9-BBN (*syn*:*anti*)
$COCH_3$	2.7:1	1:7.5
$COCF_3$	7.5:1	1:14
CO^tBu	6.5:1	1:15

Similarly, *Evans et al* [36] have also found that several classes of allylic alcohols on rhodium-catalyzed hydroboration afford allylic alcohols with high diastereoselectivity-*syn* product and isomer complementing to that furnished by uncatalyzed variant of the reaction (9-BBN)—the *anti* product.

The proposed model [B] implies that *syn* selection in catalyzed hydroboration should decrease as the σ-accepting character of the *anti* substituent decreases. This explains the higher *syn* selectivity in catalyzed hydroboration of allylic trifluoroacetates as compared with reactions of allylic acetates and carbamates [25]. The less diastereoselectivity of cationic complexes in hydroboration of chiral allylic systems than in the neutral catalyst systems [29, 30] is rationalized, as the latter have more electron density to shed *via* back-bonding. The *anti* selectivity in catalyzed hydroboration of the allylic acetate and trifluoroacetate is rationalized in terms of competition between the phenyl and acetate groups for the role of σ acceptor. The further evidence is obtained in the case of pentafluorophenyl allylic acetate. The C_6H_5 lowers the σ* orbital of the aryl ring relative to the phenyl derivative, because the C_6F_5 substituent is a better σ acceptor, and *anti* selectivity predicted is confirmed experimentally (Chart 5.5) [30].

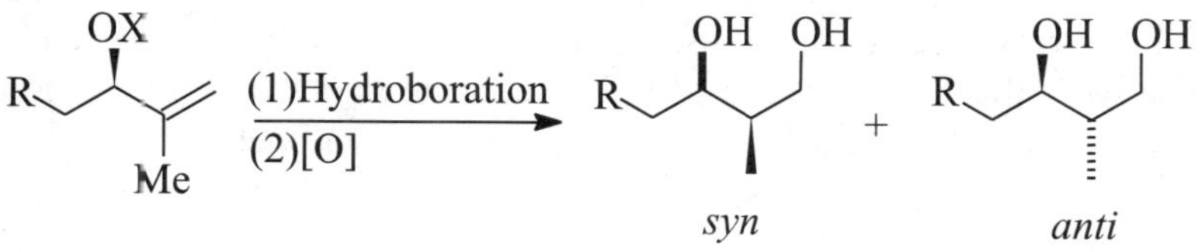

Catecholborane–rhodium complex

		syn		*anti*
(a)	R=Ph, X=$COCH_3$	1	:	3.5
(b)	R=Ph, X=$COCF_3$	1	:	1.5
(c)	R=C_6F_5, X=$COCH_3$	1	:	6.9
9-BBN				
(a)	R=C_6H_5, X=$COCH_3$	1	:	4.5
(c)	R=C_6F_5, X=$COCH_3$	1	:	3

Chart 5.5

The pentafluorophenyl substrate affords predominantly *anti* product in the catalyzed hydroboration, whereas the uncatalyzed reaction's is much less *anti* selective. It reveals that the catalyzed hydroborations are more sensitive to electronic effects than are the corresponding uncatalyzed reactions [30].

The N-protected allylic amines as phosphorated N-allylic amines also hinder N–B coordination, and clean hydroboration is achieved by 9-BBN [37] (Chart 5.6).

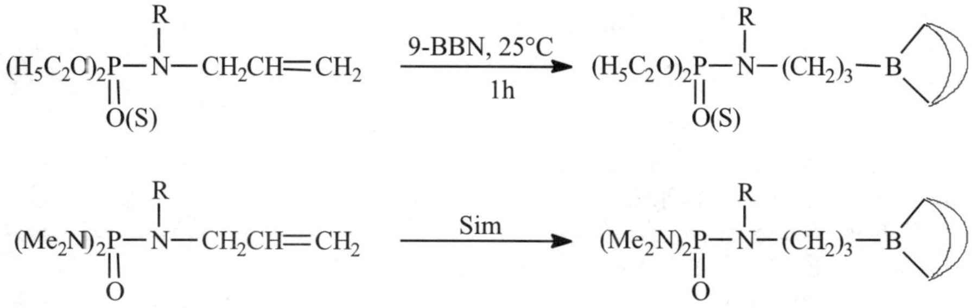

R = CH_3 or $CH_2C_6H_5$

Chart 5.6

5.1.6 Hydroboration of Alkenylheterocycles

Brown and coworkers [38] have conducted a systematic investigation of hydroboration of 2-vinylfuran, 2-vinylthiophene, and vinyl-substituted pyridines, and found that directive effects observed for 2-vinylfuran and 2-vinylthiophene are similar to those realized in styrene [38] (Chart 5.7). The vinyl substituent has strong +M effect and directs the boron to the β position, and many allylic substituents also direct the boron predominantly to the β position.

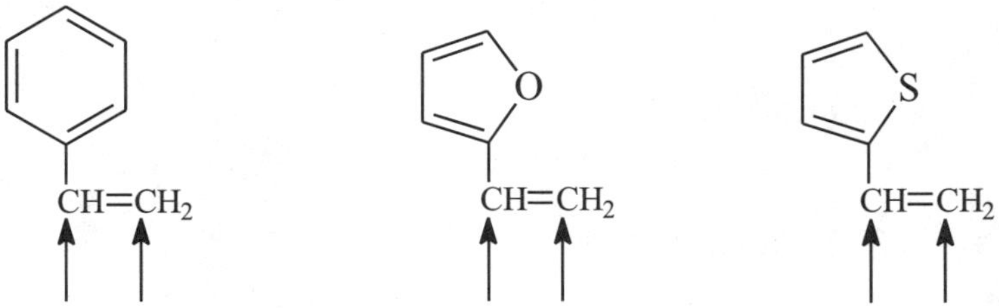

$BH_3.SMe_2$	19	81	16	84	12	88
9-BBN	2	98	3	97	2	98
Chx_2BH			0	100	0	100
Sia_2BH	2	98	0	100	0	100

Chart 5.7

These results in Chart 5.7 explain that as the bulkiness of the dialkylborane increases, for example, from 9-BBN to Chx_2BH to Sia_2BH, the boron atom prefers to add to a less hindered primary carbon atom. In addition, the heteroatoms in 2-vinylthiophene and 2-vinylfuran donate electrons to the β-carbon of the vinyl group, thus making it more electronegative (Eq. 5.13) and preferred attack of the boron of the dialkylborane.

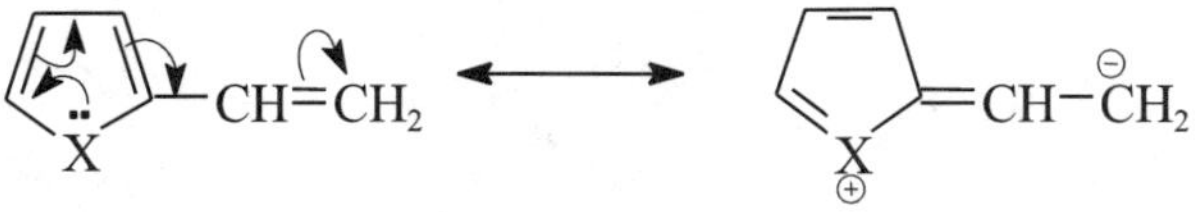

(5.13)

The results of hydroboration of 2-vinylfuran and 2-vinylthiophene are summarized in the Table 5.8 [38].

Hydroboration of *trans*-2-(1-propenyl)furan and *trans*-2-(1-propenyl) thiophene with BMS, 9-BBN, Chx_2BH, and Sia_2BH proceeds smoothly, and the organoborane thus obtained on oxidation with alkaline hydrogen peroxide yields the main product 1-(2-furanyl)-propan-1-ol, in about 90% yield arising from

Table 5.8 Hydroboration of 2-vinylfuran and 2-vinylthiophene [38]

Olefin	Hydroborating agent	Ratio of olefin to hydroborating agent	Reaction temperature (°C)	Reaction time (h)	Total yield (%)	Product distribution (%)		Boron distribution (%)	
								α	β
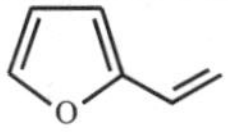						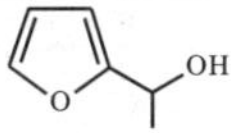	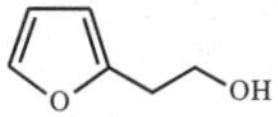		
	BMS	3:1	25	2	94	13	87	13	87
	9-BBN	1:1	25	2	93	6	94	6	94
		1:1	65	1	97	3	97	3	97
	Chx_2BH	1:1	25	4	99	0	100	0	100
		1:1	25	4	98	0	100	0	100
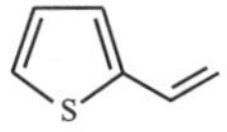						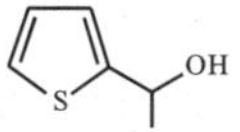	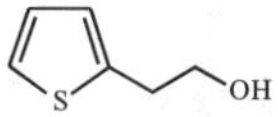		
	BMS	3.1	25	2	86			7	93
	9-BBN	1.1	25	2	100			4	96
		1.1	65	1	98			2	98
	Chx_2BH	1.1	25	4	100			0	100
	Sia_2BH	1.1	25	4	100			0	100
						7	93		
						4	96		
						2	98		
						0	100		
						0	100		

α-C–B bond. Similar results are obtained from 2-(1-propenyl)thiophene (Table 5.9) [38].

2-Vinylpyridine affords different products distribution with different molar ratios of hydroborating reagents. 2-Vinylpyridine on treatment with 9-BBN (1:1 molar ratio) also gives a 2-vinylpyridine–9-BBN complex. This complex has lose association with 9-BBN, and dissociation occurs with concurrent hydroboration. The strong electron-withdrawing power of the pyridine ring greatly reduces the electron density at the β-carbon atom as compared with five-membered heterocycles. However, addition of boron to the β-carbon is still favored due to the powerful regioselective properties of 9-BBN, and hydroboration proceeds smoothly at 25 or at 65 °C. Oxidation of the organoborane gives 2-ethylpyridine and 2-(β-hydroxyethyl)-pyridine in a 3:7 ratio in good yields (Table 5.10). The formation of ethylpyridine is explained by the organoboranes of the benzylic type being susceptible to hydrolysis under the influence of alkali under very mild conditions [39b]. Consequently, due to the strong electron-withdrawing properties of the pyridine nitrogen, the C–B bond of the α-boron derivative is prone to alkaline hydrolysis, as shown in Scheme 5.3.

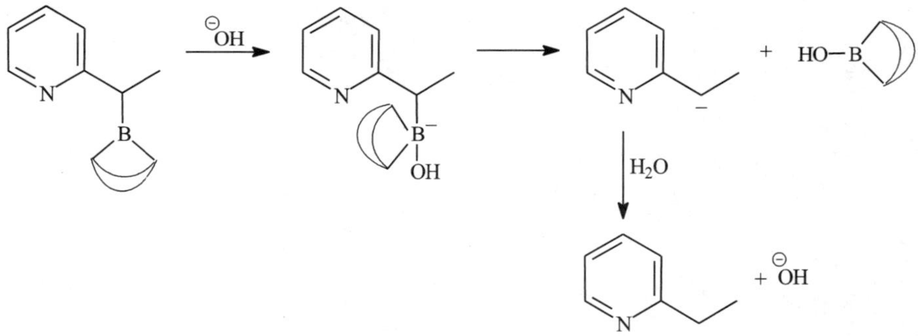

Scheme 5.3

4-Vinylpyridine, on the other hand, is strongly complexed with 9-BBN and no hydroboration has been observed. Moreover, the unreacted starting material cannot be recovered following usual alkaline hydrogen peroxide oxidation. 4-Vinylpyridine is known to undergo polymerization [40] readily.

The complex formed by treatment of 2-methyl-5-vinylpyridine with 9-BBN (1:1 molar ratio) gets dissociated readily at 25 or 65 °C due to 2-methylsubstituent, and hydroboration proceeds smoothly to place the boron atom predominantly at the β-carbon [38]. The electron-withdrawing property of pyridine nitrogen here is transmitted weakly as compared with vinylpyridine. The comparable hydroboration with BMS, 9-BBN, Chx_2BH, and Sia_2BH for their steric and electronic consideration along with their molar ratio studies are presented in Table 5.10 [38].

Table 5.9 Hydroboration of 2-propenylfuran and 2-propenylthiophene [38]

Olefin	Hydroborating agent	Ratio of olefin to hydroborating agent	Reaction temperature (°C)	Reaction time (h)	Total yield (%)	Product distribution		Boron distribution	
								α	β
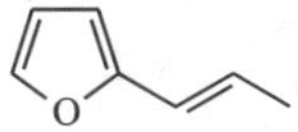						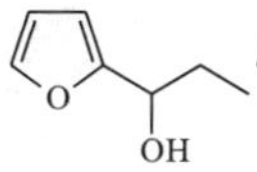	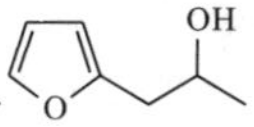		
	BMS	3:1	25	2	90	96	4	96	4
	9-BBN	1:1	25	3	85	88	12	88	12
		1:1	65	2	90	96	4	96	4
	Chx_2BH	1:1	25	6	99	98	2	98	2
	Sia_2BH	1:1	25	6	98	95	5	95	5
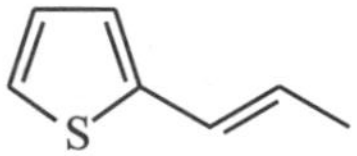						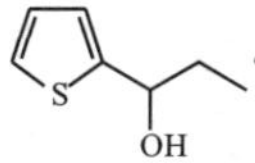	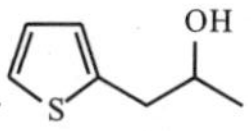		
	BMS	3:1	25	2	72	89	11	89	11
	9-BBN	1:1	25	5	85	94	6	94	6
		1:1	65	3	98	91	9	91	9
	Chx_2BH	1:1	25	10	95	95	5	95	5
	Sia_2BH	1:1	25	10	97	96	4	96	4

Table 5.10 Hydroboration of vinylpyridines [38]

Olefin	Hydroborating agent	Ratio of olefin to hydroborating agent	Reaction temperature (°C)	Reaction time (h)	Total yield (%)	Product distribution			Boron distribution (%)	
									α	β
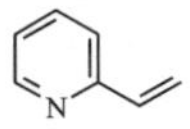						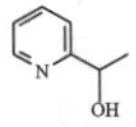	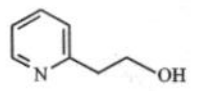	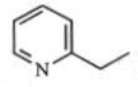		
	BMS	3:1	25	96	5	0	0	100	100	0
		3:4	25	1	61	0	33	67	67	33
	9-BBN	1:1	25	24	79	0	68	32	32	68
		1:1	65	24	97	0	71	29	29	71
		1:2	25	18	100	0	63	31	37	63
		1:2	65	10	87	0	64	36	36	64
		1:1.2	65	24	99	0	73	27	27	73
	Chx_2BH	1:1	65	24	28	0	0	100	100	0
		1:2	25	12	10	0	17	83	83	17
	Sia_2BH	1:1	25	24	30	0	0	100	100	0
		1:2	25	12	8	0	15	85	85	15

Table 5.10 (*continued*) Hydroboration of vinylpyridines [38]

Olefin	Hydroborating agent	Ratio of olefin to hydroborating agent	Reaction temperature (°C)	Reaction time (h)	Total yield (%)	Product distribution			Boron distribution (%)	
									α	β
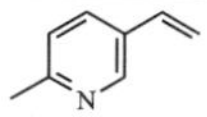						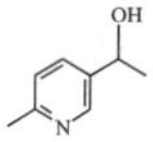	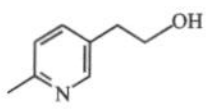	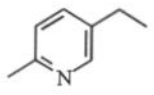		
	BMS	3:1	25	44	13	0	0	100	100	0
		3:1	65	16	9	0	0	100	100	0
		3:4	25	24	92	16	50	34	50	50
		3:4	65	9	98	28	47	25	53	47
	9-BBN	1:1	25	44	91	3	90	7	10	90
		1:1	65	17	97	2	92	6	8	92
		1:1.2	65	8	99	3	92	5	8	92
	Chx_2BH	1:1	25	24	49	0	85	15	15	85
		1:2	25	6	100	0	77	23	23	77
	Sia_2BH	1:1	25	24	39	0	88	12	12	88
		1:2	25	6	100	0	83	17	17	83

Table 5.10 *(continued)* Hydroboration of vinylpyridines [38]

Olefin	Hydroborating agent	Ratio of olefin to hydroborating agent	Reaction temperature (°C)	Reaction time (h)	Total yield (%)	Product distribution			Boron distribution (%)	
									α	β
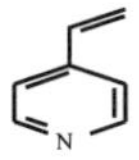						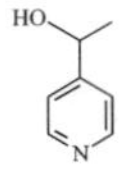	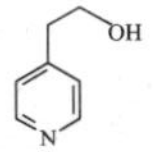	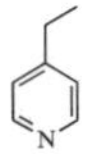		
	BMS	3:1	25	19	5	0	0	100	100	0
		3:4	25	19	52	0	0	100	100	0
	9-BBN	1:1	25	8	3	0	0	100	100	0
		1:1	65	7	0					
		1:2	25	22	36	0	31	69	69	31
		1:2	65	22	30	0	37	63	63	37
	Chx_2BH	1:1	25	24	44	0	0	100	100	0
		1:2	25	6	86	0	58	42	42	58
	Sia_2BH	1:1	25	24	46	0	0	100	100	0
		1:2	25	6	65	0	43	57	57	43

The complexation adducts of vinylpyridines and BF_3 can be hydroborated with a 1:1 molar ratio of 9-BBN, but without much success (Table 5.11) [38].

trans-2-(1-Propenyl)pyridine on hydroboration with 9-BBN (1:2 molar ratio) gives α-organoborane in more than 92% yield. In an alternative procedure *trans*-2-(1-propenyl)pyridine is complexed with borontrifluoride-etherate, and then the hydroboration is achieved in a 1:1 molar ratio with identical success. After the completion of the hydroboration, the boron trifluoride is recovered by treatment with *N,N,N′,N′*-tetramethylethylenediamine (TMEDA) (Scheme 5.4). The products formed are identified after the alkaline hydrogen peroxide oxidation. The comparative data with BMS, Chx_2BH, and Sia_2BH are presented in Table 5.12. This remarkable increase in the formation of α-organoboranes compared with that of α-vinylpyridine is attributed to the terminal methylsubstituent on the olefinic double bond.

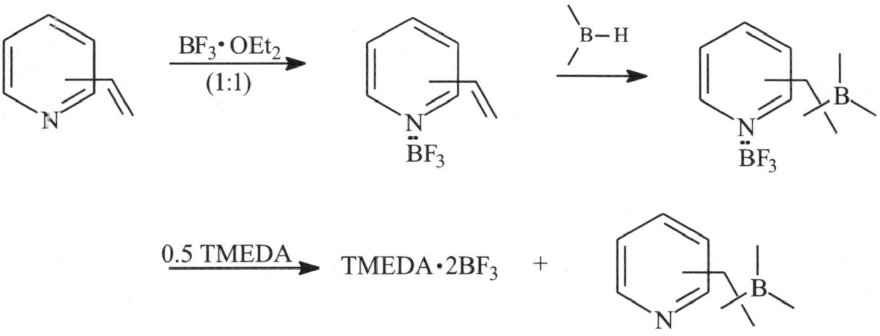

Scheme 5.4

Table 5.11 Hydroboration of vinylpyridine–borontrifluoride complexes [38]

Olefin	Hydroborating agent	Ratio of olefin to hydroborating agent	Reaction temperature (°C)	Reaction time (h)	Total yield (%)	Product distribution			Boron distribution (%)	
						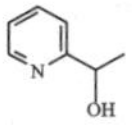	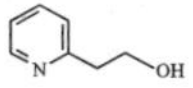	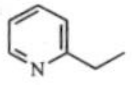	α	β
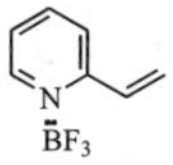										
	BMS	3:1	65	10	32	30	17	53	83	17
	9-BBN	1:1	25	42	74	0	56	44	44	56
		1:1	65	12	84	0	64	36	36	64
	Chx_2BH	1:1	25	24	20	0	16	84	84	16
	Sia_2BH	1:1	25	24	38	0	12	88	88	12

Table 5.11 *(continued)* Hydroboration of vinylpyridine–borontrifluoride complexes [38]

Olefin	Hydroborating agent	Ratio of olefin to hydroborating agent	Reaction temperature (°C)	Reaction time (h)	Total yield (%)	Product distribution			Boron distribution (%)	
						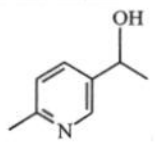	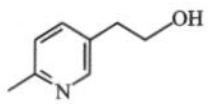	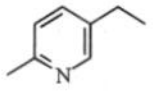	α	β
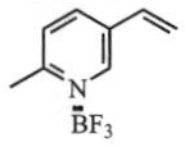										
	BMS	3:1	65	24	75	20	60	20	40	60
	9-BBN	1:1	25	6	92	2	86	12	14	86
		1:1	65	6	96	3	86	11	14	86
	Chx_2BH	1:1	25	6	91	0	71	29	29	71
	Sia_2BH	1:1	25	6	63	0	87	13	13	87

Table 5.11 *(continued)* Hydroboration of vinylpyridine–borontrifluoride complexes [38]

Olefin	Hydroborating agent	Ratio of olefin to hydroborating agent	Reaction temperature (°C)	Reaction time (h)	Total yield (%)	Product distribution			Boron distribution (%)	
						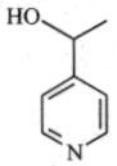	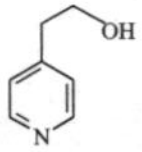	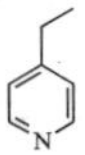	α	β
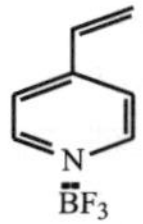										
	BMS	1:1	65	10	26	0	15	85	85	15
	9-BBN	1:1	25	80	68	0	26	74	74	26
		1:1	65	16	73	0	27	73	73	27
	Chx_2BH	1:1	25	6	59	0	41	59	59	41
	Sia_2BH	1:1	25	6	53	0	47	53	53	47

Table 5.12 Hydroboration of propenylpyridine–boron trifluoride etherate complex [38]

Olefin	Hydroborating agent	Ratio of olefin to hydroborating agent	Reaction temperature (°C)	Reaction time (h)	Total yield (%)	Product distribution			Boron distribution (%)	
									α	β
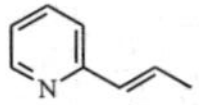						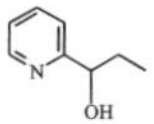	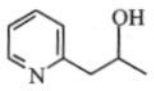	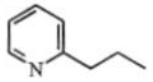		
	BMS	3:4	25	24	63	5	4	91	96	4
		3:4	65	12	99	0	4	96	96	4
	9-BBN	1:2	25	24	54	0	7	93	93	7
		1:2	65	24	88	3	5	92	95	5
	Chx_2BH	1:2	25	24	54	20	5	75	95	5
	Sia_2BH	1:2	25	36	86	3	1	96	99	1
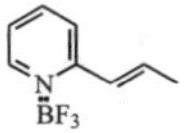										
	BMS	3:1	25	12	72	0	1	99	99	1
	9-BBN	1:1	25	24	34	0	8	92	92	8
		1:1	65	18	99	0	1	99	99	1
	Chx_2BH	1:1	25	36	86	3	1	96	99	1
	Sia_2BH	1:1	25	60	78	6	Traces	94	100	Traces

The comparative distribution of boron in the hydroboration of 2-propenylheterocycles, compared to that of *trans*-1-propenylbenzene [9] is shown in Chart 5.8 [38].

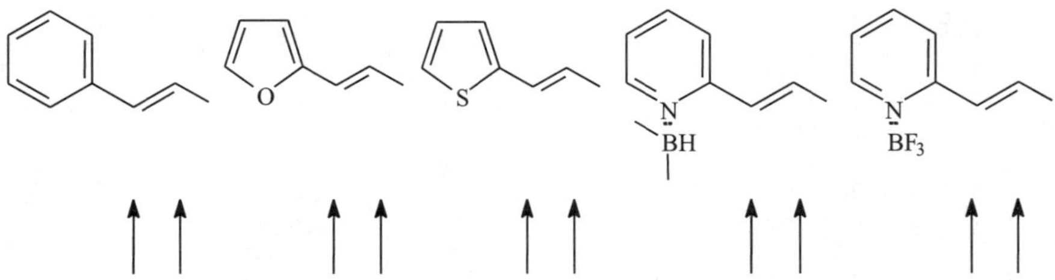

BH_3,SMe_2	85	15	91	9	89	11	96	4	99	1
9-BBN	97	3	96	4	94	6	93	7	99	1
Chx_2BH			98	2	95	5	95	5	99	1
Sia_2BH			95	5	96	4	99	1	100	0

Chart 5.8

5.1.7 Hydroboration of Enamines

Singaram and coworkers [41, 42] have conducted a comprehensive study of the hydroboration of β,β-disubstituted enamines with BMS and 9-BBN. The hydroboration studies with BMS reveal that cyclic six-membered secondary amines such as morpholine and piperidine direct initially the addition of the boron atom of BMS to α positions, which then rearrange mainly to β-isomers while all other amines direct the boron either predominantly or exclusively to the β position (Eq. 5.14; Table 5.13). The unexpected α-directing effect of a six-membered cyclic moiety in β,β-disubstituted enamines is explained in the same way that the rate difference in the reduction of cyclopentanone and cyclohexanone are explained [43] (see Fig. 5.5). In pyrrolidino enamines, the delocalization of the nitrogen lone pair into an exocyclic double bond, with concomitant concentration of the electron density on the β-carbon, is favored due to the absence of eclipsing interactions in the iminium structure [44], whereas the opposite is true for the six-membered cyclic amine moieties; the delocalization of the nitrogen lone pair onto the adjacent sp^2 carbon results an increase of eclipsing interactions and is highly disfavored. Consequently, the directive effect of the six-membered cyclic amino group is weak, and the parent isoalkyl moiety exerts a dominant directive effect, which leads predominantly to the α-adduct. The steric interactions in β,β-disubstituted enamines favor the α-borane adducts, whereas the resonance effect shown in Fig. 5.5 clearly outweighs the steric considerations.

Table 5.13 Directive effect in the hydroboration of enamines derived from 2-methylpentanal [41].

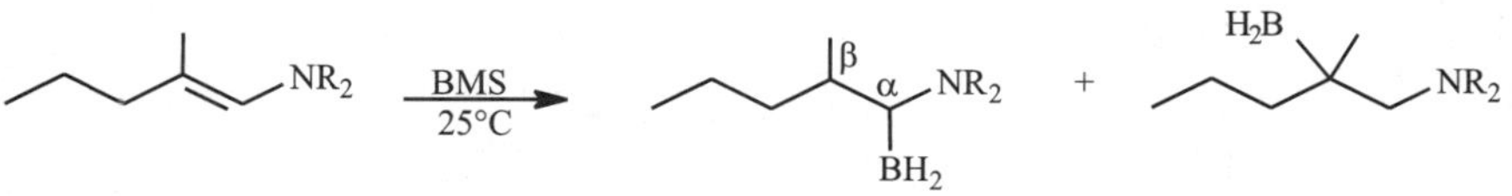

(5.14)

Amine (NR_2)	α-RBH_2	β-RBH_2
Diethylamine	20	80
Diisopropylamine	0	100
Pyrrolidine	0	100
Piperidine	80	20
Morpholine	80	20
Hexamethyleneimine	10	90

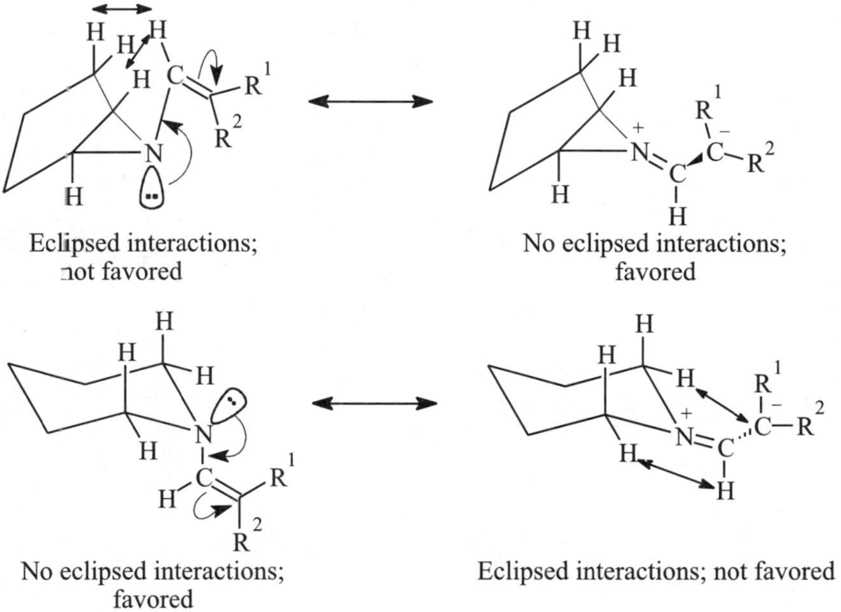

Fig. 5.5 Effect of ring size on the regiochemistry of hydroboration of cyclic amine enamines

The normal electronic effect of the dialkylamino group strongly directs the boron to the β-carbon of the enamine. However, when the hydroborating agent initially complexes with nitrogen atom, the influence of the dialkylamino group is reversed (Fig. 5.6).

Fig. 5.6 Reversal of the normal electronic effect of the enamines dialkylamino group due to coordination of the hydroboration reagent with nitrogen

Aromatic β,β-disubstituted enamines do not undergo hydroboration with 9-BBN. However, their counterpart aliphatic enamines are hydroborated at 25 °C within 12 h and also place the boron exclusively at the β position (Chart 5.9) [41, 42], regardless of the greater steric requirement of 9-BBN and the size of the amine moiety of the enamine. Thus it is explained that 9-BBN attacks the enamine double bond without prior coordination with nitrogen atom, while BMS initially coordinates to the nitrogen atom to a varying degree prior to hydroboration reaction.

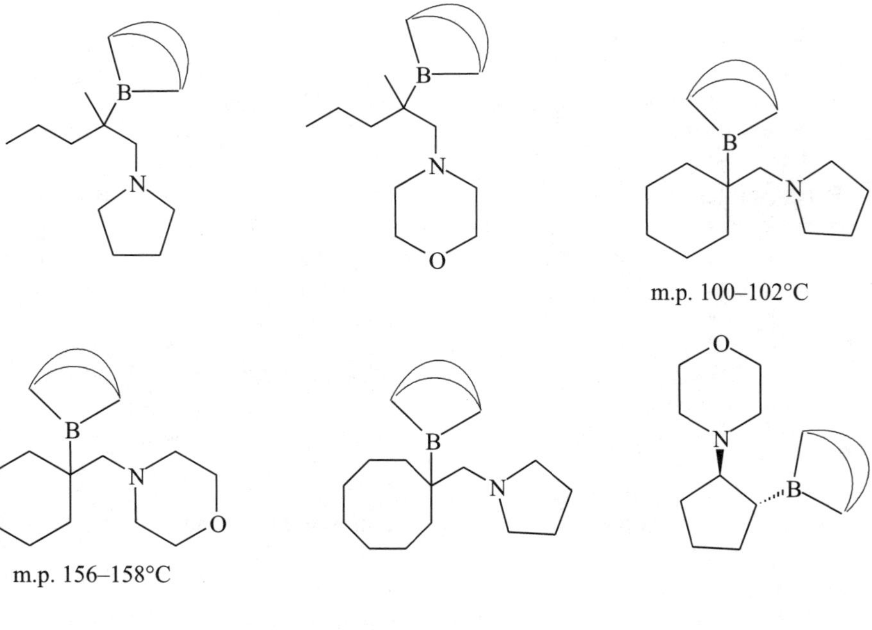

m.p. 100–102°C

m.p. 156–158°C

Chart 5.9

These trialkylboranes of 9-BBN are stable compounds when stored under nitrogen. Even the cyclic derivatives do not undergo appreciable isomerization [45]. The ^{11}B NMR chemical shifts of these are in the range of δ 12–15, indicating that their stability is attributed to the strong coordination of boron to the nitrogen atom.

The morpholinenamines of heterocyclic ketones such as tetrahydro-4*H*-pyran-4-one, tetrahydro-4*H*-thiopyran-4-one, and 1-benzyl-4-piperidone undergo hydroboration, cleanly, with 1 equiv of 9-BBN to furnish the corresponding trialkylboranes (Chart 5.10) [8].

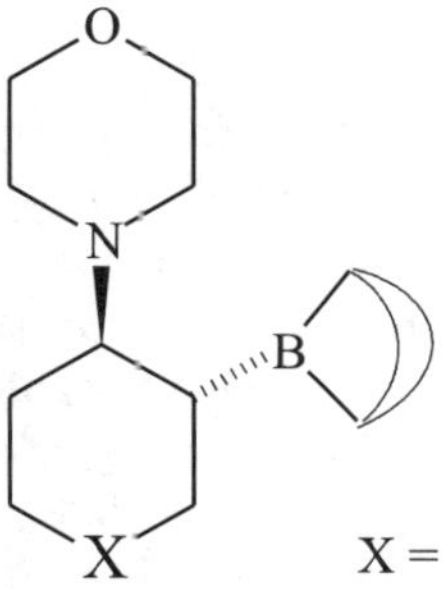

X = O, S, $NCH_2C_6H_5$

Chart 5.10

5.1.8 Hydroboration of Heterocyclic Olefins

Brown and his coworkers have systematically studied the hydroboration of 3-butenyl [3a], 2-butenyl [3b], and 1-butenyl [4] derivatives containing representative substituents. It was found that organoboranes containing a heteroatom at the β position tend to undergo 1,2-elimination (Eq. 5.15) [3].

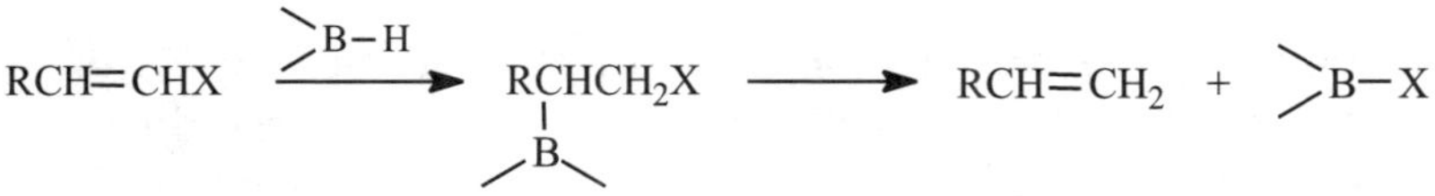

(5.15)

The extent of the leaving group depends on the nature of leaving group, the temperature, and the solvent. For good leaving group like OTs, Cl, or Ac, elimination occurs rapidly [3, 4], whereas it is minimized for poor leaving groups such as OR and OAr. Moreover, the hydroboration proceeds regioselectively, placing essentially all of the boron atoms at the β position [4, 48–50]. Thus

1-ethoxy-2-methyl-1-propene yields 88% of 1-ethoxy-2-methyl-2-propanol on hydroboration–oxidation, indicating the selectivity of boron for β-carbon even though it is tertiary (Eq. 5.16) [4].

$$H_3C{-}\underset{}{\overset{CH_3}{\overset{|}{C}}}{=}CHOC_2H_5 \xrightarrow[(2)[O]]{(1)HB} (CH_3)_2\overset{OH}{\overset{|}{C}}H{-}CH_2OC_2H_5 \quad 88\% \tag{5.16}$$

The hydroboration of heterocyclic olefins having oxygen [50–56], sulfur [57], and nitrogen [58–68] are reported. Brown and coworkers [69] have conducted detailed hydroboration studies of many heterocyclics with an endocyclic double bond with borane-methylsulfide BMS, 9-BBN, Chx_2BH, and Sia_2BH and have established the optimum conditions for clean and quantitative hydroboration. The hydroboration of 2,3- and 2,5-dihydrofuran with BMS (3:1 molar ratio) at 25 °C for 1 h affords trialkylborane, readily oxidized to 3-hydroxytetrahydrofuran, in excellent yield. But the synthesis of dialkylboranes from these olefins using olefins, BMS in 2:1 ratio is not possible at 0 °C. However, hydroboration of 2,3-dihydrofuran proceeds cleanly with 9-BBN in 1:1 ratio and oxidation affords the desired 3-hydroxytetrahydrofuran, in excellent yield. Similar results are obtained with dicyclohexylborane and disiamylborane (Table 5.14).

At higher temperatures, the ring ruptures to afford 3-buten-1-ol or diol. At 25 °C 2,5-dihydrofuran has also been hydroborated, and comparative results are summarized [69] in Table 5.14.

The experimental conditions for the hydroboration of 2-methyl-4,5-dihydrofuran, 2,3-dihydrothiophene and *N*-(benzyloxycarbonyl)-3-pyrroline have been optimized (Table 5.15) [69], and no ring cleavage is observed under these experimental conditions.

The hydroboration of 3,4-dihydropyran with BMS at 25 °C affords the product also arising from the ring cleavage, while good yield of 3-hydroxytetrahydropyran is obtained with $BH_3{\cdot}THF$ in 2-olefin–$BH_3{\cdot}THF$. However, 9-BBN, Chx_2BH, and Sia_2BH afford 3-hydroxytetrahydropyran in excellent yield (Table 5.16) [69].

Hydroboration of 2-methoxy-3,4-dihydro-2*H*-pyran and, 2-ethoxy-3,4-dihydro-2*H*-pyran have been studied, which afford the corresponding 5-hydroxy derivatives in excellent yields but with poor diastereoselectivity (Table 5.17) [69].

Table 5.14 Hydroboration of 2,3- and 2,5-dihydrofurans [69]

Olefin	Hydroborating agent	Ratio of olefin to hydroborating agent	Reaction temp (°C)	Reaction time (h)	Product distribution (mol%) 3-Hydroxy-tetrahydrofuran	3-Buten-1-ol	Butanediols 1,3-	1,4-
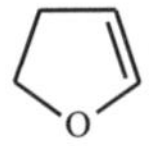								
	BH_3-SMe_2	3:1	0	2	88	4	2	4
	BH_3-SMe_2	3:1	25	1	98	Trace	0	Trace
	BH_3-SMe_2	3:1	25	4	65	27	0	4
	BH_3-SMe_2	2:1	0	8	54	19	2	3
	BH_3-SMe_2	2:1	0	4	62	19	5	7
	BH_3-SMe_2	2:1	0	0.5	54	19	2	3
	9-BBN	1:1	25	1	100	0	0	0
	9-BBN	1:1	25	4	94	6	0	0
	9-BBN	1:1	25	22	28	70	0	0
	9-BBN	1:1	65	4	31	27	0	21
	Chx_2BH	1:1	25	1	100	0	0	0
	Sia_2BH	1:1	0	2	63	0	0	0

Table 5.14 *(continued)* Hydroboration of 2,3- and 2,5-dihydrofurans [69]

Olefin	Hydroborating agent	Ratio of olefin to hydroborating agent	Reaction temp (°C)	Reaction time (h)	Product distribution (mol%)			
					3-Hydroxy-tetrahydrofuran	3-Buten-1-ol	Butanediols 1,3-	1,4-

	BH_3-SMe_2	3:1	0	10	63	7	0	0
	BH_3-SMe_2	3:1	25	1	95	0	0	0
	BH_3-SMe_2	3:1	25	2	76	3	7	13
	9-BBN	1:1	25	1.5	82	5	0	0
	9-BBN	1:1	25	4	65	22	0	6
	Chx_2BH	1:1	25	1	92	2	0	Trace
	Sia_2BH	1:1	0	2	96	0	0	0

Table 5.15 Hydroboration of 2-methyl-4,5-dihydrofuran, 2,3-dihydrothiophene and *N*-(benzyloxycarbonyl)-3-pyrroline [69]

Olefin	Hydroborating agent	Ratio of olefin to hydroborating agent	Reaction temperature (°C)	Reaction time (h)	Product	Yield (%)
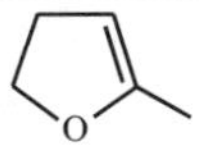	$BH_3 \cdot SMe_2$	3:1	25	1	*trans*-2-Methyl-3-hydroxytetra hydrofuran	100
	9-BBN	1:1	25	1	*trans*-2-Methyl-3-hydroxytetra hydrofuran	93
	Chx_2BH	1:1	25	1	*trans*-2-Methyl-3-hydroxytetra hydrofuran	98
	Sia_2BH	1:1	0	2	*trans*-2-Methyl-3-hydroxytetrahydro furan	97
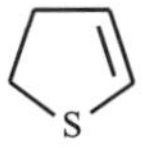	$BH_3 \cdot SMe_2$	3:1	25	1	3-Hydroxytetrahydrothiophene	94
	9-BBN	1:1	25	1	3-Hydroxytetrahydrothiophene	98
	Chx_2BH	1:1	25	1	3-Hydroxytetrahydrothiophene	100
	Sia_2BH	1:1	0	2	3-Hydroxytetrahydrothiophene	100
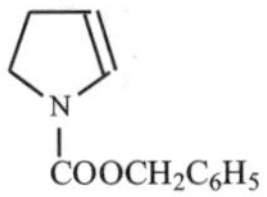	$BH_3 \cdot SMe_2$	3:1	25	1	*N*-(Benzyloxycarbonyl)-3-pyrrolidinol	88
	9-BBN	1:1	25	1	*N*-(Benzyloxycarbonyl)-3-pyrrolidinol	92
	Chx_2BH	1:1	25	1	*N*-(Benzyloxycarbonyl)-3-pyrrolidinol	82
	Sia_2BH	1:1	0	3	*N*-(Benzyloxycarbonyl)-3-pyrrolidinol	84

Table 5.16 Hydroboration of 3,4-dihydropyran [69]

Hydroborating agent	Ratio of olefin to hydroborating agent	Reaction temperature °C	Reaction time (h)	Product distribution (mol%)		
				3-Hydroxy tetrahydropyran	4-Penten-1-ol	1,5- and 1,4-Pentanediols
BH_3-SMe_2	3:1	25	1	82	15	3
BH_3·THF	2:1	0	3	90		
9-BBN	1:1	25	2	92		
Chx_2BH	1:1	25	1.5	98		2
Sia_2BH	1:1	0	18	96	2	Trace

Table 5.17 Hydroboration of 2-methoxy-3,4-dihydro-2*H*-pyran and 2-ethoxy-3,4-dihydro-2*H*-pyran [69]

Olefin	Hydroborating agent	Ratio of olefin to hydroborating agent	Reaction temperature (°C)	Reaction time (h)	Total yield (%)	Percentage of alcohols	
						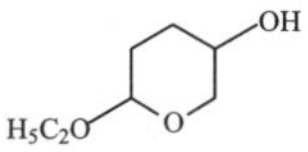	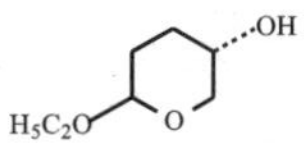
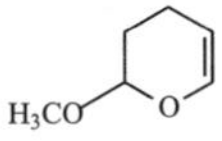	BH_3·THF	3:1	25	3	80	27	73
	$BH_3·SMe_2$	3:1	25	3	90	32	68
	9-BBN	1:1	25	12	96	49	51
	9-BBN	1:1	0	48	92	53	47
	Chx_2BH	1:1	25	2	96	47	53
	Chx_2BH	1:1	0	24	88	34	66
	Sia_2BH	1:1	0	48	98	21	79
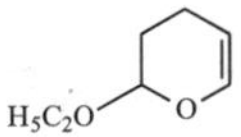	BH_3·THF	3:1	25	5	70	28	72
	$BH_3·SMe_2$	3:1	25	1	94	32	68
	9-BBN	1:1	25	12	100	56	44
	Chx_2BH	1:1	25	2	96	48	52
	Sia_2BH	1:1	0	48	96	21	79
	Sia_2BH[b]	1:1	0		92	20	66

All attempts to hydroborate five-membered nitrogen heterocyclics, 3-pyrroline, and some of its derivatives, with various hydroborating reagents including 9-BBN and Sia_2BH have failed, even in refluxing THF, probably due to complex

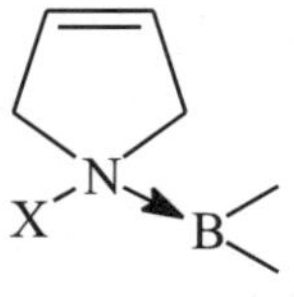

X = H, $SiMe_3$, CH_3

formation with nitrogen [69]. However, *N*-(benzyloxycarbonyl)-3-pyrroline is successfully hydroborated with 9-BBN, Chx_2BH, and Sia_2BH and affords *N*-(benzyloxycarbonyl)-3-pyrrolidinol in excellent yield, after oxidation employing usual alkaline hydrogen peroxide oxidation (Eq. 5.17).

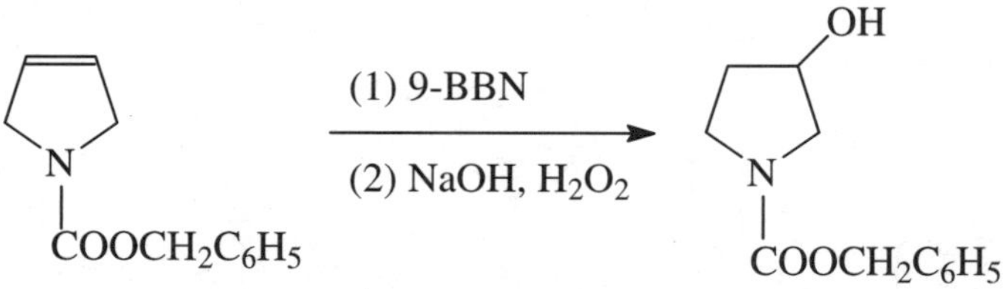

(5.17)

1,2,3,6-Tetrahydropyridine behaves in a manner similar to five-membered heterocycles except in the case of benzyloxycarbonyl derivative, which deactivates the amino moiety. The hydroboration products of BMS, 9-BBN, and Sia_2BH are readily oxidized to give good yields of *N*-(benzyloxycarbonyl)-3- and 4-piperidinols (Eq. 5.18; Table 5.18) [69].

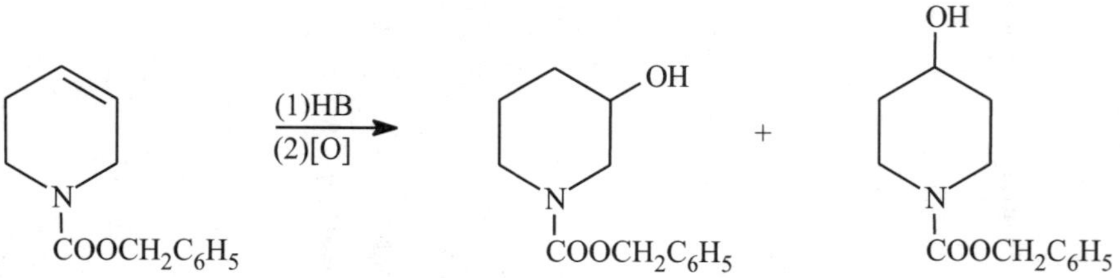

(5.18)

It is observed that strongly basic groups in a heterocyclic ring can greatly reduce the ease of hydroboration, and the introduction of boron β to the heteroatom can lead to elimination.

Table 5.18 Hydroboration of *N*-(benzyloxycarbonyl)-1,2,3,6-tetrahydropyridine [69]

					Product distribution (%)	
Hydroborating agent	Ratio of olefin to hydroborating	Reaction temperature (°C)	Reaction time (h)	Total yield (%)	*N*-(Benzyloxycarbonyl)-3-piperidinol	*N*-(Benzyloxycarbonyl)-4-piperidinol
BH_3-SMe_2	3:1	25	1	80	85	15
9-BBN	1:1	25	24	75	85	15
Chx_2BH	1:1	25	6	84	75	25
Sia_2BH	1:1	0	96	69	75	25

However, there is a report for the successful hydroboration of *N*-benzyloxy-3-pyrroline with BH_3·THF by Joulliё and coworkers [70]. This led Brown to re-investigate [71] the studies of pyrrolines, and he found that excess of BMS indeed leads to the desired alcohols, and similar results with equimolar and excess hydroborating reagents have been compiled [71] in Table 5.19.

The relative rates of hydroboration of representative heterocyclic olefins with 9-BBN have been studied [72] at temperature 25 °C, as most of the large amount of data for relative reactivities with 9-BBN is available at this temperature. These rates of reaction for heterocyclic olefins are compared with Sia_2BH as the hydroborating agent at 0 °C, which is described by Zweifel and Plamondon [50].

Table 5.19 Hydroboration of *N*-substituted-3-pyrrolines at 25 °C [71]

Olefin	Product	Hydroborating agent	Ratio of olefin to hydroborating agent	Reaction time (h)	Yield (%)
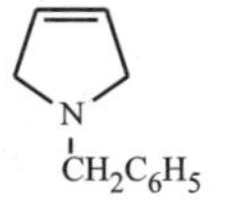	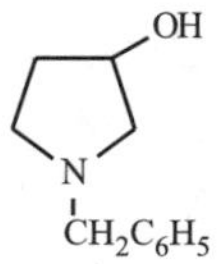	BH_3·SMe_2	1:1	24	0
			1:1.33	2	97
			1:1.66	2	97
			1:2	1	100
		9-BBN	1:1	1	96
			1:2	0.5	98
		Sia_2BH	1:1	3	99
			1:2	1	100
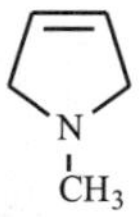	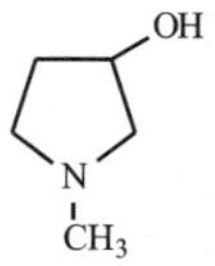	BH_3·SMe_2	1:1	48	0
			1:1.33	36	94
		9-BBN	1:2	24	97
		Sia_2BH	1:2	24	98
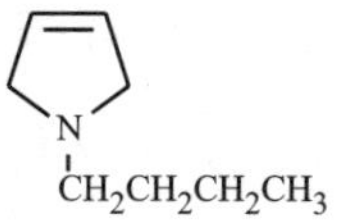	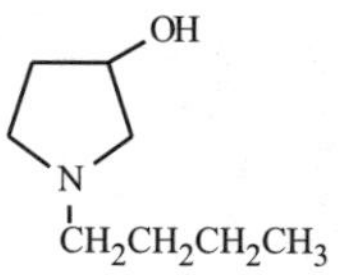	BH_3·SMe_2	1:1	48	0
			1:1.33	36	95
		9-BBN	1:2	24	93
		Sia_2BH	1:2	24	99
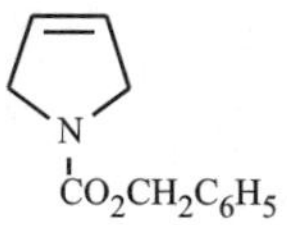	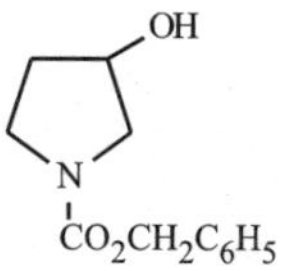	BH_3·SMe_2	1:1.33	1	88
		9-BBN	1:1	1	92
		Sia_2BH	1:1	3	84

Mesomeric effect in 2,3-dihydrofuran makes this to react 106 times faster than does cyclopentene, and moving the double bond one carbon atom away as in 2,5-dihydrofuran reduces the rate drastically, and 2,5-dihydrofuran reacts 4.2 times slower than does cyclopentene (Chart 5.11) [72] with 9-BBN.

Relative reaction	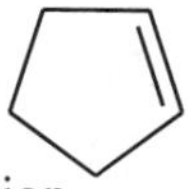		
9-BBN 25 °C	1.0	106	0.24
Sia_2BH 0 °C	1.0	21.8	1.26

Chart 5.11

Since the mesomeric effect is absent in the 2,5-isomer, the inductive effect of oxygen atom deactivates the C=C toward hydroboration. Similar deactivation is observed in an open-chain system as well:

		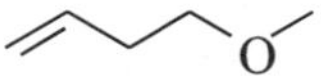
9-BBN, 25 °C	1.0	0.69

In 2-methyl-4,5-dihydrofuran both mesomeric and inductive effects operate in the same direction, and a significant acceleration in the rate of hydroboration is observed (Chart 5.12) [72].

	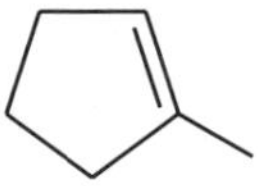	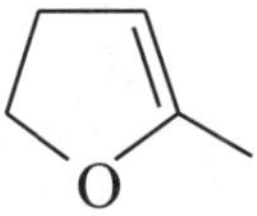
9-BBN, 25 °C	1.0	631
Sia_2BH, 0 °C	1.0	45.4

Chart 5.12

In contrast to 1-methylcyclopentene, which is less reactive than cyclopentene [11], the trend is opposite in case of heterocyclic system (Chart 5.13).

9-BBN, 25°C	1.0	0.25	1.00	1.5
Sia_2BH, 0°C			1.00	2.08

Chart 5.13

However, with a six-membered heterocycle, dihydropyran (DHP) when compared with cyclohexene, the effect in the rate is surprisingly marginal, both for 9-BBN and Sia_2BH.

9-BBN, 25°C	1.00	6.4
Sia_2BH, 0°C	1.00	1.15

However, 2,3,4,5-tetrahydrooxepin reacts 30 times slower than does cycloheptene.

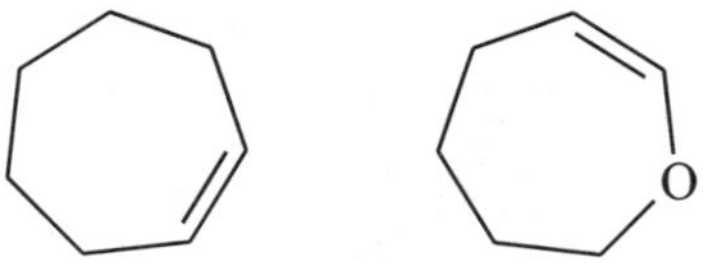

9-BBN, 25°C	1.00	0.034

Zweifel and Plamondon [50] have described that this lower reactivity is due to reduced mesomeric interaction of the nonbonded pair of electrons on the oxygen with the double bond in the less planar DHP besides the increased steric hindrance factor. The same explanation is extended to account for the behavior of 2,3,4,5-tetrahydrooxepin.

The investigation of the corresponding sulfur analogs toward hydroboration with 9-BBN shows a vast difference in the behavior of rate as compared with oxygen heterocycles, though regioselectivity toward the 3 position is same for both the oxygen and sulfur heterocycles (Chart 5.14).

9-BBN, 1.00 0.02 1.00 0.10

Chart 5.14

This behavior is explained by Brown [72] due to (1) poor mesomeric contribution from sulfur and (2) different conformation of sulfur system from those of oxygen. The C–S–C and C–O–C bond-angles and C–S and C–O bond-lengths vary considerably [73].

5.2 Hydroboration of Alkynes

The monohydroboration of terminal alkynes with dicyclohexylborane, disiamylborane, catecholborane, and dimesitylborane, and complexes of dihaloboranes and 9-BBN provides the simple and efficient route to *trans*-vinylboranes [1, 2]. Unlike other dialkyl(vinyl)boranes, vinyl-9-BBN derivatives exhibit remarkable stability [3, 4]. However, with 1:1 stoichiometry of terminal alkyne:9-BBN, 9-BBN adds twice to give a significant amount of diboryl adduct (Eq. 5.19) from which boron stabilized carbanion can be made [5, 6]. The partial solution to dihydroboration is achieved either by employing 100% excess of the alkyne [3] or using silylated derivatives [4e, h].

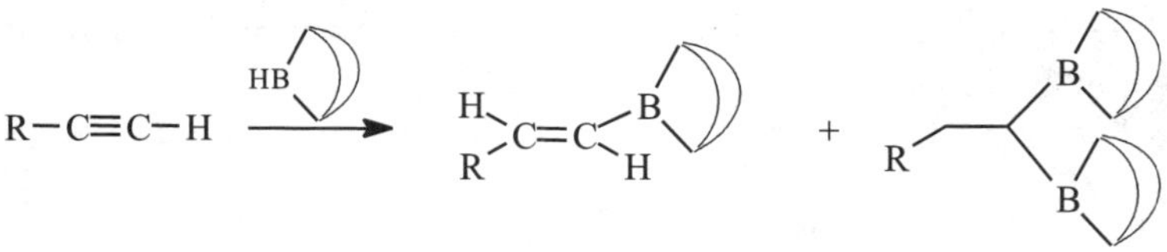

(5.19)

1 2 3

The monohydroboration of internal alkynes, on the other hand, occurs with an equimolar amount of 9-BBN in THF at 0 °C, to give 90–95% yield of *B*-alkenyl-9-BBN derivatives (Eq. 5.20) [5]. The hydroboration of the internal alkynes with 9-BBN is greatly influenced by the steric factors to afford the least hindered product in major quantity [7].

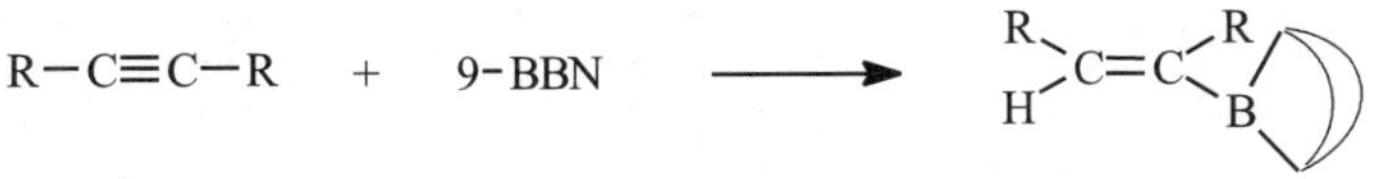

(5.20)

In the case of terminal alkynes, the sacrifice of alkyne, or to add extra reaction steps are highly unsatisfactory for many applications.

5.2.1 Hydroboration of 1-Alkynes

Soderquist and coworkers [8] reexamined the extent of monohydroboration of 1-alkynes (**1**) (Table 5.20) based on the NMR studies conducted on vinylboranes (**2**) and diboryl adduct (**3**) in THF. It is discovered that because **2** but not **3** is partially complexed by THF and thus, its ^{11}B signal in NMR is significantly shifted upfield (Table 5.21) in THF and product distribution has been assessed directly, avoiding extra steps associated with GC analyses of protonolysis and oxidation products. The data (Table 5.20) clearly reveals for the first time that the relative amount of **2** versus **3** generally increases with size of the R group in **1**, with $SiMe_3$ exhibiting its expected exceptional behavior [4h, 9].

Midland reduction [10] of a carbonyl compound with trialkylborane, involves dehydroborylation of trialkylborane in a concerted manner to produce an alkene. However, *s*-Bu-9-BBN reacts sluggishly with aldehydes to produce 2-butene as a *cis*:*trans* mixture (35:65) [10b]. On the other hand, Soderquist [11] observed that **3** accepts only one hydride from KH, a fact which suggests that the two boron atoms may act in concert to result in unusual reactivity with aldehydes compared with that with simple trialkylboranes. Indeed, it is found that when 1 equiv of PhCHO is added to **3a** at 25 °C, *B*-$PhCH_2O$-9-BBN (**4**) and **2a** are quantitatively formed in 2 h (exclusively with *trans* geometry). Moreover, the **3** → **2** conversion is quite general occurring smoothly in all representative systems to afford **2** quantitatively, as the *trans* isomer together with the formation of equal quantity of **4** (Table 5.22; Eq. 5.21) [8]. With the added aldehyde, when changed from PhCHO to 1-naphthaldehyde, the distillation of **2** in most cases become greatly simplified.

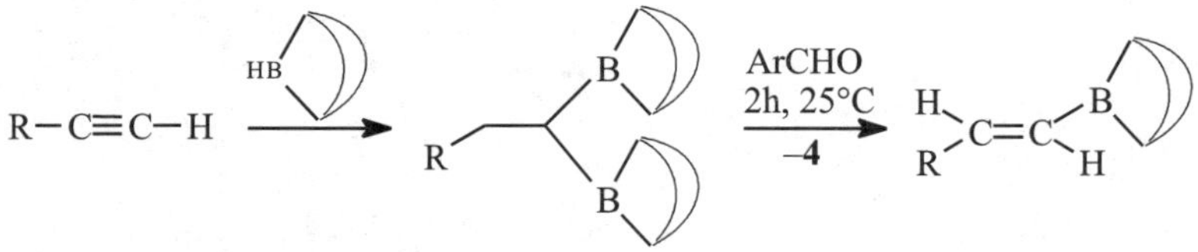

(5.21)

1 **3** **2**

Consequently, a new and completely stereoselective route to *trans* **2** from **3** is available from terminal alkynes by the dehydroborylation procedure [8].

Table 5.20 Hydroboration of 1-alkynes with 9-BBN [8]

Series	R	Ratio[a]	**2** (%)	**3** (%)
a	Me	1:1	20	40
		2:1	56	22
b	*n*-Bu	1:1	56	22
		2:1	80	10
c	Ph	1:1	92	4
		2:1	96	2
d	*i*-Pr	1:1	88	6
		2:1	94	3
e	$SiMe_3$	1:1	38	31
		2:1	74	13

Reactions are carried out at 0 °C (0.5 M 9-BBN in THF).
[a] Molar ratio of 1-alkyne:9-BBN is employed.

Table 5.21 Effect of THF on the ^{11}B NMR chemical shift of 2 [8]

R	δ ($CDCl_3$)	δ (THF)[a]
Me	77.2	74.5
n-Bu	77	73.1
i-Pr	77.7	73
Ph	78.5	70
$SiMe_3$	77.6	67.4

[a] There is no significant effect on δ for **3** with this change in solvent systems.

Table 5.22 *trans*-Vinylboranes (**2**) *via* dehydroborylation of 3 [8]

Series	R	Ar	Isolated yield of **2** (%)
a	Me	Ph	89
		1-Naph	81
b	*n*-Bu	Ph	100
		1-Naph	86
c	Ph	Ph	100
		1-Naph	100
		o-$C_6H_4(CHO)_2$	68
d	*i*-Pr	1-Naph	85
e	$SiMe_3$	1-Naph	82

The earlier concept of syntheses of vinylborane, with excess of alkyne is completely unworkable for an α,ω-diyne if both the carbon–carbon triple bonds are to be monohydroborated. However, a 4:1 ratio of 9-BBN and 1,8-nonadiyne give exclusively the corresponding 1,1,9,9-*tetra*-9-BBN adduct, which on treatment with 2 equiv of PhCHO results in clean formation of *trans,trans*-1,9-*di*-9-BBN-1,8-nonadiene (Eq. 5.22) [8].

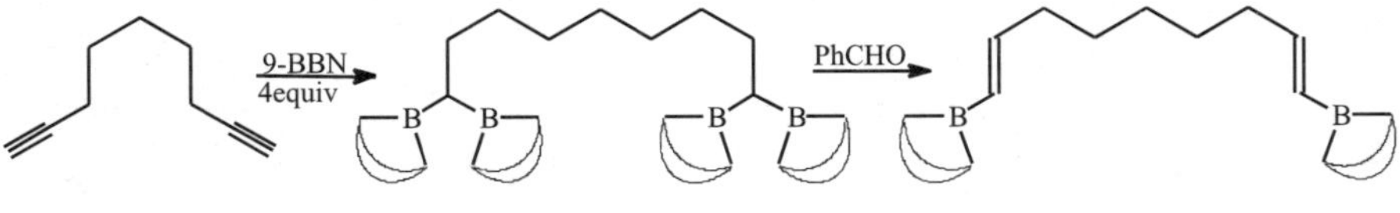

(5.22)

The new dehydroborylation route to *trans*-vinyl-9-BBN demonstrates its remarkable versatility as it undergoes deuterolysis, thermal isomerization, aldehyde insertion, oxidation, and Pd-catalyzed Suzuki's cross-coupling reactions (*vide infra*).

Comparing the MMX-minimized conformations [12] of **3a** with *n*-Pr-CH(9-BBN)CH_3 and *i*-PrCH(9-BBN)CH_3, the least hindered *gauche* conformation **3a′** has energetic preference (Fig. 5.7) over the alternative *gauche* conformation, with the β-methyl group bisecting the two α-substituents. (1,3-repulsion for the Me–C–C–9-BBN array are greater than for Me–C–C–Me).

3a′ **3a** (ΔE = + 2.6 kcal/mol)

Fig. 5.7 Energy difference of conformers

The studies reveal [8] that the delivery of 9-BBN from **3a** to an aldehyde shows a greater *trans*-selectivity regardless of the actual structural arrangement of H–C–C–B array (i.e. *syn* periplanar or *gauche*).

The dehydroborylation process has been proposed [8] to occur through the following transition state shown in Fig. 5.8.

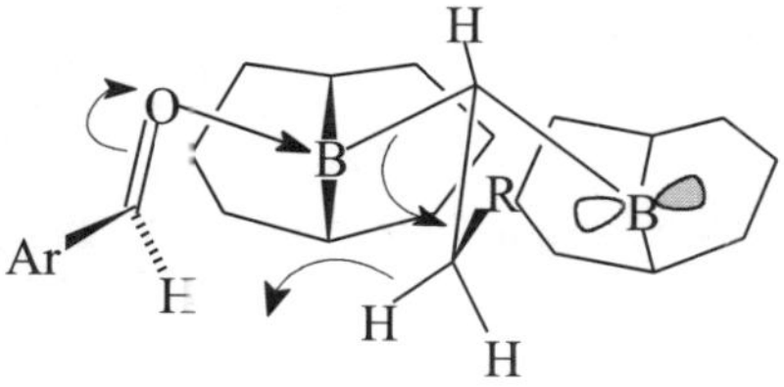

Fig. 5.8 Proposed transition state for the dehydroborylation of 3

The fast rate of reduction of ArCHO by **3** as compared with simple *B*-alkyl-9-BBN derivatives is because of added stability imparted to the transition state of the empty *p* orbital on the boron atom in the additional 9-BBN substituent [10, 13]. The stabilization also leads the reaction to occur at lower temperature than for example, with *B*-s-Bu-9-BBN, a feature that favors a more selective process. As noted above 1,3-repulsions for the Me–C–C–(9-BBN) array are greater than for Me–C–C–Me, and this aspect disfavors the alternating transition state, leading to *cis*-vinylborane. Consequently, the combination of stereo- and electronic factors favor the transition state, and the whole process makes the preparation of *trans* **2** from **3** a highly efficient and selective method [8].

The regioselectivity and stoichiometry of hydroboration of simple 1-silylacetylenes and 1-alkynes with 9-BBN are examined [14] by high-field ^{13}C NMR. The hydroboration of 1-hexyne, 3,3-dimethyl-1-butyne, and 1-decyne in dilute THF solution with 100% excess of alkynes give 94, 96, and 94% monohydroboration, respectively.

The hydroboration with 1:1 stoichiometry reduces the percentage of monohydroboration, drastically. However, for comparison the hydroboration of alkynes with 1:1 alkyne–9-BBN under essentially neat condition provides better line shapes for the carbon α to boron and no interference from solvent absorptions. (Trimethylsilyl)acetylene (**1e**) results in only β-mono- (β to TMS) and dihydroboration products (Eq. 5.23), the behavior common to nonsilylated terminal acetylenes, but differing in the extent of mono- and dihydroboration. Moreover, it is found that this silylalkyne is less reactive toward 9-BBN as compared with its alkyl counterpart. In addition, the steric repulsions between the substituted boron atom, and the α-TMS groups are sufficiently large so as to prevent their site from competing with the alternative, unencumbered position.

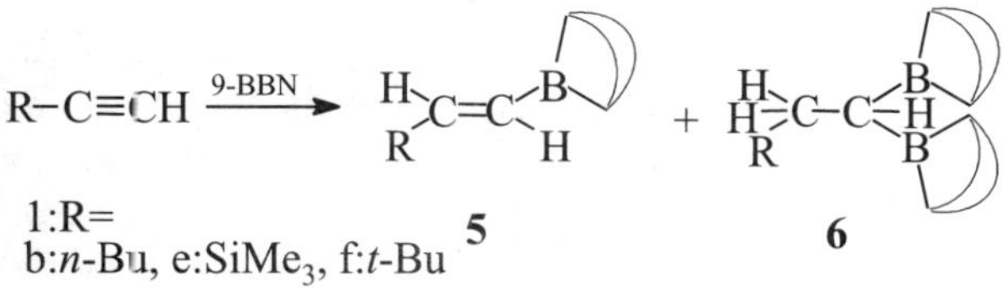

5 **6** (5.23)

The results describing the extent of hydroboration are summarized in Table 5.23 [14].

Table 5.23 Hydroboration of terminal alkynes with 9-BBN under neat conditions at 25 °C [14]

Compound	R	Yield (%)			Time (h)
		5α/5β	**5**	**6**	
1b	*n*-Bu	0/100	32	34	5
1e	Me_3Si	0/100	9	47	8
1f	*t*-Bu	0/100	79	10	7

The neat hydroborations of **1b**, **1e**, and **1f** with 9-BBN with 1:2 stoichiometry yields the corresponding *gem*-diboryl adducts (**6**), and this is the first time that β-*di*-boryl products are prepared via hydroboration of silylacetylenes.

The effect of alkyl versus silyl substitution is examined in a 1:1:1 mixture of **1e**, **1f**, and 9-BBN. The **1f** undergoes hydroboration (>95%) to provide **5f** and **6f** in an 88:12 ratio. On the other hand, hydroboration of **1e** is <5% and provides **5e** and **6e** in a 4:1 ratio. This clearly shows that *t*-Bu group is more effective in activating the triple bond to place the boron at the β position than is the TMS group (Eq. 5.24).

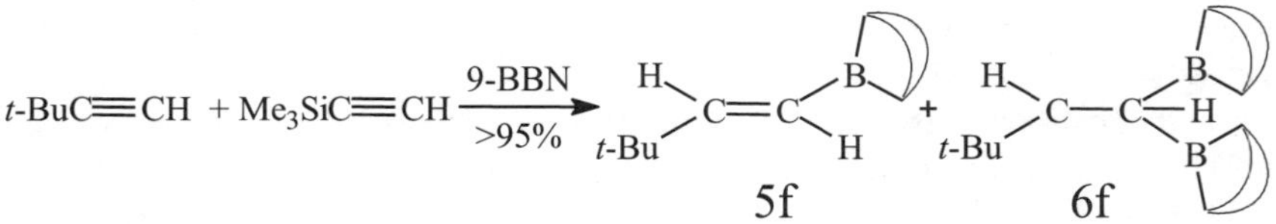

5f 88 : 6f 12

(5.24)

5.2.2 Hydroboration of 1-Silyl-2-Alkyl-1-Alkynes

In comparison to other dialkylboranes, 9-BBN is more sensitive to electronic factors (in the absence of steric factors). The above data indicate that the TMS group provides an electronically based preference for the placement of boron to the α position, a phenomenon consistently observed in the ionic additions to unsaturated organosilanes [16]. However, because of the limited ionic nature of the hydroboration process, this effect is less important in these reactions. Steric effects easily overcome the weak electronic effects of the TMS group so that β-boron placement becomes the dominant process.

In order to establish the effect of TMS, 1-hexyne (**1b**), 1-(trimethylsilyl)-1-hexyne (**7**), and 9-BBN in 1:1:1 ratio are reacted at 25 °C for 5 h (Eq. 5.25). It is found that **1b** undergoes hydroboration (>98%) to produce **5b** and **6b** in essentially equal amounts with the corresponding quantity of unreacted **1b**. The silylacetylene (**7**) remains essentially unchanged with only a traces amount of monohydroboration to vinylborane (**8**).

$$\underset{\mathbf{7}}{n\text{-BuC}{\equiv}\text{CSiMe}_3} + \underset{\mathbf{1b}}{n\text{-BuC}{\equiv}\text{CH}} \xrightarrow[>98\%]{\text{9-BBN}} \mathbf{5b} + \mathbf{6b} \tag{5.25}$$

The above study concludes that TMS group does not result in any net activation of any position in an alkyne relative to an alkyl group. In fact the addition of boron of 9-BBN is slowed by the steric bulk of a TMS group and also from adding to that position. However, the hydroboration of **7** (10% excess) with 9-BBN gives the expected vinylborane **8** (Eq. 5.26).

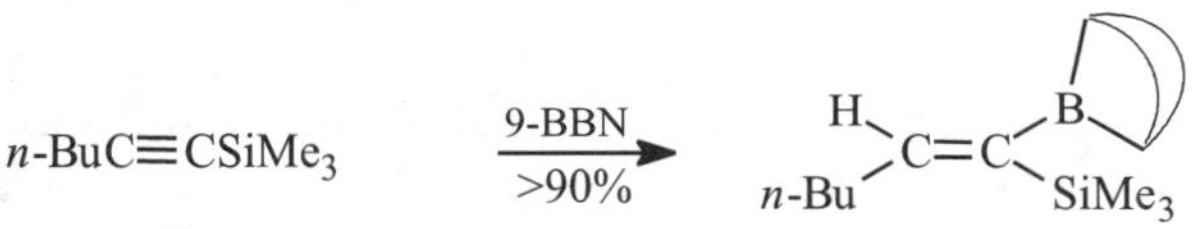

7 8 (5.26)

It is known that relative hindered vinylboranes such as **5f** (R = *t*-Bu) have less tendency to undergo competitive hydroboration with their alkyne progenitors in comparison to less hindered products such as **5b** (R = *n*-Bu). In a mixture of **7** and **1b** (R = *n*-Bu), the TMS group essentially prevents the hydroboration of **7** with 1 equiv of 9-BBN.

The hydroboration studies on **1f** (R = *t*-Bu) and **7** conclude that both bulky alkyl group and 1-trimethylsilyl substitution block the placement of second boron on that carbon. Consequently, the clean monohydroboration is achieved [14] with 9-BBN, under stoichiometric conditions when C≡C terminals possess these substituents. The reaction is highly regioselective (Eq. 5.27) and places the boron on the carbon bearing silyl substituent to afford the product in quantitative yields.

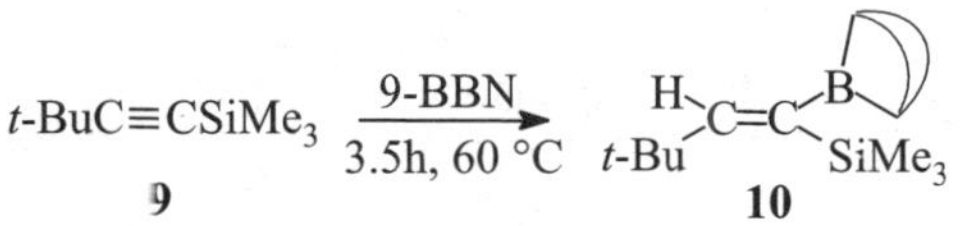

9 10 (5.27)

Thus, it is concluded that for 1-alkynes the 1-trimethylsilyl group blocks the placement of a second boron on that carbon. The large alkyl group on the β-carbon also effectively blocks this position. Therefore, clean hydroboration is achieved with 9-BBN under stoichiometric conditions.

Soderquist and colleagues [14] have further analyzed the behavior of alkynylsilanes (**11**) with 9-BBN with the help of ^{13}C NMR. The 1:1 stoichiometry of 1-(trimethylsilyl)-propyne and 9-BBN gives 54% of the (α-borylvinyl)silane **12aα** and 23% of the 1,2-diboryl adduct **13a** and is devoid of any other positional isomers. The regiochemistry of the initial monohydroboration is retained because of the alkyl group's β-directive effect (Eq. 5.28).

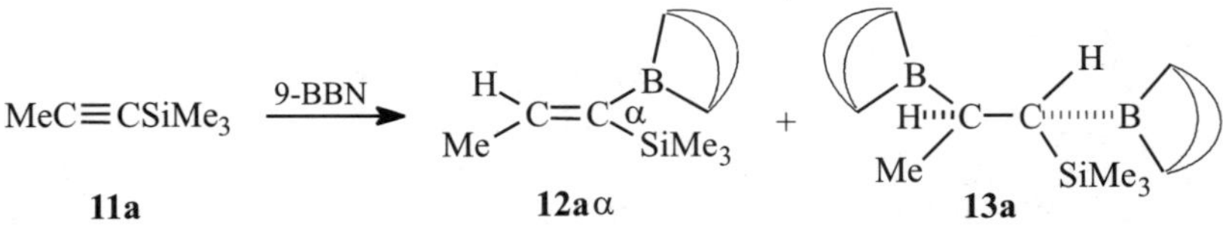

(5.28)

However, the regiochemistry and configuration of their dihydroboration products (i.e., 1,1 versus 1,2-diboryl) are altered. Thus, for alkyl-substituted silylacetylenes, TMS group alters the regioselectivity of the process at the vinylborane stage. The dominant stereoelectronic β-directing effect of the alkyl group is overcome in the case of **12aα**, which has both the boryl and silyl group at the same carbon. Steric repulsion between these two large groups forces the second hydroboration to place the boron atom at the 2 position. Unlike *t*-Bu group, the methyl group is too small to stop dihydroboration and afford 1,2-diboryl with configurations [4b, 16] as depicted in **13a**.

(β-Borylvinyl)silane adducts (**12β**), though difficult to prepare, are important intermediates, as these have two versatile functional groups [15, 17] attached to two different carbons. Moreover, since the silyl groups are normally replaced stereoselectively with hydrogen atom, the whole process becomes equivalent to the unknown Markovnikov hydroboration of terminal alkyne. Soderquist and coworkers [14] have conducted a detailed study (Table 5.24) of different silyl-substituted alkynes (**11**). The data reveal that groups smaller than triisopropylsilyl (TIPS) give initially the α-adducts (**12α**) in high regioisomeric purity. These vinyl adducts compete with **11** for 9-BBN to afford minor amounts of 1,2-diboryl products (**13**, Eq. 5.29).

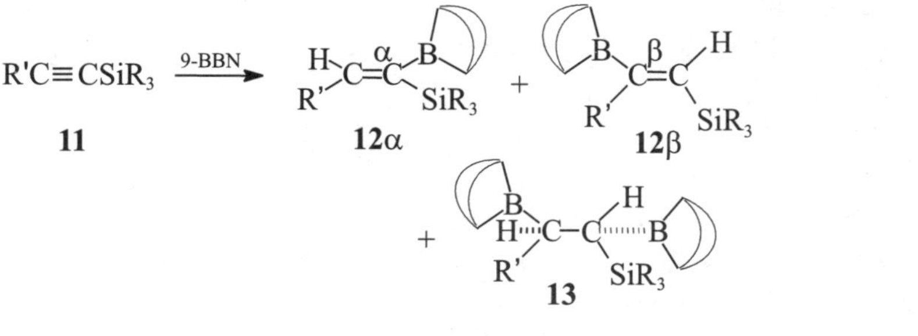

(5.29)

Table 5.24 Influence of the silyl substitution on the boron placement in different alkynylsilanes (**11**) with 9-BBN [14]

Series	R_3	R′	Yield (%)			Temperature °C and time (h)
			12α/12β	**12**	**13** (solvent)	
a	Me_3	Me	100/0	54	23 ($CDCl_3$)	25 (48)
b	Et_3	Me	97/3	70	15 (neat)	25 (36)
c	$(t\text{-Bu})Me_2$	Me	95/5	71	14 (neat)	25 (48)
d	$(i\text{-Pr})_3$	Me	0/100	100	0 (neat)	85 (1)
e	$(i\text{-Pr})_3$	*n*-Pr	0/100	100	0 (neat)	85 (3)

In contrast, the TIPS substitution not only provides the *Z*-β product cleanly, but also completely suppresses the formation of **13**. Consequently, the 1,2,2-substitution pattern (**12β**) with large groups on both the vinylic carbons, unlike the 1,1,2 (**12α**) pattern having both the large groups on a *geminal* position, prevents the dihydroboration of **11** with 9-BBN.

It is noteworthy to mention that alkyl groups principally activate the carbon β to themselves in the hydroboration process. However, when this site is blocked by steric bulk as in case of TIPS, the boron is forced to accept the internal position of the alkyne as a kinetically less preferred process.

It is reported [14] that alkynylsilanes (**11**, R′ = Me) undergo dihydroboration with 9-BBN under neat conditions to afford 1,2-dibora adducts. The normal 1,1-diboryl compounds, on heating, isomerize to mixtures of 1,2- and 1,*n*-diboryl products [17b]. However, 1,2-diboryl adducts (**14b–14d**) on heating above 160 °C undergo thermal arrangement to place both the boryl groups away from the bulky silyl group. Consequently, this "tandem walk" of two boryl groups to give 1,1-diboryl products is a fascinating synthetically useful process (Eq. 5.30).

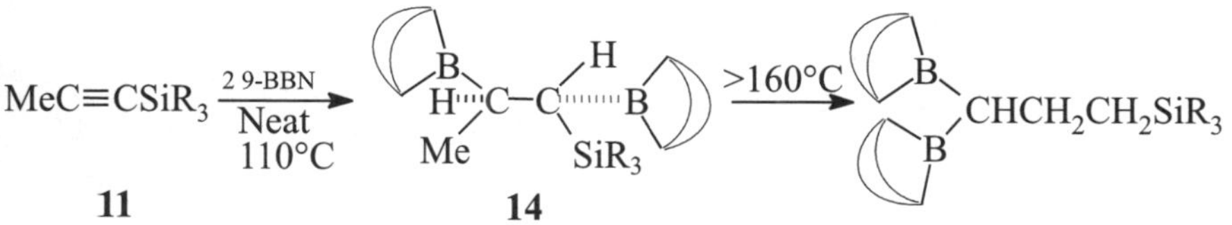

11 **14**

R_3=
b=Et_3, c=$(t\text{-Bu})Me_2$, d=$(i\text{-Pr})_3$

(5.30)

5.2.3
Hydroboration of 1-Halo-1-Alkynes

The hydroboration of 1-halo-1-alkynes is a key step in the synthesis of *trans*-alkenes [18], *trans,trans*-dienes [19], and *cis*-1-halo-1-alkenes [20]. In contrast to the hydroboration of 1-alkynes, 9-BBN reacts with 1-halo-1-alkyne in an equimolar ratio, with monohydroboration occurring at the 1 position (Eq. 5.31) [21].

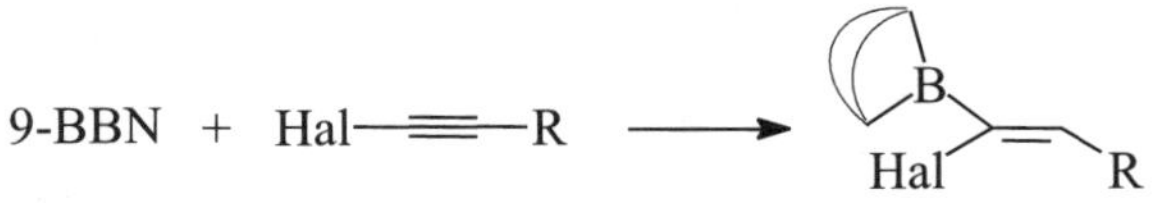

(5.31)

The relative reactivities of 1-halo-1-alkynes are summarized in Table 5.25 [21].

Table 5.25 reveals that replacing a hydrogen or alkyl group attached to triple bond with a group that reduces electron availability such as halogen or phenyl decreases the rate of the reaction.

The decrease in the rate is the order of Cl > Br > I, an expected induction trend of halogens. As the rate reduction in the largest halogen substituent, iodine, is the smallest, the electronic effects must outweigh the steric effects in the case of bromine and chlorine substituents. The results are summarized in Table 5.26 [21] and Chart 5.15.

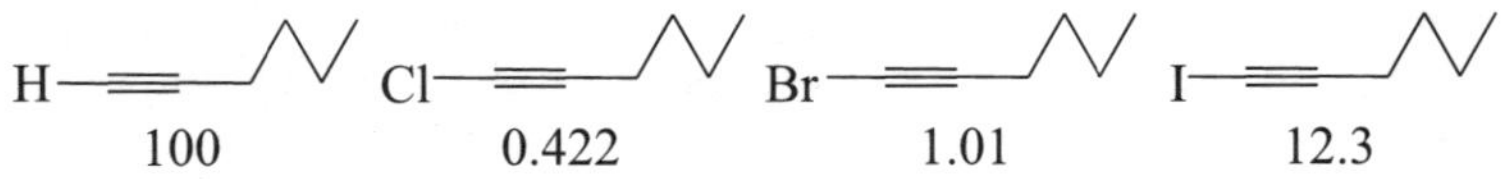

Chart 5.15

Table 5.25 Relative reactivities for the hydroboration with $(9\text{-BBN})_2$ of 1-halo-1-alkynes and their parent compounds in CCl_4 at 25 °C [21]

Compound no.	Halo alkynes	Relative reactivities	
		 = 100	 = 100
1	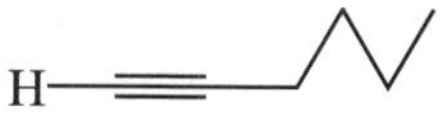	15.3	100
2	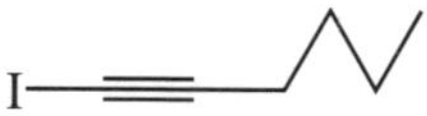	1.88	12.3
3	H—≡—Ph	1.41	9.22
4		1.31	8.56
5	I—≡—Ph	0.515	3.36
6	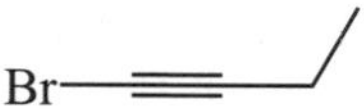	0.304	1.99
7	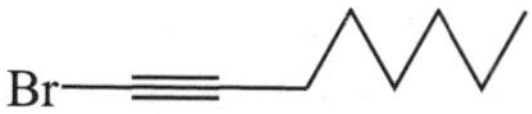	0.303	1.98
8	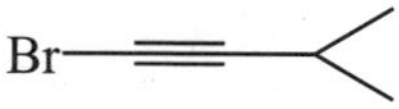	0.250	1.63
9	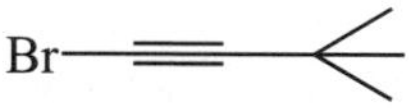	0.196	1.28
10	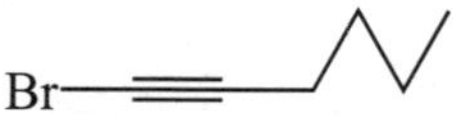	0.155	1.01
11	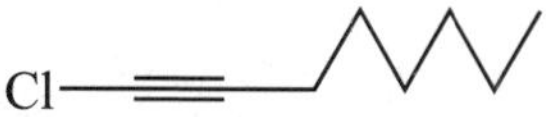	0.110	0.721
12	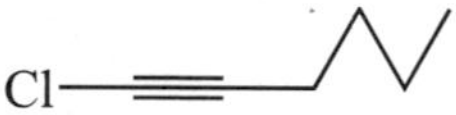	0.0645	0.422
13	Br—≡—Ph	0.0189	0.124
14	Cl—≡—Ph	0.00838	0.0548

Table 5.26 The effects of halogen substitution on the reactivity of various alkynes toward 9-BBN [21]

Haloalkyne	Parent	Factor by which reactivity is lowered		
		Cl	Br	I
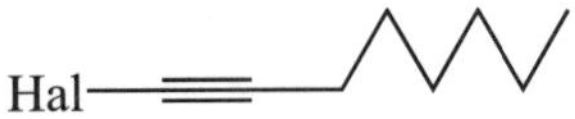		164	59.6	13.7
	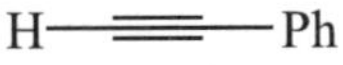	168	74.4	2.74
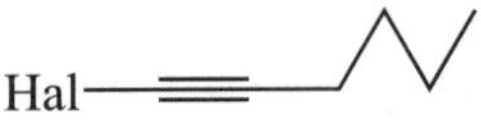	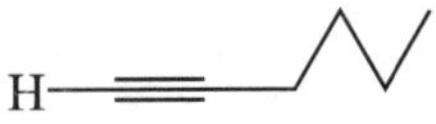	237	99	8.13
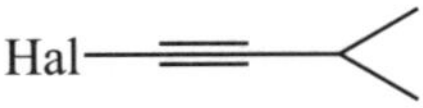	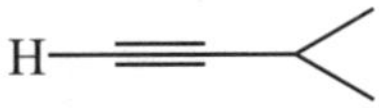		194	
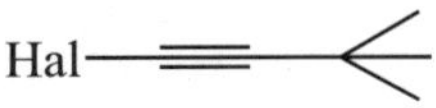	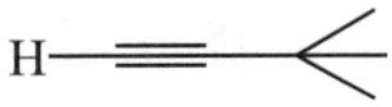		439	

The change in charge polarization across the triple bond has been found to decrease in the order I > Br > Cl. This, thus, corresponds to an increase in the amount of negative charge or electron density at C-2 and decrease in that at C-1 by proceeding from the iodo- to the bromo- to the chloro-substituted compounds. These observations correlate with the rate of hydroboration with 9-BBN at C-1 in the order of I > Br > Cl.

Another significant phenomenon in bromoalkynes is the rate-retarding effect on branching as compared with the other monohalogenated alkynes (Chart 5.16) and is explained due to greater steric requirements for hydroboration with 9-BBN. Thus, rate-retarding steric effects of methyl groups outweigh their rate-increasing inductive effects in methyl-substituted bromobutynes.

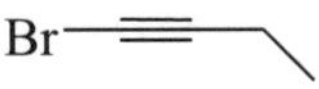

100

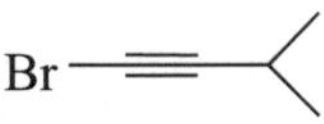

82

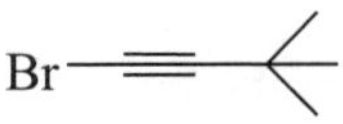

64

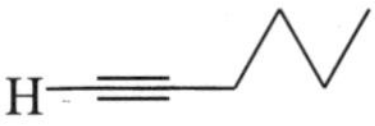

100

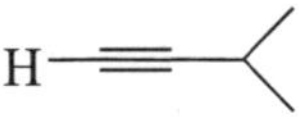

317

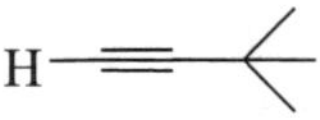

562

Chart 5.16

The sensitivity to a change in alkyl group is the lowest of iodoalkynes among its counterparts, bromoalkynes, and chloroalkynes (Chart 5.17).

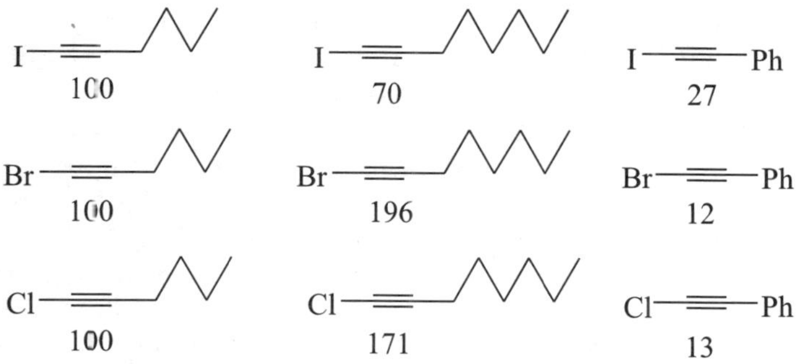

Chart 5.17

It is interesting to note that in 1-halo-1-alkynes, electronic and steric effects act together to make the first hydroboration of the alkyne occur at the terminal position. However, the second hydroboration also places the boron at the same carbon atom, yielding the *gem*-dibora derivative. As the *gem* derivative is also obtained in the hydroboration of 3-hexyne, it must be electronic rather than steric effects causing the *gem* product to be preferred. This is easily explained in terms of charge alternation effect, which would be the result of the combined inductive and mesomeric effects. The effect of the boron substituent on the direction of hydroboration of an alkene is the opposite of that of halogens. A vinyl halide favors placement of boron at the 2 position, whereas vinyl borane favors it at the 1 position. This is because the inductive and conjugative effects of halogen and boron act in opposite manners. Thus, the hydroboration in these alkenes occurs at the site of greatest electronic availability, determined primarily by the mesomeric effect.

Chlorine substitution in 1-hexyne reduces the C-1–C-2 ^{13}C NMR shift difference from 15.75 in the parent to 12.51 ppm. This reveals that the effect of chlorine on 1-hexyne makes C-1 more positive and C-2 more negative. This effect of chlorine on the alkynes agrees with the +M effect analyzed in alkenes. An identical situation is observed in 1-chloro-1-octyne and in chlorophenylacetylene.

The contrasting regiochemistry of hydroboration for 1-chloroalkynes and 1-chloroalkenes is explained. +M effect in the chloroalkynes is a smallone relative to the stronger charge pattern of the alkyne moiety, which places more electron density at the C-1 position. It is to be mentioned that chlorine, bromine and iodine are capable of exerting a –M effect as well as +M effect in these compounds [22a], and this is more apparent in the NMR data (Table 5.27) [21] of bromo- and iodoalkynes than that of chloroalkynes. However, the possibility of heavy halogen effects [22b] by bromine or iodine makes analysis of NMR data on the compounds containing these groups difficult. Consequently, in the chloroalkynes, hydroboration at C-1 is linked to a strong charge pattern set up by the alkyne moiety, which places a higher electron density at C-1 than at C-2.

The change in charge polarization across the C≡C reflected in the ^{13}C NMR data for haloalkynes indicates that polarization decreases in the order I > Br > Cl. This results in an increase in the amount of negative charge or electron density at C-2 and decrease in that at C-1, as one proceeds from the iodo- to bromo- to the chloro-substituted compounds. This thus correlates with the rate of hydroboration (which occurs at C-1) in the order I > Br > Cl. Further, the reason for the nucleophilic attack at the C-2 position [23] in 1-chloro-1-alkynes is the decrease in the electron density at that position caused by the –M effect of chlorine. Both regioselectivity and rate of hydroboration depend on the electron availability at specific sites in the molecule, and the above reason explains why hydroboration does not occur at the 2 position.

Table 5.27 ^{13}C NMR data for the 1-haloalkynes, X-C-1≡C-2-R, and their parent compounds [21]

^{13}C Shifts (±0.05 ppm)[a]			
X	R	C-1	C-2
H	*n*-Bu	67.31	83.06
Cl	*n*-Bu	56.71	69.22
Br	*n*-Bu	37.36	79.88
I	*n*-Bu	–5.86	94.89
H	*n*-Hex	67.48	83.16
Cl	*n*-Hex	56.75	69.33
Br	*n*-Hex	37.36	79.85
I	*n*-Hex	– 6.22	94.73
H	Ph	77.14	83.24
Cl	Ph	67.83	69.4
Br	Ph	50.18	80.39
I	Ph	8.96	94.85
H	Et	67.3	85
H	*i*-Pr	66.83	89.68
H	*t*-Bu	66.14	92.38
Br	Et	37.04	81.15
Br	*i*-Pr	37.38	85.08
Br	*t*-Bu	37.11	87.64

[a] Shifts are relative to the external $CDCl_3$.

The comparative data of ^{13}C NMR for 1-hexyne and the 1-halo-1-hexyne describe the importance of the –M effect, which in haloalkynes decreases in the order I > Br > Cl. In fluoroalkynes, no –M effect is possible as no *d* orbitals are available. Only a +M interaction is possible in the 1-fluoro-1-alkynes, and one expects their hydroboration at the C-1 position would (1) proceed at a slower rate than the corresponding chloro analogs, or (2) switch to C-2 position in case the electron density at C-1 decreases enough. Indeed, in 1-fluoro-1-alkynes the nucleophilic attack occurs at the C-1 position, whereas the same occurs at C-2 position in 1-chloro-1alkynes [22a]. Their comparative hydroboration studies are not conducted [21], as fluoroalkynes are difficult to prepare and handle, and often very unstable [22].

As boron is attached at specific site where electron availability is higher, the same reasoning is applicable to the question of why hydroboration does not occur at the C-2 position. Both –M and +M effects are explained in Chart 5.18.

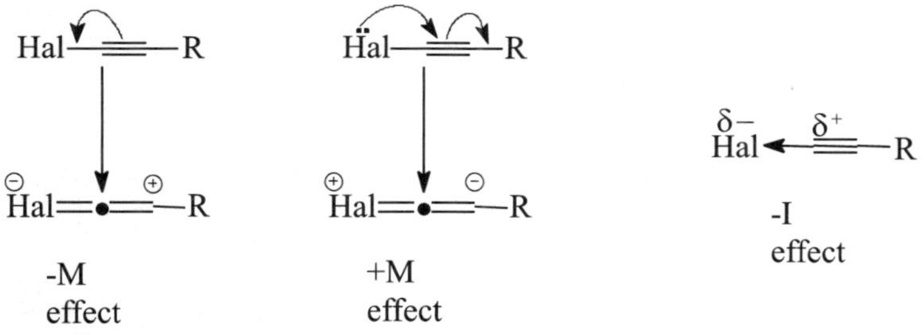

Chart 5.18

5.3 Hydroboration of Dienes

The rate of hydroboration coupled with relative reactivities of alkenes toward hydroboration with 9-BBN are used directly to predict hydroboration of unsymmetrical nonconjugated dienes. 1-Hexene is about 100 times more reactive than is *cis*-2-pentene and 116 times more reactive than is 2-methyl-2-butene. Similarly, 2-methyl-1-pentene is 194 times more reactive toward 9-BBN than *cis*-2-pentene is. The data, thus, reveal that selective monohydroboration of terminal C=C of nonconjugated dienes can be achieved, as are illustrated in Chart 5.19 [1–3].

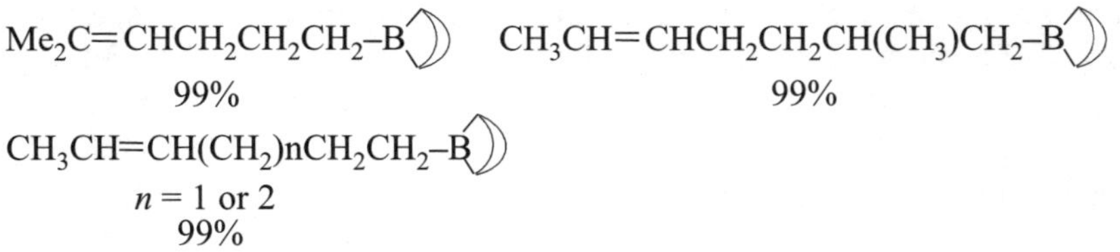

Chart 5.19

In comparison, disiamylborane is more sensitive to steric effects than 9-BBN is, and the latter more sensitive to electronic contributions as is revealed in the hydroboration of 2-methyl-1,5-hexadiene (Scheme 5.5) [1].

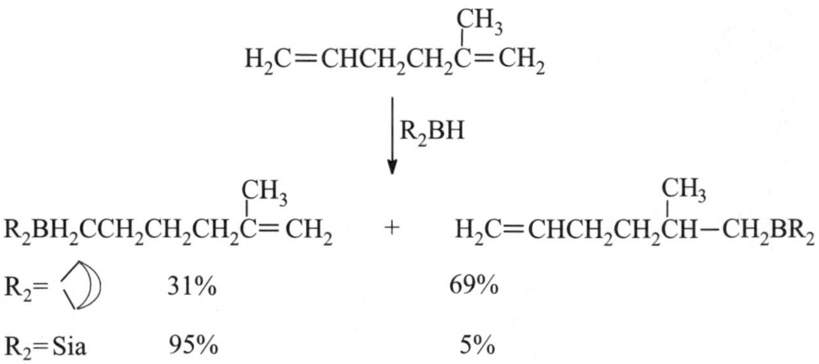

Scheme 5.5

Monohydroboration achieved in other dienes is shown in Chart 5.20.

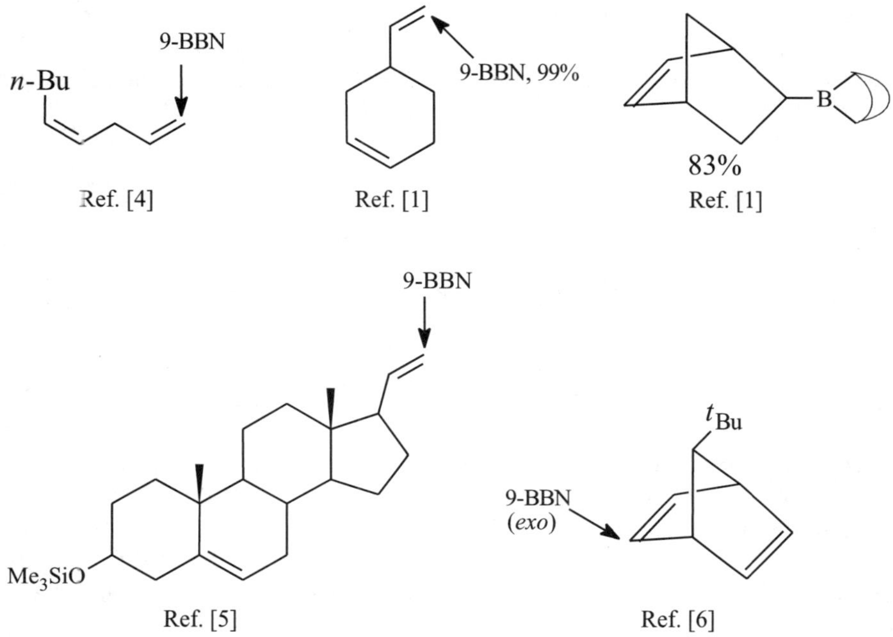

Chart 5.20

The hydroboration of symmetrical nonconjugated dienes, such as 1,5-hexadiene with 9-BBN, in a 1:1 molar ratio proceeds in an essentially statistical manner, giving approximately 25% residual diene, 50% monohydroboration product, and 25% dihydroboration product. On the other hand, conjugated dienes, for example, 1,3-butadiene, behaves differently and affords equal amounts of residual diene and 1,4-dihydroboration product. Conjugation markedly decreases the reactivity of diene toward hydroboration. Consequently, nonconjugated dienes like 1,4-hexadiene are selectively hydroborated with 9-BBN in the presence of 1,3-pentadiene (Scheme 5.6) [3].

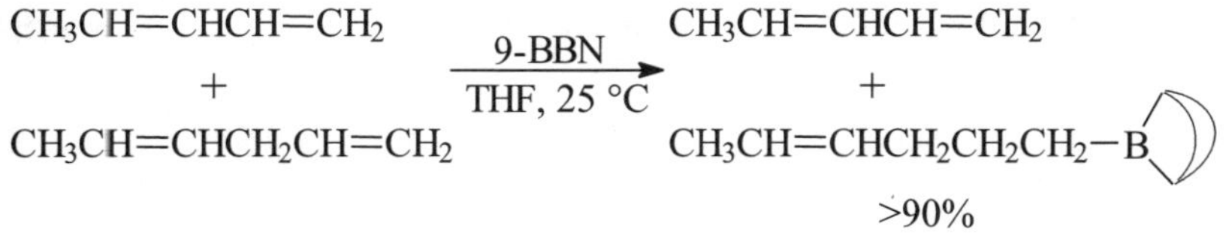

Scheme 5.6

The conjugation causes the deactivation toward the first hydroboration step. However, the formation of a homoallylic double bond in the hydroboration of 1,3-butadiene produces *B*-(3-buten-1-yl)-9-BBN, whose homoallylic double bond has reactivity comparable to that of 1-butene and second hydroboration, is much faster and produces 1,4-dibora product (Eq. 5.32) [3].

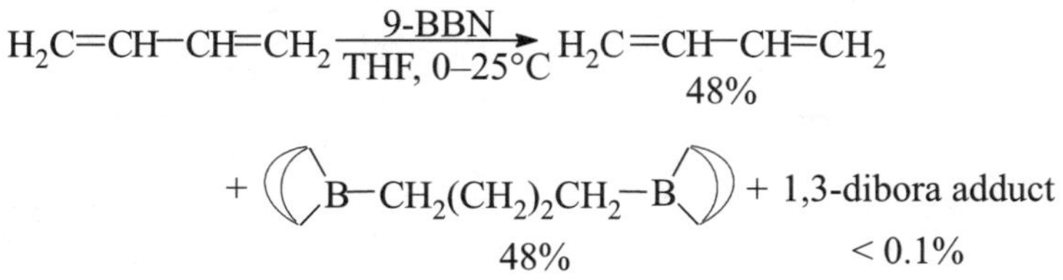

(5.32)

On the other hand, in cases where the two olefinic centers of the diene possess structures that differ considerably in reactivity toward 9-BBN, clean monohydroboration [3] is achieved.

Thus, the hydroboration of 2,5-dimethyl-2,4-hexadiene with 9-BBN proceeds rapidly to the monohydroboration product. The *geminal* methyl groups direct the boron atom to the allylic position in contrast to homoallylic position.

The large steric requirements of 9-BBN resist the second hydroboration and also its placement at the allylic position (Eq. 5.33).

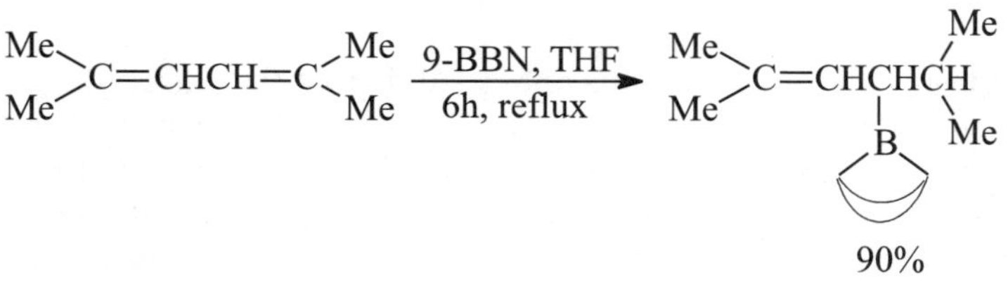

(5.33)

The product is in equilibrium with the isomer, but prefer to be attached to the secondary rather than *tertiary* center as the oxidation gives, only 2,5-diemthyl-4-hexen-3-ol. On the other hand, acetone reacts with allylic rearrangement to afford 2,3,3,6-tetramethyl-4-hepten-2-ol, in 80% yield.

Unlike 1,3-butadiene, *cis*- and *trans*-1,3-pentadienes react readily to give the corresponding monohydroboration derivatives (Scheme 5.7).

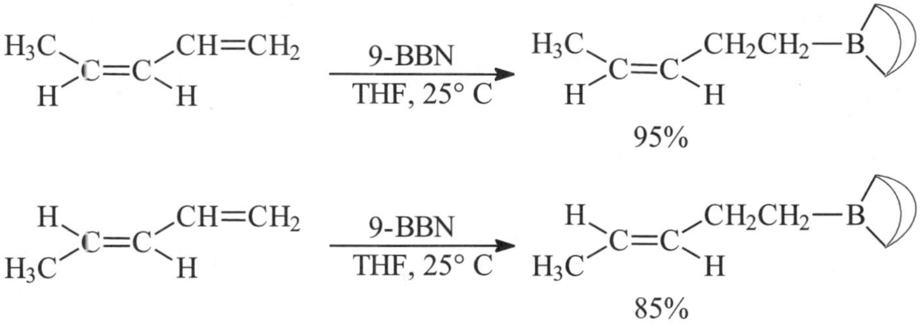

Scheme 5.7

Similarly, 5,5-dimethyl-1,3-hexadiene and 2,4-dimethyl-1,3-pentadiene afford the monohydroboration product [3].

99% 99%

Isoprene, which undergoes significant monohydroboration with disiamylborane, however, produces 1,4-dibora product (46%) with 9-BBN along with the unreacted residual isoprene (50%).

Hydroboration of cyclic 1,3-dienes also results into three major types of organoboranes [4]: allylboranes, homoallylboranes, and dibora species. As usual oxidation of cyclic allylboranes also produces olefins rather than allylic alcohols, the allylboranes are indirectly estimated by reacting them with acetaldehyde to derivatized product, before oxidation. Homoallylic and dibora species are inert to acetaldehyde. The oxidation of derivatized product affords homoallylic alcohols, which estimate the amount of allylborane. The amount of underivatized homoallylic alcohol indicates the percentage of homoallylborane, and the amount of unreacted dienes gives the amount of dihydroboration (0% diene = 0% dihydroboration, 50% diene = 100% dihydroboration).

The 1,3-cyclopentadiene reacts slowly with 9-BBN, and its dimerization competes seriously, resulting in a complex mixture of products [3, 4]. The high yield of monohydroboration is realized only in the six-membered ring diene system. As the ring size increases, monohydroboration decreases and dihydroboration increases (Chart 5.21). The high yield of monohydroboration realized in the six-membered ring system is the result of exceptional inertness due to steric requirements of the puckered six-membered ring [5].

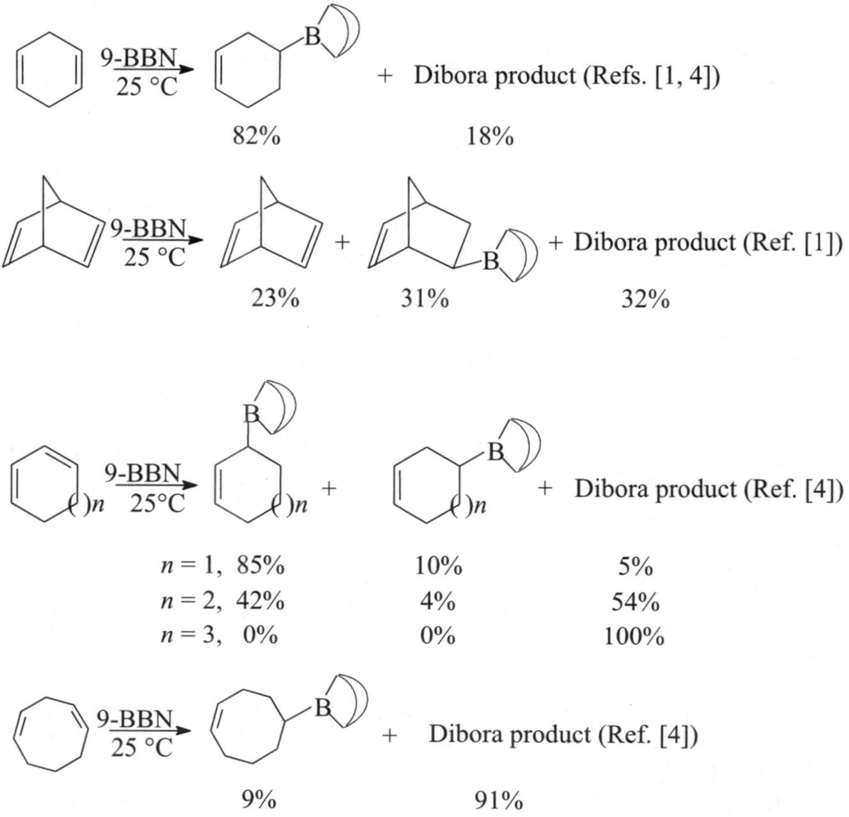

Chart 5.21

The 1,5-cyclooctadiene gives variable results [4] on hydroboration with 9-BBN.

5.4 Hydroboration of Allenes

The hydroboration of allenes or certain conjugated dienes leads to the synthesis of allylic boranes that have immense synthetic importance [1–3]. The chemistry of allylic boranes differs markedly from that of saturated analogs. Both vinylic and allylic boranes react readily with many substrates toward which trialkylboranes are inert [4]. Mikhailov in his book [5] and review [6] has documented the synthetic applicability of allylic boranes, with cautions of high thermal reactivity with respect to allylic rearrangement, e.g., 1-methyl-2-propenyl dialkylborane rearranges spontaneously to the 2-butenyl isomer even at –78 °C (Eq. 5.34).

$$R_2B\text{–}\overset{CH_3}{\overset{|}{C}}H\text{–}CH{=}CH_2 \xrightarrow{-78°C} CH_3\text{–}CH{=}CH\text{–}CH_2\text{–}BR_2 \quad (5.34)$$

In the case of hydroboration of allenes with 9-BBN, four types of organoboranes: vinylboranes (I) allylboranes (II), 1,2-dibora (III), and 1,3-dibora (IV) species are expected (Chart 5.22) [7].

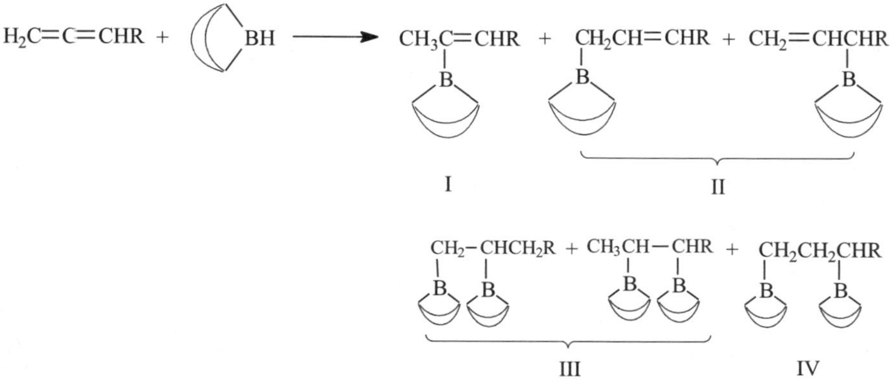

Chart 5.22

Among these species, Mikhailov [2] and Brown [8] have shown that allylboranes, especially those with low-molecular-weight, nonsterically hindered allyl groups are readily hydrolyzed and produce olefins rather than expected allylic alcohols under usual alkaline hydrogen peroxide oxidation conditions. Consequently, it is appropriate to convert the allylborane to a derivative that would give an unambiguous product on oxidation. Such a derivative is prepared by the reaction of allylborane with ketone, such as acetone, prior to oxidation [9]. The other potential adducts (I, III, and IV) are inert to acetone under these conditions. After oxidation of this derivatized mixture, the quantity of homoallylic alcohol determines the amount of the attack at the terminal carbon of the allenic system (allylborane), the quantity of ketone determines the extent of internal attack (vinylborane), and the quantity of unreacted allene reveals the amount of dihydroboration (0% allene = 0% dihydroboration, 50% allene = 100% dihydroboration). The positions of the diol reveal the point of attack in dihydroboration.

Hydroboration of propadiene leads to dihydroboration, and oxidation of the reaction mixture produces 1,3-propanediol (48%) and residual propadiene (50%) [7, 10]. It has been observed that rate of addition of the second 9-BBN is faster than is the rate of initial hydroboration. Moreover, this reveals that the attack of boron is exclusively at the both ends of the unsaturated chain in both steps, since no trace of 1,2-propanediol or acetone is detected.

The presence of an alkyl group at one end of allene chain, significantly, alters the course of reaction as hydroboration of 1,2-butadiene with 9-BBN results in the formation of mainly allylboranes (Eq. 5.35) [7].

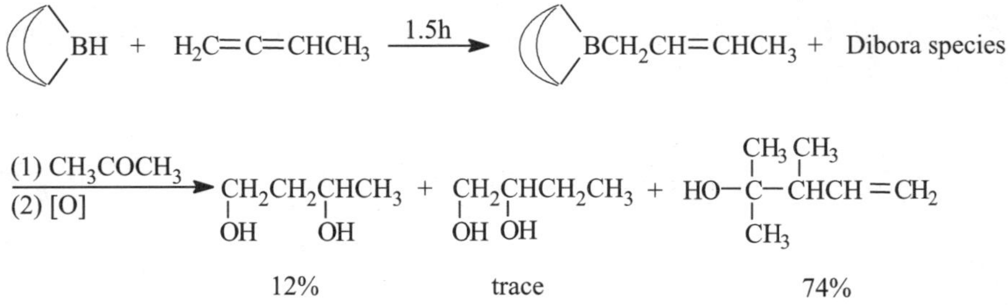

(5.35)

The studies reveal [7] that the initial attack of 9-BBN is at the terminal of the allene chain. However, allylboranes undergo a permanent allylic rearrangement [2, 8] whereby the boron easily migrates to the less substituted site. For example, if any hydroboration occurs at the carbon of the allene bearing the methyl group, the resulting α-alkyl allylborane rapidly rearranges to place the boron at the terminal carbon (Chart 5.23) [7]. The permanent allylic rearrangement also scrambles the stereochemistry of the remaining double bond. Mikhailov [2] reported that tricrotylborane exists at room temperature as a mixture of *cis* and *trans* isomers with *trans*-crotylborane predominating (70%).

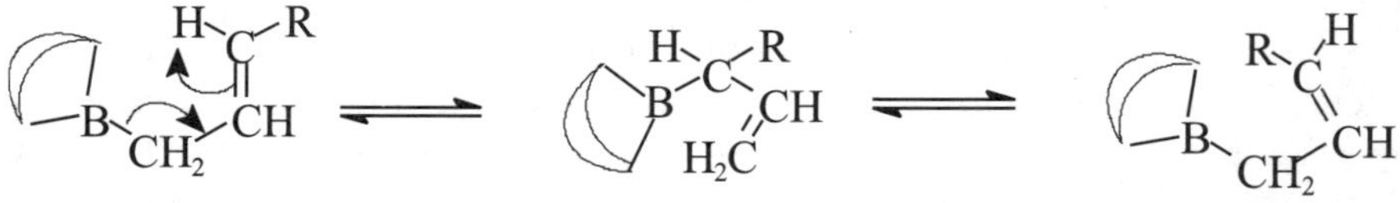

trans-crotyl α-alkylborane *cis*-crotyl

Chart 5.23

The allylboration of ketones occurs with complete allylic rearrangement [2, 7–9] (Eq. 5.36).

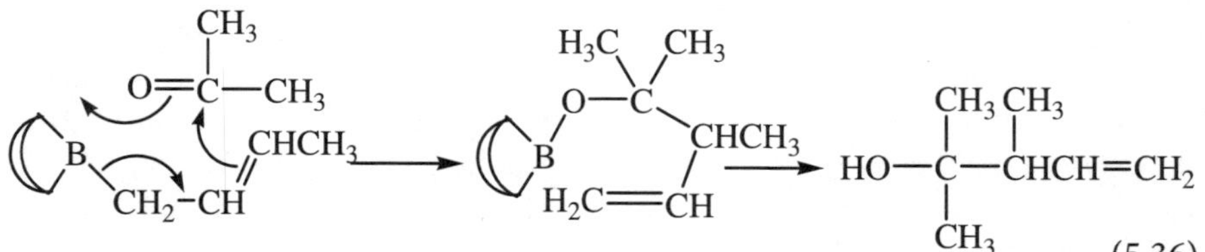

(5.36)

Brown [7, 8] have studied the directive effect of addition of 9-BBN (1 equiv) toward *B*-allyl-9-BBN. Oxidation of the bis adduct, formed after rapid hydroboration (<30 min), gives 91% of 1,3-propanediol, and no 1,2-propanediol is detected. This reveals that adduct formed is 1,3-bis-boron derivative.

Similarly, *B*-crotyl-9-BBN reacts with 9-BBN in less than 30 min at room temperature and affords after oxidation 1,3-butanediol (95%), a small amount of 1,2-isomer (3%), and traces of 2,3- and 1,4-diols (Eq. 5.37) [7].

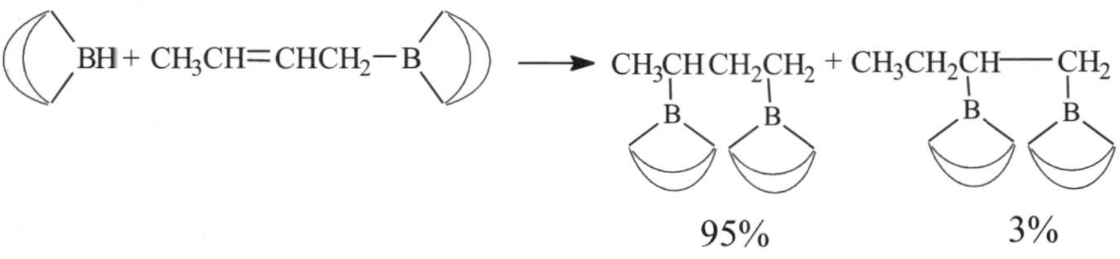

(5.37)

The above results indicate that the directivity of the attack of the second boron is due to combination of both steric and electronic factors. The attack at the C-3 position is favored on steric grounds, and the boron is located next to the smaller methyl group. As the hydroboration is also prone to electronic effects, boron is thus directed toward electronegative substituents and away from electropositive groups. Because boron is electropositive than carbon, the electronic directive effect thus also places the boron toward the 3 position [7].

Further, the addition of a single methylene unit, as in the case of 1,2-pentadiene, lowers the amount of dihydroboration by 50% (as compared with 1,2-butadiene), and increase of steric bulk at one end of the allene further decreases the amount of dihydroboration. For example, 3-methyl-1,2-butadiene gives nearly quantitative yield of the corresponding allylborane (Eq. 5.38; Table 5.28) [7].

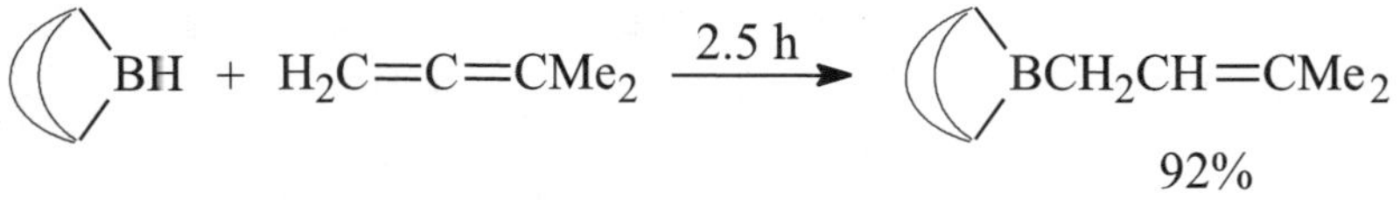

(5.38)

Replacement of the alkyl substituent with aryl moiety does not alter the regiochemistry, as phenylallene gives almost the same product distribution as does ethylallene (Table 5.28) [7]. In addition, the electronic nature of aryl groups does not influence the reaction course since phenylallene and *p*-anisylallene give almost similar results (Table 5.28) [7].

It is significant to mention that hydroboration of terminal allenes with 9-BBN gives much a cleaner reaction and affords allylboranes and dihydroboration products. On the other hand, disiamylborane gives 17% of the vinylborane derivatives [11].

Internal allenes on hydroboration with 9-BBN again give the allylboranes as the major product, as is seen in the case of 2,3-pentadiene (Eq. 5.39) [7]. Dicyclohexylborane, on the other hand, gives mixture of both vinyl- and allylboranes.

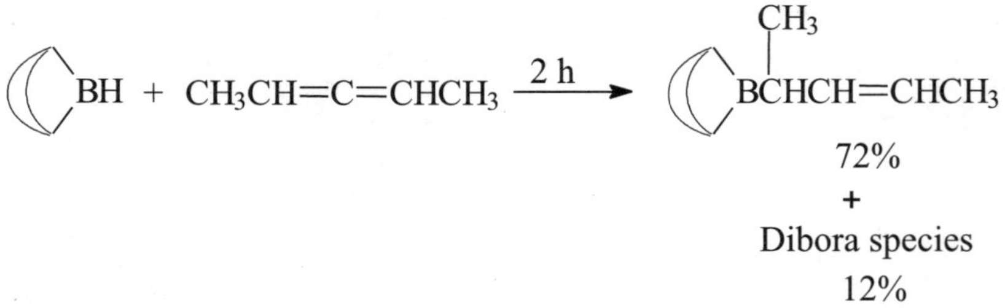

(5.39)

Similarly, trisubstituted allene, 2-methyl-2,3-pentadiene with 9-BBN affords exclusively the allylborane (Eq. 5.40; Table 5.28) [7].

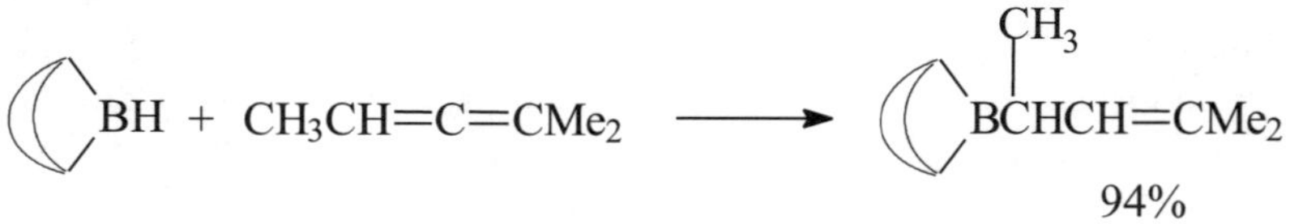

(5.40)

In sharp contrast to the behavior of 9-BBN, hydroboration of an internal allene with disiamylborane and dicyclohexylborane reveals preferential attack of the boron at the central carbon of the allenic chain, forming vinylboranes. For example, 2,4-dimethyl-2,3-pentadiene is converted [7] essentially, quantitatively, to the vinylborane with dicyclohexyl- and disiamylboranes, but exclusively to the allylborane with 9-BBN (Chart 5.24). In sharp contrast to acyclic allylborane, these allylborane products of 9-BBN do not react with acetone; however, it undergoes usual alkaline hydrogen peroxide oxidation to afford the corresponding allylic alcohol. No protonolysis is observed on oxidation. This anomalous behavior of the allylborane is due to the steric bulk surrounding the boron atom, which prevents the coordination of oxygen of the acetone or water, necessary for allylboration or protonolysis [2, 8].

The *B*-allyl-9-BBN derivatives formed by hydroboration of allenes are utilized for addition of the allyl group to the carbonyl moiety of aldehydes, ketones, and other carbonyl derivatives (allylboration). Since allenes are easily prepared from corresponding olefins [12, 13], the route the olefin–allene–allylborane–carbonyl addition adduct is a general allylation procedure, which provides an attractive alternative to Grignard-based sequences for the syntheses of complex structures.

Table 5.28 Hydroboration–derivatization–oxidation product distribution of reaction of allenes with 9-BBN and Chx_2BH [7]

Allene	Hydroborating agent	Reaction time (h)	Residual allene (%)	Homoallylic alcohol (%)	Diol (%)	Ketone (%)
Propadiene	9-BBN	2.5	50	0	1,3-diol (48)	0
1,2-Butadiene	9-BBN	1.5	7	74	1,3-diol (12) 1,2-diol (trace)	Trace
1,2-Pentadiene	9-BBN	1.5	4	82	1,2- and 1,3-diols (5)	2
3-Methyl-1,2-butadiene	9-BBN	2.5	Trace	92	–	Trace
Phenylpropadiene	9-BBN	4	11.5	86	1,2- and 1,3-diols (6)	Trace
p-Anisylpropadiene	9-BBN	4	4	87	1,2- and 1,3-diols (6)	Trace
3-Phenyl-1,2-butadiene	9-BBN	4	Trace	97	–	–
2,3-Pentadiene	9-BBN Chx_2BH	2 3	5 2	72 26	2,3- and 2,4-diols (12) (20)	– 28
4-Methyl-2,3-pentadiene	9-BBN	–	–	94	–	–
2,4-Dimethyl-2,3-pentadiene	9-BBN Chx_2BH Sia_2BH	>108 3 14	8 3 3	86 – –	(0) (5) (5)	Trace 90 90
1,2-Cyclononadiene	9-BBN	1	Trace	83	(0)	17

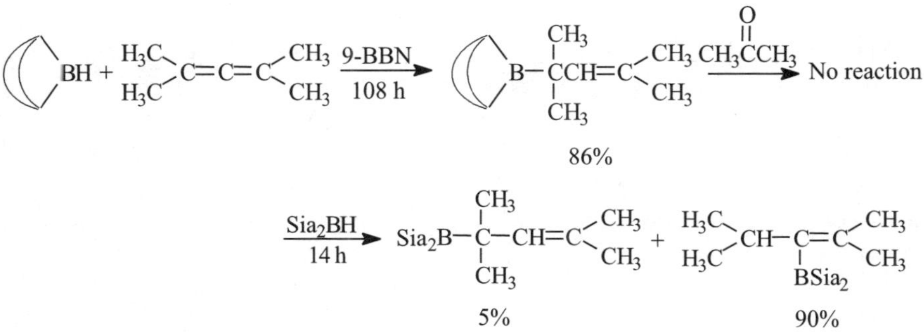

Chart 5.24

Cyclic allenes also afford the corresponding allyl derivatives (Chart 5.25) [7] with 9-BBN and vinylboranes with Sia_2BH.

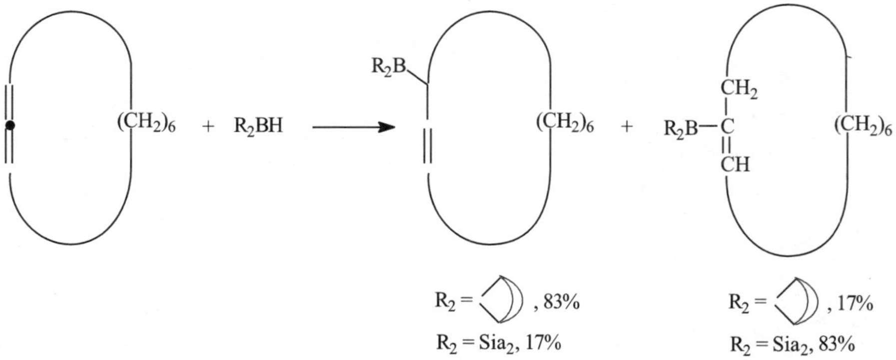

Chart 5.25

5.5 Hydroboration of Enynes and Diynes

With the availability of wide variety of hydroborating agents, it is now possible to hydroborate either of the internal or the terminal unsaturation. In a mixture of an alkene and an alkyne, 9-BBN preferentially attacks the carbon–carbon double bond only (Chart 5.26) [1].

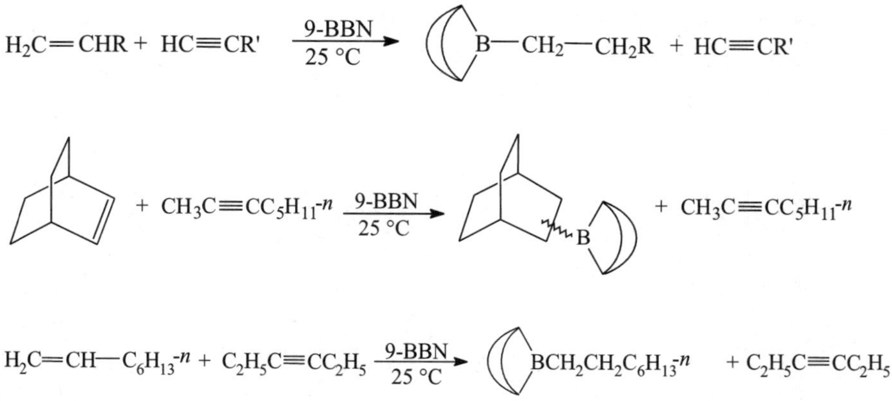

Chart 5.26

Consequently, hydroboration of enyne with 9-BBN [1] exclusively yields a product resulting from the terminal double bond, while on the other hand, dibromoborane [2] reacts only on the internal triple bond (Scheme 5.8).

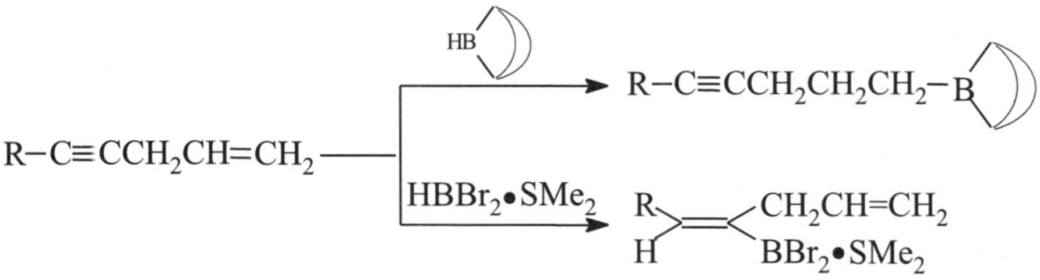

Scheme 5.8

Similarly 9-BBN differentiates between the internal and the terminal triple bonds (Eq. 5.41) of diynes, hydroborating only the terminal carbon–carbon triple bond.

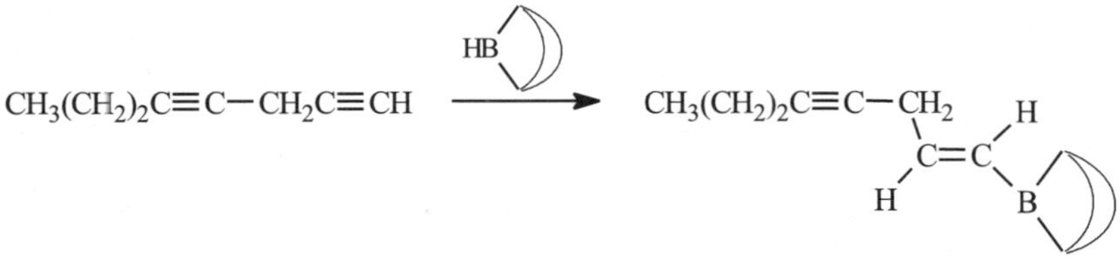

(5.41)

References

Section 5.1

1. Brown HC, Zweifel G (1960) J Am Chem Soc 82:4708
2. (a) Brown HC (1975) Organic syntheses via boranes. Wiley, New York (Reprinted as Vol 1 by Aldrich Chemical Co., 1999 catalog no. Z40:094-7); (b) Brown HC, Zaidlewicz M, Negishi E (1982) Comprehensive organometallic chemistry, vol. 7. Pergamon, Oxford; (c) Petler A, Smith K (1979) Comprehensive organic chemistry, vol. 3. Pergamon, Oxford
3. (a) Brown HC, Unni MK (1968) J Am Chem Soc 90:2902; (b) Brown HC, Gallivan RM Jr (1968) ibid. 90:2906; (c) Brown HC, Sharp RL (1968) ibid. 90:2915
4. Brown HC, Knights EF, Scouten CG (1974) J Am Chem Soc 96:7765
5. Scouten CG, Brown HC (1973) J Org Chem 38:4092
6. Taniguchi H, Brener L, Brown HC (1976) J Am Chem Soc 98:7107
7. Brown HC, Liotta R, Scouten CG (1976) J Am Chem Soc 98:5297

8. Brown HC, Nelson DJ, Scouten CG (1983) J Org Chem 48:641
9. Vishwakarma LC, Fry A (1980) J Org Chem 45:5306
10. Nelson DJ, Brown HC (1982) J Am Chem Soc 104:4907
11. Turner RB, Meador WR (1957) J Am Chem Soc 79:4133
12. Lide DR Jr, Christensen D (1961) J Chem Phys 35:1374
13. Wang KK, Scouten CG, Brown HC (1982) J Am Chem Soc 104:531
14. Brown HC, Kawakami JH, Liu K-T (1973) J Am Chem Soc 95:2209
15. Brown HC, Liotta R, Brener L (1977) J Am Chem Soc 99:3427
16. Brown HC, Klimisch RL (1966) J Am Chem Soc 88:1430
17. Berner L, Brown HC (1977) J Org Chem 42:2702
18. Brown HC, Krishnamurthy S, Yoon NM (1976) J Org Chem 41:1778
19. Brown HC, Chen J (1981) J Org Chem 46:3978
20. Fleming I, Lawrence NJ (1988) Tetrahedron Lett 29:2073
21. Knights EF, Brown HC (1968) J Am Chem Soc 90:5281
22. Fleming I, Henning R, Plaut H (1984) J Chem Soc Chem Commun 29
23. Fleming I, Sanderson PEJ (1987) Tetrahedron Lett 28:4229
24. Fleming I, Lawrence NJ (1988) Tetrahedron Lett 29:2077
25. Burgess K, Ohlmeyer MJ (1989) Tetrahedron Lett 30:395
26. Evans DA, Fu GC, Hoveyda AH (1988) J Am Chem Soc 110:6917
27. Burgess K, Cassidy J, Ohlmeyer MJ (1991) J Org Chem 56:1020
28. Burgess K, Ohlmeyer MJ (1991) J Org Chem 56:1027
29. Burgess K, Ohlmeyer MJ (1989) Tetrahedron Lett 30:5857
30. Burgess K, Ohlmeyer MJ (1989) Tetrahedron Lett 30:5861
31. Hibino J-I, Okazoe T, Takai K, Nozaki H (1985) Tetrahedron Lett 26:5579
32. Houk KN, Rondan NG, Wu Y-D, Metz JT, Paddon-Row MN (1984) Tetrahedron 40:2257
33. Still WC, Barrish JC (1983) J Am Chem Soc 105:2487
34. Houk KN, Paddon-Row MN, Rondan NG, Wu Y-D, Brown FK, Spellmeyer DC, Metz JT, Li Y, Loncharich RJ (1986) Science 231:1108
35. Burgess K, Van der Donk WA, Jarstfer MB, Ohlmeyer MJ (1991) J Am Chem Soc 113:6139
36. Evans DA, Fu GC, Hoveyda AM (1992) J Am Chem Soc 114:6671
37. Benmaarouf-Khallaayoun Z, Baboulene M, Speziale V, Lattes A (1985) J Organomet Chem 289:309
38. Brown HC, Vara Prasad JVN, Zee SH (1986) J Org Chem 51:439
39. (a) Brown HC, Zweifel G (1961) J Am Chem Soc 83:1241; (b) Brown HC, Sharp RL (1966) J Am Chem Soc 88:5851 and references cited therein
40. Saunders KJ (1973) Organic polymer chemistry. Chapman and Hall, London
41. Singaram B, Goralski CT, Fisher GB (1991) J Org Chem 56:5691
42. Singaram B, Rangaishenvi MV, Brown HC, Garalski CT, Hasha DL (1991) J Org Chem 56:1543
43. Brown HC, Ichikawa K (1957) Tetrahedron 1:221
44. Brown HC, Brewster JH, Shechter H (1954) J Am Chem Soc 76:467. For a review of the structure and reactivity of enamines, see Hickmott PW (1982) Tetrahedron 38:1975
45. (a) de Jesus M, Rosario O, Larson GL (1977) J Organomet Chem 132:301; (b) Brown HC, Racherla US (1982) Organometallics 1:765

46. Pasto DJ, Snyder R Sr (1966) J Org Chem 31:2773:2777
47. Pasto DJ, Hickman J (1968) J Am Chem Soc 90:4445
48. Mikhailov BM, Shchegoleva TA (1959) Izvest Akad Nauk SSSR 546
49. Pasto DJ, Cumbo CC (1964) J Am Chem Soc 86:4343
50. Zweifel G, Plamondon J (1970) J Org Chem 35:898
51. Srivastava RM, Brown RK (1970) Can J Chem 48:2334
52. Clark-Lewis JW, McGarry EJ (1973) Aust J Chem 26:809
53. Still WC, Goldsmith DJ Jr (1970) J Org Chem 35:2282
54. Kirkiacharian BS, Garnier M (1973) CR Acad Sci Ser C 277:1037
55. Kirkiacharian BS (1975) J Chem Soc Chem Commun 162
56. Kirkiacharian BS, Chidiac H (1975) CR Hebd Seances Acad Sci Ser C 280:775
57. Krug RC, Boswell DE (1962) J Org Chem 27:95
58. Lyle RE, Spicer CK (1970) Tetrahedron Lett 11:1133
59. Polivka Z, Ferles M (1970) Collect Czech Chem Commun 35:2392
60. Ferles M, Hauer J, Kolar J, Polivka Z, Stern P (1972) Collect Czech Chem Commun 37:2464
61. Ferles M, Stern P, Trska P, Vysata F (1973) Collect Czech Chem Commun 38:1206
62. Ferles M, Stern P, Trska P (1974) Collect Czech Chem Commun 39:3317
63. Iorio MA, Barrios GN, Menichini E, Mazzeo-Farina A (1975) Tetrahedron 31:1959
64. Irreverre F, Morita K, Robertson AV, Witkop B (1963) J Am Chem Soc 85:2824
65. Fujita Y, Irreverre F, Witkop B (1964) J Am Chem Soc 86:1844
66. Lyle RE, Carle KR, Ellefson CR, Spicer CK (1970) J Org Chem 35:802
67. Caron-Sigaut C, Le Men-Oliver L, Hugel G, Lévy J, Le Men J (1979) Tetrahedron 35:957
68. Shono T, Matsumura Y, Tsubata K, Sugihara Y, Yamane S, Kanazawa T, Aoki T (1982) J Am Chem Soc 104:6697
69. Brown HC, Vara Prasad JVN, Zee SH (1985) J Org Chem 50:1582
70. Bowers Nemia MM, Lee J, Joullie MM (1983) Synth Commun 13:1117
71. Brown HC, Vara Prasad JVN, Gupta AK (1986) J Org Chem 51:4296
72. Brown HC, Ramachandran PV, Vara Prasad JVN (1985) J Org Chem 50:5583
73. Smith MB, March J (2001) March's advanced organic chemistry reactions mechanisms and structure, 5th edn. Wiley, New York, p 20

Section 5.2

1. (a) Pelter A, Smith K, Brown HC (1988) Borane reagents. Academic, London; (b) Soderquist JA, Brown HC (1981) J Org Chem 46:4559; (c) Soderquist JA, Negron A (1991) Org Synth 70:169; (d) Köster R, Yalpani M (1991) Pure Appl Chem 63:387
2. (a) Brown HC, Campbell JB Jr (1981) Aldrichim Acta 14:3; (b) Miyaura N, Yamada K, Suginome H, Suzuki A (1985) J Am Chem Soc 107:972; (c) Martinez-Fresneda P, Vaultier M (1989) Tetrahedron Lett 30:2929; (d) Roush WR, Moriarty KJ, Brown BB (1990) ibid. 31:6509; (e) Cole TE, Quintanilla R, Rodewald S (1991) Organometallics 10:3777; (f) Oppolzer W, Radinov RN (1993) J Am Chem Soc 115:1593
3. (a) Brown HC, Scouten CG, Liotta R (1979) J Am Chem Soc 101:96; (b) Wang KK, Scouten CG, Brown HC (1982) ibid. 104:531

4. (a) Brown HC, Soderquist JA (1980) J Org Chem 45:846; (b) Soderquist JA, Brown HC (1980) ibid. 45:3571; (c) Soderquist JA, Hassner A (1983) ibid. 48:1801; (d) Soderquist JA, Shiau F-Y, Lemesh RA (1984) ibid. 49:2565; (e) Soderquist JA, Najafi MR (1986) ibid. 51:1330; (f) Soderquist JA, Negron A (1987) ibid. 52:3441; (g) Soderquist JA, Negron A (1989) ibid. 54:2462; (h) Soderquist JA, Colberg JC, Del Valle L (1989) J Am Chem Soc 111:4873; (i) Soderquist JA, Rivera I, Negron A (1989) J Org Chem Soc 54:4051; (j) Soderquist JA, Rivera I (1989) Tetrahedron Lett 30:3919; (k) Soderquist JA, Santiago B (1990) ibid. 31:5113; (l) Soderquist JA, Vaquer J (1990) ibid. 31:4545; (m) Soderquist JA, Santiago B, Rivera I (1990) ibid. 31:4981; (n) Soderquist JA, Santiago B (1990) ibid. 31:5541; (o) Santiago B, Soderquist JA (1992) J Org Chem 57:5844; (p) Rivera I, Colberg JC, Soderquist JA (1992) Tetrahedron Lett 33:6919
5. Zweifel G, Fisher RP, Horng A (1973) Synthesis 37
6. Brown HC, Rhodes SP (1969) J Am Chem Soc 91:4306
7. Knight EF, Brown HC (1968) J Am Chem Soc 90:5281
8. Colberg JC, Rane A, Vaquer J, Soderquist JA (1993) J Am Chem Soc 115:6065
9. Singleton DA, Martinez JP (1991) Tetrahedron Lett 32:7365
10. (a) Midland MM, Tramontano A, Zderic SA (1977) J Organomet Chem 134:C17; (b) Midland MM, Tramontano A, Zderic SA (1978) ibid. 156:203; (c) Midland MM, Greer S, Tramontano A, Zderic SA (1979) J Am Chem Soc 101:2352; (d) Midland MM, Zderic SA (1982) ibid. 104:525; (e) Midland MM, McLoughlin JI (1984) J Org Chem 49:1317; (f) Midland MM (1989) Chem Rev 89:1553; (g) Brown HC, Ramachandran PV, Weissman SA, Swaminathan S (1990) J Org Chem 55:6328; (h) Soderquist JA, Anderson CL, Miranda EI, Rivera I, Kabalka GW (1990) Tetrahedron Lett 31:4677; (i) Brown HC, Ramachandran PV (1992) Acc Chem Res 25:16
11. Soderquist JA, Rivera I (1988) Tetrahedron Lett 29:3195
12. PC Model: available from Serena Software, Bloomington IN 47402-3076
13. (a) Gassman PG, Singleton DA (1987) Tetrahedron Lett 28:5969; (b) Gassman PG, Gorman DB (1990) J Am Chem Soc 112:8623
14. Soderquist JA, Colberg JC, Valle LD (1989) J Am Chem Soc 111:4873
15. (a) Fleming I (1979) In: Barton DHR, Ollis WD (eds) Comprehensive organic chemistry, vol 3. Pergamon, Oxford p 647; (b) Colvin E (1981) Silicon in organic synthesis. Butterworths London; (c) Weber WP (1983) Silicon reagents on organic synthesis. Springer, Berlin Heidelberg New York; (d) Armitage DA (1982) In: Wilkinson G, Stone FGA, Abel EW (eds) Comprehensive organometallic chemistry, vol. 2. Pergamon Oxford, p 1; (e) Davies AG, Smith PVJ (1982) ibid. p 519; (f) Soderquist JA, Anderson CL Tetrahedron Lett 29 (1988) 2777
16. For the ^{13}C NMR of TMS derivatives see: Levy GC, White DM, Cargioli JD (1972) J Magn Reson 8:280
17. Zaidlewicz M (1982) In: Wilkinson G, Stone FGA, Abel EW (eds) Comprehensive organometallic chemistry, vol. 7. Pergamon, Oxford, p 143, 161
18. Negishi E, Katz J-J, Brown HC (1972) Synthesis 555
19. Negishi E, Yoshida TJ (1973) Chem Soc Chem Commun 606
20. Zweifel G, Arzoumanian H (1967) J Am Chem Soc 89:5086
21. Nelson DJ, Blue CD, Brown HC (1982) J Am Chem Soc 104:4913
22. (a) Delavarenne SU, Viehe HG (1969) Chemistry of acetylenes. Dekker, New York; (b) Maciel GE (1974) Topics in carbon-13 NMR spectroscopy. Wiley, New York
23. Dale J (1969) Chemistry of acetylenes. Dekker, New York

Section 5.3

1. Liotta R, Brown HC (1977) J Org Chem 42:2836
2. Marfat A, McGuirk PR, Helquist P (1979) J Org Chem 44:3888
3. Brown HC, Liotta R, Kramer GW (1978) J Org Chem 43:1058
4. Brown HC, Bhat KS (1986) J Org Chem 51:445
5. Zweifel G, Nagase K, Brown HC (1962) J Am Chem Soc 84:183, 190

Section 5.4

1. Mikhailov BM (1973) Intra-Sci Chem Rep 7:191
2. Mikhailov BM (1972) Organometal Chem Rev A 8:1
3. Mikhailov BM (1974) Pure Appl Chem 39:505
4. Onak T (1975) Organoborane chemistry. Academic, New York
5. Mikhailov BM, Bubnov YN (1984) Organoboron compounds in organic synthesis. Harwood, Chur, Switzerland
6. Mikhailov BM (1976) Ups Khim 45:1102
7. Brown HC, Liotta R, Kramer GW (1979) J Am Chem Soc 101:2966
8. Kramer GW, Brown HC (1977) J Organomet Chem 132:9
9. Kramer GW, Brown HC (1977) J Org Chem 42:2292
10. Chevolot L, Soulié J, Cadiot P (1974) Tetrahedron Lett 15:3435
11. (a) Sethi DS, Joshi GC, Devoprabhakara D (1969) Can J Chem 47:1083; (b) Mehrotra I, Devaprabhakara D (1971) J Organomet Chem 33:287
12. Skattabol L (1963) Acta Chem Scond 17:1683; Moore WR, Ward HR (1962) J Org Chem 27:4179
13. Makosza M, Wawrzyniewicz M (1969) Tetrahedron Lett10:4659

Section 5.5

1. Brown CA, Coleman RA (1979) J Org Chem 44:2328
2. Brown HC, Campbell JB Jr (1980) J Org Chem 45:389

6 Synthesis of Alcohols

6.1 Synthesis of Saturated Alcohols

Hydroboration-oxidation is now a standard method for *anti*-Markovnikov's *cis*-hydration [1] of alkenes from the less hindered side of the double bond. The hydroboration of olefins, mainly in THF solvent, affords conveniently the corresponding organoboranes. Oxidation of the intermediate organoboranes is most conveniently carried out by successive addition of 3 M NaOH and 30% hydrogen peroxide (depicted as [O], Eq. 6.1). Other oxidation procedures, e.g., H_2O_2 in buffer solutions, H_2O_2 in NaOAc, per acids, and trimethyl-*N*-oxides are used in certain cases when the organoboranes are unstable to water or contain functionalities sensitive to alkaline medium.

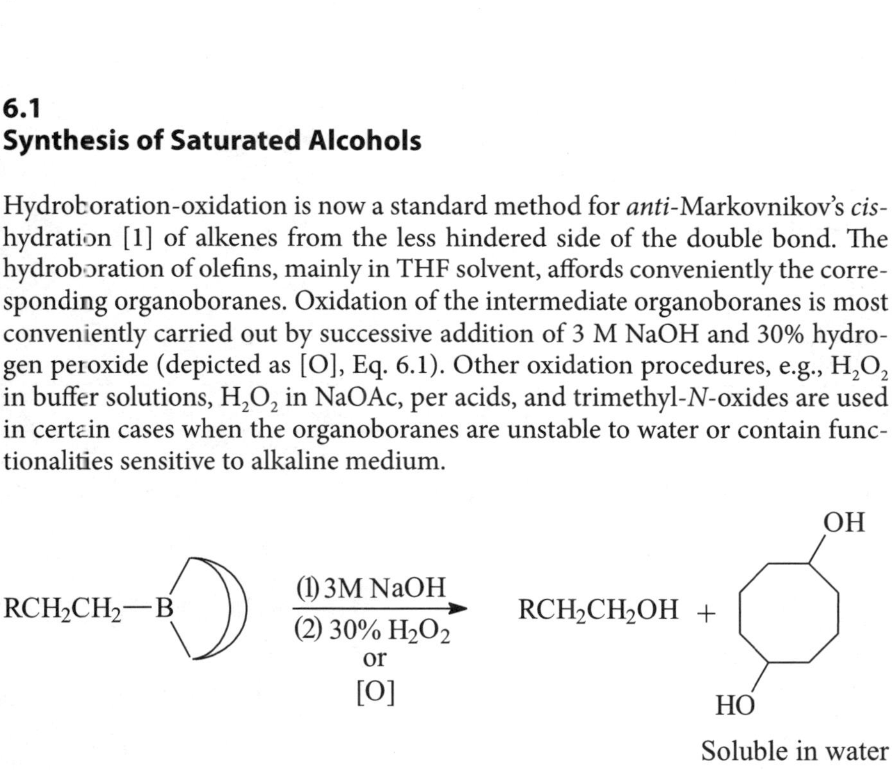

(6.1)

Among all boranes, 9-BBN shows the highest sensitivity to subtle differences in steric environment. Though, disiamylborane is sterically more hindered than 9-BBN, but surprisingly, the hydroboration of 1-hexene with 9-BBN in THF at 25 °C is more regioselective than is disiamylborane. The comparative data are given in Table 6.1 [2].

In general, the intermediate organoboranes on usual oxidation afford the corresponding alcohols, essentially in quantitative yields. Only in the case of 2,3-dimethyl-2-butene is the primary alcohol (<0.5%) derived from an isomerized organoborane. The results are summarized in Table 6.2 [2].

Table 6.1. Hydroboration-oxidation products in the reaction of unsymmetrical olefins with diborane, disiamylborane and 9-boracyclo[3.3.1]nonane (9-BBN) [2]

Olefin	Hydroborating agent	Product distribution (%)[a]		
		1-ol	2-ol	3-ol
1-Hexene	BH_3	94	6	
	Sia_2BH	99	1	
	9-BBN	>99.9		
Styrene	BH_3	80	20	
	Sia_2BH	98	2	
	9-BBN	98.5	1.5	
cis-4-Methyl-2-pentene	BH_3		57	43
	Sia_2BH		97	3
	9-BBN		99.8	0.2

[a] Total yields of products are 95±5%.

Table 6.2. Products from the hydroboration–oxidation of representative olefins with 9-BBN in THF [2]

Olefin	Time (h)	Temp (°C)	Product distribution (%)[a]
1-Hexene	2	25	1-Hexanol, >99.9
3,3-Dimethyl-1-butene	2	25	3,3-Dimethyl-1-butenol, >99.7 3,3-Dimethyl-2-butanol, <0.3
Styrene	4	25	1-Phenylethanol, 98.5 2-Phenylethanol, 1.5
2-Methyl-1-pentene	2	25	2-Methyl-1pentanol, >99.8
cis-3-Hexene	1	65	3-Hexanol, 100
cis-4-Methyl-2-pentene	1	65	4-Methyl-2-pentanol, 99.8 2-Methyl-3-pentanol, 0.2
cis-4,4-Dimethyl-2-pentene	1	65	4,4-Dimethyl-2-pentanol, 99.9 2,2-Dimethyl-3-pentanol, 0.1
2-Methyl-2-butene	1	65	3-Methyl-2-butanol, >99.8
Cyclopentene	2	25	Cyclopentanol, 100
Cycloheptene	2	25	Cycloheptanol, 100
Cyclooctene	2	25	Cyclooctanol, 100
Norbornene	2	25	*exo*-Norborneol, 99.5 *endo*-Norborneol, 0.5
Cyclohexene	1	65	Cyclohexanol
1-Methylcyclohexene	8	65	*trans*-2-Methylcyclohexanol, >99.8
2,3-Dimethyl-2-butene	8	65	2,3-Dimethyl-2-butanol, 99.5 2,3-Dimethyl-1-butanol, 0.5

[a] The combined yield of alcohols is >97%, except for 2,3-dimethyl-2-butene, where it is 95%.

Table 6.3 Products from the hydroboration-oxidation of representative olefins with 9-BBN in refluxing benzene and hexane solvents [2]

Olefins	Product	Yield (%)[a,b] Benzene	Hexane
1-Hexene	1-Hexanol	>99.5	>99.5
	2-Hexanol	<0.5	Trace
3,3-Dimethyl-1-butene	3,3-Dimethyl-1-butanol	>99.5	>99.5
	3,3-Dimethyl-2-butanol	Trace	Trace
Styrene	2-Phenylethanol	98	98
	1-Phenylethanol	2	2
cis-3-Hexene	3-Hexanol	100	100
2,3-Dimethyl-2-butene[c]	2,3-Dimethyl-2-butanol	99.5	99.5
		0.5	0.5
Cyclopentene	Cyclopentanol	100	100
Cyclohexene	Cyclohexanol	100	100
1-Methylcyclohexene[d]	*trans*-2-Methyl-cyclohexanol	>99.5	>99.5
	Cyclohexanemethanol	Trace	Trace
Norbornene	*exo*-Norborneol	99.5	99.5
	endo-Norborneol	0.5	0.5

[a] Reaction times are 8 h.
[b] Total yields of 93–100% are obtained.
[c] Reaction time of 12 h. Total yield of alcohols is 71% in benzene and 40% in hexane.
[d] Reaction time of 12 h.

The results of hydroboration-oxidation in solvents benzene and hexane are summarized in Table 6.3 [2]. Hydroboration is slower in non-ether solvents.

The *cis*-4-methyl-2-pentene reacts to place 99.8% (Chart 6.1) of the boron of 9-BBN at the less hindered of the two carbon atoms of the double bond. It is significant to mention that no selectivity is observed with either thexylborane or diborane.

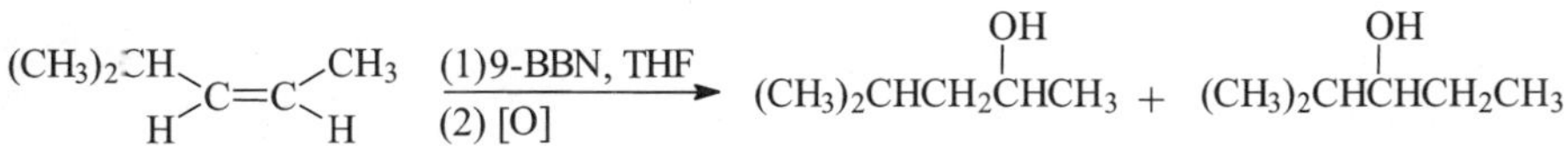

9-BBN	99.8%	0.2%
Sia_2BH	97%	3%
Chx_2BH	97%	3%

Chart 6.1

Hydroboration of the *trans* isomer is much less regioselective.

The hydroboration of substituted cycloalkenes proceeds with exceptionally high regio- and stereoselectivity. The addition of B–H occurs *cis*. Consequently, hydroboration of 1-substituted cycloalkenes cleanly produce the *trans*-2-alkyl-cyclo-9-BBN, which on oxidation produces the corresponding *trans*-alcohols (Chart 6.2) [3].

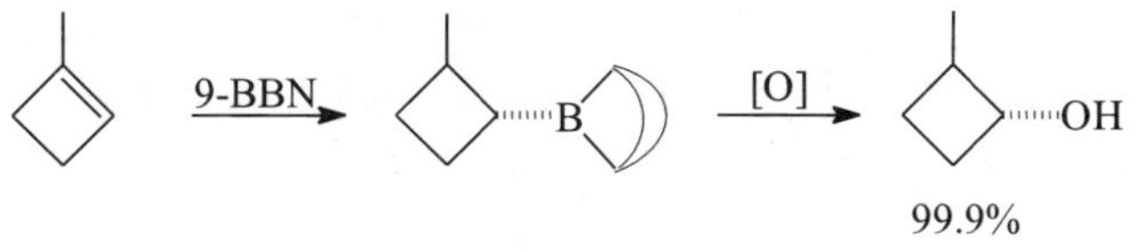

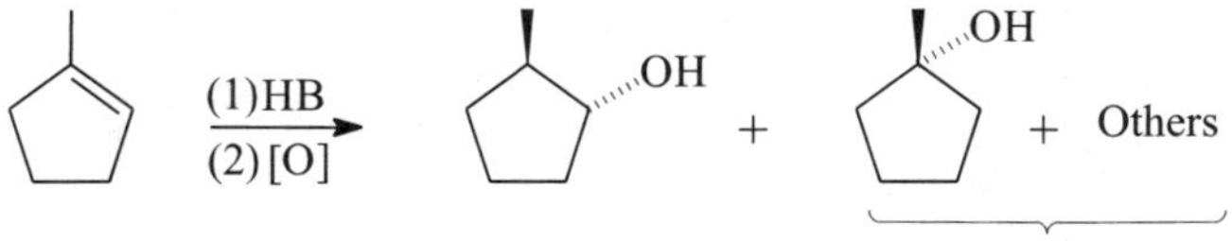

9-BBN	>99.9%	~0%
BH_3	98.5%	1.5%
$HBCl_2$	>99.8%	<0.2%

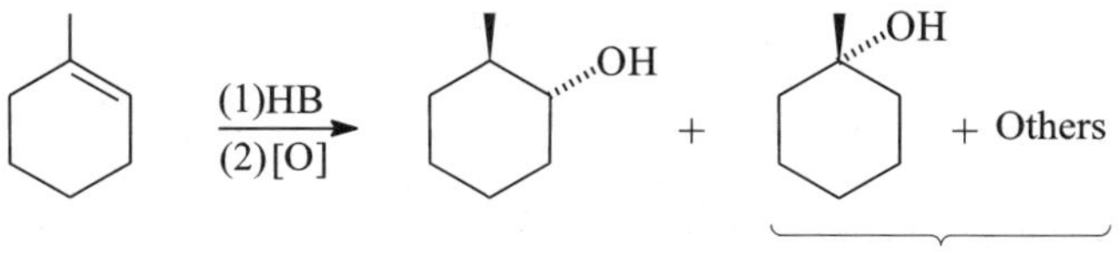

9-BBN	>99.9%	~0%
BH_3	98%	<2%

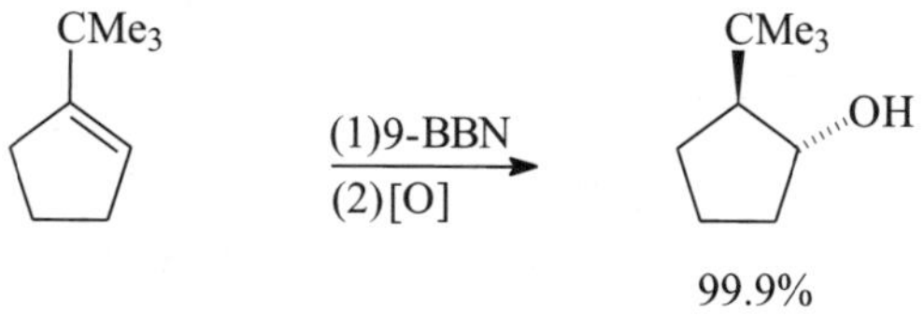

Chart 6.2

The stereoselectivities of 9-BBN, IPC_2BH, Sia_2BH, and BH_3 toward 3-substituted cycloalkenes are of little significant, as the addition of B–H occurs from both the faces. Among these hydroborating agents 9-BBN, however, gives the highest regioselectivity (Chart 6.3) [3].

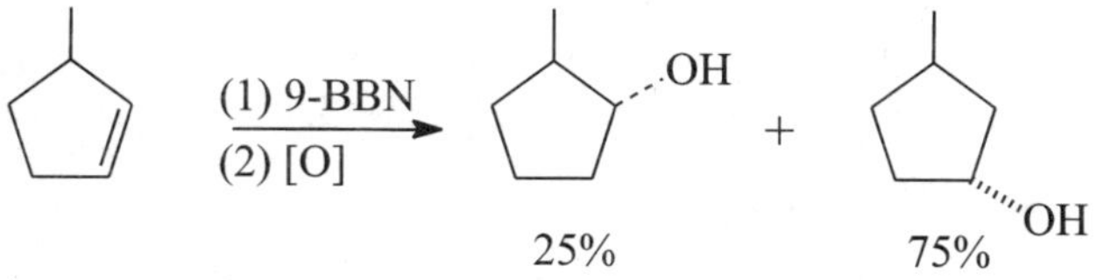

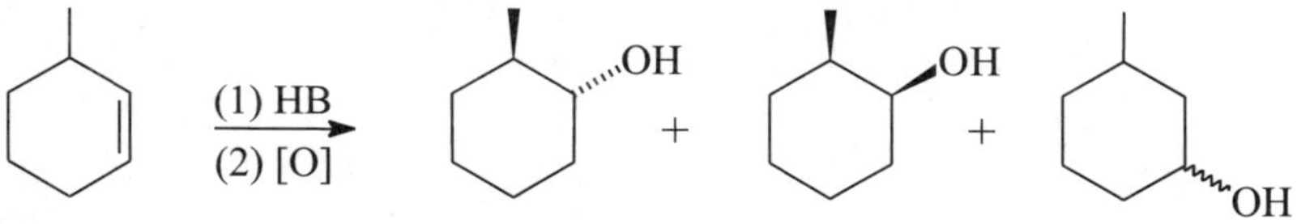

	trans	*cis*	
9-BBN	20%	0%	80%
Sia_2BH	30%	18%	52%
BH_3	34%	16%	50%

Chart 6.3

The regioselectivity of 9-BBN and BH_3 toward 4-methylcyclohexene is neither stereoselective *nor* regioselective. 4-Methylcyclopentene, however, undergoes hydroboration with 9-BBN, with high stereo- and regioselectivities (Chart 6.4) [3].

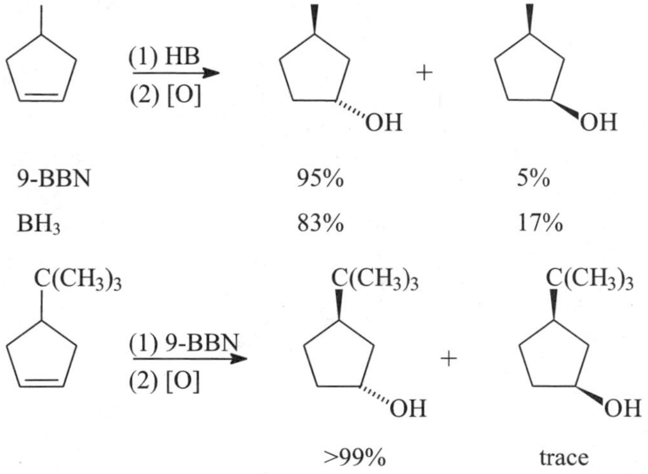

Chart 6.4

Stereoisomeric purities from 97–99.9% have been realized in bicyclic systems where the attack of 9-BBN occurs from the less hindered side (Chart 6.5) [3, 4].

9-BBN is a reagent of choice for high stereoselective hydroboration of 1,3-dimethlcycloalkenes, affording only 2,ω-dimethylcycloalkyl-9-BBN, with both methyl groups *trans* to the 9-BBN moiety (Chart 6.6) [5].

B-Alkyl-9-BBN derivatives are exceptionally resistant to thermal isomerization. *B*-3-Hexyl-9-BBN requires heating at 150 °C for 168 h to attain the equilibrium distribution of the boron along the hexyl chain, whereas isomerization of tri-3-hexylborane is complete in 1 h at the same temperature. The sluggish migration of 9-BBN along the carbon skeleton makes possible the successful hydroboration of certain labile system, not readily handled by other hydroborating reagents. For example, 1-methylcyclooctene with borane-THF produces mixtures of products arising from the facile isomerization of the organoborane intermediates. However, 9-BBN hydroborates [1] these alkenes cleanly (Chart 6.7) [3].

The stereocontrolled addition of 9-BBN has been elegantly provided in the synthesis of the steroidal alcohols [6], where the attack occurs from the less hindered α-face. Aldehydes derived from these alcohols are assigned the stereostructures (Scheme 6.1) as the proton of equatorial -CHO absorbs at δ 9.73, whereas axial -CHO appears at δ 9.96 in the ^1H NMR spectrum [7].

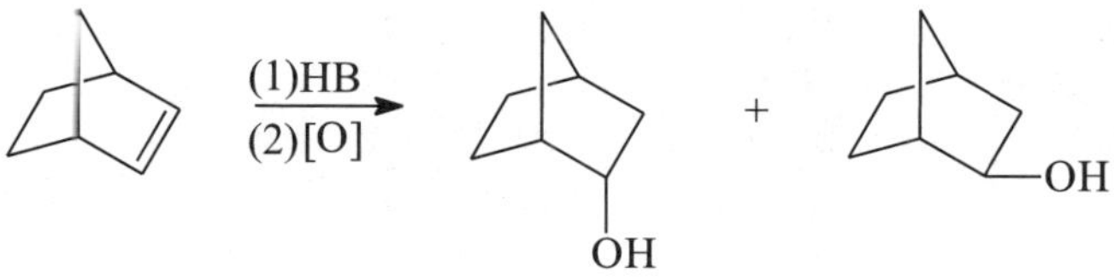

BH_3	87%	13%
9-BBN	99.5%	0.5%
BH_2Cl	99.8%	0.2%

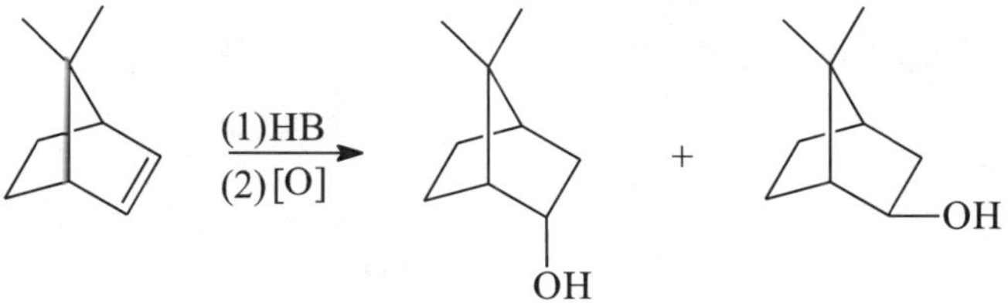

BH_3	79%	21%
9-BBN	97%	3%

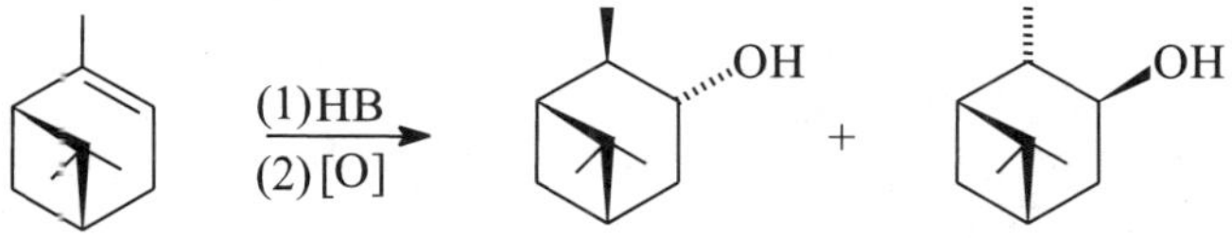

BH_3	98%	2%
9-BBN	>99.9%	–

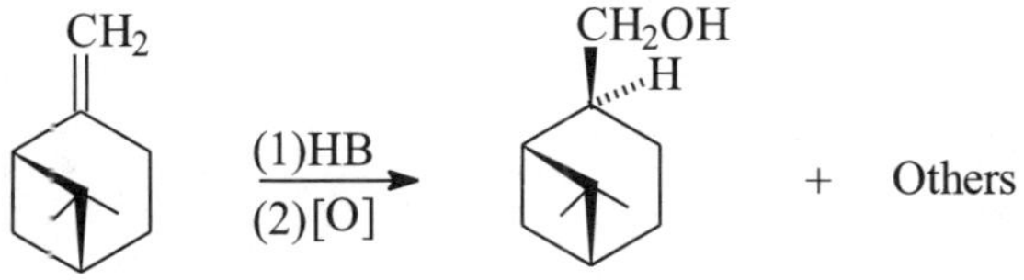

BH_3	98%	2%
9-BBN	99.9%	–

Chart 6.5

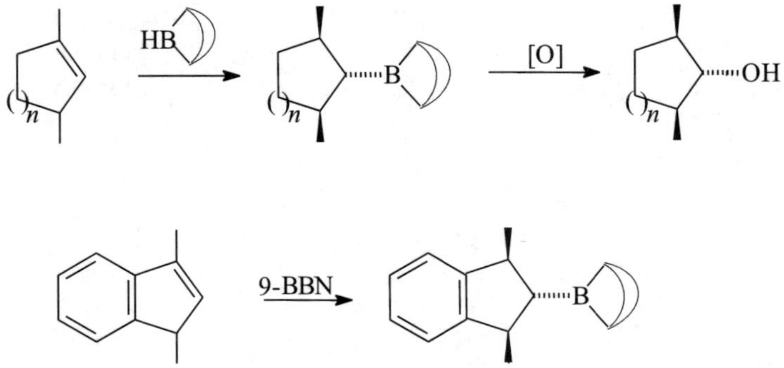

Chart 6.6

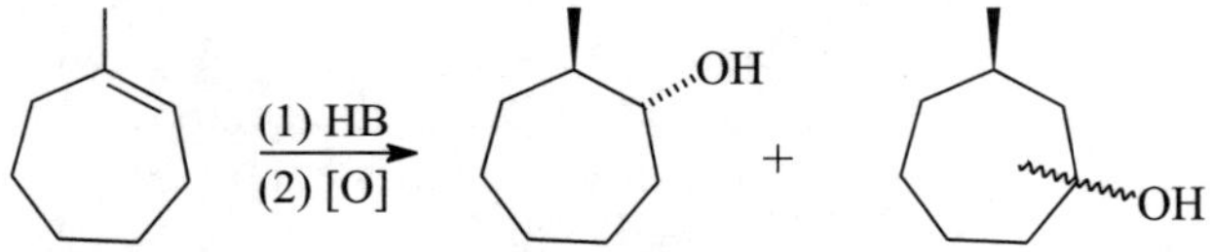

9-BBN	98%	2%
BH_3	97%	3%

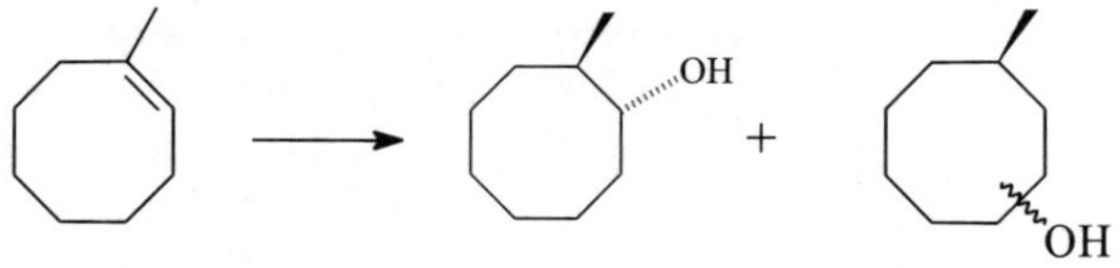

BH_3, 0.5 h, 0 °C	60%	40%
Sia_2BH, 2 h, 0 °C	60%	40%
9-BBN, 2 h, 25 °C	98%	2%

Chart 6.7

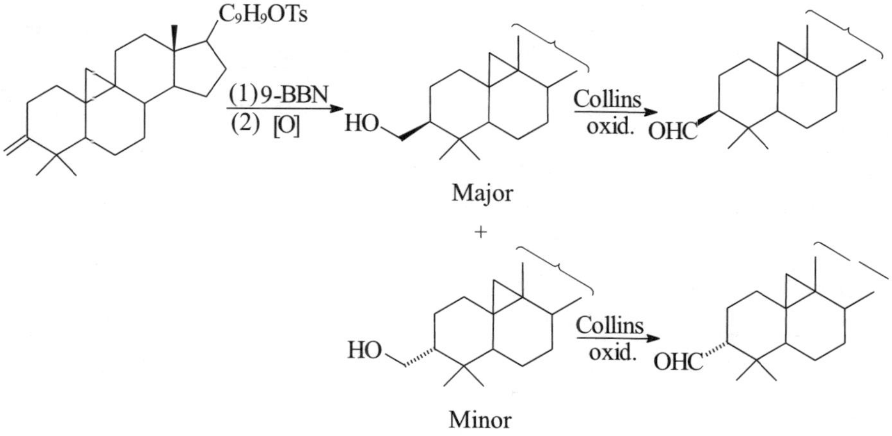

Scheme 6.1

Brown *et al* [8] have devised a general, convenient, and simple synthesis of straight-chain alcohols from internal alkynes. Long-chain internal alkynes, prepared by Eiter's procedure [9] by metalating 1-alkynes, followed by treatment with alkyl halides, are isomerized to 1-alkynes on treatment with potassium-3-aminopropylamide (KAPA) [10] in 1,3-diaminopropane (APA). KAPA is prepared by the quantitative reaction of potassium hydride with excess of (APA) [10]. This difunctional "superbase" produces exceptionally rapid migration of internal C≡C to the terminal C≡C position. The terminal alkynes thus obtained are subjected to dihydroboration with 2 equiv of 9-BBN. The dibora intermediate on alkaline hydrogen peroxide oxidation provides 61–80% yield (Table 6.4) [8] of the corresponding alcohols (Eq. 6.2).

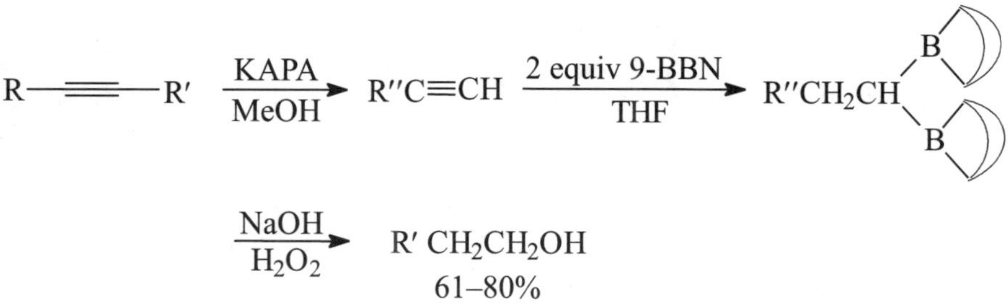

(6.2)

Table 6.4 Long-chain alcohols *via* KAPA method [8]

Alcohols	Number of carbon atoms	Yield (%)	m.p. (°C)
1-Tetracosanol	24	80	74–76
1-Triacontanol	30	76	86–88
1-Hentriacontanol	31	74	84–86
1-Tritriacontanol	33	73	86–88
1-Hexatriacontanol	36	76	90–92
1-Dotetracontanol	42	68	96–98
1-Octatetracontanol	48	61	102–104

The 1,2 transposition of oxygen functionality is an important reaction methodology in the synthesis of various organic compounds. 9-BBN enables the clean achievement of this transposition [11] in several cases, as illustrated in Scheme 6.2.

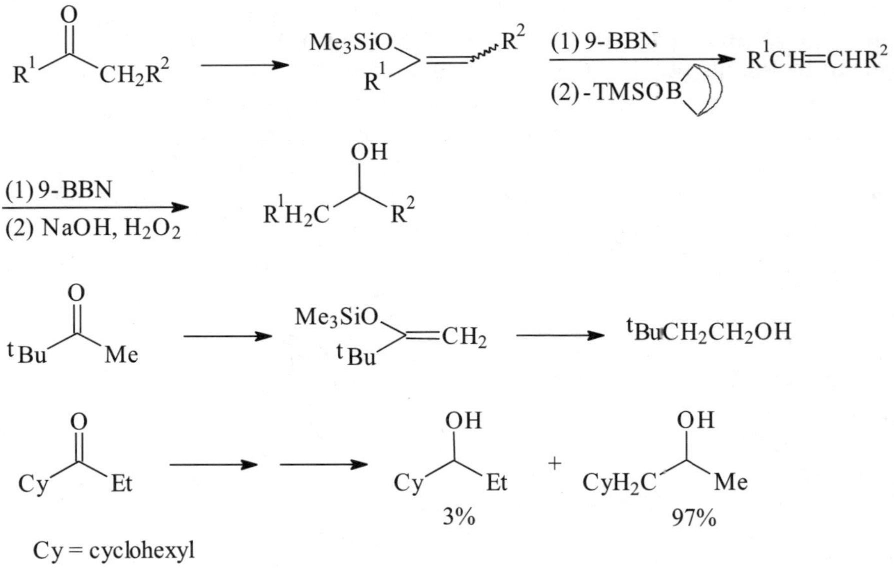

Scheme 6.2

6.2
Synthesis of Allylic Alcohols

B-Alkenyl-9-borabicyclo[3.3.1]nonane (*B*-alkenyl-9-BBN), unlike their saturated counter parts, adds across the carbonyl group of simple aldehydes and affords in good yields the corresponding allylic alcohols [1]. This Grignard-like addition to aldehyde is unique to vinyl-9-BBN derivatives due to their Lewis acidity and is in contrast to alkylboranes. No competitive *B*-hydride reductive processes are observed. The reaction proceeds with complete retention of vinyl borane stereochemistry, resulting exclusively in the *trans*-disubstituted olefin linkage in the final product. The reaction tolerates functionalities, which is not possible in Grignard synthesis of alcohols. Thus, 5-chloro-1-pentyne is hydroborated with 9-BBN and then reacted with propionaldehyde to give 8-chloro-*trans*-4-octen-3-ol (Eq. 6.3).

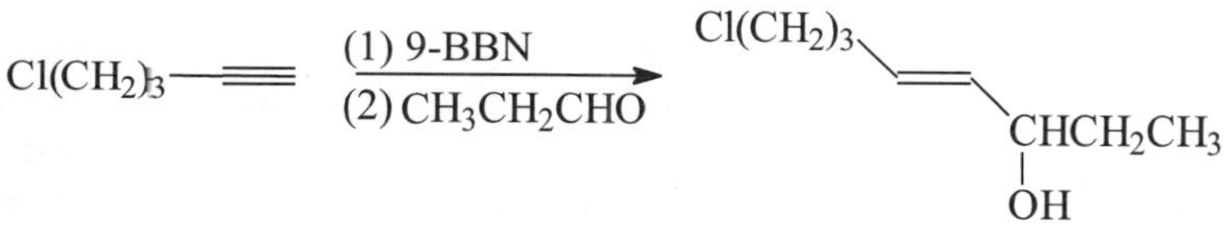

(6.3)

The results are summarized in Table 6.5 [1].

Now, the dehydroborylation process of Soderquist [2] has made easy accessibility of *trans*-vinyl-9-BBN, which is cleanly converted to *trans* allylic alcohols in good yields (Eq. 6.4).

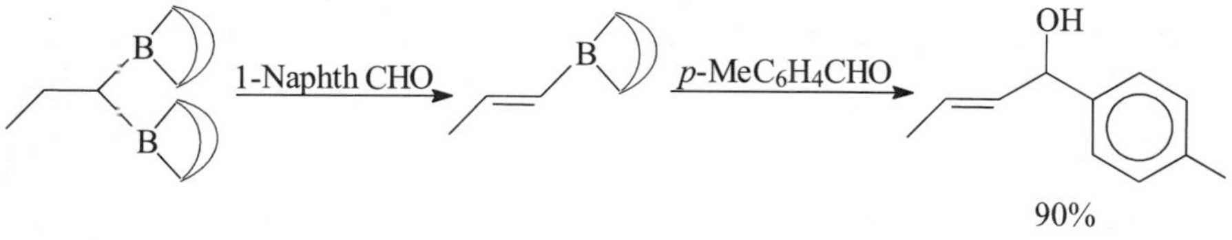

(6.4)

Evidently, this procedure provides a remarkably simple, stereospecific synthesis of allylic alcohols that was previously possible only with more reactive organometallic compounds.

The allylboranes are prepared (for the synthesis of allylalcohols) from unsymmetrical allenes and certain conjugated dienes, but allylboranes are protonolyzed readily by reagents containing labile hydrogens. However, the sterically hindered conjugated diene, 2,5-dimethyl-2,4-hexadiene reacts rapidly with 9-BBN to afford the corresponding allylborane, which on alkaline hydrogen peroxide oxidation gives 2,5-dimethyl-4-hexen-3-ol [4] (Eq. 6.5).

Table 6.5 Conversion of alkynes into allylic alcohols [1]

Alkyne	Aldehyde	Product	Yield (%)
Propyne	Formaldehyde[a]	*trans*-Crotylalcohol	48
1-Pentyne	Formaldehyde[a]	*trans*-2-Hexen-1-ol	36
1-Hexyne	Benzaldehyde	1-Phenyl-*trans*-2-hepten-1-ol	86
3,3-Dimethyl-1-butyne	Butyraldehyde	7,7-Dimethyl-*trans*-5-octen-4-ol	55
4,4-Dimethyl-2-pentyne	Acetaldehyde	3,5,5-Trimethyl-*trans*-3-hexen-2-ol	69
5-Chloro-1-pentyne	Propionaldehyde	8-Chloro-*trans*-4-octen-3-ol	47

[a] Monomeric formaldehyde.

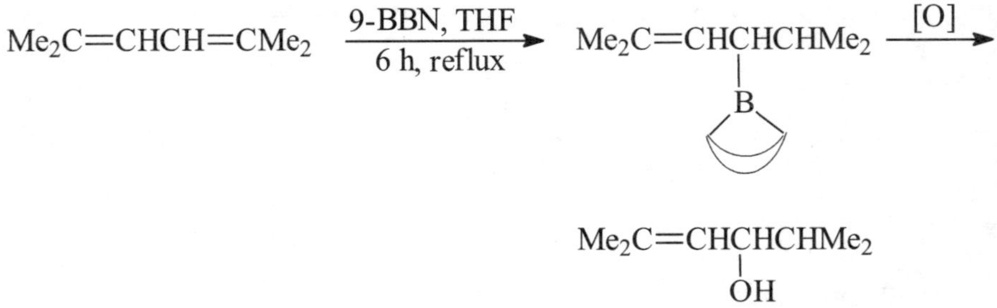

(6.5)

The *cis*-vinyl-9-BBN, prepared from (*Z*)-1-iodo-3-methyl-1-butene *via* lithiation, followed by treatment with *B*-OMe-9-BBN undergoes condensation with high stereospecificity with benzaldehyde to give the corresponding *cis*-allylic alcohol in moderate yield (Eq. 6.6) [5].

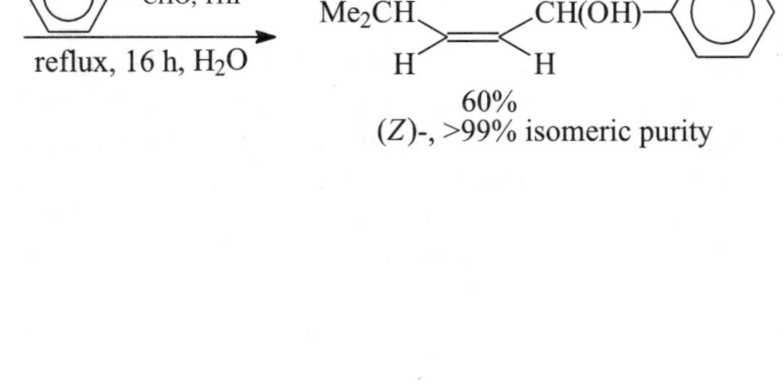

(6.6)

6.3 Synthesis of Homoallylic Alcohols

The *B*-allyl-9-BBN derivatives react with carbonyl compounds to afford borinate esters, which on protonolysis using triethanolamine afford the corresponding homoallylic alcohols. Consequently, the allylboration sequence provides a synthetically useful alternative to the familiar Grignard's synthesis of these alcohols. However, the protonolysis by triethanolamine causes a problem in the isolation of homoallylic alcohol from the thick, air-sensitive, boron-containing mixture. Fortunately, treatment of pentane solution of borinate esters of 9-BBN with 1 equiv of ethanolamine results in the rapid formation of a fluffy white precipitate of ethanolamine 9-BBN adduct, which can be easily removed (Eq. 6.7) [1].

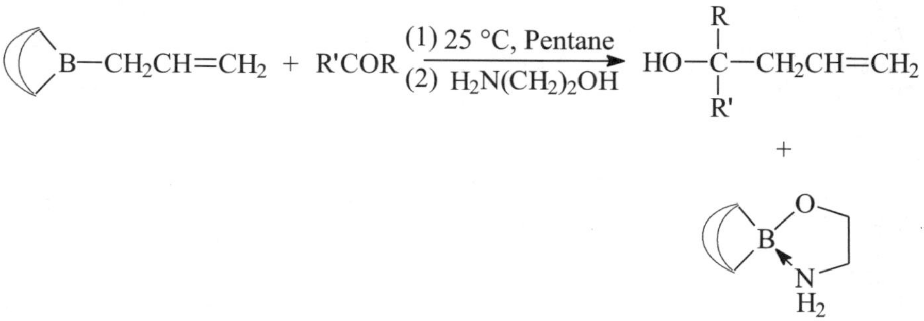

(6.7)

The stoichiometry of the allylboration of aldehydes and ketones with *B*-allyl-9-BBN is 1:1 and undergoes rapid reaction with aldehydes and ketones, being complete within a few minutes at room temperature. However, as the steric crowding around the carbonyl group increases, the reaction rate decreases, as is evident from di-*tert*-butyl ketone, which affords less than 25% of the expected product even after 5 days in refluxing *n*-octane. The steric sensitivity of the allylboration is in sharp contrast with the corresponding allylation using Grignard reagent [2]. Even, di-*tert*-butylketone is completely allylated by allylmagnesium bromide within 6 h in refluxing ether. Thus, this remarkable selectivity of allylboration offers an advantage to allylate selectively the unhindered carbonyl moiety present in the same molecule.

The results are summarized in Table 6.6 [1].

Allylboration of aldehydes and ketones with *B*-crotyl-9-BBN affords only the rearranged product (Eq. 6.8; Table 6.7) [1]. Monomeric formaldehyde gives 2-methyl-3-buten-1-ol.

(6.8)

Table 6.6 Yields of homoallylic alcohol ($HOCRR'CH_2CH{=}CH_2$) [1]

R	R′	Percentage yield (GLC)	R	R′	Percentage yield (GLC)
n-C_5H_{11}	H	92	*t*-Bu	Ph	82
t-Bu	H	97	*t*-Bu	*t*-Bu	<25
Ph	H	96	Cyclopentanone		102
$CH_3CH{=}CH$	H	91	Cyclohexanone		100
Me	Me	99	Cycloheptanone		94
Me	*t*-Bu	101	4-*tert*-Butylcyclohexanone		100[a]
Me	Ph	101	*nor*Camphor		100[b]
Ph	Ph	100	Bicyclo[3.3.1]nonan-9-one		85
i-Pr	*i*-Pr	98	$H_2C{=}CH$	Me	94
t-Bu	*n*-Pr	97	$CH_3CH{=}CH$	Me	96
t-Bu	*i*-Pr	74	$(CH_3)_2C{=}CH$	Me	94

[a] Axial:equatorial, 54.8:45.2.
[b] *endo*:*exo*, 95.8:4.2.

Table 6.7 Yields of homoallylic alcohol from *B*-crotyl-9-BBN [1]

R	R′	Percentage yield (GLC)	R	R′	Percentage yield (GLC)
H	H	88	Me	Me	100
n-C_5H_{11}	H	89[a]	Ph	Me	97
Ph	H	87	Ph	Ph	96

[a] *threo*:*erythro*, 60:40.

Unsymmetrical allylboranes (prepared from allenes or 1,3-dienes) when treated with acetone give only the rearranged homoallylic alcohol (Table 6.8) [1].

Table 6.8 Allylboration of acetone with unsymmetrical *B*-allyl-9-BBN [1]

B-Allyl-9-BBN derivative	Product	% yield (GLC)
2-Methylallyl	2,4-Dimethyl-4-penten-2-ol	95
Crotyl	2,3-Dimethyl-4-penten-2-ol	100
3,3-Dimethylallyl	2,3,3-Trimethyl-4-penten-2-ol	93
3,3-Dimethyl-1-isopropylallyl	2,3,3,6-Tetramethyl-4-hepten-2-ol	85[b]
(2,5-Dimethyl-2,4-hexadiene + 9-BBN)[a]	2,3,3,6-Tetramethyl-4-hepten-2-ol	80
2-Cyclohexen-1-yl	2-(2′-Cyclohexenyl)-2-propanol	86
(1,3-Cyclohexadiene + 9-BBN)[a]	2-(2′-Cyclohexenyl)-2-propanol	68

[a] Prepared in situ.
[b] Mainly *trans* product.

The reaction proceeds through the six-centered mechanism (Scheme 6.3). The enhanced Lewis acidity of allylboranes accounts for their ability to add easily to carbonyl derivatives whereas normal trialkylboranes do not.

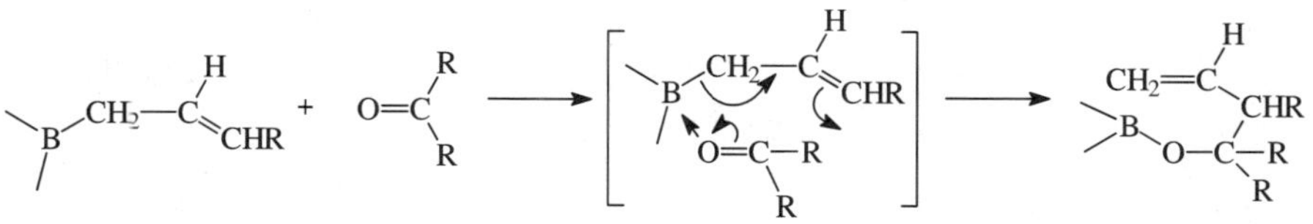

Scheme 6.3

The allylboration offers certain advantages over the familiar Grignard route for preparation of homoallylic alcohols:

1. Certain functionalities like halides can be tolerated; allylboration with unhindered aldehydes and ketones are far faster than those with esters and amides. The former groups are readily allylborated in the presence of latter.
2. The intermediate borinic ester is not strongly basic as are intermediates in the lithium and magnesium allylation.
3. The workup conditions are extremely mild. There are no problems with gels, emulsions, or extraction from aqueous media.
4. Sensitivity to the steric requirement makes possible for selective reaction.

B-Allyl-9-BBN reacts vigorously with acid chlorides (Scheme 6.4) and acid anhydrides (Scheme 6.5), slowly with carboxylic acid esters (Scheme 6.6), and *N,N*-dimethylamide (Scheme 6.7). Acid chlorides, esters, and *N,N*-dimethylamides react with 2 equiv of the allylborane; acid anhydrides utilize 4 equiv (Table 6.9) [1]. In case of acid chloride, sticky gel is formed due to the generation of hydrogen chloride resulting from the protonolysis of chloroborane. This difficulty is circumvented by the addition of 1 equiv of lithium isopropoxide before adding 2 equiv of ethanolamine.

Acid chloride:

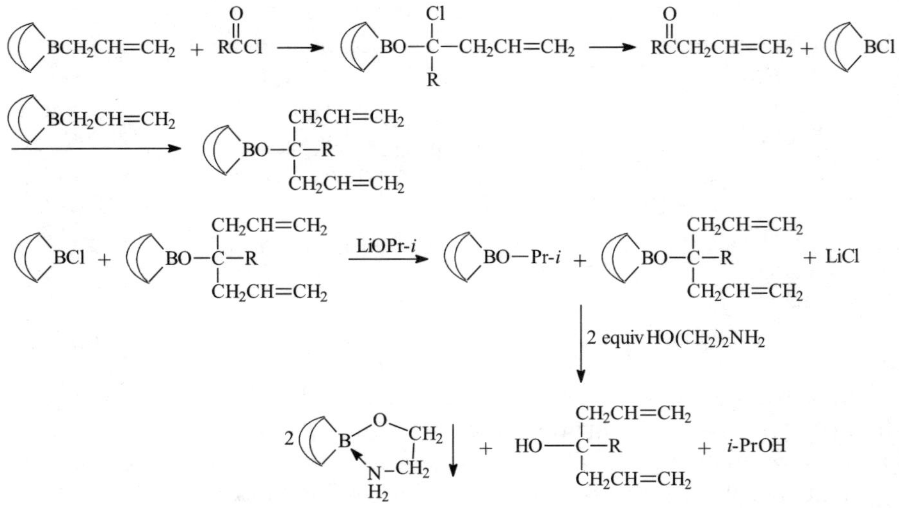

Scheme 6.4

Acid anhydride:

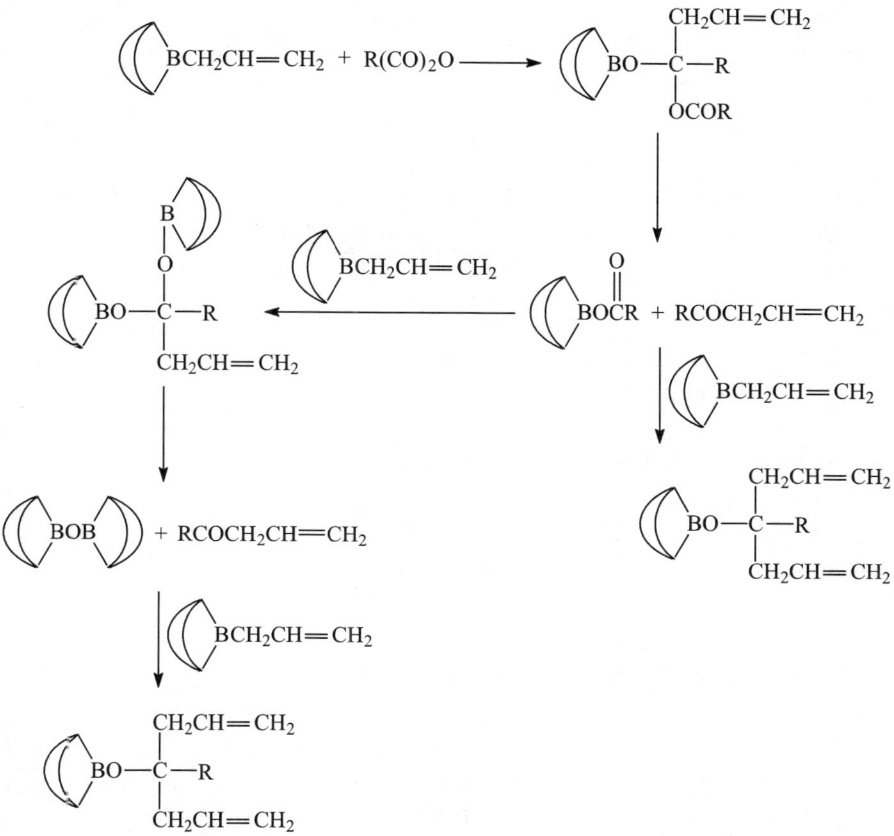

Scheme 6.5

Esters:

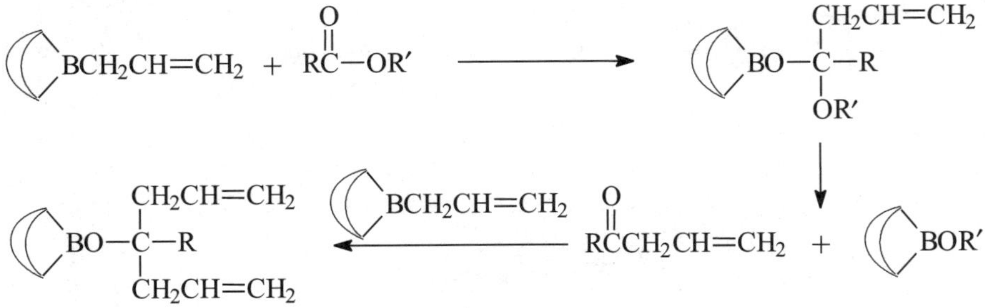

Scheme 6.6

***N,N*-Amides:**

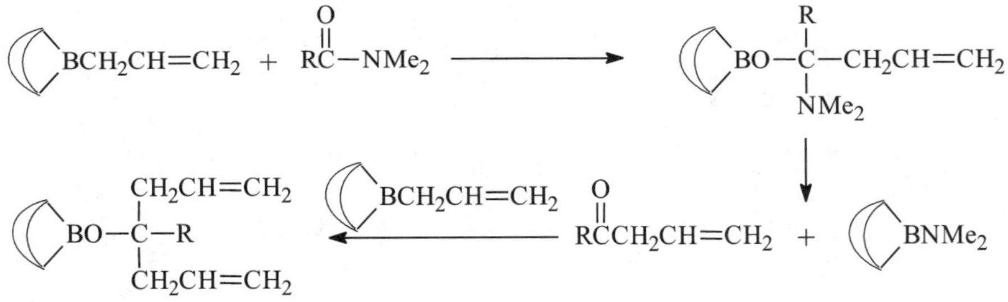

Scheme 6.7

The results are summarized in Table 6.9 [1].

The complete allylic rearrangement again occurs with the acid chloride and anhydride.

A series of homoallylic tertiary alcohols [3] with structure 1-aryl-1-(Δ^2-cyclohexenyl)ethanol are synthesized by treating *p*-haloacetophenone with *B*-2-cyclohexen-1-yl-9-BBN (Eq. 6.9). Apparently, it is not feasible to convert 1-chloro- or 1-bromo-2-cyclohexene into the corresponding Grignard reagent.

Table 6.9 Allylboration of carbonyl derivatives with *B*-allyl- and *B*-crotyl-9-BBN [1]

B-R-9-BBN R	Substrate	Stoichiometry[a]	Product[b]	Yield (%)
Allyl	Acetylchloride	2:1	A	95
	Benzoylchloride	2:1	B	89
	Acetic anhydride	4:1	A	95
	Benzoic anhydride	4:1	B	82
	Ethyl acetate	2:1	A	90
	Ethylbenzoate	2:1	B	50
	N,N-Dimethylacetamide	2:1	A	50–57
	N,N-Dimethylbenzamide	2:1	B	91
	B-Acetoxy-9-BBN	2:1	A	84
Crotyl	Acetylchloride	2:1	C	93
	Acetic anhydride	4:1	C	97

[a] Allylborane:substrate.
[b] A = 4-methyl-1,6-heptadien-4-ol; B = 4-phenyl-1,6-heptadien-4-ol; C = 3,4,5-trimethyl-1,6-heptadien-4-ol (mixture of diastereoisomers).

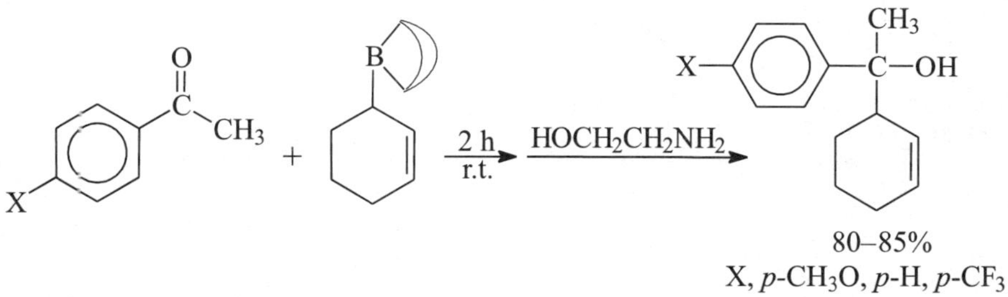

(6.9)

B-2-Cyclohexen-1-yl-9-BBN reacts with acetaldehyde, which on alkaline hydrogen peroxide oxidation affords the corresponding homoallylic alcohol (Eq. 6.10) [4].

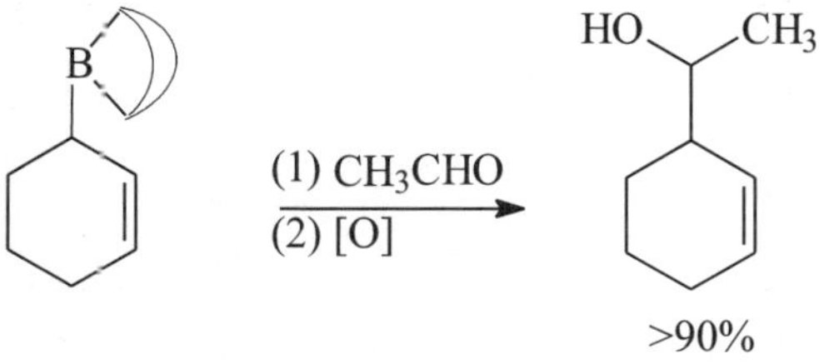

(6.10)

B-3-Cyclohexen-1-yl-9-BBN prepared from 1,4-cyclohexadiene on alkaline hydrogen peroxide oxidation yields the cyclohex-3-en-1-ol in good yield (Eq. 6.11) [4].

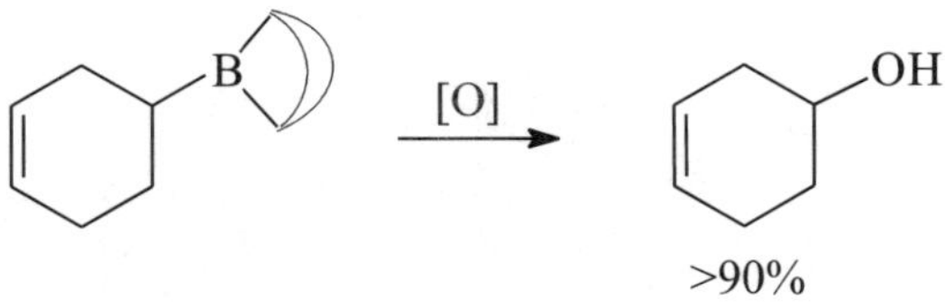

(6.11)

The oxidation process is also employed to the derivatized product of allylborane and acetone which affords the corresponding homoallylic alcohol (Table 6.10; Eq. 6.12) [5] in excellent yields.

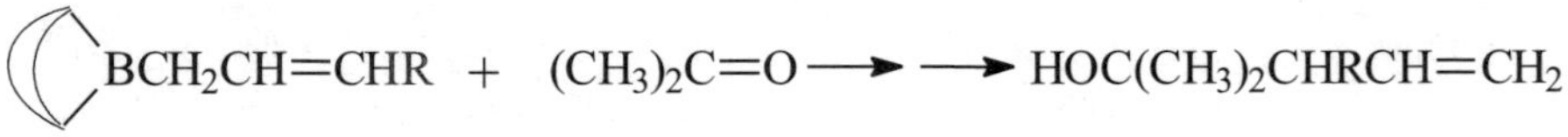

(6.12)

Table 6.10 Hydroboration–derivatization–oxidation product distribution of allenes with 1 equiv of 9-BBN (0.5 M) in THF [5]

Allene	Reaction time (h)	Residual allene (%)	Homoallylic alcohol (%)	Diols (%)	Ketone (%)
Propadiene	2.5	50	0	1,3-diol, 48	0
1,2-Butadiene	1.5	7	74	1,3-diol, 12 1,2-diol, trace	Trace
1,2-Pentadiene	1.5	4	82	1,2- and 1,3-diols, 5	2
3-Mehyl-1,2-butadiene	2.5	Trace	92	–	Trace
Phenylpropadiene	4	11.5	86	1,2- and 1,3-diols, 6	Trace
p-Anisylpropadiene	4	4	87	1,2- and 1,3-diols, 6	Trace
3-Phenyl-1,2-butadiene	4	Trace	97	–	–
2,3-Pentadiene	2	5	72	2,3- and 2,4-diols, 12	0
4-Methyl-2,3-pentadiene		–	94[a]	–	–
2,4-Dimethyl-2,3-pentadiene	>108	8	86	0	Trace
1,2-Cyclononadiene	1	Trace	83	0	17

[a] Mixture of *cis* and *trans* homoallylic alcohols.

Since allenes are easily obtained from the corresponding olefins [6, 7], the route olefin–allene–allylborane–carbonyl addition is general sequence and is a viable alternative to Grignard-based process for the synthesis of complex structures.

6.4 Synthesis of Propargylic Alcohols

Propargylic alcohols are valuable intermediates in the synthesis of many natural products like prostaglandins [1], steroids [2], carotenoids [3], and leukotrienes [4]. These alcohols are generally prepared by addition of alkynyl metals (Li, Na, K, Mg, Zn, and Al) to aldehyde [5]. However, many times it leads to mixtures of products, as they are highly nucleophilic and react with variety of functional groups, and their strong basicity also causes based-induced eliminations.

B-1-Alkynyl-9-BBN compounds prepared [6–8] from the corresponding 1-alkynes and *B*-MeO-9-BBN, readily undergo addition to aldehydes and ketones, and afford the propargylic alcohols in very high isolated yields (Scheme 6.8) [9]. They are not basic and are safely used in the presence of compounds such as sulfoxides and diethylmalonate.

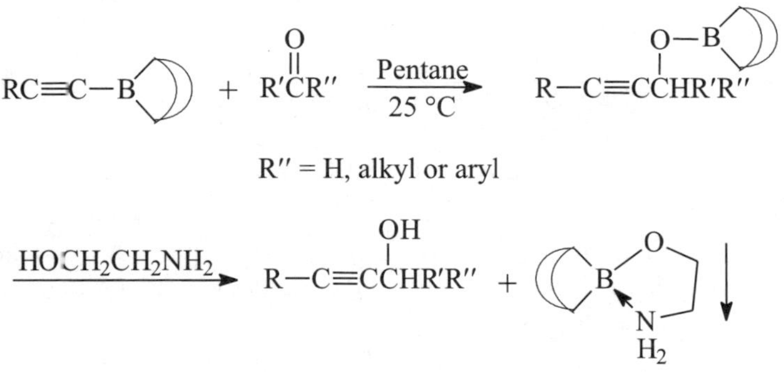

Scheme 6.8

The reaction of *B*-alkynyl-9-BBN compounds with propionaldehyde are summarized in Table 6.11 [6].

Table 6.11 Reaction of *B*-1-alkynyl-9-BBN compounds with propionaldehyde at 25 °C in pentane for 0.5 h [6]

Alkyne	Product	Yield (%)
1-Hexyne	4-Nonyn-3-ol	94
3,3-Dimethyl-1-butyne	6,6-Dimethyl-4-heptyn-3-ol	83
Phenylethyne	1-Phenyl-1-pentyn-3-ol	68
5-Chloro-1-pentyne	8-Chloro-4-octyn-3-ol	78
3-Methyl-3-buten-1-yne	6-Methyl-6-hepten-4-yn-3-ol	81

Unlike alkynyllithium, *B*-alkynyl-9-BBN compounds are prepared and stored until needed. These compounds thus add to carbonyl moiety in a Grignard-like fashion.

Further, their generality is established as *B*-1-(3-dimethylbutynyl)-9-BBN reacts with a variety of aldehydes and ketones (Table 6.12) [6] to afford the corresponding proporgylic alcohols in high yields.

Table 6.12 Reaction of *B*-1-(3,3-dimethylbutynyl)-9-BBN with aldehydes and ketones [6]

Carbonyl compound	Reaction conditions	Product	Yield (%)
Formaldehyde	25 °C, 5 min	4,4-Dimethyl-2-pentyn-1-ol	66
Propionaldehyde	25 °C, 10 min	6,6-Dimethyl-4-heptyn-3-ol	83
3-Cyclohexenecarboxaldehyde	25 °C, 1.5 days	1-(3-Cyclohexenyl)-4,4-dimethyl-2-pentyn-1-ol	76
Benzaldehyde	25 °C, 1.5 days	1-Phenyl-4,4-dimethyl-2-pentyn-1-ol	63
Pivalaldehyde	25 °C, 5 days	2,2,6,6-Tetramethyl-4-heptyn-3-ol	89
Cyclohexanone	25 °C, 10 h	1-(3,3-Dimethyl-1-butynyl)cyclohexanol	97
Cyclopentanone	65 °C, 16 h	1-(3,3-Dimethyl-1-butynyl)cyclopentanol	56
Acetone	65 °C, 16 h	2,5,5-Trimethyl-3-hexyn-2-ol	85

In addition, *B*-alkynyl-9-BBN compounds exhibit high chemospecificity between less and more sterically hindered aldehydes. *B*-1-(3,3-dimethylbutynyl)-9-BBN reacts with propionaldehyde in less than 15 min, whereas its reaction with pivalaldehyde requires 5 days for completion under identical conditions. Similarly, its reaction with cyclohexanone is complete in 10 h at room temperature, whereas cyclopentanone requires 16 h at 65 °C. The introduction of Pitzer strain in going from sp^2 to sp^3 carbon center plays an important role in the rate of reaction of *B*-alkynyl-9-BBN and cyclic ketones. Of special significance is the chemoselectivity of *B*-1-(3,3-dimethylbutynyl)-9-BBN with relatively reactive ketone, cyclohexanone (10 h, 25 °C), compared with simple aldehyde propionaldehyde (<15 min, 25 °C). As is apparent from Table 6.12, *B*-alkynyl-9-BBN compounds react with unhindered aldehydes with remarkable selectively as compared to ketones.

The chemoselectivity exhibited by *B*-1-hexynyl-9-BBN toward an equimolar mixture of propionaldehyde and cyclohexanone (–78 → 25 °C) and toward cyclopentanone and cyclohexanone (1:1 mixture; 0 → 25 °C) is depicted in Chart 6.8.

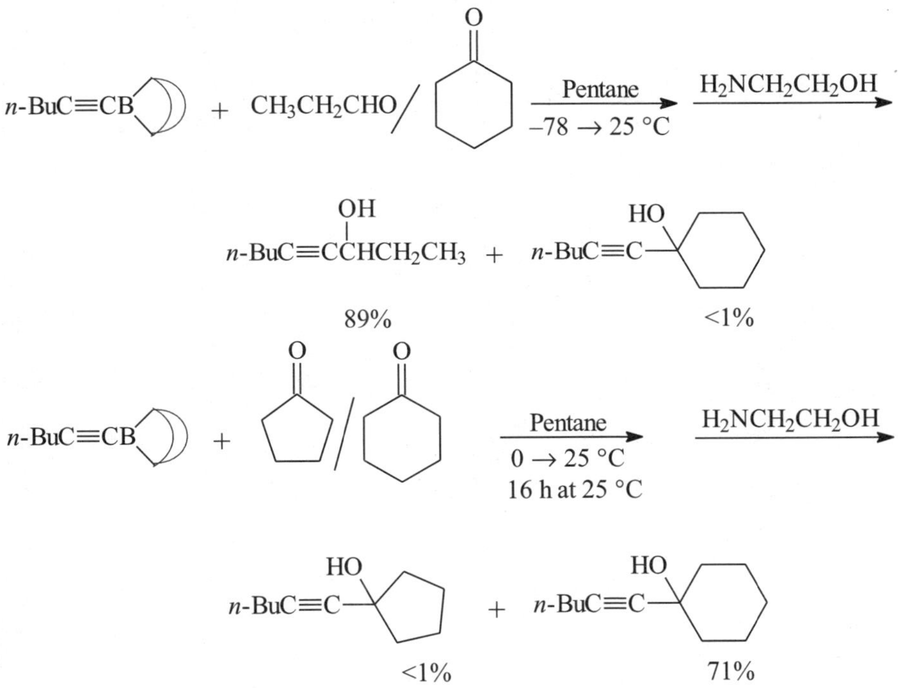

Chart 6.8

On the other hand, 1-hexynyllithium exhibits very poor selectivity under comparable conditions.

B-1-alkynyl-9-BBN compounds are exceptionally mild reagents and are essentially inert at 25 °C toward alkyl halides, acid chlorides, amides, anhydrides, esters, nitriles, acetals, and ketals.

Unfortunately, *B*-ethynyl-9-BBN decomposes on working from –78 °C to room temperature. Evans *et al* [9] prepared *B*-[2-(trimethylsilyl)ethynyl]-9-BBN (Scheme 6.9) from (trimethylsilyl)acetylene and *B*-methoxy-9-BBN. The reagent is isolated under an atmosphere of nitrogen as a solid 1:1 complex with THF in 90% yield. Similar to reagents reported by Brown *et al* [6, 7], *B*-[2-(trimethylsilyl)ethynyl]-9-BBN reacts with a variety of aldehydes and ketones

(Scheme 6.10) under very mild conditions to provide 2-(trimethylsilyl)ethynyl alcohols in excellent yields (Table 6.13). Unlike the alkynyl metals, this reagent reacts much faster with aldehydes than with ketones (Table 6.13) [9], thus providing the synthetic chemists a chemoselective tool.

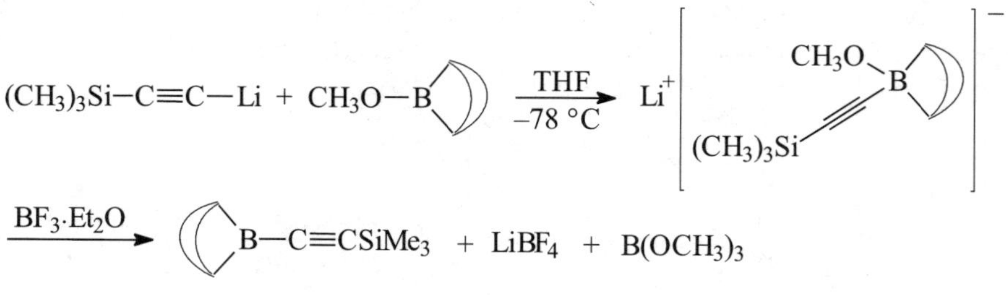

Scheme 6.9

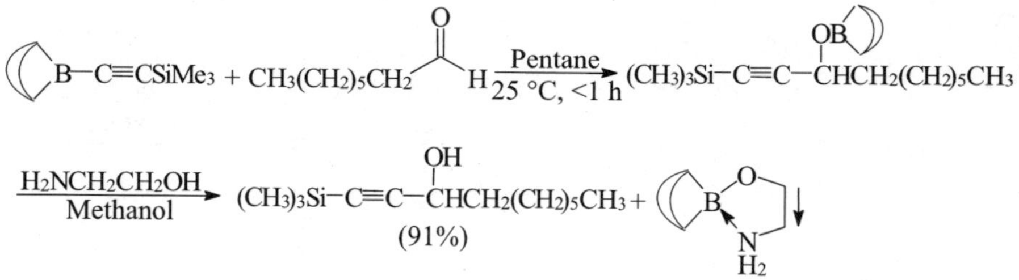

Scheme 6.10

6.5 Synthesis of Homopropargylic Alcohols

Homopropargylic alcohols are important intermediates, as these structural units are present in a variety of natural products and biologically active compounds [1]. Synthesis of these compounds is generally accomplished by reactions of their organometallic equivalents using an array of metals (Mg, Li, Ti, Zn, Al, Sn, Si) [2]. However, their utility is limited due to their ambident nucleophilic nature, which makes them to react with electrophiles unselectively to produce a mixture of products.

Brown and coworkers [3, 4] discovered that *B*-allenyl-9-BBN [3] readily prepared from *B*-chloro-9-BBN [5] and allenylmagnesium bromide [6] (Eq. 6.13) undergoes a facile condensation with aldehydes and ketones to yield exclusively the corresponding homopropargylic alcohols in excellent yields [3, 4]. *B*-Alle-

Table 6.13 Reaction of *B*-[2-(trimethylsilyl)ethynyl]-9-BBN with aldehydes and ketones at 25 °C in pentane [9]

Carbonyl compound	Reaction time	Product(s)	Yield (%)
1-Octanal	< 1 h	1-(trimethylsilyl)-1-decyn-3-ol	91
Hydrocinnamaldehyde	6 h	5-phenyl-1-(trimethylsilyl)-1-pentyn-3-ol	89
Pivalaldehyde	5 d	4,4-dimethyl-1-(trimethylsilyl)-1-pentyn-3-ol	93
(2*R*)-2-Methoxy-2-methylhexanal	3 d	(3*S*,4*R*)-4-methoxy-4-methyl-1-(trimethylsilyl)-1-octyn-3-ol (3*R*,4*R*)-4-methoxy-4-methyl-1-(trimethylsilyl)-1-octyn-3-ol	51 10
Cyclohexanone	16 h	1-[2-(trimethylsilyl)ethynyl]-cyclohexanol	88
2-Octanone	2 d[a]	1-(trimethylsilyl)-3-methyl-1-nonyn-3-ol	71

[a] Reaction temperature is 60 °C.

nyl-9-BBN exhibits excellent selectivity and reacts with aldehydes in the presence of ketones and esters and also distinguishes the less hindered aldehydes or ketones from the more sterically hindered.

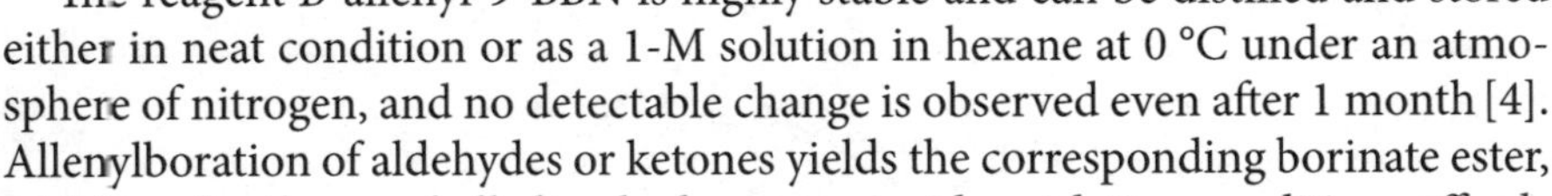

(6.13)

The reagent *B*-allenyl-9-BBN is highly stable and can be distilled and stored either in neat condition or as a 1-M solution in hexane at 0 °C under an atmosphere of nitrogen, and no detectable change is observed even after 1 month [4]. Allenylboration of aldehydes or ketones yields the corresponding borinate ester, which under the usual alkaline hydrogen peroxide oxidation conditions affords homopropargylic alcohols in excellent yields (Scheme 6.11.; Table 6.14) [3, 4].

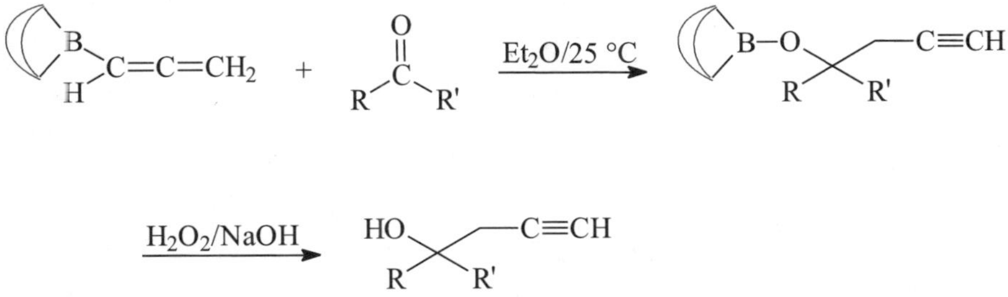

Scheme 6.11

Table 6.14 Allenylboration of representative carbonyl compounds with *B*-allenyl-9-BBN at 25 °C in Et_2O [3]

Carbonyl compound	Homopropargylic alcohol	Yield (%)
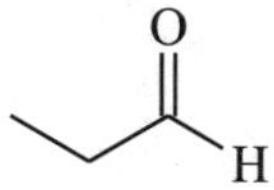	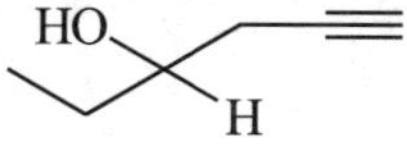	82
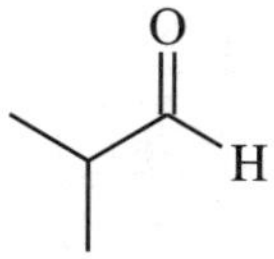	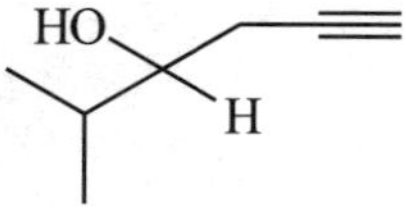	88
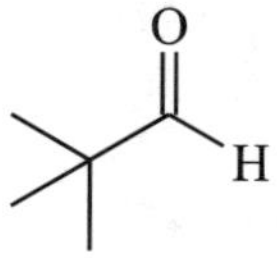	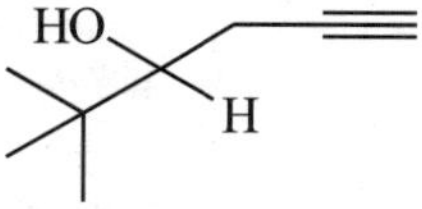	89
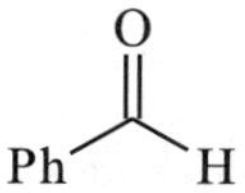	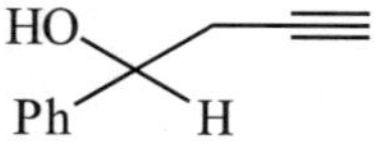	82
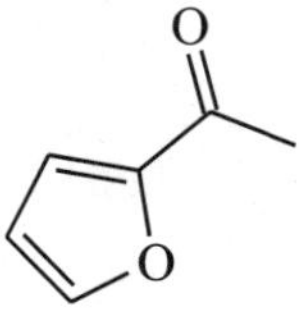	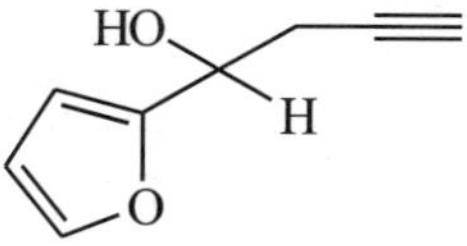	79
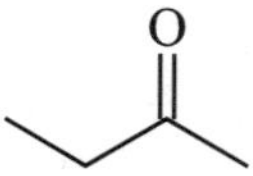	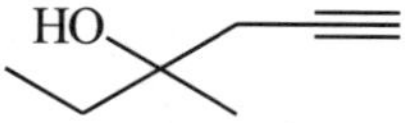	89
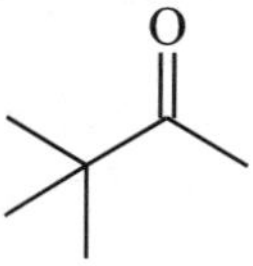	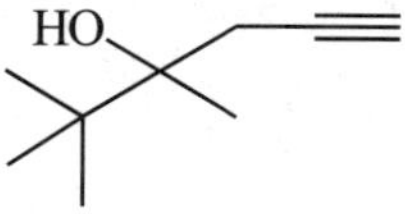	88[b]
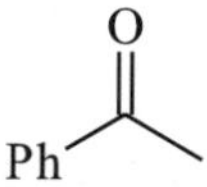	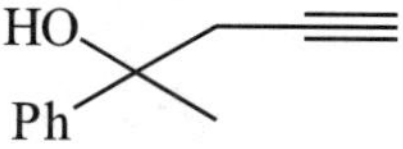	86

[a] Reactions are complete in <5 min.
[b] Require 90 min for completion.

Table 6.14 *(continued)* Allenylboration of representative carbonyl compounds with *B*-allenyl-9-BBN at 25 °C in Et_2O [3]

Carbonyl compound	Homopropargylic alcohol	Yield (%)
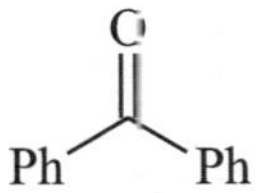	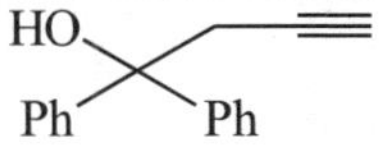	90
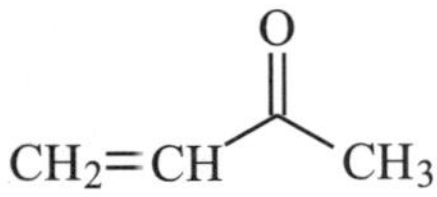	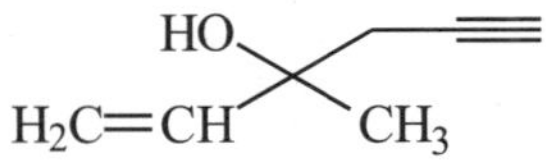	71
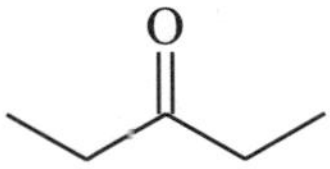	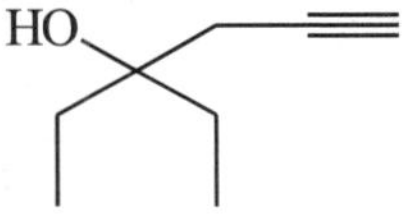	87
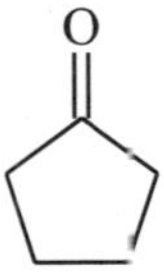	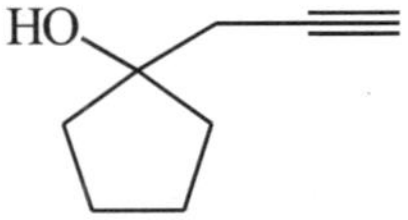	88
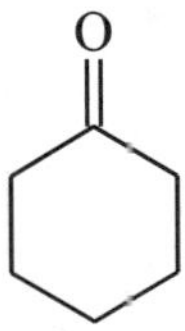	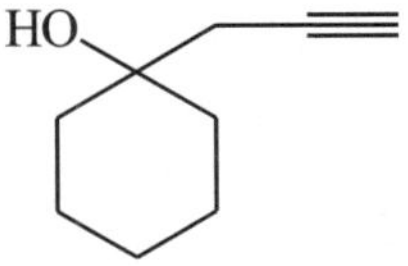	87

[a] Reactions are complete in <5 min.
[b] Require 90 min for completion.

Unlike trialkylboranes, *B*-allenyl-9-BBN reacts with α,β-unsaturated carbonyl compounds in 1,2-fashion to give, exclusively, the olefinic homopropargylic alcohols (Table 6.14) [3].

Both acetyl and benzoyl chlorides react with 2 equiv of *B*-allenyl-9-BBN (Scheme 6.12) in ether to afford the corresponding tertiary homopropargylic alcohols cleanly and in excellent yields [3] and without the contamination of allenic side products.

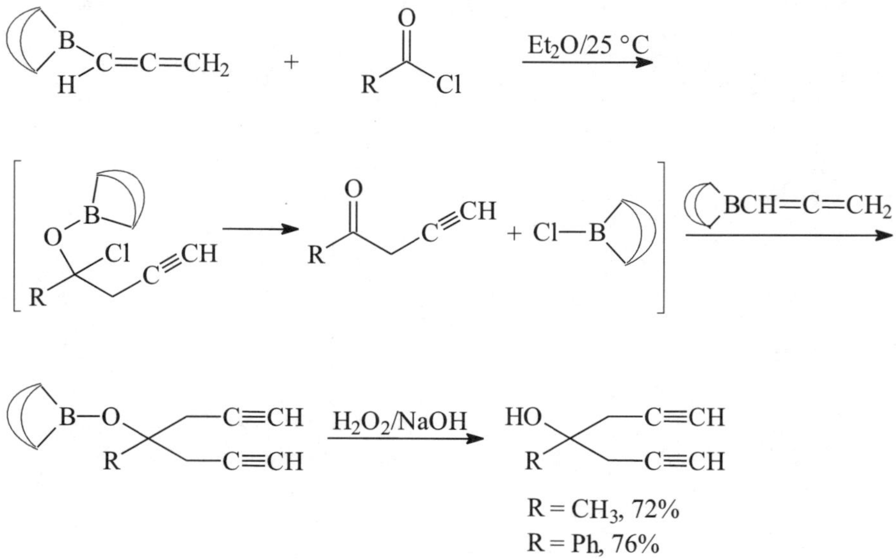

Scheme 6.12

The carboxylic acid esters also react (Scheme 6.13) [3] at reflux, but side products (15%) are formed.

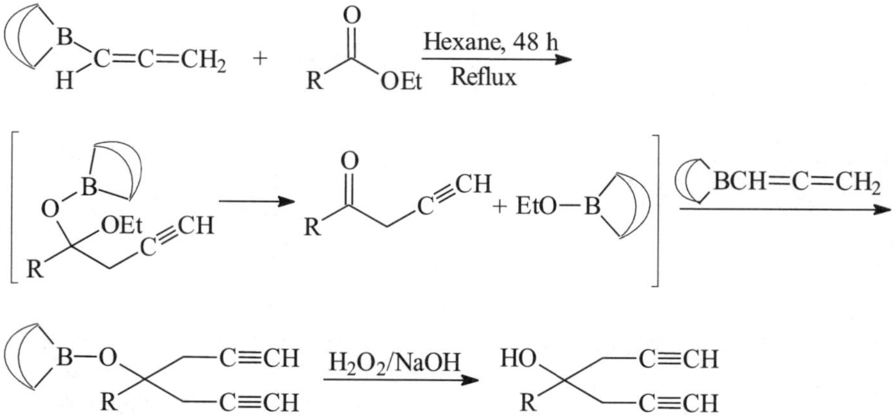

Scheme 6.13

Chemoselective studies [3] reveal that aldehydes react much faster than do ketones (Scheme 6.14) [3].

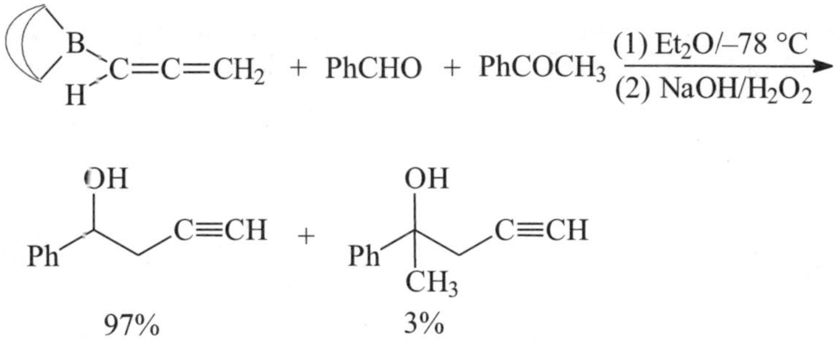

Scheme 6.14

Carboxylic acid esters exhibit no reactivity in presence of benzaldehyde at room temperature. The comparative reactive data are summarized in Table 6.15.

Comparative studies of these pairs with allenylmagnesium bromide (Table 6.15) [3] establishes that the Grignard reagent is much less chemoselective.

Table 6.15 Reactivities of different carbonyl compounds toward *B*-allenyl-9-BBN in Et_2O [3]

Carbonyl compound	Temp (°C) (reaction time, h)	Product with *B*-allenyl-9-BBN (%)	Product with allenyl magnesium bromide (%)
Benzaldehyde	−78 (4)	97	40
Acetophenone		3	60
Pivalaldehyde	0 (0.5)	98	
tert-Butylmethyl ketone		2	
Benzaldehyde	25 (0.5)	100	81
Ethylbenzoate		0	9
Cyclohexanone	−78 (4)	87	65
Cyclopentanone		13	35
Diethylketone	−78 (4)	99	48
Cyclohexanone		1	52
Cyclohexanone	−78 (4)	96	45
tert-Butylmethyl ketone		4	55
Cyclopentanone	−25 (4)	79	51
tert-Butylmethyl ketone		21	49
Acetophenone	−78 (4)	100	80
Benzophenone		0	20

In addition, its high regioselective allenylboration toward diethylketone [A] and 4-*t*-butylmethylketone [B] as compared with the behavior of allenylmagnesium bromide and di(*n*-butyl)allenylboronate makes *B*-allenyl-9-BBN, a valuable reagent [3].

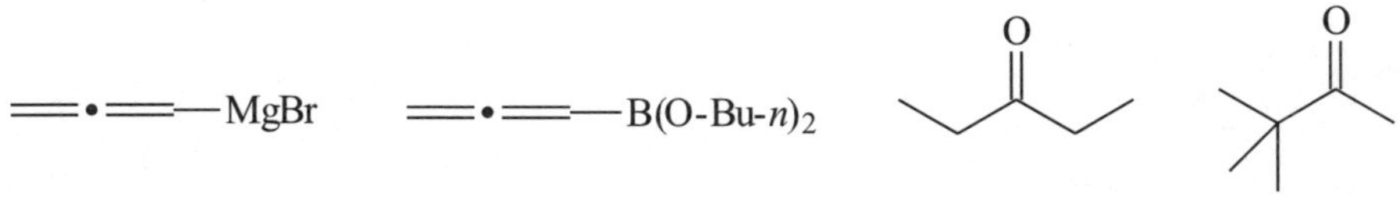

[A] [B]

The results of allenylboration are summarized in Table 6.16 [4].

Table 6.16 Comparison of the allenylboration of [A] and [B] with allenylmagnesium bromide, di(*n*-butyl)-allenylboronate and *B*-allenyl-9-BBN [4]

		Product alcohol (%)	
Ketone	Reagent	Homopropargylic	Allenic
[A]	Allenylmagnesium bromide	88	12
[A]		41	59
[A]	*B*-Allenyl-9-BBN	100	0
[B]	Allenylmagnesium bromide	73	27
[B]	Allenylboronate	40	60
[B]	*B*-Allenyl-9-BBN	100	0

6.6 Synthesis of Silyl Alcohols

In this section the unsaturated silyl alcohols are classified based on the relative position of hydroxyl group to mono unsaturation and the total sum of unsaturations. For example enyne is with C=C and C≡C and allenene is C=C=C and C=C in a silyl alcohol molecule.

6.6.1 Saturated Alcohols

Soderquist *et al* [1] have observed that hydroboration of vinylsilanes with 9-BBN gives (*B*-borylethyl)silanes in quantitative yields, which on oxidation provides isomerically pure 2-silylethanols (Eq. 6.14). The same reaction with BH_3 affords the 1- and 2-silylethanols mixture in 60:40.

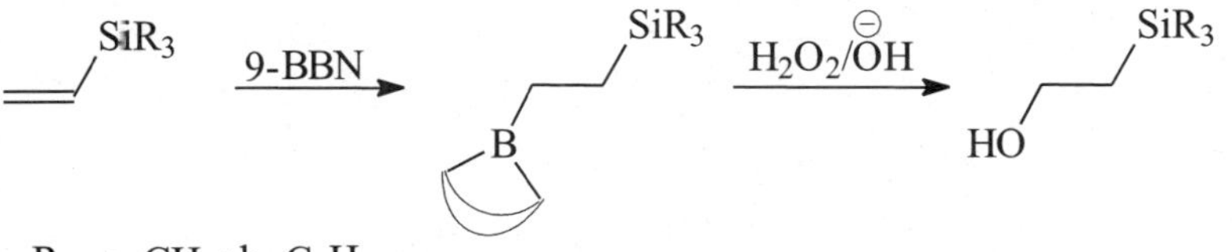

R = a, CH_3; b, C_2H_5 (6.14)

Similarly, allyltrimethylsilane, 3-buten-1-yltrimethylsilane and propen-2-yl-trimethylsilane on hydroboration with 1 equiv of 9-BBN, followed by alkaline hydrogen peroxide oxidation give, in quantitative yields, the isomerically pure terminal alcohols (Chart 6.9) [1b]. However, *cis*-propen-1-yltrimethylsilane with 9-BBN followed by oxidation gives a 56:44 mixture of 1-(trimethylsilyl)-1- and -2-propanols, in quantitative yield.

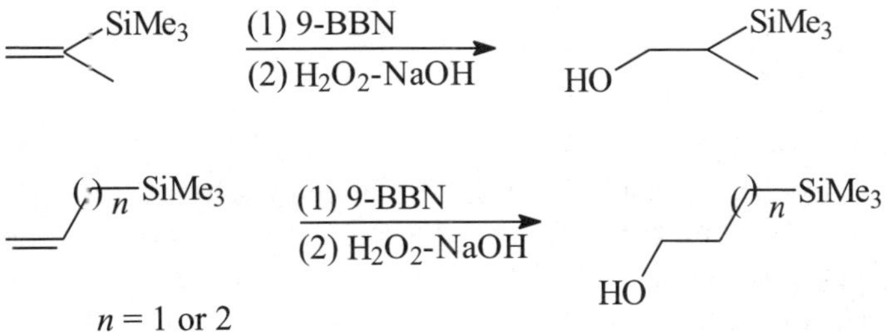

n = 1 or 2

Chart 6.9

On the other hand, the bulkier 2-(triisopropylsilyl)ethanol is prepared only from triisopropyl(α-methoxyvinyl)silane. This is based on Larson's findings that silyl enol ethers [2] undergo hydroboration with 9-BBN, followed by β-elimination of the *B*-methoxy-9-BBN and subsequent rehydroboration. Consequently, this one-pot procedure [1] of hydroboration–dehydroboration–hydroboration gives the desired alcohol, 2-(triisopropylsilyl)ethanol (Chart 6.10) in 89% isolated yield. As evident in Chart 6.10, (α-methoxyvinyl)silane acts as the key intermediate for the preparation of either 1- or 2-(trialkylsilyl)ethanols.

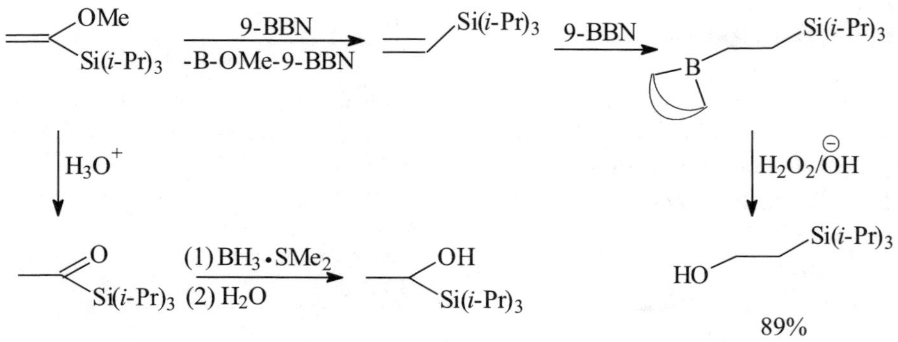

Chart 6.10

Soderquist *et al* have further, reported [3] the hydroboration of 1-trimethylsilyl

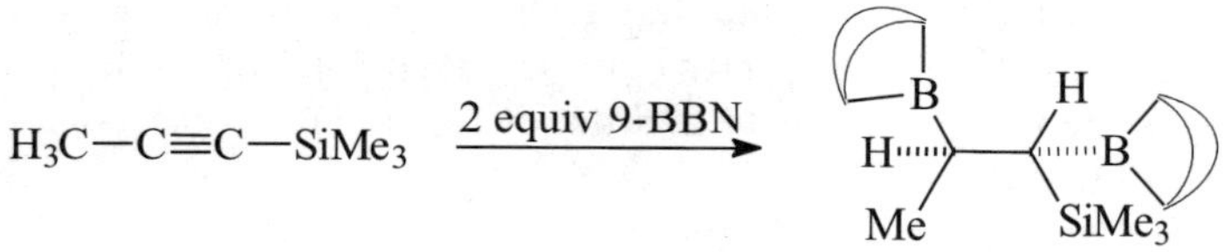

(6.15)

propyne with 2 equiv of 9-BBN under neat conditions (4.5 h, 110 °C) to 1,2-diboryladduct (Eq. 6.15). The dibora adduct is isomerized to the 3,3-diboryl-product (1 h, >160 °C). This "tandem walk" of two 9-BBN groups away from the bulky silyl group is a fascinating process that makes these adducts, synthetically useful. Oxidation of 3,3-diboryladduct (R = Et) affords the 3-(triethylsilyl)-1-propanol as the sole product (Eq. 6.16).

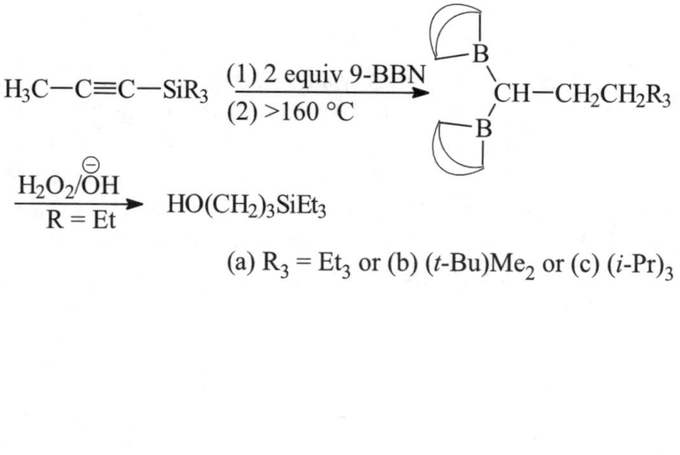

(a) R_3 = Et_3 or (b) (*t*-Bu)Me_2 or (c) (*i*-Pr)$_3$ (6.16)

6.6.2
Allylic Alcohols

The monohydroboration of internal allenes with 9-BBN affords the corresponding allylboranes [1]. Unlike the normal allylboranes, oxidation of silylallylboranes with alkaline hydrogen peroxide produces [2] the *E*- and *Z*-geometric isomers [3, 4] predominating in *E* isomers (R = CH_3, *n*-Bu, *i*-Pr) of allyl alcohols. Increase in size of the R group, however, results in the reversal of selectivity (Chart 6.11) [4].

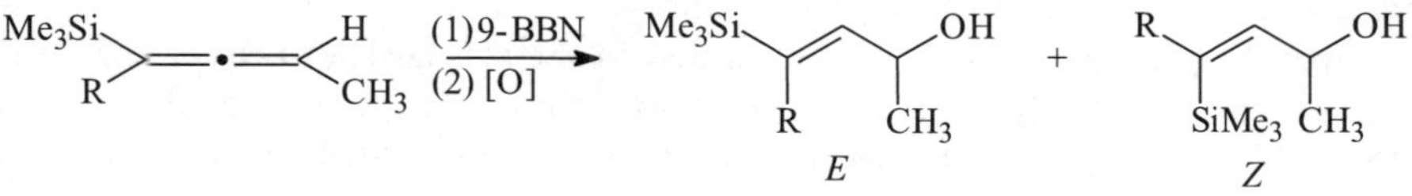

R = CH_3	98 : 2
R = *n*-Bu	94 : 6
R = *i*-Pr	82 : 18
R = *t*-Bu	28 : 72

Chart 6.11

Hydroboration of 1-(trimethylsilyl)allenes occurs at the terminal positions and afford after oxidation the corresponding allylic alcohols (Table 6.17; Chart 6.12) in good yield and in high isomeric purity [5].

Chart 6.12

Table 6.17 The isomeric ratio of allyl alcohols [5]

Allyl alcohol R	Ratio *E*:*Z*	Allyl alcohol R	Ratio *E*:*Z*
Me	>99:1	Ph	98:2
n-Bu	98:2	*i*-Pr	92:8
Me_3SiCH_2	>99:1	*t*-Bu	42:58

Unique to *B*-vinyl derivatives of 9-BBN is their Grignard-like addition to aldehydes [6] to afford stereodefined alcohols. *trans*-*B*-Vinyl-9-BBN synthesis via dehydroborylation [7] process is extended to the synthesis of *trans*-2-trimethylsilyl vinylborane [8]. *trans*-2-Trimethylsilylvinyl-9-BBN adds cleanly to aldehydes (Eq. 6.17) to provide pure *trans*-3-trimethylsilyl(TMS) allylic alcohols in excellent yields (Table 6.18) [8].

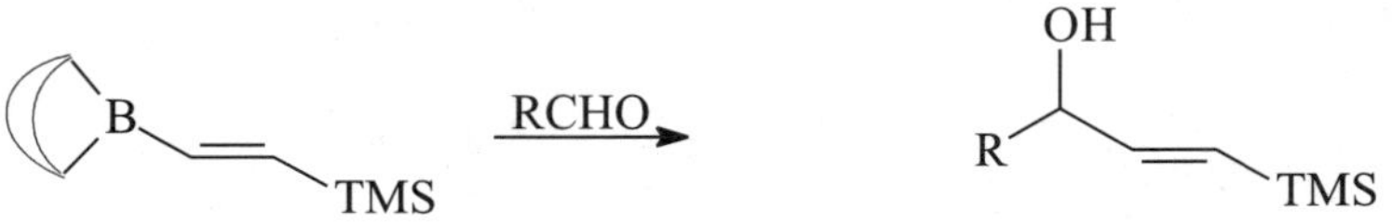

(6.17)

Table 6.18 *trans*-3-TMS Allylic alcohols *via* the Brown's vinylation [8]

R	Yield[a] (%)	R	Yield (%)
C_6H_5	82	*m*-Cl-C_6H_4	71
p-MeO-C_6H_4	75	*o*-CH_2OMe-C_6H_4	81
o-MeO-C_6H_4	88	Me_2CH	70
p-Cl-C_6H_4	75		

[a] Reactions are carried out neat, 120 °C, 1 h.

Enones derived from these alcohols are particularly useful for cyclopentenone annulation through the silicon-directed Nozarov cyclization [9].

The aliphatic aldehydes, except isobutyraldehyde, generally provide poor yields.

The dehydroborylation and addition processes have been utilized, in a one-pot sequence, to generate primary and allylic alcohol functionalities (Chart 6.13).

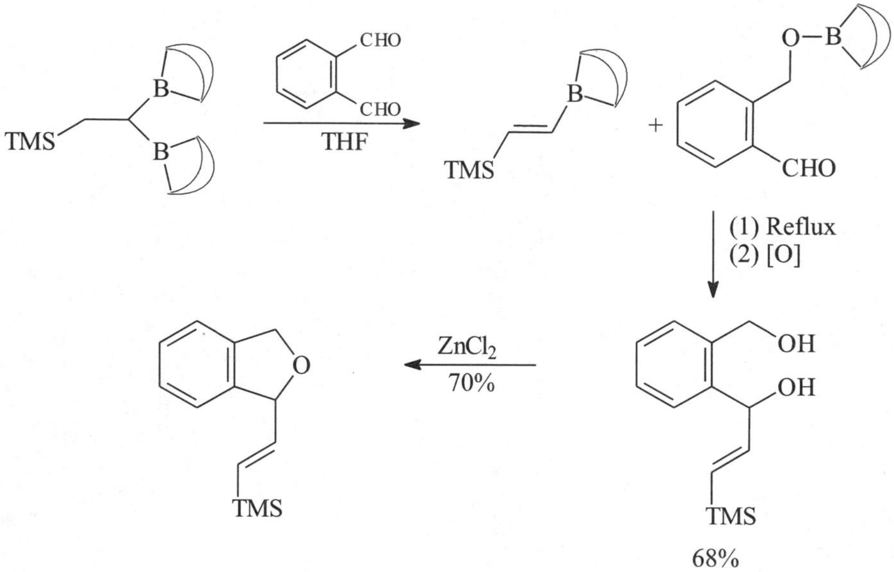

Chart 6.13

The efficient dehydration of the diol with $ZnCl_2$ [10] gives the novel dihydroisobenzofuran in 70% isolated yield [8].

6.6.3 Homoallylic Alcohols

Hydroboration of terminal allenylsilanes with 9-BBN occurs from the less hindered side to afford [1] the kinetic products with *Z* geometry, which undergo fast 1,3-sigmatropic rearrangements to thermodynamically more stable (*E*)-2-(trimethylsilyl) allylboranes. Condensation of *E* isomer with aldehydes proceeds through six-membered transition state and on treatment with 2-aminoethanol [2] gives the desired silated homoallylic alcohols with high diastereomeric purity (Scheme 6.15).

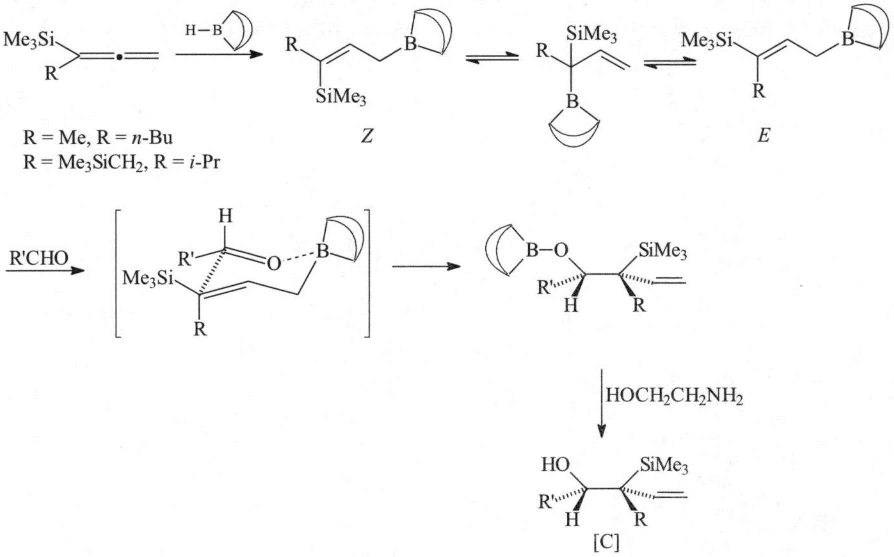

Scheme 6.15

The allylic rearrangement is stopped by the use of sterically more demanding Chx_2BH, at low temperature or bulkier R group of an allene. In the six-membered chair-like transition state trimethylsilyl group now is forced to assume the axial position by *Z* geometry of the double bond, and both R and R′ occupy the equatorial positions. The corresponding diastereomer is obtained by treatment with 2-aminoethanol (Scheme 6.16) [2].

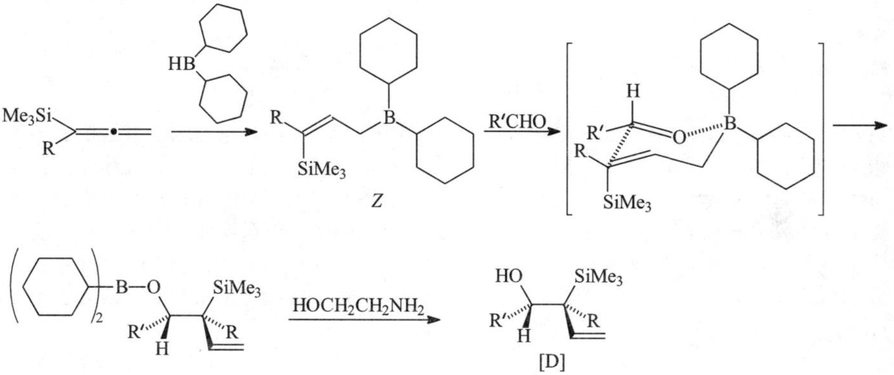

Scheme 6.16

The results are summarized in Table 6.19 [2].

Table 6.19 Stereoselective synthesis of C and D [2]

Hydroborating agent	Allene R	Hydroboration temperature (time)	Yield %	Isomeric ratio of C:D
9-BBN	Me	r.t. (3 h)	89	98:2
	n-Bu	r.t. (4 h)	89	97:3
	Me_3SiCH_2	r.t. (4 h)	93	>99:1
	i-Pr	r.t. (24 h)	75	88:12
$Cl_{x2}BH$	Me	r.t. (3 h)	93	97:3
	n-Bu	−15 °C (6 h)	90	4:96
		r.t. (3 h)	86	30:70
		50 °C (7 h)	91	91:9
	Me_3SiCH_2	0 °C (4 h)	81	6:94
		50 °C (20 h)	78	83:17
		Reflux THF (52 h)	73	94:6
	i-Pr	r.t. (4 h)	92	3:97

6.6.4 Homopropargylic Alcohols

The borane derivative derived from readily available 3-(*tert*-butyldimethylsilyl)-1-(trimethylsilyl)-1-propyne [1] condenses with hexanal and benzaldehyde. The reaction proceeds through six-membered transition state to furnish after workup with 2-aminoethanol the homoprogylic alcohols with high diastereoselectivities. The *SR*/*RS*:*RR*/*SS* ratio is >98:2.

The high diastereoselectivity is attributed to the preferential adoption of the *tert*-butyldimethylsilyl group and the alkyl group of the aldehydes on the opposite side of the transition state (Scheme 6.17) [2].

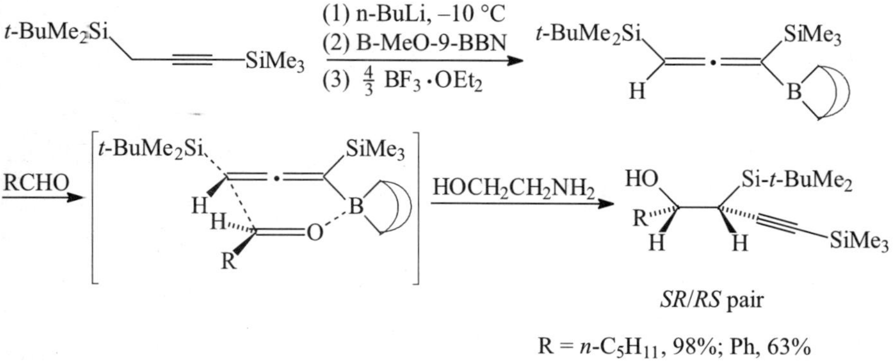

Scheme 6.17

The high diastereoselectivity also results on condensation of hexanal and benzaldehyde with γ-(trimethylsilyl)allenylborane. The *RR*/*SS*:*SR*/*RS* for hexanal is 85:15 and for benzaldehyde is 95:5 (Scheme 6.18) [2].

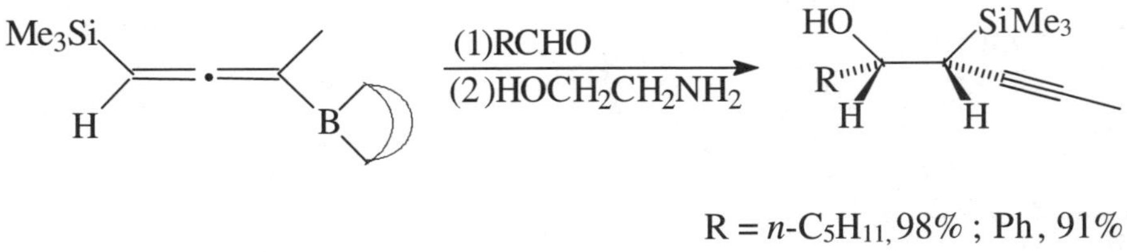

Scheme 6.18

6.6.5 Diyne Alcohols

In a protocol, similar to synthesis of homopropargylic alcohol, allenylborane derived from 3-(*tert*-butyldimethylsilyl)-1-(trimethylsilyl)-1-propyne affords, on condensation with conjugated acetylenic aldehydes (R = *n*-BuC≡C and Ph-C≡C), the diynic alcohols with high diastereoselectivities. The *SR*/*RS*:*RR*/*SS* ratio is >98:2 (Scheme 6.19) [1] with both aldehydes.

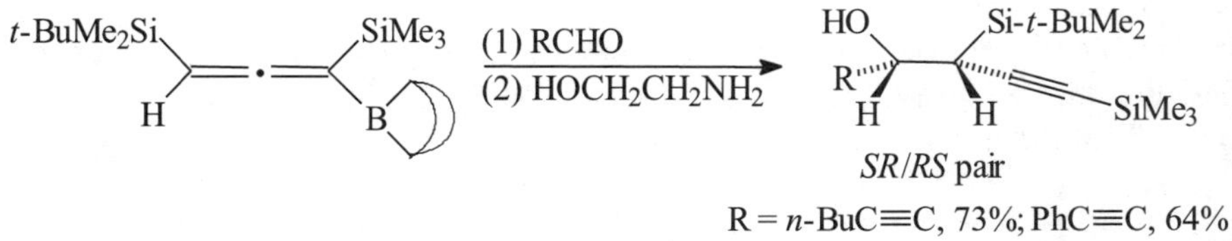

Scheme 6.19

On the other hand, γ-(trimethylsilyl)allyl gives poor diastereoselectivity, and *RR*/*SS*:*SR*/*RS* is 70:30 with both aldehydes (Scheme 6.20) [1].

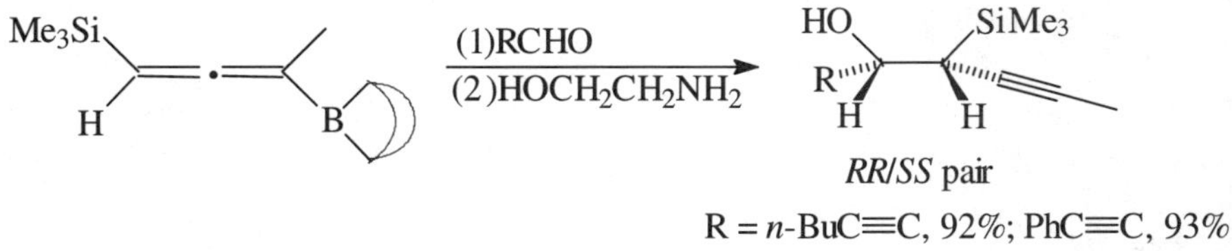

Scheme 6.20

6.6.6
Endiyne Alcohols

(*Z*)-3-Methyl-2-penten-4-ynal reacts with allenylborane to afford after workup the corresponding open-chain endiyne silyl alcohol. The terminal alkynyl group undergoes $Pd(PPh_3)_3$-catalyzed cross-coupling with 1 equiv of iodobenzenes to give the condensation products (Chart 6.14) [1].

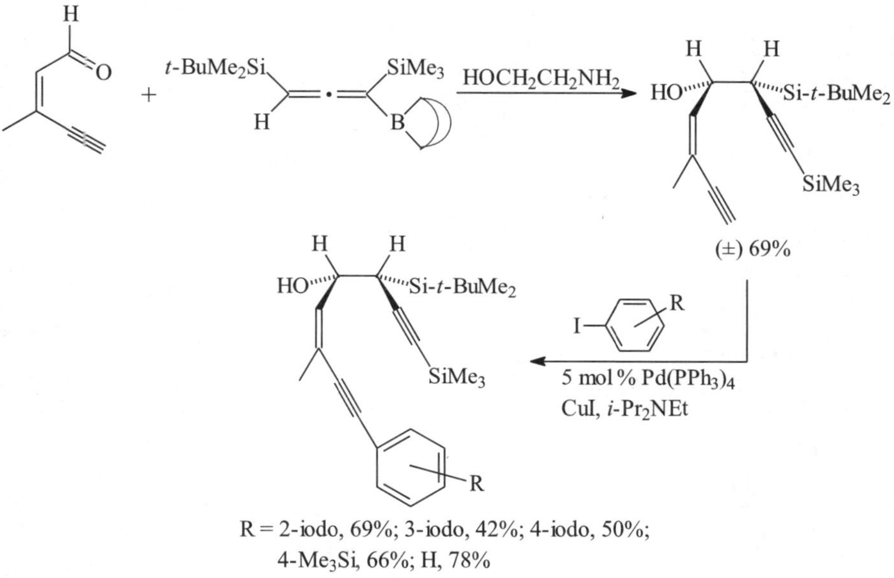

Chart 6.14

The methodology is extended where 2 equiv of endiyne alcohol undergoes cross-coupling with variety of aryliodides. The sequences enhance the versatility and flexibility of this synthetic methodology for preparation of an array of condensation adducts (Chart 6.15) [1].

The cyclopentene derivatives are prepared by reacting the conjugated cyclopentenal, as depicted in Chart 6.16 [2].

6.6.7
Diendiyne Alcohols

In a similar manner, allenylborane derived from 3-(*tert*-butyldimethylsilyl)-1-(trimethylsilyl)-propyne gives mainly (1*S*,2*R*)-2-(*tert*-butyldimethylsilyl)-1-[2-(3-methyl-3-buten-1-ynyl)-1-cyclopentenyl]-4-(trimethylsilyl)-3-butyn-1-ol (54%) and (1*S*,2*R*)-2-(*tert*-butyldimethylsilyl)-1-[2-1-cyclohexenylethynyl)-1-cyclopentenyl]-4-(trimethyl silyl)-3-butyn-1-ol (55%) (Chart 6.17) [1].

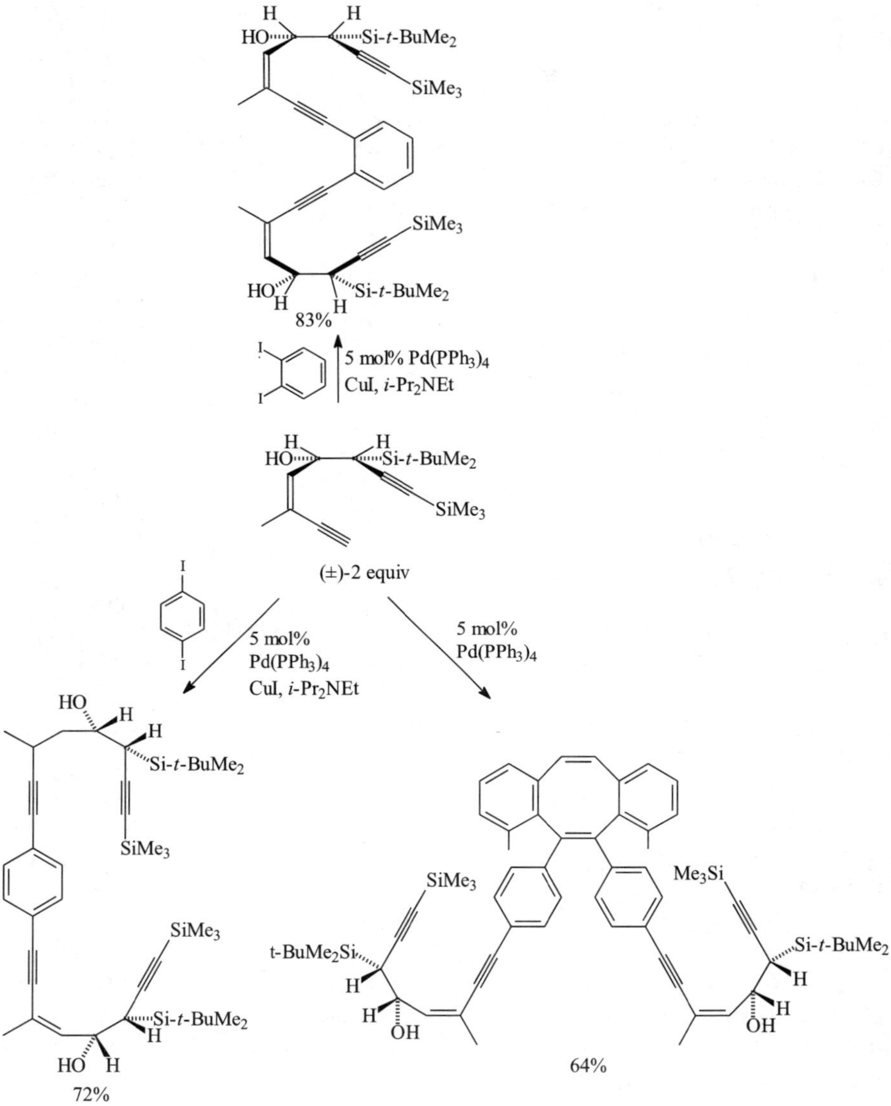

Chart 6.15

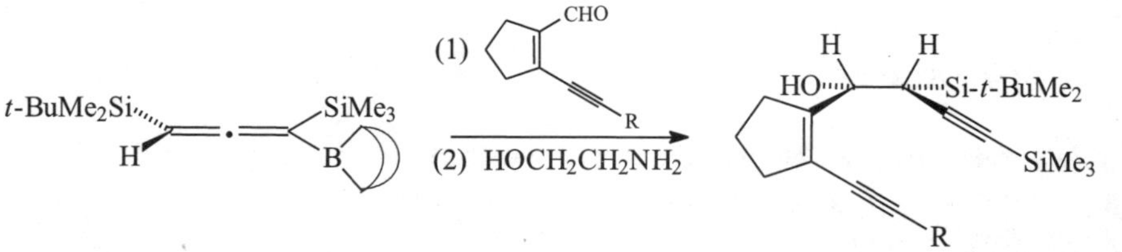

R = $SiMe_3$, 61%; *n*-Bu, 65%; Ph, 56%

Chart 6.16

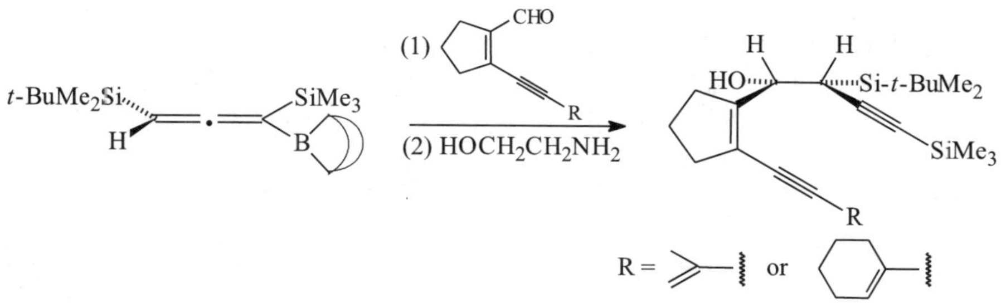

Chart 6.17

6.6.8 Allenic Alcohols

Wang *et al* [1] have reported that propargylic organoborane intermediates derived from the corresponding lithium reagents react with aldehydes and certain ketones with high regioselectivity to form the corresponding trimethylsilyl-substituted α-allenic alcohols.

1-(Trimethylsilyl)-propyne gives lithium reagent on treatment with *tert*-butyllithium. The subsequent reactions with *B*-OMe-9-BBN at 0 °C and with $BF_3{\cdot}OEt_2$ [2] afford the corresponding propargylic organoborane. To the reaction mixture is then added an aldehyde or ketone, and after oxidative workup provides the corresponding trimethylsilyl-substituted α-allenic alcohol in excellent isolated yield (Scheme 6.21). None of the corresponding *B*-acetylenic alcohols is detected.

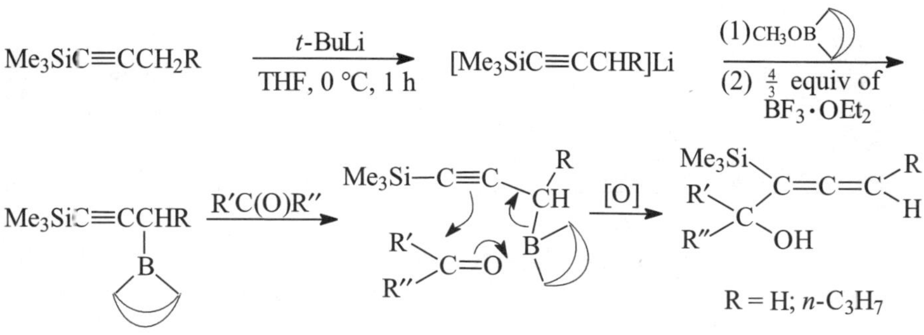

Scheme 6.21

The reaction of aldehydes or ketones proceed through the six-center electronic transfer with propargylic–allenic rearrangement [3].

The reaction of propargylic borane (R = H), however, is dramatically affected by the reaction temperature, and its reaction is carried out at –78 °C and slowly warmed to room temperature. α-Allenic alcohols are obtained as the predominant products from all the aldehydes and certain ketones (Table 6.20) [1].

Table 6.20 Reactions of representative aldehydes and ketones with propargylic borane [1]

R	R′	R″	Yield (%)
H	*n*-C_5H_{11}	H	82
	i-C_3H_7	H	82
	t-C_4H_9	H	85
	C_6H_5	H	88
	(*E*)-$CH_3CH{=}CH$	H	79
	CH_3	CH_3	86
	C_2H_5	C_2H_5	91
		-$(CH_2)_5$-	91
	CH_3	C_6H_5	93
n-C_3H_7	*n*-C_5H_{11}	H	78 (88:12)
	i-C_3H_7	H	74 (88:12)
	C_6H_5	H	72 (87:13)
	CH_3	CH_3	71 (91:9)
	C_2H_5	C_2H_5	75 (83:17)
		-$(CH_2)_5$-	76 (91:9)
	CH_3	C_6H_5	88 (54:46)

The relative reactivities of representative aldehydes or ketones toward propargylic borane (R = H) are summarized in Table 6.21 [4].

Table 6.21 Relative reactivities of representative aldehydes and ketones toward trimethylsilyl-substituted propargylic borane (R = H) in THF at 25 °C [4]

Aldehyde or ketone	Relative reactivity	Aldehyde or ketone	Relative reactivity
Hexanal	100	Acetone	4.4×10^{-2}
iso-Butyraldehyde	87	2-Pentanone	2.1×10^{-2}
Benzaldehyde	37	Acetophenone	1.8×10^{-2}
Crotonaldehyde	31	Cyclopentanone	2.5×10^{-3}
Pivaldehyde	3	3-Pentanone	2.4×10^{-3}
Cyclohexanone	7.3×10^{-2}	2-Methylcyclopentanone	10^{-3}
2-Methylcyclohexanone	5.2×10^{-2}	Butyrophenone	5.9×10^{-4}

The relative reactivities [4] of $Me_3SiC{\equiv}CCH_2B$⟩ and $Me_3SiC{\equiv}CCH_2BChx_2$ are almost the same toward ketones, but $Me_3SiC{\equiv}CCH_2BChx_2$ is ~2.7 more reactive toward cyclopentanone as compared with $Me_3SiC{\equiv}CCH_2B$⟩

Aldehydes are much more reactive than ketones are; cyclohexanone is about 30 times more reactive than is cyclopentanone; 2-pentanone, (a methylketone) is about 9 times more reactive than are 3-pentanone and cyclopentanone. This selectivity is utilized (Chart 6.18) to transform one of the carbonyl moieties to the corresponding allenic alcohols. The alcohol and ketone react intramolecularly to give the corresponding hemiketal. The hemiketal as the preferred tautomer has also been observed for 2-(3-hydroxypropyl) cyclohexanone [5].

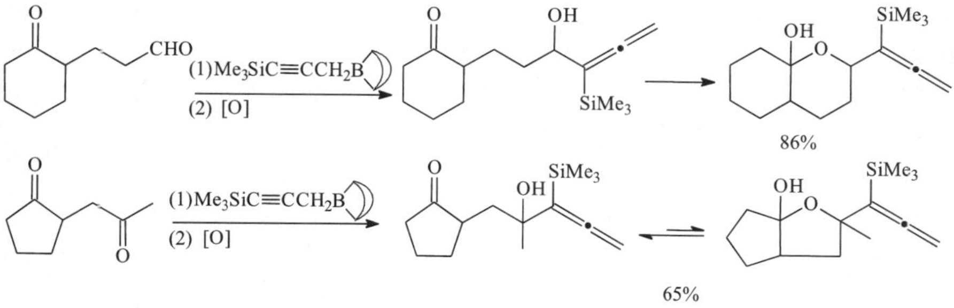

Chart 6.18

The carboxylic esters also do not exhibit reactivity, thus, the ketone group is selectively utilized for condensation (Chart 6.19) [4].

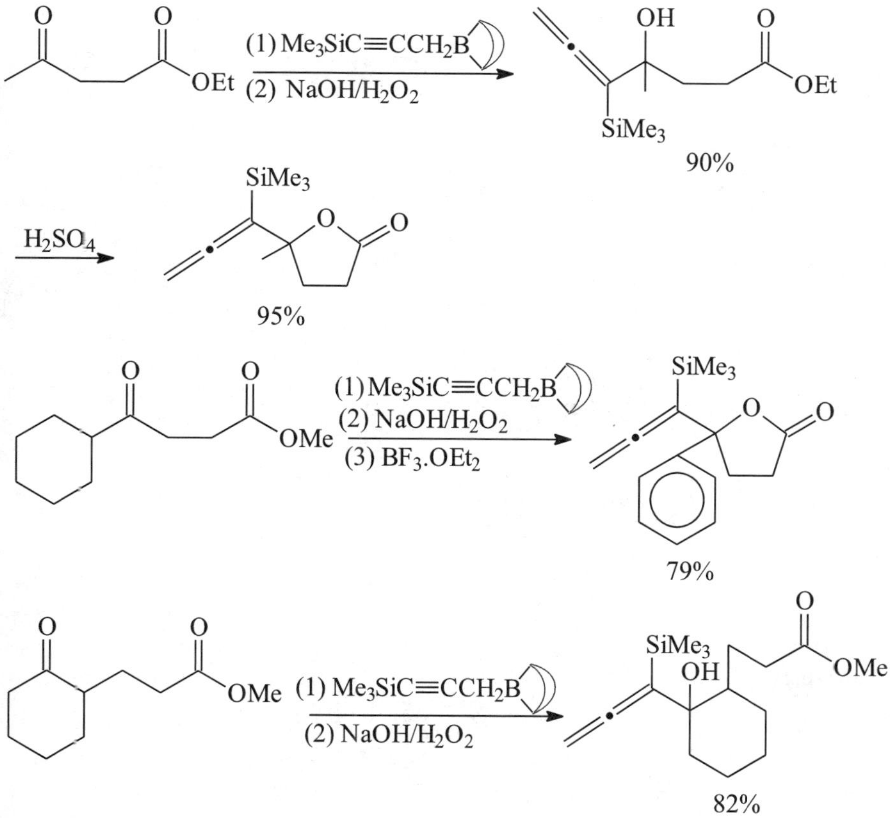

Chart 6.19

6.6.9 Allenene Alcohols

The isomeric silyl alcohols having three carbon–carbon double bonds have been prepared by the following sequence with *Z* geometry of the α-double bond of triene, when 9-BBN is utilized as the hydroborating agent (Scheme 6.22) [6]. Alternatively, α-carbon–carbon double bond with *E* geometry is obtained using dicyclohexylborane as the hydroborating agent (Scheme 6.23) [1].

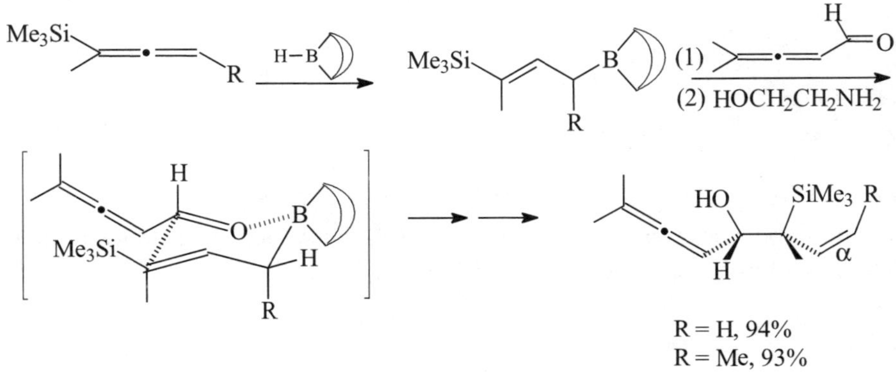

Scheme 6.22

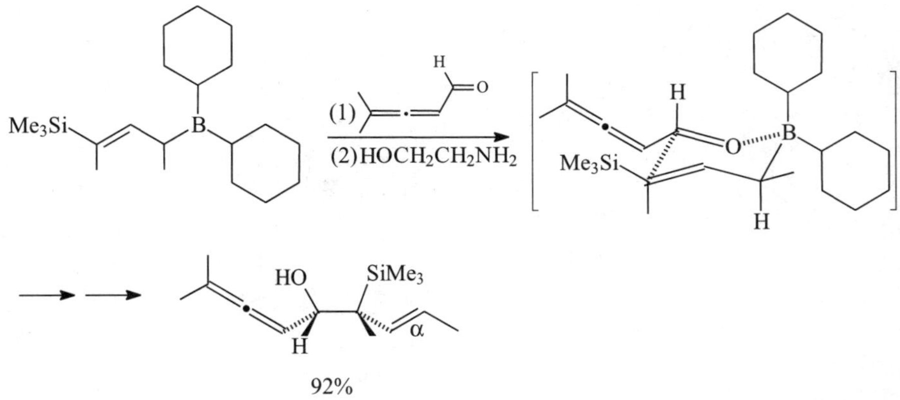

Scheme 6.23

6.6.10
Allenyne Alcohols

The alcohols having allene and enyne moieties are prepared as depicted in Scheme 6.24 [1] and Scheme 6.25 [2].

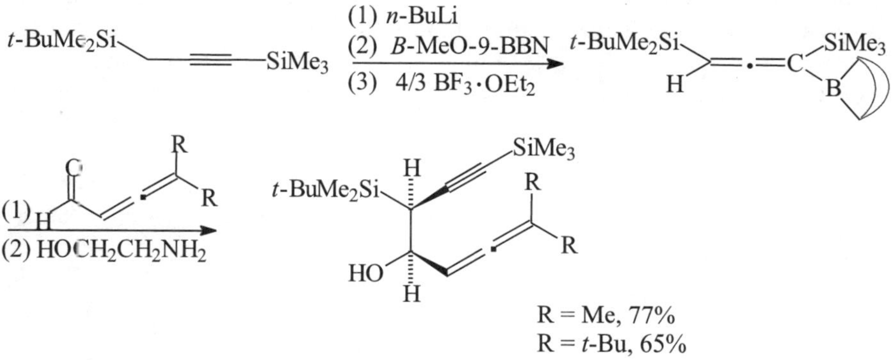

Scheme 6.24

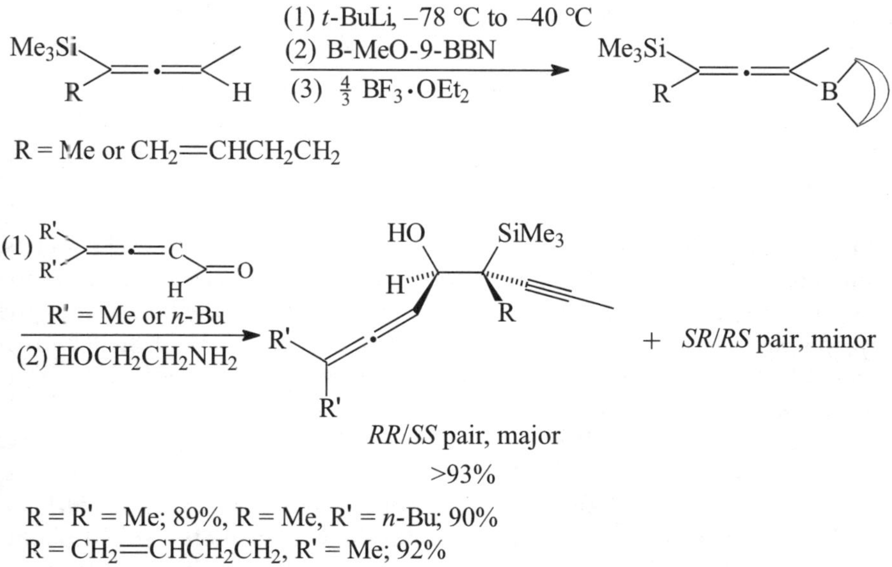

Scheme 6.25

6.6.11
Allendienyne Alcohols

Wang *et al* [1] have synthesized acyclic dienyne–allene alcohols, a precursor required for the construction of tetracyclic steroidal skeleton. The conjugated allenic aldehyde, 2,3,9-decatrienal, prepared [2] from readily available 1,2,8-nonatriene, undergoes condensation with allenylborane [3] smoothly and affords after treatment with 2-aminoethanol, hydroxyallendienyne in excellent isolated yield (Chart 6.20) [1]. Similar result is obtained with allenylborane ($R = CH_3$). The diastereoselectivity of the two new formed stereocenters is high ($de = 94\%$).

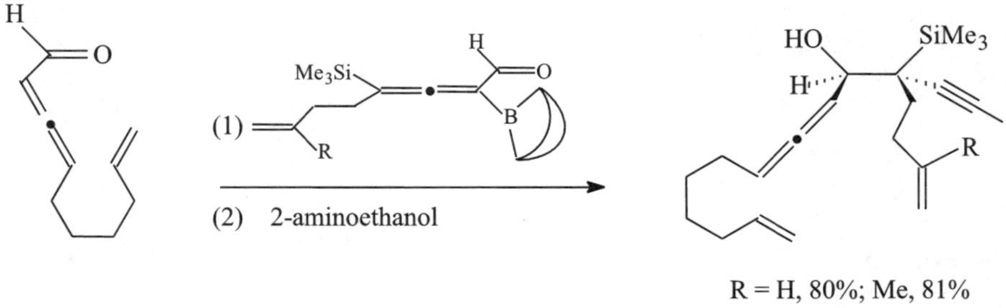

Chart 6.20

6.7
Synthesis of Heterocyclic Alcohols

Brown *et al* [1] have elegantly hydroborated a variety of heterocyclic olefins, which on oxidation yield the corresponding alcohols in excellent yields. In the case of heterocyclic olefin containing a double bond α to the heteroatom, the hydroboration reaction is highly regioselective, placing boron at the β-carbon atom. The use of 9-BBN for the synthesis of variety of heterocyclic alcohols is given in Chart 6.21.

In the case of 1,2-dihydrofuran, the hydroboration with 9-BBN is complete in 1 h, the oxidation of which afforded the corresponding alcohol in quantitative yield. However, the alkylborane, is not very stable at 25 °C in THF and gradually rearranges to the open-chain unsaturated alcohol. The hydroboration of the intermediate with 9-BBN affords the corresponding diol (Scheme 6.26).

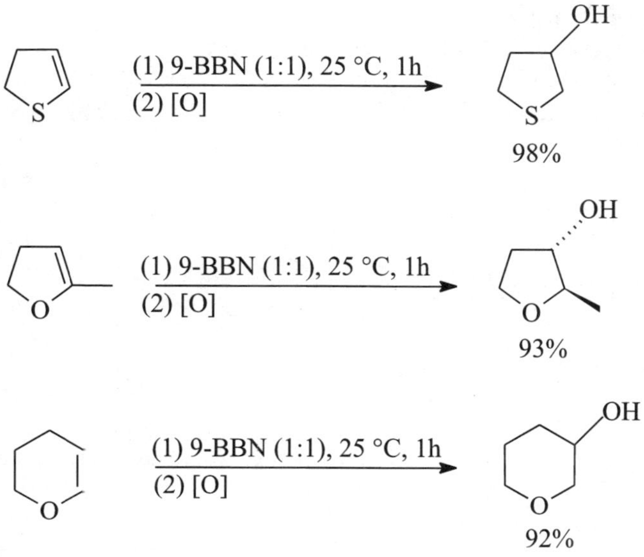

Chart 6.21

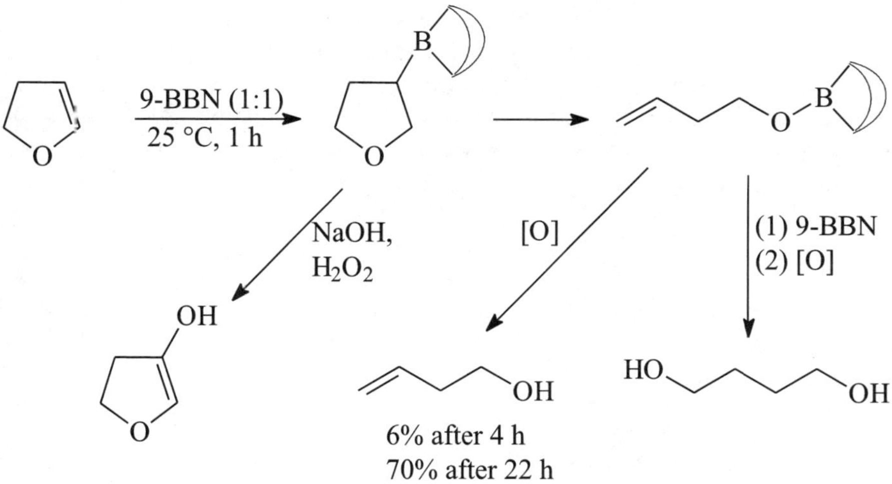

Scheme 6.26

The hydroboration of pyridine derivative having a double bond β to the nitrogen affords the corresponding 3-hydroxy and 4-hydroxy derivatives in an 85:15 ratio (Eq. 6.18), whereas the product distribution ratio is 75:25 in the case of dicyclohexylborane and disiamylborane as the hydroborating reagents.

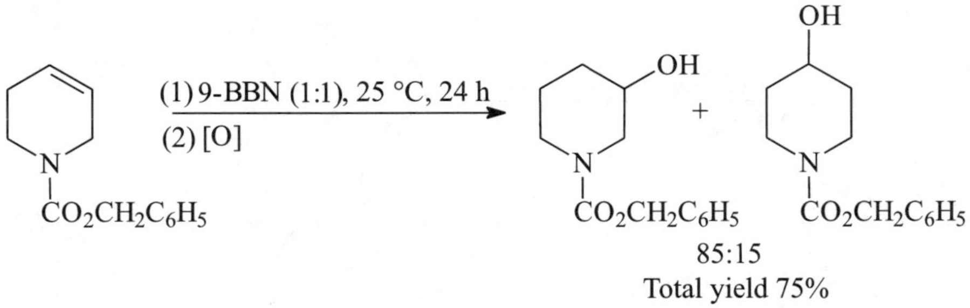

(6.18)

Brown and coworkers [2] studied the hydroboration of N-substituted 3-pyrrolines and the intermediate organoboranes on oxidation afford the corresponding 3-hydroxy pyrrolines in good yield (Table 6.22) [2].

The relative reactivity of heterocyclic and carbocyclic olefins toward 9-BBN are depicted in Chart 6.22 [3].

The *exo*-methylene sugars, which are easily prepared [4] from readily available sugar lactones, by Tebbe's reagent [5] are useful precursors for *C*-glycoside synthesis. The stereoselective hydroboration of *exo*-methylene sugar with 9-BBN followed by alkaline H_2O_2 oxidation afford *C*-(hydroxymethyl)-2,3,4,6-tetra-*O*-benzyl-β-D-glucopyranoxide in 94% yield [4] (Chart 6.23). On the other hand borane–THF complex gives 1:1 mixture of α- and β-hydroxymethyl glucoxides.

Table 6.22 Hydroboration-oxidation of N-substituted 3-pyrrolines with 9-BBN [2]

Olefin	Alcohol	Olefin: 9-BBN ratio	Reaction time (h)	Yield (%)
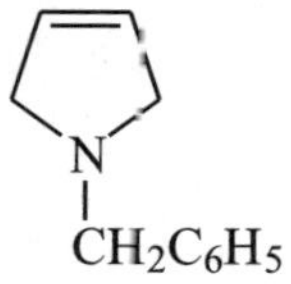	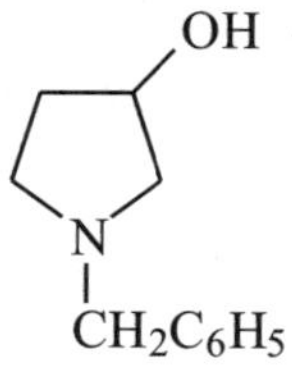	1:2 1:1	1 1	100 96
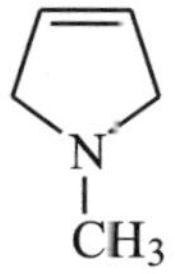	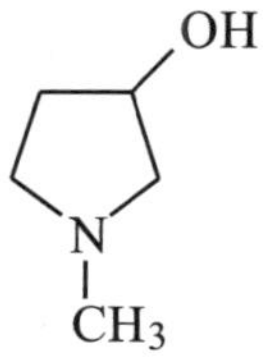	1:2	24	97
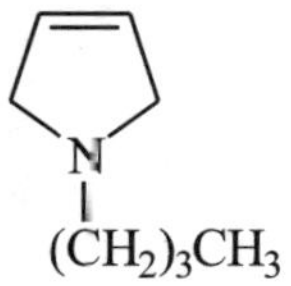	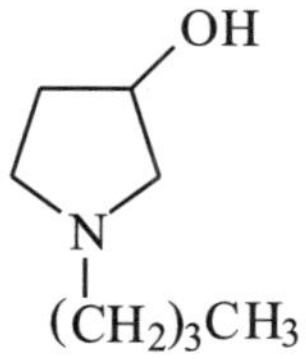	1:2	24	93
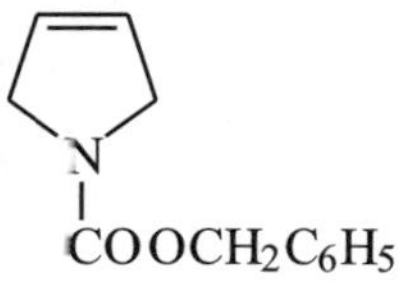	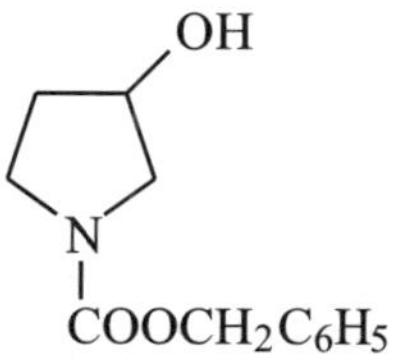	1:1	1	92

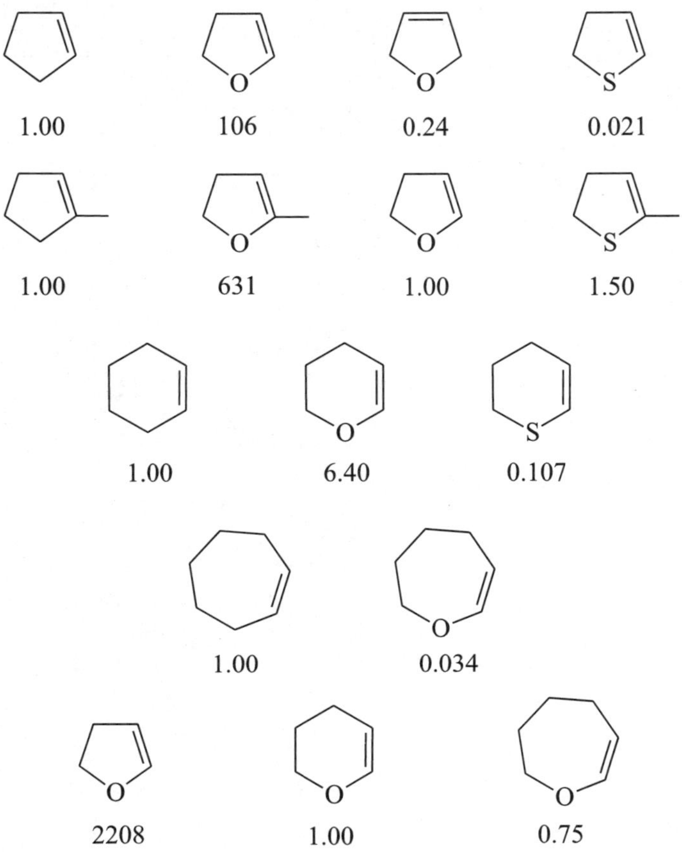

Chart 6.22

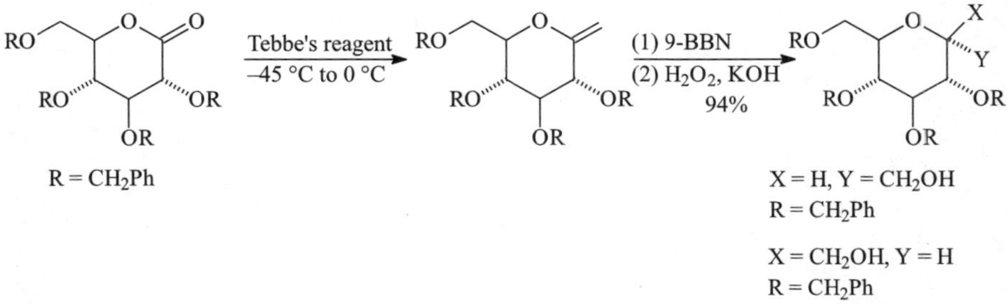

Chart 6.23

6.8 Synthesis of Amino Alcohols

Many amino alcohols are important therapeutic agents for the treatment of a wide variety of human diseases [1] and are also extraordinarily important as chiral auxiliaries in organic synthesis [2]. Singaram and coworkers [3] prepared various trialkylboranes by hydroboration of β,β-disubstituted enamines with 9-BBN. Among the three oxidizing agents; trimethylamine *N*-oxide, sodium perborate, and 30% hydrogen peroxide–solid sodium hydroxide for oxidation of the intermediate organoborane; 30% hydrogen peroxide–solid sodium hydroxide is the best reagent for such oxidation (Eq. 6.19) to the corresponding amino alcohols (Table 6.23) [3].

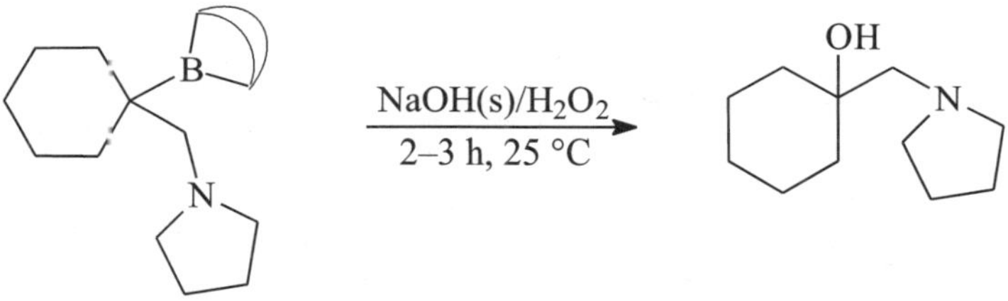

(6.19)

Table 6.23 Synthesis of amino alcohols from β,β-disubstituted enamines [3]

Enamine	Amino alcohol	Yield (%)
2-Methyl-1-pyrrolidino-1-pentene	2-methyl-1-pyrrolidino-pentan-2-ol	64
2-Methyl-1-morpholino-1-pentene	2-methyl-1-morpholino-pentan-2-ol	84
1-Diethylamino-2-methyl-1-pentene	1-diethylamino-2-methylpentan-2-ol[a]	62
1-(Cyclohexylidenemethyl)pyrrolidine	1-(1-pyrrolidinomethyl)cyclohexanol	95
4-(Cyclohexylidenemethyl)morpholine	1-(4-morpholinomethyl)cyclohexanol	91

[a] Hydroboration with BMS.

N-Tosyl-5-aminocyclooctanol is prepared [4] in excellent yield by reacting *B*-octyl-9-BBN with hypervalent iodine compound [*N*-(*p*-toluenesulfonyl)imino] phenyliodinane at room temperature (Eq. 6.20).

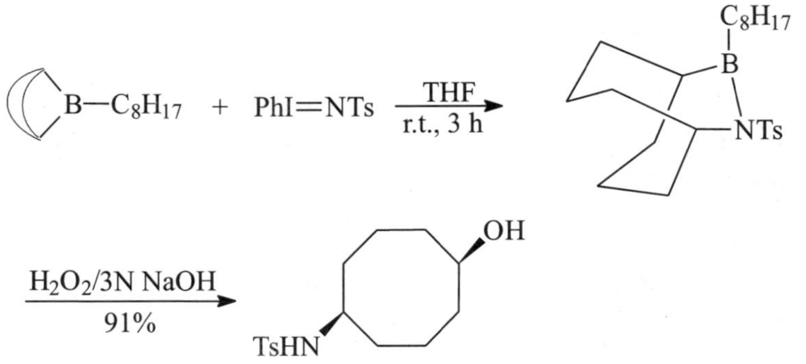

(6.20)

6.9 Asymmetric Synthesis of Alcohols

Yamamoto and coworkers [1] have developed a methodology for the preparation of stereochemically homogeneous allylic metals (Mg and Ba) from the corresponding allylic chlorides. These allylic metals react in high diastereoselective-γ-allylation with aldehydes in the presence of *B*-methoxy-9-BBN as an additive. With this method *threo* homoallylic alcohols are selectively obtained from (*E*)-allylic metal compounds and *erythro* isomers from (*Z*)-allylic metal compounds (Scheme 6.27) in excellent yields [1].

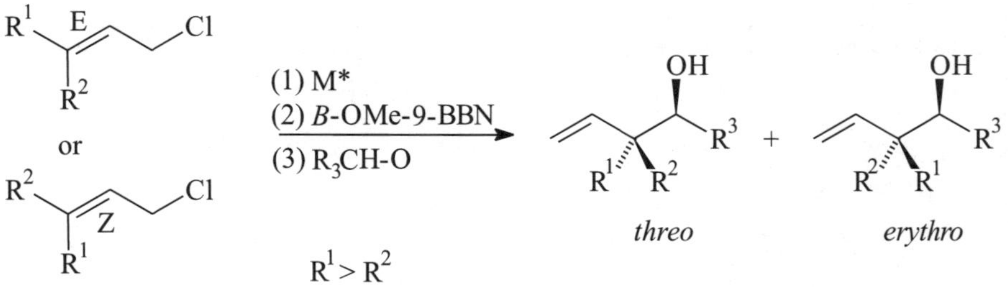

Scheme 16.27

Consequently, diastereoselective allylation reactions of stereochemically homogeneous geranyl and neryl Grignard reagents with hexanal are achieved using *B*-OMe-9-BBN as an additive: (< −95 °C) RMgCl; *threo:erythro*, yield; R = geranyl, 97:3, 82% R = neryl, 6:94, 83%. The comparative effects of additives are summarized in Table 6.24 [1].

Table 6.24 Effect of additives in the diastereoselective allylation using stereochemically fixed allylic Grignard reagent [1]

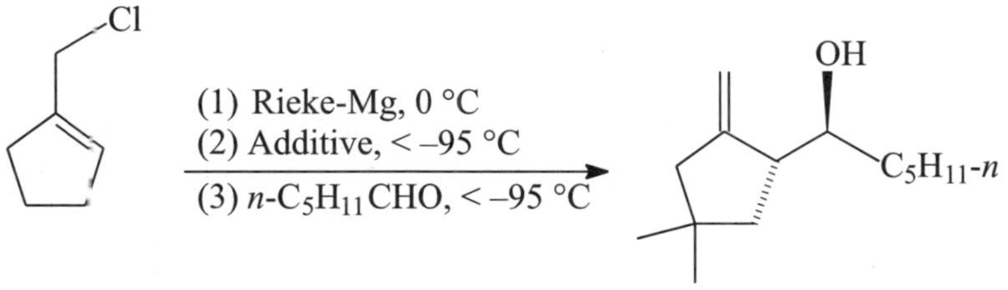

Threo

Additive	*threo:erythro*	Yield (%)
–	69:31	>99
$Ti(O^iPr)_4$	81:19	99
$B(OEt)_3$	62:38	57
B-OTf-9-BBN	77:23	>99
B-OMe-9-BBN	97:3	71
Me_3Al	68:32	92
Et_2Zn	61:39	87

Allylation is carried out using an allylic chloride (1 equiv) magnesium (3 equiv), additive (1 equiv) and hexanal (0.4 equiv).

Similarly, the reaction of monosubstituted allylic barium reagents give rise to exceedingly high diastereoselectivity (Table 6.25) [1].

These results support the intermediate allylic boron ate complex formed from stereochemically homogeneous allylmetal and *B*-OMe-9-BBN, with a plausible chair-like transition state T_1. This transition state is further stabilized by the adjacent four-membered ring formed by the second metal [2]. The alternative transition state T_2 is destabilized by the 1,3-diaxial repulsion of the bulky cyclooctyl ring of 9-BBN (Fig. 6.1). *erythro-threo* Assignments are made

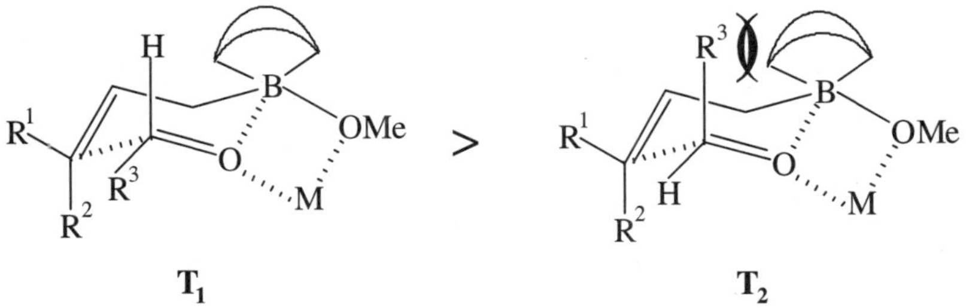

Fig. 6.1 Effect of 1,3-diaxial interaction on stability of bicyclic transition states

Table 6.25 Diastereoselective γ-allylation of various aldehydes by allylic barium reagents using *B*-OMe-9-BBN as an additive [1]

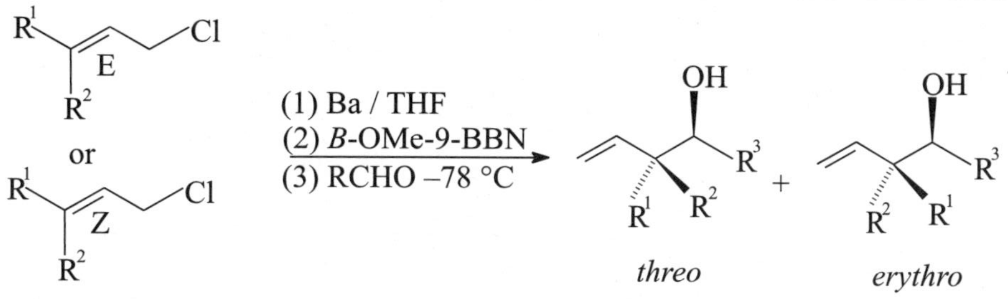

$R^1 > R^2$

Allylic-Cl	*threo*:*erythro* yield (%) n-$C_5H_{11}CHO$	PhCHO	(*E*)-PhCH=CHCHO
(*E*)-2-Decenyl-Cl	97:3 (82)	96:4 (96)	97:3 (85)
(*Z*)-2-Decenyl-Cl	4:96 (77)	2:98 (98)	2:98 (96)
Geranyl-Cl	97:3 (91)	97:3 (96)	98:2 (96)
Neryl-Cl	1:99 (70)	3:97 (67)	2:98 (61)

Allylation is carried out using allylic chloride (1 equiv), barium (1.1 equiv), *B*-OMe-9-BBN (1 equiv) and aldehyde (0.4 equiv).

by transforming the diastereomeric isomers to their corresponding aldols by ozonolysis. The stereochemical assignments are based on the magnitude of the vicinal coupling [3] constant; J_{ab} ~ 3 Hz for *erythro*, whereas J_{ab} is much higher (~ 8 Hz) for *threo*.

Yamamoto *et al* have further reported [4] a one-pot process for the stereoselective synthesis of (*Z*)-2-alkenylsilanes and -tins in high chemical and isomeric yields (Chart 6.24). The (*Z*)-2-alkenyltin prepared as illustrated in Chart 6.25 adds to aldehyde and complete *erythro*-selective coupling is realized *via* the thermal [5] or Lewis acid–mediated reaction [6] (Chart 6.25) to give homoallylic alcohols.

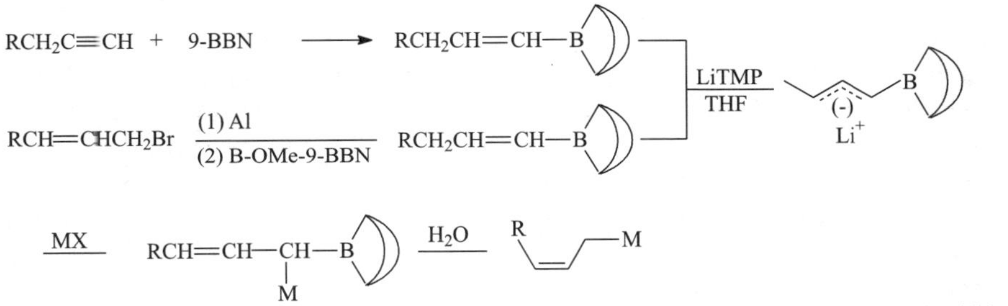

R = *n*-Pr or Me; LiTMP = Lithium 2,2,6,6-tetramethylpiperidide

M = $SiMe_3$, $SnMe_3$ or Sn(*n*-Bu)$_3$

Chart 6.24

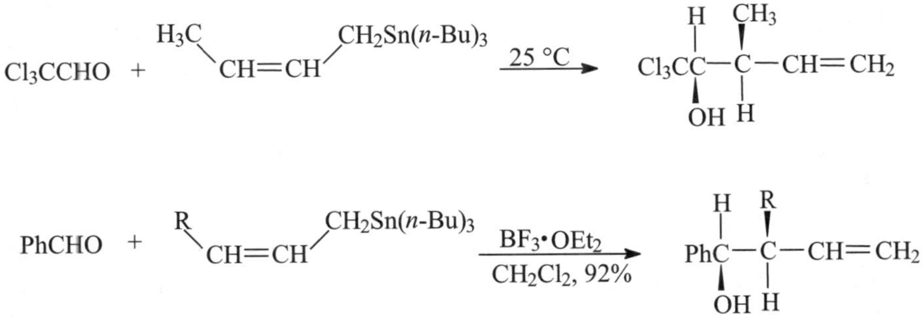

R = *n*-Pr or Me

Chart 6.25

The Wacker-type oxidation [7] of homoallylic alcohols affords the ketoalcohol in high yields (Eq. 6.21).

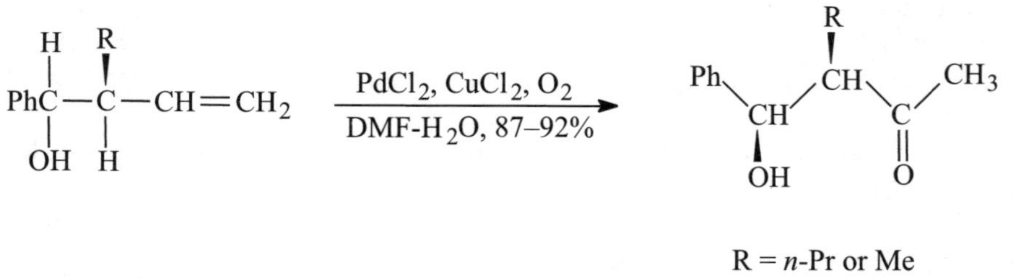

(6.21)

The utility of α-silyl- and α-stannyl-substituted crotyl-9-BBN [4] has been extended to realize the stereoregulated synthesis of four [8a] carbon units (Table 6.26). Consequently, the reaction of α-silyl- or -stannyl-substituted crotyl-9-BBN with aldehydes in the presence of certain bases (such as pyridine, *n*-butyllithium, or *sec*-butyllithium) provides the high regulation of the stereochemistries over four consecutive acylic carbon atoms; the *threo* relation between C-1 and C-2 and the *cis*-configuration at C-3 and C-4 (Scheme 16.28) [8a].

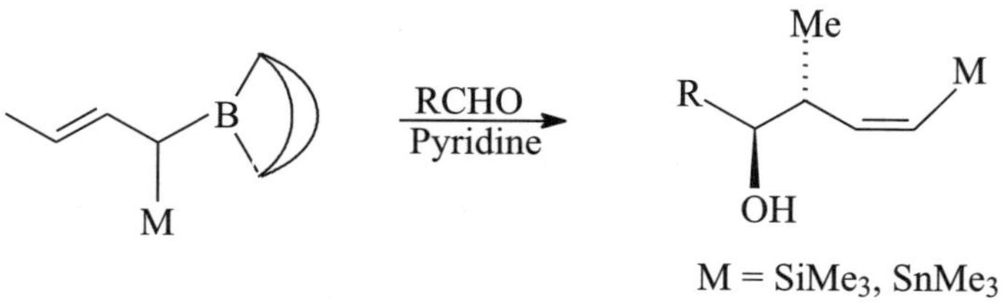

Scheme 6.28

It is significant to mention that use of base is essential to realize the high stereoregulation. The base probably forms the ate complex and influences the reactivity and selectivity of crotyl-9-BBN derivatives. *n*-Butyl or *sec*-butyl bases, sometimes, cause the migration of the butyl groups.

The inspection of the model reveals that the Me_3Si or Me_3Sn group occupies the axial position of the six-membered cyclic transition state because of the steric repulsion by the protons of 9-BBN ring (Fig. 6.2). This leads to the selective formation of *threo* isomer among eight possible combinations. Due to rapid allylic rearrangement, the crotyl derivative consists of a mixture of *cis* and *trans* isomers. However, the formation of ate complexes prevents the rearrangement, and the geometry of the crotyl unit is fixed to *trans*.

Table 6.26 Stereoregulated reaction of α-silyl or α-stannyl substituted crotyl-9-BBN with aldehydes [8a]

Aldehydes	Product	Yield (%)	Isomer ratio
PhCHO	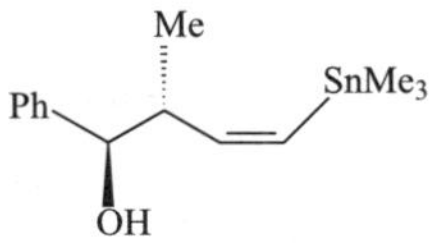	75[a]	(*threo*,*Z*)-(*erythro*,*Z*)-(*E*)-others = 88:5:5:trace
PhCHO	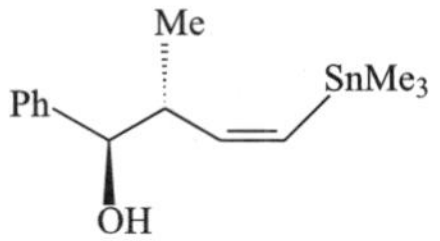	90[b]	(*threo*,*Z*)-(*erythro*,*Z*)-(*E*)-others =92:trace:1:6
PhCHO	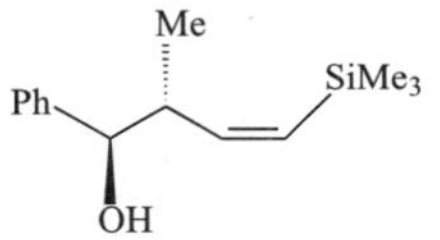	50[c]	(*threo*, *Z*)-others = >98:<2
	PhCH(OH)-*n*-Bu	45	
PhCHO	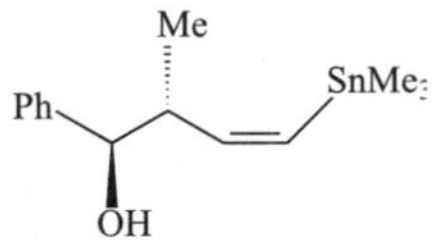	56[d]	(*threo*, *Z*)-others = >98:<2
	PhCH(OH)-*sec*-Bu	40	
PhCHO	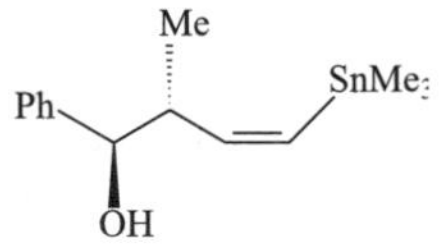	90[b]	Not analyzed
n-BuCHO	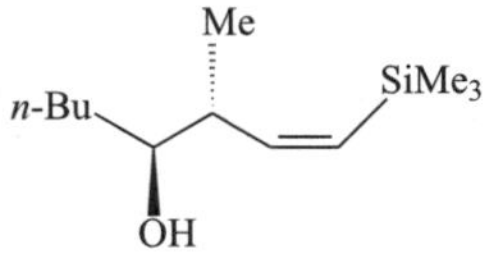	70[a]	(*threo*, *Z*)-α isomer-others = 90:6:4

[a] 1 equiv of pyridine is used.
[b] 2 equiv of pyridine is used.
[c] 1 equiv of *n*-BuLi is used.
[d] 1equiv of *sec*-BuLi is used.

Table 6.26 *(continued)* Stereoregulated reaction of α-silyl or α-stannyl substituted crotyl-9-BBN with aldehydes [8a]

Aldehydes	Product	Yield (%)	Isomer ratio
n-BuCHO	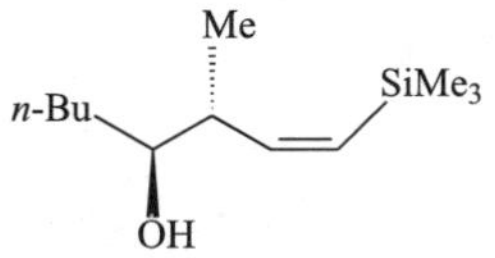	62[c]	(*threo*, *Z*)-others = >98:<2
n-BuCHO	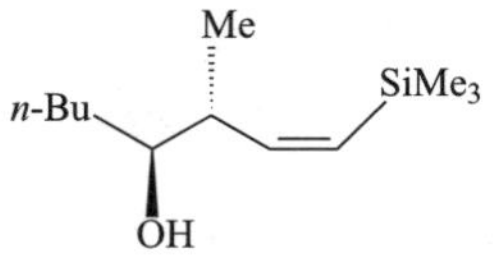	70[b]	Not analyzed
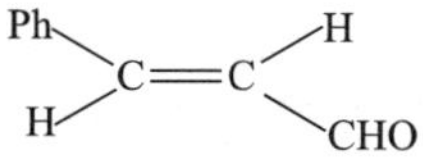	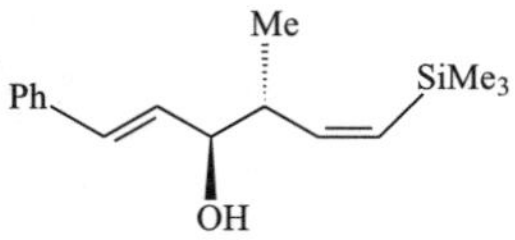	85[b]	Not analyzed
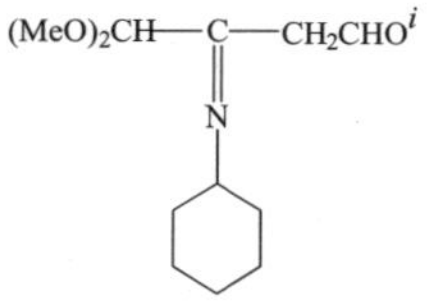	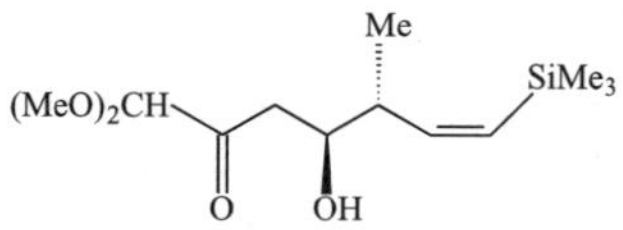	40[b]	Not analyzed

[a] 1 equiv of pyridine is used.
[b] 2 equiv of pyridine is used.
[c] 1 equiv of *n*-BuLi is used.
[d] 1equiv of *sec*-BuLi is used.

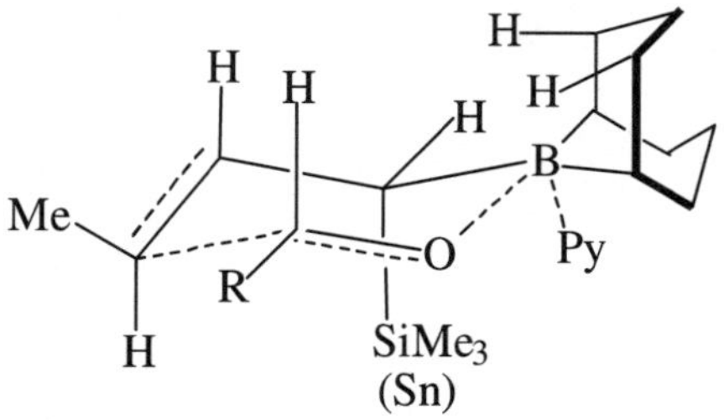

Fig. 6.2 Steric repulsion by 9-BBN protons places Si(Sn)Me_3 to axial position in chair-like transition state

Yamamoto and coworkers [8b] have reported the diastereoselective C–C bond formation for the synthesis of tertiary alcohols. The reaction of *B*-crotyl-9-BBN with pyruvates produces the *threo* (*anti*) isomer as the major product. By increasing the steric bulk of the ester groups of pyruvate, the *threo* (*anti*) isomer is obtained predominantly or exclusively (Eq. 6.22; Table 6.27) [8b]. On the other hand, reaction of allenic organometallics gives mainly the *erythro* (*syn*) isomer.

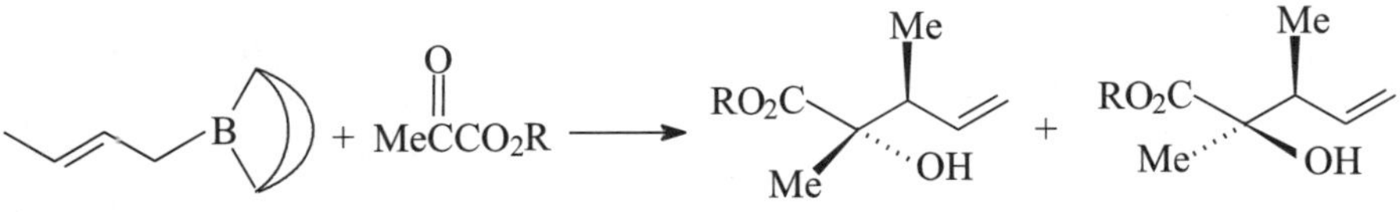

threo (*anti*) major product *erythro* (*syn*)

(6.22)

Table 6.27 Reaction *B*-crotyl-9-BBN with pyruvate [8b]

	Enol ester			Enol ester	
Pyruvate R	Yield (%)	*threo*: *erythro*	Pyruvate R	Yield (%)	*threo*: *erythro*
Me	96	73:27	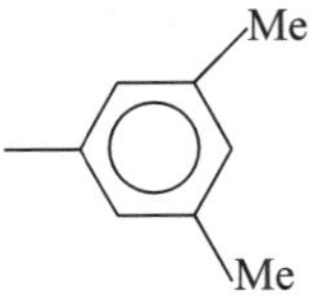	88	65:35
CH_2CMe_3	90	85:15			
CH_2Ph	94	80:20	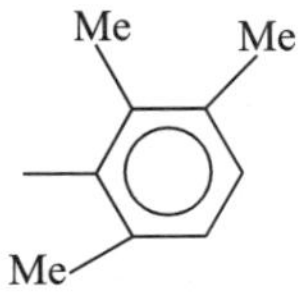	80	70:30
Ph	89	80:20			
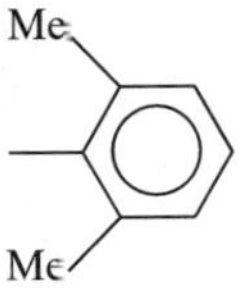	79	75:25	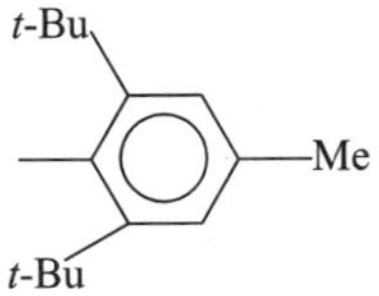	90	100:0

α-Stannyl substituted crotyl 9-BBN undergoes condensation with methylpyruvate and affords exclusively the *threo–cis* isomer as a single product in 60% yield (Eq. 6.23) [8b]. The silyl derivative, however, gives also the α adduct (Eq. 6.24) [8b].

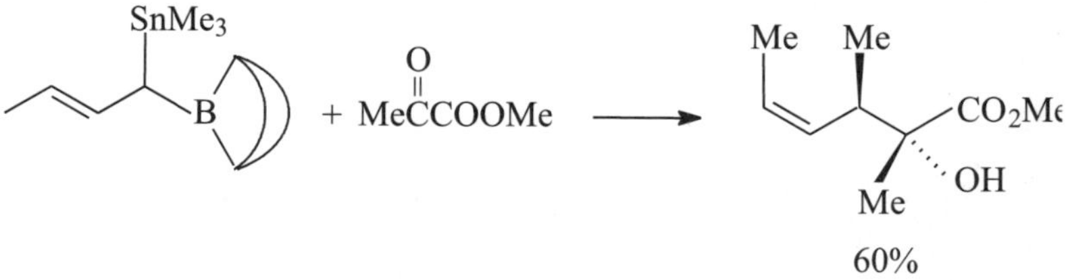

(6.23)

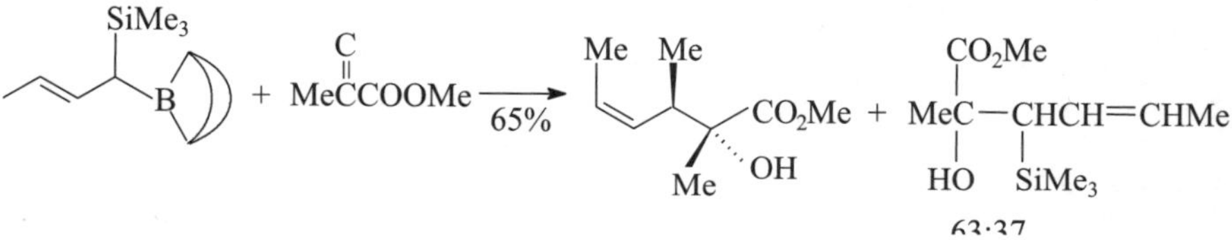

(6.24)

Brown *et al* [9] have synthesized optically active *B*-R*-9-BBN by an indirect method. The asymmetric hydroboration of prochiral olefins are achieved by monoisopinocampheylborane, $IPCBH_2$ (100% *ee*) [10], prepared from (+)-α-pinene to afford IpcR*BH in essentially 100% optically purity. By a series of sequences involving elimination of the chiral auxiliary unit with acetaldehyde under mild conditions yields corresponding boronic ester in very high enantiomeric purity. The optical active 2-alkyl-1,3,2-dioxaborinanes are then prepared by esterification of the corresponding boronic acid with propanediol (Scheme 6.29). Brown *et al* [11] in a real breakthrough have discovered that $LiAlH_4$ readily converts these relatively inert boronic esters to R^*BH_3Li intermediates, which are very stable and can be stored under nitrogen even at 25 °C, without any hydride loss, redistribution, isomerization, or racemization of the alkyl groups. The optically active monoalkylboranes R^*BH_2 reagents are generated as and when needed by a convenient, simple reaction with trimethylsilylchloride [12]. The chiral R^*BH_2 is then used for the hydroboration of 1,5-cyclooctadiene, which on reflux is converted to 1,5-isomer to afford the optically pure *B*-R*-9-BBN.

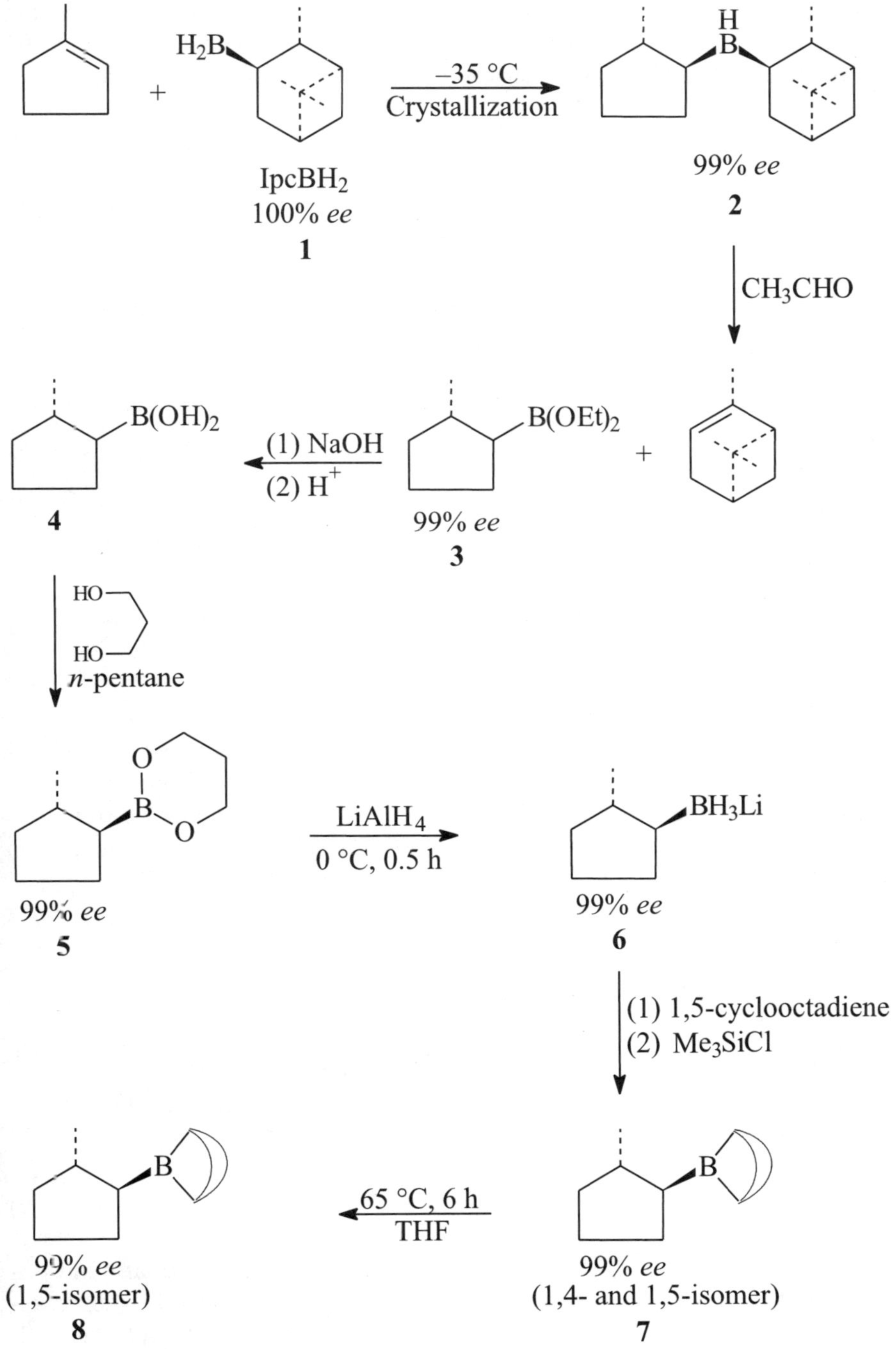

H₂B
+
IpcBH₂
100% ee
1
–35 °C
Crystallization
99% ee
2
CH₃CHO
B(OEt)₂
+
99% ee
3
(1) NaOH
(2) H⁺
B(OH)₂
4
HO
HO
n-pentane
99% ee
5
LiAlH₄
0 °C, 0.5 h
BH₃Li
99% ee
6
(1) 1,5-cyclooctadiene
(2) Me₃SiCl
99% ee
(1,4- and 1,5-isomer)
7
65 °C, 6 h
THF
99% ee
(1,5-isomer)
8

Scheme 6.29

Consequently, the following representative *B*-alkyl-9-borabicyclononanes (Chart 6.26) are prepared in very high optical purity, which yield the corresponding chiral alcohols following alkaline hydrogen peroxide oxidation.

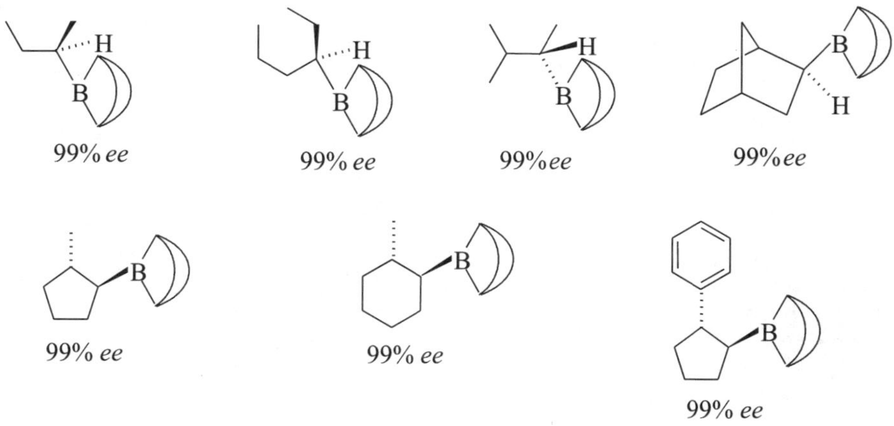

Chart 6.26

It is pertinent to mention that addition of 1,5-cyclooctadiene to R^*BH_2 affords a mixture both of 1,4- and 1,5-isomers of the cyclooctyl moiety of *B*-alkyl-9-borabicyclononanes, and there is no need to isomerize to 1,5-isomer as mixture affords the good yields of alcohol (99% ee) on oxidation along with 70:30 mixture of *cis*-1,4- and 1,5-cyclooctanediols (Eq. 6.25).

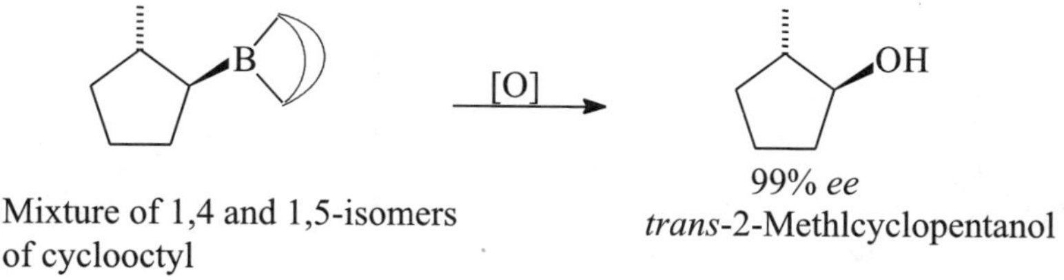

(6.25)

The boron azaenolate with an oxazoline ring as the chiral auxiliary unit undergoes condensation [13] with aldehydes to give mainly *erythro*-β-hydroxy esters (Eq. 6.26). The reaction proceeds with high diastereoselection, but enantioselection is moderate.

The results are summarized in Table 6.28 [13].

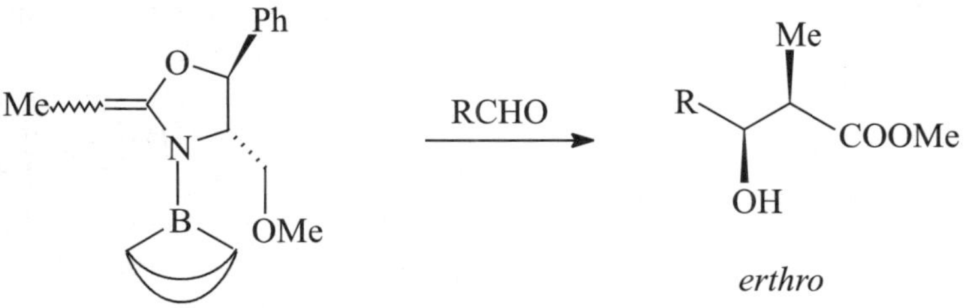

(6.26)

Table 6.28 Synthesis of *erythro*-β-hydroxyesters from boron azaenolate [13]

RCHO	*erythro*:*threo* (%)	*erythro* (% ee)	Configuration	Yield (%)
C_2H_5CHO	98:2	40	2*S*,3*R*	44
Me_2CHCHO	98:2	41	2*S*,3*R*	42
t-BuCHO	97:3	60	2*S*,3*R*	50

The boron (*Z*)-enolates (>99%, Z), prepared by 1,4-addition of 9-BBN or catecholborane to phenyl-1-alkenylketones, are converted to trimethylsilylenol ether by treatment with methylmagnesiumbromide at −78 °C and then with chlorotrimethylsilane. On the other hand, protonolysis of (*Z*)-enolates afford 1-phenyl-1-alkanone (Scheme 6.30) [14]. The high selectivity–forming boron enolate is not observed in the reaction of catecholborane or 9-BBN with alkyl alkenyl ketones [15].

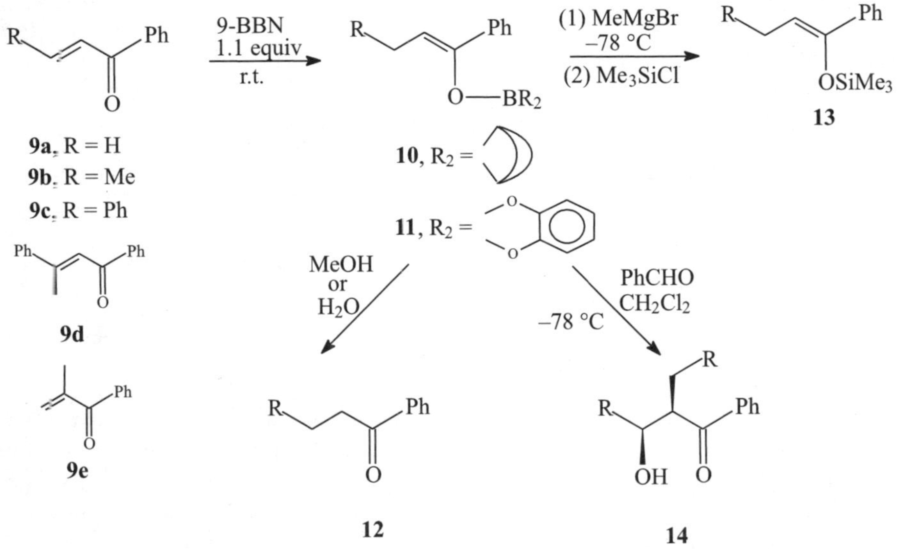

Scheme 6.30

The results of hydroboration and hydroboration protonolysis are summarized in Table 6.29 [14].

Table 6.29 Hydroboration of phenyl 1-alkenyl ketones (**1**) with 9-BBN and catecholborane [14]

Hydroboration of phenyl 1-alkenyl ketones **9**							
Ketone 9	Borane	Reaction time (h)	Boron enolate	Yield (%)	Z/E	Hydroboration and protonolysis product yield (%)	
9a	9-BBN	4	**10a**	>95	>99/1	**12a**	85
9a	Catechol-borane	1.5	**11a**	75	>99/1	**12a**	54
9b	9-BBN	8	**10b**	>95(99)	>99/1	**12b**	96
						12b	94
						12b	95
						12b	98
9b	Catechol-borane	<0.1	**11b**	>95	>99/1	**12b**	94
9c	9-BBN	3	**10c**	>95	>99/1	**12c**	95
9c	Catechol-borane	<0.5	**11c**	>95	>99/1	**12c**	97
9d	9-BBN	4		>95	>99/1	**12d**	92
9d	Catechol-borane	<0.5		>95	>99/1	**12d**	99
9e	9-BBN	10		>95	–	**12e**	88

(*Z*)-Enolates of 9-BBN, on treatment with benzaldehyde in chloroform at −78 °C, give high yields of aldol products with over 96% *syn* selectivity (Scheme 6.30; Table 6.30). The (*Z*)-enolates derived from catecholborane afford *syn*/*anti* aldols in poor ratio as compared to when derived from 9-BBN.

Table 6.30 Aldol reaction of boroenolate of 9-BBN with benzaldehyde [14]

Ketone	Enolate	Product	Yield (%)	*syn*/*anti*
9a	**10a**	**14a**	97	96/4
9b	**10b**	**14b**	99	99/1
9c	**10c**	**14c**	92	97/3
9e		**14e**	88	Not determined

A simple and efficient method for the stereochemical control of consecutive stereogenic center construction has been realized [16] by utilizing hydroboration of dialkenyl carbinol derivative. Consequently, hydroboration of **15** with 9-BBN, followed by alkaline H_2O_2 oxidation yields 90% of *anti*, *anti*-isomer **16**, whereas $ThxBH_2$ affords *syn*, *anti*-isomer **17** in 62% yield (Chart 6.27) [16].

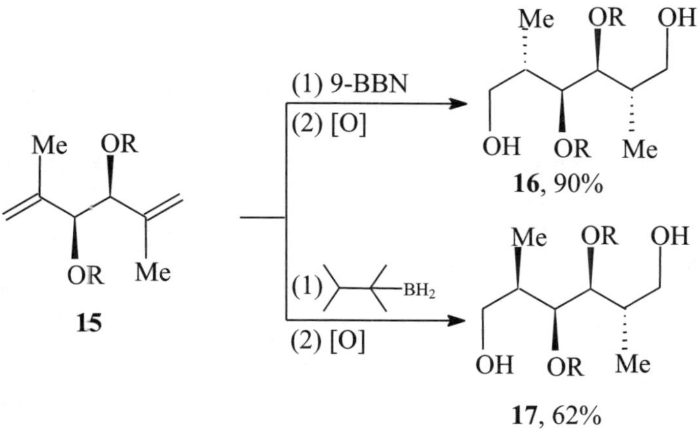

Chart 6.27

The light induced bromination of *B*-OMe-9-BBN [17] in presence of water provides *cis*-bicyclo[3.3.0]octane-1-boronic acid, which readily undergoes oxidation to *cis*-bicyclo[3.3.0]octan-1-ol in 65% yield, accompanied by cyclooctane-1,5-epoxide in 22% yield [18]. *B*-Me-9-BBN under identical conditions affords only the bicyclic alcohol (Chart 6.28) [19].

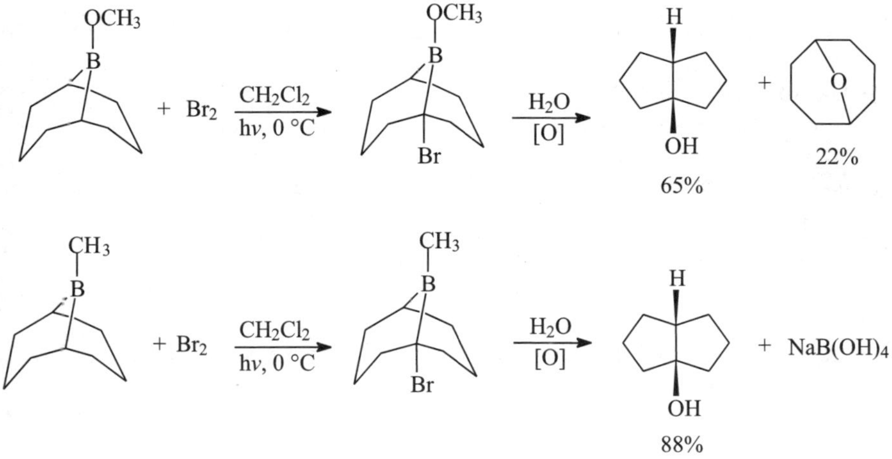

Chart 6.28

References

Section 6.1

1. Brown HC, Kramer GW, Levy AB, Midland MM (1975) Organic synthesis via boranes. Wiley InterScience, New York
2. (a) Brown HC, Knights EF, Scouten CG (1974) J Am Chem Soc 96:7765; (b) Knights EF, Brown HC (1968) ibid. 90:5280
3. Brown, HC, Liotta R, Brener L (1977) J Am Chem Soc 99:3427
4. Wang KK, Scouten CG, Brown HC (1982) J Am Chem Soc 104:531
5. Brener L, Brown HC (1977) J Org Chem 42:2702
6. Desai MC, Chawla HPS, Dev S (1982) Tetrahedron 38:379
7. Buchanan GW, Stothers JB, Wu S-T (1967) Can J Chem 45:2955
8. Brown HC, Iyer RR, Bhat NG, Racherla US, Brown CA (1992) Tetrahedron 48:9187
9. Eiter K (1972) Naturwissenshaften 59:468
10. Brown CA, Yamashita A (1975) J Am Chem Soc 97:891

Section 6.2

1. Jacob P III, Brown HC (1977) J Org Chem 42:579
2. Colberg JC, Rane A, Vaquer J, Soderquist JA (1993) J Am Chem Soc 115:6065
3. Kramer GW, Brown HC (1977) J Organomet Chem 132:9
4. Brown HC, Liotta R, Kramer GW (1978) J Org Chem 43:1058
5. Brown HC, Bhat NG, Rajagopalan S (1986) Organometallics 5:816

Section 6.3

1. Kramer GW, Brown HC (1977) J Org Chem 42:2292
2. (a) Benkeser RA (1971) Synthesis 7:347; (b) Courtois G, Miginiac L (1974) J Organomet Chem 69:1
3. Brown HC, Ravindranathan M, Rao CG (1978) J Am Chem Soc 100:1218
4. Brown HC, Bhat KS (1986) J Org Chem 51:445
5. Brown HC, Liotta R, Kramer GW, (1979) J Am Chem Soc 101:2966
6. (a) Skattebol L (1963) Acta Chem Scand 7:1683; (b) Moore WR, Ward HR (1962) J Org Chem 27:4179
7. Makosza M, Wawrzyniewicz M (1969) Tetrahedron Lett 10:4659

Section 6.4

1. (a) Fried J, Lin C, Mehra M, Kao W, Dahren P (1971) Ann NY Acad Sci 180:38; (b) Bagli JF, Bogri T (1975) Prostaglandins 10:503
2. Johnson WS, Brinkmeyer RS, Kapoor VM, Yarnell TM (1977) J Am Chem Soc 99:8341
3. Chan K-K, Cohen N, De Noble JP, Specian AC Jr, Saucy G (1976) J Org Chem 41:3497
4. Nicolaou KC, Zipkin RE, Dolle RE, Harris BD (1984) J Am Chem Soc 106:3548
5. (a) Demare H, Cadiot P (1968) Bull Soc Chim Fr 205; (b) Cadiot P, Chodkiewicz W (1968) Chemistry of alkylenes. Dekker, New York; (c) Golse MR, Liermain A (1962) Bull Soc Pharma Bordeaux 101:3; (1968) Chem Abstr 58:4451; (d) Smith WN, Kuehn ED (1973) J Org Chem 38:3588; (e) Pittman CU Jr, Olah GA (1965) J Am Chem Soc 87:5632
6. Brown HC, Molander GA, Singh SM, Racherla US (1985) J Org Chem 50:1577

7. Brown HC, Sinclair JA (1977) J Organomet Chem 131:163
8. Molander GA, Thesis Ph.D. (1979) Purdue University
9. Evans JC, Goralski CT, Hasha DL (1992) J Org Chem 57:2941

Section 6.5

1. (a) Fryhle CB, Williard PG, Rybak CM (1992) Tetrahedron Lett 33:2327; (b) Nicolaou KC, Skokotas G, Furuya S, Suemune H, Nicolaou D (1990) Angew Chem Int Ed Engl 29:1064; (c) Hirama M, Tokuda M, Fujiwara K (1990) Synlett 9:651; (d) Chan TH, Arya P (1989) Tetrahedron Lett 30:4065; (e) Friesen RW, Blouin M (1993) J Org Chem 58:1653; (f) Montellano PRO (1973) J Chem Soc Chem Commun 79 and reference therein; (g) Galantay F, Bacso I, Coombs RV (1974) Synthesis 5:344 and references therein
2. For reviews see (a) Yamamoto H (1991) In: Heathcock CH (ed) Comprehensive organic synthesis, vol. 2. Pergamon, pp 81–87; (b) Epaztein R (1984) In: Buncel E, Durst T (eds) Comprehensive carbanion chemistry, part B. Elsevier, Amsterdam, pp 107–175; (c) Moreau J-L (1980) In: Patai S (ed) The chemistry of ketenes, allenes and related compounds, part 1. Wiley, New York, pp 363–413; (d) Klein J (1978) In: Patai S (ed) The chemistry of the carbon–carbon triple bond, part 1. Wiley, New York, pp 343–379; (e) Brandama L, Verkruijee HD (1981) Synthesis of acetylenes, allenes and cumelenes. Elsevier, Amsterdam
3. Brown HC, Khire UR, Narla G, Racherla US (1995) J Org Chem 60:544
4. Brown HC, Khire UR, Racherla US (1993) Tetrahedron Lett 34:15
5. Brown HC, Dhar RK, Ganesan K, Singaram B (1992) J Org Chem 57:499
6. Hopf H, Bohm I, Kleinschroth J (1982) Org Synth 60:41

Section 6.6.1

1. (a) Soderquist JA, Rivera I, Negron A (1989) J Org Chem 54:4051; (b) Soderquist JA, Brown HC (1980) ibid. 45:3571
2. Larson GL, Fuentes LM (1979) Synth Commun 9:841
3. Soderquist JA, Colberg JC, Del Valle L (1989) J Am Chem Soc 111:4873

Section 6.6.2

1. (a) Kramer GW, Brown HC (1977) J Organomet Chem 132:9; (b) Brown HC, Liotta R, Kramer GW (1979) J Am Chem Soc 101:2966
2. Wang KK, Gu YG, Liu C (1990) J Am Chem Soc 112:4424
3. Stork G, Jung ME, Colvin E, Noel Y (1974) J Am Chem Soc 96:3684
4. Marshall JA, Shearer BG, Crooks SL, (1987) J Org Chem 52:1236
5. Wang KK, Liu C, Gu YG, Burnett FN, Sattsangi PD (1991) J Org Chem 56:1914
6. (a) Jacob PJ III, Brown HC (1977) J Org Chem 42:579; (b) Soderquist JA, Santiago B (1990) Tetrahedron Lett 31:5541
7. Colberg JC, Rane A, Vaquer J, Soderquist JA, (1993) J Am Chem Soc 115:6065
8. Rane AM, Vaquer J, Colberg JC, Soderquist JA (1995) Tetrahedron Lett 36:987
9. (a) Denmark SE, Habermas KL, Hite GA, Jones TK (1986) Tetrahedron 42:2821; (b) Jones TK, Denmark SE (1983) Helv Chim Acta 66:2377; (c) Denmark SE, Wallace MA, Walker CB Jr (1990) J Org Chem 55:5543; (d) Fleming I, Dunogues J (1989) Org React 37:57
10. Kim S, Chung KN, Yang S (1987) J Org Chem 52:3917

Section 6.6.3

1. Krishnamurthy S, Brown HC (1975) J Org Chem 40:1864
2. Gu YG, Wang KK (1991) Tetrahedron Lett 32:3029

Section 6.6.4

1. Furuta K, Ishiguro M, Haruta R, Ikeda N, Yamamoto H (1984) Bull Chem Soc Jpn 57:2768; (b) Yamakado Y, Ishiguro M, Ikeda N, Yamamoto H (1981) J Am Chem Soc 103:5568
2. Wang KK, Wang Z, Gu YG (1993) Tetrahedron Lett 34:8391

Section 6.6.5

1. Wang KK, Wang Z, Gu YG (1993) Tetrahedron Lett 34:8391

Section 6.6.6

1. Wang KK, Shi C, Petersen JL (1998) J Org Chem 63:4413
2. Wang KK, Liu B, Petersen JL (1996) J Am Chem Soc 118:6860

Section 6.6.7

1. Wang KK, Liu B, Petersen JL (1996) J Am Chem Soc 118:6860

Section 6.6.8

1. Wang KK, Nikam SS, Ho CD (1983) J Org Chem 48:5376
2. Brown HC, Sinclair JA (1977) J Organomet Chem 131:163
3. Favre E, Gaudemar M, (1974) J Organomet Chem 76:297–305
4. Wang KK, Liu C (1985) J Org Chem 50:2578
5. Cazaux M, De Jeso BCR (1980) Hebd Seances Acad Sci Ser C 290:49

Section 6.6.9

1. Andemichael YW, Wang KK (1992) J Org Chem 57:796

Section 6.6.10

1. Wang KK, Wang Z, Sattsangi PD (1996) J Org Chem 61:1516
2. Andemichael YW, Gu YG, Wang KK (1992) J Org Chem 57:794

Section 6.6.11

1. Andemichael YW, Huang Y, Wang KK (1993) J Org Chem 58:1651
2. (a) Skattebøl L (1966) J Org Chem 31:2789; (b) Clinet JC, Linstrumelle G (1977) Nouv J Chem 1:373
3. Andemichael YW, Gu YG, Wang KK (1992) J Org Chem 57:794

Section 6.7

1. Brown HC, Vara Prasad JVN, Zee SH (1985) J Org Chem 50:1582
2. Brown HC, Vara Prasad, JVN, Gupta AK (1986) J Org Chem 51:4296
3. Brown HC, Ramachandran PV, Vara Prasad JVN (1985) J Org Chem 50:5583
4. RajanNabu TV, Reddy GS (1986) J Org Chem 51:5458
5. (a) Tebbe FN, Parshall GW, Reddy GS (1978) J Am Chem Soc 100:3611, for an in situ preparation of this reagent; (b) Cannizzo LF, Grubbs RH (1985) J Org Chem 50:2386

Section 6.8

1. Grayson M (ed) (1982) Kirk-Othmer encyclopedia of chemical technology, vol. 17, 3rd ed. Wiley InterScience, New York, pp 311–345
2. Tomicka K (1990) Synthesis 541
3. Singaram B, Goralski CT, Fisher GB (1991) J Org Chem 56:5691
4. Yang R-Y, Dai L-X (1992) Synthesis 481

Section 6.9

1. Habaue S, Yasue K, Yanagisawa A, Yamamoto H (1993) Synlet 788
2. Lee D, Martin JC (1984) The penta-coordinate hypervalent boron compounds are actually isolated. J Am Chem Soc 106:5745
3. Evans DA, Nelson JV, Taber TR (1982) In: Allinger NL, Eliel EL, Wilen SH (eds) Topics in stereochemistry, vol 3. Wiley, New York, p 5
4. Yatagai H, Yamamoto Y, Maruyama K (1980) J Am Chem Soc 102:4548
5. Servens C, Pereyre M (1972) J Organomet Chem 35:C20
6. Maruyama K, Narula Y (1978) J Org Chem 43:3796; Maruyama K, Naruta Y (1978) Chem Lett 431; Naruta Y, Maruyama K (1979) ibid. 881; Naruta Y, Maruyama K (1979) ibid. 885; Naruta Y, Oshida S, Maruyama K (1979) ibid. 917
7. Tsuji J, Shimizu I, Yamamoto K (1976) Tetrahedron Lett 2975
8. (a) Yamamoto Y, Yatagai H, Maruyama K (1981) J Am Chem Soc 103:3229; (b) Yamamoto Y, Maruyama K, Komatsu T, Ito W (1986) J Org Chem 51:886
9. Brown HC, Joshi NN, Pyun C, Singaram B (1989) J Am Chem Soc 111:1754
10. Brown HC, Schwier JR, Singaram B (1978) J Org Chem 43:4395
11. Brown HC, Singaram B, Cole TE (1985) J Am Chem Soc 107:460
12. Cole TE, Bakshi RK, Srebnik M, Singram B, Brown HC (1986) Organometallics 5:2303
13. Meyers AI, Yamamoto Y (1981) J Am Chem Soc 103:4278
14. Matsumoto Y, Hayashi T (1991) Synlett 349
15. (a) Boldrini GP, Bortolotti M, Tagliavini E, Trombini C, Umani-Ronchi A (1991) Tetrahedron Lett 32:1229; (b) Boldrini GP, Moneini F, Tagliavini E, Trombini C, Umani-Ronchi A (1990) J Chem Soc Chem Commun 1680
16. Harada T, Matsuda Y, Wada I, Uchimura J (1990) J Chem Soc Chem Commun 21
17. Kramer GW, Brown HC (1974) J Organometal Chem 73:1
18. Yamamoto Y, Brown HC (1974) J Org Chem 39:861
19. Yamamoto Y, Brown HC (1973) Chem Commun 80

7 Synthesis of Aldehydes and Ketones

Carbonyl compounds are an important class among organic molecules. Literature records several methods for their synthesis. However, there are very few methods to convert carbon–carbon unsaturation to carbonyl compounds. Hydroboration of acetylenes, followed by oxidation provides a novel method for carbonyl synthesis. It has been noted that regioselectivities achieved in the monohydroboration of internal acetylenes with thexylborane [1], disiamylborane [1], dicyclohexylborane [1], and catecholborane [2] are similar to, but less pronounced than, that realized by 9-BBN [3]. The *B*-alkenyl-9-BBN derivatives undergo oxidation to the corresponding ketones or aldehydes under aprotic conditions with trimethylamine *N*-oxide [4, 5] or under protic conditions by inverse addition to buffered hydrogen peroxide [3]. The inverse addition, i.e., the slow addition of the *B*-alkenyl-9-BBN in THF to the buffered H_2O_2, suppresses the otherwise undesirable protonolysis reaction and favors the oxidation pathway to the desired aldehyde or ketone.

7.1 Synthesis of Aldehydes

The monohydroboration of 1-alkyne (1 mole) with 9-BBN (0.5 mole), followed by oxidation affords the corresponding aldehyde (Eq. 7.1) [3] with no evidence of the presence of the isomeric-2-alkanones.

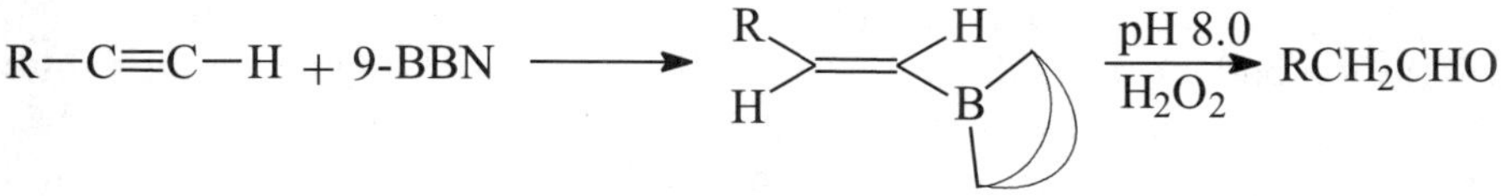

(7.1)

B-Alkyl-9-BBN on carbonylation with carbon monoxide in the presence of potassium triisopropoxide affords the corresponding aldehyde or alcohol (Scheme 7.1) [6] in excellent yields.

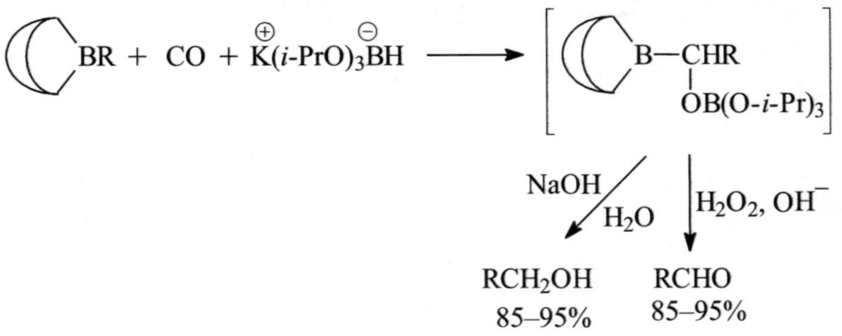

Scheme 7.1

Similar reactions of 9-BBN with alkenes and dienes, followed by carbonylation in the presence of lithium trimethoxyaluminohydride afford the corresponding alcohols or aldehydes (Chart 7.1) [7a].

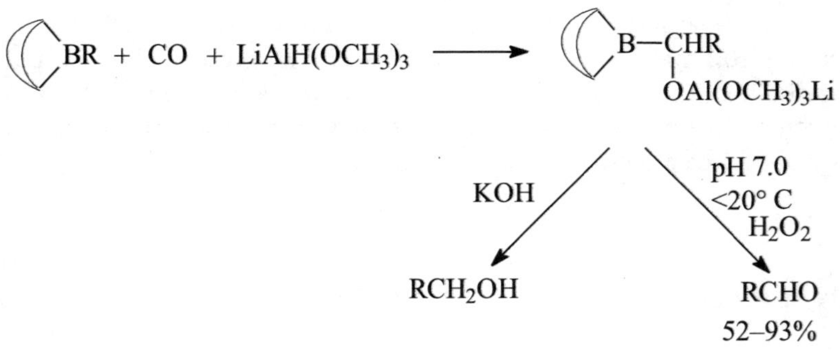

Chart 7.1

The results are summarized in Table 7.1 [7a].

The method is extended to introduce aldehyde groups [7b] into the molecules containing other functional groups. The mild reducing agent lithium-tri-*t*-butoxyaluminohydride reduces the carbonyl intermediate from *B*-R-9-BBN and does not reduce the esters or nitriles (Chart 7.2; Table 7.2) [7b].

Table 7.1 Aldehydes produced *via* the hydroboration–carbonylation of representative olefins using 9-BBN [7a]

Olefin	Aldehyde	Yield (%)
1-Hexene	*n*-Heptaldehyde	93
Isobutylene	Isovaleraldehyde	52
2-Methyl-2-butene	2,3-Dimethylbutanal	60
2,3-Dimethyl-2-butene	No reaction	
Cyclopentene	Cyclopentanecarboxaldehyde	79
Cyclohexene	Cyclohexanecarboxaldehyde	81
Styrene	Hydrocinnamaldehyde	84
Norbornene	Norbornane-2-carboxaldehyde	59
1,5-Hexadiene	Suberaldehyde	78
4-Vinylcyclohexene	3-(3-Cyclohexenyl)propanal	93

Table 7.2 Conversion of functionally substituted olefins into aldehydes *via* hydroboration with 9-BBN and carbonylation in the presence of lithium tri-*t*-butoxyaluminohydride [7b]

Olefin	Product	Yield (%)
Methyl-10-undecenoate	10-Carbomethoxyundecanal	99
10-Undecenenitrile	10-Cyanododecanal	99
10-Undecen-1-ylacetate	12-Acetoxydodecanal	99
Allyl benzoate	4-Benzoxybutanal	89
Ethyl vinylacetate	4-Carbethoxybutanal	83
Allyl cyanide	4-Cyanobutanal	85
3-Buten-1-ylacetate	5-Acetoxypentanal	92

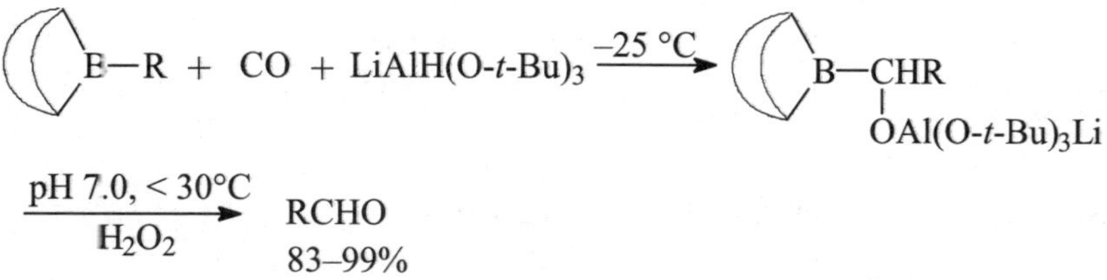

Chart 7.2

Cha and coworkers have described one-pot conversion of carboxylic acids to aldehydes by treatment of acyloxy-9-borabicyclo[3.3.1]nonane with either lithium 9-boratabicyclo[3.3.1]nonane (Scheme 7.2; Table 7.3) [8] or by stepwise treatment with *tert*-butyllithium and 9-BBN (Scheme 7.3; Table 7.4) [9].

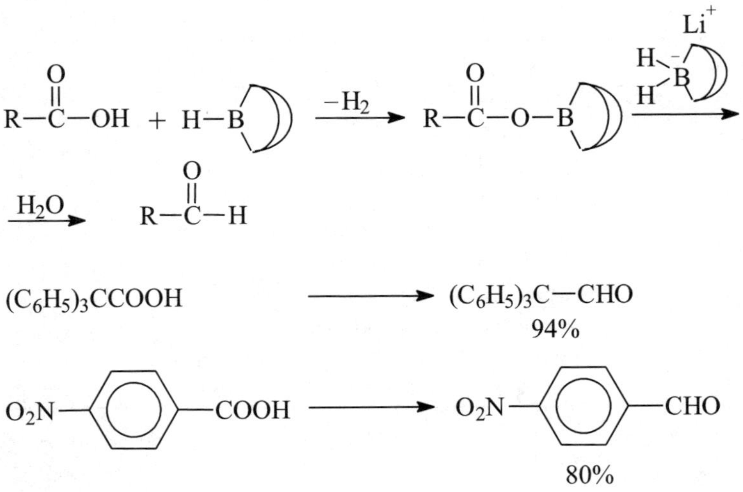

Scheme 7.2

Li 9-BBNH can be readily prepared by hydridating 9-BBN with lithium hydride [10] or *tert*-butyllithium [11] in THF.

Table 7.3 Yields of aldehydes in the reduction of representative carboxylic acids through treatment of acyloxy-9-borabicyclo[3.3.1]nonanes with lithium-9-boratabicyclo[3.3.1]nonane (Li 9-BBNH) in tetrahydrofuran at room temperature [8]

Acid	Yield of aldehyde (%)	Acid	Yield of aldehyde (%)
Butyric	90	Triphenylacetic	94
Hexanoic	92	6-Bromohexanoic	87
Decanoic	92	1,10-Decanedicarboxylic	99 (Dialdehyde)
Stearic	99	Benzoic	78
Isobutyric	94	α-Naphthoic	80
Isopentanoic	89	*p*-Methoxybenzoic	79
Pivalic	85	*p*-Chlorobenzoic	76
Cyclopropanecarboxylic	85	*p*-Aminobenzoic	79
Cyclohexanecarboxylic	88	*p*-Nitrobenzoic	80
Diphenylacetic	96	Terephthalic	92

Reaction with 50% excess Li 9-BBNH (1.05 equiv for monocarboxylic and 2.1 equiv for dicarboxylic acid) for 1 h with aliphatic and for 6 h with aromatic carboxylic acids, both at room temperature.

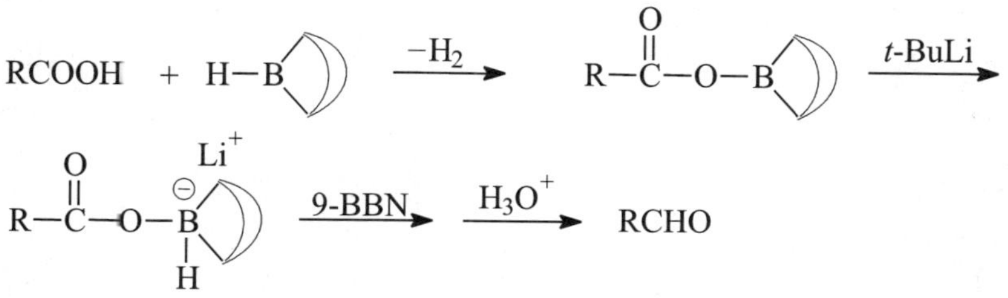

Scheme 7.3

Table 7.4 Yields of aldehydes in the reduction of representative carboxylic acids through treatment of acyloxy-9-borabicyclo[3.3.1]nonanes with *tert*-butyllithium and 9-BBN in tetrahydrofuran at room temperature [9]

Acid	Yield of aldehyde (%)	Acid	Yield of aldehyde (%)
Butyric	92	Triphenylacetic	93
Hexanoic	93	6-Bromohexanoic	87
Decanoic	91	1,10-Decanedicarboxylic	99 (Dialdehyde)
Stearic	99	Benzoic	81
Isobutyric	98	α-Naphthoic	79
Isopentanoic	96	*p*-Methoxybenzoic	72
Pivalic	92	*p*-Chlorobezoic	78
Cyclopropanecarboxylic	84	*p*-Aminobenzoic	82
Cyclohexanecarboxylic	88	*p*-Nitrobenzoic	83
Diphenylacetic	95	Terephthalic	98

Reaction with 5% excess 9-BBN (1.05 equiv for monocarboxylic and 2.1 equiv for dicarboxylic acid) for 1 h with aliphatic and for 6 h with aromatic carboxylic acids both at room temperature, after addition of 10% excess *tert*-butyllithium at −20 °C.

Cha *et al* [12] have reported that acyloxy-9-BBN derivatives are also readily reduced by lithium aluminum hydride in the presence of pyridine, and the reduction stops at the aldehyde stage. Further, Cha and coworkers have found that lithium tris(diethylamino) aluminum hydride (LTDEA), readily prepared [13] from the reaction of $LiAlH_4$ and 3 equiv of diethylamine in THF (Eq. 7.2), reduces the acyloxy group to the corresponding aldehyde in fair yield. However, in the presence of 2 equiv of pyridine the reduction stops at the aldehyde stage, and hydrolysis affords excellent yields of aldehydes (Eq. 7.3, Table 7.5) [13].

$$LiAlH_4 + 3\,Et_2NH \xrightarrow{0\,^\circ C} \underset{\text{LTDEA}}{Li(Et_2N)_3AlH} + 3\,H_2\uparrow \quad (7.2)$$

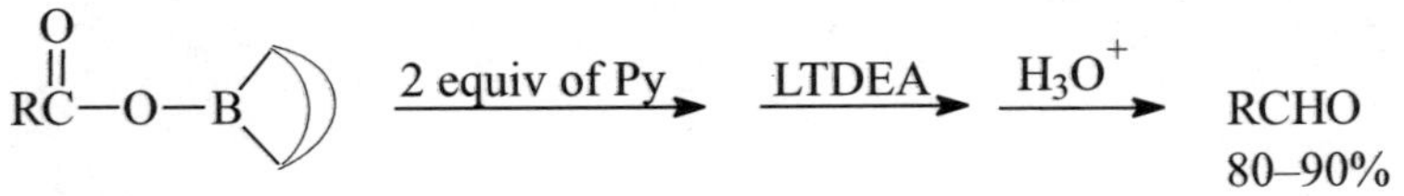

(7.3)

Table 7.5 Yields of aldehydes by LTDEA reduction of carboxylic acids [13]

Acid	Yield (%)	Acid	Yield (%)
Acetic	92	Maleic	91
Butyric	93	Benzoic	96
Caproic	91	*o*-Toluic	92
Decanoic	90	*m*-Toluic	90
Pentadecanoic	93	*p*-Toluic	94
Palmitic	92	*o*-Anisic	92
Stearic	93	*m*-Anisic	91
Isobutyric	90	*p*-Anisic	98
Isopentanoic	90	*o*-Chlorobenzoic	92
Cyclopropanecarboxylic	90	*m*-Chlorobenzoic	94
Cyclohexanecarboxylic	91	*p*-Chlorobenzoic	96
Phenylacetic	93	*o*-Nitrobenzoic	81
Diphenylacetic	94	*m*-Nitrobenzoic	80
Triphenylacetic	96	*p*-Nitrobenzoic	85
Bromoacetic	90	α-Naphthoic	97
6-Bromohexanoic	87	β-Naphthoic	98
Succinic	92	Phthalic	90
Adipic	94	Terephthalic	93
1,10-Decanedicarboxylic	95		

Potassium 9-*sec*-amyl-9-boratabicyclo[3.3.1]nonane (K 9-*sec*-Am-9-BBNH) has shown excellent chemo- and stereoselectivity [14, 15] toward various functional groups. 9-*sec*-Am-9-BBN, prepared readily by hydroborating 2-methyl-2-butene with 9-BBN [15], is easily transformed to reducing agent K 9-*sec*-Am-9-BBNH by treating it with excess of KH in THF (Eq. 74) [15].

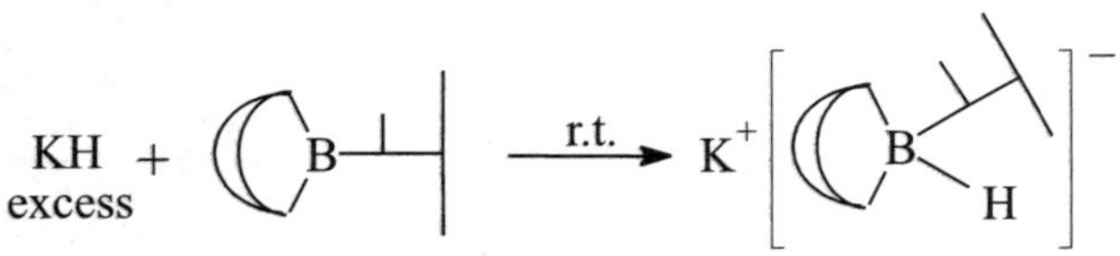

(7.4)

The agent K 9-*sec*-Am-BBNH reduces aromatic nitriles to the corresponding aldehydes in excellent yields [16]. The reagent tolerates some of the functional groups as *p*-carboxybenzonitrile is reduced to *p*-carboxybenzaldehyde in almost quantitative yield (Table 7.6) [16].

Table 7.6 Yields of aldehydes in the reduction of representative aromatic nitriles with potassium-9-*sec*-amyl-9-boratabicyclo[3.3.1]nonane in tetrahydrofuran at room temperature [16]

Compound	Yield of aldehyde (%)	Compound	Yield of aldehyde (%)
Benzonitrile	98	*o*-Tolunitrile	60
1-Naphthonitrile	91	*m*-Tolunitrile	97
Phthalonitrile	64	*p*-Tolunitrile	83
Terephthalonitrile	97	2-Cyanopyridine	73
p-Carboxybenzonitrile	98	3-Cyanopyridine	96
2,4-Dichlorobenzonitrile	80	4-Cyanopyridine	65

The reagent has been found to selectively reduce the aromatic nitriles in the presence of aliphatic nitriles (Table 7.7) [16].

Table 7.7 Selective reduction of aromatic nitriles in the presence of aliphatic nitriles with potassium-9-*sec*-amyl-9-boratabicyclo[3.3.1]nonane in THF at room temperature [16]

Compound used	Product	Mol (%)
Benzonitrile	Benzaldehyde	91
+	Benzonitrile	2
Caprylonitrile	Caprylic aldehyde	0
	Caprylonitrile	99
Terephthalonitrile	Terephthalaldehyde	93
+	Terephthalonitrile	0
Decanenitrile	Decylaldehyde	0
	Decanenitrile	98

7.2 Synthesis of Ketones

7.2.1 Synthesis of Saturated and Aromatic Ketones

Excellent yields of the ketones are achieved from internal alkynes (Eq. 7.5; Table 7.8) [1] via hydroboration with 1 equiv of 9-BBN. The boron of 9-BBN adds to the less hindered carbon of carbon–carbon triple bond, and oxidation of the resulting *B*-alkenyl-9-BBN affords the corresponding ketones.

$$RC{\equiv}CMe \xrightarrow[(2)\ H_2O_2,\ pH\ 8]{(1)\ 9\text{-BBN, THF}} RCOCH_2CH_3 + RCH_2COCH_3 \quad (7.5)$$

Table 7.8 Directive effects in the hydroboration of methyl-substituted alkynes by 9-BBN [1]

$RC{\equiv}CCH_3$	$RCOCH_2CH_3$ (%)	RCH_2COCH_3 (%)	Total yield of ketones (%)
n-Propyl	22	78	90
i-Propyl	4	96	92
t-Butyl	0	>99	>99
Cyclohexyl	4	96	96
Phenyl	65	35	90

These studies also show that 9-BBN is not only most sensitive to the steric environment, but it is also most sensitive to the electronic environment [1].

The homologated ketones are synthesized by adding the α-bromoketone and potassium *t*-butoxide simultaneously to the *B*-R-9-BBN at 0 °C. This simultaneous addition procedure provides good yields of the desired ketone with an additional alkyl group (Chart 7.3) [2a].

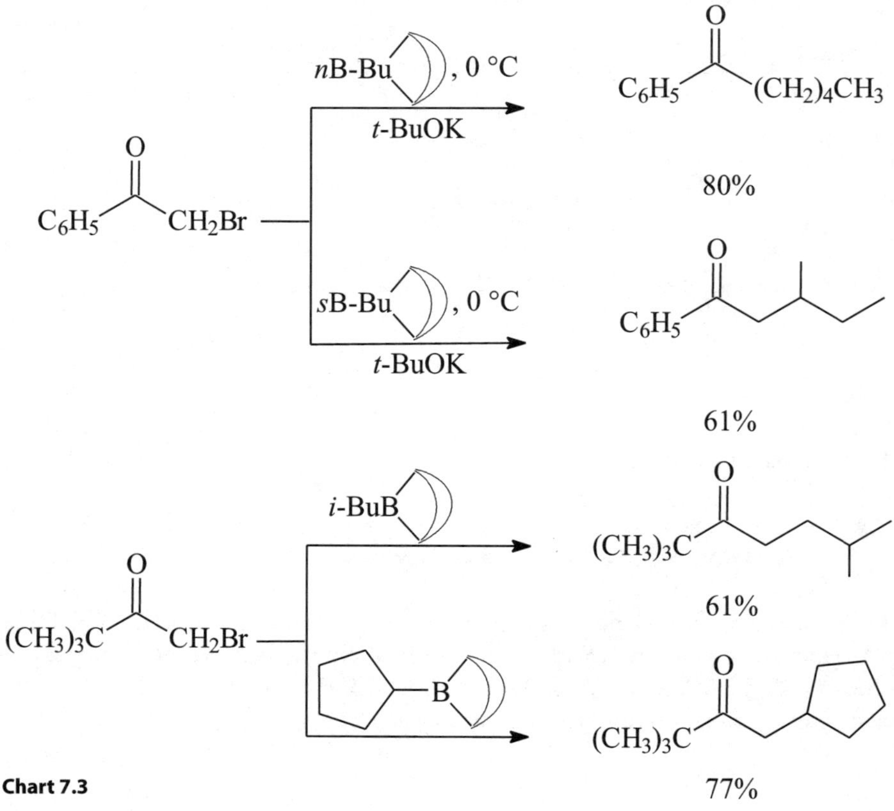

Chart 7.3

The results are summarized in Table 7.9 [2a].

The *B*-alkyl-9-BBN and *B*-aryl-9-BBN not available *via* hydroboration undergo reactions with ethyl haloacetates and α-bromoketones. The reaction sequence provides a convenient procedure for the α-arylation of ketones and esters under the influence of potassium *t*-butoxide (Eq. 7.6; Table 7.10) [2b].

Table 7.9 Alkylation of α-bromoketones with *B*-alkyl-9-borabicyclo[3.3.1] nonanes under the influence of potassium *t*-butoxide in tetrahydrofuran [2a]

Olefin	*B*-R-9-BBN (mmol)	α-Bromo-ketone[a]	Addition[b]	Product	Yield (%)
Ethene	10	A	KOR	Butyrophenone	71
	10	A	SA		82
	20	A	KOR		91
1-Butene	10	A	KOR	Hexanophenone	78
	10	A	SA		80
	20	A	KOR		83
2-Butene	10	A	KOR	β-Methylvalero-phenone	45
	10	A	SA		61
	20	A	KOR		65
Isobutylene	10	A	KOR	γ-Methylvalero-phenone	33
	10	A	SA		55
	20	A	KOR		65
Cyclopentene	10	A	SA	α-Cyclopentyl-acetophenone	20
Cyclohexene	10	A	SA	α-Cyclohexyl-acetophenone	30
4-Vinylcyclo hexene	10	A	SA	γ-(3-Cyclohexenyl)-butyrophenone	60
1-Butene	10	B	SA	2,2-Dimethyl-3-octanone	78
cis-2-Butene	10	B	SA	2,2,5-Trimethyl-3-heptanone	79
Isobutylene	10	B	SA	2,2,6-Trimethyl-3-heptanone	61
Cyclopentene	10	B	SA	2,2-Dimethyl-4-cyclopentyl-3-butanone	77
Cyclohexene	10	B	SA	2,2-Dimethyl-4-cyclohexyl-3-butanone	60
Ethene	10	C	SA	2-Ethylcyclo-hexanone	0

Ten mol of 9-R-9-BBN in tetrahydrofuran is treated with 10 mmol of the α-bromoketone and 10 mmol of potassium *t*-butoxide.

[a] *A* phenacyl bromide, *B* α-bromopinacolone, *C* α-bromocyclohexanone.

[b] *KOR* addition of potassium *t*-butoxide to the mixture of the other two reactants, *SA* simultaneous addition of base and α-bromoketone to 9-R-9-BBN. Both reactions are carried out at 0°.

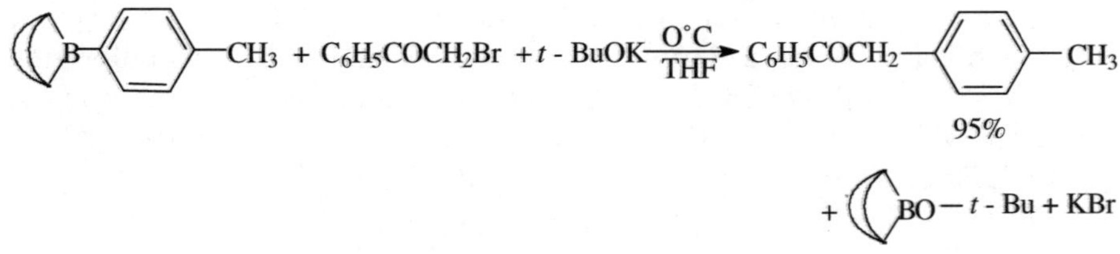

(7.6)

Table 7.10 Reaction of *B*-aryl- and *B*-methyl-9-BBN with α-bromoketones [2b]

R of *B*-R-9-BBN	α-Halo compound	Product	Yield[a] (%)
Methyl	$C_6H_5COCH_2Br$	$C_6H_5COCH_2CH_3$	40 (73)
	$(CH_3)_3CCOCH_2Br$	$(CH_3)_3CCOCH_2CH_3$	– (68)
Phenyl	$C_6H_5COCH_2Br$	$C_6H_5COCH_2C_6H_5$	93 (91)
	$(CH_3)_3CCOCH_2Br$	$(CH_3)_3CCOCH_2C_6H_5$	92 (90)
p-Tolyl	$C_6H_5COCH_2Br$	C_6H_5CO-$CH_2C_6H_4CH_3$-*p*	95

[a] The yields in *parentheses* are with base potassium 2,6-di-*t*-butylphenoxide [2c].

However, this methodology to homologate α-bromoacetone for the synthesis of methylketones fails. This ketone is extraordinarily sensitive to the action of potassium *t*-butoxide, an butoxide, an exceptionally strong base. The problem is circumvented by utilizing potassium 2,6-di-*t*-butylphenoxide. The presence of bulky substituents at *ortho* positions prevent the organoborane from coordinating with base.

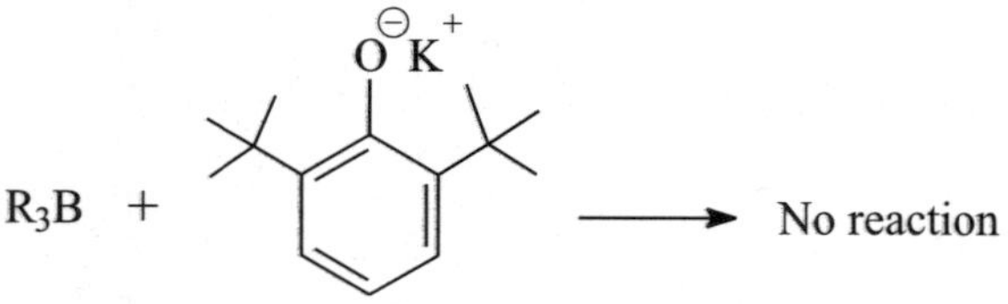

The base, however, acts to abstract the proton from the α-haloketones to produce the α-halocarbanion. The latter is immediately removed by the reaction with the free, uncomplexed organoborane, and the transfer of alkyl group from boron occurs and product methyl ketone is obtained by the hydrolysis of the intermediate (Scheme 7.4) [2c]. It is significant to mention that protonolysis of the reaction intermediate by 2,6-di-*t*-butylphenol produced in the reaction is relatively difficult. Consequently, ethyl alcohol is added to liberate ketone.

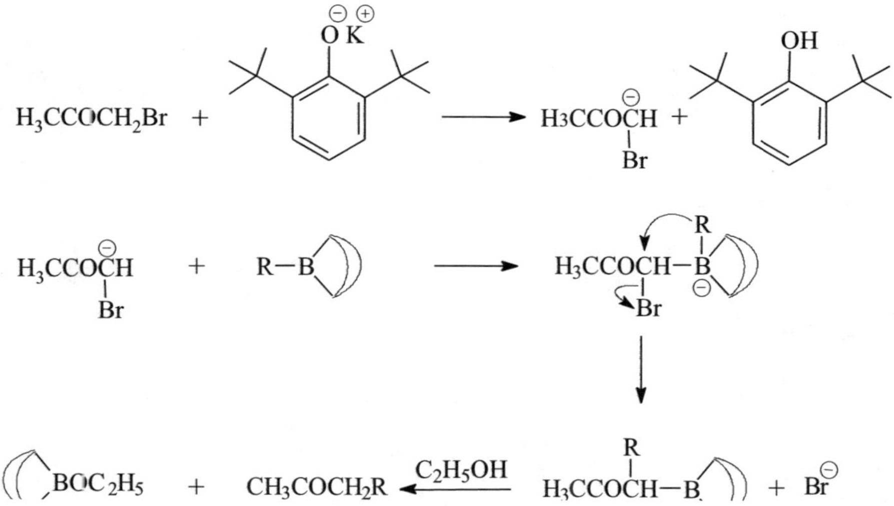

Scheme 7.4

The results are summarized in Table 7.11 [2c].

Table 7.11 α-Alkylation and α-arylation of bromoacetone with organoboranes under the influence of potassium 2,6-di-*t*-butylphenoxide [2c]

R of *B*-R-9-BBN	Product	Yield (%)
n-Butyl	2-Heptanone	80
2-Butyl	4-Methyl-2-hexanone	71
Isobutyl	5-Methyl-2-hexanone	62
Cyclopentyl	1-Cyclopentyl-2-propanone	73
Cyclohexyl	1-Cyclohexyl-2-propanone	72
exo-Norbornyl	1-(2-Norbornyl)-2-propanone	25
Phenyl	1-Phenyl-2-propanone	76

Methylketones are also easily prepared through the oxidation of *B*-alkyl-9-BBN with TMANO; the overall one-pot procedure represents the regioselective hydroacetylation of alkenes (Scheme 7.5) [3].

The hydroboration of 1,3-dimethylcycloalkenes occurs *cis* to produce with exceptionally high stereospecificity [4], affording *cis*-2, ω-dimethylcycloalkyl-9-BBN. These intermediates on treatment with alkaline hydrogen peroxide, followed by chromic acid oxidation [5] afford the corresponding *cis*-2-ω-dimethylcycloalkanones, which otherwise are difficult to prepare in pure form. Consequently, *cis*-2,5-dimethylcyclopentanone and *cis*-2,6-dimethylcyclohexanone are prepared with excellent high isomeric purities and in essentially quantitative yields (Chart 7.4) [4].

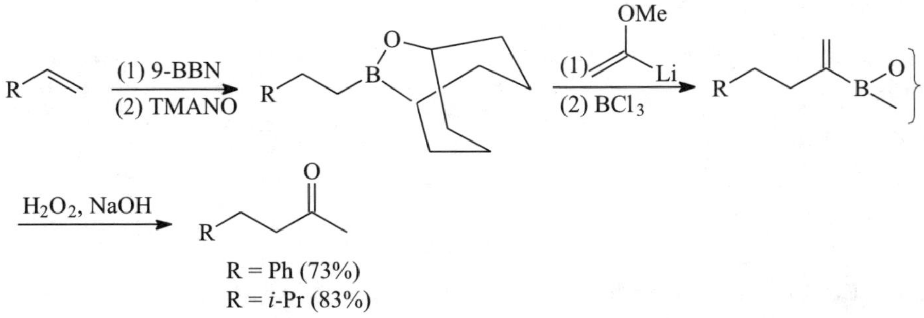

Scheme 7.5

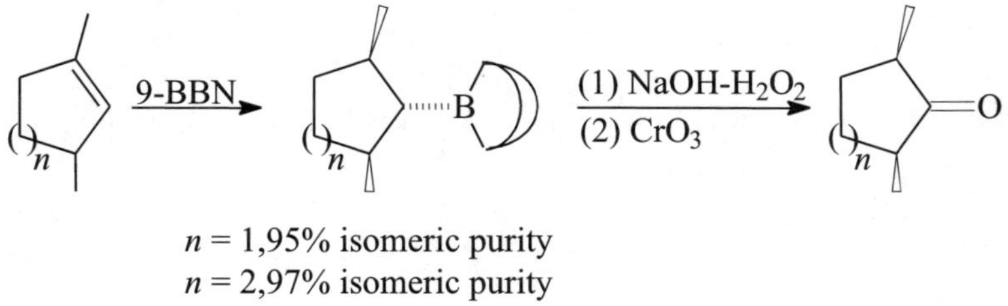

Chart 7.4

Cyclooctyl group of 9-BBN has been used to synthesize bicyclo[3.3.1]nonan-9-one (Chart 7.5) [6].

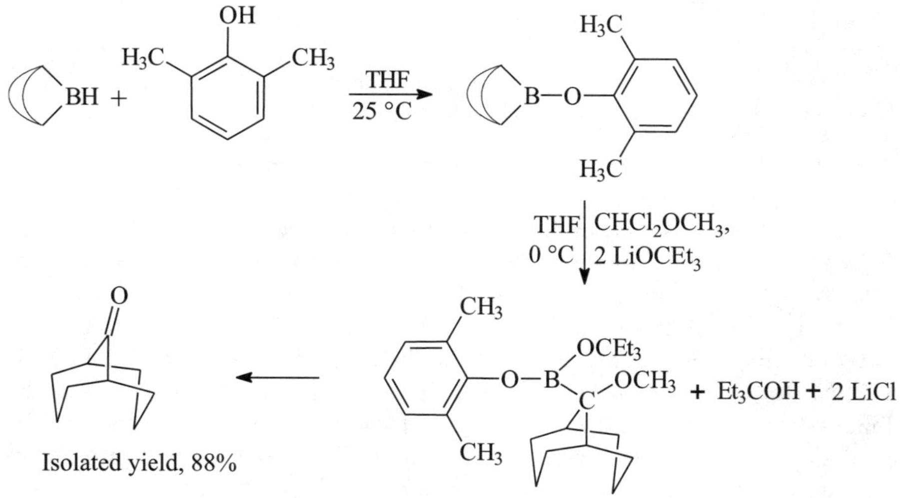

Chart 7.5

The cross-coupling reaction of organic electrophiles with organometallics in the presence of transition metal provides a synthetic method of forming carbon–carbon bonds. However, such reactions are limitedly applicable to 1-alkenyl, 1-alkynyl, aryl, allyl and benzyl halides but not extended to alkyl halides with the sp^3 carbon-containing β-hydrogen [7]. It is reported [8] that 9-R-9-BBN reacts with iodoalkanes under a carbon monoxide atmosphere in the presence of K_3PO_4 and catalytic amount of $Pd(PPh_3)_4$ and yields the unsymmetrical ketones in good yields (Chart 7.6). The reaction takes place slowly under dark conditions. The irradiation of light accelerates the rate of coupling. The irradiation of UV or a higher carbon monoxide pressure provides no satisfactory results.

The primary, secondary, and tertiary iodoalkanes are carbonylated and coupled with 9-alkyl-9-BBN. A variety of functional groups in both iodoalkanes and 9-R-9-BBN are tolerated, as is evident in the synthesis of several functionalized ketones having acetal, nitrile, and carbomethoxy groups (Table 7.12) [8]. The side reactions of alkyl halides with β-hydrogen results by β-hydrogen elimination to the alkene, and the isomerization of alkyl groups are not serious and limit to less than 10%.

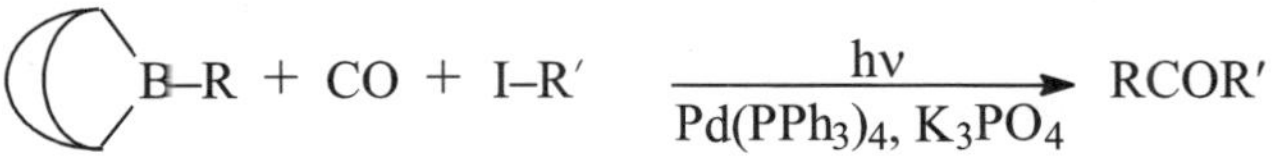

R = primary; R′ = primary, secondary, and tertiary

Chart 7.6

An interesting transformation occurs in the case of reaction with 1-iodo-5-hexene where it furnishes 1-cyclopentyl-2-decanone (Chart 7.7) [8] instead of the expected ketone.

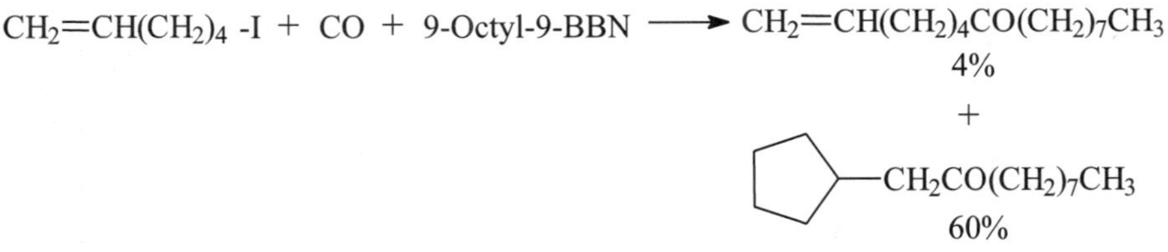

Chart 7.7

Table 7.12 Synthesis of unsymmetrical ketones: palladium-catalyzed carbonylative cross-coupling of 9-R-9-BBN with iodoalkanes [8]

Entry	Iodide	Alkene	Product	Yield (%)
1	$C_6H_{13}I$	1-Octene	C_6H_{13}-CO-C_8H_{17}	67
2	MeI	CH_2=CH$(CH_2)_8CO_2Me$	Me-CO-$(CH_2)_{10}CO_2Me$	76
3	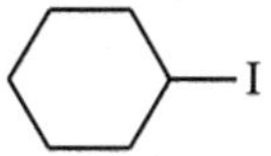	1-Octene	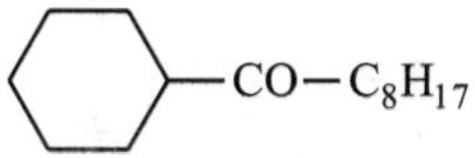	65
4	t-C_4H_9I	1-Octene	t-C_4H_9-CO-C_8H_{17}	69
5	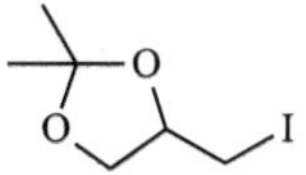	4-Allylveratrole	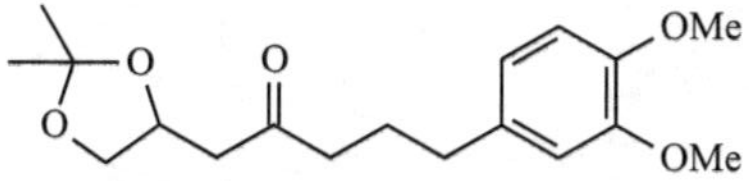	73
6	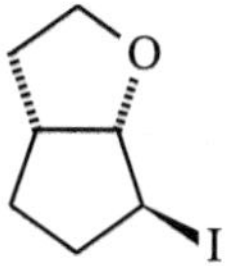		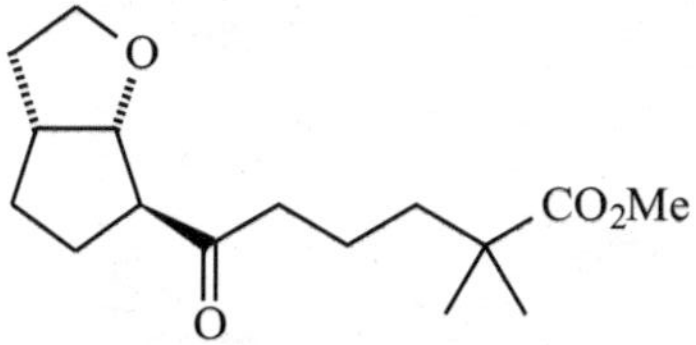	65
7	$NC(CH_2)_3I$	CH_2=CH$(CH_2)_2OCH_2Ph$	$NC(CH_2)_3$-CO-$(CH_2)_4OCH_2Ph$	70
8	$MeO_2C(CH_2)_3I$	CH_2=CH$(CH_2)_8CN$	$MeO_2C(CH_2)_3$-CO-$(CH_2)_{10}CN$	65
9	MeI	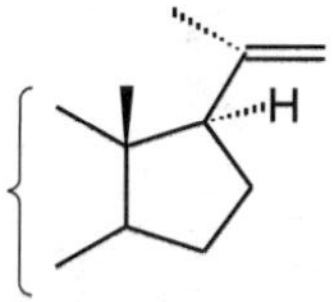	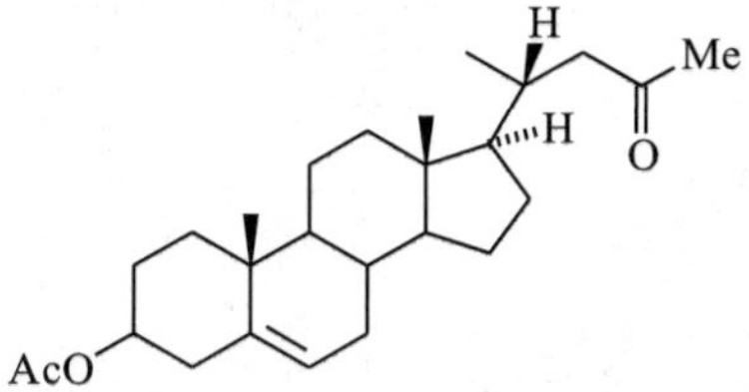	50

7.2.2 Synthesis of Enones

7.2.2.1 Synthesis of (*E*)-β-γ-Enones

B-trans-1-Alkenyls-9-BBN prepared *in situ* in THF react with α-halo carbanions generated from phenacyl bromide to afford the (*E*)-β,γ-unsaturated ketones in good yields (Eq. 7.7; Table 7.13) [9].

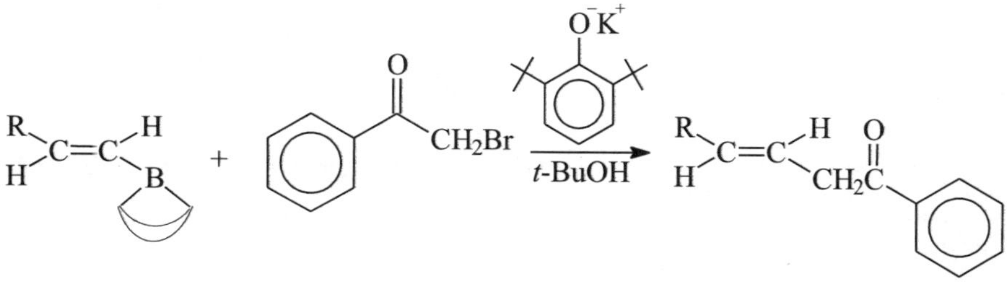

(7.7)

Table 7.13 β,γ-Unsaturated ketones from the reaction of *B-trans*-alkenyl-9-BBN derivatives with phenacyl bromide under the influence of 2,6-di-*tert*-butylphenoxide [9]

B-Alkenyl-9-BBN	β,γ-Unsaturated ketone	Yield (%)	Isomeric purity (%)
B-1-Hexenyl-9-BBN	(3*E*)-1-Phenyl-3-octen-1-one	61	96
B-3,3-Dimethyl-1-butenyl-9-BBN	(3*E*)-5,5-Dimethyl-1-phenyl-3-hexen-1-one	71	98
B-2-Cyclohexyl-1-ethenyl-9-BBN	(3*E*)-4-Cyclohexyl-1-phenyl-3-buten-1-one	60	98

7.2.2.2 Synthesis of γ,δ-Enones

B-Alkenyls-9-BBN prepared by the hydroboration of acetylenes with 9-BBN are transferred in a stereospecific manner to α,β-unsaturated carbonyl compounds in a 1,4-fashion (Scheme 7.6) [10]. The hydrolysis of the enolborinate provides the corresponding γ,δ-enones.

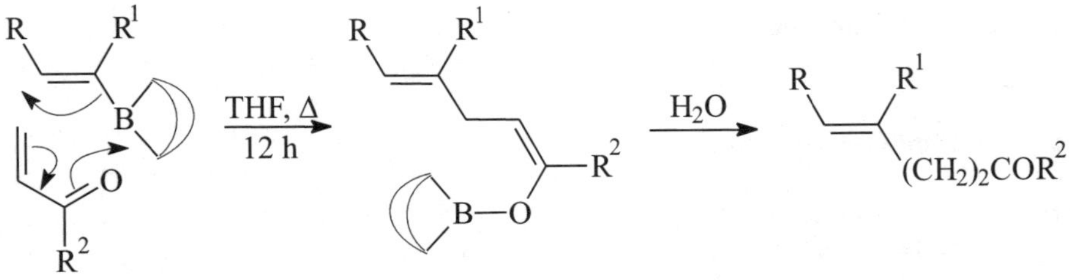

Scheme 7.6

The results are summarized in Table 7.14 [10].

Table 7.14 Conversion of alkynes into 4-alkenyl-2-butanones by the reaction of the corresponding *B*-alkenyl-9-BBN derivatives with methylvinyl ketone [10]

Alkyne	Product	Yield (%)
1-Hexyne	*trans*-5-Decen-2-one	87
3,3-Dimethyl-1-butyne	7,7-Dimethyl-*trans*-5-octen-2-one	85
Phenylethyne	*trans*-6-Phenyl-5-hexen-2-one	93
3-Hexyne	5-Ethyl-*trans*-5-octen-2-one	35
1-Phenyl-1-propyne	5-Methyl-*trans*-6-phenyl-5-hexen-2-one	62
5,5-Dimethyl-2-pentyne	5,7,7-Trimethyl-*trans*-5-octen-2-one	69
5-Chloro-1-pentyne	9-Chloro-*trans*-5-nonen-2-one	66

The conjugated *cisoid* ketones react satisfactorily with *B*-alkenyls-9-BBN, whereas *transoid* conjugated ketones, like cyclopentenone and cyclohexenone, afford a complex mixture of products. The reaction proceeds thermally and does not appear to involve radicals. The alkenyl group of borane is transformed with strict retention of configuration. As the reaction occurs only with *cisoid* enones, and thus suggesting a cyclic transition state (Chart 7.8) [10].

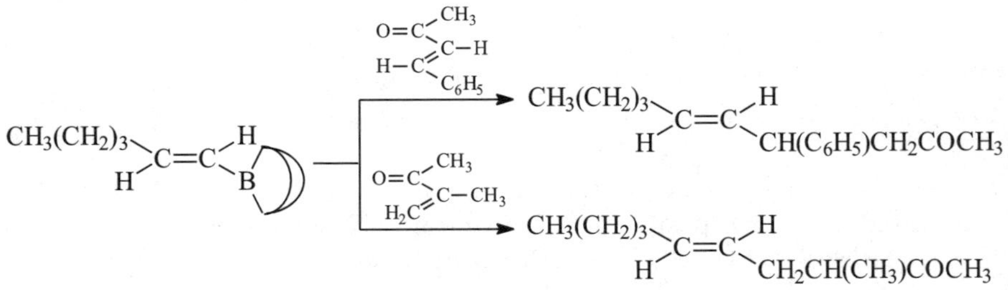

Chart 7.8

The unknown *B*-(*Z*)-1-alkenyls-9-BBN have been prepared by the reaction of (*Z*)-1-lithioalkens with *B*-OMe-9-BBN followed by the treatment of the ate complexes with borontrifluoride-diethyletherate. The *B*-(*Z*)-1-alkenyl-9-BBN reacts in a stereospecific manner with a methylvinylketone to afford the corresponding (*Z*)-γ,δ-enones in good yields (Scheme 7.7) [11].

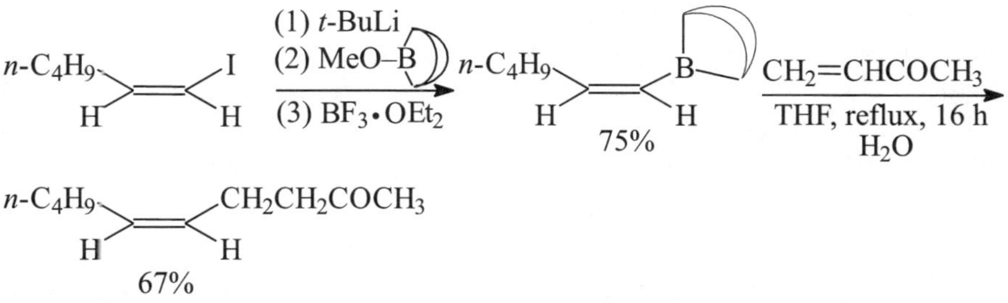

Scheme 7.7

7.2.3 Synthesis of γ,δ-Ynones

B-1-Alkynyl-9-BBN, easily and quantitatively prepared by the reaction of boron trifluoride-diethyletherate with the corresponding lithium methyl alkynyldialkylborinate [12], undergoes a smooth 1,4-addition in pentane at room temperature to methyl vinyl ketone (MVK) and ketones capable of adopting *cisoid* conformation [13]. The reaction proceeds through six-membered cyclic transition state to afford the enol borinate intermediate, which on hydrolysis afford γ,δ-acetylenic ketones, in high yields (Scheme 7.8).

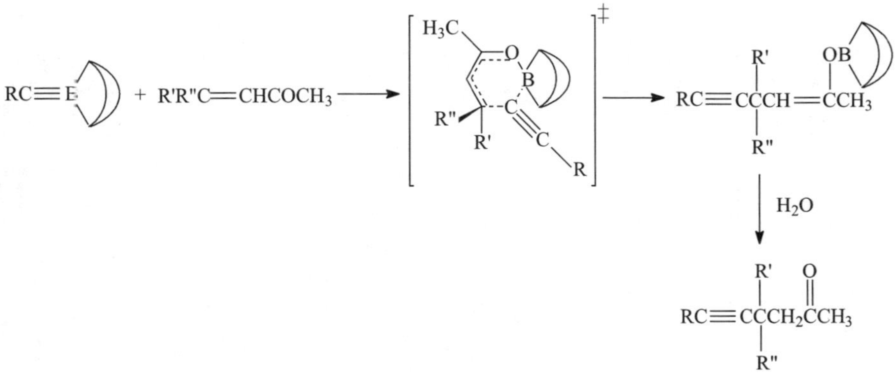

Scheme 7.8

The results are summarized in Table 7.15 [13].

Table 7.15 Conversion of alkynes into 4-alkynyl-2-butanones by the reaction of the corresponding *B*-1-alkynyl-9-BBN derivatives with MVK [13]

Alkyne	Product	Yield (%)
3,3-Dimethyl-1-butyne	7,7-Dimethyl-5-octyn-2-one	100
1-Hexyne	5-Decyn-2-one	96
Phenylethyne	6-Phenyl-5-hexyn-2-one	80
Cyclohexylethyne	6-Cyclohexyl-5-hexyn-2-one	97
2-Methyl-1-buten-3-yne	7-Methyl-7-octen-5-yn-2-one	81
5-Chloro-1-pentyne	9-Chloro-5-nonyn-2-one	87

Unlike *cisoid* ketones (entries 1–4, Table 7.16), *transoid* ketones (entries 5 and 6, Table 7.16) [13] do not react to give the desired product.

Table 7.16 Synthesis of γ,δ-acetylenic ketones by the conjugate addition of *B*-1-hexynyl-9-BBN to α,β-unsaturated ketones [13]

Entry	Enone	Reaction time	Yield of 1,4-addition product (%)
1	$CH_2{=}CHCOCH_3$	10 min	96
2	$CH_3CH{=}CHCOCH_3$	1 h	100
3	$(CH_3)_2CH{=}CHCOCH_3$	5 days	70
4	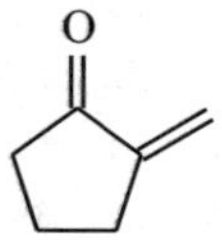	10 min	70
5	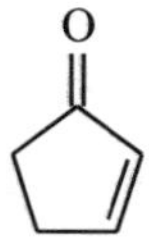	3 days	0
6	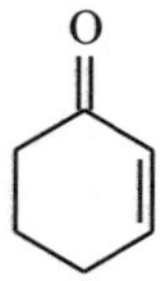	3 days	0

7.2.4
Synthesis of Dienones

Conjugated dienones are found in nature [14], and both conjugated *trans, trans*-dienoes [15] and conjugated *cis–trans* dienones [16] are prepared easily in high stereospecific manner but in modest yields. Brown *et al* [17] have developed the conjugate addition–elimination reaction of *B*-1-alkenyl-9-BBN with the commercially available 4-methoxy-3-butene-2-one to provide the corresponding conjugated *trans*, *trans*-dienones in essentially the quantitative yields (Eq. 7.8; Table 7.17) [17].

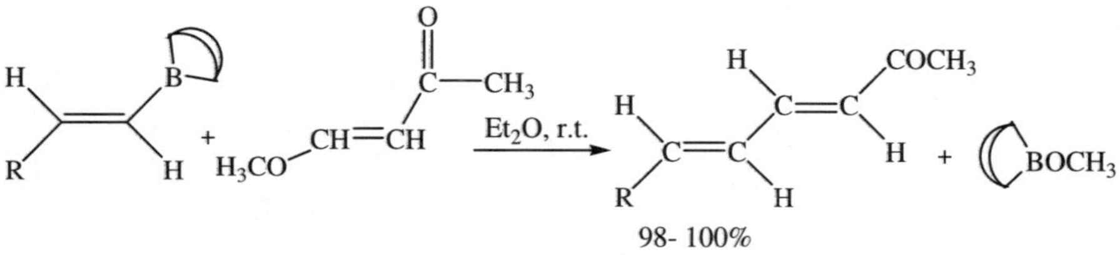

(7.8)

Table 7.17 Preparation of conjugated *trans,trans*-dienones [17]

Alkenylborane	Product	Yield (%)
B-trans-1-nonen-1-yl-9-BBN	*trans,trans*-3,5-tridecadien-2-one	100
B-trans-1-octen-1-yl-9-BBN	*trans,trans*-3,5-dodecadien-2-one	98
B-trans-3,3-dimethyl-1-buten-1-yl-9-BBN	*trans,trans*-7,7-dimethyl-3,5-octadien-2-one	99
B-trans-1-(5-chloropenten)-1-yl-9-BBN	*trans,trans*-9-chloro-3,5-nonadien-2-one	100

7.2.5
Synthesis of Conjugated Enynones

B-1-Alkynyls-9-BBN easily and quantitatively prepared from lithium methyl alkynyldialkylborinate [12] undergo a facile condensation with readily available 4-methoxy-3-buten-2-one in hexane at room temperature to provide, in excellent yield, conjugated enynones (Eq. 7.9) [18].

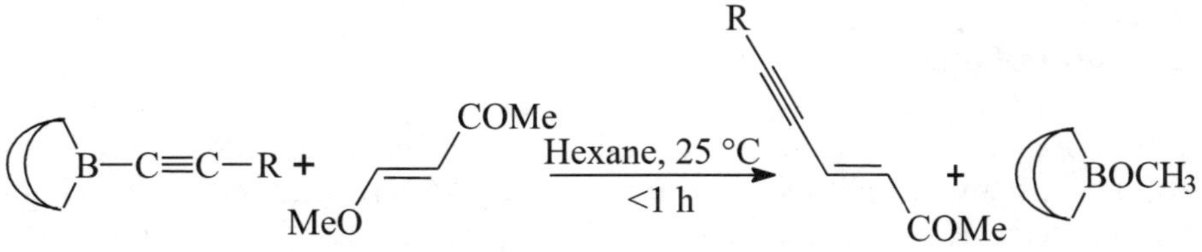

(7.9)

The reaction mechanism involves the initial coordination of the carbonyl group at the ketone with boron atom of 9-BBN, followed by the formation of cyclic transition state as sketched in Scheme 7.9.

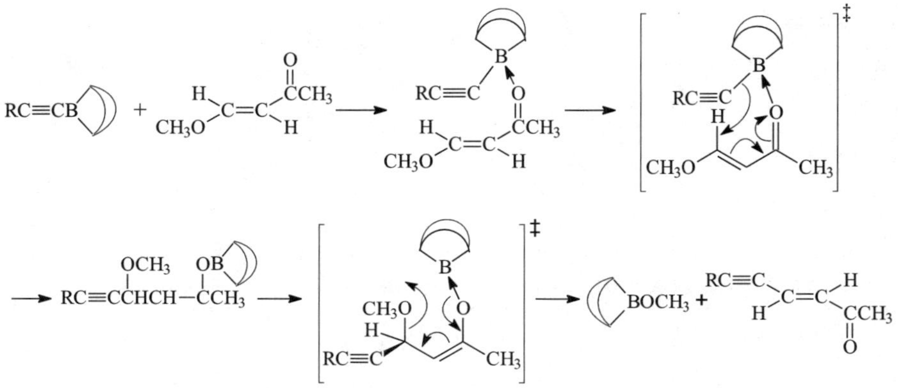

Scheme 7.9

The results are summarized in Table 7.18 [18].

Table 7.18 Conversion of alkynes into 4-alkynyl-3-buten-2-ones by the reaction of the corresponding *B*-1-alkynyl-9-BBN [18] derivatives with 4-methoxy-3-buten-2-one [18]

Alkyne	Product	Yield (%)
1-Hexyne	*trans*-3-Decen-5-yn-2-one	85
1-Decyne	*trans*-3-Tetradecen-5-yn-2-one	81
Phenylethyne	*trans*-6-Phenyl-3-hexen-5-yn-2-one	88
3,3-Dimethyl-1-butyne	*trans*-7,7-Dimethyl-3-octen-5-yn-2-one	98
Cyclohexylethyne	*trans*-6-Cyclohexyl-3-hexen-5-yn-2-one	72
5-Chloro-1-pentyne	*trans*-9-Chloro-3-nonen-5-yn-2-one	74
2-Methyl-1-buten-3-yne	*trans*-7-Methyl-3,7-octadien-5-yn-2-one	90

A 20% excess of 4-methoxy-3-buten-2-one is employed in all reactions.

The simple workup procedure provides highly pure products (≥97%) without the need for further purification of the crude material. The small excess of 4-methoxy-3-buten-2-one utilized in the reaction is conveniently hydrolyzed to water-soluble and/or highly volatile products (Eq. 7.10). The *B*-methoxy-9-BBN byproduct is oxidized to water-soluble *cis*-1,5-cyclopentanediol and boric acid. The whole sequence thus provides [18] clean products after simple extraction of the reaction mixture.

$$CH_3OCH{=}CHCOCH_3 + NaOH\ aq \longrightarrow CH_3COCH_3 + CH_3OH + HCOONa \quad (7.10)$$

7.2.6
Synthesis of β-Ketosilanes

Soderquist and coworkers have reported [19] that hydroboration with 9-BBN of 1-silylacetylenes, having bulkier groups on the Si, place the boron exclusively at the internal position of the alkyne as a kinetically less preferred process. Consequently, triisopropylsilyl-substituted alkynes afford the novel β-silylated vinylboranes, which on oxidation with alkaline hydrogen peroxide affords the corresponding ketones (Eq. 7.11) in excellent yields, a truly noteworthy example of the stability imparted to this functionality by the triisopropyl substitution on silicon [20]. It should be noted that earlier studies have shown failures to obtain β-ketosilanes from suitable organoprecursors due to their known sensitivity to bases and nucleophiles [21].

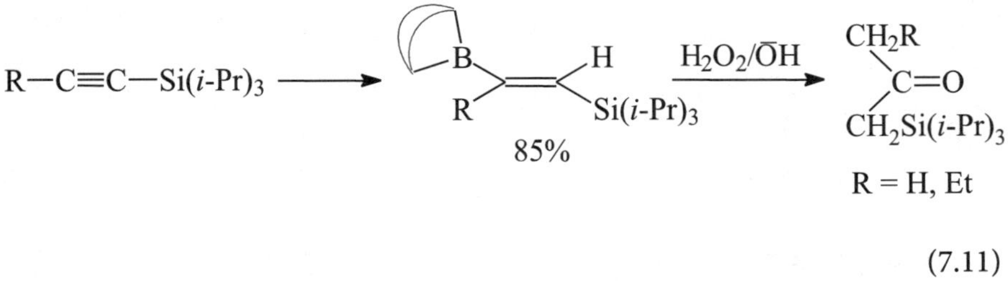

(7.11)

7.2.7
Synthesis of 1-Metallacyclohexan-4-ones

The 1-metallacyclohexan-4-ones are efficiently synthesized [22] from the corresponding divinyl compounds. The intermediate borinanes are prepared in a one-pot method which involves the hydroboration of divinyls with 9-BBN to get 1,5-diborayl adduct, and the regiospecific exchange of these adducts with borane–dimethylsulfide complex (BMS) leads to the formation of isolable metallaborinanes. The byproduct *B*-MeO-9-BBN can be recycled back to 9-BBN in >90% yields [23]. This avoids the necessity to selectively crystallize a portion of the 9-BBN prior to methanolysis. The conversion of borinanes to 1-metallacyclohexan-4-ones is achieved *via* the DCME procedure using 5 equiv of LiO(*t*-Bu), as illustrated in Scheme 7.10 [22].

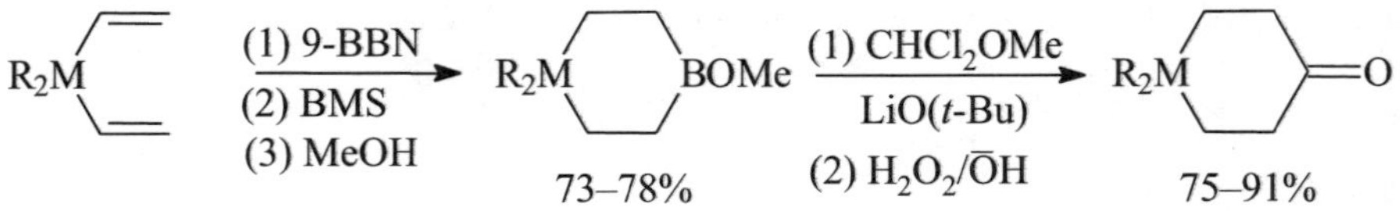

Scheme 7.10

The syntheses of 1-sila- and 1-germacyclohexan-4-ones are summarized in Table 7.19 [22].

Table 7.19 Divinylsilanes and divinylgermanes to metallacyclohexanones [22]

Series	Yield (%)	
	Borinanes	Metallacyclohexanones
Me_2Si	76	88
Et_2Si	78	75
Et_2Ge	73	91
Pr_2Ge	76	88

7.2.8 Synthesis of Chiral Ketones

Brown *et al* [24] have reported the preparation of *B*-R*-9-BBN with high optical purity (*vide infra*). An equimolar quantities of *B*-R*-9-BBN and α-bromoketones react very rapidly in the presence of equimolar quantity of potassium *tert*-butoxide. The moderate yield is realized of the chiral ketones, after the alkylation of α-bromoketones. In this reaction, both *B*-alkyl and *B*-cyclooctyl groups compete for migration. The results are summarized in Scheme 7.11 [24].

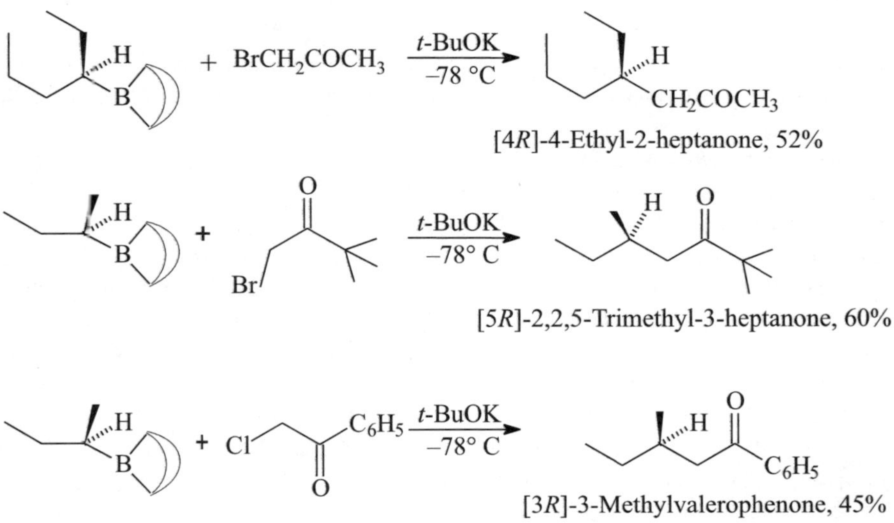

Scheme 7.11

References

Section 7.1

1. Zweifel G, Clark GM, Polston NL (1971) J Am Chem Soc 93:3395
2. Brown HC, Gupta SK (1975) J Am Chem Soc 97:5249
3. Brown HC, Scouten CG, Liotta R (1979) J Am Chem Soc 101:96
4. Kabalka GW, Hedgecock HC Jr (1975) J Chem Educ 52:745
5. Köster R, Morita Y (1967) Justus Liebig Ann Chem 704:70
6. Brown HC, Hubbard JL, Smith K (1979) Synthesis 701
7. (a) Brown HC, Knights EF, Coleman RA (1969) J Am Chem Soc 91:144; (b) Brown HC, Coleman RA (1969) ibid. 91:4606
8. Cha JS, Kim JE, Oh SY, Kim JD (1987) Tetrahedron Lett 28:4575
9. Cha JS, Kim JE, Yoon MS, Kim YS (1987) Tetrahedron Lett 28:6231
10. Brown HC, Mathew, CP, Pyun C, Son JC, Yoon NM (1984) J Org Chem 49:3091
11. (a) Krishnamurthy S (1974) Aldrichimica Acta 7:55; (b) Krishnamurthy S, Brown HC (1976) J Org Chem 41:3064
12. Cha JS, Lee JC, Yoon MS, Seo JB, Kim JM (1987) Bull Korean Chem Soc 11:76
13. Cha JS, Lee JC, Lee HS, Lee SE (1992) Org Prep Proced Int 24:327
14. Cha JS, Yoon MS, Lee KW, Lee JC (1989) Bull Korean Chem Soc 10:75
15. (a) Cha JS, Yoon MS, Kim YS, Lee KW (1988) Tetrahedron Lett 29:1069; (b) Cha JS, Yoon MS, Lee KW, Lee JC (1988) Heterocycles 27:1455
16. Cha JS, Yoon MS (1989) Tetrahedron Lett 30:3677

Section 7.2

1. Brown HC, Scouten CG, Liotta R (1979) J Am Chem Soc 101:96
2. (a) Brown HC, Rogić MM, Nambu H, Rathke MW (1969) J Am Chem Soc 91:2147; (b) Brown HC, Rogić MM (1969) ibid. 91:4304; (c) Brown HC, Nambu H, Rogić MM (1969) ibid. 91:6852
3. Soderquist JA, Ramos J, Matos K (1997) Tetrahedron Lett 38:6639
4. Brener L, Brown, HC (1977) J Org Chem 42:2702
5. Brown HC, Garg CP, Liu K-T (1971) J Org Chem 36:387
6. Carlson BA, Brown HC (1993) Synth 776
7. (a) Castle PL, Widdowson DA (1986) Tetrahedron Lett 27:6013; (b) Yuan K, Scott WJ (1989) ibid. 30:4779; (c) Kobayashi T, Tanaka M (1982) J Organomet Chem 231:C12; Hashen KE, Woell JB, Alper H (1984) Tetrahedron Lett 24:4879; Woel JB, Ferguson SB, Alper H (1985) J Org Chem 50:2134; Takeuchi R, Tsuji Y, Fujita M, Kondo T, Watanabe Y (1989) ibid. 54:1831
8. Ishiyama T, Miyaura N, Suzuki A (1991) Tetrahedron Lett 32:6923
9. Brown HC, Bhat NG, Campbell JB Jr (1986) J Org Chem 51:3398
10. Jacob P III, Brown HC (1976) J Am Chem Soc 98:7832
11. Brown HC, Bhat NG, Rajagopalan S (1986) Organometallics 5:816
12. Sinclair JA, Brown HC (1976) J Org Chem 41:1078; (1977) J Organomet Chem 131:163
13. Sinclair JA, Molander GA, Brown HC (1977) J Am Chem Soc 99:954
14. Renold W, Näf-Muller R, Keller U, Willhalm B, Ohloff G (1974) Helv Chim Acta 57:1301
15. (a) Surber W, Thens V, Colombi L, Schinz H (1956) Helv Chim Acta 39:1299; (b) Grünanger P, Grieco P (1958) Gazz Chim Ital 88:296
16. Näf F, Decorzant R (1974) Helv Chim Acta 57:1309
17. Molander GA, Singaram B, Brown HC (1984) J Org Chem 49:5024
18. Molander GA, Brown HC (1977) J Org Chem 42:3106
19. Soderquist JA, Colberg JC, Del Valle L (1989) J Am Chem Soc 111:4873
20. (a) Cunico RF, Bedell L (1980) J Org Chem 45:4797; (b) Muchowski JM, Naef R, Maddox ML (1985) Tetrahedron Lett 26:5375; (c) Corey EJ, Rucker C (1984) ibid. 25:4345
21. (a) Soderquist JA, Hassner A (1983) J Org Chem 48:1801; for preparation and properties of β-ketosilanes see (b) Colvin EW (1988) Silicon reagents in organic synthesis. Academic, London (1988)
22. Soderquist JA, Negron A, (1989) J Org Chem 54:2462
23. Soderquist JA, Negron A (1987) J Org Chem 52:3441
24. Brown HC, Joshi NN, Pyun C, Singaram B (1989) J Am Chem Soc 111:1754

8 Synthesis of Carboxylic Acids

8.1 Synthesis of Unsaturated Acids

Saturated, unsaturated, and functionalized carboxylic acids are important classes of natural products. Moreover, optically active hydroxy carboxylic acids are important biological molecules [1] as well as intermediates for organic synthesis [2].

The synthesis of wide varieties of functionalized [3] and unsaturated carboxylic acids [4] have been achieved through organoboranes. The reaction sequence involves the attack of organoboranes on the dianion of phenoxy acetic acid followed by the alkyl migration as illustrated in Eq. 8.1.

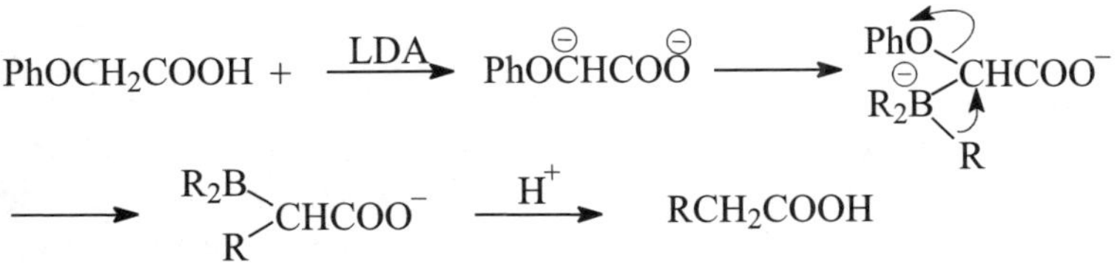

(8.1)

9-BBN has been elegantly employed for the selective hydroboration to afford the corresponding alkyl boranes, which on treatment with dianion of phenoxy acetic acid, followed by alkaline H_2O_2 oxidation produce in good to moderate yields the corresponding carboxylic acids with two carbon homologation (Eqs. 8.2, 8.3).

8.2 Synthesis of β-Amino Acids

β-Amino acids are valuable for the synthesis of peptidomimetics [5], functionalized β-lactams and some naturally occurring materials [6–8]. Burgess and Ohlmeyer [9] have reported that the allylic amines on hydroboration afford

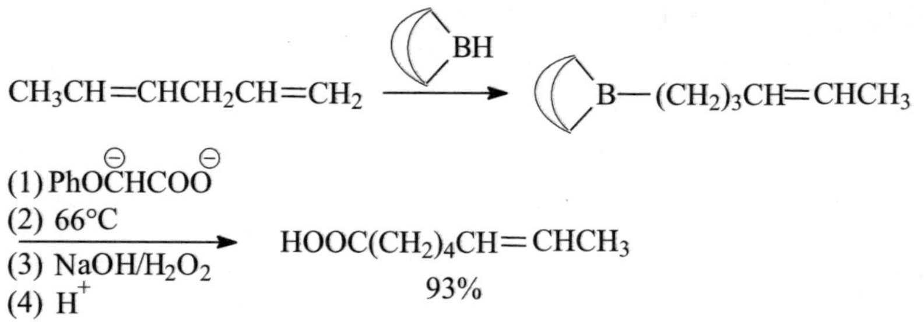

(8.2)

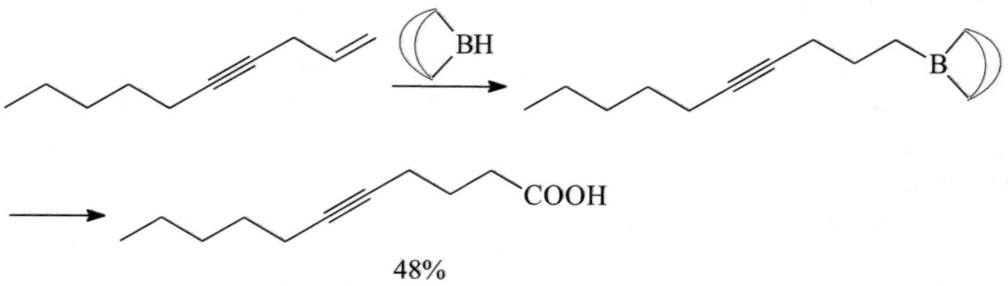

(8.3)

the *anti*-selective products when BH_3 is used as the hydroborating agent, and *syn*-selective if it is 9-BBN or catecholborane–rhodium-catalyzed hydroboration (Table 8.1) [10]. The stereoselection is enhanced when two N-protecting groups are used [5]. This protocol is employed [10] for the synthesis of α,β-disubstituted β-amino acids with high relative and absolute stereoselectivity from naturally occurring α-amino acids (Scheme 8.1).

Table 8.1 Diastereoselectivities for hydroboration–oxidation of allylamine derivatives [10]

R^1	R^2	Hydroborating agent	*Syn*:*anti*	Yield of major isomer (%)
$CH_2{}^iPr$	Me	BH_3	1:>20	60
$CH_2{}^iPr$	Me	9-BBN	>20:1	85
n-Bu	*n*-Bu	BH_3	1:16	77
n-Bu	*n*-Bu	9-BBN	1.3:1	86
CH_2Ph	Ph	BH_3	1:>20	80
CH_2Ph	Ph	9-BBN	No reaction	0

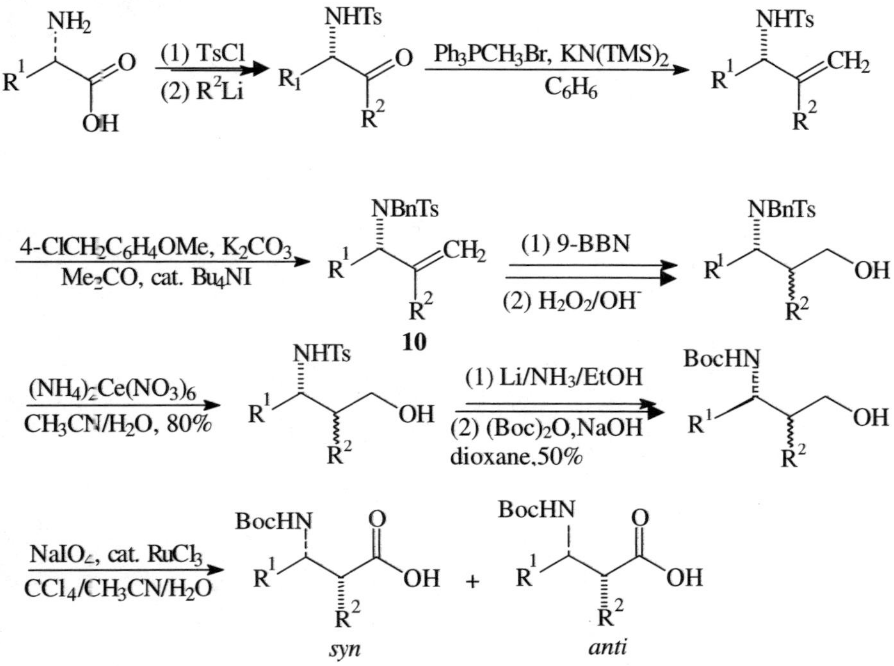

a; $R^1 = CH_2{}^iPr$, $R^2 = Me$; b; $R^1 = R^2 = Bu$-*n*

c; $R^1 = CH_2Ph$, $R^2 = Ph$; d; $R^1 = CH_2Ph$, $R^2 = CH_2Ph$

Scheme 8.1

References

1. Tahara S, Hosokawa K, Mizutani J (1980) Agric Biol Chem 44:193
2. Midland MM, Lee PE (1981) J Org Chem 46:3933
3. Hara S, Kishimura K, Suzuki A, Dhillon RS (1990) J Org Chem 55:6356
4. Hara S, Kishimura K, Suzuki A (1978) Tetrahedron Lett 2891
5. Xie J, Soleilhac J, Schmidt C, Peyroux J, Roques BP, Fournie-Zaluski M (1989) J Med Chem 32 1497
6. Namikoshi M, Rinehart KL, Dahlem AM, Beasley VR (1989) Tetrahedron Lett 30:4349
7. Beatty MF, Jennings-White C, Avery MA (1991) J Chem Soc Chem Commun 351
8. Beatty MF, Jennings-White C, Avery MA (1992) J Chem Soc Perkin Trans I 1637
9. Burgess K, Ohlmeyer MJ (1991) J Org Chem 56:1027
10. Burgess K, Liu LT, Pal B (1993) J Org Chem 58:4758

9 Synthesis of Esters

9.1 Synthesis of Achiral Esters

B-Alkyl-9-BBN reacts with ethyl bromoacetate under the influence of potassium *t*-butoxide to afford [1] the corresponding ethyl ester with the addition of two carbon moiety (Eq. 9.1).

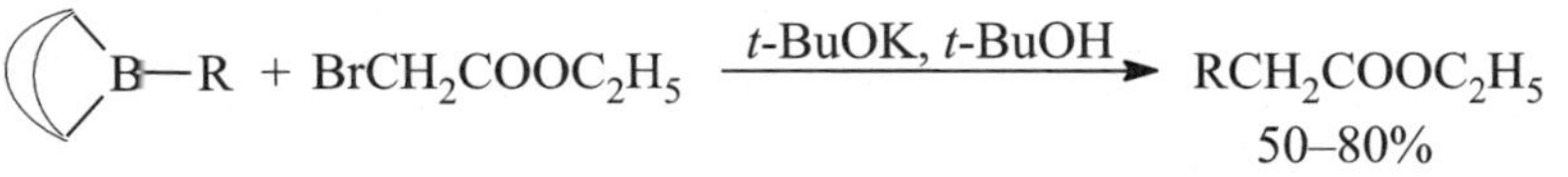

(9.1)

The lower yield in the range of 50–80% is because of some competition between migration of the *B*-alkyl group and boroncyclooctyl bond. The reaction of *B*-alkyl-9-BBN with ethyldiazoacetate [2] also results in the preferential migration of cyclooctylboron bond [3]. However, the preferential migration of alkyl groups are realized in the reactions of ethyldibromoacetate and ethyldichloroacetate, which provide the corresponding α-halocarboxylic acid esters in yields of 70–90% (Eq. 9.2) [1].

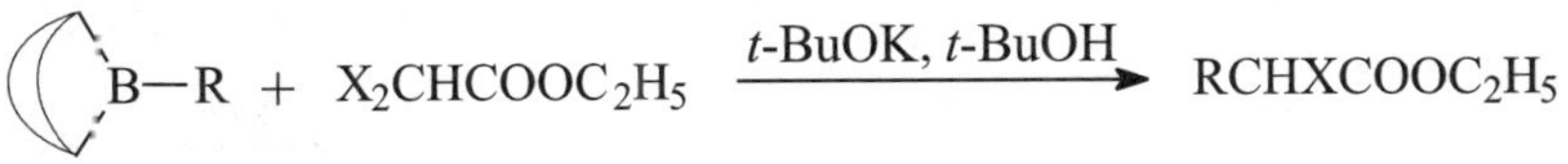

(9.2)

The results are summarized in Table 9.1[1].

The use of milder base potassium 2,6-di-*t*-butoxide to alkylate ethylbromoacetate in THF results in the drop of yield to 0%. This is reasoned owing to the slow protonolysis of the rearranged boron intermediate, so that it is capable of reacting competitively with freshly formed α-bromo carbanion.

The problem is overcome by making 1-M solution of the organoborane in THF and then adding an equimolar quantity of 2,6-di-*t*-butylphenol, followed

Table 9.1 Conversion of olefins into ethylalkanoates and 2-haloalkanoates by the reaction of the corresponding *B*-alkyl-9-borabicyclo[3.3.1]nonanes with ethyl α-haloacetates under the influence of potassium *t*-butoxide [1]

Olefin	Ethyl α-haloacetate	Product	Yield (%)
Ethene	Br	Ethyl *n*-butyrate	51
1-Butene	Br	Ethyl hexanoate	59
2-Butene	Br	Ethyl 3-methylpentanoate	68
Isobutylene	Br	Ethyl 4-methylpentanoate	53
1-Hexene	Br	Ethyl octanoate	74
1-Hexene	Cl	Ethyl octanoate	74
Cyclopentene	Br	Ethyl cyclopentylacetate	63
1-Methylcyclopentene	Br	Ethyl(*trans*-2-methylcyclopentyl) acetate	57
Cyclohexene	Br	Ethylcyclohexylacetate	62
Cyclohexene	Br_2	Ethyl α-bromocyclohexylacetate	68
Cyclohexene	Cl_2	Ethyl α-chlorocyclohexylacetate	88
Cyclopentene	Cl_2	Ethyl α-chlorocyclopentylacetate	90

by addition of 1-M solution of potassium *t*-butoxide in *t*-butylalcohol. The reaction mixture is thus roughly 50:50 in THF and *t*-butylalcohol. To this reaction mixture at 25 °C is added the bromoester. The *B*-alkyl-9-BBN reagents react satisfactorily to afford the corresponding esters in high yields (Table 9.2) [4]. Consequently, *t*-butylalcohol facilitates the protonolysis of the boron intermediate.

Table 9.2 α-Alkylation of ethylbromoacetate and dibromoacetate with *B*-R-9-BBN under the influence of potassium 2,6-di-*t*-butylphenoxide [4]

R	Bromo ester	Product	Yield (%)
Ethyl	Br	Ethyl butyrate	70
Isobutyl	Br	Ethyl isovalerate	56
Cyclopentyl	Br	Ethyl cyclopentylacetate	57
Ethyl	Br_2	Ethyl α-bromobutyrate	83
Isobutyl	Br_2	Ethyl α-bromoisovalerate	81
Cyclopentyl	Br_2	Ethyl α-bromocyclopentylacetate	78

9.2 Synthesis of Chiral Esters

The use of 9-BBN, affords a stereodefined alkylation of ethyl bromoacetate. Tris(*trans*-2-methylcyclopentyl)borane fails to react with ethyl bromoacetate in the presence of potassium *t*-butoxide, whereas *B*-*trans*-2-methylcyclopentyl-9-BBN affords ethyl(*trans*-2-methylcyclopentyl)acetate in >98% purity (Eq. 9.3) [3].

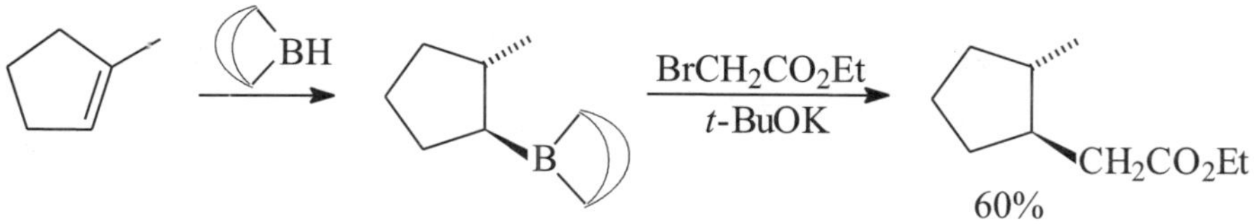

(9.3)

The lack of asymmetric hydroboration for prochiral alkene to synthesize *B*-R*-9-BBN (R* = chiral R) as reagent resulted into the handicap of the utility of this methodology for the synthesis of chiral esters. This is, however, circumvented elegantly by Brown and his coworkers [5, 6], whereby *B*-R*-9-BBN was synthesized indirectly in essentially 100% optical purity, first by asymmetric hydroboration of prochiral olefins with diisopinocampheylborane Ipc_2BH (99% *ee*) [6] or monoisopinocampheylborane $IpcBH_2$ (100% ee) [7] prepared from (+)-α-pinene. Consequently, asymmetric hydroboration of prochiral olefin with $IpcBH_2$ in the molar ratio of 1:1, followed by crystallization provides chiral isopinocampheylalkylboranes, IpcR*BH (**1**) in essentially 100% optical purity. Treatment of **1**ith acetaldehyde results in the selective facile elimination of chiral auxiliary, leading to the formation of the corresponding boronic ester (**2**) in very high enantiomeric purity. Reaction of **2** with NaOH followed by protonolysis affords the corresponding boronic acid (**3**) which on esterification with 1,3-propanediol yields optically active 2-alkyl-1,3,2-dioxaborinane (**4**).

These chiral alkylboronic esters are exceptionally promising intermediates for C–C bond formation reaction in the synthesis [8, 9] of α-chiral aldehydes, β-chiral alcohols, α-chiral acids, and α-chiral amines. Brown *et al* [10], in a real breakthrough, discovered that $LiAlH_4$ readily converts these relatively inert boronic esters to a very high reactive lithium monoalkylborohydrides R^*BH_3Li (**5**) of very high optical purity. The optically active monoalkylborane (R^*BH_2) is generated, when required, by a convenient reaction with trimethylsilyl chloride [6]. Consequently, the desired *B*-R*-9-BBN is prepared conveniently by hydroboration of 1,5-cyclooctadiene with RBH_2 (prepared *in situ*), and the desired stable 1,5-isomer is obtained by thermal isomerization. The whole sequence is illustrated in Scheme 9.1.

The optically active 9-R*-9-BBNs prepared by this method are shown in Chart 9.1.

Alkali metal *tert*-butoxide in THF is used for their homologation-reaction. All three alkali metal *tert*-butoxide in THF work equally well for the α-alkylation (Eq. 9.4) of ethyl bromoacetate.

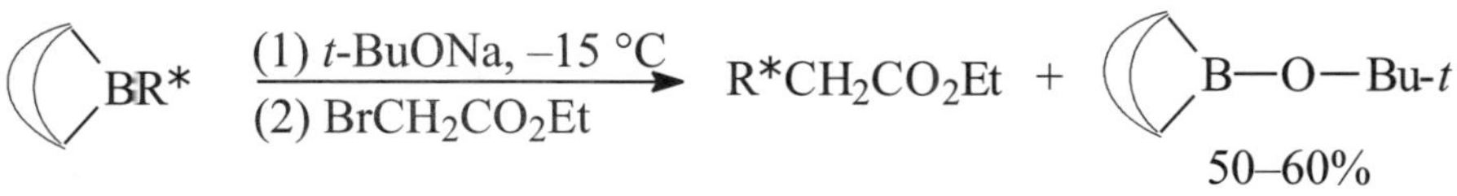

(9.4)

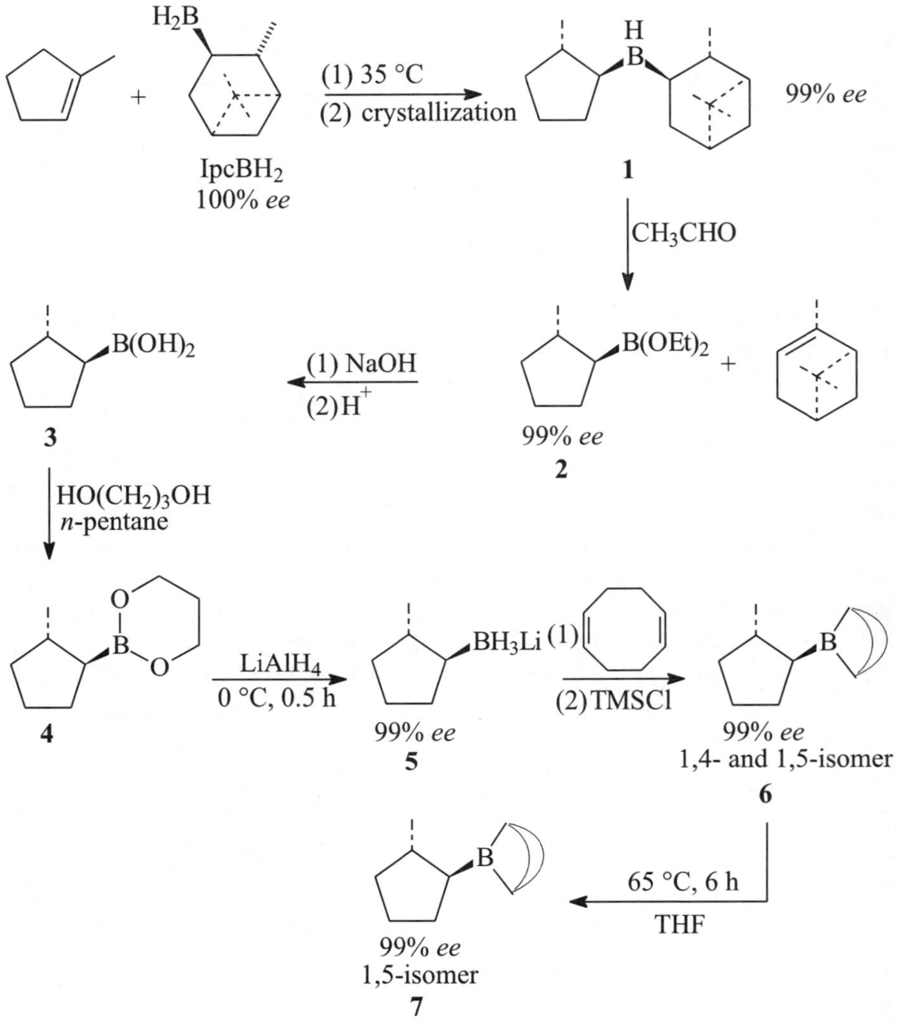
H2B
IpcBH2
100% ee
(1) 35 °C
(2) crystallization
H
B
99% ee
1
CH3CHO
B(OEt)2
99% ee
2
(1) NaOH
(2) H+
B(OH)2
3
HO(CH2)3OH
n-pentane
O
B
O
4
LiAlH4
0 °C, 0.5 h
BH3Li
99% ee
5
(1)
(2) TMSCl
B
99% ee
1,4- and 1,5-isomer
6
65 °C, 6 h
THF
B
99% ee
1,5-isomer
7

Scheme 9.1

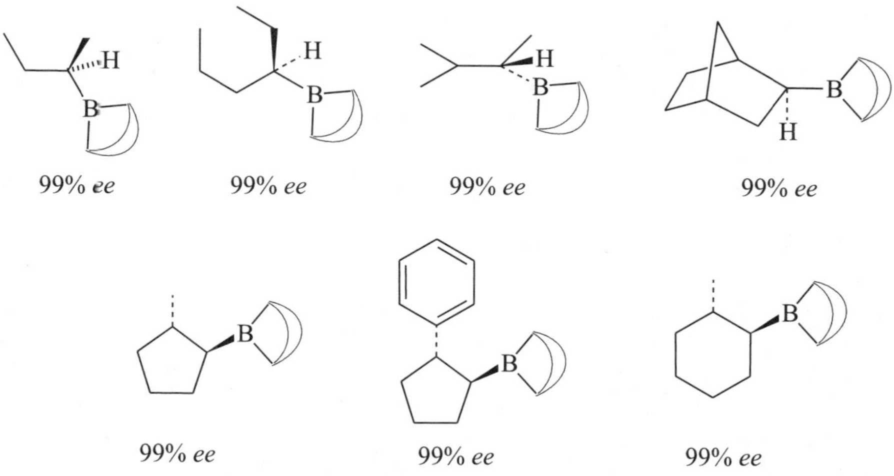

Chart 9.1

These reactions proceed through the following mechanism (Scheme 9.2).

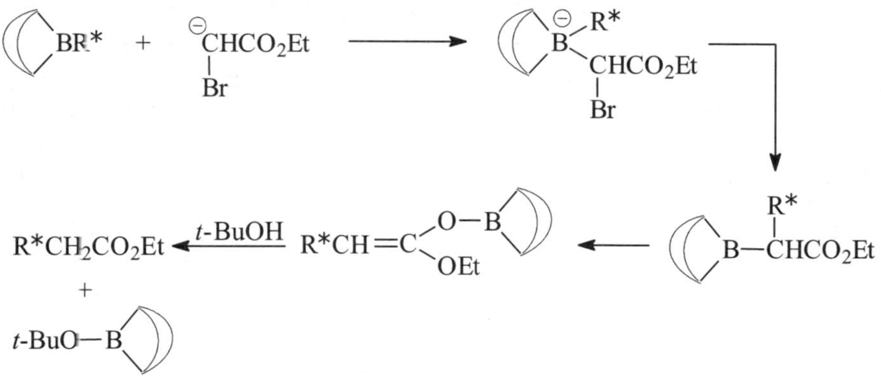

Scheme 9.2

The chiral 9-R*-BBNs are utilized for the synthesis of various homologated esters, in very high optical purities (Table 9.3) [5].

Table 9.3 Alkylation of ethyl bromoacetate with optically active *B*-R*-9-BBN [5]

*R**	Product	Yield (%)	ee (%)
[*R*]-2-Butyl	[3*R*]-Ethyl-3-methyl pentanoate	60	99
[*S*]-3-Methyl-2-butyl	[3*R*]-Ethyl-3,4-dimethyl pentanoate	55	99
[1*S*,2*S*]-*trans*-2-Methyl cyclopentyl	[1*S*, 2*S*]-Ethyl *trans*-2-methyl cyclopentyl acetate	50	99

Both (+) and (–) α-pinenes are now readily available in pure form. Consequently, it is possible by this remarkable simple procedure to achieve the synthesis of both enantiomers of functional derivatives involving carbon-atom homologation from olefins via asymmetric hydroboration.

9.3 Synthesis of (*E*)-β,γ-Unsaturated Esters

B-trans-1-Alkenyls-9-BBN obtained conveniently by the hydroboration of 1-alkynes with 9-BBN in THF undergo facile reaction with and α-halo carbanion generated from ethylbromoacetate with the hindered base potassium-2,6-di-*tert*-butylphenoate to afford, in 57–65% yields, the corresponding (*E*)-β,γ-unsaturated esters (Eq. 9.5; Table 9.4) [11]. However, the reaction gives a 1:1 mixture of *E* and *Z* isomers in the case of internal alkenyl-9-BBN compounds [11]. The other method for preparing β,γ-unsaturated esters is the deprotonation and reprotonation of α,β-unsaturated esters [12, 13], but this method usually affords a mixture of (*E*) and (*Z*)-β,γ-unsaturated esters. It is reported that both terminal and internal *B*-alkenyl-9-BBN react in a stereodefined manner [14] with ethyl-(dimethylsulfuranylidene)-acetate to give the corresponding (*E*)-β,γ-unsaturated esters in 60–80% yields (Scheme 9.3). (Carbethoxymethyl)dimethylsulfon ium bromide is easily generated from dimethylsulfide and ethyl-α-bromoacetate [15], and stereodefined (*E*)-*B*-alkenyls-9-BBN are obtained [16] by the hydroboration of internal and terminal alkynes. The *B*-alkenyl-9-BBN derivative [16] generated *in situ* readily reacts with [14] (carbethoxymethyl)dimethylsulfonium bromide and sodium hydride in THF, which followed by oxidation specifically with H_2O_2/ŌAc affords (*E*)-β,γ-unsaturated esters (Table 9.5) [14].

(9.5)

Table 9.4 β,γ-Unsaturated esters from a base-promoted α-alkenylation of ethyl bromoacetate with *B-trans*-1-alkenyl-9-BBN derivatives [11]

B-Alkenyl-9-BBN	β,γ-Unsaturated ester	Yield (%)	Isomeric purity (%)
B-1-hexenyl-9-BBN	Ethyl-(3*E*)-3-octenoate	63	96
B-3,3-dimethyl-1-butenyl-9-BBN	Ethyl-(3*E*)-5,5-dimethyl-3-hexenoate	57	98
B-2-cyclohexyl-1-ethenyl-9-BBN	Ethyl-(3*E*)-4-cyclohexyl-3-butenoate	65	98
B-1-octenyl-9-BBN	Ethyl-(3*E*)-3-decenoate	65	95

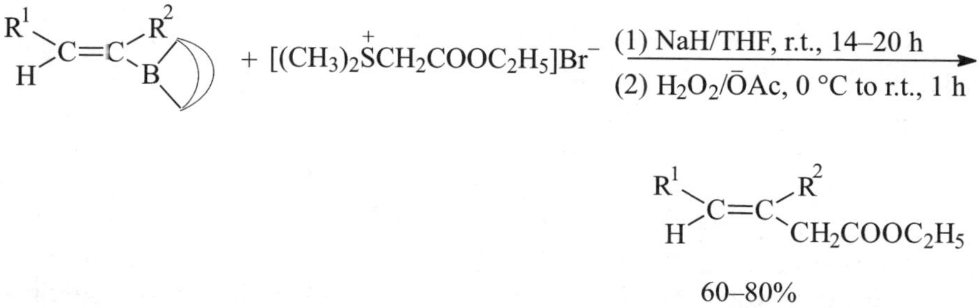

Scheme 9.3

The mechanism of the reaction is depicted in Scheme 9.4 [14].

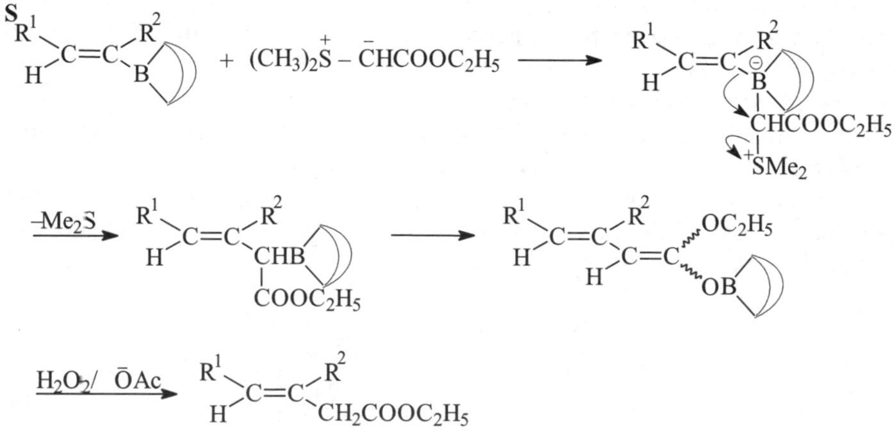

Scheme 9.4

Table 9.5 Yields of (*E*)-β,γ-unsaturated esters [14]

R^1	R^2	Yield (%)
n-C_3H_7	H	76
n-C_4H_9	H	63
n-C_5H_{11}	H	64
n-C_6H_{13}	H	62
n-C_8H_{17}	H	80
Ph	H	60
n-C_3H_7	n-C_3H_7	65
n-C_4H_9	n-C_4H_9	73
n-C_5H_{11}	n-C_5H_{11}	80
n-C_6H_{13}	n-C_6H_{13}	71

The reaction conditions are important for the preparation of pure (*E*)-β,γ-unsaturated esters. No *Z* isomer is found in the product when the reaction mixture is oxidized with $H_2O_2/\bar{O}Ac$ at 0 °C. However, when the reaction mixture is oxidized with $H_2O_2/\bar{O}Ac$ at room temperature or $H_2O_2/\bar{O}H$ at 0 °C, a small amount of *Z* isomer is formed in the product.

9.4 Synthesis of α-Amino Acid Esters

Very high enantio- and diastereoselective (up to 96% *ee*) synthesis of some amino acid esters have been achieved by the reaction of 9-allyl- and 9-but-2-enyl-9-BBN with α-imino esters (Eq. 9.6) [17].

The reaction of allyl-9-BBN with α-iminoesters in which (*S*)-(−)-α-methylbenzylamine is utilized as the chiral source produced the corresponding amino acid in high chemical (92%) and optical (92% *ee*) yield. These are increased to 96% when (−)-1-cyclohexylethylamine is employed. The reaction of allylzinc or magnesium or titanium reagents, on the other hand, has resulted in lower or negligible *ee*'s.

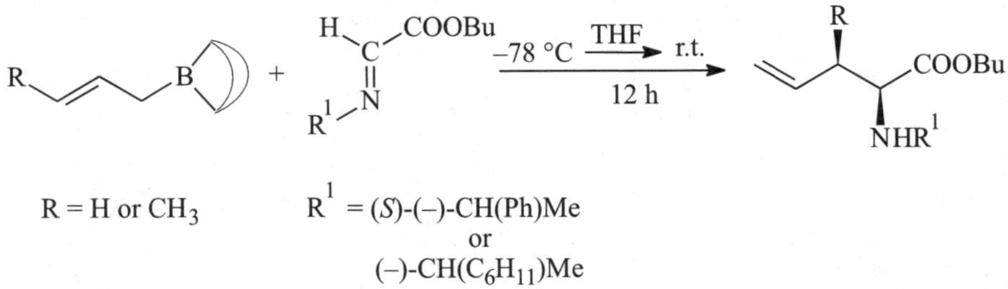

(9.6)

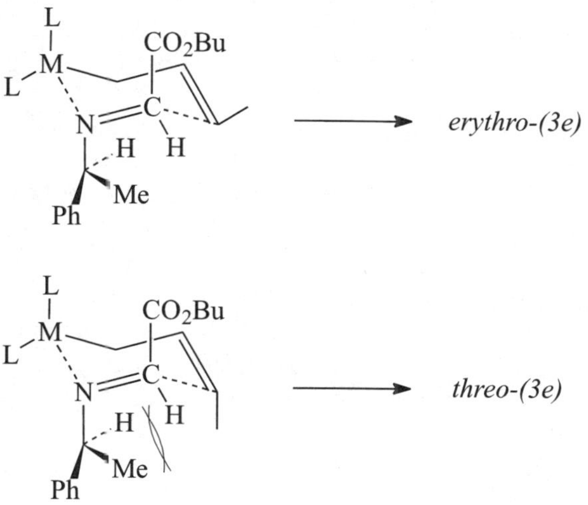

Scheme 9.5

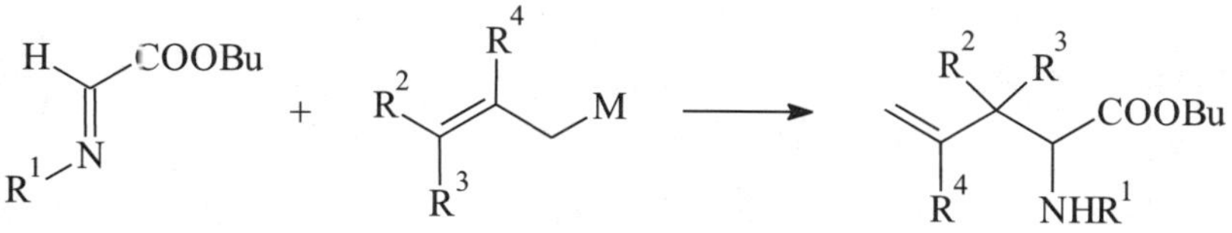

a: $R^1 = PhCH(CH_3)$

b: $R^1 = (C_6H_{11})CH(CH_3)$

c: $R^1 = SO_2C_6H_4$-*p*-CH_3

a: $R^1 = PhCH(CH_3)$, $R^2 = R^3 = R^4 = H$

b: $R^1 = (C_6H_{11})CH(CH_3)$, $R^2 = R^3 = R^4 = H$

c: $R^1 = PhCH(CH_3)$, $R^2 = R^3 = H$, $R^4 = CH_3$

d: $R^1 = (C_6H_{11})CH(CH_3)$, $R^2 = R^3 = H$, $R^4 =$

CH_3

e: $R^1 = PhCH(CH_3)$, $R^2 = R^3 = CH_3$, $R^4 = H$

f: $R^1 = SO_2C_6H_4$-*p*-CH_3, $R^2 = CH_3$, $R^3 = R^4 = H$

Chart 9.2

The high enantio- and diastereoselectivity of but-2-enyl-9-BBN are due to the attack of boron reagent from the *si* face [18] of iminoester. The *si*-face attack results from the steric and stereoelectronic effects of the imine and BBN groups. The transition state leading to *threo* is highly destabilized as compared with the *erythro*-intermediate (Scheme 9.5). Consequently, in the present reaction the *erythro*-selectivity is more than 93%. Therefore, the α-methylbenzylamine group dictates not only the enantioselectivity, but also enhances diastereoselectivity for the synthesis of amino acids and related compounds *via* C–C bond formation.

Further, Yamamoto *et al* [19] have recorded very high enantio- and diastereoselective synthesis of amino acid derivatives *via* the reaction of allylic 9-BBN and α-iminoesters having a chiral auxiliary at the R′ group. Earlier, Kagan [20] reported reactions of α-iminoesters with Grignard reagent. However, the major drawback is the lack of regioselectivity, due to the nucleophilic attack of all the three possible electrophilic centers; the imine carbon, ester carbon, and nitrogen atom. The reaction of allylic 9-BBN proceeds regioselectively at the imine carbon in high yields to give the corresponding amino acid derivatives with very high enantio- and diastereoselectivity (up to 96% *ee*, Chart 9.2).

The results are summarized in Table 9.6 [19].

Table 9.6 Reaction of chiral iminoesters with allylic organometallic compounds [19]

	Allylmetal						
R′	M	R^2	R^3	R^4	Yield (%)	*de* (%)	*erythro:threo*
a	9-BBN	H	H	H	92	92	
a	ZnBr	H	H	H	94	10	
a	MgCl	H	H	H	–	0	
b	9-BBN	H	H	H	94	96	
b	ZnBr	H	H	H	53	30	
a	9-BBN	H	H	CH_3	80	90	
a	ZnBr	H	H	CH_3	83	16	
b	9-BBN	H	H	CH_3	94	78	
b	ZnBr	H	H	CH_3	90	14	
a	9-BBN	CH_3	CH_3	H	33	54	
a	ZnBr	CH_3	CH_3	H	60	16	
c	9-BBN	CH_3	H	H	75		85:15
c	$Ti(O\text{-}i\text{-}Pr)_2$	CH_3	H	H	75		60:40
c	MgCl	CH_3	H	H	78		59:41
c	$Zr(Cp)_2Cl$	CH_3	H	H	73		51:49

As depicted in Chart 9.3, the transition state [A] resulting from allyl-9-BBN and chiral iminoester (a) leads to aminoester with *SS* configurations. However, transition state [B] resulting from methallyl-9-BBN and chiral iminoester (a) possesses three 1,3-diaxial interactions as compared with only one 1,3-diaxial interaction in [A]. Thus, [B] is highly destabilized and thus methallylboration

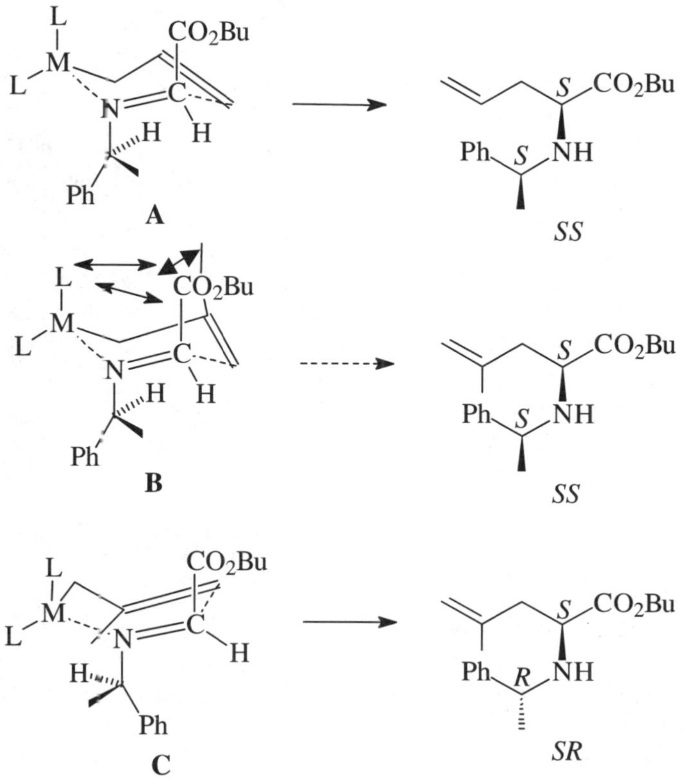

Chart 9.3

leads to an opposite chiral induction. In boat transition state [C], such bad interactions are diminished. Thus, iminoester derived from (*R*)-(+)-PhMeCHNH$_2$ reacts with methallyl-9-BBN through [C] to give (*S*,*R*)-amino acid ester with high diastereoselectivity. The direction of chiral induction in aldehydes is identical, irrespective of allylboration, methylboration and prenylboration [21].

The reaction is extended with crotyl-9-BBN [19]. The chiral iminoester reacts with crotyl-9-BBN (Eq. 9.7) in essentially quantitative yield to give Cram-*erythro* amino acid ester.

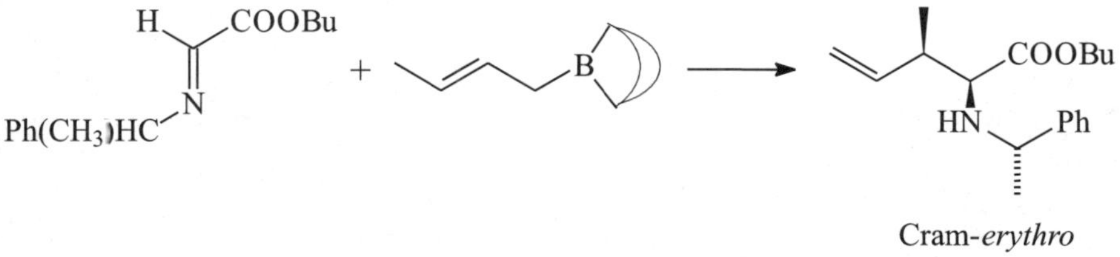

(9.7)

References

1. Brown HC, Rogić MM (1969) J Am Chem Soc 91:2146
2. Hooz J, Linke S (1968) J Am Chem Soc 90:5936, 6891
3. Even acrolein, which reacts with remarkable speed with hindered trialkylboranes, fails to react with *B*-R-9-BBN. See Brown HC, Rogić MM, Rathke MW, Kabalka GW (1969) J Am Chem Soc 91:2150
4. Brown HC, Nambu H, Rogić MM (1969) J Am Chem Soc 9:6855
5. (a) Brown HC, Joshi NN, Pyun C, Singaram B (1989) J Am Chem Soc 111:1754; (b) Brown HC, Singaram B (1984) ibid. 106:1797
6. Brown HC, Singaram B (1984) J Org Chem 49:945
7. Brown HC, Schwier JR, Singaram B (1978) J Org Chem 43:4395
8. Brown HC, Imai T, Desai MC, Singaram B (1985) J Am Chem Soc 107:4980
9. Brown HC, Kim, K-W, Cole TE, Singaram B (1986) J Am Chem Soc 108:6761
10. Brown HC, Singaram B, Cole TE (1985) J Am Chem Soc 107:460
11. Brown HC, Bhat NG, Campbell JB Jr (1986) J Org Chem 51:3398
12. (a) Krebs E-P, (1981) Helv Chim Acta 64:1023; (b) Kende AS, Toder BH (1982) J Org Chem 47:163
13. Ikeda Y, Ukai J, Ikeda N, Yamamoto H (1987) Tetrahedron 43:743
14. Deng M-Z, Li N-S, Huang Y-Z (1992) J Org Chem 57:4017
15. Speziale AS, Tung CC, Ratts KW, Yao A (1965) J Am Chem Soc 87:3460
16. (a) Jacob P III, Brown HC (1977) J Org Chem 42:579; (b) Wang KK, Scouten CG, Brown HC (1982) J Am Chem Soc 104:531
17. Yamamoto Y, Ito W, Maruyama K (1985) J Chem Soc Chem Commun 1131
18. Yamamoto Y, Komatsu T, Maruyama K (1984) J Am Chem Soc 106:5031
19. Yamamoto Y, Nishii S, Maruyama K, Komatsu T, Ito W (1986) J Am Chem Soc 108:7778
20. Fiaud J, Kagan HB (1970) Tetrahedron Lett 1813:1019
21. (a) Brown HC, Jadhav PK (1983) J Am Chem Soc 105:2029; (b) Brown HC, Jadhav PK (1984) Tetrahedron Lett 25:1215; (c) Brown HC, Jadhav PK, Perumal PT (1984) ibid. 25:5111

10 Synthesis of Nitriles

10.1
Synthesis of Achiral Nitriles

The *B*-R-9-BBN reacts with chloroacetonitrile [1] under the influence of potassium 2,6-di-*t*-butylphenoxide to produce the corresponding nitriles in high yield (Eq. 10.1). The reaction intermediate undergoes protonolysis readily and without adding ethanol or other protonolyzing agent as required, for example, in the synthesis of ketones under similar conditions.

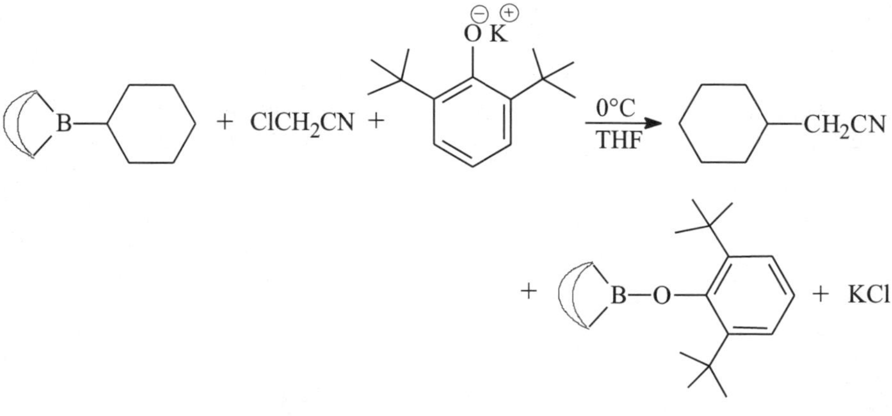

(10.1)

The results are summarized in Table 10.1 [1].

The reaction is extended to the synthesis of α-chloronitriles and dialkylacetonitriles [2]. Monoalkylation of dichloroacetonitriles with 1 equiv of *B*-R-9-BBN proceeds nicely under the influence of 2,6-di-*tert*-butylphenoxide to afford the α-chloronitriles in satisfactory yields (Eq. 10.2; Table 10.2) [2].

Table 10.1 α-alkylation and α-arylation of chloroacetonitrile with *B*-R-9-BBN under the influence of potassium 2,6-di-*t*-butylphenoxide [1]

R	Product	Yield (%)
n-Butyl	Hexanonitrile	76
2-Butyl	3-Methylpentanonitrile	65
Isobutyl	4-Methylpentanonitrile	57
Cyclopentyl	Cyclopentylacetonitrile	72
Cyclohexyl	Cyclohexylacetonitrile	77
exo-Norbornyl	2-Norbornylacetonitrile	65
Phenyl	Phenylacetonitrile	75

Table 10.2 Monoalkylation of dichloroacetonitrile under the influence of potassium 2,6-di-*tert*-butylphenoxide at 25 °C [2]

Organoborane R_3B or *B*-R-9-BBN	Product	Yield (%)
Triethyl	2-Chlorobutyronitrile	89
B-Ethyl	2-Chlorobutyronitrile	87
B-*n*-Butyl	2-Chlorohexanonitrile	75
B-Isobutyl	2-Chloro-4-methylpentanonitrile	73
B-*sec*-Butyl	2-Chloro-3-methylpentanonitrile	69
B-Cyclopentyl	2-Chloro-2-cyclopentylacetonitrile	76
B-Cyclohexyl	2-Chloro-2-cyclohexylacetonitrile	78

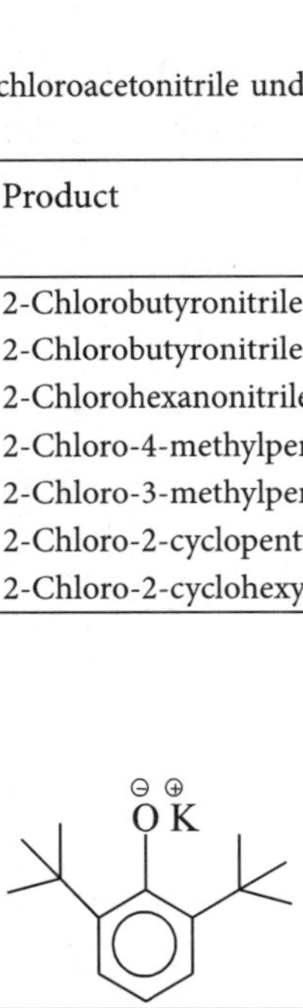

$$B\text{-R-9-BBN} + Cl_2CHCN \xrightarrow[25\,°C,\ 3\,h]{} RCHClCN \qquad (10.2)$$

The reaction is also highly satisfactory for introduction of two alkyl groups. The two identical alkyl groups are introduced by utilizing 2 mol of base and 2 mol of the organoborane. Alternatively, the dialkylacetonitriles with different alkyl groups are easily obtained in two successive monoalkylations (Chart 10.1) [2].

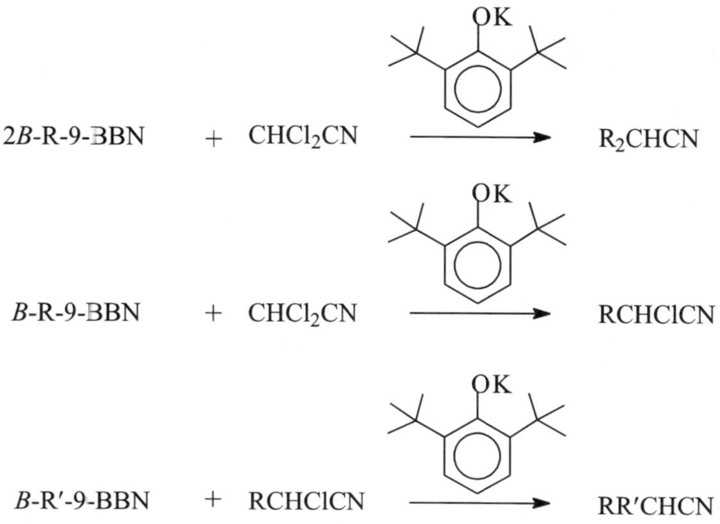

Chart 10.1

Two primary alkyl groups are readily introduced in this way, whereas introduction of more-hindered groups, such as isobutyl or cyclopentyl, is more sluggish and requires reflux in THF for an extended period. As the protonolysis of the boron intermediate in these cases is slow, a mixed solvent, *tert*-butylalcohol-THF (1:4), is used to liberate the desired product (Chart 10.2).

The introduction of two different alkyl groups is achieved by the successive alkylation of dichloroacetonitriles and without isolating the intermediate. The typical results are shown in Chart 10.2 [2].

The results of dialkylation are summarized in Table 10.3 [2].

The reaction is extended to the synthesis of dinitriles (Eq. 10.3) [3].

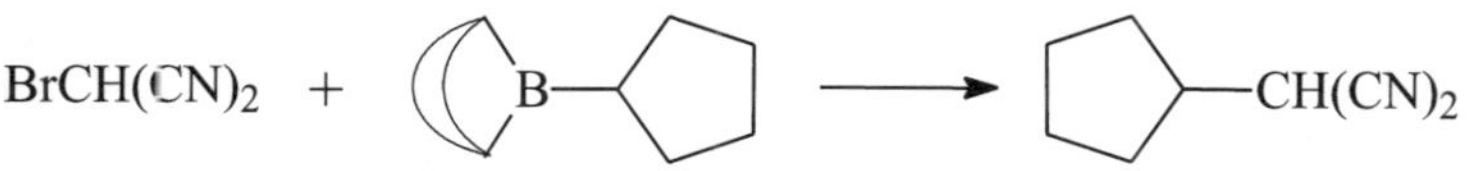

(10.3)

10.2 Synthesis of Chiral Nitriles

The chiral *B*-R*-9-BBN alkylates the chloroacetonitrile to afford in good yields the corresponding chiral nitriles. Sodium *tert*-butoxide in THF is best suited for the α-alkylation of chloroacetonitrile (Eq. 10.4) [4].

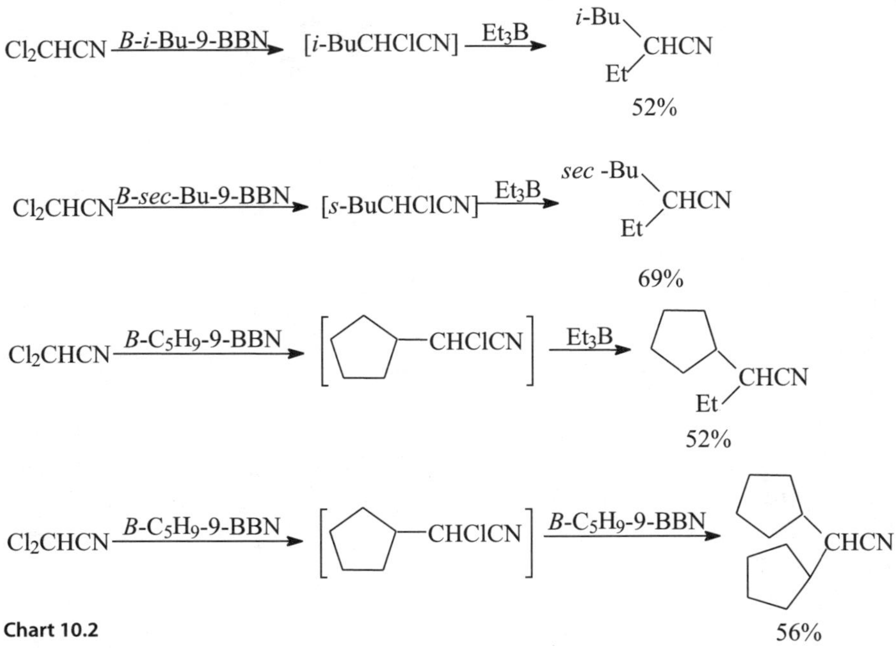

Chart 10.2

Table 10.3 Dialkylation of dichloroacetonitrile and alkylation of α-alkyl-α-chloroacetonitriles under the influence of potassium 2,6-di-*tert*-butylphenoxide [2]

Organoborane R_3B or *B*-R-9-BBN	Chloronitrile Cl_2CHCN or RCH(Cl)CN	Reaction conditions		Product	Yield (%)
		Temp (°C)	Time (h)		
Ethyl	Dichloro	25	3	Diethylacetonitrile	96
B-Ethyl	Dichloro	25	3	Diethylacetonitrile	97
B-*n*-Butyl	Dichloro	25	3	Di-*n*-butylacetonitrile	85
B-Isobutyl	Dichloro	Reflux	24	Diisobutylacetonitrile	66
B-*sec*-Butyl	Dichloro	Reflux	24	Di-*sec*-butylacetonitrile	46
Triethyl	Isobutylchloro	25	3	2-Ethyl-4-methyl-pentanonitrile	88
Triethyl	*sec*-Butylchloro	25	3	2-Ethyl-3-methyl-pentanonitrile	91
Triethyl	Cyclopen-tylchloro	25	3	2-Cyclopentylbutyronitrile	83
Triethyl	Cyclohexylchloro	25	3	2-Cyclohexylbutyronitrile	85
B-Cyclohexyl	*sec*-Butylchloro	Reflux	24	2-Cyclohexyl-3-methyl-pentanonitrile	61
B-Cyclopentyl	Cyclopen-tylchloro	Reflux	24	Dicyclopentylacetonitrile	56

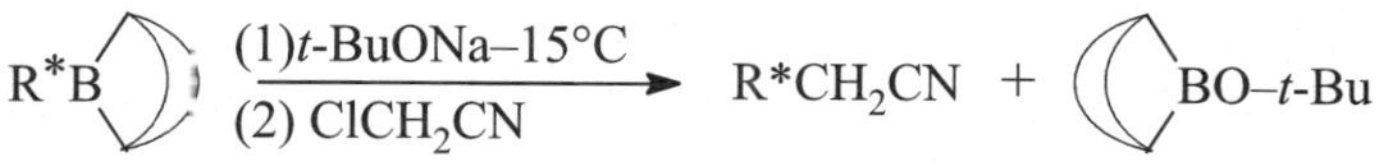

(10.4)

The results are summarized in Table 10.4 [4].

Table 10.4 Alkylation of chloroacetonitrile with optically active *B*-R*-9-BBN [4]

R*	Product	Yield (%)	ee (%)
[*R*]-3-Hexyl	[3*R*]-3-Ethylhexanenitrile	66	99
[1*S*,2*S*]-*trans*-2-Methylcyclopentyl	[1*R*,2*S*]-*trans*-2-Methylcyclopentylacetonitrile	48	99
[1*S*,2*S*,4*S*]-*exo*-2-Norbornyl	[1*S*,2*R*,4*R*]-*exo*-2-Norbornylacetonitrile	53	99

The methodology has been elegantly employed for the stereocontrolled synthesis of the side chain of steroids with natural configuration (Eq. 10.5) [5].

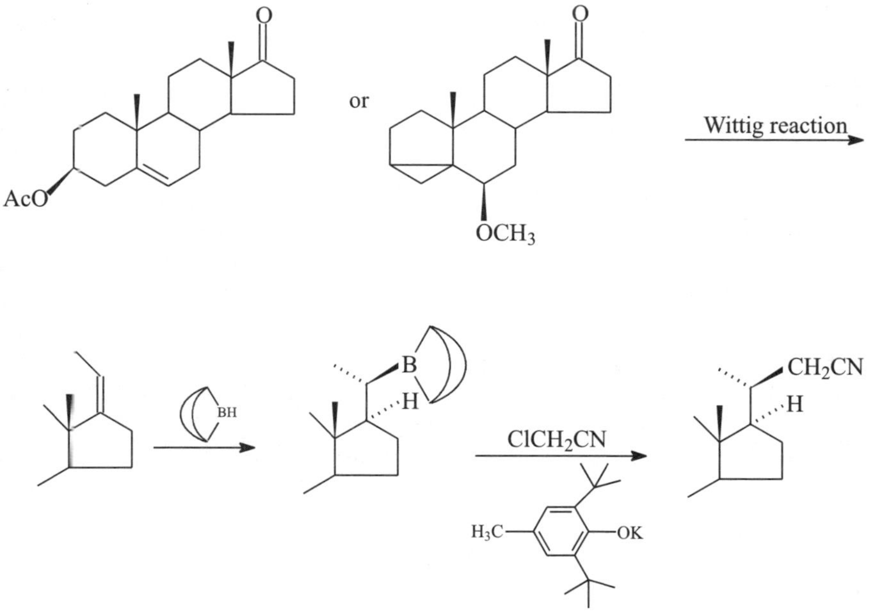

(10.5)

10.3 Synthesis of (*E*)-β,γ-Unsaturated Nitriles

B-trans-1-Alkenyl-9-BBN undergoes facile reaction with the α-halo carbanion generated from chloroacetonitrile in the presence of potassium 2,6-di-*tert*-butylphenoxide (Eq. 10.6) and yields the corresponding (*E*)-β,γ-unsaturated nitriles in good yields (Table 10.5) [6].

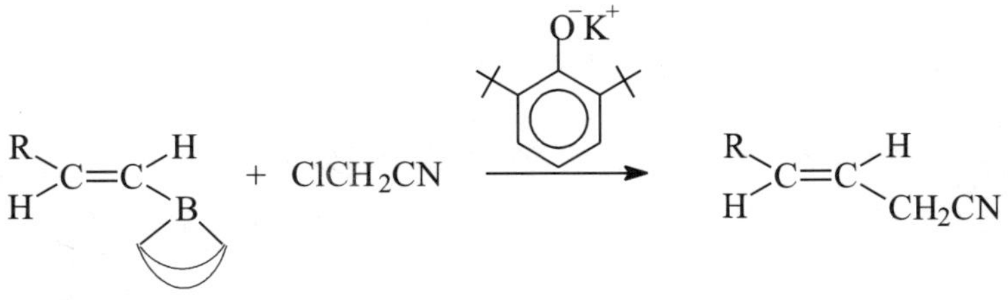
(10.6)

Table 10.5 β,γ-Unsaturated nitriles from a base-promoted α-alkenylation of chloroacetonitrile with *B*-*trans*-1-alkenyl-9-BBN derivatives [6]

B-Alkenyl-9-BBN	β,γ-Unsaturated ester	Yield (%)	Isomeric purity (%)
B-1-Hexenyl-9-BBN	(2*E*)-1-Cyano-2-heptene	70	99
B-3,3-Dimethyl-1-butenyl-9-BBN	(2*E*)-1-Cyano-4,4-dimethyl-2-pentene	69	98
B-2-Cyclohexyl-1-ethenyl-9-BBN	(2*E*)-1-Cyano-3-cyclohexyl-2-propene	65	96

References

1. Brown HC, Nambu H, Rogić MM (1969) J Am Chem Soc 91:6854
2. Nambu H, Brown HC (1970) J Am Chem Soc 92:5790
3. Nambu H, Brown HC (1970) Organomet Chem Syn 1:95
4. Brown HC, Joshi N. N, Pyun C, Singaram B (1989) J Am Chem Soc 111:1754
5. Midland MM, Kwon YC (1981) J Org Chem 46:228
6. Brown HC, Bhat NG, Campbell JB Jr (1986) J Org Chem 51:3398

11 Synthesis of (*E*)-β,γ-Unsaturated Amides

β,γ-Unsaturated amides can be prepared from allylamino [1], allylphosphate [2], or allylalcohol [3] with a one–carbon atom increase. Some of these are not stereodefined, and others require the use of valuable noble metal catalysts and carbon monoxide. However, it is possible to synthesize in a stereodefined manner (*E*)-β,γ-unsaturated amides from *E*-9-alkenyl-9-BBN with *N,N*-dialkyl(dimethylsulfuranylidene)acetamide [4] with a two–carbon increase. *N,N*,-Dialkyl-α-sulfonium substituted amides are easily prepared by the reaction of dimethylsulfide and *N,N*-dialkyl-α-bromoacetamide [5]. The terminal *E*-9-alkenyl-9-BBN generated [6] *in situ* reacts (Scheme 11.1) with *N,N*-dialkyl(dimet hylsulfuranylidene)acetamide, prepared from *N,N*-dialkyl-α-sulfonium–substituted amide and sodium hydroxide at 0 °C in THF. The reaction mixture after oxidation with H_2O_2/ŌAc affords (*E*)-β,γ-unsaturated amides in 60–70% yields (Table 11.1) [4]. The mechanism of the reaction is depicted in Scheme 11.2.

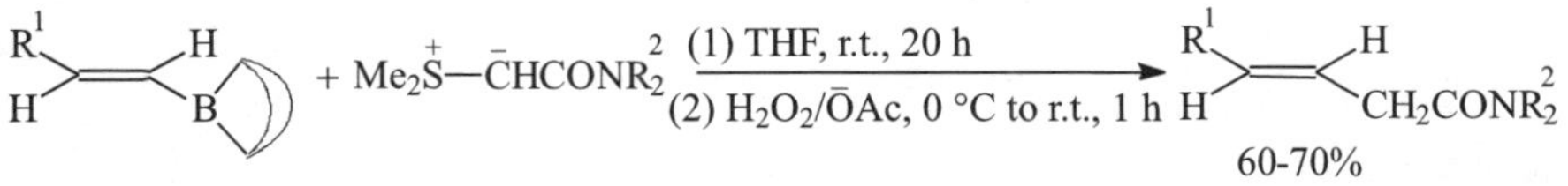

Scheme 11.1

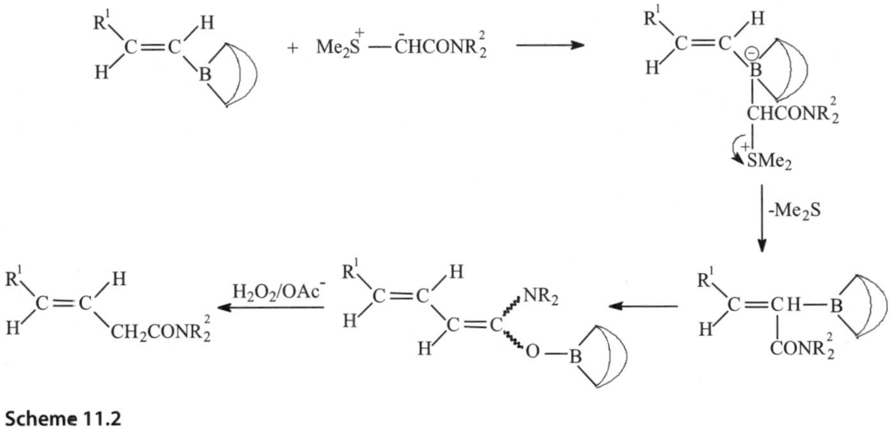

Scheme 11.2

Table 11.1 Yields of *N*,*N*-dialkyl-(*E*)-β,γ-unsaturated amides [4]

R^1	R^2	Yield (%)
Pr^n	Et	60
Pr^n	Pr^i	60
Bu^n	Et	61
Bu^n	Pr^i	61
n-Pent	Et	66
n-Pent	Pr^i	70
n-Hex	Et	64
n-Hex	Pr^i	65
n-Oct	Et	62
n-Oct	Pr^i	61
Ph	Et	69
Ph	Pr^i	67

Unlike the synthesis of (*E*)-β,γ-unsaturated esters [7], internal (*E*)-9-alkenyls-9-BBN do not react with *N*,*N*-dialkyl(dimethylsulfuranylidene)acetamide. This is attributed to the large steric bulkiness of both internal 9-alkenyl-9-BBN and *N*,*N*-dialkyl(dimethylsulfuranylidene)acetamide, which hinder the latter attack to internal 9-alkenyl-9-BBN.

References

1. Murahashi S-I, Imada Y, Nishimura K, (1988) J Chem Soc Chem Commun 1578
2. Murahashi S-I, Imada Y (1985) Chem Lett 1477
3. Buchi G, Cushman M, Wuest H (1974) J Am Chem Soc 96:5563
4. Deng M-Z, Li N-S, Huang Y-Z (1993) J Chem Soc Chem Commun 65
5. (a) Weaver WE, Whaley WM (1947) J Am Chem Soc 69:515; (b) Speziale AS, Tung CC, Ratts KW, Yao A (1965) ibid. 87:3460
6. (a) Jacob P III, Brown HC (1977) J Org Chem 42:579; (b) Wang KK, Scouten CG, Brown HC (1982) J Am Chem Soc 104:531
7. Deng M-Z, Li N-S, Huang Y-Z (1992) J Org Chem 57:4017

12 Synthesis of Amines

12.1 Synthesis of Chiral Amines

1,2- and 1,3-Asymmetric inductions in an acyclic system remain a pressing concern for synthetic organic chemists [1]. Reactions of allyl organometallic with chiral aldehydes or chiral imines yield products generally with not so high Cram selectivity [2] (Eq. 12.1).

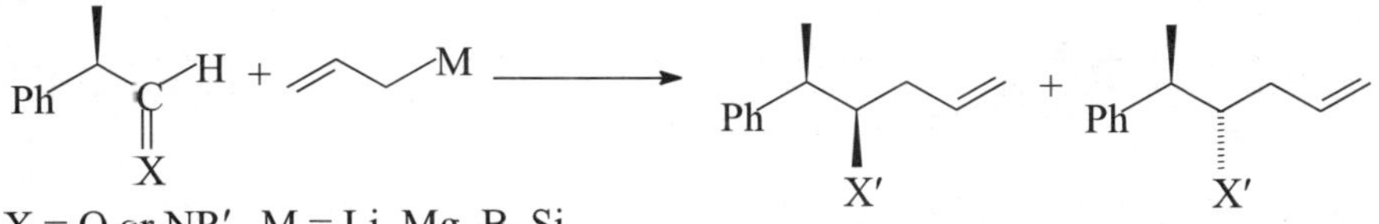

X = O or NR′ M = Li, Mg, B, Si...

1 Cram product **2** *anti*-Cram product

X′ = OH or NHR

(12.1)

Yamamoto and coworkers [3], however, have reported that reactions of allyl-9-BBN with certain chiral imines give the corresponding allylation products with very high enantiomeric excess, in essentially, the quantitative yields (Eq. 12.2). Table 12.1 summarizes these very high 1,2-asymmetric inductions. The main reason for poor selectivity of imine is the complex reactivity of imines toward other organometallic compounds [4–6].

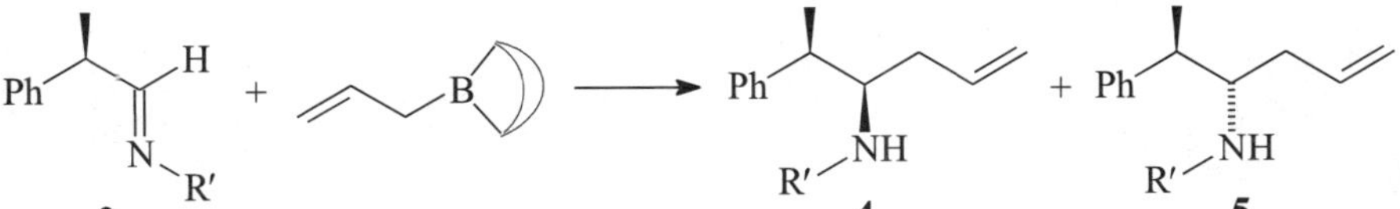

3 **4** **5**

Up to 100:0

Yield 88–98% (12.2)

Table 12.1 High Cram selectivity in 1,2-asymmetric induction [3]

Entry	Imine, **3** PhCH(CH_3)CH=NR′ R′	Allylorganometal (M)	Cram:*anti*-Cram (**4**):(**5**)
1	*n*-Pr	9-BBN	96:4
2	*i*-Pr	9-BBN	100:0
3	*n*-Pr	$SnBu_3/TiCl_4$	93:7
4	*i*-Pr	$SnBu_3/TiCl_4$	92:8
5	*n*-Pr	MgCl	84:16
6	*n*-Pr	MgBr	68:32
7	*i*-Pr	MgCl	70:30

Entry	Aldehyde	Allylorganometal	Cram:*anti*-Cram (**1**):(**2**)
8.	PhCH(CH3)CHO	9-BBN	55:45
9.		$SnBu_3/TiCl_4$	69:31
10.		MgCl	60:40

The reactions are carried out on 1-mmol scale and imine **3** and the aldehydes are used in a racemic form.

The high 1,2-asymmetric induction obtained with allyl-9-BBN is explained that the *trans* geometry [7] of aldimines necessarily forces the metal to coordinate the nitrogen atom *syn* to the alkyl group (R′), and hence the chiral center goes to the axial position of **6** (Fig. 12.1). As the conformation of the chiral center is fixed in **6** because of the steric influence of L and thus the allyl group attacks from the less hindered side (**6** and **7** or **7′**, Fig. 12.2).

The transition state **9** is highly unstable because of the steric repulsion between the methyl group and the 9-BBN ring protons. The bridged proton (H) of the 9-BBN ring covers the six-membered ring. The plane of the imine group

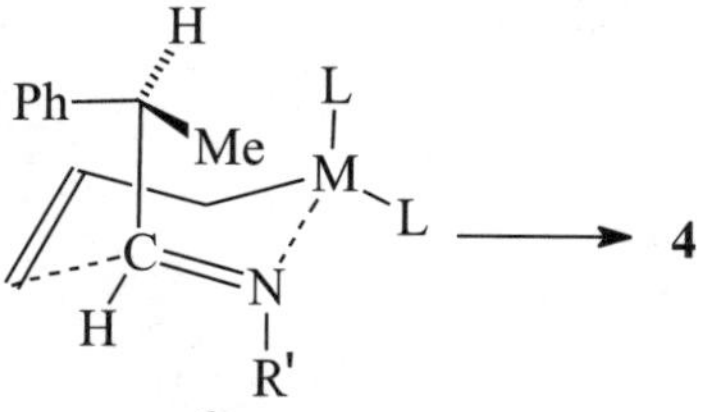

6

Fig. 12.1 Cyclic (chelate) model for 1,2-asymmetric induction – Cram product

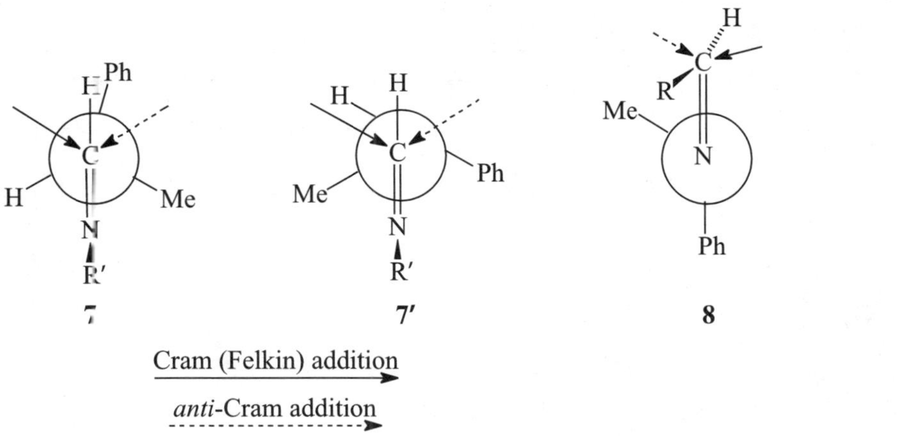

Fig. 12.2 Cram (Felkin) addition *vs anti* – Cram addition – Open chain model

(RCH=NR′) and the plane of the C–H bond intersect with nearly an orthogonal angle [8]. Thus, the *anti*-Cram attack leading to **5** becomes extremely unfavorable as depicted below:

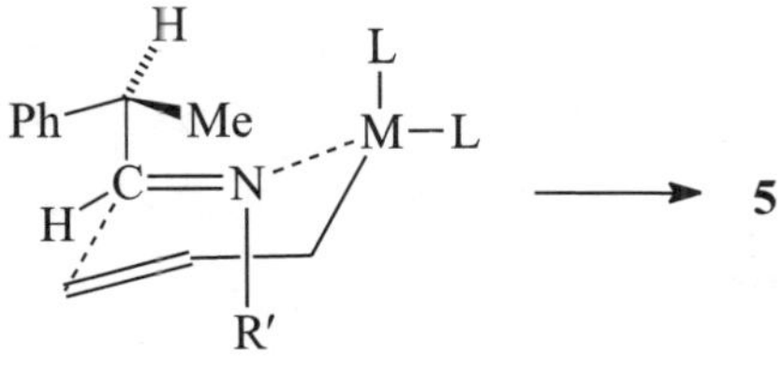

9

Yamamoto and coworkers [3] have extended this methodology for 1,3-asymmetric induction. The results of reaction of **10** with allylic organometallic compounds (Eq. 12.3) are summarized in Table 12.2 [3b].

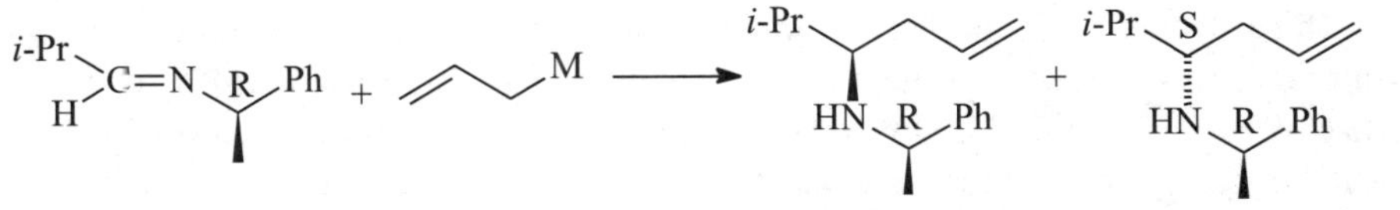

Cram product *anti*-Cram product

10 **11** **12**

(12.3)

Table 12.2 High 1,3-asymmetric[a] induction [3b]

Entry	Imine	Allylorganometal (M)[b]	Cram:*anti*-Cram (**11**):(**12**)
1	**10**	9-BBN	92:8
2	(*R*)- **10**	$B(OCH_3)_2$	60:40
3	(*R*)- **10**	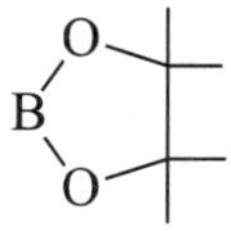	67:33
4	(*R*)- **10**	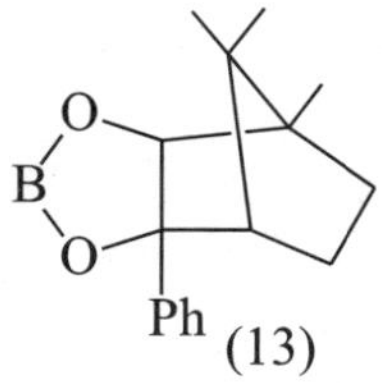 (13)	80:20
5	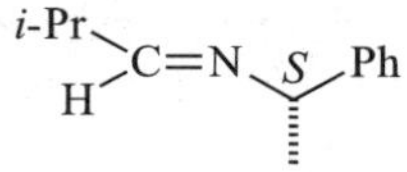	13	80:20
6	**10**	$SnBu_3/TiCl_4$	82:18
7	**10**	$SnBu_3/BF_3$	67:33
8	**10**	MgBr	80:20

[a] Normally, racemic 10 is used.
[b] Isolated yield with allyl-9-BBN and -MgBr, 88–98% and with allylstannane, 60–70%.

The reaction is reported to proceed through the conformation **14** in which the chiral center is fixed, and in this state the chiral group PhCH(Me) and L group have a set of 1,2-axial–equatorial interaction. The allyl group attacks from the less hindered side (**8** and **14**).

The transition state **15** is destabilized owing to the steric repulsion between the methyl group (and/or phenyl group) and L (Fig. 12.3).

The 1,3-diaxial interaction is normally greater than the 1,2-axial-equatorial interaction is. Apparently the selectivity of entry 2 (Table 12.1) is 100%, whereas that of entry 1 (Table 12.2) is 92%. To summarize, the modified Cram (or Felkin) model (**7** or **7′**) explains 1,2-asymmetric induction. For 1,3-asymmetric allylboration [7], the extended Cram model (**8**) is proposed.

The 1,2-asymmetric induction with crotyl organometallic compounds (M = B, Zr, and Mg) is not so high as with the allyl system. The reaction of crotylorganometal with **3** (R′ = *n*-Pr and *i*-Pr) produces all the possible four isomers.

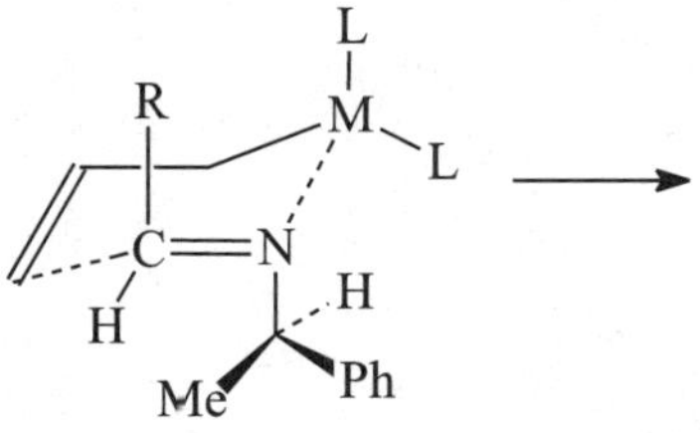

14 → **11** (Cram product)

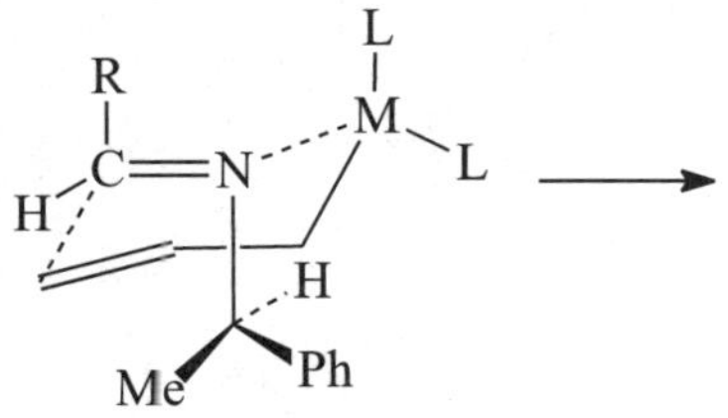

15 → **12** (*anti*-Cram product)

Fig. 12.3 Cyclic (chelate) models for 1,3- asymmetric induction

The reaction of **10** with crotyl-9-BBN is extremely sluggish. However, **16** reacts with crotyl-9-BBN to afford the Cram:*anti*-Cram in a 100:0 ratio. The *erythro* selectivity via **16**, which possesses the less bulky ethyl group, is also in good agreement with the previous result (Eq. 12.4) [9].

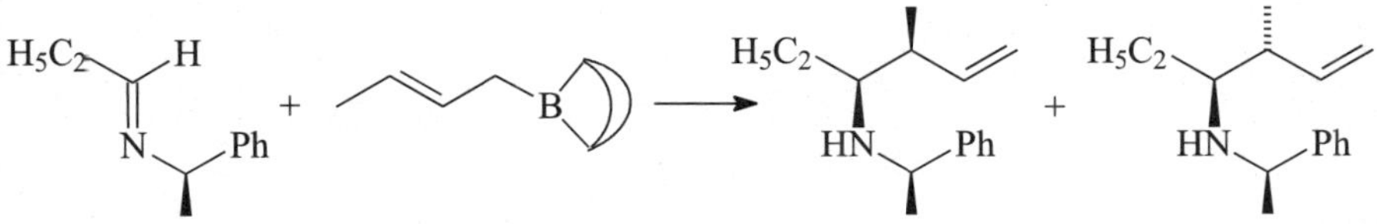

16 Cram-*erythro* Cram-*threo*

75:25

(12.4)

The reactions between the imines and the allenic organometallic compounds give the *threo* allylamines with very high stereoselectivity, while the reaction with 9-but-2-enyl-9-BBN gives the analogous *erythro* allylamines stereoselectivity (Scheme 12.1) [10].

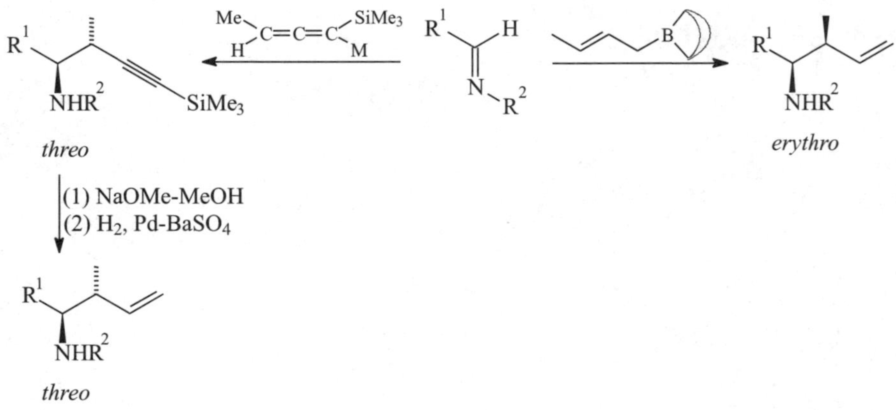

Scheme 12.1

12.2 Synthesis of Homopropargylic Amines

Brown and coworkers [11] have demonstrated that *B*-allenyl-9-BBN reacts vigorously with imines (Eq. 12.5) at room temperature and affords the corresponding homopropargylic amines in excellent yields along with a small amount of (2–4%) of the allenic amines (Table 12.3) [11].

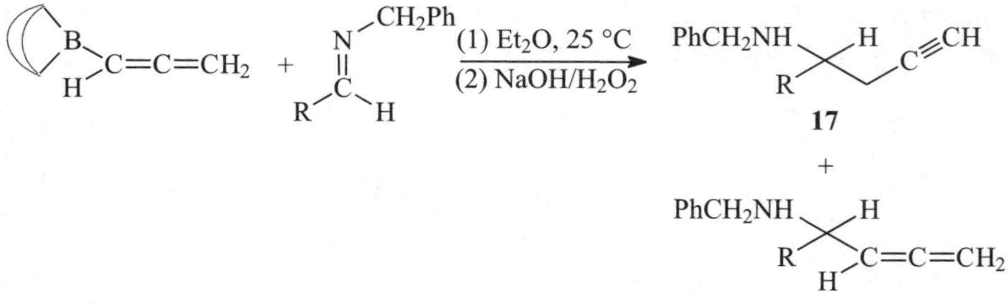

17

18

(12.5)

Table 12.3 Allenylboration of imines ($RCH{=}NCH_2Ph$) [11]

R	Product **17** (**18**)	Yield (%)
$(CH_3)_2CH$	96 (4)	90
Ph	98 (2)	84

12.3 Synthesis of Secondary Amines

2-Azanorbornenes undergo facile acid-catalyzed *retro* Diels–Alder reaction to primary amines [12] or N-methylated amines [13]. However, *retro* Diels–Alder can be interrupted to get cyclopentenylmethyl amines by complexation of 2-azanorbornenes with 9-BBN (Scheme 12.2; Table 12.4) [14].

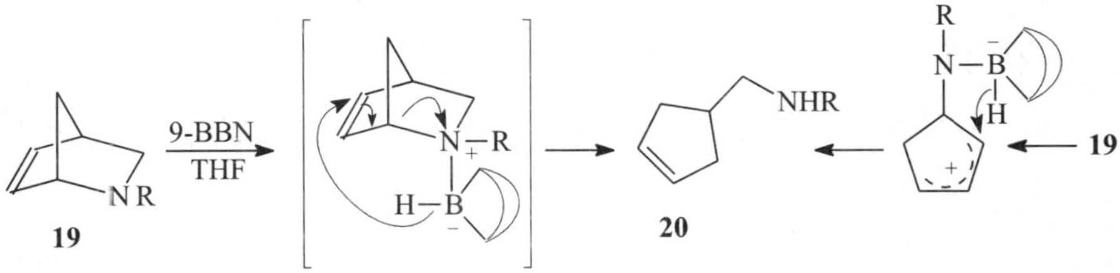

Scheme 12.2

Table 12.4 9-BBN-Mediated ring openings of azanorbornenes [14]

Entry	R	Time (h)	Yield (%)
1	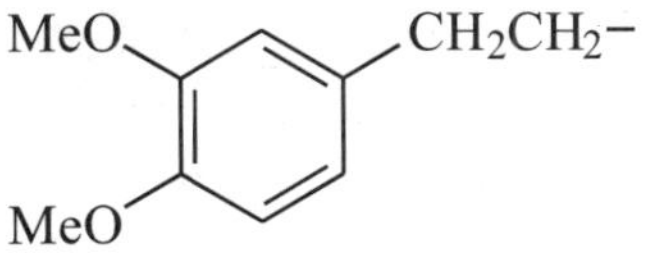	20	56
2	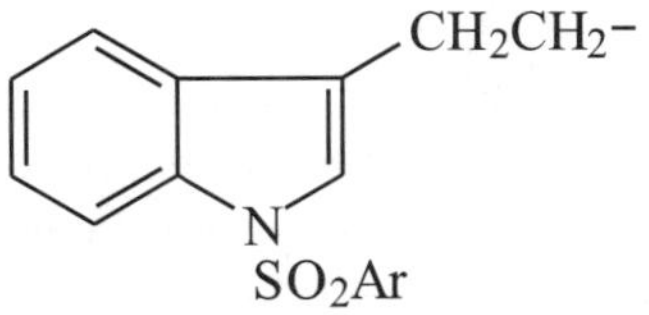	18	50
3	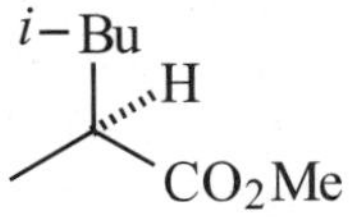	13	51

The methodology is extended to number of functionalized 3-substituted 2-azanorbornenes [15] (Table 12.5). The data reveal that *exo* substitution provides modest yield of cyclopentenylamines, whereas the *endo*-substituted azanorbornenes give poor yields of the amines. The low yield of *endo*-substituted substrate may be attributed to severe steric crowding on the *endo* face. In addition, the substituent at C-3 retains its geometry (Eq. 12.6).

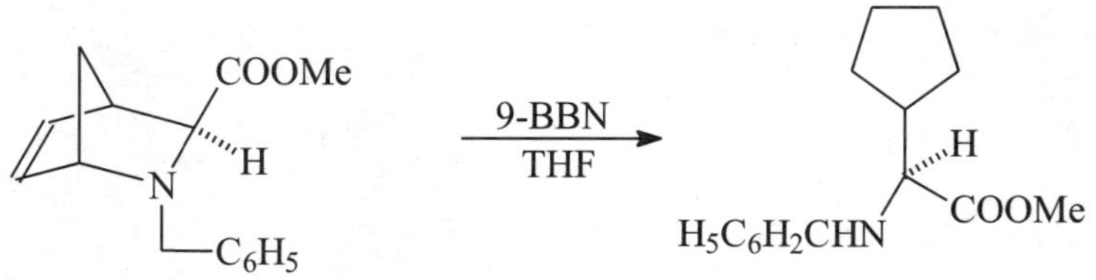

(12.6)

Table 12.5 Reaction of 3-substituted azanorbornenes with 9-BBN [15]

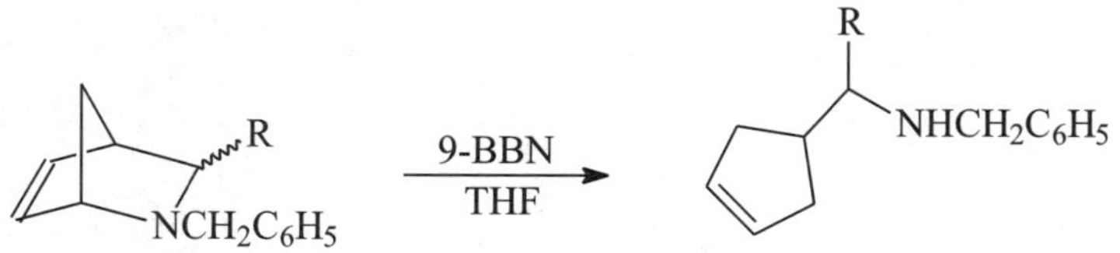

Entry	R	Time (h)	Yield (%)
1	–COOMe (*exo*)	0.5	44
2	–COOMe (*endo*)	1	16
3	$-CH_2OTBDMS$ (*exo*)	0.5	59
4	$-CH_2OMOM$ (*exo*)	0.5	55
5	$-CH_2OTBDMS$ (*endo*)	24	22

References

1. (a) Heathcock CH, (1984) In: Durst T, Buncel E (eds) Comprehensive carbanion chemistry. Elsevier, New York; (b) Evans DA, Nelson JV, Taber TR (1982) Topics in stereochemistry, vol. 1. Wiley InterScience, New York, p 1; (c) Morrison, JD (ed) (1983) Asymmetric synthesis, vol. 2. Academic, New York; (d) Masamune S (1981) In: Trost BM, Hutchinson R (eds) Organic synthesis today and tomorrow. Pergamon, New York, p 197; (e) Hoffmann RW (1982) Angew Chem Int Ed Engl 21:555; (f) Yamamoto Y, Maruyama K (1982) Heterocycles 18:357
2. Excellent 1,2- and 1,3-asymmetric induction through chelation control has been realized in the nucleophillic addition of organometallics to α- and β-alkoxy–substituted aldehydes. Reetz MT, Kesseler K, Schmidtberger S, Wenderoth B, Steinbach R (1983) Angew Chem Int Ed Engl 22:989; Reetz MT, Jung A (1983) J Am Chem Soc 105:4833; Still, WC, Schneider JA (1980) Tetrahedron Lett 21:1035; Still WC, Mcdonald JH (1980) ibid. 21:1031
3. (a) Yamamoto Y, Komatsu T, Maruyama K (1984) J Am Chem Soc 106:5031; (b) Yamamoto Y, Nishi S, Maruyama K, Komatsu T, Ito W (1986) J Am Chem Soc 108:7778
4. (a) Layer RW (1963) Chem Rev 63:489; (b) Stork G, Dowd S (1963) J Am Chem Soc 85:2178
5. Scully FE Jr (1980) J Org Chem 45:1515

6. Thies H, Schonenberger H (1956) Chem Ber 89:1918
7. McCarty CG (1970) In: Patai S (ed) The chemistry of the carbon–nitrogen double bond. InterScience, New York, p 363
8. Yamamoto Y, Yatagai H, Maruyama K (1981) J Am Chem Soc 103:3229
9. Yamamoto Y, Komatsu T, Maruyama K (1985) J Org Chem 50:3115
10. Yamamoto Y, Ito W, Maruyama K (1984) J Chem Soc Chem Commun 1004
11. Brown HC, Khire UR, Narla G, Racherla US (1995) J Org Chem 60:544
12. Grieco PA, Parker DT, Fobare WF, Ruckle R (1987) J Am Chem Soc 109:5859; Grieco PA, Clark JD (1990) J Org Chem 55:2271
13. Grieco PA, Bahsas A (1987) J Org Chem 52:5746
14. Gaul MD, Fowler KW, Grieco PA (1993) Tetrahedron Lett 34:3099
15. Grieco PA, Larsen SD, Fobare WF (1986) Tetrahedron Lett 27:1975

13 Synthesis of Halides

13.1 Synthesis of Halides *via* Hydroboration

Organic halides are important synthetic intermediates in numerous reactions. They are also important due to the variety of radiohalogen-containing pharmaceuticals that have been developed in recent years [1–3]. The trialkylboranes prepared *via* hydroboration are conveniently converted to alkyl halides when 1 equiv of sodium hydroxide or sodium methoxide in methanol is added to a mixture of 1 mol of organoborane and 1 mol of iodine. The reaction becomes instantaneous and affords the corresponding alkyl halides. A second mole of iodine and base react similarly (Eq. 13.1) [4–6], but the third alkyl resists the reaction under these conditions.

$$R_3B + 2\,I_2 + 2\,NaOH \longrightarrow 2\,RI + 2\,NaI + RB(OH)_2 \qquad (13.1)$$

These difficulties are circumvented to some extent for the synthesis of secondary alkyl bromides by the action of bromine, in dark conditions, on the *B*-alkyl-9-BBN derivatives (Eqs. 13.2, 13.3) [7].

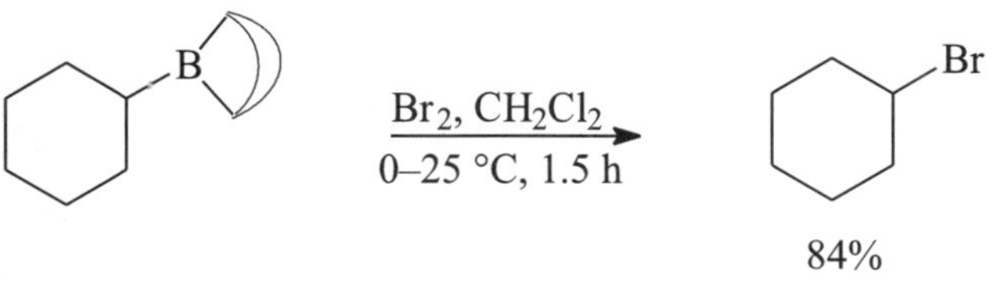

(13.2)

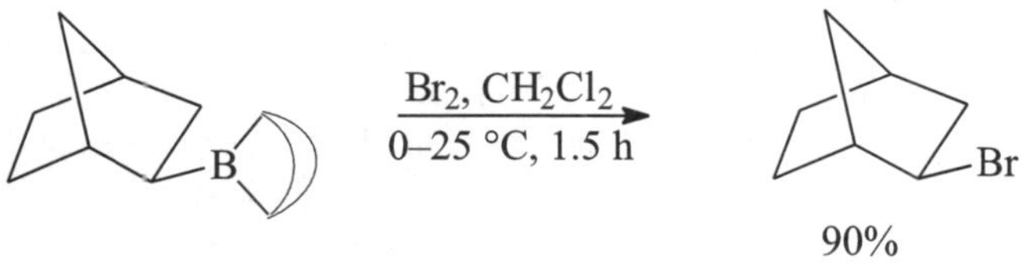

(13.3)

It is interesting to note that the reaction does not proceed through a direct rupture of the boron–carbon bond, but involves a highly selective free radical substitution of the α-hydrogen of the *sec*-alkyl group by bromine. Protonolysis of the intermediate by hydrogen bromide gives the final product (Chart 13.1) [8].

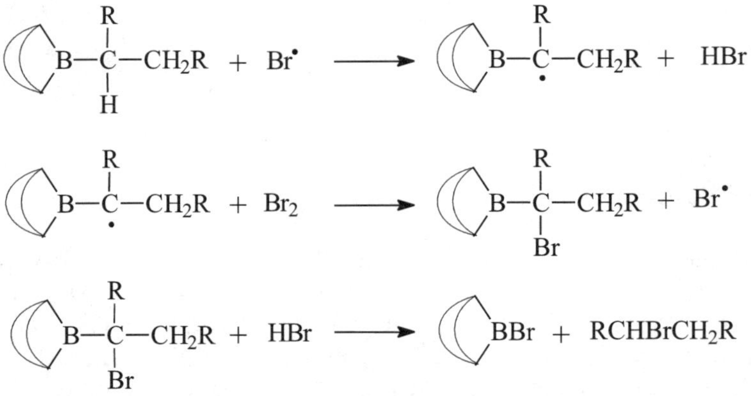

Chart 13.1

B-Isopropyl-9-BBN is 5.5 times more reactive as compared with the α position in cumene and some 660 times more reactive than is the *tertiary* position in isobutene [8].

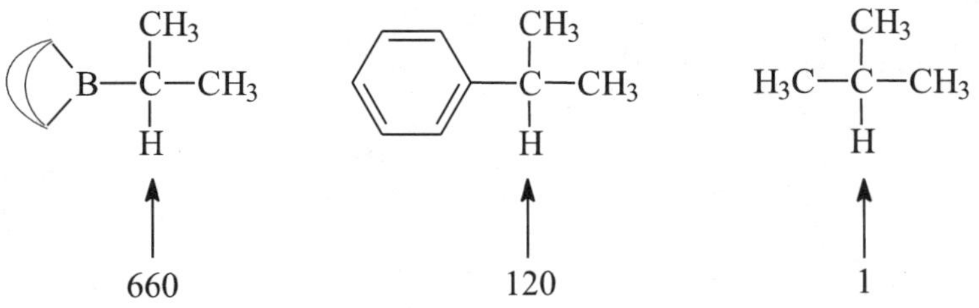

The remarkable activation of *sec*-alkyl of *B*-*sec*-alkyl-9-BBN is attributed to the stabilization of the free radical, owing to the interaction of the odd electron with the vacant *p* orbital of boron atom [8]. Attack on the two α-bridgehead positions is less facile because in these positions the odd electron would necessarily occupy an orbital that is orthogonal to the vacant *p* orbital on boron.

It has been found [9] during allylboration of acid chlorides, if the reaction mixture is not worked up immediately after its completion, it leads to the formation of *tertiary* chloride (Chart 13.2). The reaction probably occurs due to the cleavage of the borinate ester with hydrogen chloride formed by trace hydrolysis of the acid chloride or *B*-Cl-9-BBN, produced during the reaction. The *tertiary* and benzylic borinate esters are known to undergo carbon–oxygen cleavage with hydrogen halide [10].

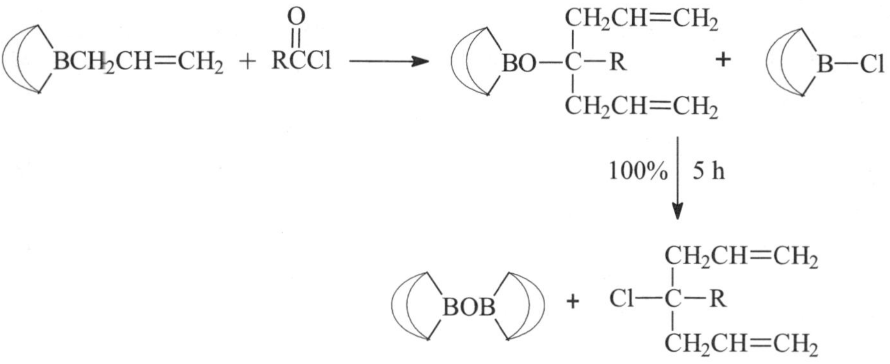

Chart 13.2

13.2 Synthesis of Halides *via* Haloboration

13.2.1 Synthesis of 2-Halo-1-Alkenes

The stereodefined alkenyl halides are of prime importance due to the developments of di- or trisubstituted alkene synthesis by cross-coupling reactions between organometallics and alkenyl halides, catalyzed by transition metal compounds [1–4]. 1-Halo-1-alkenes are conveniently synthesized *via* hydroboration–halogenation reaction [5]. However, with these methods it is not possible to synthesize 2-halo-1-alkenes. Although the halometalation reaction is considered to be a powerful tool for the synthesis of 2-halo-1-alkenes, the reaction has not been adequately developed for such purpose [6]. Lappert and coworkers [7–10] are perhaps the first to report the systematic haloboration reactions of unsaturated hydrocarbons with BX_3. However, there is no systematic approach to the haloboration with *B*-X-9-BBN to organic synthesis [11]. The desired *B*-X-9-BBN is prepared [12] by the reaction of 9-BBN with phosphorous pentachloride, bromine, or HI. Alternatively, *B*-X-9-BBN are very conveniently obtained by the hydroboration of 1,5-cyclooctadiene with stable methylsulfide complexes of monochloro-, monobromo-, and monoiodoboranes as ($H_2BX{\cdot}SMe_2$) [13]. This process, however, affords a mixture of *B*-halo-9-borabicyclo[3.3.1]nonanes (*B*-X-9-BBN) and a thermodynamically less stable [4.2.1]isomer, which predominates in the mixture. The less stable isomer, however, is readily converted to the more stable isomer *B*-X-9-BBN by gentle heating. The methylsulfide complexes are isolated as stable crystalline solids. Distillation, followed by addition of 1 equiv of the respective borontrihalide, provides *B*-X-9-BBN free from SMe_2 (Scheme 13.1) [13]. 9-X-9-BBNs have a great promise for synthetic use [12–14]. It has been found that *B*-Br-9-BBN·SMe_2 does not haloborate the carbon–carbon triple bond; the free *B*-Br-9-BBN generated by treatment with tribromoborane, however, reacts smoothly with 1-alkynes (Eq. 13.4).

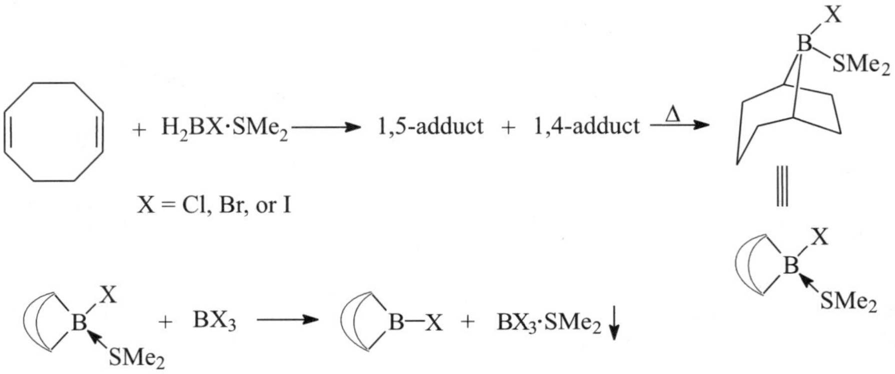

Scheme 13.1

Consequently, it is reported that *B*-bromo-9-borabicyclo[3.3.1]nonane (*B*-Br-9-BBN), and *B*-iodo-9-borabicyclo[3.3.1]nonane (*B*-I-9-BBN) haloborate the 1-alkynes, stereo-, regio-, and chemoselectively and after protonolysis; 2-halo-1-alkenes are obtained in excellent yields (Eq. 13.4; Table 13.1) [15].

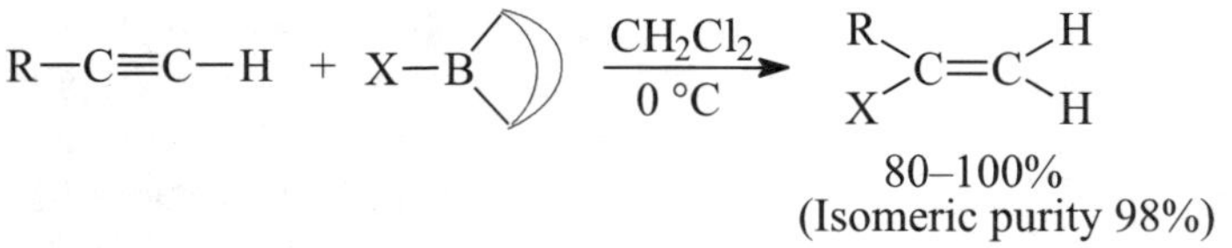

(13.4)

X = Br or I

The experiments depicted in Chart 13.3 confirm that haloboration proceeds via Markovnikov *cis* addition of halogen and boron to the terminal triple bond. Moreover, it is concluded that *B*-Br-9-BBN attacks only the terminal carbon–

carbon triple bond, while internal carbon–carbon triple bonds, and terminal and internal C=C bonds [15] withstand the bromoboration reaction conditions (Chart 13.4) [15]. However, *B*-I-9-BBN is little more reactive and less chemoselective for such multiple bonds.

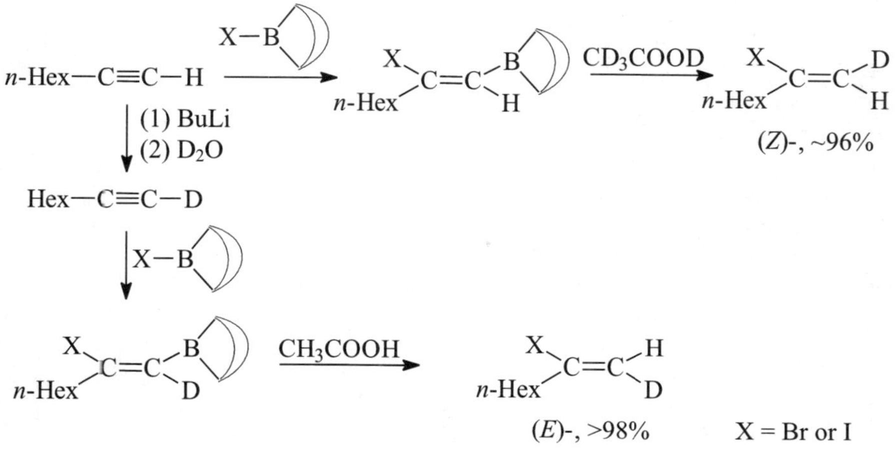

Chart 13.3

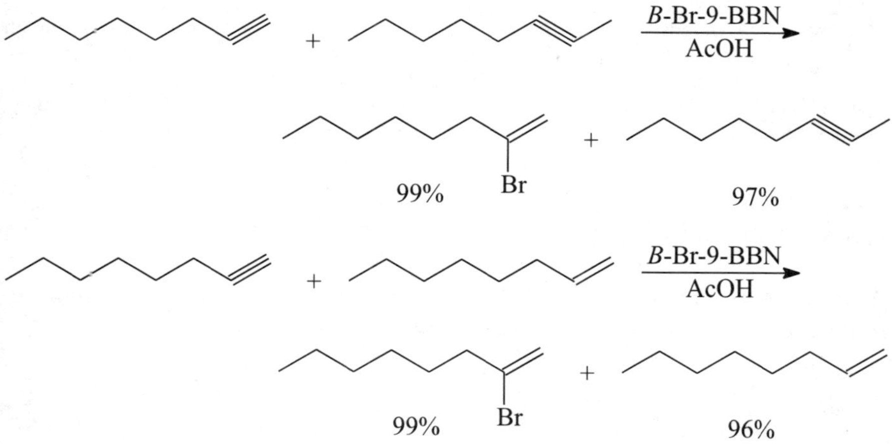

Chart 13.4

The results are summarized in Table 13.1 [15].

Table 13.1 The synthesis of 2-bromo- and 2-iodo-1-alkenes [15]

Entry	Alkyne	Method[a]	Product	Yield (%)	Regio-selectivity (%)
1	1-Hexyne	A	2-Bromo-1-hexene	99	99
2	1-Hexyne	B	2-Iodo-1-hexene	100	99
3	1-Octyne	A	2-Bromo-1-octene	95	99
4	1-Octyne	B	2-Iodo-1-octene	100	99
5	1-Decyne	A	2-Bromo-1-decene	88	99
6	1-Decyne	B	2-Iodo-1-decene	100	99
7	Phenylethyne	A	α-Bromostyrene	95	99
8	Phenylethyne	B	α-Iodostyrene	85	99
9	1,6-Heptadiyne	C	2,6-Dibromo-1,6-hepta diene	82	98
10	1,6-Heptadiyne	D	2,6-Diiodo-1,6-hepta diene	80	98
11	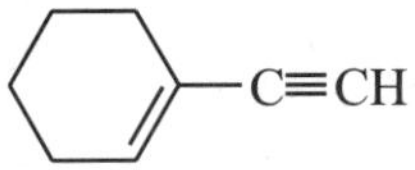	A	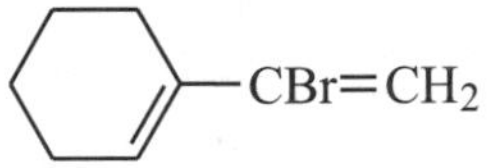	81	98
12	Cyclohexylethyne	A	2-Bromo-2-cyclo-hexylethene	93	99
13	Cyclohexylethyne	B	2-Iodo-2-cyclohexylethene	94	99
14	$HC{\equiv}C(CH_2)_8COOMe$	E	$H_2C{=}CBr(CH_2)_8COOMe$	88	99
15	Propargyl bromide	F	2,3-Dibromo-1-propene	90	98
16	Propargyl bromide	B	3-Bromo-2-iodo-1-propene	88	98
17	$HC{\equiv}C(CH_2)_3OAc$	E	$H_2C{=}CBr(CH_2)_3OAc$	82	99

[a] Method A: The amount of *B*-Br-9-BBN is 1.2 equiv, and the reaction is carried out at 0 °C for 3 h in CH_2Cl_2. Method B: *B*-I-9-BBN is 1 equiv and at –20 °C for 1 h in pentane. Method C: *B*-Br-9-BBN is 2.4 equiv and at 0 °C for 3 h in CH_2Cl_2. Method D: *B*-I-9-BBN is 2.1 equiv and at –20 °C for 1 h in pentane. Method E: *B*-Br-9-BBN is 2.2 equiv and at 0 °C for 3 h in CH_2Cl_2. Method F: *B*-Br-9-BBN is 1.5 equiv and at room temperature for 1 h in CH_2Cl_2.

13.2.2 Synthesis of (*Z*)-1-Alkynyl-2-Halo-1-Alkenes

As the haloboration occurs stereospecifically, the reaction is extended for a stereodefined synthesis of (*Z*)-1-alkynyl-2-halo-1-alkenes (Eq. 13.5) [16].

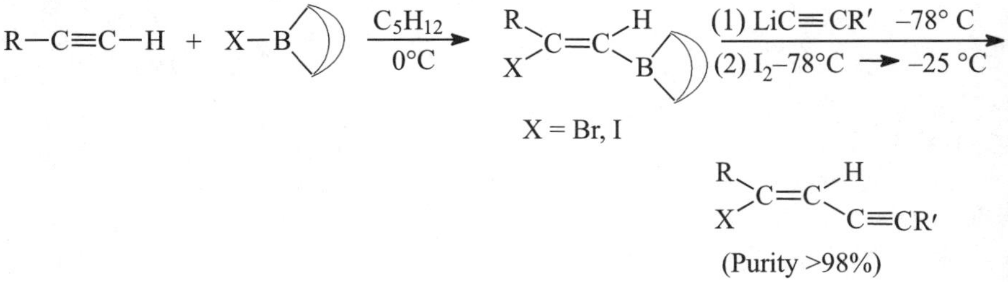

(13.5)

The reaction proceeds through the following pathway (Scheme 13.2). The results are summarized in Table 13.2 [16].

Table 13.2 The synthesis of [*Z*]-1-alkynyl-2-halo-1-alkenes from 1-alkynes [16]

R	R′	X	Yield (%)	Selectivity (%)
Hexyl	Butyl	Br	77	98
Hexyl	Butyl	I	74	99
	Butyl	Br	62	98
Butyl	Hexyl	Br	66	98
Butyl	Butyl	Br	66	98
Butyl	Butyl	I	83	98
Hexyl	Ethyl	Br	65	99
Hexyl	Ethyl	I	70	98
Ethyl	Hexyl	Br	63	98
Ethyl	Hexyl	I	67	98
Phenyl	Butyl	Br	53	99
Hexyl	$PhOCH_2$	Br	62	98
Hexyl	$PhOCH_2$	I	73	98
Cyclohexyl	Butyl	I	72	98

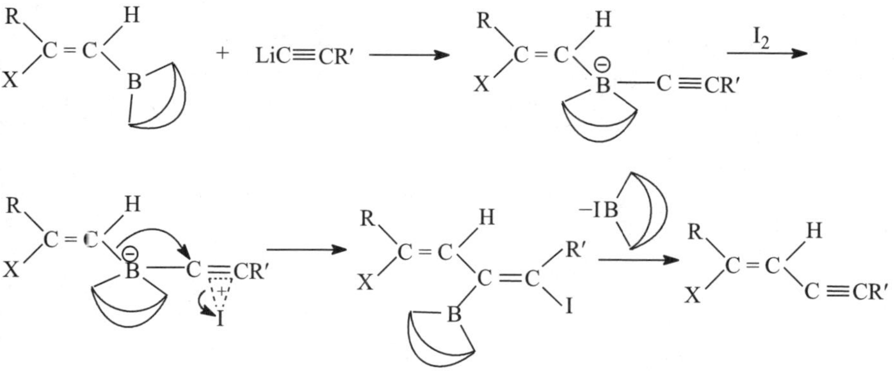

Scheme 13.2

13.2.3
Synthesis of (*Z*)-δ-Halo-γ,δ-Unsaturated Ketones

δ,δ-Disubstituted-γ,δ-unsaturated ketones are important synthetic intermediates for synthesis of natural products such as terpenoids. For their synthesis,

Michael-type additions to alkenylmetallics have serious defects because of the inapplicability due to readily-polymerizability of acyclic α,β-unsaturated ketones [17]. Organoboranes, on the other hand, give good results, even for acyclic enones [18], but there is no direct and general method for the synthesis of β,β-disubstituted alkenylboranes. However, cross-coupling reaction [1–4] between δ-halo-γ,δ-unsaturated ketones and organometallics do afford the δ,δ-disubstituted-γ,δ-unsaturated ketones.

Haloalkenylboranes are far less reactive, compared with simple alkenylboranes, presumably owing to the electronegative halogen substituent. It is found that the reaction of *B*-(*Z*)-2-bromo-1-heptenyl-9-BBN with methylvinylketone (MVK) gives the desired haloketone in poor yield (*ca.* 10%) under the known conditions [19]. In sharp contrast, the reaction takes place smoothly under Lewis acid conditions (in pentane, 100% excess of *B*-X-9-BBN). Unexpectedly, the major product isolated after the reaction with MVK under aforementioned conditions is not enone, but the aldol. All the attempts to circumvent the unexpected aldol condensation proved futile. However, enone is obtained selectively from aldol *via retro* aldol condensation on heating with a base (NaOH) (Scheme 13.3; Table 13.3) [20]. No aldol adducts are formed in the case of enones other than methylvinylketone. Consequently, this method provides

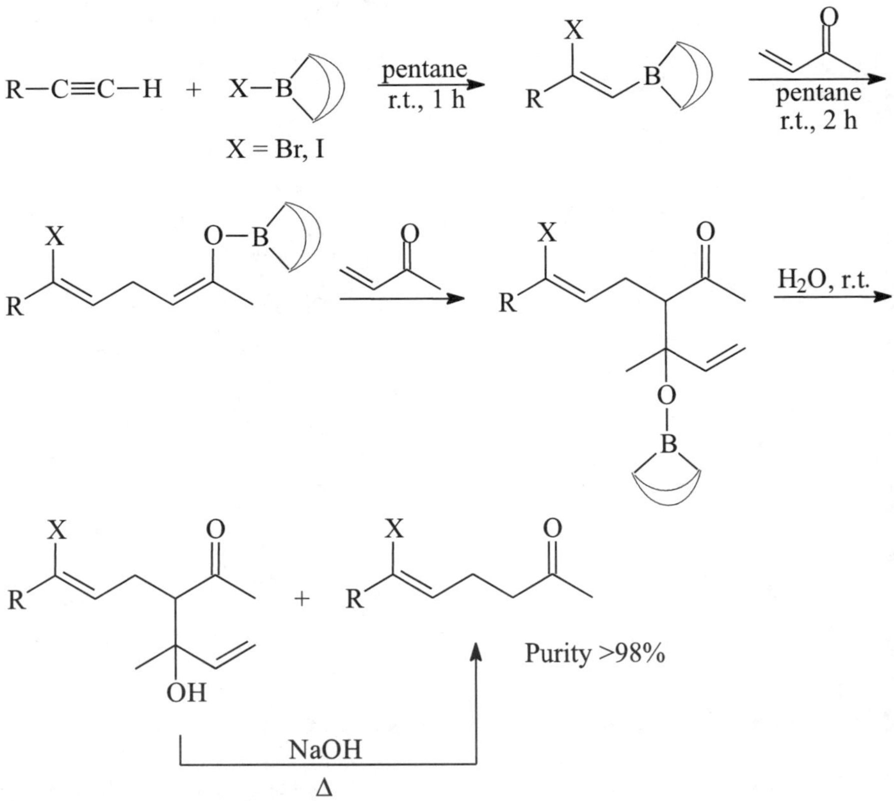

Scheme 13.3

a general route [20] to the (*Z*)-δ-halo-γ,δ-unsaturated ketones, inaccessible *via* conventional procedures. It is significant to mention that reaction with α-(trimethylsilyl)butanone [21] gives the 1,4-adduct directly without *retro*-aldol treatment (entry 9, Table 13.3).

Table 13.3 Synthesis of (*Z*)-δ-halo-γ,δ-unsaturated ketones [20]

Entry	X	Alkyne	Enone	Product[a]	Yield[b] (%)	Isomeric purity (%)
1	Br	1-Decyne	MVK[a]	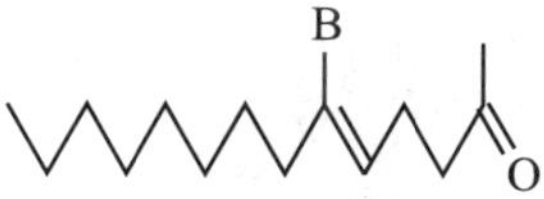	83	99
2	Br	1-Heptyne	MVK[a]	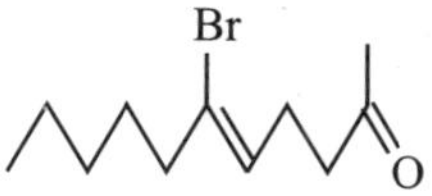	83	99
3	Br	Phenyl-ethyne	MVK[a]	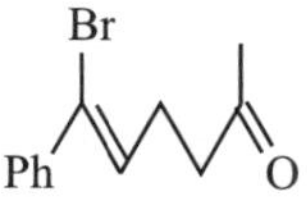	83	99
4	I	1-Octyne	MVK[a]		92	99
5	Br	1-Hexyne	PhCH=CHCOCH$_3$	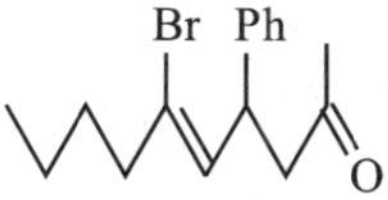	64	98
6	Br	1-Hexyne	PhCH=CHCOPh	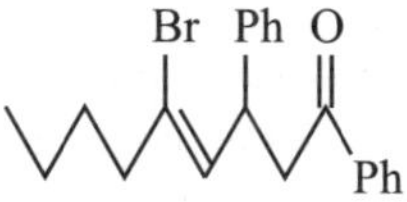	72	98
7	I	1-Hexyne	PhCH=CHCOPh	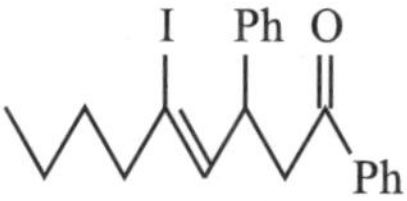	79	99
8	I	1-Octyne	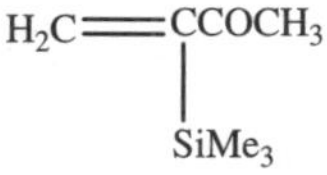		52[b]	99
9	I	BrCH$_2$C≡CH	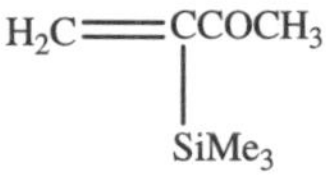		80[b]	99

[a] After the reaction with MVK, the mixture is treated with a base to effect retro-aldol condensation.
[b] The Me_3Si group is cleaved during workup.

The utility of haloboration-conjugate addition sequence has been demonstrated in the selective synthesis of several natural products such as sulcatol, *trans*-geranylacetone, and *trans*-nerolidol (Scheme 13.4) [20].

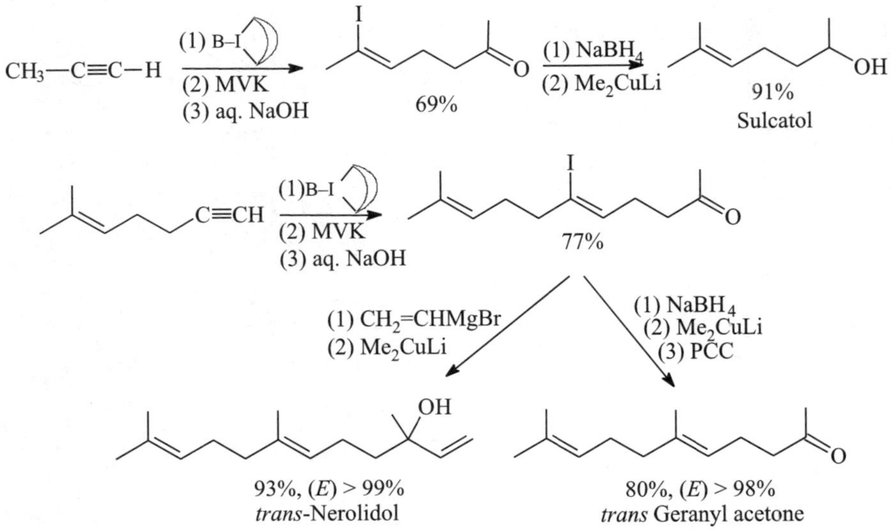

Scheme 13.4

References

Section 13.1

1. Lambrecht RM, Wolf AP (1975) In: Subramanian G, Rhodes BA, Cooper JF (eds) Radiopharmaceuticals. Society of Nuclear Medicine, New York, p 109
2. Heindel ND, Burns HD, Honda T, Brady LW (eds) (1978) The chemistry of radiopharmaceuticals. Masson, New York
3. Mazaitis JK, Gibson RE, Komai T, Eckelman WC, Francis B, Reba RC (1980) J Nucl Med 21:142
4. Brown HC, Rathke MW, Rogić MM (1968) J Am Chem Soc 90:5038
5. Brown HC, De Lue NR, Kabalka GW, Hedgecock HC Jr (1976) J Am Chem Soc 98:1290
6. De Lue NR, Brown HC (1976) Synthesis 114
7. Lane CF, Brown HC (1971) J Organomet Chem 26:C51
8. Brown HC, De Lue NR (1974) J Am Chem Soc 96:311
9. Kramer GW, Brown HC (1977) J Org Chem 42:2292
10. Abel EW, Gerrard W, Lappert MF (1957) J Chem Soc 112:3833

Section 13.2

1. Suzuki A (1982) Acc Chem Res 15:182
2. Negishi E, Matsushita H, Okukado N (1981) Tetrahedron Lett 2715
3. (a) Rand CL, Van Horn DE, Moore MW, Negishi E (1981) J Org Chem 46:4093; (b) Normant JF, Alexakis A (1981) Synthesis 841
4. (a) Collman JP, Hegedus LS (1980) Principles and applications of organotransition metal chemistry. University Science, Mill Valley, California; (b) Negishi E-I, (1982) Acc Chem Res 15:340; (c) Davies SG (1982) Organotransition metal chemistry: applications to organic synthesis. Pergamon, Oxford; (d) Hegedus LS (1983) Group VIII transition metals in organic synthesis. In: Scheffold R (ed) Modern synthetic methods, vol. 3. Otto Salle, Frankfurt, p 61; (e) Posner GH (1980) An introduction to synthesis using organocopper reagents. Wiley, New York, p 72; (f) Tsuji J (1980) Organic synthesis with palladium compounds. Springer, Berlin Heidelberg New York; (g) Jolly PW (1982) Nickel-catalyzed coupling of organic halides and related reactions. In: Wilkinson G, Stone FGA, Abel EW (eds) Comprehensive organometallic chemistry, vol. 8. Pergamon, Oxford, p 713; (h) Trost BM (1982) Organopalladium compounds in organic synthesis and in catalysis. In: Wilkinson G, Stone FGA, Abel EW (eds) Comprehensive organometallic chemistry, vol. 8. Pergamon, Oxford, p 799
5. Suzuki A, Dhillon RS (1986) Top Curr Chem 130:23
6. Negishi E (1980) Organometallics in organic syntheses, vol. 1. Wiley, New York
7. Lappert MF (1960) Angew Chem 72:36
8. Joy F, Lappert MF (1960) Proc Chem Soc 353
9. Joy F, Lappert MF, Prokai B (1966) J Organomet Chem 5:506
10. Lappert MF, Prokai B (1964) ibid. 384
11. Suzuki A (1986) Pure Appl Chem 58:629 and references cited therein
12. Brown HC, Kulkarni SU (1979) J Organomet Chem 168:281
13. Brown HC, Kulkarni SU (1979) J Org Chem 44:2422
14. Bhatt MV (1978) J Organomet Chem 156:221
15. Hara S, Dojo H, Takinami S, Suzuki A (1983) Tetrahedron Lett 24:731
16. Hara S, Satoh Y, Ishiguro H, Suzuki A (1983) Tetrahedron Lett 24:735
17. (a) Marfat A, McGuirk PR, Helquist P (1979) J Org Chem 44:3888; (b) Hooz J, Layton RB (1973) Can J Chem 51:2098
18. Negishi E (1982) In: Wilkinson G, Stone FGA, Abel EW (eds) Comprehensive organometallic chemistry, vol. 7. Pergamon, Oxford, p 291
19. Jacob P III, Brown HC (1976) J Am Chem Soc 98:7832
20. Satoh Y, Serizawa H, Hara S, Suzuki A (1985) J Am Chem Soc 107:5225
21. Boeckman RK Jr, Blum DM, Ganem M, Harvey N (1977) Org Synth 58:152

14 Synthesis of Dialkylsulfides

Organoboranes participate in either ionic or radical reactions [1]. However, there are a few reports on the applications of organoboranes as initiators for radical reactions [2, 3]. It is reported [4] that in the presence of a catalytic amount of 9-BBN or *B*-hexyl-9-BBN, or to some extent *B*-methoxy-9-BBN, initiates the radical addition of alkanethiols to alkenes under very mild conditions to provide the corresponding dialkylsulfides almost in quantitative yields (Scheme 14.1). On the other hand, the radical reactions of alkenes with thiols are generally initiated by thermal decomposition of peroxides or azo compounds, by UV radiations, or by radiolysis [5]. The reaction initiated by 9-BBN is completely inhibited in the presence of galvinoxyl, a radical trapping agent. This confirms that 9-BBN participates in the initiation of radical addition (Table 14.1) [4].

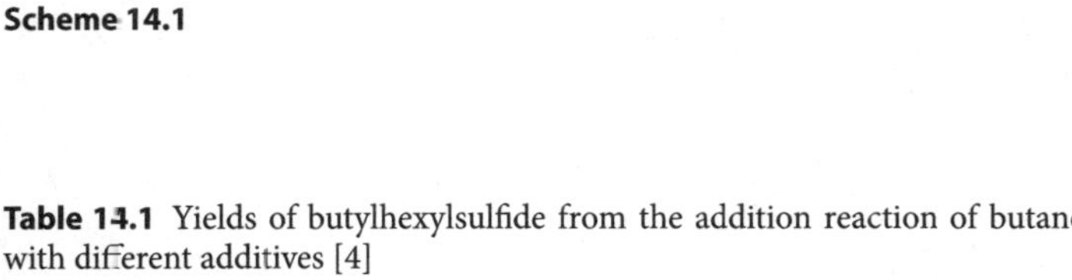

Scheme 14.1

Table 14.1 Yields of butylhexylsulfide from the addition reaction of butanethiol and hex-1-ene with different additives [4]

Additive	Reaction time (h)	Yield (%)
None	24	0.2
9-BBN (0.5%)	1	>99
Galvinoxyl (1%)	1	0
B-Hexyl-9-BBN (1%)	1	98

The initiation is not limited to primary alkanethiols and terminal alkenes. The addition of *sec*- or *tert*-alkanethiols also affords the corresponding dialkylsulfides (Table 14.2) [4] in good yields. The reactions are found to be initiated both by 9-BBN and *B*-hexyl-9-BBN (Eq. 14.1).

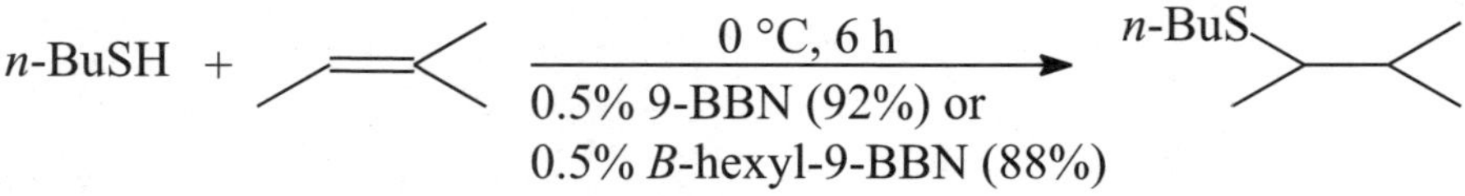

(14.1)

Table 14.2 Reactions of alkanethiols with terminal alkenes in the presence of 9-BBN at 0 °C [4]

R^1 of $R^1CH{=}CH_2$	R^2 of R^2SH	Reaction time (h)	Yield of $R^1CH_2CH_2SR^2$ (%)
Bu^n	Bu^n	1	>99
	Bu^s	3	96
	Bu^t	3	91
	c-C_6H_{11}	3	92
	CH_2Ph	3	90
cyclo-C_6H_{10}	Bu^n	1	99[a]
	Bu^s	3	97[a]
	Bu^t	3	80[a]
Me_3SiCH_2-CH_2	Bu^n	1	99
	Bu^s	3	80
	Bu^t	3	78

[a] The product is *c*-$C_6H_{11}SR^2$.

It is the first report that 9-BBN though having a B–H bond, initiates the radical reaction. It is significant to mention that no other derivative of 9-BBN yields such satisfactory results.

References

1. Pelter A, Smith K, Brown HC (1988) Borane reagents. Academic, London
2. (a) Arase A, Hoshi M, Masuda Y (1979) Chem Lett 961; (b) Hoshi M, Masuda Y, Arase A (1984) ibid. 195
3. Oshima K, Utimoto K (1989) J Synth Org Chem Jpn 47:40
4. Masuda Y, Hoshi M, Nunokawa Y, Arase A (1991) J Chem Soc Chem Commun 1444
5. See for example Abell PI (1973) In: Kochi JK (ed) Free radicals vol II. Wiley, New York, p 63

15 Synthesis of Thiophene Oligomers

Oligomers of thiophene are prepared in one-pot procedure by the reaction of *B*-OMe-9-BBN and 2-lithiothiophene. These oligomers contain two to five thiophene units linked through their C-2 and C-5 positions (Eq. 15.1) [1]. The reaction is carried out under nitrogen between –80 and 0 °C.

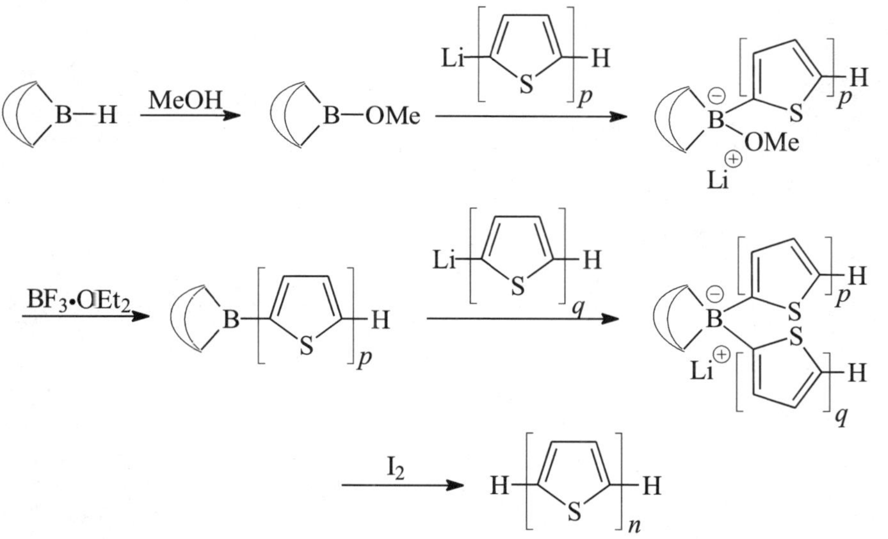

$n = p + q$

$n = 2$, 81%; $n = 3$, 37%; $n = 4$, 48%

$n = 5$, 55%; $n = 6$, 59%

(15.1)

Reference

1. Kagan J, Arora SK (1983) Tetrahedron Lett 24:4043

16 Synthesis of Cyclopropanes and Cyclobutanes

The cyclization of hydroborated allylic chlorides to the corresponding cyclopropanes discovered by Hawthorne and Dont [1] is applied to the synthesis of variety of cyclopropanes [2]. However, the directive effect of chlorine directs the diborane to give roughly 50:50 of the two isomeric boron derivatives. Consequently, in such cases the maximum yield of cyclopropane derivative is not greater than 50%.

The difficulties encountered are, however, overcome by the use of disiamyl borane and sodium triethylborohydride [3] as the base, in a nonaqueous system. However, 9-BBN is an ideal for this purpose, and a wide variety of cyclopropane derivatives (Chart 16.1) [4–6] are prepared in good yields.

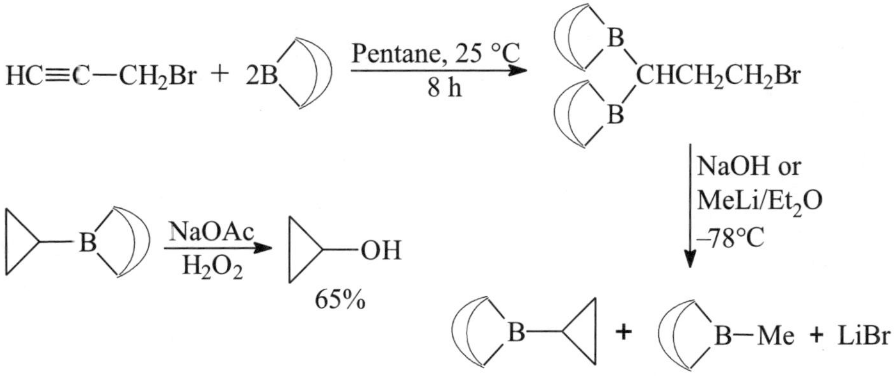

Closely related to the above reaction is the base-induced elimination from α,α-dibora-γ-chloropropane and ω-sulfonyloxy derivatives, which produce cycloalkylboranes. Oxidation of these leads to the corresponding cycloalkanols (Chart 16.2) [6, 7].

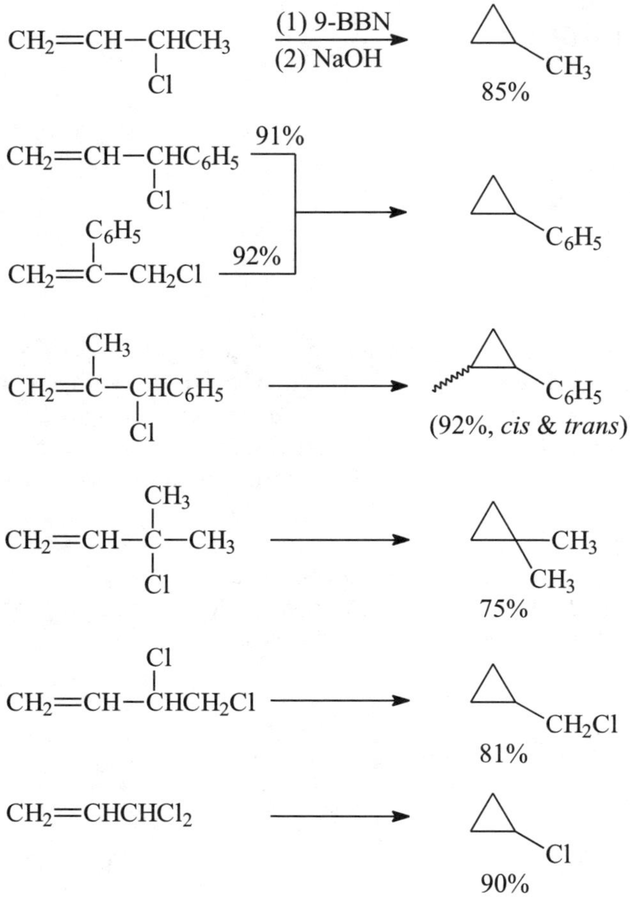

Chart 16.1

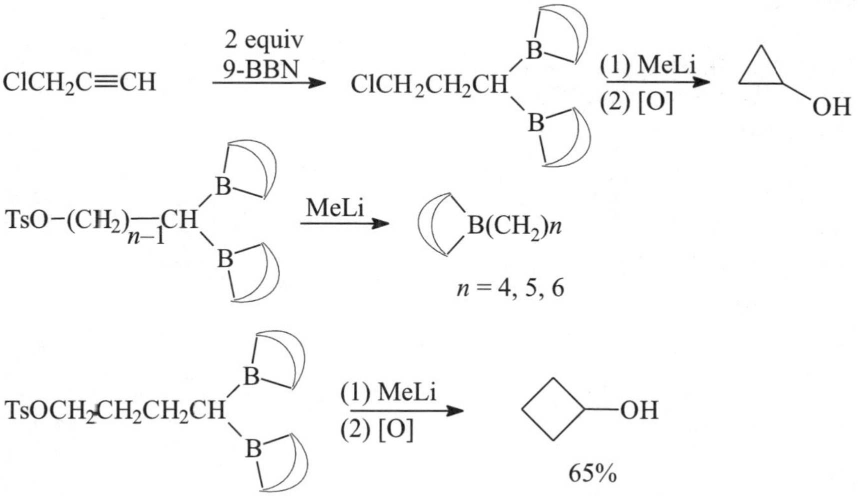

Chart 16.2

The mechanism of the reaction is given in Scheme 16.1.

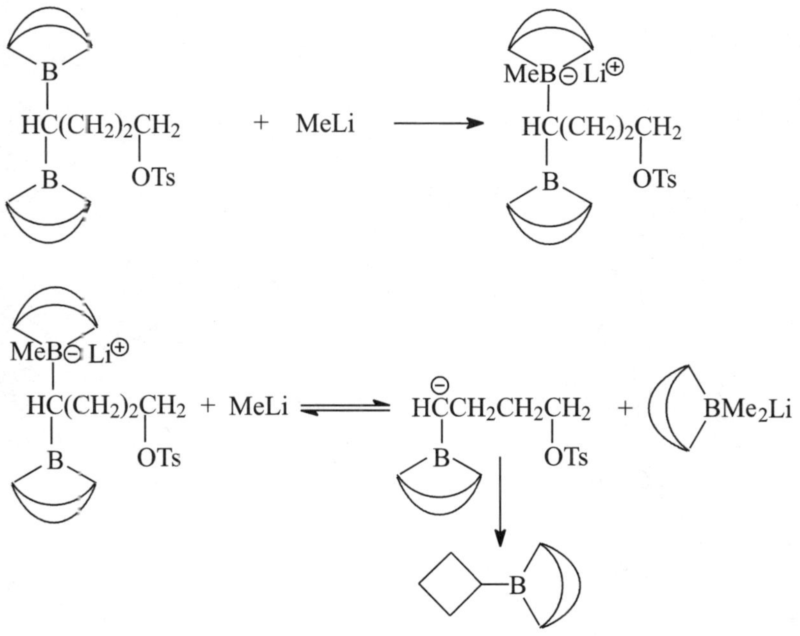

Scheme 16.1

References

1. Hawthorne MF, Dupont JA (1958) J Am Chem Soc 80:5830
2. Hawthorne MF (1960) J Am Chem Soc 82:1886
3. Binger P, Köster R (1961) Tetrahedron Lett 156
4. Brown HC, Chen JC (1981) J Org Chem 46:3978
5. Brown HC, Rhodes SP (1969) J Am Chem Soc 91:2149
6. Brown HC, Rhodes SP (1969) J Am Chem Soc 91:4306
7. Scouten CG (1974) Ph.D. Thesis, Purdue University, West Lafayette, Indiana

17 Synthesis of Borinanes

The 9-BBN has been elegantly employed for the convenient synthesis of the six-membered ring boracyclane, borinane. Boracyclanes (Chart 17.1) like borinane (**1**), 3,5-dimethyl-borinane (**2**), and 3,6-dimethyl borepane (**3**) [1] are found to be advantageous hydroborating agents over 9-BBN in certain reactions involving free radical mechanism as free radicals rapidly attack the boron atom of the 9-borabicyclononane structure.

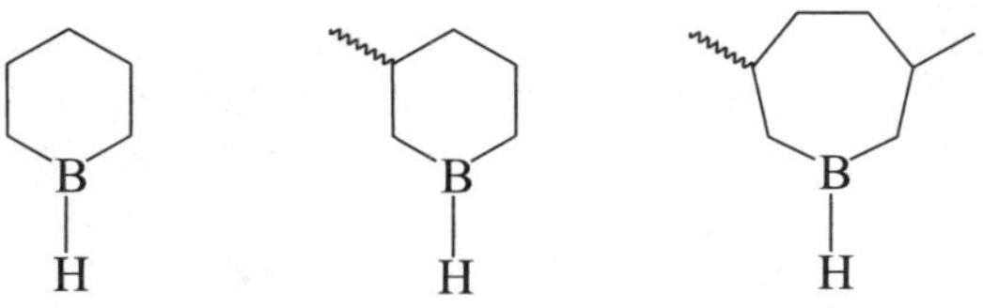

(**1**) (**2**) (**3**)

Chart 17.1

Borinane is earlier prepared *via* the cyclic hydroboration of the corresponding diene (Eq. 17.1) [2] in 40% yield.

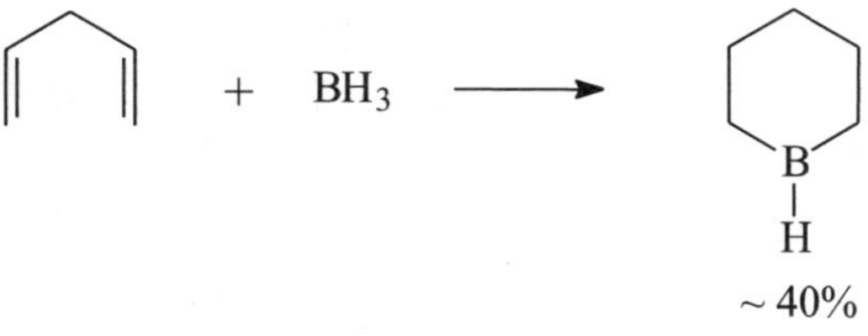

(17.1)

Köster's method is later improved by incorporating an isomerization step, as illustrated in Scheme 17.1 [3, 4].

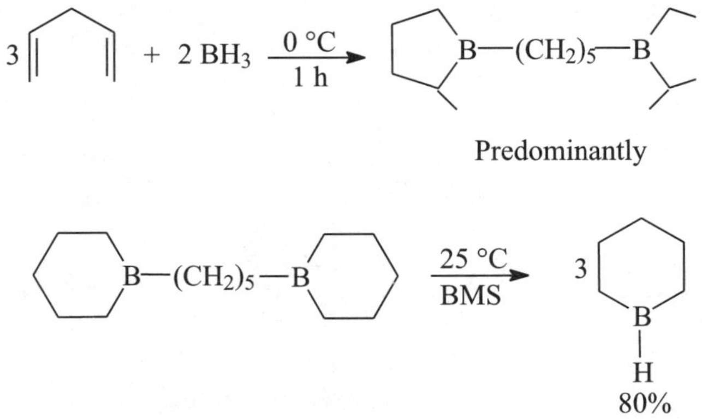

Scheme 17.1

Brown and Pai [5, 6] have elegantly hydroborated 1,4-pentadiene with 2-M equiv of 9-borabicyclo[3.3.1]nonane in *n*-hexane to afford dumbbell shaped compound (**4**), without any regioisomers. The subsequent treatment with borane-dimethylsulfide complex (BMS) leads to cyclization of pentadiene moiety forming borinane along with the regeneration of 2 M equiv of 9-BBN (Scheme 17.2). The separation of the mixture, however, has posed the problem as none of the solvents examined such as THF, ether, dimethoxyethane, dioxolane, *n*-pentane, *n*-hexane effected a complete separation of the two hydroborating agents. Cooling of the reaction mixture to −78 °C leads to the crystallization of 70–80% of pure 9-BBN. Concentration of the mother liquor followed by cooling to −78 °C or even lower results in simultaneous precipitation of both dialkylboranes.

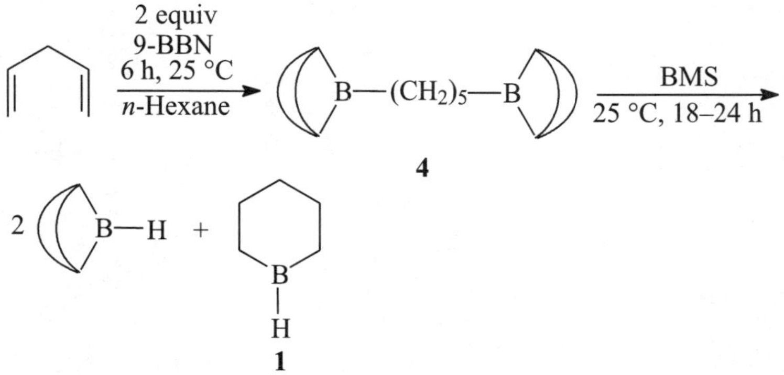

Scheme 17.2

However, it is found that 9-BBN does not form an adduct with Et_3N in cyclohexane solution [7, 8]. On the contrary, borinane is about 80% associated with Et_3N under similar conditions (Chart 17.2) [9].

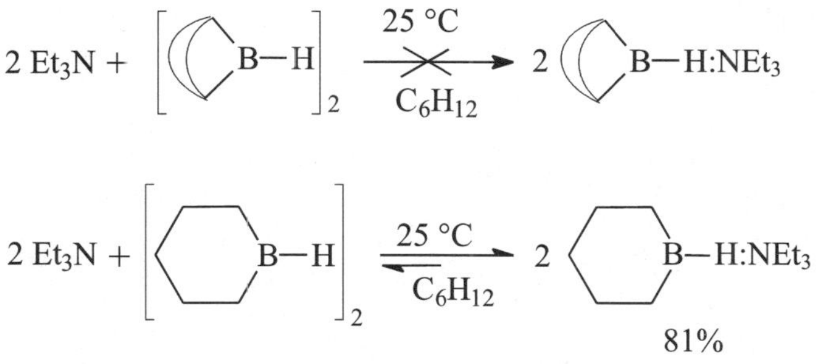

Chart 17.2

Accordingly, a hexane reaction mixture on treatment with 100% excess of Et_3N with respect to the contained borinane leads to complete conversion to its amine complex, while 9-BBN remains uncomplexed. Cooling the mixture to −78 °C results in the almost quantitative (98%) crystallization of pure 9-BBN. The mother liquor containing a borinane–triethylamine complex on treatment with $BF_3{\cdot}OEt_2$ liberates a hexane solution of free borinane (Eq. 17.2) [5, 6], and the $BF_3{\cdot}NEt_3$ complex separates out.

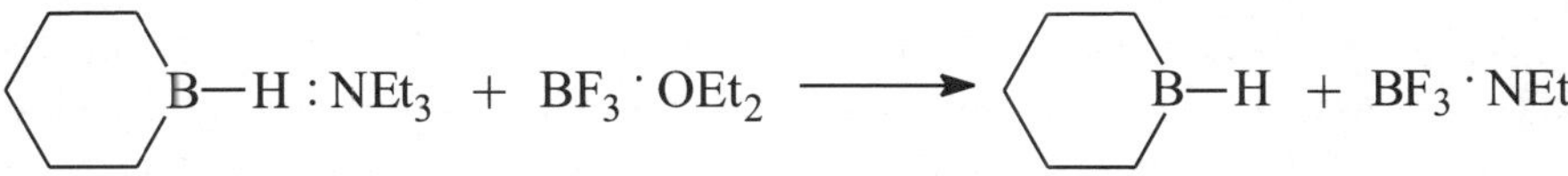

(17.2)

Alternatively, borinane is precipitated selectively from the reaction mixture as its bis adduct with either *N,N,N',N'*-tetramethylethylenediamine (TMEDA) or 1,4-diazabicyclo[2.2.2]octane (DABCO) in nonpolar solvents like hexane. However, the bis adducts of both 9-BBN and borinane precipitate, quantitatively, from the hexane solution of the boranes at 0 °C. Fortunately, there is a large kinetic difference between the rates of adduct formation: 9-BBN takes a couple of hours to react with difunctional amines, whereas borinane reacts almost instantaneously. Thus, this selective reaction affords almost quantitative precipitation of the bis adduct of borinane with TMEDA and DABCO (Scheme 17.3) [6].

In both the cases, 9-BBN remains in the solution and be recycled for subsequent synthesis. The free borinane is liberated from the amines complexes by reaction with $BF_3{\cdot}OEt_2$.

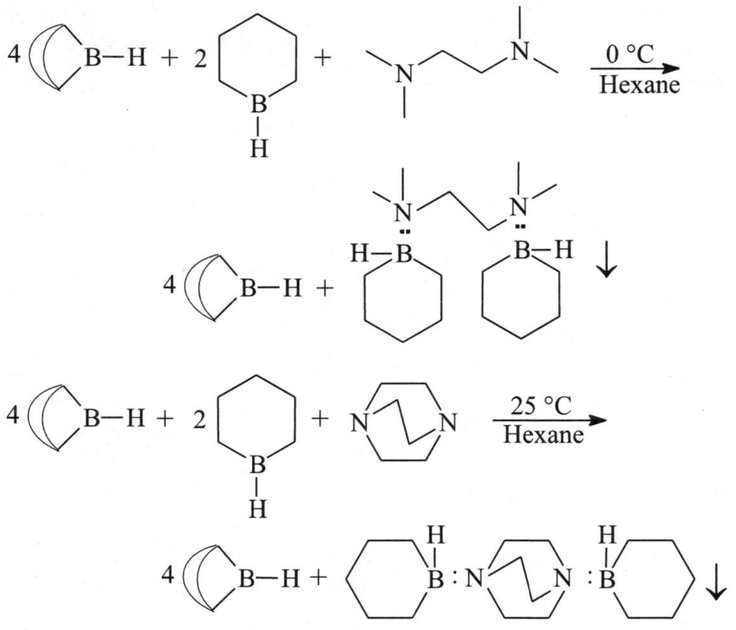

Scheme 17.3

Brown and Pai [6] have further reported that it is possible to avoid undesirable steps associated with the use of amines. Borinane is extremely air sensitive, and at 100 °C it changes into undesirable products. Consequently, by exercising due care, it is possible to distill pure borinane from 9-BBN in ~100% purity and greater than 90% yield at a pot temperature of 60–70 °C and 0.01-mm pressure. Borinane is distilled over as a low-melting solid, leaving 9-BBN as a solid pot residue.

The pure borinane prepared by this method is utilized for its further synthesis leading to 3 mol from 2 mol (Eq. 17.3), which also eliminates the problem of separation.

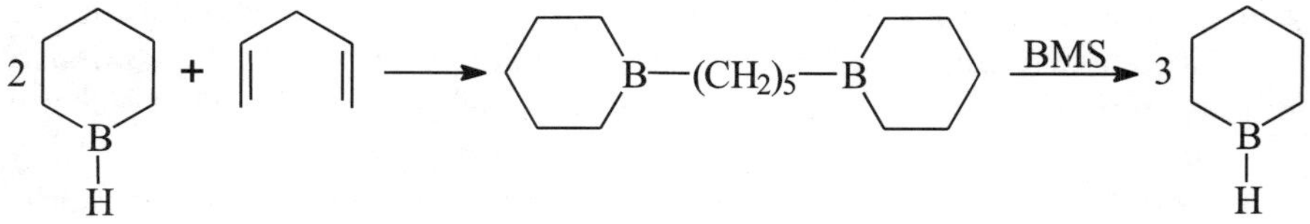

(17.3)

References

1. Negishi E, Brown HC (1973) J Am Chem Soc 95:6757
2. Köster R (1964) Advan Organometal Chem 2:257
3. Brown HC, Negishi E (1971) J Organomet Chem 26:C67
4. Negishi E, Burke PL, Brown HC (1972) J Am Chem Soc 94:7431
5. Brown HC, Pai GG (1982) Heterocycles 17:77
6. Brown HC, Pai GG (1983) J Organomet Chem 250:13
7. Brown HC, Kulkarni SU (1977) Inorg Chem 16:3090
8. Brown HC, Wang KK (1979) Rec Trav Chim Pays Bas 98:117
9. Brown HC, Pai GG (1981) J Org Chem 46:4713

18 Synthesis and Transformations of Butterflyboranes: *cis*-Bicyclo[3.3.0]oct-1-yldialkylboranes

Lithium dialkyl ate complexes of 9-BBN are conveniently prepared by the addition of organolithium to *B*-alkyl-9-BBN (Eq. 18.1) [1].

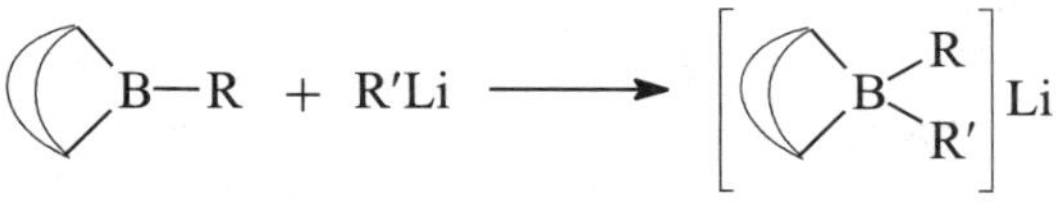

(18.1)

However, secondary and tertiary lithium reagents containing β-hydrogen undergo hydride transfer to afford lithium monoalkyl ate complexes of 9-BBN (Eq. 18.2).

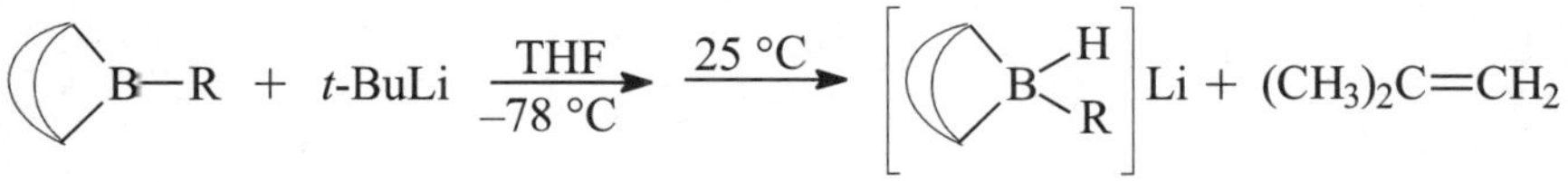

(18.2)

The transfer of hydride from the organolithium to *B*-alkyl-9-BBN is a kinetic phenomenon, since it is possible to prepare tetralakylborates having secondary and tertiary alkyl groups with β-hydrogen if they are present in the starting *B*-R-9-BBN derivatives. Consequently, secondary and tertiary alkyl groups containing ate complexes are prepared as shown in Eq. 18.3.

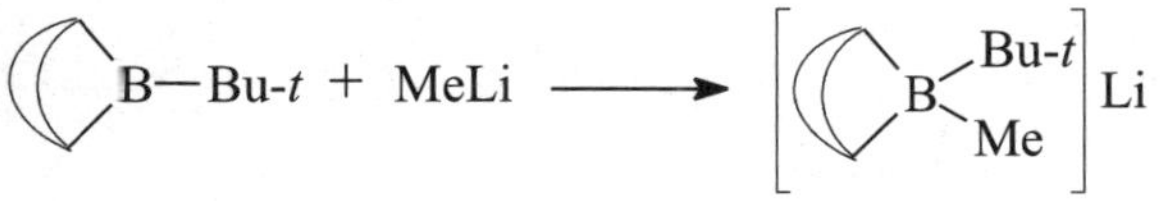

94–99%

(18.3)

The usual oxidation of these ate complexes surprisingly gives products other than the expected alcohol *cis*-1-5-cyclooctanediol (Eq. 18.4) [5]. Since no gas evolution accompanies reaction, the “hydride” is not lost as hydrogen gas; a redox reaction must occur.

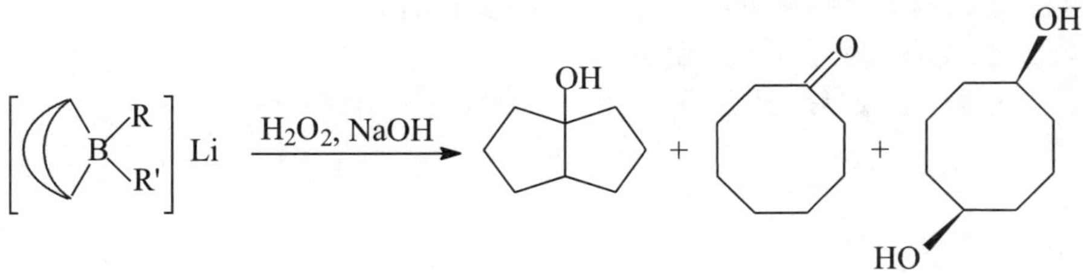

(18.4)

The formation of bicyclic alcohol is explained on the basis that dialkyl ate complexes reduce hydrogen peroxide with concurrent or subsequent migration of boron-carbon bond (Scheme 18.1).

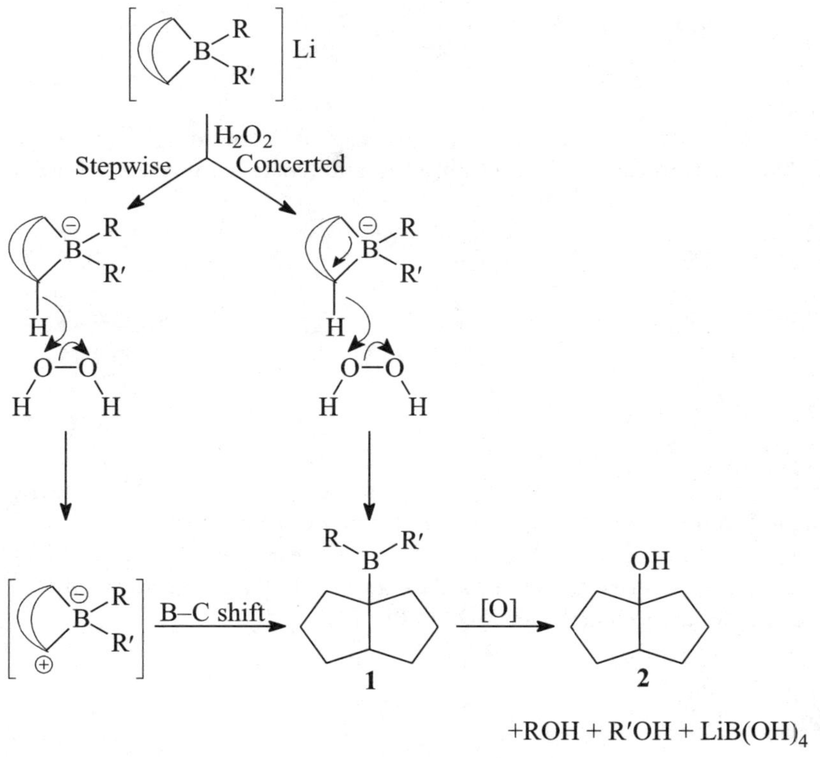

Scheme 18.1

The yield and boiling points of *cis*-bicyclo[3.3.0]oct-1-yldialkylboranes (**1**) are summarized in Table 18.1 [5].

Table 18.1 Yield and boiling points of cis-bicyclo[3.3.0]-oct-1-yldialkylboranes (**2**) [5]

Alkyl groups[a]		Yield (%)	% Purity (GLC)	b.p. (°C)	mm Hg
R	R′				
Methyl	Methyl	96	93	76–78	20
Ethyl	Methyl	99	97	28–32	0.005
Isopropyl	Methyl	97	94	39–43	0.005
Methyl	*n*-Butyl	95	99	53–56	0.005
t-Butyl	Methyl	97	86	46–49	0.005
n-Butyl	*n*-Butyl	94	[b]	65–68	0.005

[a] *B*-R-9-BBN; R′Li.
[b] The organoborane decomposes in GLC.

The rearranged intermediate (**1**), if prevented from destruction, is proven to be an interesting and useful organoborane [5]. These novel trialkylboranes (**2**) are dubbed as the "dialkyl butterfly boranes" since molecular models of these structures resemble butterflies when viewed in a certain way.

Yamamoto and coworkers [6] have reinvestigated the work of Jägar and Hesse [7]. It is found, for example, that when lithium di-*n*-butyl-9-BBN is treated with benzyl chloride for 24 h at room temperature, followed by normal alkaline hydrogen peroxide oxidation, it gives quantitative yields of toluene, 1-butanol and bicyclic alcohol (**2**). Similar studies with cyclohexanone afford a quantitative reduction to cyclohexanol.

These studies thus suggest that butterfly organoboranes can be trapped by reactive halides or carbonyl derivatives. Consequently, acid chlorides and chloral are found to be two promising materials [8]. The reduction of chloral is rapid and quantitative. However, the precipitation of lithium 2,2,2-trichloroethoxide makes difficult the isolation of the butterflyboranes (Eq. 18.5).

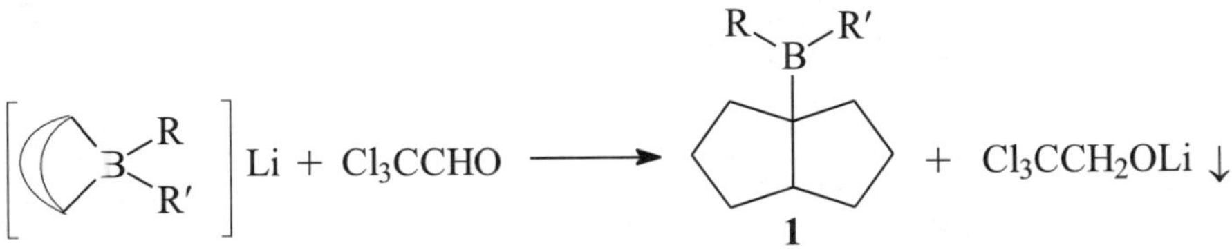

1

(18.5)

This difficulty is circumvented by the use of acid chlorides where lithium chloride solidifies and is easily separated. The other product lithium ethoxide is effectively removed by use of an another equivalent of acid chloride that converts it into an ester and lithium chloride [5, 8]. The acetyl chloride is proven to be an excellent substrate for the synthesis of butterflyborane (Eq. 18.6), where the resulting ethyl acetate is conveniently removed with solvent.

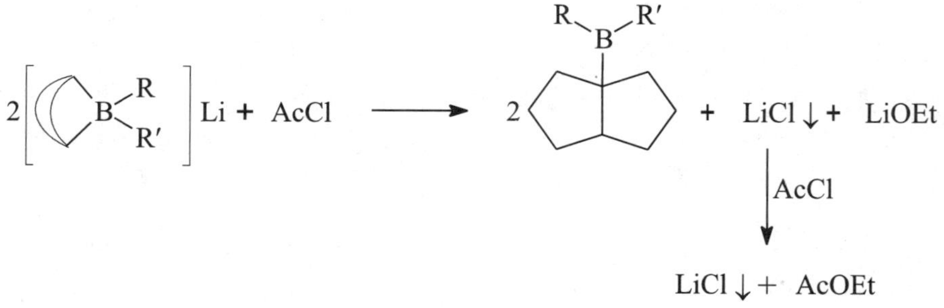

(18.6)

The butterflyorganoboranes are valuable intermediates for the preparation of a variety of 1-substituted *cis*-bicyclo[3.3.0]octane (Scheme 18.2) [5, 9], which otherwise are difficult to prepare.

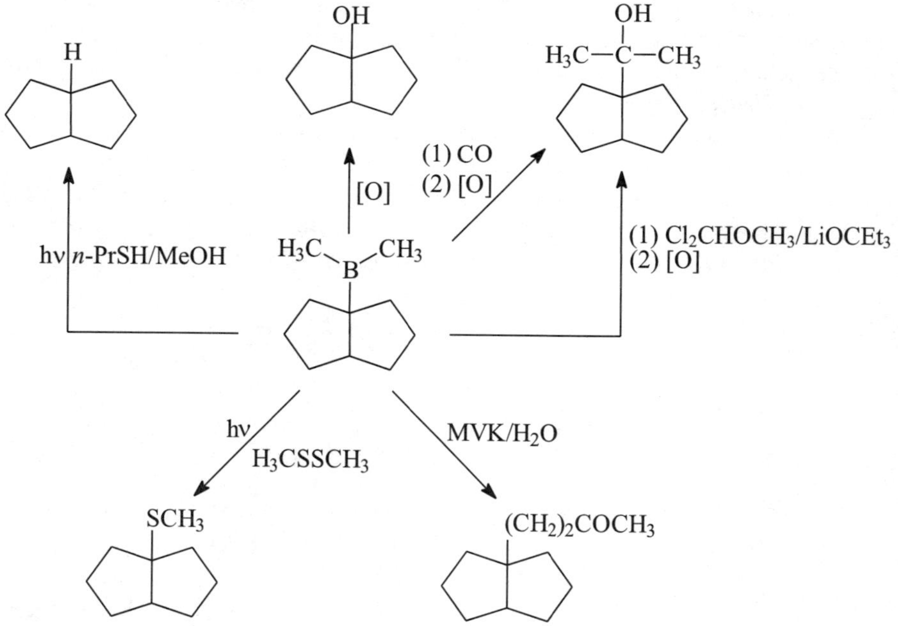

Scheme 18.2

The other *tert*-alcohols prepared via carbonylation method are *cis*-bicyclo[3.3.0]oct-1-yldi-*n*-butylcarbinol and *cis*-bicyclo[3.3.0]oct-1-ylmethyl-*t*-butylcarbinol (Chart 18.1) [9].

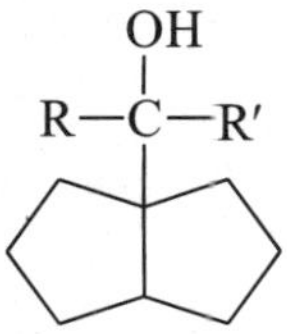

R = R′ = *n*-Bu, 87%, b.p. 97–99°C (20 Torr)
R = *t*-Bu, R′ = Me, 41%, m.p. 41°C

Chart 18.1

References

1. Kramer GW (1976) Ph.D. Thesis, Purdue University, West Lafayette, Indiana
2. Corey EJ, Becker KR, Varma RK (1972) J Am Chem Soc 94:8618
3. Negishi E, Chiu, K-W, Yoshida T (1975) J Org Chem 40:1676
4. Krishnamurthy S, Brwon HC (1976) J Am Chem Soc 98:3383
5. Kramer GW, Brown HC (1977) Heterocycles 7:487
6. Yamamoto Y, Toi H, Murahashi S-I, Moritani I (1975) J Am Chem Soc 97:2558
7. Jäger H, Hesse G (1962) Chem Ber 95:345
8. Kramer GW, Brown HC (1976) J Am Chem Soc 98:1964
9. Kramer GW Brown HC (1977) J Org Chem 42:2832

19 Synthesis of α-Bromoboranes

Organozirconium obtained [1–5] by hydrozirconation are valuable intermediates for organic synthesis. These are converted to the desired organic products by utilizing the reactivity of the zirconium–carbon bond with a variety of electrophilic reagents.

It is reported [6] that hydrozirconation of various *B*-alkenyl-9-BBN by Schwartz's reagent, H(Cl)ZrCp$_2$ [1–5, 7], proceeds smoothly in dichloromethane or THF and affords 1,1-bimetallics of boron and zirconium (Eq. 19.1). It is pertinent to mention that hydrozirconation of alkenylboranes with Schwartz's reagent does not take place in both diethyl ether and hexane, and in benzene the reaction is slow. This suggests that the electronic donor ability of solvent favors hydrozirconation process. It has also been found that cyclic alkenylboronic esters having larger groups than cyclooctyl on boron, like diisopinocampheylhexenylborane and diisocaranylhexenylborane, undergo partial or incomplete hydrozirconation with Schwartz's reagent. Moreover, cyclic alkenylboronic esters, such as hexenyl-1,2,3-benzodioxaborole, ethylene glycol hexenylboronate, and pinacol hexenylboronate, undergo partial hydrozirconation with 1 equiv of Schwartz's reagent slowly. It suggests that both electronic and steric factors greatly influence the course of hydrozirconation of *B*-alkenylboranes.

The selective cleavage of C–Zr bond of the 1,1-bimetalloalkanes with bromine results in a convenient method for the preparation of α-bromoboranes. Consequently, addition of bromine in dichloromethane *in situ* at −35 °C results in the discharge of color and formation of white precipitate within 1 h of zirconocene dihalide. The removal of CH$_2$Cl$_2$, followed by extraction with hexane of reaction mixture provides the α-bromoborane in high yield (Table 19.1) [6].

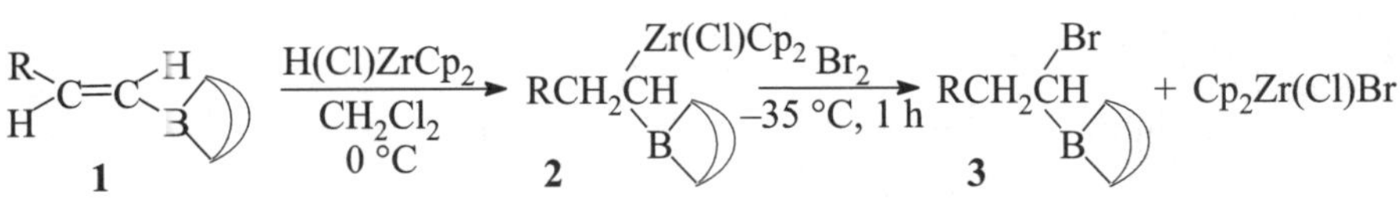

1 2 3

(19.1)

Table 19.1 Synthesis of α-bromoboranes (**3**) by bromination of 1,1-bimetallic compounds (**2**) [6]

R in vinyl-9-BBN, 1	Hydrozirconation time (h)	Bromination product	Yield %
n-Butyl	1	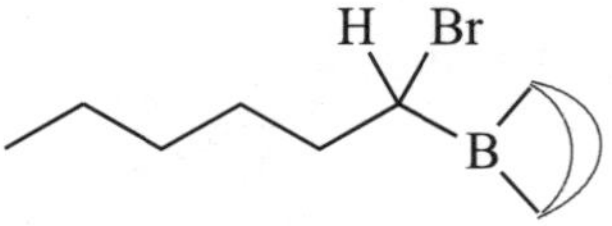	97
3-Chloropropyl	1	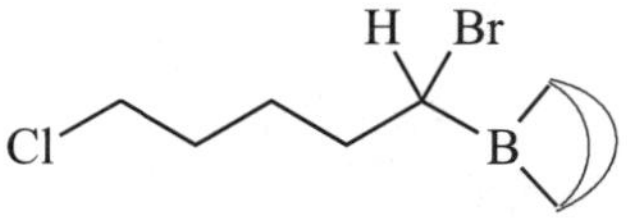	99
1-Methylpropyl	1.5	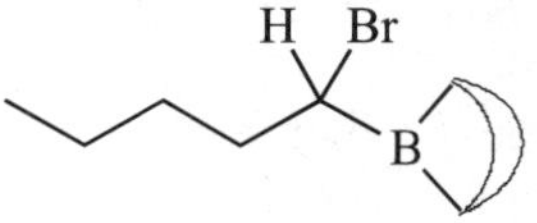	95
3-Phenylpropyl	1.5	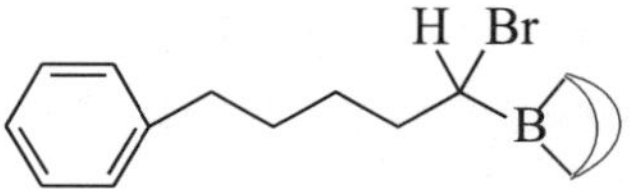	99
Cyclopentyl	2	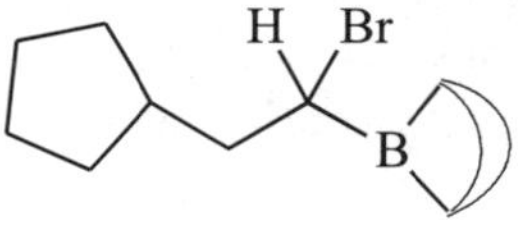	91
t-Butyl	6	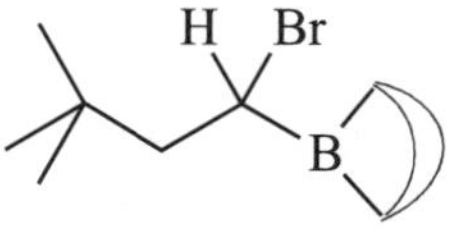	87
Phenyl	6	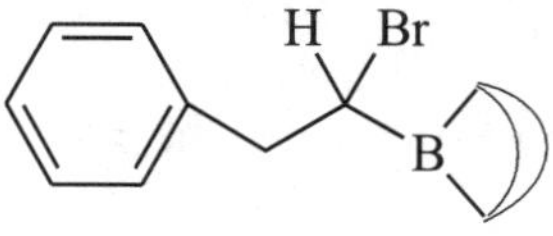	83

The α-bromoboranes are valuable intermediates that can be converted into a multitude of organic products [8–10].

Brown and De Lue have reported [11] the synthesis of α-bromoboranes of *B*-*sec*-alkyl-9-BBN. Consequently, bromination of *B*-*i*-Pr-9-BBN in methylene chloride with continuous removal of hydrogen bromide affords 2-bromo-2-propyl-9-BBN (Eq. 19.2) (*vide supra*).

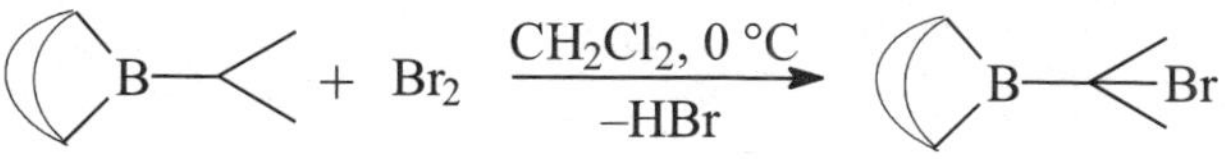

(19.2)

These bromoboranes are used for the synthesis of *B*-alkyl-9-BBN derivatives, not directly available via hydroboration (*vide supra*).

References

1. Schwartz J, Hart DW (1974) J Am Chem Soc 96:8115
2. Schwartz J, Labinger JA (1976) Angew Chem Int Ed Engl 15:333
3. Hart DW, Blackburn TF, Schwartz J (1975) J Am Chem Soc 97:680
4. Bertelo CA, Schwartz J (1976) J Am Chem Soc 98:262
5. Carr DB, Schwartz J (1979) J Am Chem Soc 101:3521
6. Zheng B, Srebnik M (1993) Tetrahedron Lett 34:4133
7. Buchwald SL, LaMaire SJ, Nielson RB, Watson BT, King SM (1987) Tetrahedron Lett 28:3895
8. Brown HC, Kramer GW, Levy AB, Midland MM (1975) Chapter 7.10. In: Organic synthesis via boranes. Wiley InterScience, New York (1975)
9. Pelter A, Smith K, Brown HC (1988) Chapter 4.3. In: Borane reagents. Academic, London
10. Matteson DS (1989) Tetrahedron 45:1859
11. Brown HC, De Lue NR (1974) J Am Chem Soc 96:311

20 Synthesis of Borinates

20.1
Enol Borinates

20.1.1
Synthesis of (*E*)- and (*Z*)-Enolborinates from Saturated Ketones

The literature describes number of methods for the generation of enolborinates [1]. Enolborinates are valuable intermediates [2] and react with high stereospecificity in aldol condensation [3–8]. Brown and coworkers [9] have carried out a detailed study of the stereochemistry of ketone enolization with *B*-Cl-9-BBN [10], *B*-OTf-9-BBN [3c], Chx_2BCl [10], and Chx_2BOTf [9] to get (*E*)- or (*Z*)-enolborinates. It is found that stereochemical outcome of the reaction varies not only with the steric requirements of R_2B and steric requirements of the amine [4, 5b, c], but also with the nature of the leaving group, Cl or OTf (Table 20.1) [9].

Table 20.1 Enolization of propiophenone and diethylketone with R_2BX and amines [9]

Reagent R_2BX	Amine R_3N	Propiophenone *Z*:*E*	Propiophenone *syn*:*anti*[a]	Diethylketone *syn*:*anti*[a]
B-Cl-9-BBN	Et_3N	52:48 65:35	60:40	>99:1
	i-Pr_2EtN	>99:1	95:5	>99:1
B-OTf-9-BBN	Et_3N	>99:1	93:7	>99:1
	i-Pr_2EtN	>99:1	95:5	>99:1
Chx_2BCl	Et_3N	>1:99	5:95	21:79
	i-Pr_2EtN	51:49		72:28
Chx_2BOTf	Et_3N	67:33		80:20
	i-Pr_2EtN	>99:1	98:2	93:7

[a] Diastereoselection of aldol product of benzoldehyde

The effects of varying the amine and alkyl groups on boron are more significant in the case of R_2BCl than in the case of R_2BOTf, and quantitative synthesis of (*Z*)-enolborinate has been achieved using *i*-Pr_2EtN and *B*-Cl-9-BBN, whereas enolization with Chx_2BCl and amine Et_3N gives (*E*)-enolborinate (Scheme 20.1; Table 20.2) [9], almost exclusively.

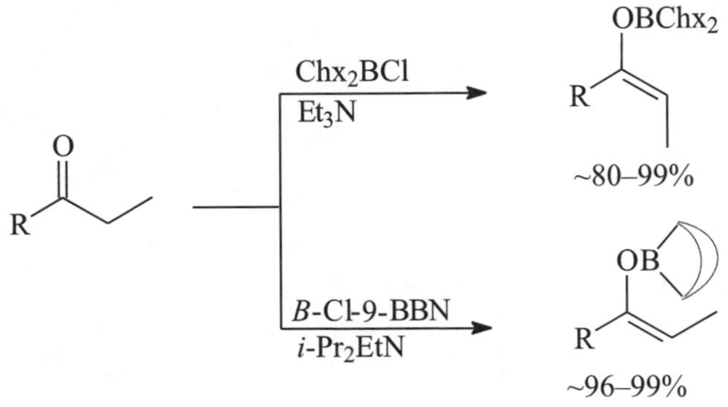

Scheme 20.1

Table 20.2 Stereoselective enolization of representative ketones with R_2BCl and amines [9]

Ketone	*B*-Cl-9-BBN[a] *Z*:*E*	Chx_2BCl[b] *Z*:*E*	Yield of *E* isomer (%)
Propiophenone	>99:1	<1:99	>99
Phenyl benzyl ketone	>99:1	15:85	85
			>99
Isopropylethyl ketone	65:35	<1:99	
	98:2		
			>99
Cyclohexylethyl ketone	60:40	<1:99	
	96:4		79
Diethyl ketone	>99:1	21:79	

[a] Enolization with *i*-Pr_2EtN at 25° C
[b] Enolization with Et_3N at 0° C

The reaction is applied for the synthesis of *syn* or *anti* aldol (Scheme 20.2) [9].

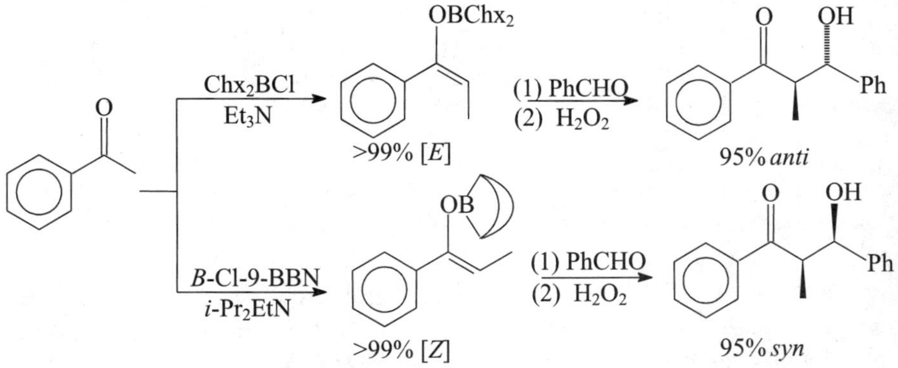

Scheme 20.2

Evans and coworkers have established that *Z*-enolborinates give *syn* aldols whereas *E*-enolborinates provide *anti* aldols, stereoselectively [4b] (Scheme 20.3). Various R_2BX reagents have been designed and used for the enolboration of ketones in the presence of various tertiary amines of different steric requirements [3, 4] (Scheme 20.4), with different leaving groups. It has been established that the R_2BX reagents with the very powerful leaving group, triflate, favor the formation of *Z*-enolborinates while with the poorer leaving group, chloride, favor the formation of *E*-enolborinate [9]. Brown et al. [11] have examined the various *B*-X-9-BBN and Chx_2BX reagents with different leaving groups such as OTf, OMs, I, Br and Cl (Chart 20.1) of variable steric and electronic requirements for their role in *Z*- or *E*-enolborinates syntheses.

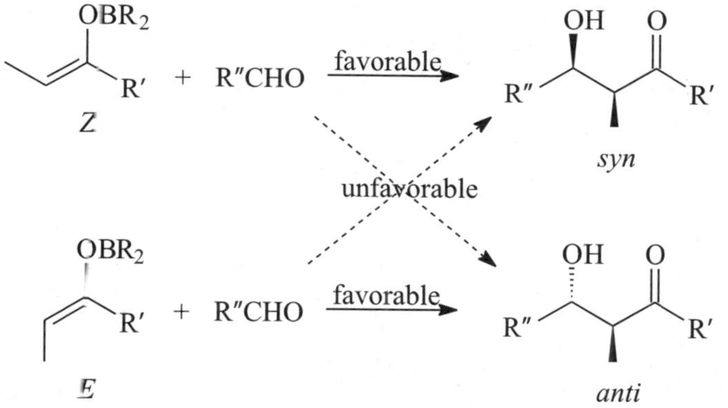

Scheme 20.3

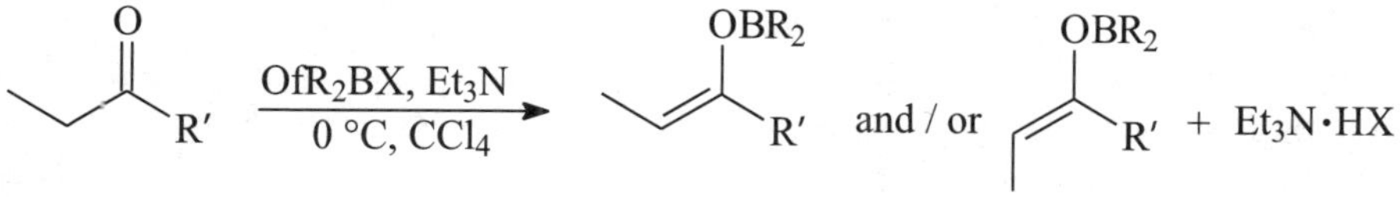

Scheme 20.4

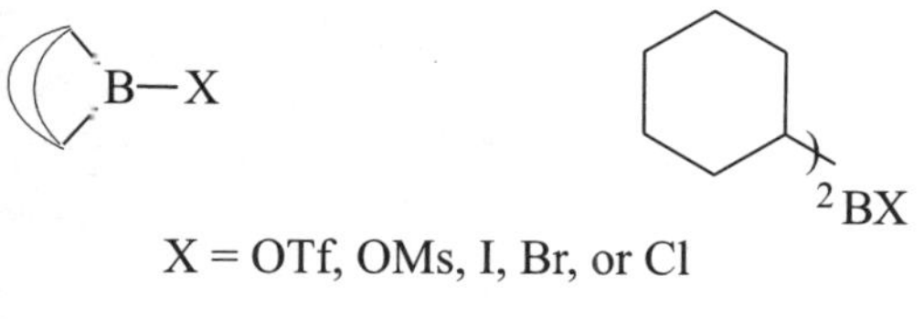

Chart 20.1

For the model ketone, ethylisopropylketone, the regiochemistry of the enolboration is always toward the less hindered ethyl side, irrespective of the R_2BX reagent. The results are summarized in Table 20.3 [11], which reveal that the powerful leaving group (OTf) favors the formation of *Z*-enolborinate, and the weaker leaving group (Cl) favors the formation of *E*-enolborinate. The comparison of *B*-X-9-BBN and Chx_2BX reagents concludes that the lower steric requirement of R of R_2BX reagents and stronger leaving effects of X favor the formation of *Z*-enolborinate, while those with relative bulkier R groups (cyclohexyl) and poor leaving groups favor the formation of *E*-enolates.

Table 20.3 Effect of the leaving group on enolate geometry in the enolboration of ethylisopropyl ketone with various R_2BX/Et_3N [11]

B-X-9-BBN (%)				Chx_2BX (%)		
X	*Z*	*E*	Yield	*Z*	*E*	Yield
OTf	88	12	96	25	75	95
OMs	82	18	94	23	77	93
I	73	27	97	32	68	98
Br	57	43	96	11	89	95
Cl	46	54	95	<3	>97	97

The stereoselective enolboration of diethylketone with *B*-X-9-BBN and Chx_2BX (Table 20.4) [11] reveals that the smaller steric requirement of the 9-BBN moiety on boron controls the stereochemistry of the enolboration process more than the corresponding leaving group does. Thus, irrespective of the nature of leaving group, all the *B*-X-9-BBN reagents yield *Z*-enolborinates selectively from diethylketone. However, the effect of leaving group is much larger with bulkier Chx_2BX reagents. The stronger Lewis acid, Chx_2BOTf with a better leaving group, favors *Z*-enolate, whereas the relatively weaker Lewis acid, Chx_2BCl with a poorer leaving group, favors the formation of *E*-enolate, with a maximum selectivity of 79%. Bco_2BCl, with higher steric requirements, is the only organoboron reagent available for the predominant formation of *E*-enolate from diethylketone [7c].

The steric requirement of the substituent R′ in EtCOR′ also plays a significant role in the control of enolate geometry. The smaller R′ groups favor the formation of *Z*-enolates, while bulkier R′ groups favor the formation of *E*-enolate. Consequently, EtCOBu-*t* with all the R_2BX (except R_2BI) favor the formation of *E*-enolborinate, either exclusively or predominantly (Table 20.5) [11].

The R_2BI reagents are highly reactive and behave in an unusual manner to give, surprisingly, the isomeric *Z*-enolate essentially exclusively from EtCOBu-*t* and EtCOPh. The high reactivity of R_2BI is attributed to the influence of the iodide leaving group.

Table 20.4 Effect of the leaving group on enolate geometry in the enolboration of diethyl ketone with various R_2BX/Et_3N [11]

B-X-9-BBN (%)				Chx_2BX (%)		
X	*Z*	*E*	Yield	*Z*	*E*	Yield
OTf	>97	<3	97	80	20	96
OMs	>97	<3	95	80	20	93
I	>97	<3	97	56	44	98
Br	>97	<3	97	30	70	96
Cl	>97	<3	95	21	79	97

Table 20.5 Effect of the leaving group on enolate geometry in the enolboration of ethyl *tert*-butyl ketone with various R_2BX/Et_3N [11]

B-X-9-BBN (%)				Chx_2BX (%)		
X	*Z*	*E*	Yield	*Z*	*E*	Yield
OTf	10	90	90	<3	>97	85
OMs	<3	>97	87	<3	>97	66
I	>97	<3	95	>97	<3	96
Br	<3	>97	94	10	90	82
Cl	<3	>97	94	<3	>97	60

The stereoselective enolboration of propiophenone (Table 20.6) [11] is also controlled by the nature of leaving groups of *B*-X-9-BBN and of Chx_2BX. Chx_2BI behaves unusually, favoring the formation of *Z*-enolate. The selective generation of *E*-enolborinate with Chx_2BCl/Et_3N from propiophenone is one of the significant achievements.

Table 20.6 Effect of the leaving group on enolate geometry in the enolboration of propiophenone with various R_2BX/Et_3N [11]

B-X-9-BBN (%)				Chx_2BX (%)		
X	*Z*	*E*	Yield	*Z*	*E*	Yield
OTf	>97	<3	97	67	33	96
OMs	>97	<3	96	62	38	95
I	>97	<3	98	>97	<3	97
Br	83	17	96	5	95	97
Cl	52	48	97	<3	>97	97

It is significant to mention that *B*-X-9-BBN (X = OTf, Cl, I) and Chx_2BX (X = OTf, Cl) do not enolize esters. Among the other reagents of Chx_2BX (X = OMs, I, Br) and of *B*-X-9-BBN (X = OMs, Br), Chx_2BI is the only preferred reagent for the enolboration of esters in terms of yields and selectivity [12].

R_2BOTf reagents are prepared [13] by the controlled addition of trifluoromethane sulfonic acid (1 equiv) to R_2BH (1 equiv). R_2BOMs reagents are obtained by the controlled addition of methanesulfonic acid (1 equiv) to R_2BH (1 equiv) [11]. Chx_2BBr is prepared by the direct hydroboration of cyclohexene with monobromoborane methylsulfide [11]. *B*-Br-9-BBN is synthesized by treating 9-BBN with HBr gas.

Both Chx_2BCl and *B*-Cl-9-BBN are obtained from the corresponding organoborane and anhyd HCl in ether [7a].

20.1.2 Stereoselective Synthesis of (*Z*)-Enol Borinates from α,β-Unsaturated Ketones

1,4-Hydroboration of α,β-unsaturated carbonyl compounds have been studied [14, 15], mainly due to the synthetic utility of boron enolates [1]. Phenyl-1-alkenylketones undergo selective 1,4-hydroboration either with 9-BBN or catecholborane to produce the corresponding (*Z*)-boron enolates (Chart 20.2) [16] in high yields in solvents like chloroform, dichloromethane, THF, or benzene and in high isomeric purity. However, this selectivity has not been observed in the reactions of 9-BBN or catecholborane with alkylalkenylketones (**3**).

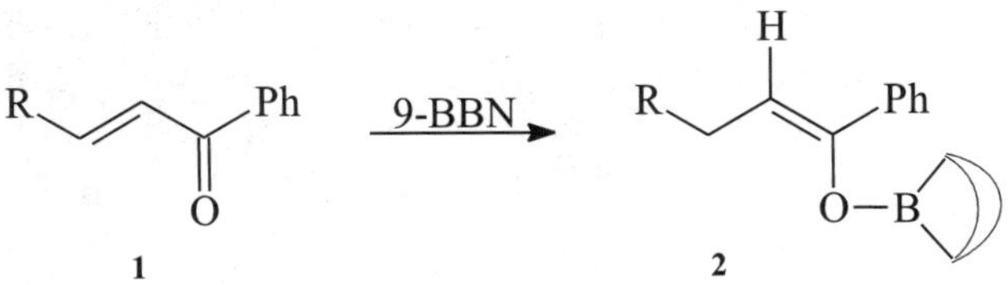

1 **2**

>95%

a, R = H; b, R = Me; c, R = Ph

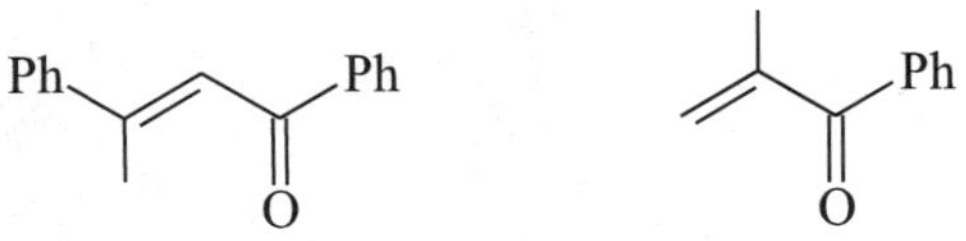

1d **1e**

Chart 20.2

The results are summarized in Table 20.7 [16].

Table 20.7 Hydroboration of phenyl-1-alkenyl ketones with 9-BBN [16]

Ketone	Reaction time (h)	Boron enolate	Yield (%)	*Z*/*E*
1a	4	**2a**	>95	>99/1
1b	8	**2b**	>95	>99/1
1c	3	**2c**	>95	>99/1
1d	4	**2d**	>95	>99/1
1e	10	**2e**	>95	–

20.1.3
Synthesis of Stable *cis*-*B*-Vinyl-9-Oxa-10-Borabicyclo[3.3.2]decane (*cis*-*B*-Vinyl-OBBD) Derivatives

Soderquist and coworkers [17] have reported the synthesis of stable *cis*-vinyl-OBBDs. The *B*-methoxy-9-BBN on selective oxidation with anhydrous trimethylamine-*N*-oxide (TMANO) (85%, $CHCl_3$, 0 °C) affords the corresponding borinic ester. The borinic ester on alkynylation, followed by demethoxylation gives the stable alkynylborinate. The hydroboration of alkynylborination with dicyclohexylborane (Chx_2BH) affords cleanly the 1:1 *gem*-diboryl adduct. This is selectively protiodeborylated with acetic acid at 0 °C and gives the corresponding *cis*-*B*-vinylborinate. The reaction sequence is outlined in Scheme 20.5 [17].

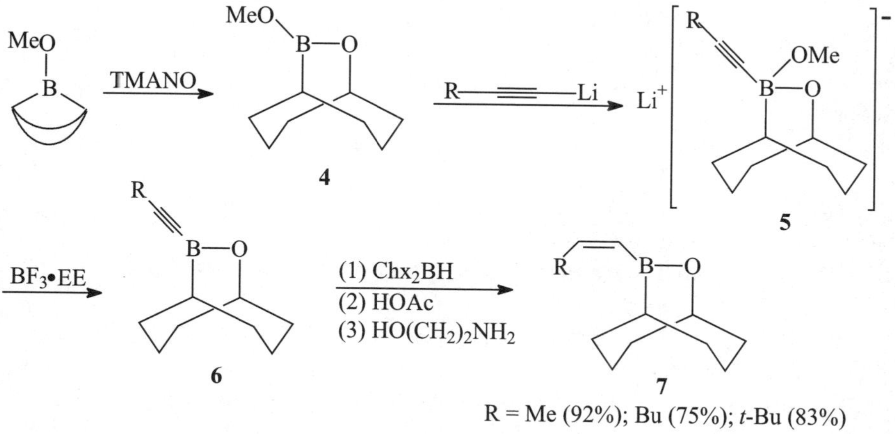

Scheme 20.5

20.1.4 Synthesis of Stable *trans-B*-Vinyl-9-Oxa-10-Borabicyclo[3.3.2]decane (*trans-B*-Vinyl-OBBD)

The terminal alkynes react with 2 equiv of 9-BBN and affords, quantitatively, the corresponding 1,1-diboraylalkanes. Soderquist has reported that this trialkylborane reacts with 1 equiv of benzaldehyde or 1-NaphCHO in 2 h at 25 °C, and quantitatively form *B*-ArCH$_2$O-9-BBN and *B*-alkenyl-9-BBN, exclusively with *trans* configuration [18]. The *trans-B*-alkenyl-9-BBN undergoes selective oxidation [18] with 1 equiv of anhydrous trimethylamine-*N*-oxide (TMANO) [19] and affords almost quantitatively the corresponding stable *trans-B*-vinyl-9-oxa-10-borabicyclo[3.3.2]decane derivatives (*trans-B*-vinyl-OBBD). *trans-B*-Vinyl-OBBD derivatives are inert to atmospheric oxygen and are unreactive toward protonolysis (HOAc, 25 °C, 8 h) or insertion process (PhCHO, neat, 80 °C, 6 h). The reaction sequence for the synthesis of *trans-B*-vinyl-OBBD is outlined in Scheme 20.6 [18].

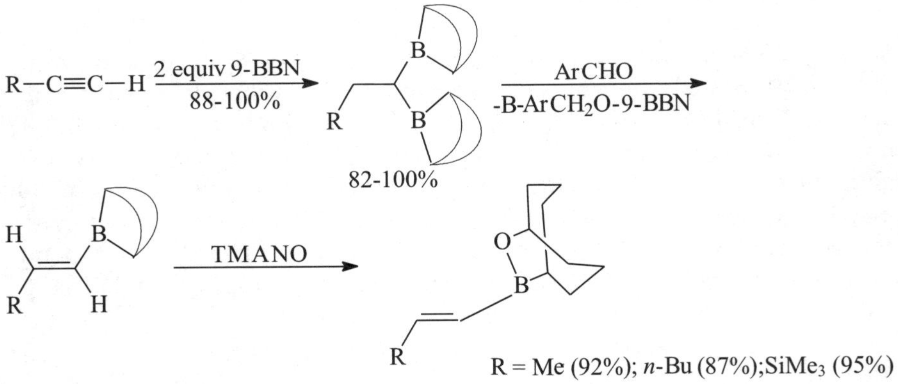

Scheme 20.6

20.1.5 Markovnikov Vinylborinates: Synthesis of *B*-Vinyl-9-Oxa-10-Borabicyclo[3.3.2]decane (Vinyl-10-OBBD)

Efficient routes to both *cis*- [17] and *trans*- [18] *B*-vinyl-9-BBN are available. However, Markovnikov vinylboranes ($R_2BCHR{=}CH_2$) are prepared from trialkyl(α-methoxyvinyl)borate complexes ($Li[R_3BCH(OMe){=}CH_2]$) and chlorotrimethylsilane (TMSCl) [19]. This process is limited to the availability of BR_3 (e.g., R = 1° and 2° but not 3°) and also produces unwanted regioisomeric byproducts. In addition, two of these three R groups are wasted and also compete with vinyl moiety to produce side products. The process also fails for 9-BBN derivatives where one of the bridgehead C–B bonds migrates rather than the *B*-alkyl group [19]. This problem is overcome by Soderquist [20]; and Markovnikov vinylborinates are prepared from 9-oxa-10-borabicyclo[3.3.2]decanes.

9-Oxa-10-borabicyclo[3.3.2]decyl (OBBD) derivatives [20] provide superior stationary ligation for many of the known conversions. These bicyclic intermediates are thermally stable and selectively transfer the *B*-substituent, avoiding the wastage of valuable groups as is encountered with tri- and dialkylboranes and also often with *B*-R-9-BBN.

Soderquist and coworkers [21] have reported the facile oxidation of *B*-R-9-BBN with anhydrous trimethyl-*N*-oxide (TMANO), which is both highly selective and efficient. The resulting borinates resist further oxidation with this reagent and are also remarkably stable toward further oxidation in the open air. These alkylborinates react [20] with α-methoxyvinyllithium (LiAMV, 1.64 equiv) and lead to the quantitative formation of the stable ate complexes. The smooth migration of the alkyl group takes place when ate complexes are treatment with borontrichloride (1.3 equiv) and produce the corresponding Markovnikov vinylborinates (Scheme 20.7) [20].

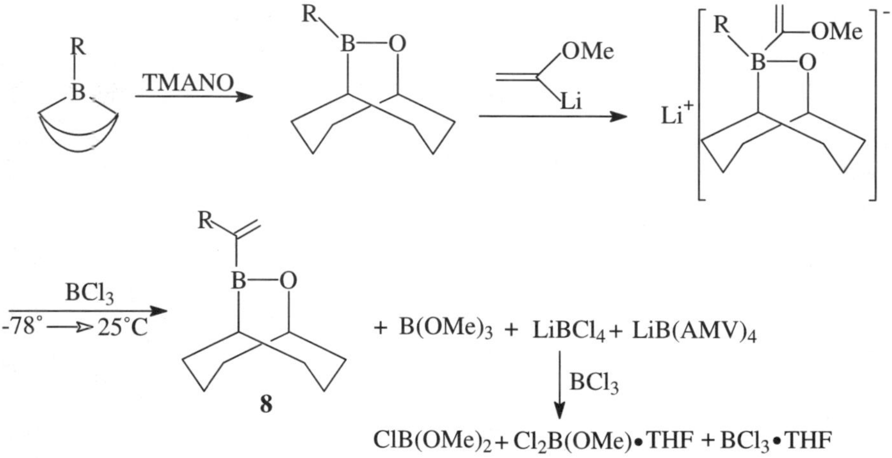

Scheme 20.7

The results are summarized in Table 20.8 [20].

These vinylborinates are remarkably air stable and are easily purified by column chromatography (SiO_2).

Unlike the other vinylborinates, the vinylborinate that contains the allylsilane moiety is very sensitive to the amount of boron trichloride used. The use of 1.3 equiv of borontrichloride also provides an allyldichloroborane (it undergoes

Table 20.8 Preparation of Markovnikov vinylborinates (**8**) from *B*-alkyl-OBBD derivatives [20]

8	R	Yield (%)
a	Me	57
b	*n*-Hx	87
c	$TMSCH_2$	81
d	$TMSCH_2CH_2$	83
e	*c*-Hx	83
f	*t*-Bu	90

hydrolysis on silica), which is formed through the electrophilic conversion of allylsilane (Chart 20.3). However, this is avoided by the use of 1 equiv of borontrichloride [20].

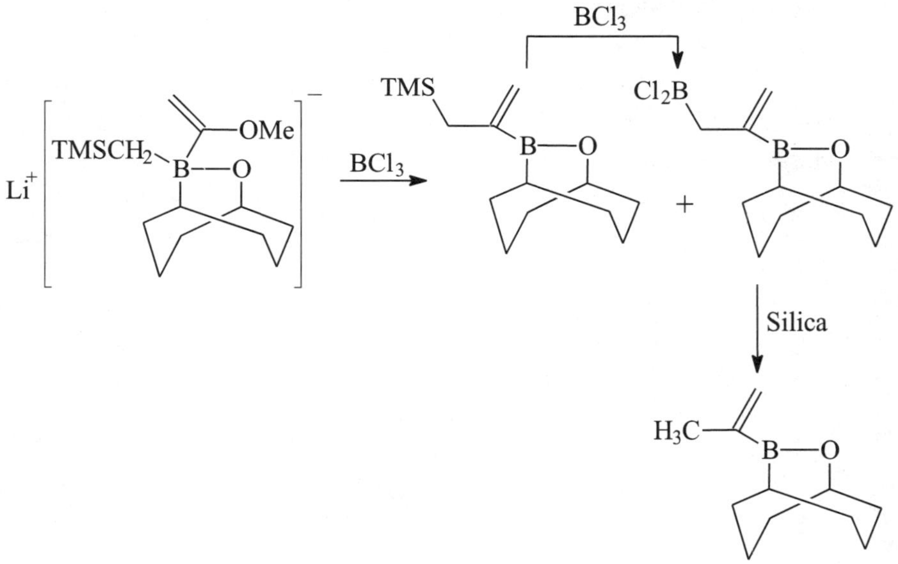

Chart 20.3

20.2 Synthesis of Alkynylborinates

Alkynylboranes are valuable chemical synthones [1] and are employed in enantioselective additions to aldehydes [2], as acetylene equivalent in Diels–Alder cycloaddition [3], as effective precursors to *cis*-vinylboronates [4], and as alkylating agents in Suzuki coupling [5]. These intermediates are normally prepared either through the acid-mediated (BF_3-Et_2O, HCl) demethoxylation of their lithium ate complexes [6] or alternatively, through the transmetalation of alkynyltins with bromoboranes [2, 3]. However, difficulties are encountered with the isolation of the alkynylboranes product in pure form due to THF complexation in the first case, and with the separation of organotin byproduct in the second case. Unlike their dialkyl counterparts, alkynylboronates [RC≡CB(OR′)$_2$] have lower boron Lewis acidities and are isolable free of coordinated solvents [6b]. This clue led Soderquist *et al* [7] to the efficient preparation of alkynylborinates [(RC≡CBR″(OR′)] in pure, uncomplexed form while still retaining the much of the desirable Lewis acid chemistry of dialkyl derivatives. These versatile alkynylborinates possess remarkable oxidation and thermal stabilities.

The stable boronic ester 10-methoxy-9-oxa-10-borabicyclo[3.3.2]decane (OBBD) is readily prepared from *B*-OMe-9-BBN through its selective oxidation [8] with anhydrous trimethyl *N*-oxide (TMANO), (85%, $CHCl_3$, 0 °C). The OBBD undergoes alkylation and demethoxylation cleanly and alkynylborinate, 10-alkynyl-9-oxa-10-borabicyclo[3.3.2]decane, free of coordinated THF, is isolated.

The process is one-pot procedure and 1-alkynes react smoothly to give the corresponding alkynylborinates in high yields and chemical purities (Scheme 20.8) [7].

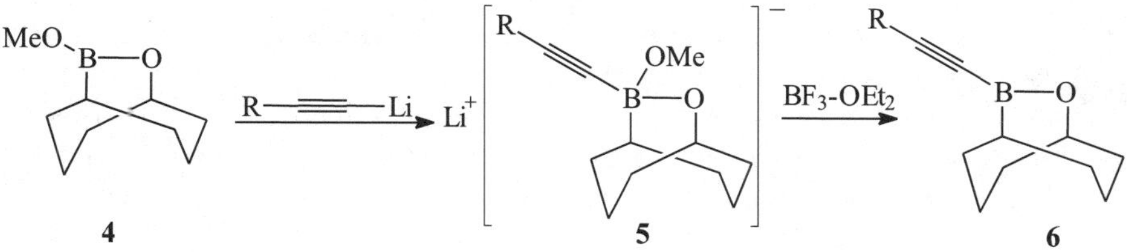

Scheme 20.8

The results are summarized in Table 20.9 [7].

Table 20.9 Alkynylborinates from boronic ester; 10-methoxy-9-oxa-10-borabicyclo[3.3.2]decane [7]

R	Product	Yield (%)
Me	**6a**	71
Et	**6b**	89
Bu	**6c**	86
t-Bu	**6d**	96
TMS	**6e**	81

The alkynylborinates are easier to handle than their unoxidized counterpart **9** [6a]. Moreover, their preparation from **4** is a much simpler overall operation than it is from **9** (Eq. 20.1).

(20.1)

References

Section 20.1

1. (a) Hooz J, Unke S (1969) J Am Chem Soc 90:5936; (b) Hooz J, Gunn DM (1969) ibid. 91:6195; (c) Kabalka GW, Brown HC, Suzuki A, Honma S, Arase A, Itoh M (1970) ibid. 92:712
2. (a) Evans DA (1984) Chapter 1. In: Morrison JD (ed) Asymmetric synthesis, vol. 3. Academic, New York; (b) Heathcock CH (1984) Chapter 2. In: Morrison JD (ed) Asymmetric synthesis, vol. 3. Academic, New York; (c) Evans DA, Nelson JV, Taber TR (1982) Top Stereochem 13:1
3. (a) Mukaiyama T, Inoue T (1976) Chem Lett 559; (b) Inoue T, Uchimaru T, Mukaiyama T (1977) ibid. 153; (c) Inoue T, Mukaiyama T (1980) Bull Chem Soc Jpn 53:174
4. (a) Evans DA, Vogel E, Nelson JV (1979) J Am Chem Soc 101:6120; (b) Evans DA, Nelson JV, Vogel E, Taber TR (1981) ibid. 103:3099
5. (a) Masamune S, Mori S, Van Horn D, Brooks DW (1979) Tetrahedron Lett 19:1665; (b) Hirama M, Masamune S (1979) ibid. 24:2225; (c) Van Horn DE, Masamune S (1979) ibid. 24:2229; (d) Hirama M, Garvey DS, Lu LDL, Masamune S (1979) ibid. 41:3937
6. d'Angelo J (1976) Tetrahedron 32:2979
7. (a) Brown HC, Dhar RK, Ganesan K, Singaram B (1992) J Org Chem 57:499; (b) Brown HC, Dhar RK, Ganesan K, Singaram B (1992) ibid. 57:2716; (c) Brown HC, Ganesan K, Dhar RK (1992) ibid. 57:3767
8. (a) Reetz MT, Kunish F, Heitmann P (1986) Tetrahedron Lett 27:4721; (b) Paterson I, Lister MA, McClure CK (1986) ibid. 27:4787
9. Brown HC, Dhar RK, Bakshi RK, Pandiarajan PK, Singaram B (1989) J Am Chem Soc 111:3441
10. (a) Brown HC, Ravindran N, Kulkarni SU (1979) J Org Chem 44:2417; (b) Kramer GW, Brown HC (1974) J Organomet Chem 73:1
11. Brown HC, Ganesan K, Dhar RK (1993) J Org Chem 58:147
12. Ganesan K, Brown HC (1994) J Org Chem 59:2336
13. Paterson I, Goodman JM, Lister MA, Schumann RC, McClure CK, Norcross RD (1990) Tetrahedron 46:4663
14. Evans DA, Fu GC (1990) J Org Chem 55:5678
15. (a) Boldrini GP, Mancini F, Tagliavini E, Trombini C, Umani-Ronchi A (1990) J Chem Soc Chem Commun 1680; (b) Boldrini GP, Bortolotti M, Tagliavini E, Trombini C, Umani-Ronchi A (1991) Tetrahedron Lett 32:1229

16. Matsumoto Y, Hayashi T (1991) Synlett 349
17. Soderquist JA, Rane AM, Matos K, Ramos J (1995) Tetrahedron Lett 36:6847
18. Colberg JC, Rane AM, Vaquer J, Soderquist JA (1993) J Am Chem Soc 115:6065
19. Soderquist JA, Rivera I (1989) Tetrahedron Lett 30:3919
20. Soderquist JA, Ramos J, Matos K (1997) Tetrahedron Lett 38:6639
21. (a) Soderquist JA, Najafi MR (1986) J Org Chem 51:1330; (b) Soderquist JA, Anderson CL (1986) Tetrahedron Lett 27:3961

Section 20.2

1. Pelter A, Smith K, Brown HC (1988) Borane reagents. Academic, London
2. Corey EJ, Cimprich KA (1994) J Am Chem Soc 116:3151
3. Singleton DA, Leung S-W (1992) J Org Chem 57:4796
4. (a) Srebnik M, Bhat NG, Brown HC (1988) Tetrahedron Lett 29:2635; (b) Gano JE, Srebnik M (1993) ibid. 34:4889; (c) Deloux L, Srebnik M (1994) J Org Chem 59:6871
5. Soderquist JA, Matos K, Rane A, Ramos J (1995) Tetrahedron Lett 36:2401
6. (a) Brown HC, Sinclair JA (1977) J Organomet Chem 131:163; (b) Brown HC, Bhat NG, Srebnik M (1988) Tetrahedron Lett 29:2631
7. Soderquist JA, Rane AM, Matos K, Ramos J (1995) Tetrahedron Lett 36:6847
8. (a) Soderquist JA, Najafi MR (1986) J Org Chem 51:1330; (b) Soderquist JA, Anderson CL (1986) Tetrahedron Lett 27:3961

21 Synthesis and Transformation of Polymers

Polymers have intruded into our daily life, and it looks impossible to remain isolated without these molecules. Consequently, a lot of work is being carried out in various research laboratories to synthesize new types of polymers with better physical and chemical properties. The development of different organoboranes is also contributing significantly for their preparation. Chung *et al* [1] have prepared functionalized copolymer by Ziegler-Natta process, employing 9-(5-hexenyl)-9-BBN and 1-octene to yield octane–hexenol copolymer.

Diblock copolymers having both polydiene and polyalcoholic blocks have been synthesized by the hydroboration of butadiene–isoprene diblock copolymer, followed by H_2O_2 oxidation [2]. Hydroxyl groups containing polymers have been prepared by the hydroboration of 2,5-norbornadiene with 9-BBN in the presence of a WCl_6-Me_4Sn catalyst, followed by alkaline H_2O_2 to afford poly(*exo*-5-hydroxy norbornene) (Eq. 21.1) [3].

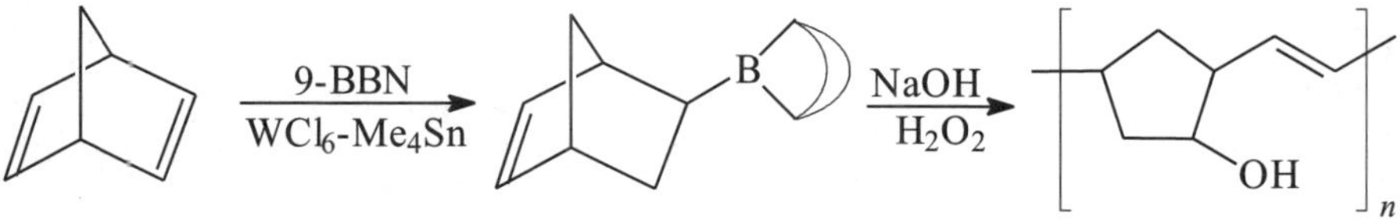

(21.1)

The selective hydroboration of block copolymers containing 1,2-polybutadiene blocks and polyisoprene poly 2-methyl-1,3-pentadine, 1,3-polybutadiene blocks, on 1,2-polybutadiene blocks have been achieved using 9-BBN as the hydroborating agent [3]. The resulting hydroborated copolymers on oxidation afford the hydroxylated copolymers having 2 glass temperature, which show the phase-separated nature of the copolymers.

Copolymers with amorphous A block and liquid crystalline B block have been synthesized. The hydroboration of polybutadiene or butadiene styrene block copolymer with 9-BBN yields the intermediate, which on oxidation, followed by esterification with cholesteryl chloroformate, leads to the formation [4] of the said polymer.

Surface modifications of polymers is brought about by the introduction of alcohol functionality, e.g., poly(tetrafluoroethylene-co-hexafluoropropylene) on reduction with sodium naphthalide in THF results in an unsaturated modified surface layer, the thickness of which is controlled with reaction time and temperature. The air sensitive surface contains alcohols, ketones, aliphatic C–H bonds in addition to C=C and C≡C. The more alcoholic groups are introduced by hydroboration–oxidation, but the esterification leads to the formation of ester in lower yield. This reveals that the reactivity of OH group is similar to hindered alcohols. The reactivity of the surface can be enhanced by chain extension of secondary surface alcohols with ethylene oxide to form a surface containing primary alcohols groups separated from the polymer backbone by C-2 spacer. On the other hand, primary alcohols are directly introduced to the surface by reaction of the reduced layer with 9-BBN, followed by carbonylation and reduction [5].

The polymerization sequence of *B*-R-9-BBN prepared from 1,7-octadiene is employed [6] for intermolecular cross-coupling with 1,4-dibromobenzene or 1,4-diiodobenzene in the presence of dichloro[1,1′-bis(diphenylphosphino) ferrocene] palladium (II) [$PdCl_2$(dppf)]. Similarly, hydroboration of diolefin with 9-BBN, followed by the intermolecular cross-coupling of the resulting α,ω-bis(*B*-alkanediyl-9-BBN) with dihaloarenes are performed in the presence of $PdCl_2$(dppf), a base, and phase-transfer catalyst [7]. Both the steps are performed in the same flask.

The ultrafiltration membrane with high permselectivity and solvent resistance has been prepared from the butadiene–styrene copolymer. The copolymer on hydroboration with 9-BBN, followed by NaOH-H_2O_2 oxidation affords ultrafiltration membrane [8] with good permeation property and separation behavior for several solutes.

Liquid-crystalline side block copolymer with an amorphous A block and liquid crystalline B block has been synthesized from polybutadiene or butadiene–styrene copolymer on hydroboration with 9-BBN, followed by oxidation and esterification with cholesteryl chloroformate [4].

References

1. Ramakrishnan S, Berluche E, Chung TC (1990) Macromolecules 23:378
2. Chung TC (1989) J Polym Sci Polym Chem 27:3251
3. Chung TC, Ramakrishnan S (1989) Polymer Prepr 30:128
4. Adams J, Gronski W (1989) Makromol Chem Rapid Commun 10:553
5. Bening RC, McCarthy TJ (1990) Macromolecules 23:2648
6. Cramer E, Percec V (1990) Polym Prepr 31:516
7. Cramer E, Percec V (1990) J Polym Sci Polym Chem 28:3029
8. Amano T, Koyama K, Hashida I (1991) Kagaku to Kogyo (Osaka) 65:129

22 Synthesis of Alkali Metal 9-Boratabicyclo[3.3.1] nonane (Li, K, and Na 9-BBNH)

Alkali metal trialkylborohydrides are attractive reducing agents in addition to other numerous synthetic applications [1]. Brown and coworkers [2] have also synthesized alkali metal di- and monoalkylborohydrides. Consequently, the reaction of lithium, sodium, and potassium hydrides with 9-BBN, Chx_2BH, Sia_2BH, IPC_2BH, $ThxBH_2$, and $IPCBH_2$ afford the corresponding hydrides, which are examined with respect to rate, stoichiometry, and products.

The reactions are generally carried out with metal hydride (50% excess) suspended in a 0.5-M solution of an alkylborane in THF. When alkylborane is soluble in THF, a standard solution of the alkylborane is introduced to the solid metal hydride. When alkylborane is insoluble in THF, a suspension of metal hydride is introduced to the alkylborane. The reaction mixture, at room temperature, is vigorously stirred or carried out under reflux conditions, depending on the reactivity of the borane. The reaction course is monitored by measuring the hydrogen evolved or by determining the ^{11}B NMR spectrum. The results are summarized in Table 22.1 [2].

The Table 22.1 reveals that there is a remarkable influence of steric effects, and increasing the size of the alkyl group on boron, markedly decreases the reaction rate.

Lithium hydride reacts completely with 9-BBN and $ThxBH_2$ at room temperature after 48 h (Eq. 22.1), whereas Chx_2BH, Sia_2BH, and IPC_2BH remain inert. In addition, the reactions of 9-BBN and $ThxBH_2$ in THF under reflux condition are completed in 3 h. However, the other organoboranes at reflux undergo rapid redistribution and/or elimination of alkyl group.

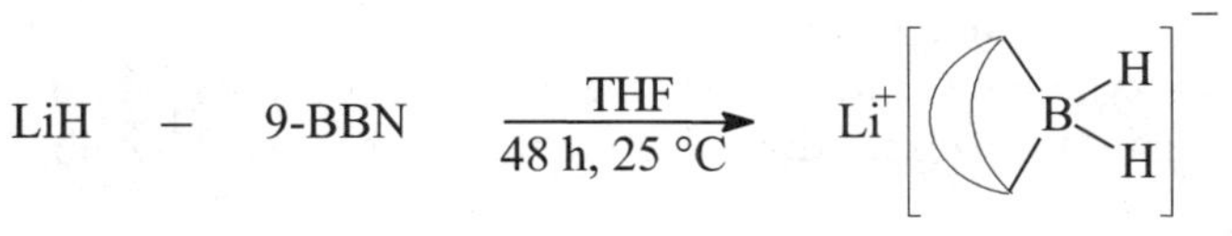

(22.1)

The reaction of all dialkylboranes, on the other hand, is more facile with sodium hydride at 25 °C and follows the order 9-BBN > Chx_2BH > Sia_2BH > IPC_2BH.

Potassium hydride reacts instantly and quantitatively with all the alkylboranes [2].

Table 22.1 Reaction of saline hydrides with representative mono- and dialkylboranes [2] in THF

			Mono- or dialkylborohydride (%)							
Mono- or dialkylborane	Metal hydride	Temp (°C)	15 min	30 min	1 h	2 h	3 h	6 h	12 h	24 h
9-Borabicyclo[3.3.1]-nonane	LiH	25	0	0	0	0	0	0	30	81
	LiH	65	0	4	30	38	54	97	97	
	NaH	25	0	0	0	56	100	100		
	KH	25	100	100						
Dicyclohexylborane	LiH	25	0	0	0	0	0	0	0	20
	LiH	65	0	24	42	58	65	70		
	NaH	25	0	0	0	4	29	98	98	
	KH	25	100	100						
Disiamylborane	LiH	25	0	0	0	0	0	0	12	49
	LiH	65	0	8	73	98	98			
	NaH	25	0	14	27	47	63	89	100	100
	KH	25	100	100						
Diisopinocampheylborane	LiH	25	0	0	14	25	34	49	68	89
	LiH	65	0	12	26	61	95	100		
	NaH	25	0	0	0	12	28	70	99	
	KH	0	0	65	100	100				
Thexylborane	LiH	25	0	0	0	0	0	0	0	0
	LiH	65	0	0	22	78	98	98		
	NaH	25	0	0	0	20	40	100	100	
	KH	25	100	100						
Monoisopinocampheylborane	NaH	25	0	5	10	18	27	55	99	99
	KH	25	80	100	100					

Solutions are 0.5 M in organoborane, and ~50% excess of metal hydride is used.

The order of reactivity of metal hydrides decreases in the following order: KH >> NaH > LiH. The higher reactivity of potassium hydride is attributed to its lower crystal-lattice energy that facilitates the reaction with alkylboranes [3]. The reaction of three hydrides with 9-BBN at 25 °C shows that potassium hydride is exceptionally reactive: lithium hydride reacts 30% in 12 h; sodium hydride, 100% in 3h; and potassium hydride, 100% in <10 min.

Both mono- and dialkylboranes exhibit a simple acid–base reaction in a 1:1 stoichiometry, producing the corresponding alkali metal mono- and dialkylborohydrides (Chart 22.1).

$$R_2BH \quad + \quad MH_{solid} \longrightarrow MR_2BH_{2soln.}$$

$$RBH_2 \quad + \quad MH_{solid} \longrightarrow MRBH_{3soln.}$$

Chart 22.1

The ratio of M:B:H in solution is 1:1:2 for 9-BBN derivatives and 1:1:3 for thexylborane derivatives (Table 22.2) [2].

Table 22.2 Analysis of alkali metal mono- and dialkylborohydride solutions in THF [2]

MH	RBH_2/ R_2BH	Time (h)	Temp (°C)	$[M^+]$ (M)	$[RBH_2]$/ $[R_2BH]$ (M)	$[H^-]$ (M)	Ratio M	B	H
LiH	9-BBN	6	65	0.46	0.5	0.96	1	1.09	2.07
NaH	9-BBN	3	25	0.46	0.5	0.95	1	1.09	2.06
KH	9-BBN	<0.1	25	0.48	0.5	0.98	1	1.04	2.04
LiH	$ThxBH_2$	3	65	0.48	0.5	1.49	1	1.04	3.1
NaH	$ThxBH_2$	6	25	0.5	0.5	1.47	1	1	2.94
KH	$ThxBH_2$	<0.1	25	0.5	0.5	1.5	1	1	3

The alkali metal is estimated by hydrolyzing an aliquot of the reaction mixture with water and titrating the base formed against a standard acid. Boron is determined by GC analysis of the alcohol obtained by oxidation of the alkylboranes. Hydride is estimated by measuring the hydrogen evolved after hydrolysis of the solution.

In the IR of trialkylborohydride [4], a characteristic absorption of B–H stretch occurs around 2,000 cm^{-1}. The exact frequency depends mainly on the solvent used, nature of the alkyl group, and the cation [5]. Dialkylborohydride and monoalkylborohydride exhibit B–H absorption around 2,100 and 2,200 cm^{-1}, respectively. The absorption in sodium borohydride occurs at 2,300 cm^{-1}. The change in the B–H stretching frequency is attributed to the inductive effect of the alkyl substituent (Table 22.3) [2].

Table 22.3 IR and ^{11}B NMR spectral data of alkali metal alkylborohydrides in THF [2]

Mono- and dialkylborohydrides	ν_{B-H} (cm^{-1})	Chemical shift, δ (multiplicity)	J_{B-H} (Hz)
Lithium 9-boratabicyclo[3.3.1]nonane	2,150	−17.4 (t)	72.1
Lithium disiamylborohydride	2,100	−14.8 (t)	69
Lithium thexylborohydride	2,200	−24.47 (q)	71
Sodium 9-boratabicyclo[3.3.1]nonane	2,125	−18.37 (t)	72
Sodium disiamylborohydride	2,100	−14.95 (t)	71
Sodium dicyclohexylborohydride		−11.82 (t)	71.8
Sodium thexylborohydride	2,225	−25.69 (q)	72
Potassium 9-boratabicyclo[3.3.1]nonane	2,145	−16.27 (t)	69
Potassium disiamylborohydride	2,125	−13.06 (t)	71
Potassium diisopinocampheylborohydride		− 4.84 (t)	69
Potassium thexylborohydride	2,200	−21.74 (q)	71
Potassiummonoisopinocampheylboro hydride		−21.20 (q)	71

The progress of the reaction between metalhydride and alkylborane is monitored by ^{11}B NMR spectroscopy, and chemical shifts in δ values are relative to $Et_2O{\cdot}BF_3$. Mono- and dialkylboranes exhibit signals between δ +20 to +35, while the corresponding hydrides show signals between δ –5 to –22. In the 9-BBN and $ThxBH_2$ systems, even in the presence of unreacted alkylboranes, the borohydrides show sharp triplets and quartets, respectively. On the other hand, Chx_2BH, Sia_2BH, and IPC_2BH exhibit broad singlets in the presence of unreacted dialkylboranes, which get resolved into sharp triplets on utilization of the whole of the dialkylborane. This results from the exchange phenomenon, which is greatest in the potassium derivatives and is similar to the KR_3BH-R_3B system [6]. The exchange phenomenon is the least in lithium derivatives.

These derivatives are very stable and can be stored at 25 °C under nitrogen without any hydride loss, redistribution, or isomerization of the alkyl groups for at least 3 months.

Further, methyl iodide readily and quantitatively removes metal hydride from these adducts, thus regenerating the free mono- and dialkylboranes (Chart 22.2).

$$MR_2BH_2 + MeI \xrightarrow[0.5\ h,\ 25\ °C]{THF} R_2BH + MI\downarrow + CH_4\uparrow$$

$$MRBH_3 + MeI \xrightarrow[0.5\ h,\ 25\ °C]{THF} RBH_2 + MI\downarrow + CH_4\uparrow$$

$$M = Li, Na, K$$

Chart 22.2

This system, thus, provides a method to store the otherwise unstable hydroborating agents.

References

1. (a) Krishnamurthy S (1974) Aldrichimica Acta 7:55; (b) Brown HC, Dickason WC (1970) J Am Chem Soc 92:709; (c) Krishnamurthy S, Brown HC (1976) ibid. 98:3383; (d) Krishnamurthy S, Brown HC (1976) J Org Chem 41:3064; (e) Krishnamurthy S, Vogel F, Brown HC (1977) ibid. 42:2534
2. Brown HC, Singaram B, Mathew CP (1981) J Org Chem 46:2712
3. Brown CA (1974) J Org Chem 39:3913
4. Brown HC, Krishnamurthy S, Hubbard JL (1978) J Am Chem Soc 100:3343
5. Binger P, Benedikt G, Rotermund GW, Köster R (1968) Justus Liebigs Ann Chem 717:21
6. Brown CA (1978) J Organomet Chem 156:C17

23 Synthesis of *B*-R-9-BBN Not Available *via* Hydroboration

The *B*-alkyl, alkenyl, alkynyl, or aryl-9-BBN derivatives not available *via* hydroboration are obtained by alkylation of 9-BBN or *B*-Cl-9-BBN or *B*-MeO-9-BBN with organolithium, Grignard reagent, organocadmium, lithium dialkylcuprates, (phenylthio)alkylcuprates, and hydroboration–carbonylation–reduction.

In THF or hydrocarbon solvents, 9-BBN reacts rapidly with 1 equiv of organolithium reagents at –78 °C. The intermediate formed is oxidized with methyl iodide [1] or protonolyzed with methanesulfonic acid [2] to afford excellent yields of *B*-R-9-BBN (Chart 23.1; Table 23.1) [1].

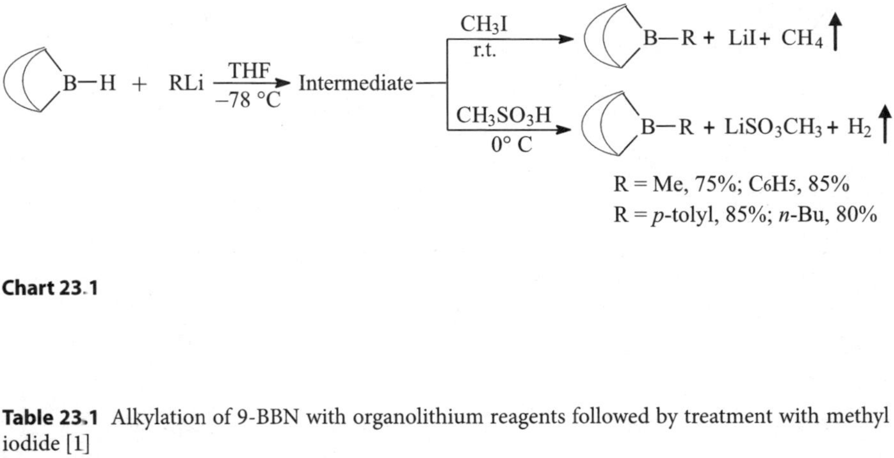

Chart 23.1

Table 23.1 Alkylation of 9-BBN with organolithium reagents followed by treatment with methyl iodide [1]

RLi	Yield (%)	RLi	Yield (%)
Methyl	98	*tert*-Butyl	98
iso-Propyl	82	*neo*-Pentyl	[a]
n-Butyl	99	Phenyl	94
sec-Butyl	94	*p*-Tolyl	97

[a] *neo*-Pentylalcohol is obtained in 100% yield after oxidation of the intermediate.

B-Chloro-9-BBN, prepared readily from 9-BBN by treatment with dry hydrogen chloride in ether, reacts rapidly with organolithiums at –78 °C in *n*-pentane. The intermediate ate complex is not stable and breaks down to organoborane and lithium chloride even at –78 °C (Chart 23.2) [1].

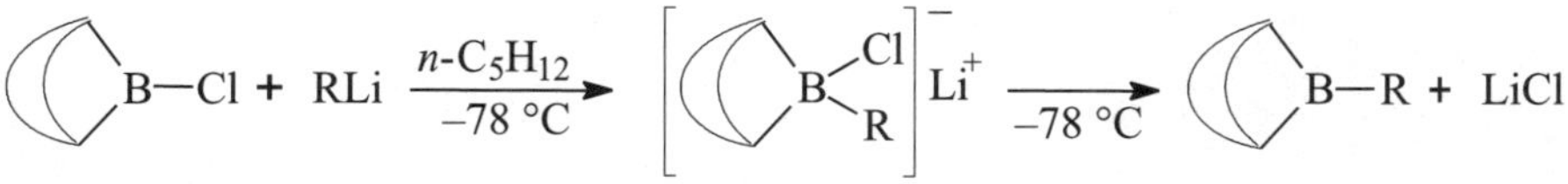

Chart 23.2

The lithium salt precipitates out quantitatively from pentane solution. Consequently, the pure *B*-R-9-BBN is isolated by decantation from the salt and removal of the solvent. It is necessary that the stoichiometry of the reaction is closely controlled in order to achieve high yields. The excess organolithium reagent reacts to give the insoluble ate complex.

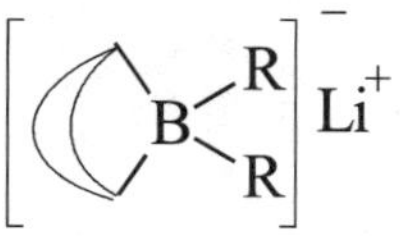

The results of alkylation of *B*-Cl-9-BBN with organolithiums are summarized in Table 23.2 [1].

Table 23.2 Alkylation of *B*-chloro-9-BBN with organolithium reagents [1]

RLi	Yield (%)	RLi	Yield (%)
Methyl	97	*tert*-Butyl	99
Ethyl	97	Cyclopentyl	84
iso-Propyl	97	*neo*-Pentyl	94
n-Butyl	96	Phenyl	100
sec-Butyl	99		

The dialkylborinate, *B*-OMe-9-BBN, prepared by methanolysis of 9-BBN is remarkably stable toward redistribution, as heating it for 1 week in refluxing *n*-decane produces no change in composition.

B-OMe-9-BBN on treatment with an organolithium in *n*-pentane at –78 °C results in the formation of white precipitate. On warming to room temperature

(25 °C) this white precipitate dissolves with slow formation of another white precipitate of lithium methoxide. The pure organoborane is isolated by decantation of the liquid from the insoluble lithium methoxide and removal of the solvent (Chart 23.3) [1]. It is necessary that the stoichiometry of the *B*-OMe-9-BBN is carefully controlled to achieve high yields.

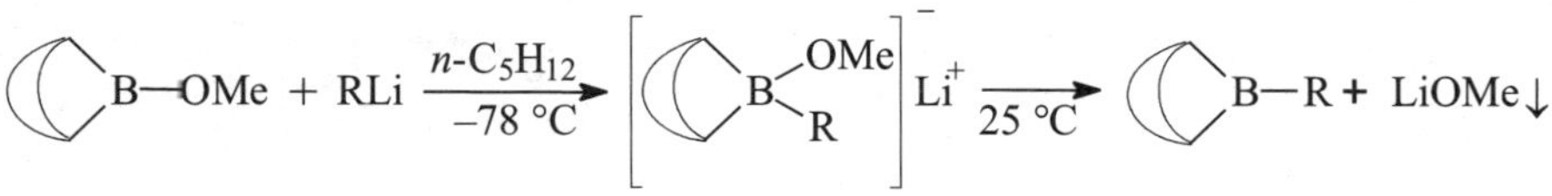

Chart 23.3

The results are summarized in Table 23.3 [1].

Table 23.3 Alkylation of *B*-OMe-9-BBN with organolithium reagents [1]

RLi	Yield (%)	RLi	Yield (%)
Methyl	94	*sec*-Butyl	87
Ethyl	93	*tert*-Butyl	90
Allyl	95	*neo*-Pentyl	97
iso-Propyl	90	Phenyl	95
n-Butyl	89	Benzyl	97

The synthesis of *B*-R-9-BBN using Grignard reagent is fruitful only in the case of methylmagnesium compounds. Consequently, methyl Grignard reagent and dimethylmagnesium give good yield of *B*-Me-9-BBN (Chart 23.4) [1].

Reagent	Yield (%)		
	X = I	X = Br	X = Cl
CH_3MgX	93	101	95
$CH_2{=}CHCH_2MgX$	-	66	95
Me_2Mg	77		

Chart 23.4

The boiling points of some of the *B*-R-9-BBN are summarized in Table 23.4.

Table 23.4 Boiling points of *B*-R-9-BBN [1]

R	B.P. (°C)/mm	R	B.P. (°C)/mm
Methyl	23–24/0.5	*neo*-Pentyl	68–69/0.7
Ethyl	32–34/0.2	Cyclcopentyl	68–70/0.2
iso-Propyl	28–30/0.05 35.5–37/0.25	Allyl	41–42/0.05
n-Butyl	58–59/0.5	Benzyl	105–125/0.5
sec-Butyl	56–58/0.2	Phenyl	90/0.4
t-Butyl	27–28/0.02	*p*-Tolyl	155–160/5.5

The *B-tert*-butyl-9-BBN can also be obtained from *B*-isopropyl-9-BBN. The *B*-isopropyl-9-BBN on bromination in methylene chloride with continuous removal of HBr gives *B*-2′-bromopropyl-9-BBN. The bromoderivative on treatment with methyllithium provides *B-tert*-butyl-9-BBN (Chart 23.5) [3], in quantitative yield.

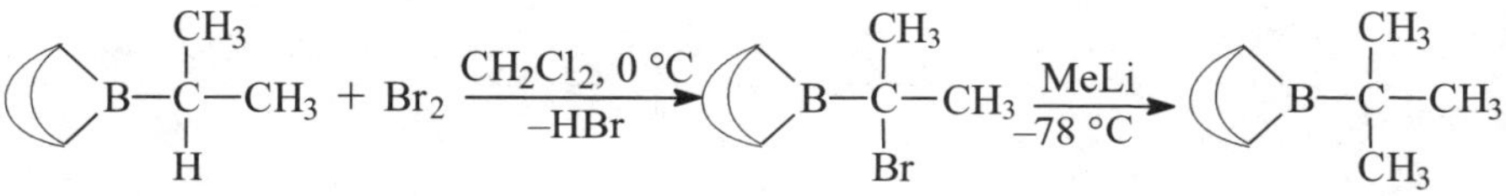

Chart 23.5

The *B*-benzyl-9-BBN and *B-p*-$CH_3OC_6H_4CH_2$-9-BBN are obtained in good yields (Chart 23.6) [4] on treatment of *B*-C_6H_5-9-BBN and *B-p*-methoxyphenyl-9-BBN, respectively, with thiomethoxymethyllithium-TMEDA [5].

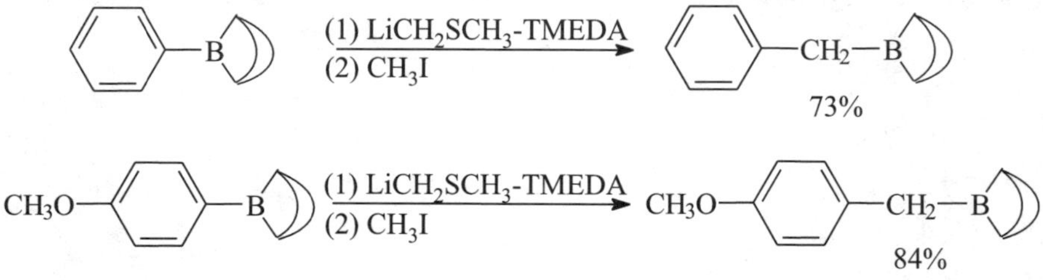

Chart 23.6

B-Alkyl-9-BBN derivatives have been synthesized, in a single stage approach to their preparation, by the alkylation of the electron deficient boron atom of 9-BBN (Scheme 23.1) with readily available lithium dialkylcuprates [7].

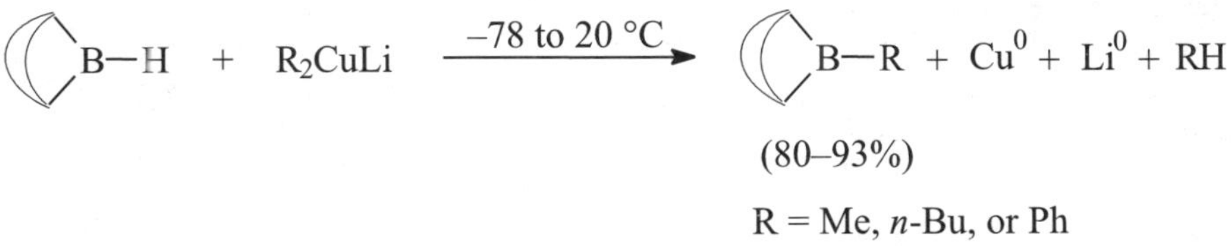

Scheme 23.1

The reaction has an obvious disadvantage, as only one of the alkyl groups of cuprate is utilized, a problem if such cuprates are difficult to prepare or involve expensive starting material. A systematic investigation, however, solved the problem by using heterocuprates [8]. The results show high-yield transfer of the alkyl group with lithium (phenylthio) alkycuprates (Scheme 23.2; Table 23.5) [8], moderate-yield with lithium phenoxyalkylcuprates, and disappointing results with lithium *tert*-bitoxyalkyl cuprate [8].

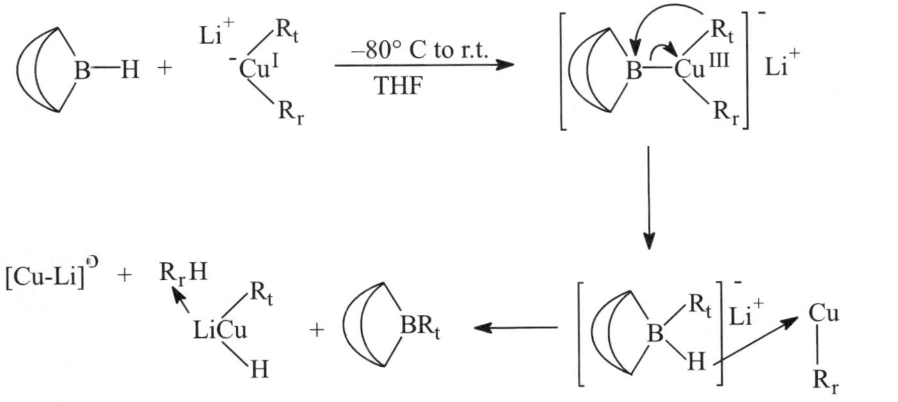

R_t= valuable (transferable) alkyl group; R_r= dispensable, residue alkyl group=PhS

Scheme 23.2

Thus, the reaction of mixed (phenylthio)alkylcuprates with 9-BBN overcomes the problem associated with (1) the thermal instability of phenoxy- and *tert*-butoxyalkylcuprates, (2) the low reactivity of mixed acetylenic cuprates, and (3) the wasting of one R group of the homocuprates.

The stereoisomer, not obtained by direct hydroboration, has been prepared by an addition step of carbonylation of *B*-R-9-BBN. The hydroboration of 2-methylene*nor* bornene is expected to proceed predominantly from the *exo* face to give the *endo* derivative. However, the *exo* derivative is prepared *via* hydrobo-

Table 23.5 Reaction of (phenylthio)alkylcuprates, PhS(R_t)CuM with 9-BBN [8]

R_t	M	Oxidative product	Yield (%)
Me	Li	MeOH	92
Ph	MgBr	PhOH	77
$PhCH_2$	MgBr	$PhCH_2OH$	67
n-Bu	Li	*n*-BuOH	86
t-Bu	Li	*t*-BuOH	83
n-Pr	MgBr	*n*-PrOH	90

ration–carbonylation of *nor*bornene [9]. The stereochemistry is determined by the hydroboration step, and carbonylation proceeds with retention of configuration [10]. Consequently, reduction by lithium aluminum hydride of the intermediate formed in the hydride-induced carbonylation of *B*-R-9-BBN provides a valuable method for the stereospecific homologation of *B*-R-9-BBN to *B*-RCH_2-9-BBN. The results are shown in Chart 23.7 [9].

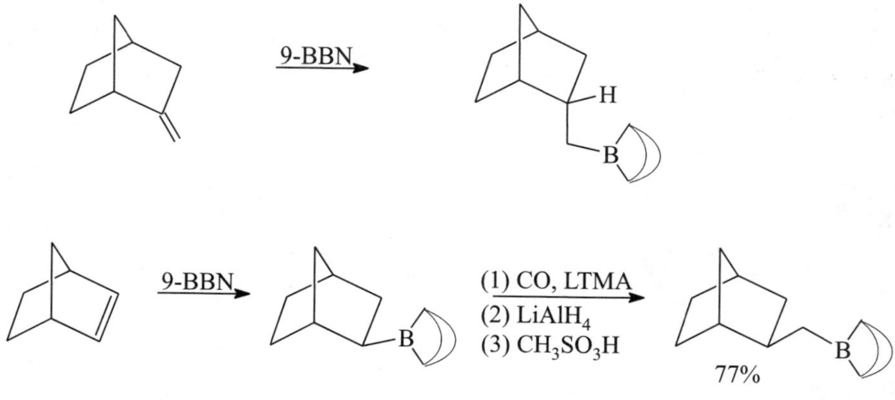

Chart 23.7

Strict temperature control during carbonylation and reduction is crucial for success. Thus, carbonylation at –20 °C in the presence of 1.3 equiv of freshly prepared lithium trimethoxyaluminum hydride (LTMA) and reduction at –20 °C with 1 molar equiv of lithium aluminum hydride are carried out.

The reaction is extended, for example, in the synthesis of *trans* isomer from 1-methylcyclopentene. The hydroboration of 2-methylmethylenecyclopentane yields a mixture predominanting in the *cis* isomer. On the other hand, hydroboration-carbonylation of 1-methylcyclopentene yields only the *trans* isomer (Chart 23 8) [9].

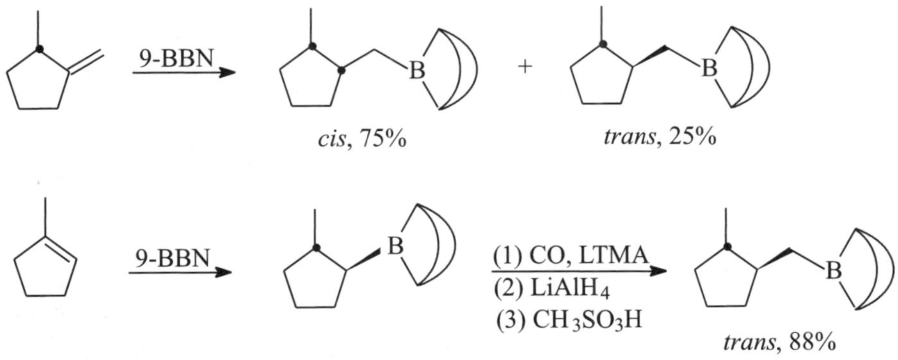

Chart 23.8

B-1-Alkynyl-9-BBN derivatives are prepared as 1:1 THF complexes in excellent yields. *B*-Methoxy-9-BBN in THF on treatment with an alkynyllithium reagent at –78 °C results in the formation of an adduct. This adduct is stable and may be isolated as a complex with THF. However, the treatment of this ate complex with 1.3 equiv of $BF_3{\cdot}OEt_2$ at –78 °C, followed by warming to 25 °C results in the formation of trimethylborate, lithium tetrafluoroborate, and the desired *B*-1-alkynyl–9-BBN·THF complex. This complex is easily isolated by evaporation of the THF solvent, extraction with pentane, and crystallization. The *B*-1-alkynyl–9-BBN·THF complexes are stable crystalline solids that can be stored at room temperature for up to 1 year with no apparent decomposition (Chart 23.9; Table 23.6) [11].

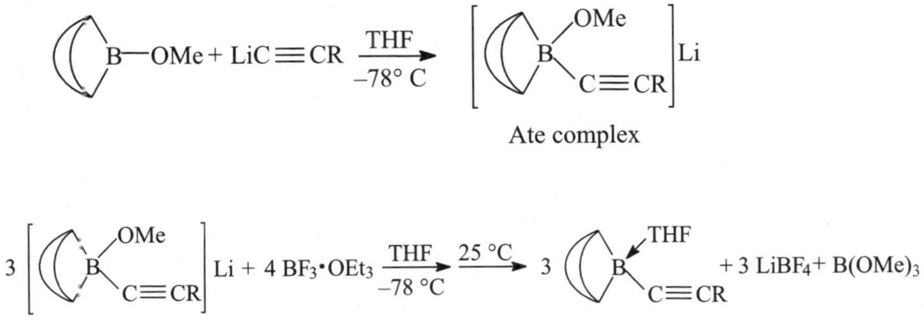

Chart 23.9

Table 23.6 Preparation of *B*-1-alkynyl-9-BBN·THF derivatives [11]

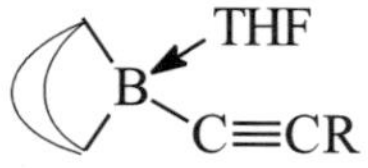
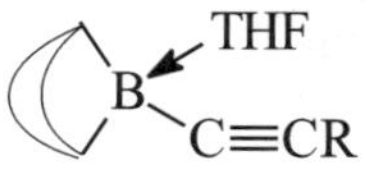

R	Yield (%)	R	Yield (%)
H	[a]	*n*-Hexyl	100
Ethyl	88	Phenyl	90
n-Butyl	86	3-Chloro-*n*-propyl	95
t-Butyl	94		

[a] Product is decomposed on warming from −78 °C.

The *B*-(*E*)-1-alkenyl-9-BBN derivatives exhibit exceptional stereospecific reactivity [12] and are prepared via the regio- and stereospecific monohydroboration [13] of 1-alkynes with 9-BBN or dehydroboration [14] of the 1,1-diboryl adduct. For the synthesis of *B*-(*Z*)-1-alkenyls-9-BBN, the hydroborylation of 1-haloalkynes with 9-BBN, followed by treatment with *tert*-butyllithium afford [15] a mixture of two products, due to indiscriminate migration of both the hydride and cyclooctyl ring. The controlled catalytic hydrogenation of *B*-1-alkynyl-9-BBN derivatives under a variety of experimental conditions also results in a complex mixture of products [16].

The *B*-(*Z*)-1-alkenyl-9-BBN, however, are prepared from (*Z*)-1-lithio-1-alkene [17] and *B*-OMe-9-BBN. (*Z*)-1-Iodo-1-alkene [18] on reaction with *t*-butyllithium in pentane affords the corresponding (*Z*)-1-lithio-1-alkene. The treatment of a solution of (*Z*)-1-lithio-1-alkene in diethylether-*n*-pentane mixture at −78 °C with *B*-OMe-9-BBN in THF affords the ate complex. Treatment of the ate complex with 1.33 equiv of borontrifluoride-diethyl etherate at −78 °C, followed by warming results in the formation of the desired *B*-(*Z*)-1-alkenyl-9-BBN (Chart 23.10; Table 23.7) [19].

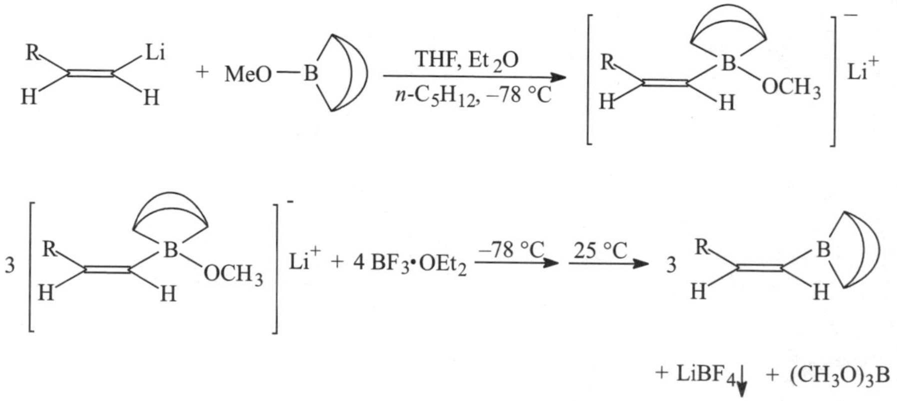

Chart 23.10

Table 23.7 Preparation of *B*-(*Z*)-1-alkenyl-9-BBN [19]

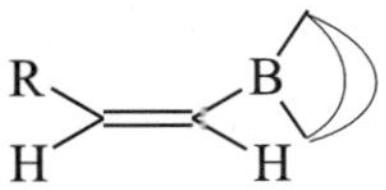

R	Yield (%)	b.p. (°C [*p*, Pa])
n-C_4H_9	75	80–82 (13.33)
n-C_5H_{11}	70	104–106 (66.67)
i-C_3H_7	72	54–56 (40)
c-C_6H_{11}	74	88–90 (4)
t-C_4H_9	70	60–62 (26.67)

References

1. Kramer GW, Brown HC (1974) J Organomet Chem 73:1
2. Brown HC, Rogić MM (1969) J Am Chem Soc 91:4304
3. Brown HC, De Lue NR (1974) J Am Chem Soc 96:311
4. Negishi E, Yoshida T, Silveira A Jr, Chiou BL (1975) J Org Chem 40:814
5. Peterson DJ (1967) J Org Chem 32:1717
6. (a) Whitley CG (1981) J Am Chem Soc Chem Commun 5; (b) Whitley CG (1982) S Afr J Chem 35:9
7. (a) Posner GH (1972; 1975) Org React 19:1 and 22:253; (b) Posner GH, Whitten CE, Sterling JJ (1973) J Am Chem Soc 95:7788
8. Whitley CG, Zwane I (1985) J Org Chem 50:1969
9. Brown HC, Ford TM, Hubbard JL (1980) J Org Chem 45:4067
10. Brown HC, Knights EF, Coleman RA (1969) J Am Chem Soc 91:2144
11. Brown HC, Sinclair JA (1977) J Organomet Chem 131:163
12. (a) Jacob P III, Brown HC (1976) J Am Chem Soc 98:7832; (b) Jacob P III, Brown HC (1977) J Org Chem 42:579; (c) Brown HC, Molander GA (1981) ibid. 46:645; (d) Molander GA, Singaram B, Brown HC (1984) ibid. 49:5024
13. Brown HC, Scouten CG, Liotta R (1979) J Am Chem Soc 101:96
14. Colberg JC, Rane A, Vaquer J, Soderquist JA (1993) J Am Chem Soc 115:6065
15. Campbell JB Jr, Molander GA (1978) J Organomet Chem 156:71
16. Molander GA (1979) Ph.D. Thesis, Purdue University, West Lafayette, Indiana
17. Corey EJ, Beames DJ (1972) J Am Chem Soc 94:7210
18. Zweifel G, Arzoumanian H (1967) J Am Chem Soc 89:5086
19. Brown HC, Bhat NG, Rajagopalan S (1986) Organometallics 5:816

24 Synthesis of Unsaturated Compounds

24.1
Synthesis of Alkenes

24.1.1
Synthesis of Acyclic, Cyclic, and Heterocyclic Alkenes

As already mentioned, the internal triple bond is hydroborated (90–96%) with equimolar quantity of 9-BBN, while twofold excess of terminal alkyne is used for monohydroboration (90–100%) with 9-BBN [1]. Protonolysis of *B*-vinyl-9-BBN derivatives provides the corresponding alkenes. Thus, protonolysis of the product from 5-decyne and 9-BBN yields only *cis*-5-decene (Eq. 24.1).

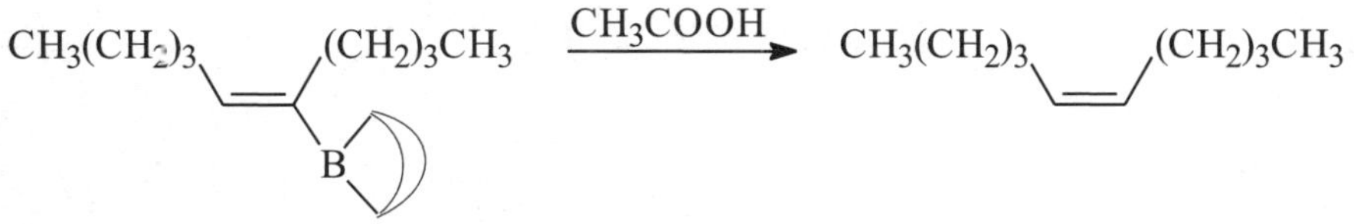

(24.1)

This method is utilized [1] for the synthesis of deuterated alkenes (Eq. 24.2) with *trans* geometry, which otherwise are prepared only by tedious methods.

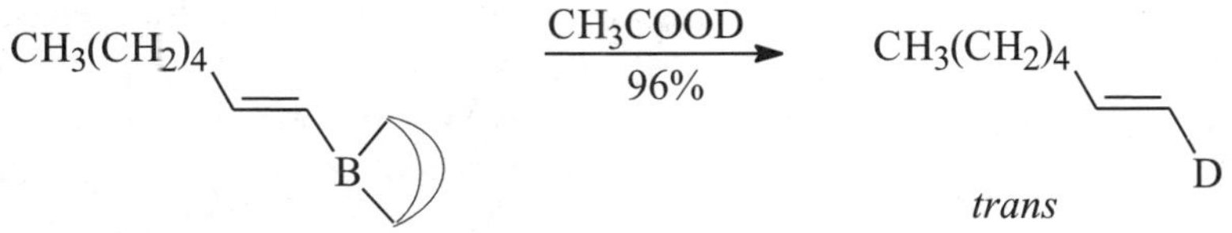

(24.2)

An alternative and more efficient method for synthesis of *trans*-vinyl-9-BBN was developed by Soderquist [2], where 1 equiv of 9-BBN is sacrificed instead of the terminal alkyne. The process involves the dihydroboration of terminal alkyne (1 equiv) with 2 equiv of 9-BBN. The 1,1-dibora product is then treated with 1 equiv of PhCHO for 2 h at 25 °C, and *B*-$PhCH_2O$-9-BBN and vinyl-9-

BBN are quantitatively formed, the latter with exclusively the *trans* configuration (Scheme 24.1) [2].

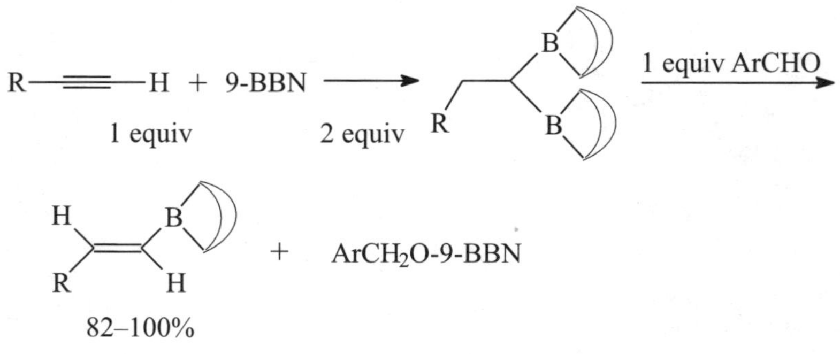

Scheme 24.1

The use of 1-naphthaldehyde instead of benzaldehyde renders the distillative separation of *trans*-vinylboranes, a greatly simplified process in most cases.

The vinylboranes that are not formed efficiently even with large excess of terminal alkynes, however, are prepared conveniently from 1,1-dibora adducts and are smoothly converted to *trans*-1-deuterio-1-alkenes with CD_3COOD or AcOD at 0 °C (Eq. 24.3).

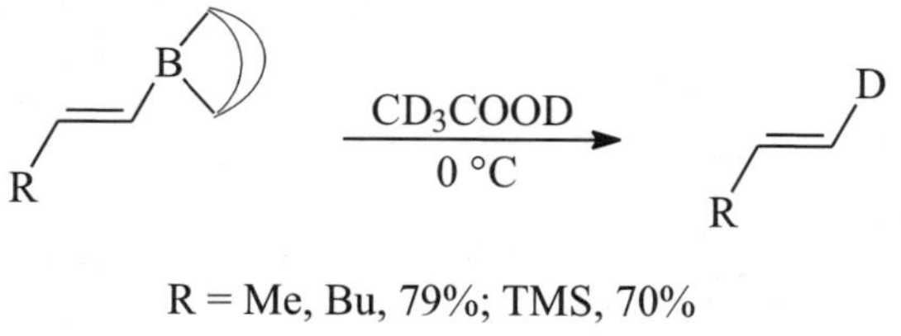

(24.3)

The conversion of carbonyl compounds to alkenes has always attracted considerable interest, and numerous methods have been developed for their conversions [3–10].

It has been found [4b] that carbonyl derivatives as enamines *via* the hydroboration process give the corresponding alkenes in excellent yield. However, the drastic reaction conditions make the isolation of the highly volatile alkenes difficult. Moreover, no stereochemistry has been assigned to starting acyclic enamines or the resulting product alkenes. Brown and coworkers have found [11, 12] that 1-morpholinocyclopentene with 9-BBN in THF affords the corresponding trialkylborane cleanly. This trialkylborane slowly (in 192 h) undergoes elimination at 25 °C to give cyclopentene (Eq. 24.4).

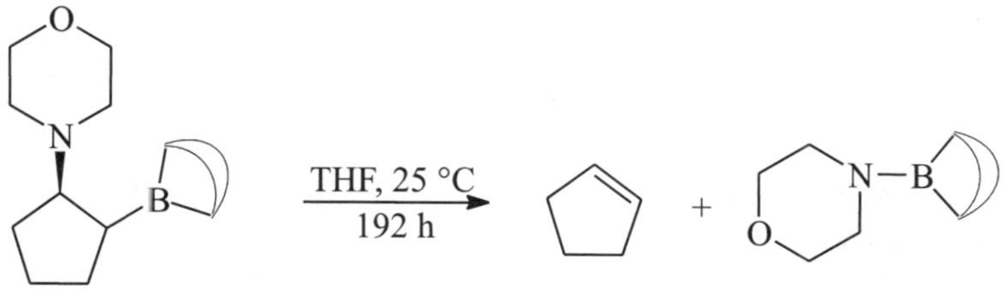

(24.4)

In addition, trialkylborane obtained from 1-morpholinocyclohexene and 9-BBN on alkaline hydrogen peroxide oxidation does not yield any aminoalcohol. In fact, the elimination takes place after the addition of sodium hydroxide at 25 °C to give cyclohexene. It is further found [13] that even water and methanol also facilitate this elimination reaction.

The hydroboration of enamine is remarkable regio- and chemoselective as the boron adds to the electron-rich enamine double bond. Consequently, hydroboration of enamines and the treatment of the resulting trialkylboranes with methanol afford the corresponding alkenes in excellent yield [13]. The process is a general procedure for the synthesis of terminal alkenes from aldehydes and internal or cyclic alkenes from ketones (Eq. 24.5).

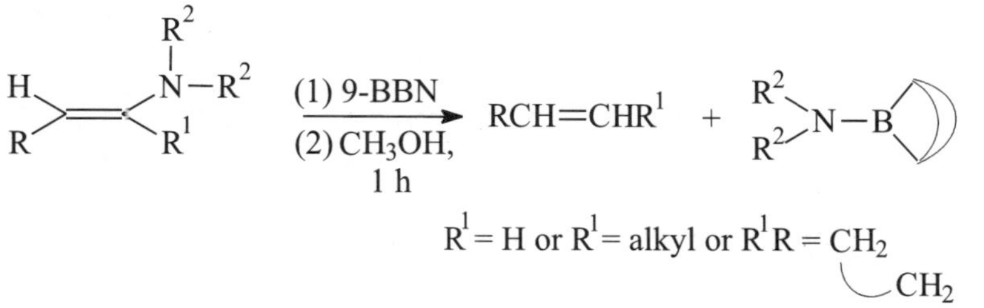

(24.5)

The results of formation of terminal monoene are summarized in Table 24.1 [13a].

This general procedure for the synthesis of alkenes from the enamines, when applied to the acyclic enamines derived from the acyclic ketones by modification of hydroboration-elimination procedure, permits a facile, diastereospecific conversion to either (*Z*)- or (*E*)-alkene at will: (A) The hydroboration of enamine with 9-BBN, followed by treatment with methanol gives (*Z*)-alkenes of 99% isomeric purity. (B) The hydroboration of the same enamine with borane–methylsulfide, followed by methanolysis and oxidation with neutral hydrogen peroxide gives (*E*)-alkenes of 99% isomeric purity (Scheme 24.2) [13a].

The results are summarized in Table 24.2 [13a].

Table 24.1 Alkenes from aldehyde enamines [13a]

Enamine	Alkene	Yield (%)
(*E*)-1-Morpholino-1-octene	1-octene	80
(*E*)-1-Morpholino-3-phenyl-1-propene	3-phenyl-1-propene	82
(*E*)-1-Phenyl-2-pyrrolidino- ethene	1-phenylethene	76

Table 24.2 (*Z*)- and (*E*)-Alkenes from acyclic ketone enamines [13a]

Enamine	Procedure	Alkene	Yield %
(*E*)-1-Morpholino-1-phenyl-1-propene	A	(*Z*)-1-Phenyl-1-propene	80
	B	(*E*)-1-Phenyl-1-propene	75
(*E*)-1,2-Diphenyl-1-morpholinoethene	A	(*Z*)-1,2-Diphenylethene	65
	B	(*E*)-1,2-Diphenylethene	80
(*E*)-1-Morpholino-1-(4-pyridyl)-1-propene	A	(*Z*)-1-(4-Pyridyl)-1-propene	60
	B	(*E*)-1-(4-Pyridyl)-1-propene	30
(*E*)-1-Morpholino-1-(2-thienyl)-1-propene	A	(*Z*)-1-(2-Thienyl)-1-propene	68
	B	(*E*)-1-(2-Thienyl)-1-propene	77
(*E*)-4-Pyrrolidino-3-heptene	A	(*Z*)-3-Heptene	69
	B	(*E*)-3-Heptene	50
(*E*)-5-Morpholino-4-nonene	A	(*Z*)-4-Nonene	75
	B	(*E*)-4-Nonene	80
(*E*)-6-Morpholino-5-undecene	A	(*Z*)-5-Undecene	90
	B	(*E*)-5-Undecene	85

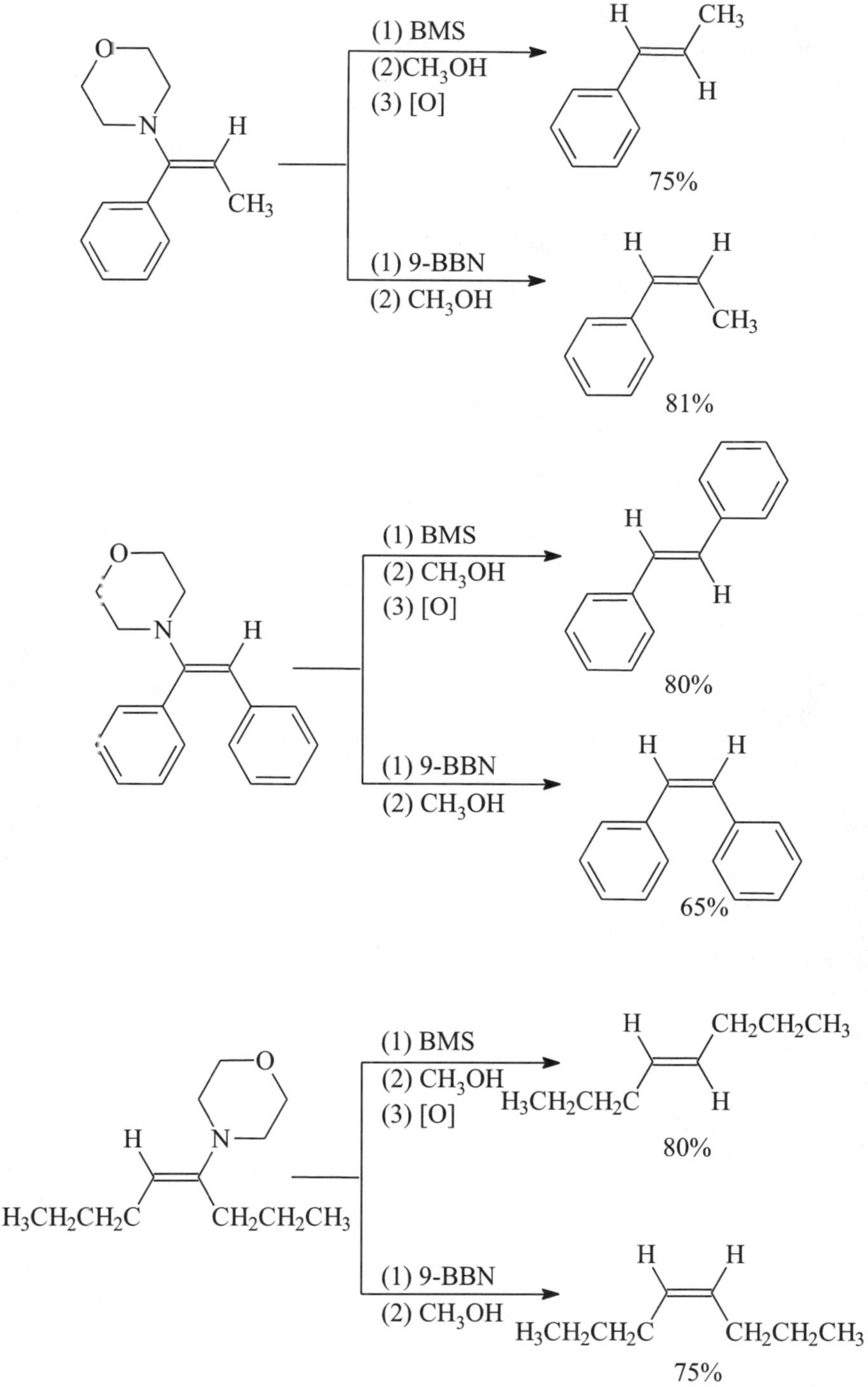
(1) BMS
(2)CH3OH
(3) [O]
75%
(1) 9-BBN
(2) CH3OH
81%
(1) BMS
(2) CH3OH
(3) [O]
80%
(1) 9-BBN
(2) CH3OH
65%
(1) BMS
(2) CH3OH
(3) [O]
H3CH2CH2C
CH2CH2CH3
80%
(1) 9-BBN
(2) CH3OH
75%

Scheme 24.2

The elimination of BMS derivative takes place via *syn*-elimination, while of trialkylborane of 9-BBN proceeds through *anti*-elimination (Chart 24.1) [13].

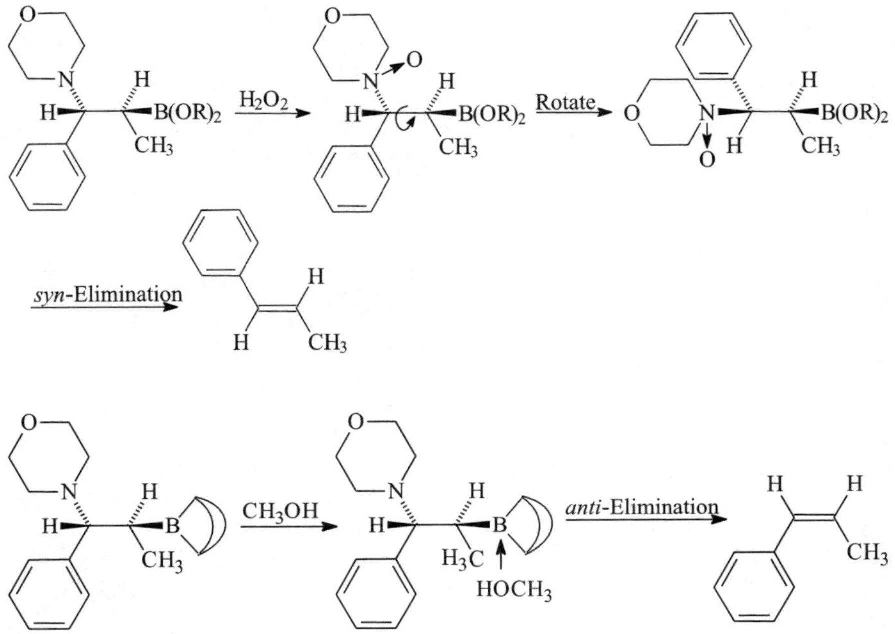

Chart 24.1

The reaction is also general procedure for the conversion of cyclic ketones through the hydroboration–elimination sequence (Eq. 24.6) to corresponding cycloalkenes [13].

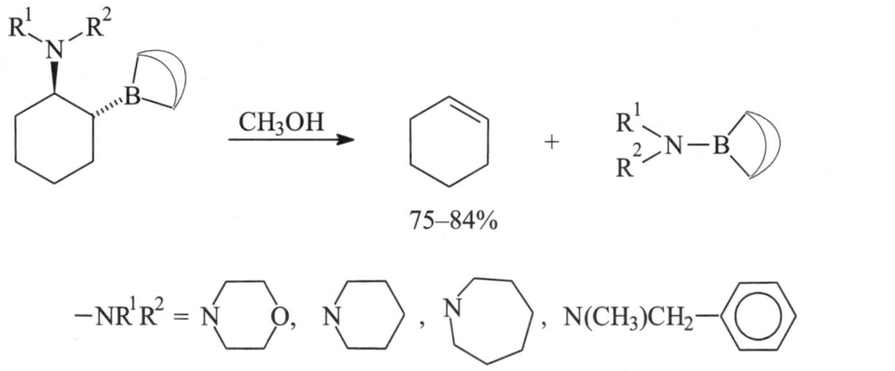

(24.6)

The formation of pyrrolidine enamines from 2-alkylcycloalkenones is highly regiospecific [14] and affords the less substituted enamines. Thus, 2-methylcyclohexanone yields a mixture of 6-methyl-1-pyrrolidinocyclohexene (90%) and 2-methyl-1-pyrrolidinocyclohexene (10%), which is then converted to 3-methylcyclohexene (Eq. 24.7) [13].

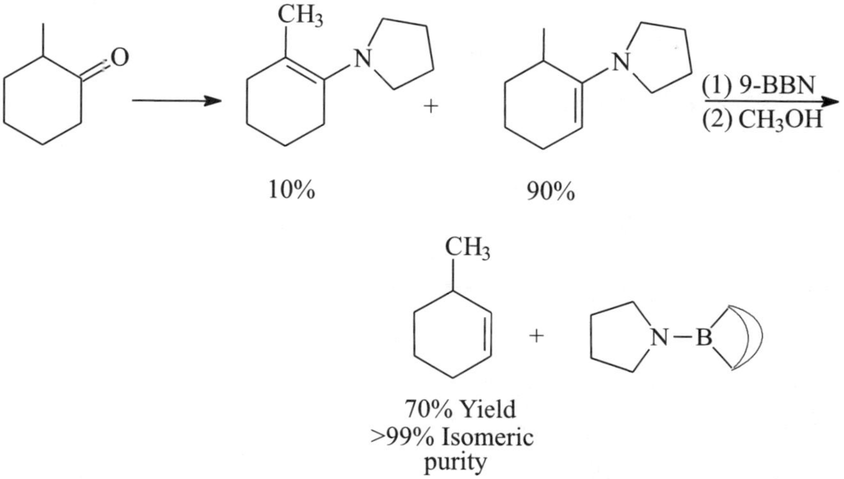

(24.7)

The results of unsubstituted cycloalkanones and substituted cyclohexanones are summarized in Table 24.3 [13a].

Table 24.3 Cycloalkenes and substituted cycloalkenes from ketone enamines [13a]

Enamine	Cycloalkene	Yield (%)
1-Morpholinocyclopentene	Cyclopentene	82
1-Morpholinocyclohexene	Cyclohexene	84
1-Piperidinocyclohexene	Cyclohexene	80
1-(Benzylmethylamino)cyclohexene	Cyclohexene	75
6-Methyl-1-pyrrolidinocyclohexene	3-Methylcyclohexene	70
6-Cyclohexyl-1-pyrrolidinocyclohexene	3-Cyclohexylcyclohexene	75
6-Phenyl-1-pyrrolidinocyclohexene	3-Phenylcyclohexene	69
1-Morpholino-4-*tert*-butylcyclohexene	4-*tert*-Butylcyclohexene	72
3,3,5,5-Tetramethyl-1-piperidino-cyclohexene	3,3,5,5-Tetramethylcyclo-hexene	68

The generality of hydroboration–elimination is further extended to the synthesis of heterocyclic alkenes. The results are presented in Table 24.4 [13a]. Even though each of these trialkylboranes contains two leaving groups β to the boron atom, only *B*-morpholino-9-BBN is eliminated, furnishing the corresponding heterocyclic alkenes (Eq. 24.8) [13].

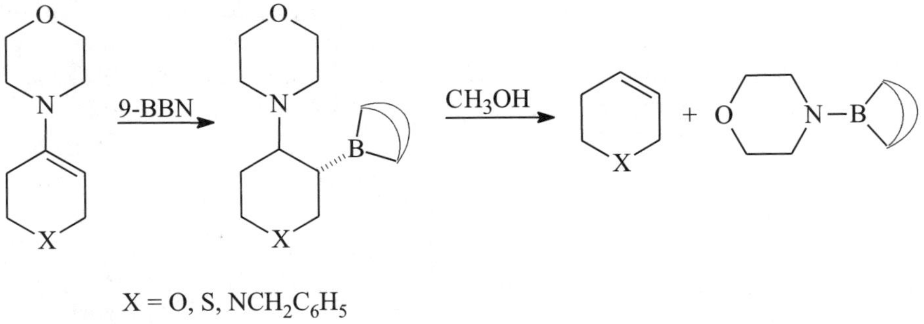

(24.8)

Table 24.4 Heterocyclic alkenes from the corresponding ketone enamines [13a]

Enamine	Heterocyclic alkene	Yield (%)
4-(1,4-Dioxaspiro[4,5]dec-7-en-8-yl)-morpholine	3-Cyclohexen-1-one ethylene ketal	75
4-(3,6-Dihydro-2*H*-pyran-4-yl)-morpholine	5,6-Dihydro-2*H*-pyran	55
4-(3,6-Dihydro-2*H*-thiopyran-4-yl)morpholine	5,6-Dihydro-2*H*-thiopyran	65
4-(1,2,3,6-Tetrahydro-1-benzyl-4-pyridinyl)-morpholine	1,2,3,6-Tetrahydro-1-benzylpyridine	80

Trisubstituted olefins are prepared from vinyliodide on treatment with BuLi and alkylated 9-BBN at –80 °C, followed by stirring with iodine in THF (Eq. 24.9) [15].

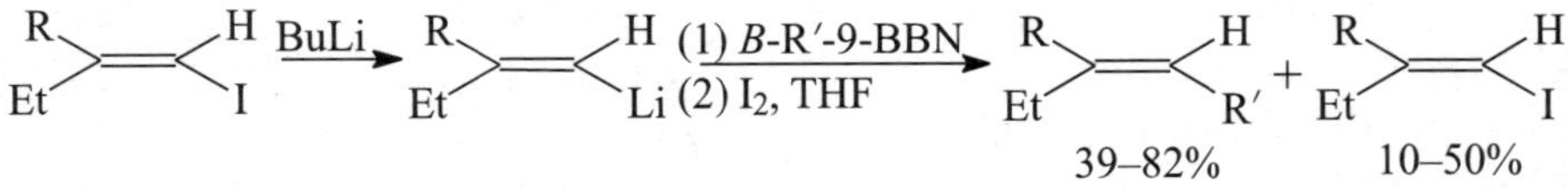

(24.9)

The reaction assumes importance as the incoming alkyl group attaches to the position of the iodine in the parent compound. The results are summarized in Table 24.5 [15].

Table 24.5 Synthesis of trisubstituted alkenes from vinyliodides and *B*-R-9-BBN [15]

R	R′	Yield (%)	R	R′	Yield (%)
n-C_6H_{13}	*n*-C_4H_9	63	*n*-C_6H_{13}	*c*-C_5H_9	82
n-C_6H_{13}	*i*-C_4H_9	66	*n*-C_6H_{13}	*c*-C_6H_{11}	74
n-C_6H_{13}	*s*-C_4H_9	76	C_6H_5	*n*-C_4H_9	39

The yield of trisubstituted olefins is controlled by the iodination step, as it leads to the formation of the parent vinyl iodide as the side product, which is ~50% in phenyl substituted and ~10–15%, in other cases.

The low yield of phenyl substituted olefin is due to the competition between ipso substitution and attack at the carbon β to boron. The result of substitution depends on the stabilization of the positive charge on the iodonium ion. When C-1 stabilizes a positive charge best, the major reaction path is the desired α-transfer. Alternatively, as C-2 is better able to support a positive charge, the iodonium ion opens, leaving a positive charge on C-2, and *B*-R′-9-BBN is either simultaneously or sequentially lost (Chart 24.2) [15].

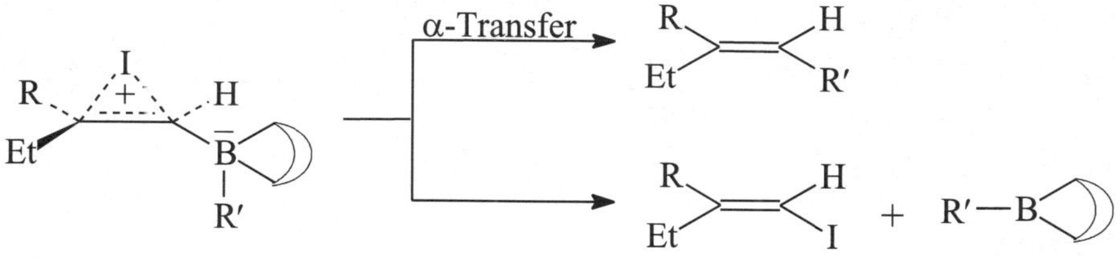

Chart 24.2

The *B*-alkyl-9-BBN undergoes an interesting reverse reaction to afford the parent alkene when treated with benzaldehyde. Consequently, the reaction is uniquely employed for the synthesis of exocyclic olefins (Chart 24.3). The hydroboration of cyclic olefins with an internal double bond, followed by homologation with carbon monoxide in the presence of lithium trimethoxyaluminum hydride afford *B*-(cycloalkylmethyl)-9-BBN. This intermediate on treatment with benzaldehyde leads to an exocyclic methylene compound (Chart 24.3) [16]. Since the synthesis proceeds from the cycloalkene, thus it provides a valuable alternative to the customary methylenation of carbonyl compounds by Wittig and related procedures. The method also provides a clean synthesis of deuterium-labeled compounds (Eq. 24.10) [16], without positional scrambling or loss of label. Consequently, methylmethylene-d$_2$-cyclopentane in 52% isolated yield is obtained.

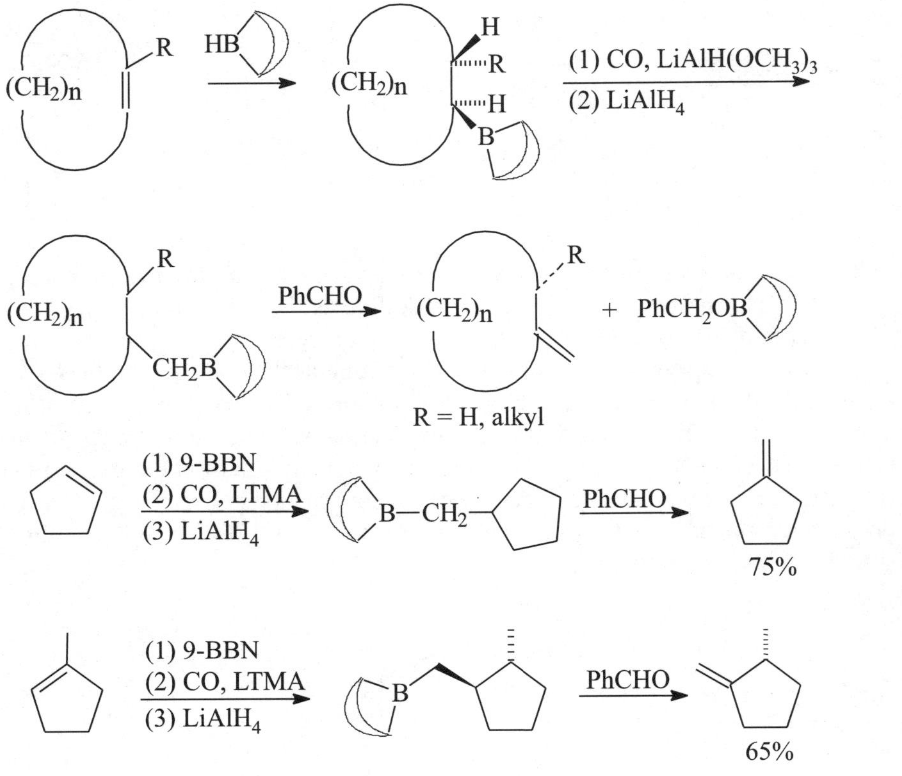

Chart 24.3

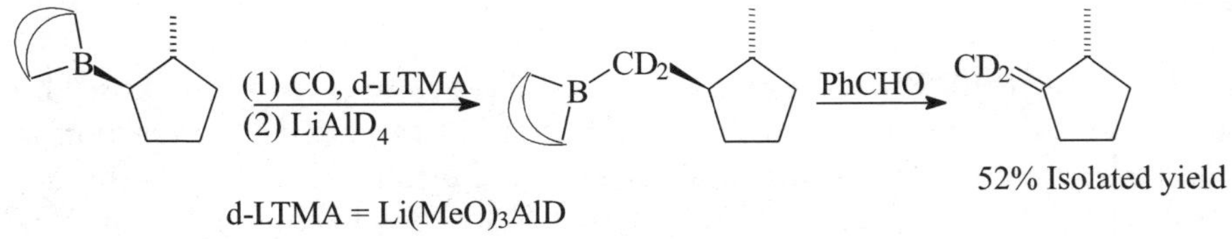

(24.10)

Trialkylboranes, obtained after the hydroboration of β,β-disubstituted enamines on hydroboration [17] with 9-BBN, do not react with methanol under conditions that work well for α,β-disubstituted enamine [13]. These trialkylboranes are even inert at 65 °C with methanol for 12 h and are recovered unchanged. However, thermal decomposition of these trialkylboranes affords the corresponding alkenes (Eq. 24.11). Consequently, when neat trialkylboranes are maintained at 200 °C for 6 h, the alkenes are recovered in moderate to excellent yield by a simple distillation (Table 24.6) [17].

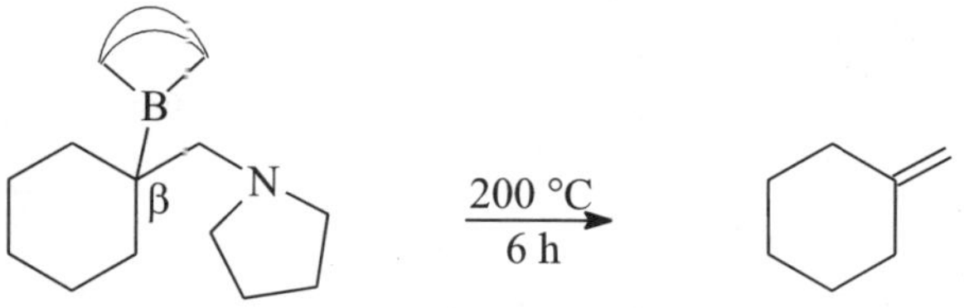

(24.11)

Table 24.6 Synthesis of alkenes from β,β-disubstituted enamines [17]

Enamine	Alkene	Yield (%)
2-Phenyl-1-pyrrolidino-1-propene	2-Phenyl-1-propene[a]	84
1,1-Diphenyl-2-pyrrolidinoethene	1,1-Diphenylethene[a]	80
2-Methyl-1-pyrrolidino-1-pentene	2-Methyl-1-pentene[b]	66
2-Methyl-1-pyrrolidino-1-undecene	2-Methyl-1-undecene[b]	80
1-(Cyclohexylidene-methyl)-pyrrolidine	Methylenecyclohexane[b]	82

[a] 1 M BMS in THF; oxidation–elimination with neutral H_2O_2.
[b] 10 M 9-BBN in THF; elimination by pyrolysis at 200 °C.

24.1.2 Synthesis of (*Z*)-Alkenylsilane and (*Z*)-Alkenyltin

Yamamoto reported a convenient one-pot procedure for the synthesis of (*Z*)-2-alkenylsilanes and -tins (Scheme 24.3) [18]. The boron stabilized allyl carbanions are prepared *in situ* by treatment of 1- or 2-alkenyl-9-BBN [19] with lithium 2,2,6,6-tetramethyl piperidide (LiTMP). The carbanion reacts with trialkylsilyl and -tin halide exclusively at the α position to give 1,1-borasilyl or 1,1-boratin derivatives, which are subsequently protonolyzed with H_2O. The results are summarized in Table 24.7 [18].

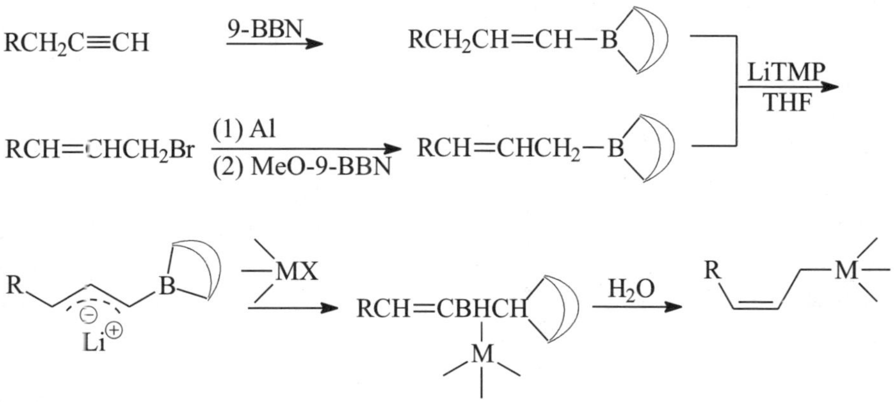

Scheme 24.3

Table 24.7 Stereoselective synthesis of (Z)-2-alkenylsilanes and -tins [18]

9-BBN derivative	Electro-phile	Product	Yield (%)	Isomeric purity (%)
n-BuCH=CH–B	Me_3SiCl	n-PrCH=CHCH_2SiMe_3	72	~100
n-BuCH=CH–B	n-Bu_3SnCl	n-PrCH=CHCH_2Sn(n-Bu)$_3$	76	~100
n-BuCH=CH–B	Me_3SnBr	n-PrCH=CHCH_2SnMe_3	70	~100
CH_3CH=CHCH_2–B	Me_3SiCl	CH_3CH=CHCH_2SiMe_3	40	~100
CH_3CH=CHCH_2–B	Me_3SnBr	CH_3CH=CHCH_2SnMe_3	72	70[a]

[a] Crotyltins are known to undergo facile isomerization in contrast to other derivatives [20].

The hydroboration of internal allenes with 9-BBN affords the corresponding allylboranes [21]. These allylboranes when treated with water, leading to the formation of predominantly of (Z)-4-(trimethylsilyl)-2-alkenes (Chart 24.4) [22].

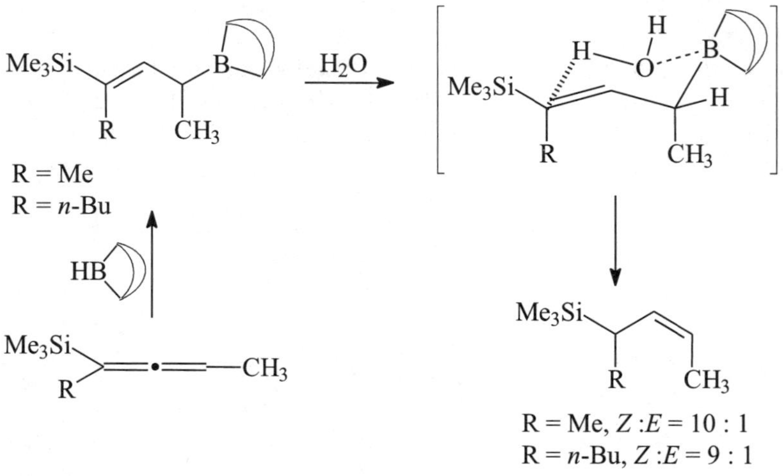

Chart 24.4

Soderquist has reported [23] that bulky alkylsilyl substituted 1-alkynes place the 9-BBN β to the silyl group to give (2-borylvinyl)silanes. The protolysis of these intermediates affords the *cis* vinyl products cleanly, in quantitative yield (Eq. 24.12).

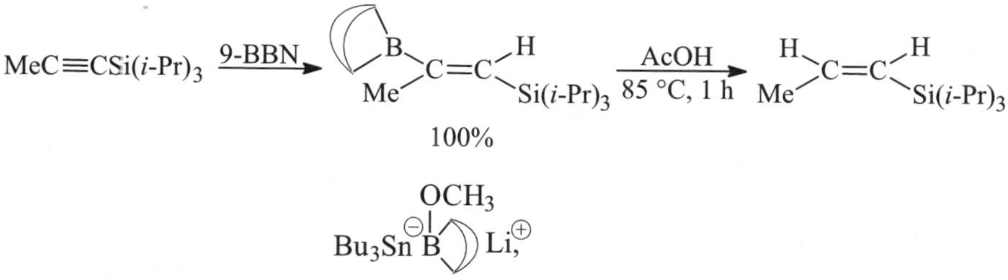

(24.12)

The ate complex prepared from *B*-methoxy-9-BBN and Bu_3SnLi [24] adds under copper(I)-catalyzed reaction to 1-alkyne to yield exclusively or predominantly lithium [2-tri-*n*-butylstannyl-(*Z*)-1-alkenyl]-1-borates [25]. These borates are selectively coupled in the presence of organopalladium or organocuprate with a variety of electrophiles (Chart 24.5), exclusively at the vinylcarbon–boron bond to form a carbon–carbon bond (Table 24.8) [25]. The process is extremely versatile, as various functional groups are tolerated. However, an alkene with opposite regiochemistry (entry K, Table 24.8) [25] is obtained when $BF_3 \cdot Et_2O$ is added to the initial reaction, which is then refluxed.

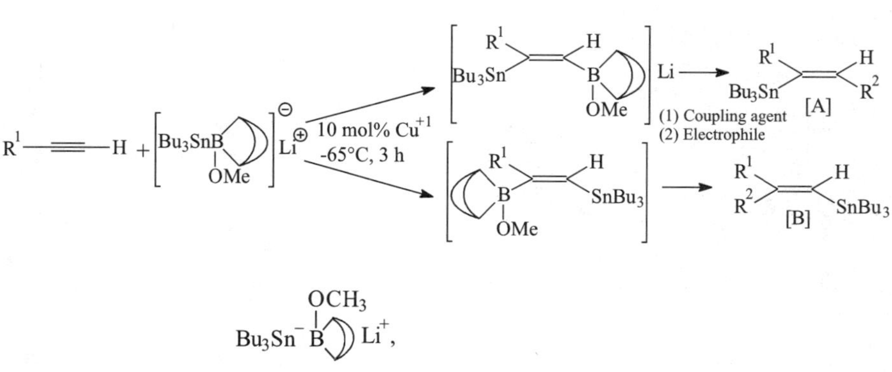

Chart 24.5

Table 24.8 Addition of to 1-alkynes and further reactions with electrophiles [25]

Entry	R^1	Coupling agent and conditions	Electrophile	R^2, yield (A+B %)	A:B
A	HOC_4H_8	1 eqiv $CuBr{\cdot}Me_2S$ –78 °C, 1 h	HCl/MeOH	H, 68	100:0
B	$THPOC_4H_8$	1 eqiv $CuBr{\cdot}Me_2S$	HCl/MeOH	H, 73	100:0
C	CNC_4H_8	–78 °C, 1 h	HCl/MeOH	H, 81	100:0
D	$AcOC_4H_8$	1 eqiv $CuBr{\cdot}Me_2S$	HCl/MeOH	H, 77	100:0
E	BrC_4H_8	–78 °C, 1 h	HCl/MeOH	H, 54	70:30
F	C_6H_5	1 eqiv $CuBr{\cdot}Me_2S$	HCl/MeOH	H, 82	46:54
G	C_7H_{15}	–78 °C, 1 h	a, HCl/MeOH b,	H, 92	98:2 96:4
			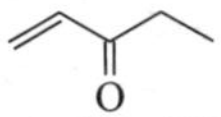 1 eq. $BF_3{\cdot}OEt_2$	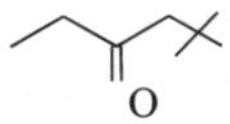 : 53%	
H	C_7H_{15}	1 equiv CuCN –30 °C, 1h	HCl/MeOH	H, 80	66:34
I	C_7H_{15}	$Pd(Ph_3P)_4$			
J	C_7H_{15}	$Pd(Ph_3P)_2^{+2}$ NaOMe, RT	PhI	Ph, 63	96:4
K	C_7H_{15}	1.3 eqiv $BF_3{\cdot}Et_2O$ –78 °C → reflux	AcOH	H, 62	9:91

The 9-BBN ring possesses remarkable thermal stability, and this feature is utilized in the one-pot conversion of 1-triethylsilyl-1-propyne to *trans*-(3-deuterio-2-propen-1-yl)triethylsilane, as illustrated in Scheme 24.4 [2].

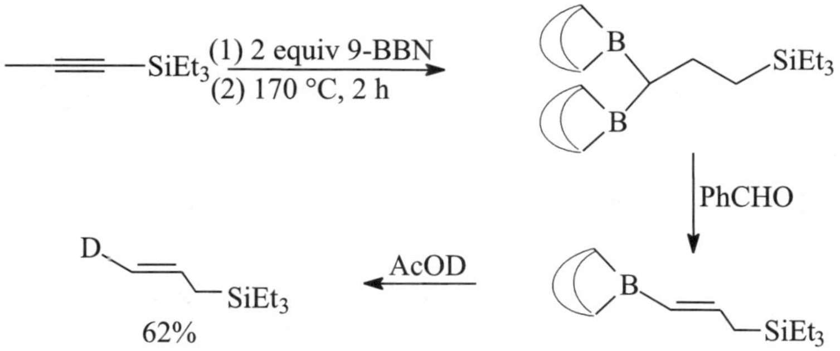

Scheme 24.4

24.1.3
Synthesis of (*Z*)-1-Halo-1-Alkenes

Unlike 1-alkynes, the hydroboration of 1-haloalkynes with 1 equiv of 9-BBN or Chx_2BH does not produce any dibora adducts [26]. Moreover, the hydroboration of 1-halo-1-alkynes with these reagents yields exclusively C-1 monohydroboration products. The C-1 monohydroboration products are easily protonolyzed [26] with AcOH to give (*Z*)-1-halo-1-alkene (Eq. 24.13).

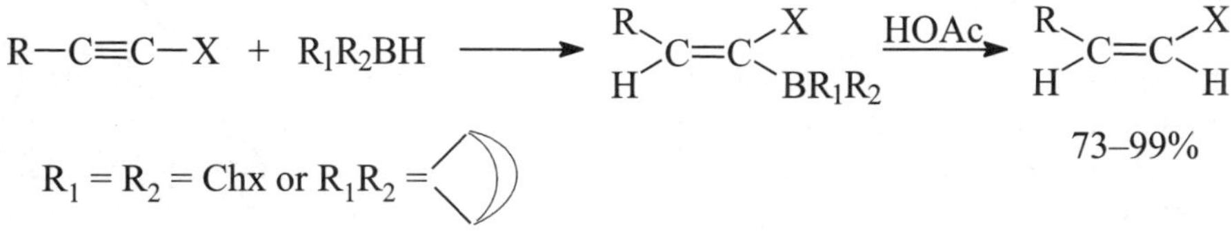

(24.13)

It is important to mention that to avoid the interference of cyclohexanol in the distillation of the product that is formed if oxidation is employed, the borane residue is removed *via* the adduct with ethanolamine when Chx_2BH is employed. In the case of 9-BBN, the hydroboration is performed in THF, and then to significantly reduce the time for protonolysis, THF is removed and replaced with pentane. The ethanolamine adduct of 9-BBN is very sticky and more difficult to handle than in the synthesis with Chx_2BH. The distillation of some of (*Z*)-1-halo-1-alkenes leads to some isomerization (≤20%) to the *trans* product. However, the isomerization is reduced to less than 2% by the addition of $NaHCO_3$ and 2,6-di-*tert*-butyl-*p*-cresol (BHT) to the mixture before distillation [27]. The results are summarized in Table 24.9 [26].

Table 24.9 Synthesis of (*Z*)-1-halo-1-alkenes from 1-halo-1-alkynes [26]

Compound	Hydroborating agent	Yield (%)	Isomeric purity (%)
(*Z*)-1-Chloro-1-octene	Chx_2BH	93	>99
(*Z*)-1-Chloro-1-octene	9-BBN	73	>99
(*Z*)-1-Bromo-1-hexene	Chx_2BH	85	98
(*Z*)-1-Bromo-1-octene	Chx_2BH	91	98
(*Z*)-1-Bromo-1-octene	9-BBN	92	98
(*Z*)-1-Bromo-2-cyclopentyl-1-ethene	9-BBN	82	>98
(*Z*)-1-Bromo-2-phenylethene	Chx_2BH	84	>98
(*Z*)-1-Iodo-1-hexene	Chx_2BH	94	>99
(*Z*)-1-Iodo-1-hexene	9-BBN	84	>99
(*Z*)-1-Iodo-1-octene	Chx_2BH	98	>99
(*Z*)-1-Iodo-3-methyl-1-butene	Chx_2BH	75	>98
(*Z*)-1-Iodo-3,3-dimethyl-1-butene	Chx_2BH	73	>98
(*Z*)-1-Iodo-2-cyclohexyl-1-ethene	Chx_2BH	80	>98

24.2 Synthesis of Dienes

24.2.1 Synthesis of 1,ω-Dienes

The dienamines derived from enals undergo hydroboration with 9-BBN, with high regio- and chemoselectivity at the enamine carbon–carbon double bond and place the boron atom on a carbon having higher electron density. The high selectivity is confirmed by the hydroboration of a 1:1 mixture of 1-hexene and 1-morpholino-1-octene with 1 equiv of 9-BBN. The reaction mixture affords, after the addition of methanol, a 1:1 mixture of 1-hexene and 1-octene, and no 1-hexanol is detected. Consequently, this process is used to convert unsaturated aldehydes to dienes. Thus, enamine of citronellal on hydroboration–elimination affords a chiral non-conjugated diene, β-citronellene (Scheme 24.5) [1].

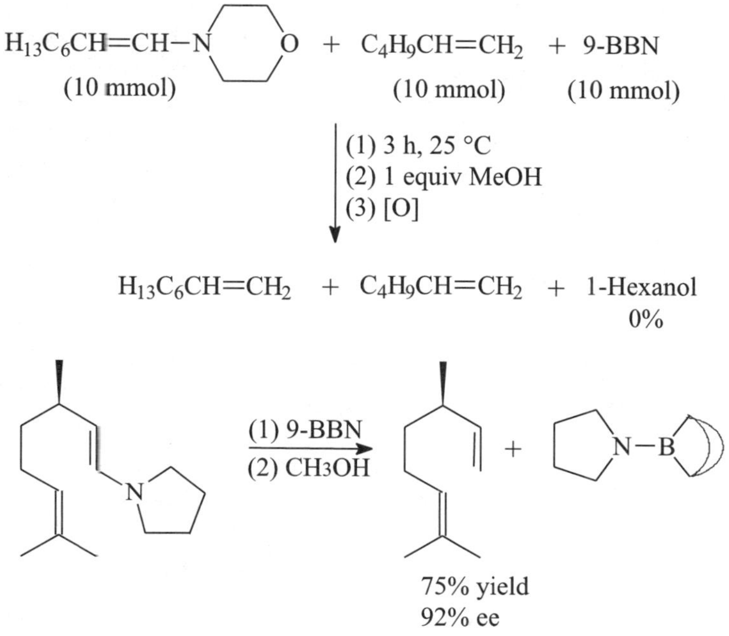

Scheme 24.5

The results of this process for the synthesis of dienes are summarized in Table 24.10 [1].

Table 24.10 The dienes from aldehyde enamines [1]

Dienamines	Dienes	Yield (%)
(*R*)-(−)-(*E*)-1-Pyrrolidino-3,7-dimethyl-1,6-octadiene	(*R*)-(−)-3,7-Dimethyl-1,6-octadiene	75
(1*E*,4*E*)-1-Morpholino-1,4-decadiene	(4*E*)-1,4-Decadiene	82
(1*E*,4*Z*)-1-Morpholino-1,4-decadiene	(4*Z*)-1,4-Decadiene	89
(*E*)-1-(Hexamethyleneimino)-1,10-undecadiene	1,10-Undecadiene	72

In an another method, dialkyl chloroboranes undergo methylcopper-induced coupling to produce symmetrical (*E*,*E*)-1,3-dienes in excellent yields and high stereochemical purity [2, 3]. In this reaction it is essential to use 3 equiv of methylcopper in order to achieve the effective conversion of organoboranes into dienes.

A more milder procedure is where sodium methoxyalkenyl dialkylborates obtained by the simple treatment of alkenyldialkylboranes with sodium methoxide react readily with cuprous bromide–methyl sulfide at 0 °C to afford symmetrical conjugated *E,E*-dienes (Chart 24.6; Table 24.11) [4]. However, if the temperature is lowered to −15 °C, the dark blue–black complex formed is stable and can be trapped by allylbromide, which afford a stereochemically defined synthesis of (4*E*)-1,4-dienes [5] (Chart 24.6; Table 24.12).

Table 24.11 Synthesis of 1,3-dienes [4]

Alkyne	Hydroborating agent	Product	Yield (%)
1-Hexyne	9-BBN, Chx_2BH	(5*E*,7*E*)-5,7-Dodecadiene (5*E*,7*E*)-5,7-Dodecadiene	95 97
3-Hexyne	Chx_2BH	(3*E*,5*E*)-4,5-diethyl-3,5-octadiene	99
3,3-Dimethyl-1-butyne	9-BBN	(3*E*,5*E*)-2,2,7,7-Tetramethyl-3,5-octadiene	98
4-Acetoxy-1-butyne	Chx_2BH	(3*E*,5*E*)-1,8-Diacetoxy-3,5-octadiene	80
Phenylethyne	Chx_2BH	(1*E*,3*E*)-1,4-Diphenyl-1,3-butadiene	79
5-Chloro-1-pentyne	Chx_2BH	(4*E*,6*E*)-1,10-Dichloro-4,6-decadiene	95
1-Iodo-1-hexyne[a]	Chx_2BH	(5*Z*,7*Z*)-5,7-Dodecadiene	85

[a] The intermediate borane is treated with *t*-BuLi to get (1*Z*)-1-1-hexen-1- yldicyclohexylborane.

Table 24.12 Synthesis of 1,4-dienes [5]

Alkyne	Hydroborating agent	Product	Yield (%)
1- Hexyne	9-BBN	(4*E*)-1,4-Nonadiene	92
1- Hexyne	Chx_2BH	(4*E*)-1,4-Nonadiene	93
3- Hexyne	9-BBN Chx_2BH	(4*E*)-4-Ethyl-1,4-heptadiene (4*E*)-4-Ethyl-1,4-heptadiene	95 96
4-Acetoxy-1-butyne	Chx_2BH	(4*E*)-7-Acetoxy-1,4-heptadiene	73
1-Iodo-1-Hexyne[a]	Chx_2BH	(4*Z*)-1,4-Nonadiene	76
1-(Trimethylsilyl)-1-Nonyne	Chx_2BH	(4*Z*)-4-(Trimethylsilyl)-1,4-dodecadiene	83

[a] The intermediate borane is treated with *t*-BuLi.

1,3-Butadienes are important starting material for the Diels–Alder reaction [6]. Moreover, many naturally occurring biologically active organic molecules possess conjugate diene functionality [7, 8]. Matteson [9], Yamamoto [10], and Wang [11] have reported that condensations of trimethylsilyl-substituted allylboronates or allylboranes with aldehydes proceed with diastereoselectivity, and the resultant β-hydroxysilanes undergo loss of trimethylsilanol to give dienes [12, 13]. Under basic workup conditions, a *syn*-elimination of trimethylsilanoxide affords one geometry of the carbon–carbon double bond, while acid work up procedures gives the complimentary isomer [14].

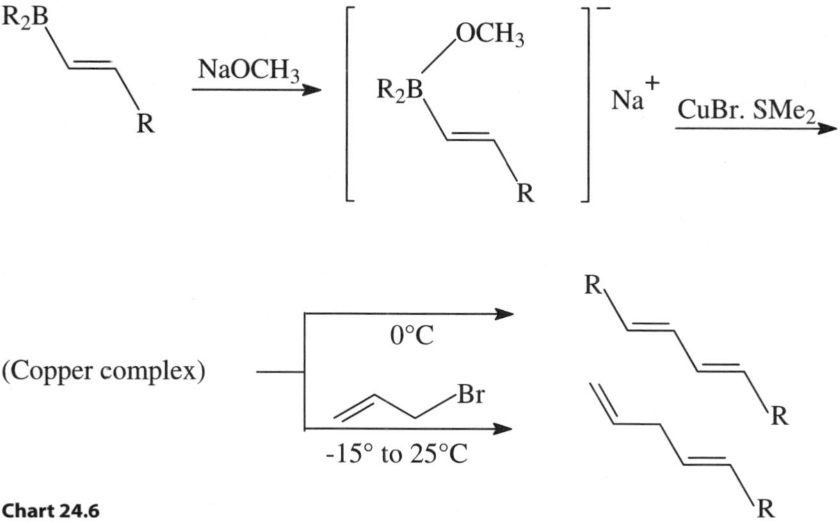

Chart 24.6

Trimethylsilyl-substituted allenes [15, 16] undergo ready hydroboration with 9-BBN to afford the corresponding allylboranes. For example, hydroboration of 1,2-bis(trimethylsilyl)-2,3-butadiene (**2**), readily prepared from **1** (40% yield) [15], with 9-BBN [17] affords the corresponding allylborane (**3**). Condensation of **3** takes place smoothly with aldehydes and ketones, and after basic or acid work up affords the [*E*]- or [*Z*]-2-[(trimethylsilyl) methyl]-1,3-butadienes, respectively (Scheme 24.6 Table 24.13) [18].

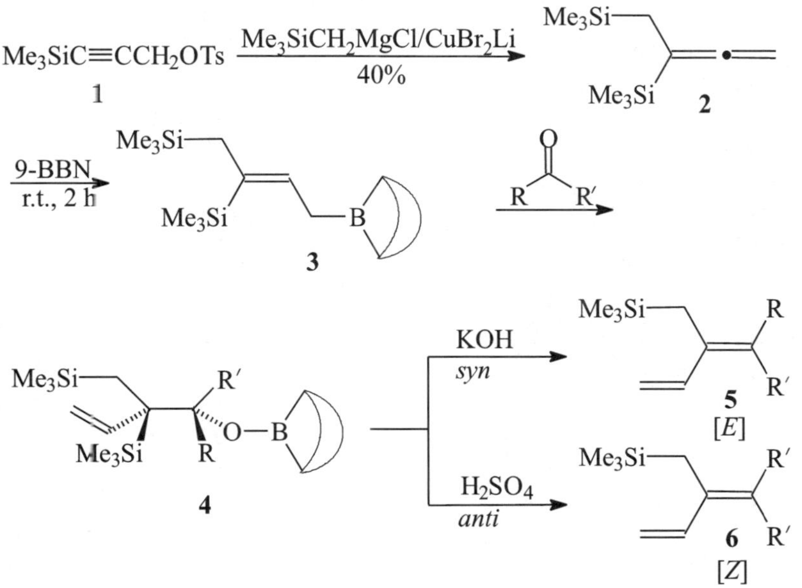

Scheme 24.6

Table 24.13 Stereoselective synthesis of 2-[(trimethylsilyl) methyl]-1,3-butadienes [18]

Diene	R	R′	Yield (%)	**5**:**6**
5a	H	CH_3	61	98:20
6a	H	CH_3	50	2:98
5b	H	n-C_5H_{11}	82	97:30
6b	H	n-C_5H_{11}	87	0.5:99.5
5c	H	C_6H_5	88	98:20
6c	H	C_6H_5	83	2:98
5d	H	(E)-$CH_3CH{=}CH$	59	98:20
5e		-$(CH_2)_5$-	58	
5e		-$(CH_2)_5$-	61	
5f	CH_3	n-C_3H_7	68	55:45

Wang and coworkers [19] extended this methodology for the synthesis of various terminal 1,3-butadienes from the readily available variety of trimethylsilyl-substituted terminal allenes [15, 16]. Hydroboration of allenes **7–12** with 9-BBN gives the corresponding allylboranes **13–18** (Scheme 24.7). Allylboranes **13–17** are predominantly E isomers (E:Z ≥ 91:9), while **18** favors Z geometry. (E:Z = 28:72). Subsequent condensation of these allylboranes with hexanal, benzaldehyde, acetaldehyde, or crotonaldehyde takes place smoothly to yield the corresponding 1,3-butadienes, after either basic or acidic workup (Table 24.14) [19a].

Table 24.14 Stereoselective synthesis of terminal 1,3-butadienes [19a]

Diene	Workup (time)	R	R[1]	Yield (%)	Isomer ratio (**21**:**22**)
21	NaOH (2 h)	Me	n-C_5H_{11}	73	95:5
21	NaOEt (30 min)			72	98:2
22	H_2SO_4 (1 h)			65	3:97
21	NaOH (40 min)	Me	Ph	71	98:2
22	H_2SO_4 (2 h)			69	2:98
21	NaOH (2h)	n-Bu	n-C_5H_{11}	81	96:4
22	H_2SO_4 (2 h)			70	3:97
21	NaOH (1 h)	n-Bu	Ph	81	98:2
22	H_2SO_4 (2 h)			75	2:98
21	NaOH (2 h)	n-Bu	(E)-$CH_3CH{=}CH$	53	93:7
21	NaOH (30 min)	Me_3SiCH_2	Me	61	98:2
22	H_2SO_4 (30 min)			50	2:98
21	NaOH (30 min)	Me_3SiCH_2	n-C_5H_{11}	82	97:3
21	NaOEt (30 min)			85	97:3
22	H_2SO_4 (30 min)			87	0.5:99.5
21	NaOH (30 min)	Me_3SiCH_2	Ph	88	98:2
22	H_2SO_4 (40 min)			83	2:98
21	NaOH (2h)	Me_3SiCH_2	(E)-$CH_3CH{=}CH$	59	98:2
21	NaOH (30 min)	Ph	n-C_5H_{11}	67	96:4

Table 24.14 *(continued)* Stereoselective synthesis of terminal 1,3-butadienes [19a]

Diene	Workup (time)	R	R^1	Yield (%)	Isomer ratio (**21**:**22**)
21	NaOEt (15 min)			80	95:5
22	H_2SO_4 (2 h)			84	5:95
21	NaOH (15 min)	Ph	Ph	58	>99.5:0.5
22	H_2SO_4 (2 h)			56	<0.5:99.5
21	NaOH (30 min)	*i*-Pr	*n*-C_5H_{11}	77	88:12
21	NaOEt (30 min)			82	88:12
22	H_2SO_4 (2 h)			65	11:89
21	NaOH (30 min)	*i*-Pr	Ph	85	91:9
22	H_2SO_4 (2 h)			84	11:89
21	NaOH (2 h)	*t*-Bu	*n*-C_5H_{11}	55	41:59
21	NaOEt (1 h)			79	29:71
22	H_2SO_4 (5 h)			69	72:28
21	NaOH (30 min)	*t*-Bu	Ph	70	41:59
21	NaOEt (30 min)			83	29:71
22	H_2SO_4 (1.5 h)			68	62:38

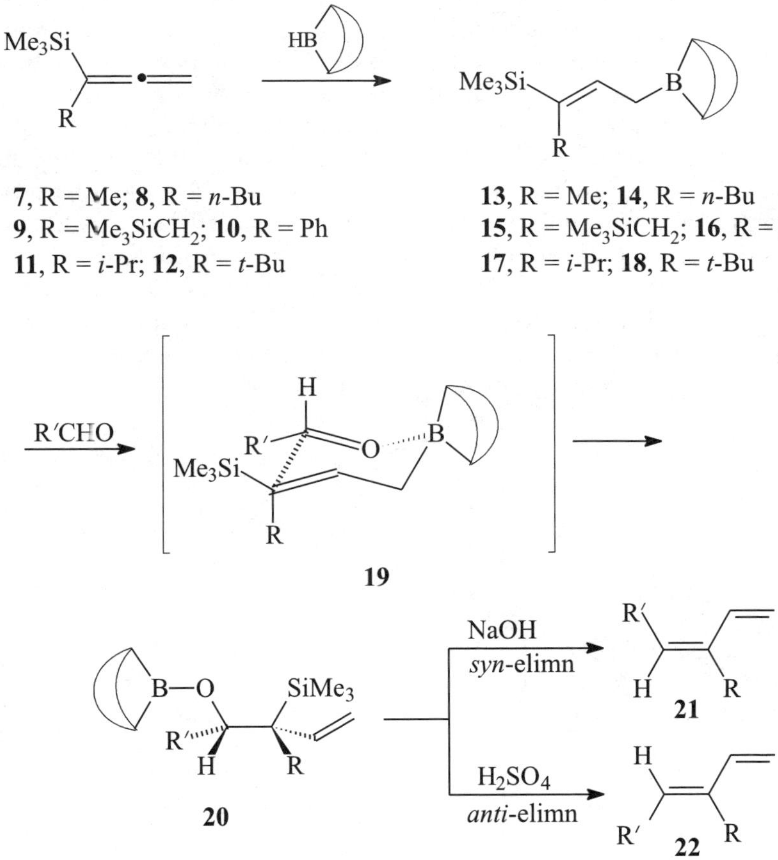

Scheme 24.7

The dienes formed (Table 24.14) have high isomer purity; thus, the condensation step is highly diastereoselective, with **13–17** to afford predominantly **20** (Scheme 24.7). Subsequent treatment with NaOH or NaOEt to promote Peterson olefination [14] affords diene **21** as the major isomer after *syn* elimination of trimethylsilyloxide. On the other hand, diene **22** is obtained preferentially by reaction with concentrated H_2SO_4, which induces *anti*-elimination [14].

The *E* geometry of the C=C in **13–18** is assigned on the basis of its ^{1}H NMR study and by oxidizing the allylboranes to the corresponding allylic alcohols by alkaline H_2O_2 (Table 24.15) [19a].

Table 24.15 The geometry of the double bond of allyboranes **13–18** [19a]

Allylborane	*E*:*Z* ratio[a]	Allylborane	*E*:*Z* ratio[a]
13	98:2 (>99:1)	**16**	99:1 (98:2)
14	97:3 (98:2)	**17**	91:9 (92:8)
15	97:3 (>99:1)	**18**	28:72 (42:58)

[a] The numbers in *parentheses* refer to the *E*:*Z* ratio of the corresponding allylic alcohols.

When glanced, it appears that hydroboration of **1+2** occurs preferentially from the side of the bulkier trimethylsilyl group [20], and in the case of **12** from the *tert*-butyl side. This is in contradiction to the general behavior of the hydroboration, which prefers the less hindered side of a carbon–carbon double bond [21]. However, allylic boranes are known to undergo rapid [1,3]sigmatropic rearrangement, which then leads to interconversion of the geometry of double bond. Thus, the product distribution described in Table 24.15 reflects only the thermodynamic equilibrium of *E* and *Z* isomers of allylboranes **13–18** rather than the kinetic preference of the hydroboration step. This is explained based on large $A^{(1,3)}$ allylic interaction [22] with the sterically bulkier trimethylsilyl group in **13b–17b** and is responsible for shifting the equilibrium toward **13a–17a**. On the other hand, interaction with *t*-butyl group is more severe, and **18b** becomes the preferred isomer (Chart 24.7) [19a].

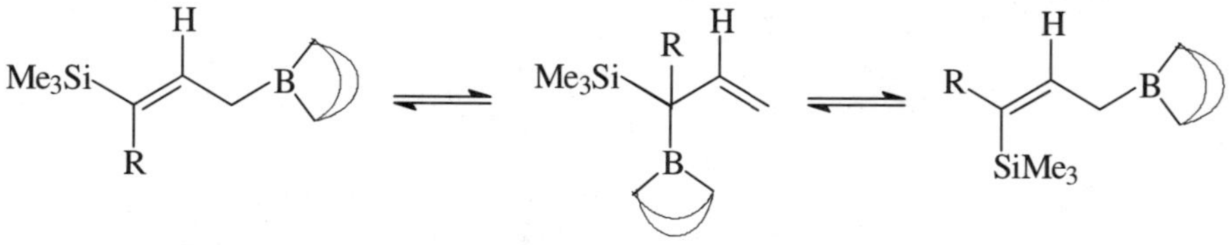

13a, R = Me; **14a**, R = n-Bu
15a, R = Me3SiCH2; **16a**, R = Ph
17a, R = i-Pr; **18a**, R = t-Bu

13b, R = Me; **14b**, R = n-Bu
15b, R = Me3SiCH2; **16b**, R = Ph
17b, R = i-Pr; **18b**, R = t-Bu

Chart 24.7

The ketones are reported [19] to react very slowly, for example, cyclohexanone reacts with **15** at reflux in THF and forms diene, after basic workup, in 58% yield; 2-pentanone does not react at all. However, the addition of $BF_3{\cdot}OEt_2$ (ca. 0.2 ml for a 3-mmol reaction), makes the reaction fast both with cyclohexanone and 2-pentanone (Eqs. 24.14, 24.15) [19]. The reaction in presence of $BF_3{\cdot}OEt_2$ proceeds through an acyclic transition state with enhanced reaction rate [23].

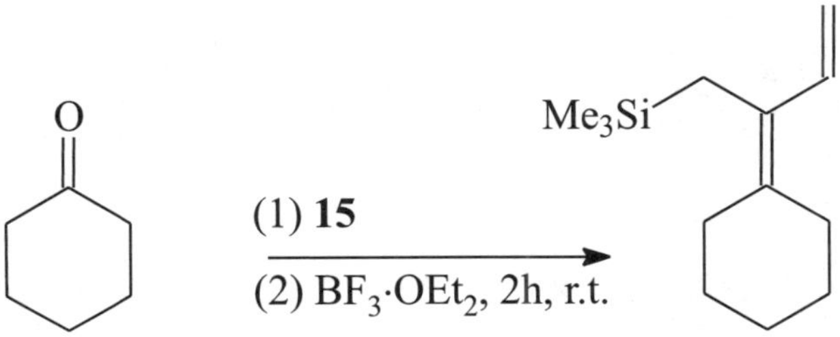

23, 61%

(24.14)

24

(24.15)

The presence of $BF_3{\cdot}OEt_2$ also simultaneously promotes Peterson olefination [14, 24]. The corresponding dienes are produced [19] without treating with NaOH or H_2SO_4.

The lower reactivity of allyboranes **14** or **15** toward the keto group is exploited [18, 19a] for selective condensation with the aldehyde group of **25** [25] to form **26** (Eq. 24.16).

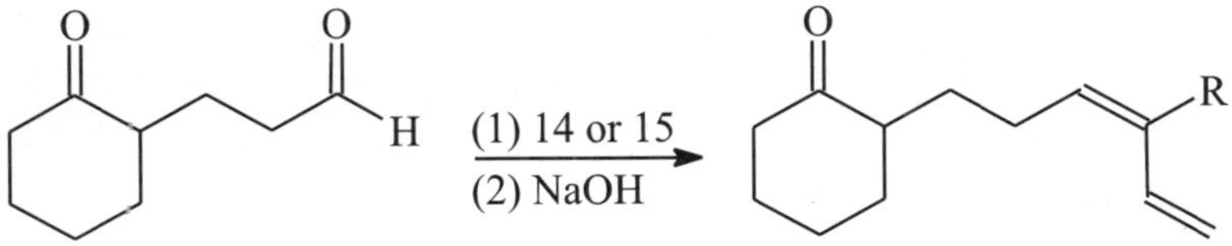

25

26a, R = *n*-Bu, 73% ($E : Z = 4 : 96$)
26b, R = Me_3SiCH_2, 87% ($E : Z = 95 : 5$)

(24.16)

The ester moiety is also much less reactive, and selective condensation is achieved [19] with **27** [26] (Eq. 24.17). Hydrolysis of **28a** affords the corresponding (6*E*)-7-methyl-6,8-nonadienoic acid (*E*:*Z* = 96:4) [19a].

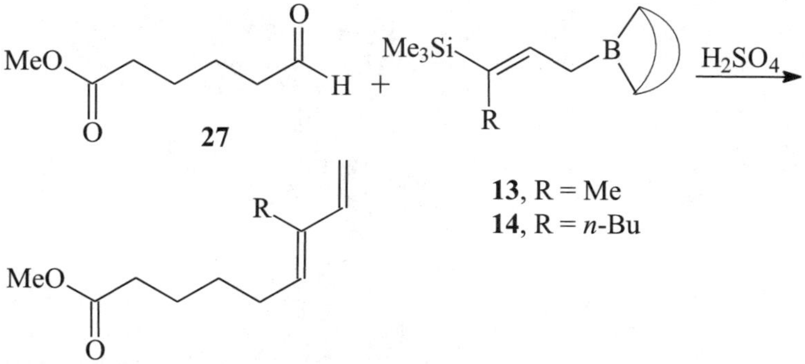

13, R = Me
14, R = *n*-Bu

28a, R = Me, 81% (*E* :*Z* = 96 : 4)
28b, R = *n*-Bu, 59% (*E* :*Z* = 96 : 4)

(24.17)

The concept of using an excess of the alkyne to enhance vinylborane formation is not workable for an α,ω-diyne if both alkyne moieties are to be hydroborated. However, for example with 4:1 9-BBN:1,8-nonadiyne stoichiometry, the 1,1,9,9-tetra-9-BBN adduct is cleanly formed. The treatment of tetraborane adduct with 2 equiv of PhCHO results in the clean formation of *trans,trans*-1,9-di-9-BBN-1,8-nonadiene which on deuterolysis with AcOD affords the pure *trans,trans*-dideuterated diene in 64% overall yield (Scheme 24.8) [27].

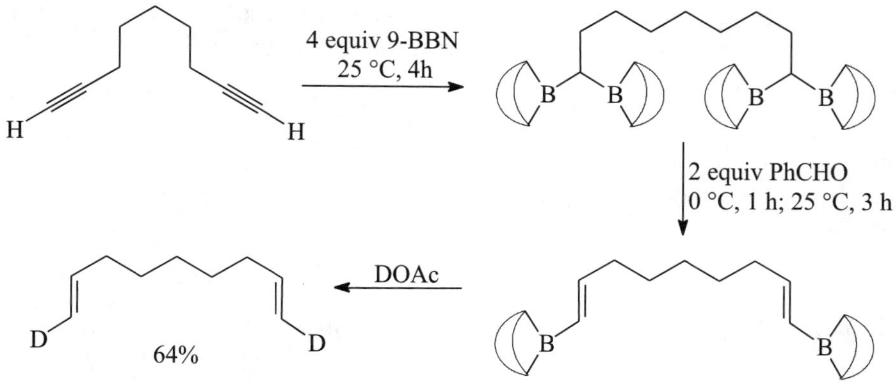

Scheme 24.8

24.2.2
Synthesis of Internal 1,3-Dienes

In the stereoselective synthesis [27–29] of internal 1,3-butadienes (Chart 24.6, Table 24.11) with high isomeric purity, the alkenyl reagents with predetermined geometry must be used. However, it is not always an easy task to synthesize certain alkenyl reagents with required geometry. The condensation reaction between aldehydes and allylic organometallics containing a γ-trimethylsilyl or γ-phosphorus substituent constitutes one of the most promising direct routes to 1,3-dienes [28]. Wang and coworkers [30] have reported the condensation of γ-trimethylsilyl substituted allylboranes and aldehydes to get the stereodefined 1,3-dienes.

The reaction sequence involves: (1) the hydroboration of internal allenes **29** and **30** with 9-BBN [17], (2) condensation with aldehyde followed by (3) elimination of hydroxy trimethylsilane by either (a) basic (*syn*-elimination) or (b) acidic (*anti*-elimination) workup conditions [14] to afford the Peterson olefination reaction. The reaction sequence is sketched in Scheme 24.9 [30].

It is postulated that the attack of 9-BBN occurs initially from the less hindered allyl side of **29** or **30** to produce the *Z* isomer [A], which then rapidly rearranges through [B] to the thermodynamically more stable *E* isomer (**31** or **32**, Eq. 24.18).

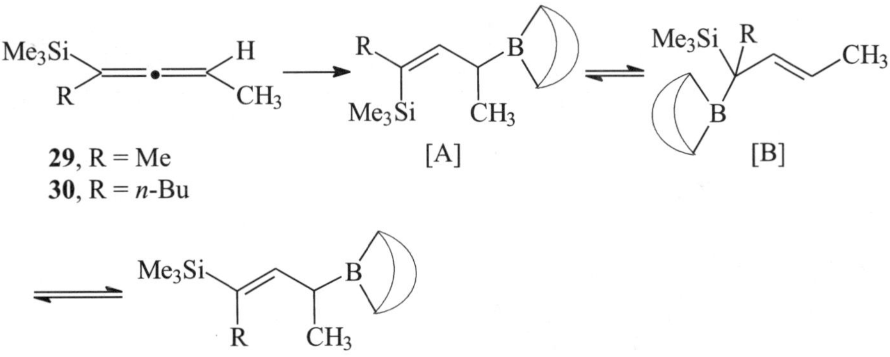

(24.18)

The formation of *Z* geometry for α–double bond indicates a large preference for the allylic methyl group on the α-carbon to adopt the axial position in the **33** and **34** transition states, in spite of the presence of the 1,3-diaxial interaction between the allylic methyl group and the R group. However, if the methyl group assumes the equatorial position as in **39**, it would have suffered from a large nonbonded interaction with the rigid bicyclic framework on the boron atom. Apparently, this repulsion is severe enough to override the 1,3-diaxial interaction.

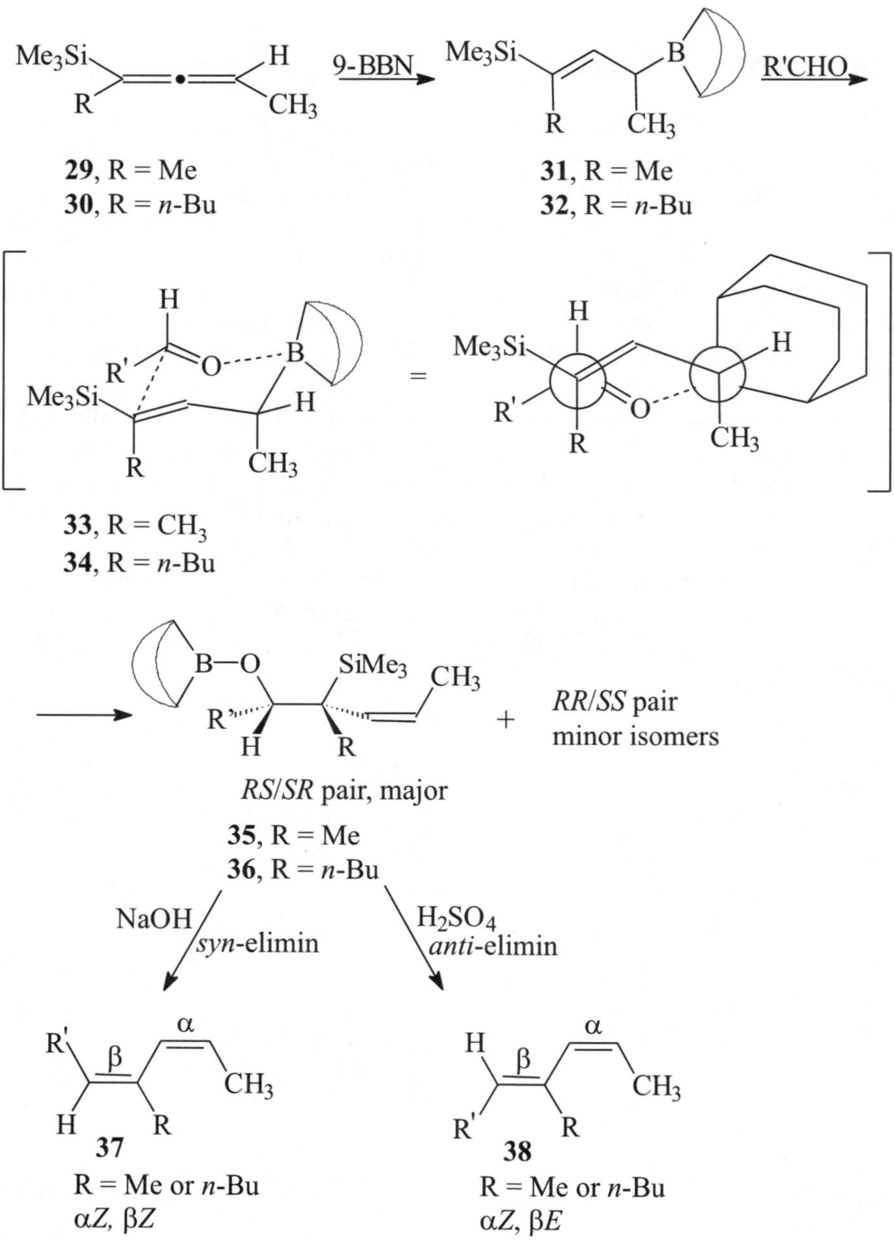
Me3Si
R
H
CH3
9-BBN
B
R'CHO
29, R = Me
30, R = n-Bu
31, R = Me
32, R = n-Bu
R'
O
33, R = CH3
34, R = n-Bu
B—O
SiMe3
RR/SS pair
minor isomers
RS/SR pair, major
35, R = Me
36, R = n-Bu
NaOH
syn-elimin
H2SO4
anti-elimin
α
β
37
R = Me or n-Bu
αZ, βZ
38
R = Me or n-Bu
αZ, βE

Scheme 24.9

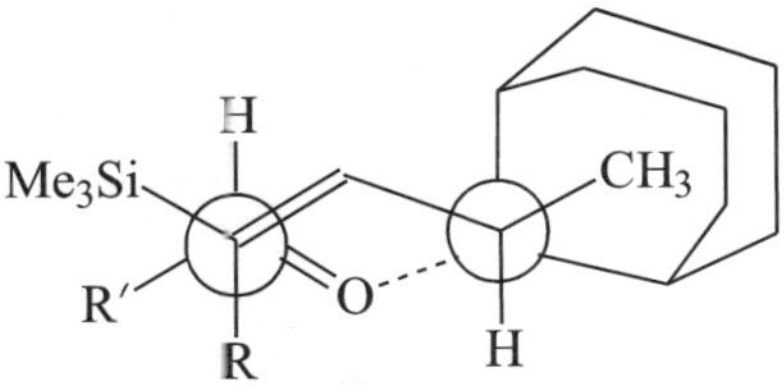

39

The use of dicyclohexylborane [31] as the hydroborating agent for allenes, however, affords the 1,3-dienes with preference for α*E* geometry [32] of dienes (Scheme 24.10). Apparently the less rigid cyclohexyl group could rotate away to avoid excessive nonbonded interaction with the equatorial allylic methyl in the transition state.

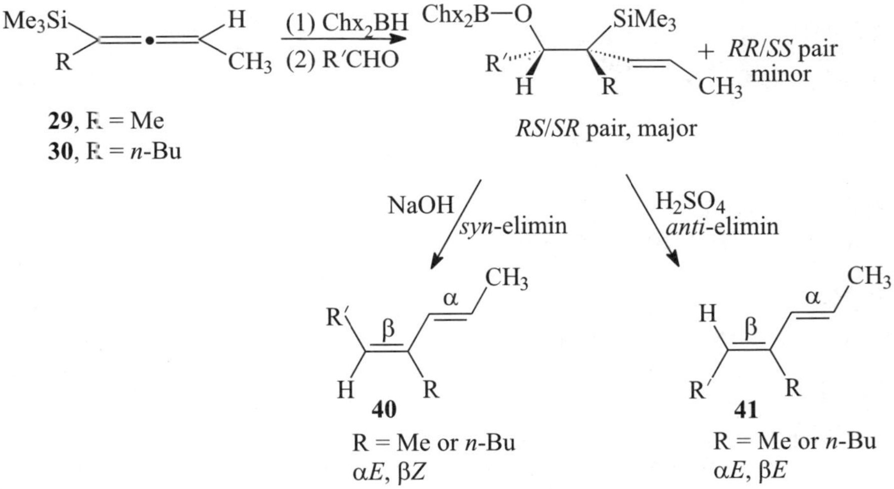

Scheme 24.10

Consequently, the four possible geometric isomeric of several representative internal 1,3-dienes are synthesized (Table 24.16) [30], with high isomeric purity.

Table 24.16 Stereoselective synthesis of internal 1,3-dienes [30]

Borane	Workup	Diene	R	R^1	Yield (%)	Isomer ratio (**37**:**38**:**40**:**41**)
9-BBN	NaOH	**37a**	Me	n-C_5H_{11}	77	94:1:4:1
	H_2SO_4	**38a**			80	1:90:3:6
Chx_2BH	NaOH	**40a**			78	0:0:98:2
	H_2SO_4	**41a**			70	0:0:08:92
9-BBN	NaOH	**37b**	Me	C_6H_5	86	92:1:5:2
	H_2SO_4	**38b**			87	1:91:2:6
Chx_2BH	NaOH	**40b**			83	0:0:97:3
	H_2SO_4	**41b**			82	0:0:5:95
9-BBN	NaOH	**37c**	n-Bu	n-C_5H_{11}	68	97:1:2:0
	H_2SO_4	**38c**			65	1:93:2:4
Chx_2BH	NaOH	**40c**			77	0:0:97:3
	H_2SO_4	**41c**			73	0:0:9:91
9-BBN	NaOH	**37d**	n-Bu	C_6H_5	83	92:1:4:2
	H_2SO_4	**38d**			86	1:92:2:5
Chx_2BH	NaOH	**40d**			85	0:0:97:3
	H_2SO_4	**41d**			79	0:0:3:97

24.2.3 Synthesis of Nitrogen-Containing Heterocyclic (*Z*)-1,3-Dienes

The reaction is extended [33] to synthesize 1,3-butadienes having a nitrogen containing heterocyclic substituents. Consequently, γ-(trimethylsilyl)allylboran e **13** is condensed with 2-, 3-, and 4-pyridinecarboxaldehydes (**42**) and affords after base-induced Peterson olefination reaction the corresponding (*Z*)-1,3-butadienes (**44a–c**, Scheme 24.11; Table 24.17) [33]. The condensation reaction tolerates the presence of an N–H group without the undesired protonolysis of allylboranes (**13**), as in the case of dienes **44d–f** (Table 24.17) [33].

The poor diastereoselectivity in **44e** is not due to deterioration of diastereoselectivity during condensation step. Instead the *Z* isomer of **44e** is isomerized to more stable *E* isomer during treatment with NaOEt and during purification by column chromatography. The rate of isomerization is much faster in imidazole derivative **44f**, and it is attributed to the more acidic character of the hydrogen atom of N–H bond [34], which results in rapid deprotonation of **44f** by NaOEt and consequently, an accelerated rate of isomerization.

The limited success to synthesize *E* isomers by Peterson olefination employing concentrated sulfuric acid is attributed to the formation of pyridinium salt in **44a–c**, which results in reduced rate and formation of undesirable side product. The nitrogen ring system of **43d–f** can not tolerate the presence of sulfuric acid and condensation adducts decompose.

Table 24.17 Synthesis of 1,3-butadienes having a nitrogen-containing heterocyclic substituent [33]

Diene	Base (time)	Yield (%)	Isomer ratio (*Z*:*E*)
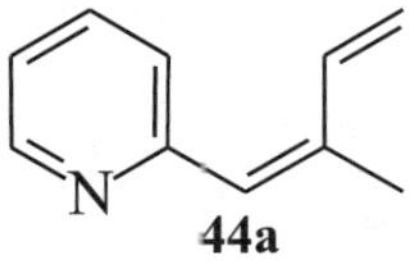 **44a**	NaOH (1 h)	92	98:2
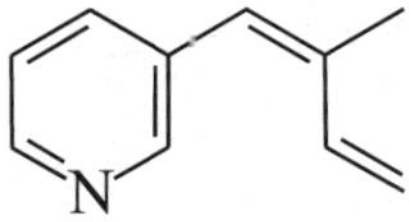 **44b**	NaOH (1 h)	83	98:2
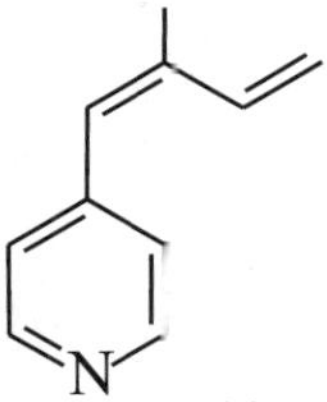 **44c**	NaOH (1 h)	91	98:2
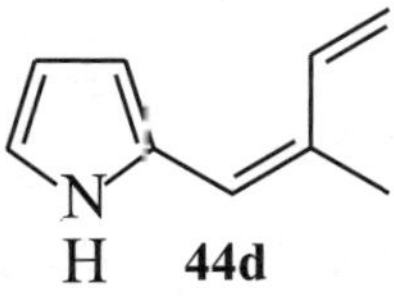 **44d**	NaOEt (15 h)	67	98:2 (98:2)[a]
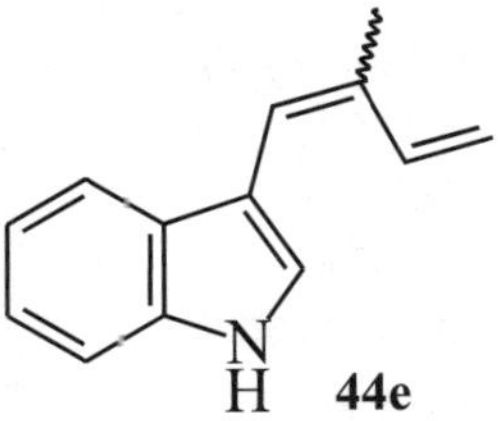 **44e**	NaOEt (15 h)	88	3:1 (9:1)
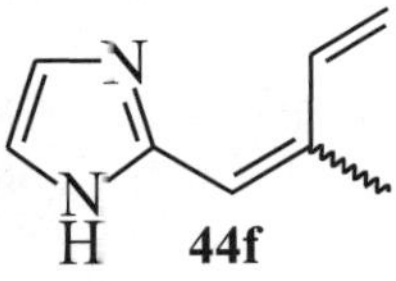 **44f**	NaOEt (15 h)	63	2:3 (2:3)

[a] The ratio in *parentheses* is the isomer ratio of the crude product before purification.

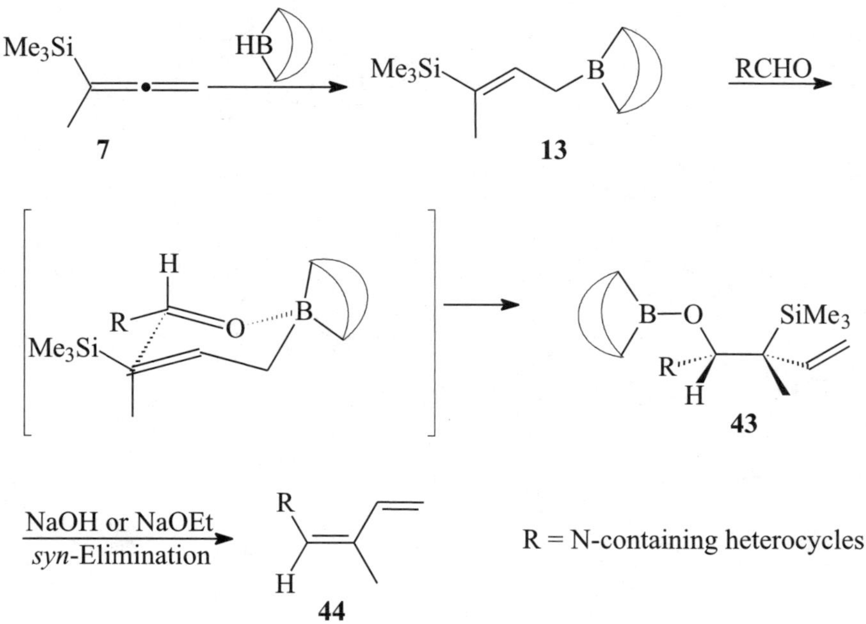

Scheme 24.11

24.2.4 Synthesis of Sulfur-Substituted 1,3-Dienes

Sulfur substituted 1,3-butadienes are valuable synthons, particularly in Diels-Alder reactions, where they impart an added level of reactivity and regioselectivity [35]. Pearson *et al* [36] have reported the stereoselective synthesis of 2-(phenylthio)-1,3-butadiene by the protocol delineated in Scheme 24.12.

Allene **46** is prepared from 1-(phenylthio)-1-propyne **45** [37a] by deprotonation [37b] with LDA, followed by quenching with chlorotrimethylsilane (TMSCl). Hydroboration of **46** with 9-BBN leads to allylborane. The facile allylic rearrangement of allylic dialkylboranes allows the formation of more stable *Z* isomer **47**. Condensation of **47** with aldehydes proceeds smoothly at room temperature, and the intermediate **48**, on treatment with sulfuric acid gives the *Z*-diene **49** and with 4 N sodium hydroxide affords the *E*-diene **50** (Table 24.18) [36].

The reaction proceeds with good chemoselectivity and tolerates the presence of other electrophilic and/or Lewis basic groups (bromides, esters, azides, Table 24.18; Eq. 24.19).

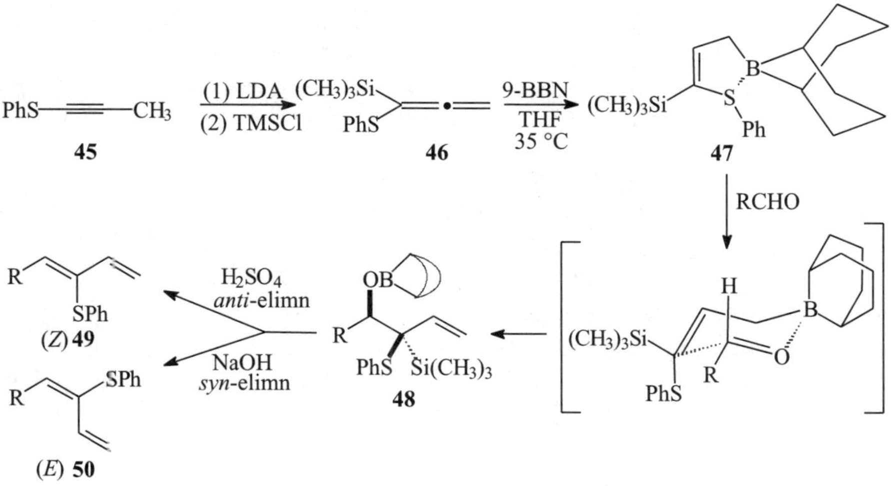

Scheme 24.12

Table 24.13 Stereoselective synthesis of 2-(phenylthio)-1,3-butadienes from aldehydes (RCHO) and allylborane 47 [36]

R	Diene **49**	Yield (%)	Diene **50**	Yield (%)
n-C_5H_{11}	**a**	80	**a**	92
n-C_8H_{17}	**b**	78	**b**	74
i-C_3H_7	**c**	76	**c**	83
c-C_6H_{11}	**d**	82	**d**	87
Ph	**e**	58	**e**	80
CH_2Ph	**f**	68	**f**	84
$(CH_2)_3CH_2Br$	**g**	73	**g**	82
$(CH_2)_2CH_2Br$	**h**	78	**h**	86
$(CH_2)_3CH_2N_3$	**i**	79	**i**	86
$(CH_2)_2CH_2N_3$	**j**	63	**j**	80
(E)-CH=CHPh	**k**	66	**k**	83

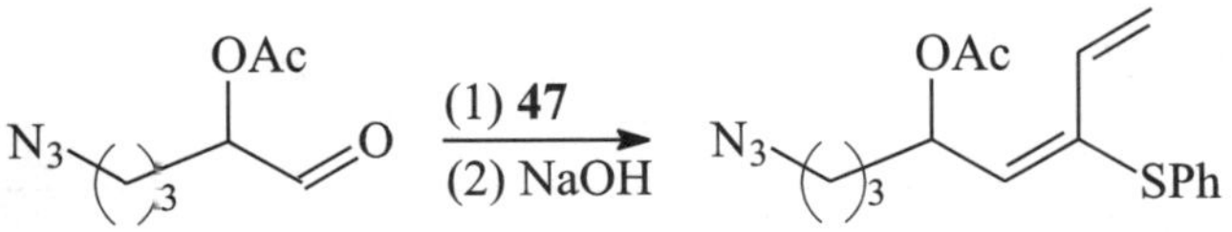

93% (E/Z = > 97/3)

(24.19)

The dienes **49** and **50** isomerize slowly on standing or extended contact with silica gel and the actual isolated *E*:*Z* ratio depends on the method of purification. Using flash-chromatography the diene isomerizes with 0–5% change in the values reported in Table 24.18.

24.2.5 Synthesis of Stannyl Dienes

The lithium[2-tri-*n*-butylstannyl-*Z*-1-alkenyl]-1-borates, prepared by copper catalyzed addition of Bu_3SnB-(OCH_3)-9-BBNLi$^+$ to 1-alkynes (*vide supra*), are selectively coupled *via* organopalladium or organocuprate with allylbromide, exclusively at the vinyl boron bond. The process results into the exclusive or predominant formation of 5-tributylstannyl-1,4-(*E*)-dienes (Chart 24.8) [38].

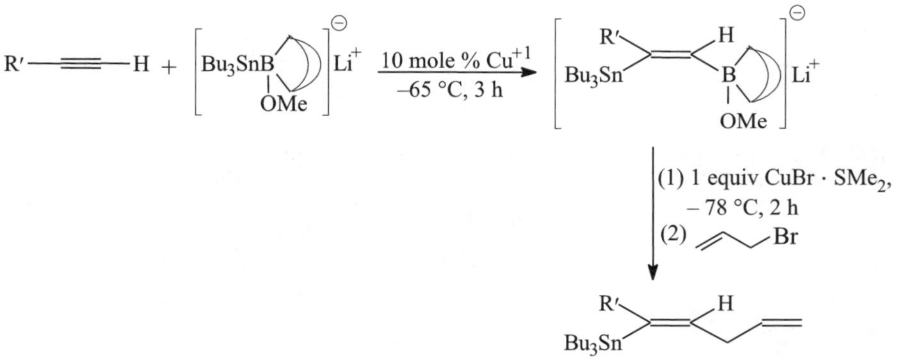

Chart 24.8

The results are summarized in Table 24.19 [38].

Table 24.19 Synthesis of 5-tributylstannyl-1,4(*E*)-dienes [38]

R′	Yield (%)[a]	R′	Yield (%)
HOC_4H_8	57	BrC_4H_8	53
$THPOC_4H_8$	62	C_6H_5	80
CNC_4H_8	78 (96:4)	C_7H_{15}	89 (92:8)
$AcOC_4H_8$	73	C_7H_{15}	78 (92:8)[b]

[a] A ratio of 1,4(*E*)-diene:other isomers.
[b] $Pd(Ph_3P)_4$ is used instead of $CuBr{\cdot}SMe_2$.

24.2.6 Synthesis of *B*-Isoprenyl Derivatives

The functionalized 1,3-butadiene moiety is an important structural feature of many natural products. One of the methodologies to achieve the synthesis of this structural moiety is the formation of isopropenyl anion, which is subsequently incorporated into various molecules [39]. The synthesis of isopropenyl anion is, however, cumbersome. Brown and Randad [40] have achieved a convenient procedure, which is detailed as: (1) application of the Schlösser reaction [41] to 2,2,6,6-tetramethylpiperidine provides the potassium amide, (2) the potassium amide cleanly converts isoprene to isoprenyl–potassium [42], and (3) treatment of *B*-OMe-9-BBN with isoprenyl–potassium provides the corresponding ate complex. Addition of controlled amount of BF_3·EE to the ate complex yields the desired *B*-isoprenyl-9-BBN (Scheme 24.13).

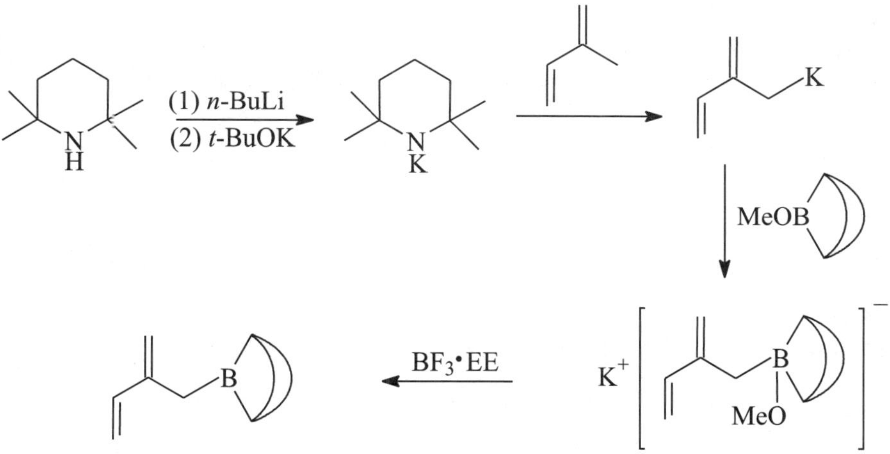

Scheme 24.13

Typical of allylboranes, *B*-isoprenyl-9-BBN reacts with acetaldehyde with allylic rearrangement to provide the borinate intermediate *via* a six-membered transition state [43]. The alkaline hydrogen peroxide oxidation of this intermediate affords the desired 4-methylene-5-hexen-2-ol (Eq. 24.20). Table 24.20 [41] summarizes the results of representative aldehydes with *B*-isoprenyl-9-BBN.

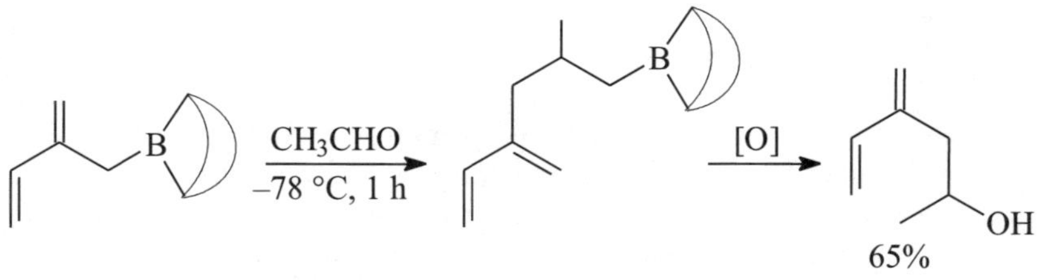

(24.20)

Table 24.20 Isoprenylation of aldehydes with *B*-isoprenyl-9-BBN [41]

Aldehyde	Product	Yield (%)
Acetaldehyde	4-Methylene-5-hexen-2-ol	65
2-Methylpropionaldehyde	2-Methyl-5-methylene-6-hepten-3-ol	65
Benzaldehyde	3-Methylene-1-phenyl-4-penten-1-ol	60
Isovaleraldehyde	2-Methyl-6-methylene-7-octen-4-ol (ipsenol)	65
β,β-Dimethylacrolein	2-Methyl-6-methylene-2,7-octadien-4-ol (ipsdienol)	60

24.3 Synthesis of Enynes

24.3.1 Stereoselective Synthesis of Conjugated (*E*)- and (*Z*)-Enynes

Brown and Molander [1] have reported the synthesis of conjugated enynes. The reaction involves the synthesis of alkenylcopper intermediates from alkenylboron derivatives of 9-BBN, which undergo coupling with 1-halo-1-alkynes in a stereodefined manner and afford the conjugated *trans* enynes (>95% pure) in almost quantitative yields. The cuprous iodide is the reagent of choice as compared to cuprous bromide-dimethylsulfide (Eq. 24.21; Table 24.21) [1].

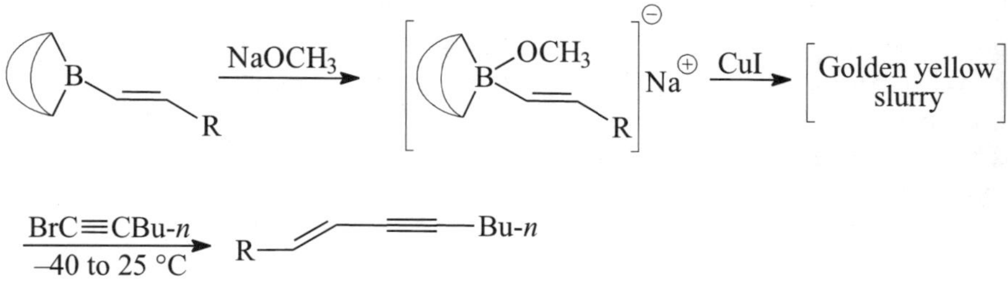

(24.21)

Table 24.21 Synthesis of conjugated enynes [1]

Alkenylborane	Product	Yield (%)
B-[(1*E*)-1-Hexene-1-yl]-9-BBN	(5*E*)-5-Dodecen-7-yne	93
B-[(1*E*)-3,3-Dimethyl-1-buten-1-yl]-9-BBN	(3*E*)-2,2-Dimethyl-3-decen-5-yne	90
B-[(1*E*)-1-(5-Chloropenten-1-yl)-9-BBN	(4*E*)-1-Chloro-4-undecen-6-yne	98

In another method, γ-(trimethylsilyl)-allenylborane undergoes condensation with hexanal and benzaldehyde with high diastereoselectivity [2] and elimination (Peterson olefination) [3] with KH (*syn*-elimination) or acid (*anti*-elimination) yields the corresponding enynes in high isomeric purity (Scheme 24.14) [2].

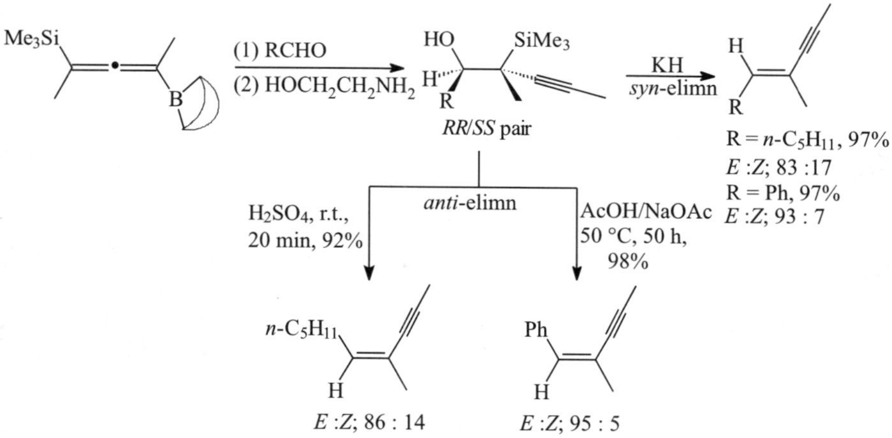

Scheme 24.14

24.3.2 Stereoselective Synthesis of Silylated (*E*)- and (*Z*)-Enynes

Wang *et al* [2] have reported that treatment of the readily available 3-(*tert*-butyldimethylsilyl)-1-(trimethylsilyl)-1-propyne [4] with *n*-butyllithium, followed by *B*-MeO-9-BBN and 4/3 $BF_3{\cdot}OEt_2$ [5] yields *B*-allenyl-9-BBN. The condensation of *B*-allenyl-9-BBN with hexanal or benzaldehyde and followed by the elimination step of the Peterson olefination [3] affords the conjugated enynes (Scheme 24.15) [2].

The results are summarized in Table 24.22 [3].

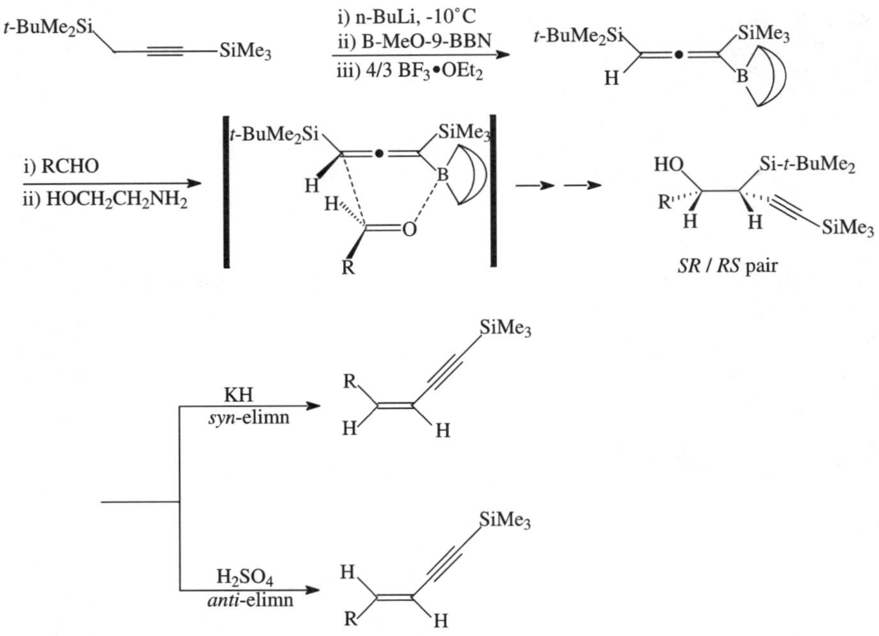

Scheme 24.15

Table 24.22 Stereoselective synthesis of silylated enynes [3]

	R and elimination conditions	Yield (%)	Isomer ratio
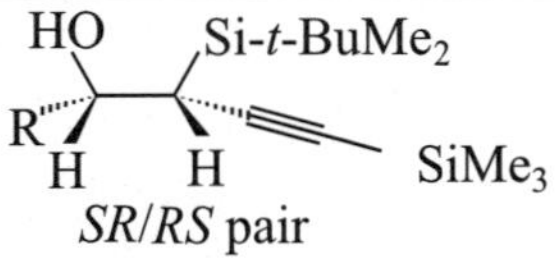			*SR*/*RS*:*RR*/*SS*
	n-C_5H_{11}	98	>98:2
	Ph	63	>98:2
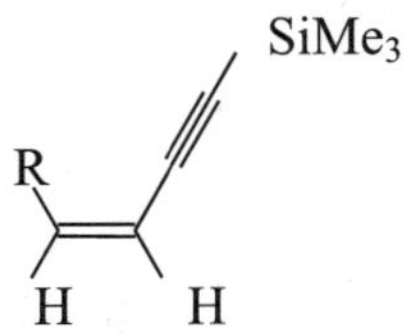			*Z*: *E*
	n-C_5H_{11}, KH(Et_2O, r.t., 1 h)	87	>99:1
	Ph, KH(Et_2O, r.t., 30 min)	94	>99:1
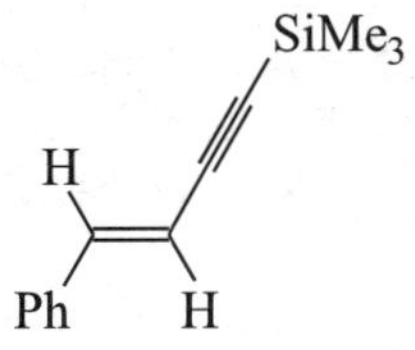	H_2SO_4 (r.t., 2 h)	88	98:2

24.4
Synthesis of Endiynes

24.4.1
Stereoselective Synthesis of (*E*)- and (*Z*)-Endiynes

γ-(Trimethylsilyl)-allenylborane undergoes diastereoselective condensation [1] with acetylenic aldehydes to produce predominantly *RR/SS* pairs of the corresponding silyl derivatives. The elimination step of Peterson's olefination [2] with KH (*syn*-elimination) provides the corresponding endiynes with moderate isomeric purity (Scheme 24.16) [1].

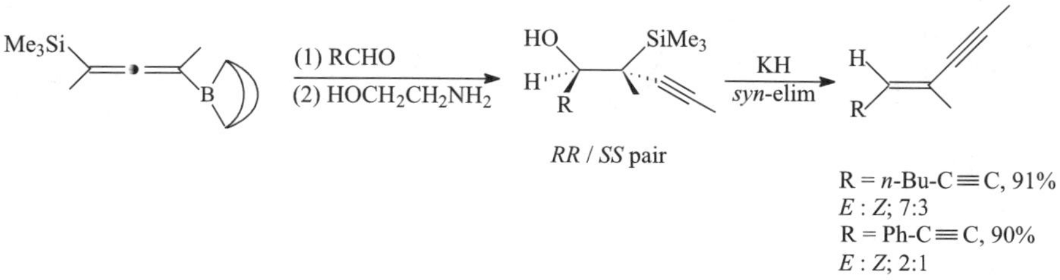

Scheme 24.16

24.4.2
Stereoselective Synthesis of (*E*)- and (*Z*)-Silylated Endiynes

The γ-(*tert*-butyldimethylsilyl)-allenylboranes, which are conveniently prepared (*vide supra*), undergo condensation [1] with conjugated acetylenic aldehydes [3] and furnish after workup with 2-aminoethanol, the corresponding adducts with high diastereoselectivity (*de* >96%). Endiynes with *Z* geometry (>99% *Z*) are obtained by treating the adducts with KH (*syn*-elimination), whereas endiynes with *E* geometry (≥98% *E*) are prepared by treating the condensation adduct with H_2SO_4 (*anti*-elimination) [2] (Scheme 24.17).
The data are summarized in Table 24.23 [1].

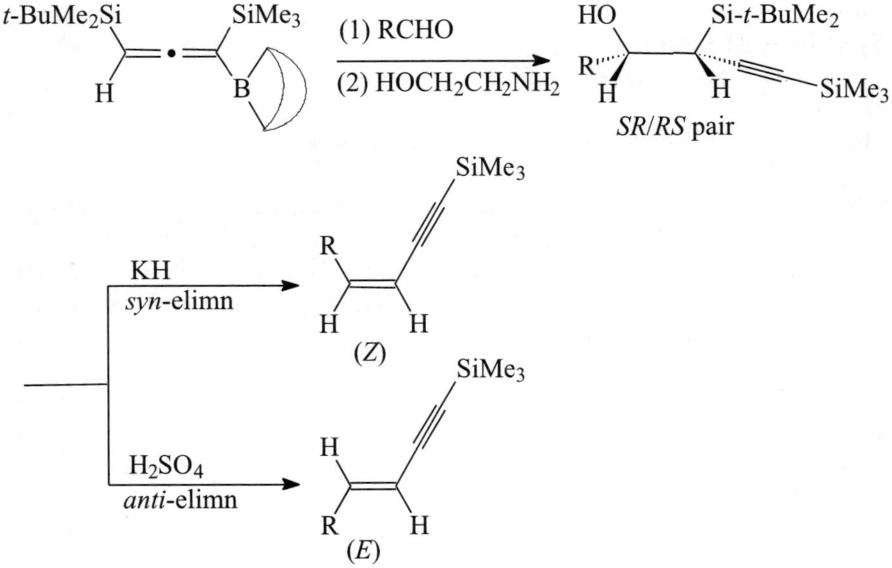

Scheme 24.17

Table 24.23 Stereoselective synthesis of silylated endiynes [1]

	R and elimination conditions	Yield (%)	Isomer ratio
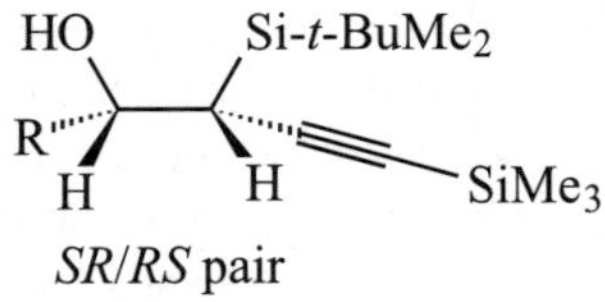			*SR*/*RS*:*RR*:*SS*
	n-Bu-C≡C	73	>98:2
	Ph-C≡C	64	>98:2
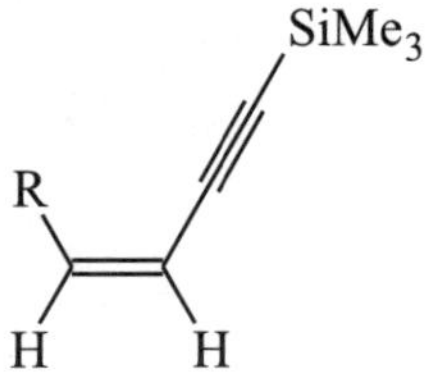			*Z*:*E*
	n-Bu-C≡C, KH(Et_2O, r.t., 30 min)	94	>99:1
	Ph-C≡C, KH(Et_2O, r.t., 1 h)	81	>99:1
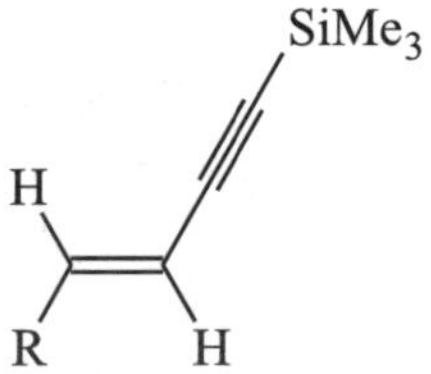			*E*:*Z*
	n-Bu-C≡C, H_2SO_4 (r.t., 5h)	85	>98:2
	Ph-C≡C, H_2SO_4 (r.t., 2h)	79	>98:2

24.5
Synthesis of 5-Methylene-1,3-Cyclohexadienes (*o*-Isotoluenes) and 1,2,4,6-Heptatetraenes: (Diene-Allenes)

Synthesis of 5-methylene-1,3-cyclohexadiene (*o*-isotoluene) was first reported by Bailey and Baylouny [1]. Several different routes to this alicyclic isomer of toluene [2] and its benzologs [3] have been reported. However, almost all of these methods are multistep synthesis and generally require gas-phase pyrolysis. Andemichael and Wang [4] have reported the condensation reaction methodology to synthesize diene-allenes [5] and *o*-isotoluenes by using the readily available conjugated allenic aldehydes [6] and γ-(trimethylsilyl)-allylboranes. Consequently, condensation of allylboranes with 4-methyl-2,3-pentadienal [6], followed by treatment with 2-aminoethanol produces allenic alcohols with high diastereomeric purity (*de* >90%) (Scheme 24.18) [4].

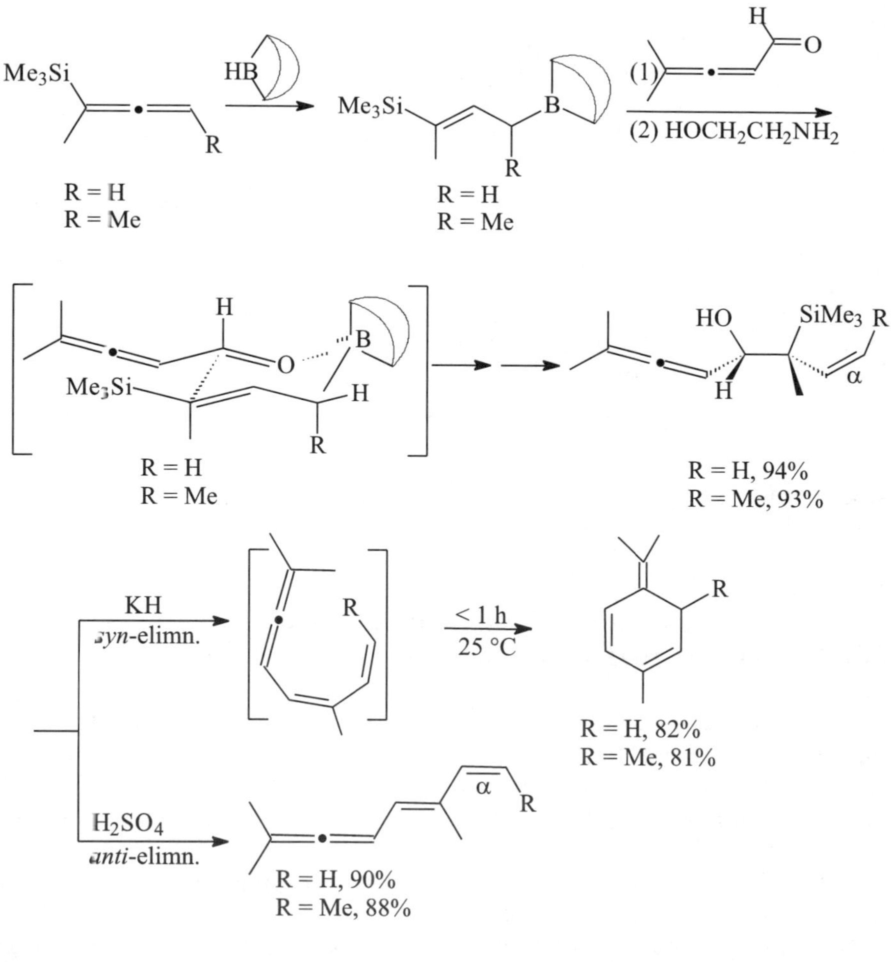

Scheme 24.18

To promote Peterson olefination, the use of KH results in the formation of *o*-isotoluene derivatives [2d, 7]. On the other hand, treatment of allenic alcohols with concentrated sulfuric acid produces the diene-allenes. The preference for *Z* geometry of the methyl-substituted α C=C in silylated allenic alcohol is attributed to the allylic methyl group at the C-1 position, favoring the axial position in the chair-like transition state, thus avoiding the severe steric interaction between the equatorial allylic methyl and the equatorial rigid bicyclic ligand on the boron.

Similarly, condensation of allylborane formed with bicyclohexylborane with 4-methyl-2,3-pentadienal leads to *o*-isotoluene and the corresponding diene-allene with *E* geometry (>98%) of the α C=C double bond (Scheme 24.19) [4].

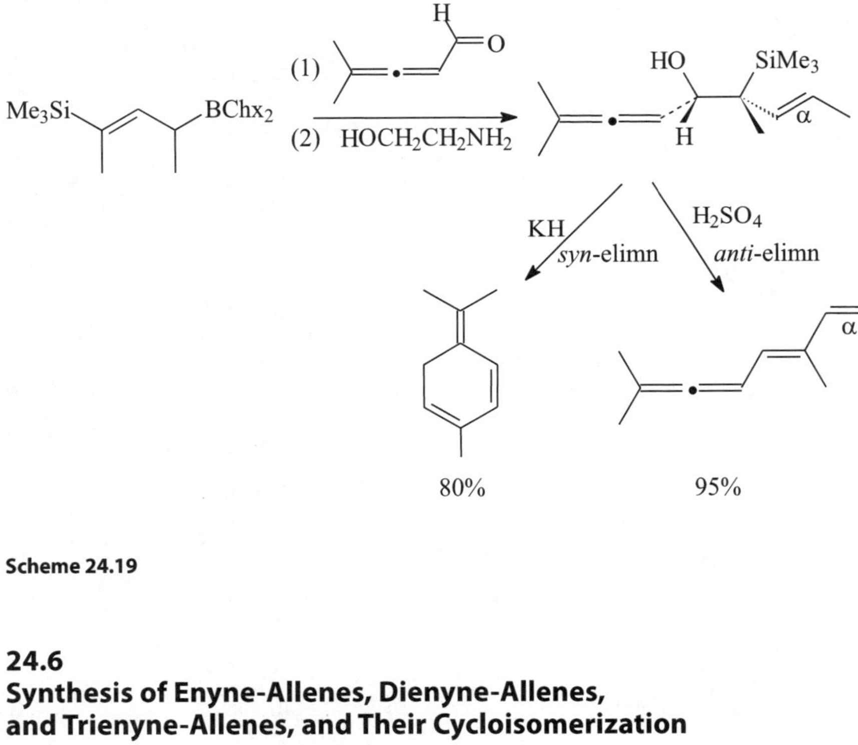

Scheme 24.19

24.6 Synthesis of Enyne-Allenes, Dienyne-Allenes, and Trienyne-Allenes, and Their Cycloisomerization

The endiyne and enyne-allene have attracted the attention, as they undergo facile cycloaromatization to produce reactive biradicals [1], which could mimic the DNA-cleaving mechanism and properties of the new class of very potent antitumor antibiotics calicheamicins [2], esperamicins [3], neocarzinostatin [4], and dynemicins [5].

9-Allenyls-9-BBN required for the synthesis of these unsaturated compounds are obtained by the following reaction sequence. Allenylsilanes [6] are lithiated with *tert*-butyllithium [7], followed by treatment with *B*-OMe-9-BBN and

4/3 $BF_3 \cdot OEt_2$ [8] to produce allenylboranes. Wang *et al* [9] have synthesized both (*E*)- and (*Z*)-1,2,4-heptatrien-6-ynes (enyne-allenes) by condensing allenylboranes with readily available conjugated allenic aldehydes [10]. The condensation affords, after treatment with 2-aminoethanol [11], hydroxypropargylsilanes with high diastereomeric purity (≥93:7) and in excellent yields. The KH-induced *syn*-elimination [12] of hydroxytrimethylsilanes give (*E*)-1,2,4-heptatrien-6-ynes as the major isomers (≥96%). The H_2SO_4- or $NaHSO_3$- induced *anti*-elimination [12] yields (*Z*)-1,2,4-heptatrien-6-ynes (enyne-allenes) and (*Z*)-dienyne-allenes. Both base and acid eliminations afford the corresponding *E* and *Z* isomers, respectively, in ≥96% isomeric purity (Scheme 24.20) [9].

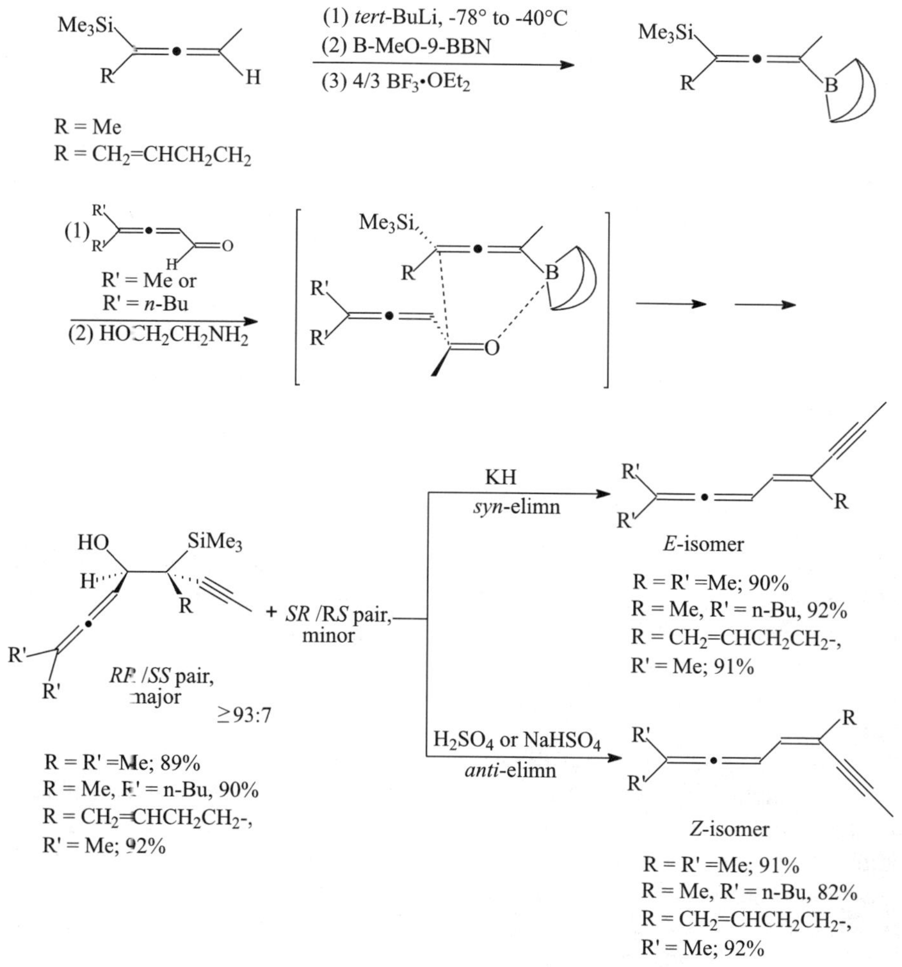

Scheme 24.20

The other enyne-allenes prepared by Wang by this method are outlined in Scheme 24.21 [13].

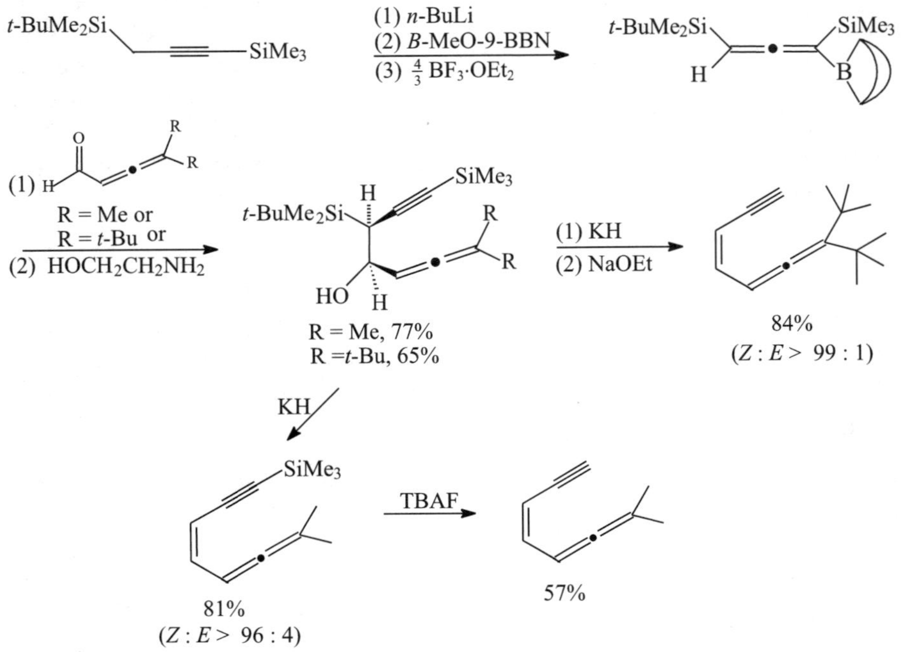

Scheme 24.21

The KH-induced *syn*-elimination of hydroxytrimethylsilane (R = Me) produces enyne-allene (*Z*:*E* = 96:4), which on treatment with tetrabutylammonium fluoride (TBAF) affords the desilylated adduct. Conversion of hydroxytrimethylsilane (R = *t*-Bu) to enyne-allene (*Z*:*E* = 99:1) is achieved in one operation by treatment with KH, followed by desilylation with sodium ethoxide.

The methodology is extended to the synthesis of triene-allene with high (*Z*)-isomeric purity of the new formed carbon–carbon double bond (Scheme 24.22) [14].

Further, it is reported that these trienyne-allenes undergo cycloaromatization in a single step. On heating, acyclic enyne-allenes undergo a sequence of intramolecular transformations with a cascade of energy, and represent a new approach to a one-step O(zero) → ABCD ring construction of the tetracyclic steroidal skeleton having an aromatic C-ring (Scheme 24.23) [14].

Similarly, dienyne-allene on refluxing in benzene undergoes cycloaromatization to afford a bicyclic aromatic adduct in 74% yield (Scheme 24.24) [9].

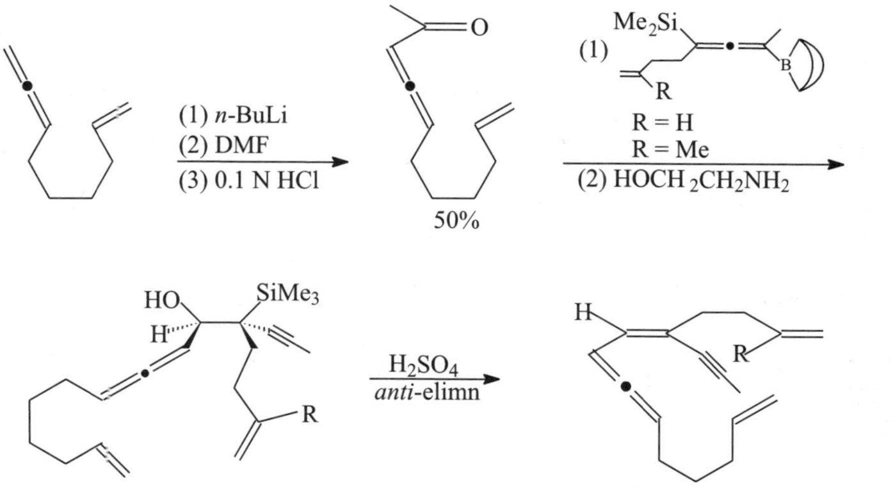

R = H, 80%
R = Me, 81%

R = H, 93%; $Z:E = 96:4$
R = Me, 86%; $Z:E = 97:3$

Scheme 24.22

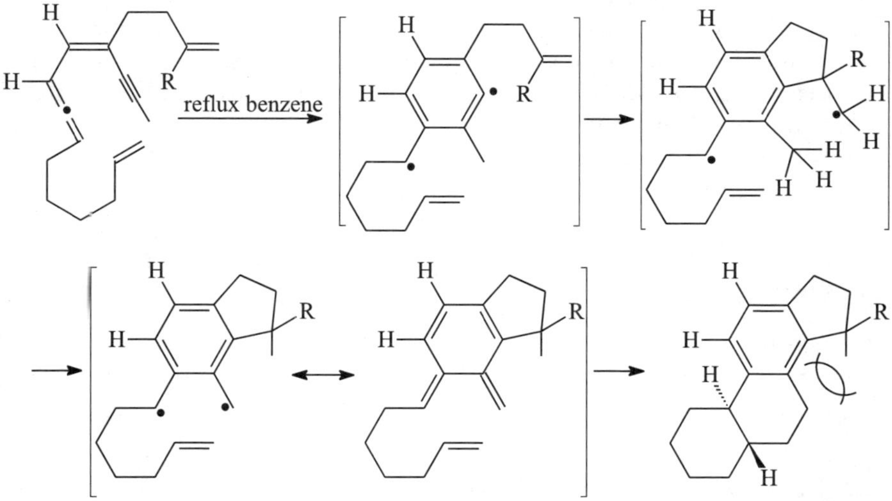

R = H, 50%
R = Me, 13%

Scheme 24.23

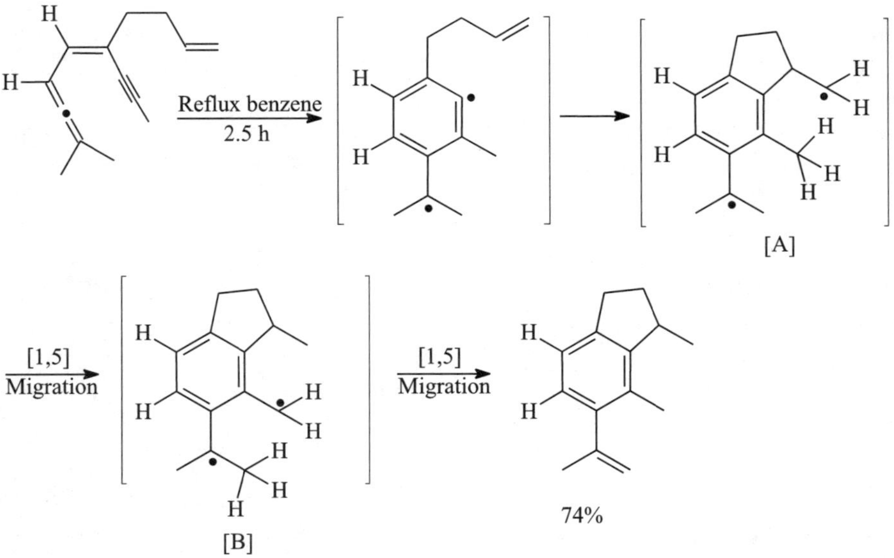

Scheme 24.24

The reaction involves two intermolecular H shifts. The biradical [A] decays with an initial [1,5]-sigmatropic H shift to give another biradical [B]. This biradical undergoes a second [1,5]-sigmatropic shift to afford the aromatic bicyclic adduct. The mechanism gets support by the synthesis of [C] from deuterated dienyne-allene, where deuterium from one of the deuterated *gem*-dimethyl groups migrates to benzylic system (Scheme 24.25) [9].

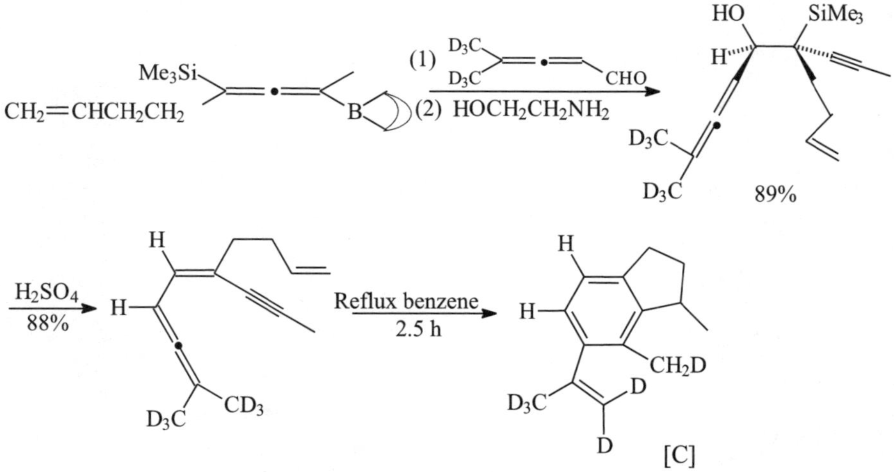

Scheme 24.25

24.7 Synthesis of Silylated *Z,Z*-Diendiynes and *Z,Z,Z,Z*-Tetraentetraynes, and Their Cycloisomerization

The synthesis of (*Z,Z*)-3,5-octadien-1,7-diynes (diendiynes) by condensation of enynyl aldehydes and *B*-allenyl-9-BBN is outlined in Scheme 24.26 [1]. The reaction sequence involves the treatment of the readily available 3-(*tert*-butyldimethylsilyl)-1-(trimethylsilyl)-1-propyne [2] with *n*-butyllithium, followed by reactions with *B*-MeO-9-BBN [3] and 4/3 BF_3·OEt_2 [4], which afford allenylboranes. The required (*Z*)-enynyl aldehydes are prepared by a $Pd(PPh_3)_4$-catalyzed cross-coupling between 2-bromo-1-cyclopentenecarboxaldehyde [5] and terminal alkynes [6]. The condensation of allenylborane, prepared *in situ* with (*Z*)-enynyl aldehydes, affords the corresponding silylalcohol with high diastereoselectivity (Table 24.24) [1]. The high diastereoselectivity in silyl alcohols is attributed to the preference of the *tert*-butyldimethylsilyl group and *Z*-enynyl group of

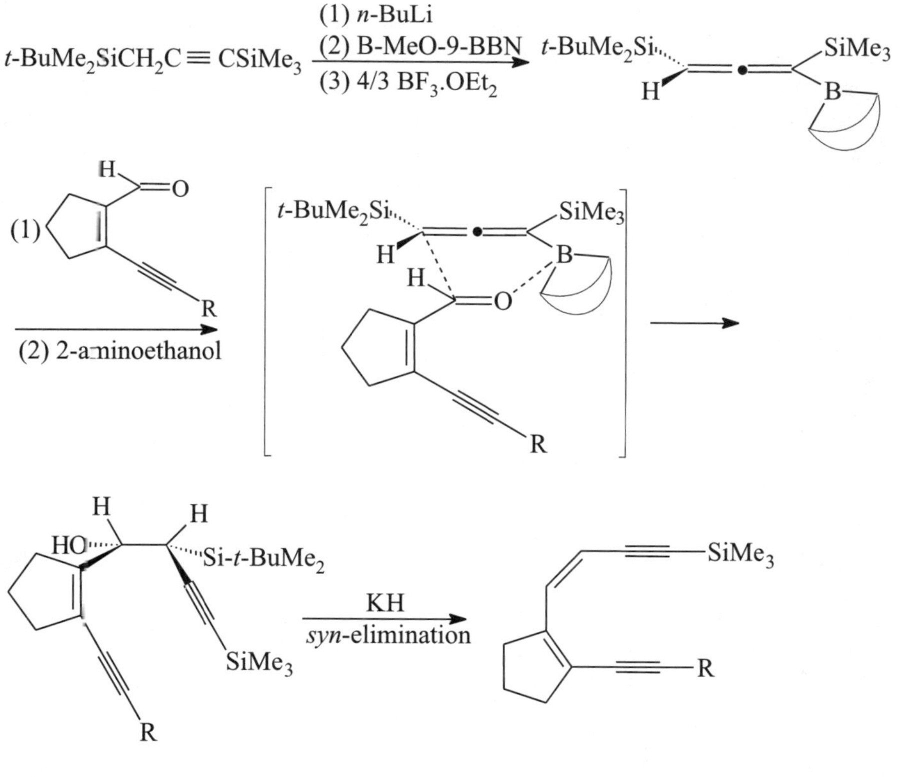

Scheme 24.26

the aldehydes to the opposite sides in the six-membered transition states in order to minimize the nonbonded steric interactions [7]. The *syn*-elimination, with potassium hydride [8], of silyl and OH group affords (*Z,Z*)-3,5-octadien-

1,7-diynes with high geometric purity (Z:$E \geq 92$:8, Table 24.24) [1]. Diendiynes, owing to the presence of two substituents at the alkynyl terminus, are stable enough to be purified by column chromatography.

Table 24.24. Stereoselective synthesis of silyl alcohols and diendiynes [1]

R	Silylalcohols	Yield (%)	Diendiynes	Yield (%) (*Z*:*E*)[a]
Me_3Si	a	61	a	78 (93:7)
Bu	b	65	b	80 (92:8)
Phenyl	c	56	c	83 (95:5)
Methylethenyl	d	54	d	81 (95:5)
1-Cyclohexenyl	e	55	e	81 (94:6)

[a] The preferential formation of the *Z* isomer by treatment of silylalcohol with potassium hydride indicates that the *SR*/*RS* pair are produced predominantly.

The desilylation of the disilyl derivative of diendiyne with tetrabutylammonium fluoride (TBAF) results in the formation of diendiyne. The diendiyne then undergoes electrocyclization reactions to benzocyclobutadiene, followed by its dimerization and yielding of all the four possible angular dimers, with nearly equal proportions in 66% isolated yield. However, in the presence of excess of cyclopentadiene, the Diels–Alder reaction occurs to give the *endo* isomer (Scheme 24.27) [1].

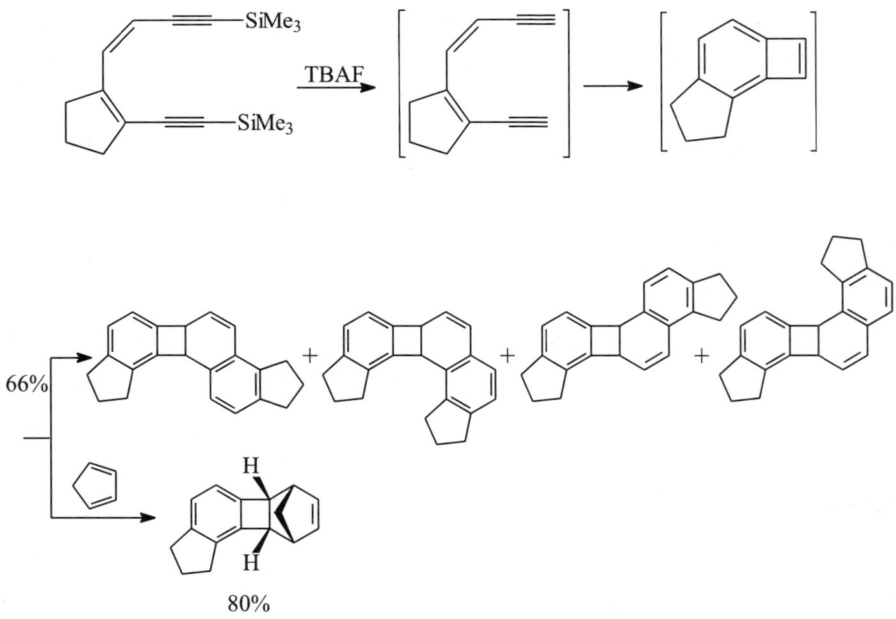

Scheme 24.27

The treatment of butyl- and trimethylsilyl-substituted terminal carbon–carbon triple-bonds derivative with TBAF leads to selective dimerization and affords only one angular dimer product in 61% yield. However, the presence of excess of cyclopentadiene leads to the Diels–Alder adduct in 50% isolated yield along with 20% of the dimer (Scheme 24.28) [1].

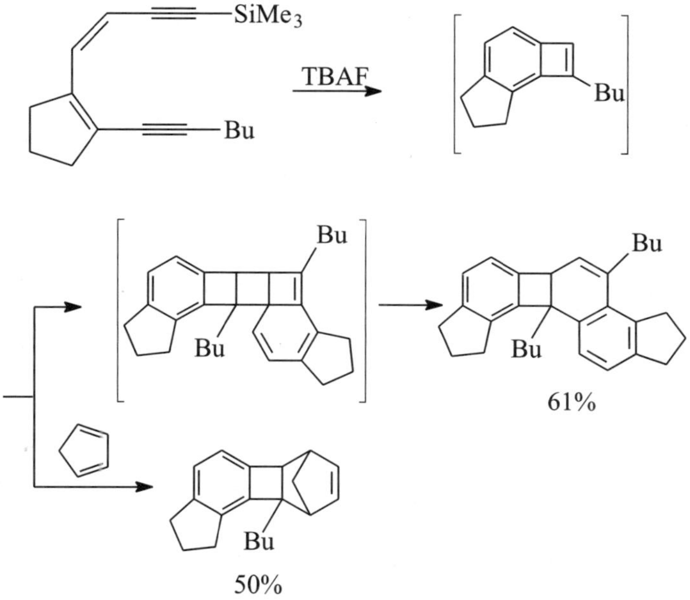

Scheme 24.28

Interestingly, when a diendiyne having a phenyl group is reacted with TBAF, the dimer formed initially is converted to rearranged linear dimer in 89% yield (Eq. 24.22), within 3 h of heating. The isomerization also occurs slowly at room temperature, which suggests that dimer is the *syn* isomer. This is consistent with the observation that the *syn* isomer is prone to thermal isomerization [9b], whereas the *anti* isomer is thermally much more stable.

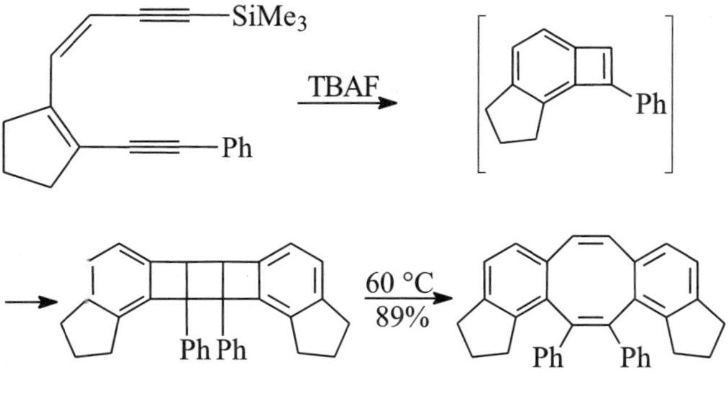

(24.22)

The treatment of isopropenyl derivative with TBAF, however, affords 1,5-cyclooctadiene with *trans* geometry. Here, formal [4+4]cycloaddition of the intermediate occurs (Scheme 24.29) [1]. The unexpected formation of *trans* geometry is

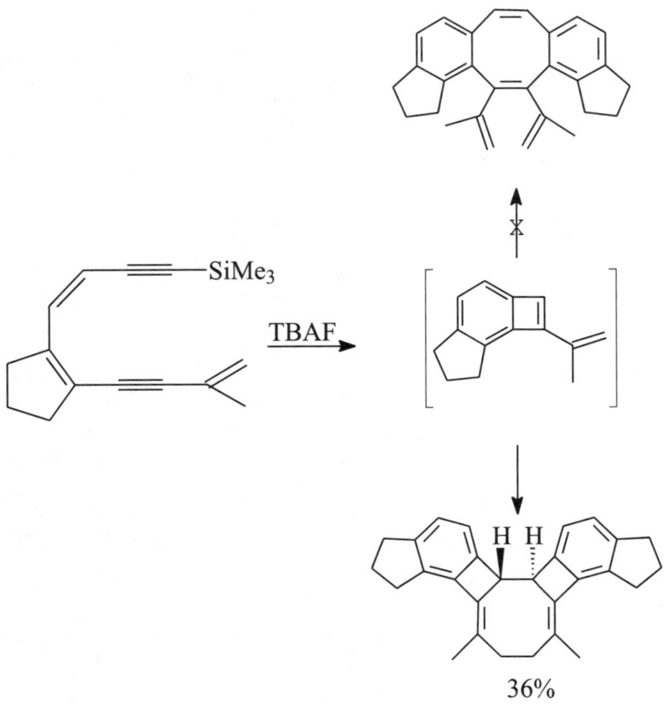

Scheme 24.29

explained *via* a two-step diradical mechanism. A radical pathway involves an initial head-to-head dimerization to form diradical, which then undergoes an intramolecular radical–radical combination to form the *trans* isomer (Eq. 24.23).

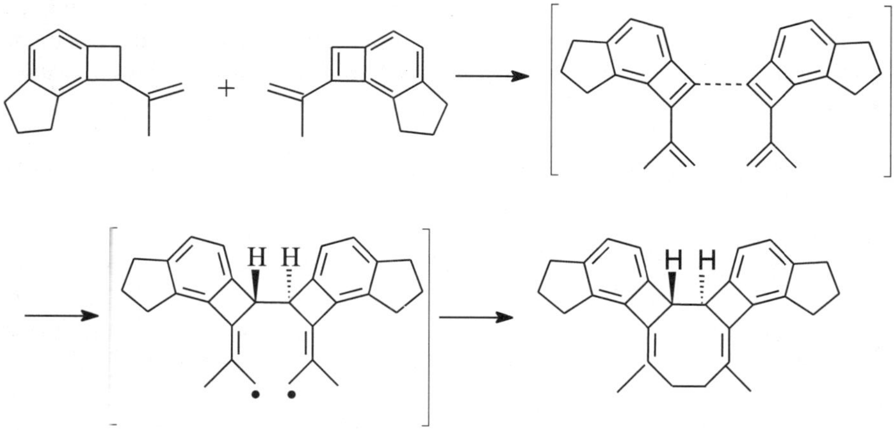

(24.23)

Similarly, cyclohexenyl derivative on treatment with TBAF undergoes dimerization to afford two *trans* isomers (Eq. 24.24).

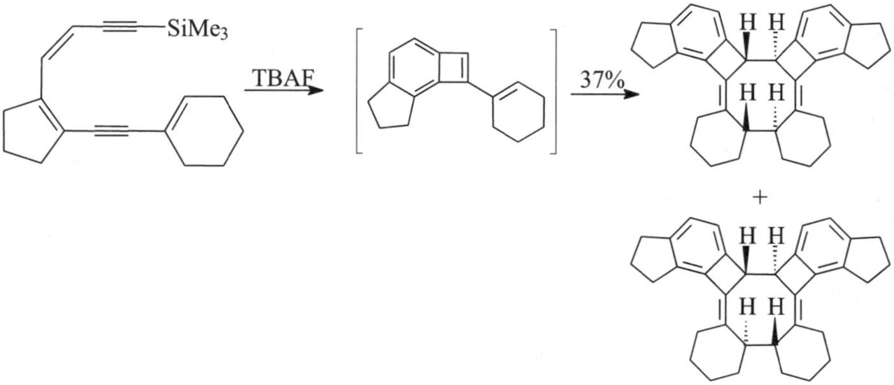

(24.24)

The condensation-Peterson olefination sequence has been further extended [10] to the synthesis of open chain (*Z*,*Z*)-diendiynes as outlined in Scheme 24.30. The reactions involve condensation of enynal, synthesized by oxidation of the commercially available (*Z*)-3-methyl-2-penten-4-yn-1-ol [11], and silyl substituted allenyl-9-BBN, prepared *in situ*, followed by treatment with 2-aminoethanol. The condensation product silyl alcohol, 6-*tert*-butyldimethylsilyl-8-trimethylsilyl-3-methyl-3-(*Z*)-octen-1,7-diyn-5-ol is formed in 69% yield with high diastereoselectivity (*de* > 99%). The $Pd(PPh_3)_4$-catalyzed cross-coupling, between terminal alkyne-silylalcohol and aryliodides, gives the aryl adducts of silylalcohols in 42–78% yields. The aryl adducts (entries c–e, Table 24.25) are obtained from monocoupling of 1,2-, 1,3-, or 1,4-diiodobenzene with 1 equiv of terminal alkyne. The conversions of silylalcohols to diendiynes are achieved with potassium hydride (*syn*-elimination) [8] in diethylether at 0 °C. The diendiynes have predominantly *Z*,*Z* geometry (>98%) and are thermally stable at room temperature (Scheme 24.30; Table 24.25) [10].

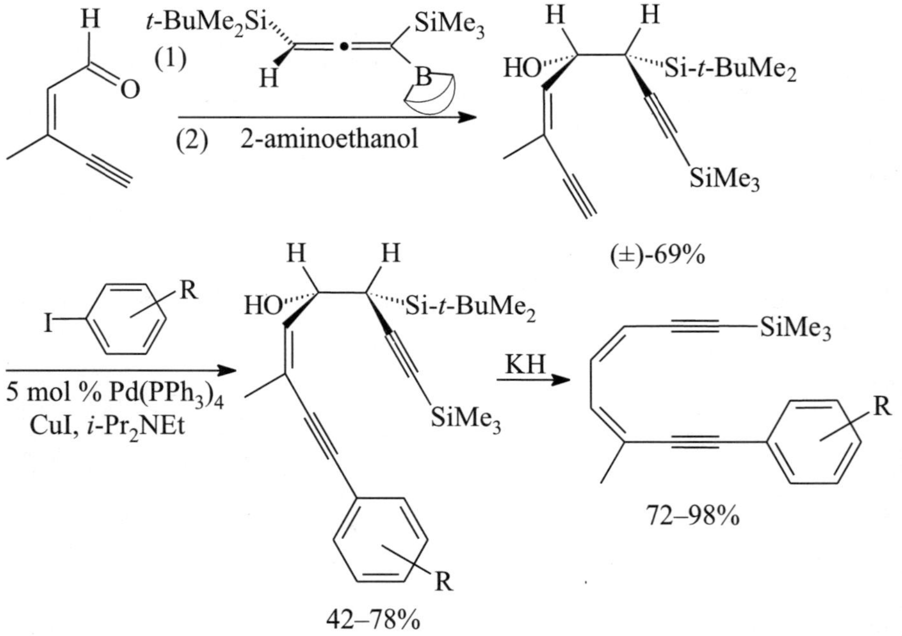

Scheme 24.30

Table 24.25 Synthesis of aryl adduct, diendiynes, and dibenzo[*a*,*e*]cyclooctenes [10]

R	Aryl adducts	Yield (%)	Diendiynes	Yield (%)	Dibenzo[*a*,*e*] cyclooctenes	Yield (%)
H	a	78	a	98	a	74
4-Me_3Si	b	66	b	98	b	54
2-Iodo	c	69	c	95	c	47
3-Iodo	d	42	d	72	d	40
4-Iodo	e	50	e	84	e	43

[a] The preferential formation of the *Z,Z* isomer (>98%) of diendiynes by treatment of aryl adduct of silyl alcohol with KH to induce *syn*-elimination indicates that the *RS/SR* pair is produced predominantly.

The treatment of silyl derivatives of diendiynes with TBAF triggers a cascade sequence, initiating from the desilylated diendiynes, and through benzocyclobutadienes afford dibenzo[*a*,*e*]cyclooctenes (*syn*-dibenzocyclooctatetraenes) (Scheme 24.31; Table 24.25) [10].

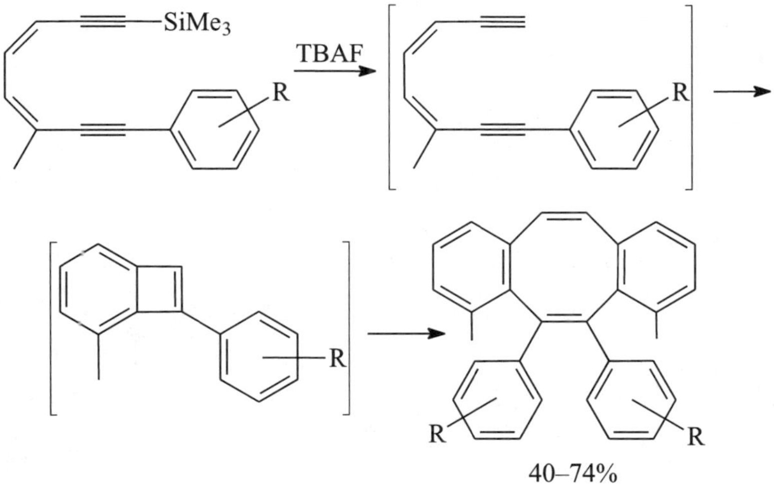

Scheme 24.31

However, when 2 equiv of silyl-substituted terminal alkyne-silylalcohol is coupled, independently, with 1 equiv of 1,2- or 1 equiv of 1,4-diodobenzene, it leads to the corresponding 1,2- or 1,4-aryl adducts. The KH-induced *syn*-elimination of silyl-alcohol moiety leads to the formation of silyl derivative of *o*- or *p*-substituted, *Z,Z,Z,Z*-tetraentetraynes (Schemes 24.32, 24.33) [10].

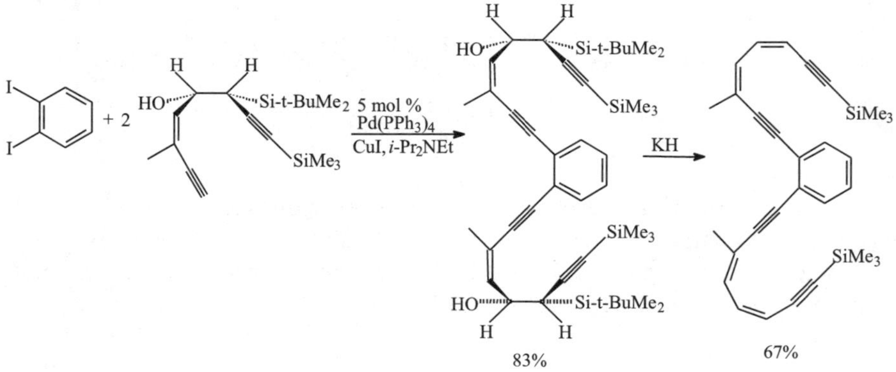

Scheme 24.32

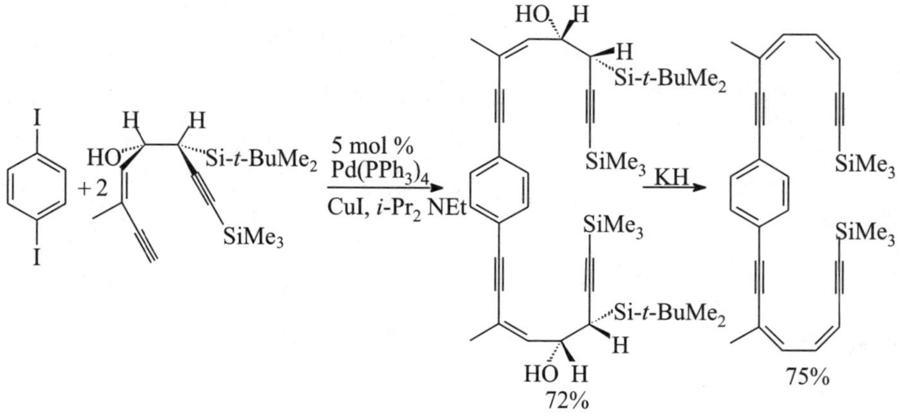

Scheme 24.33

The silyl derivatives of *o*- and *p*-substituted tetraentetraynes on treatment with TBAF afford shining black and yellow solids, respectively. These solids are mixture of oligomers.

The diiodo derivative of dibenzo[*a,e*]cyclooctene (entry e, Table 24.25) undergoes cross-coupling with 2 equiv of 6-*tert*-butyldiemthylsilyl-8-trimethylsilyl-3-methyl-3-(*Z*)-octen-1,7-diyn-5-ol (ca. Scheme 24.30) to afford the silylalcohols as a 1:1 mixture of two diastereomers in 64% isolated yield. The subsequent KH treatment gives the corresponding *Z,Z,Z,Z*-tetraentetrayne (Scheme 24.34) [10].

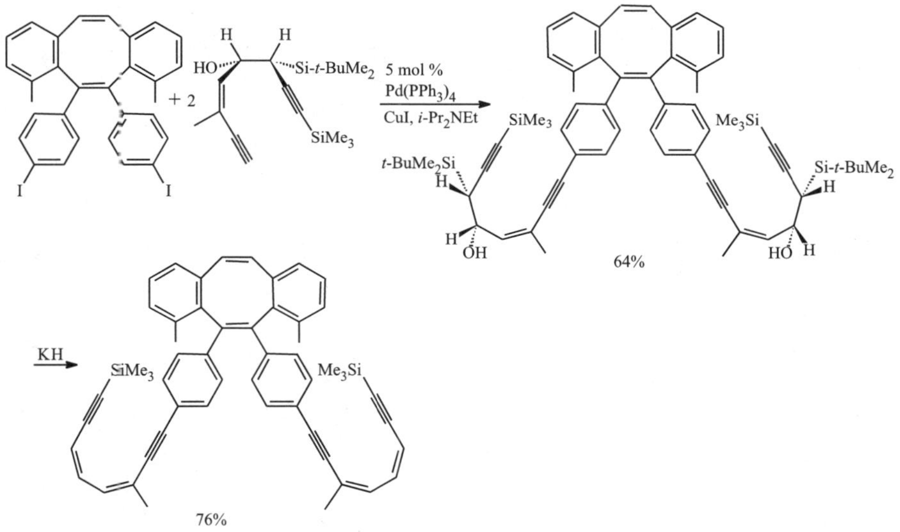

Scheme 24.34

Treatment of diiododerivative of dibenzo[*a,e*]cyclooctene (entry e, Table 24.25) with 1 equiv of 6-*tert*-butyldimethylsilyl-8-trimethylsilyl-3-methyl-3-(*Z*)-octen-1,7-diyn-5-ol (ca. Scheme 24.30) produces the monocoupling adduct, which on treatment with KH yields the corresponding *Z,Z*-diendiyne. The cascade sequence triggered by TBAF affords the diiodide derivative of tris-dibenzo[*a,e*]cyclooctene. The reactions sequence is outlined in Scheme 24.35 [10].

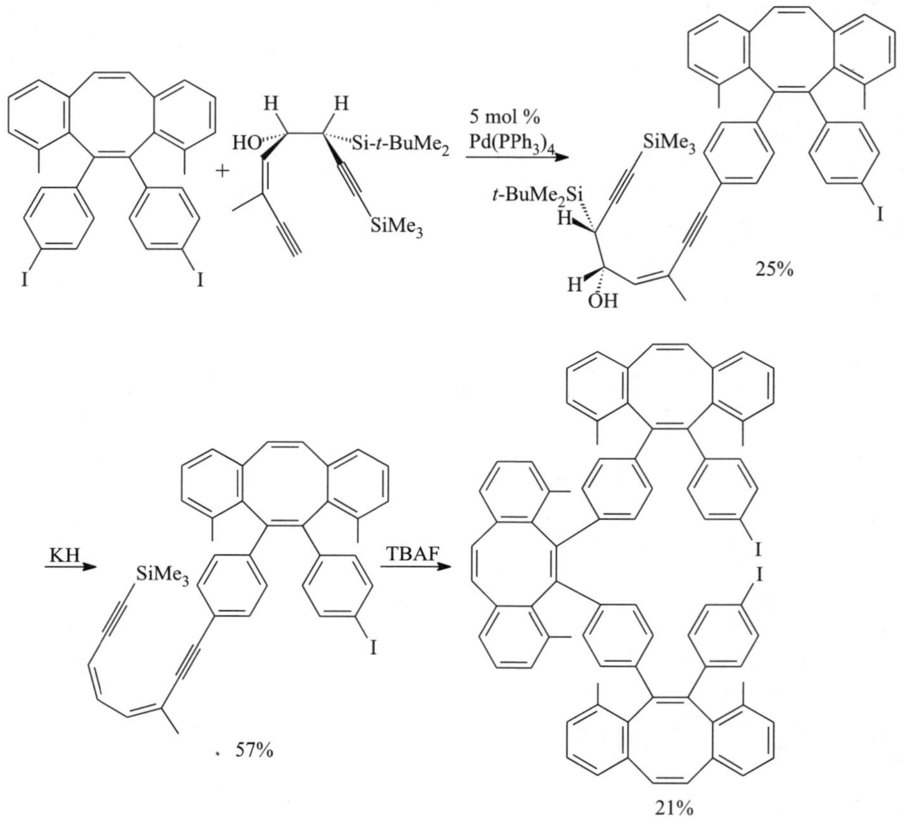

Scheme 24.35

Consequently, 6-*tert*-butyldimethylsilyl-8-trimethylsilyl-3-methyl-3-(*Z*)-octene-1,7-diyn-5-ol serves as a latent diendiyne moiety, which on cross-coupling with aryliodides provides an easy access to a variety of diendiynes, (*Z,Z*)-1-aryl-3,5-octadiene-1,7-diynes. These diendiynes are precursors of 5,6-diaryldibenzo[*a,e*] cyclooctenes. The generation of two diendiynyl moieties (tetraentetraynes) in the same molecule produces oligomers having multiple dibenzo[*a,e*]cyclooctenyl units.

References

Section 24.1

1. (a) Brown HC, Scouten CG, Liotta R (1979) J Am Chem Soc 101:96; (b) Wang KK, Scouten CG, Brown HC, (1982) J Am Chem Soc 104:531
2. Colberg JC, Rane A, Vaquer J, Soderquist JA (1993) J Am Chem Soc 115:6065
3. (a) Fetizon M, Jurion M, Anh NT (1969) J Chem Soc Chem Commun 112; (b) Ireland RE, Pfister G (1969) Tetrahedron Lett 26:2145; (c) Ireland RE, Muchmore DC, Hengartner U (1972) J Am Chem Soc 94:5098; (d) Ireland RE, Kowalski CJ, Tilley JW, Walba DM (1975) J Org Chem 40:990; (e) Welch SC, Walters ME (1978) ibid. 43:2715
4. (a) Pasto DJ, Snyder R Sr (1966) J Org Chem 31:2777; (b) Lewis JW, Pearce AA (1964) Tetrahedron Lett 2039; (c) Lewis JW, Pearce AA (1969) J Chem Soc B 863; (d) Friary RJ, Gilligan JM, Szajewski RP, Falci KJ, Frank RW (1973) J Org Chem 38:3487; (e) Froborg J, Magnusson G, Thoren S (1975) Tetrahedron Lett 1621; (f) Montury M, Gore J (1977) ibid. 219; (g) Froborg J, Magnusson G (1978) J Am Chem Soc 100:6728
5. (a) Caglioti L, Cainelli G, Maina G, Selva A (1962) Gazz Chem Ital 92:309; (b) Larson GL, Hernandez E, Alonso C, Nieves I (1975) Tetrahedron Lett 4005
6. Trost BM, Ornstein PL (1981) Tetrahedron Lett 3463
7. (a) Nelson SJ, Detre G, Tanabe M (1973) Tetrahedron Lett 447; (b) Cacchi S, Morera E, Ortar G (1984) ibid. 4821
8. (a) Shapiro RH, Heath MJ (1967) J Am Chem Soc 89:5734; (b) Shapiro RH (1975) Org; React; (N.Y.) 23:405; (c) Shapiro RH, Lipton MF, Kolonko KJ, Buswell RL, Capuano LA (1975) Tetrahedron Lett 1811; (d) Adlington RM, Barrett AGM (1983) Acc Chem Res 16:55
9. Another approach to the synthesis of (*E*)- and (*Z*)- alkenes from aldehydes based on boron intermediates has been described: Pelter A, Buss D, Colclough EJ (1987) Chem Commun 297
10. (a) Harrison IT, Harrison S (1971) Compendium of organic synthetic methods. Wiley Inter-Science, New York, pp 517–520; (b) Motherwell WB (1973) J Chem Soc Chem Commun 935
11. Goralski CT, Singaram B, Brown HC (1987) J Org Chem 52:4014
12. (a) Singaram B, Goralski CT, Rangaishenvi MV, Brown HC (1989) J Am Chem Soc 111:384; (b) Goralski CT, Singaram B, Brown HC (1989) US Patent 4,886,924; (c) Goralski CT, Singaram B, Brown HC (1990) US Patent 4,895,996
13. (a) Singaram B, Rangaishenvi MV, Brown HC, Goralski CT, Hasha DL (1991) J Org Chem 56:1543; (b) Singaram B, Goralski CT, Rangaishenvi MV, Brown HC (1989) J Am Chem Soc 111:384; (c) Goralski CT, Singram B, Brown HC (1990) US Patent 4,895,996
14. (a) Gurowitz WD, Joseph MA (1965) Tetrahedron Lett 4433; (b) Borowitz IJ, Williams GJ (1967) J Org Chem 32:4157
15. Levy AB, Angelastro R, Marinelli ER (1980) Synthesis 11:945
16. Brown HC, Ford TM (1981) J Org Chem 46:647
17. Singaram, B, Goralski CT, Fisher GB (1991) J Org Chem 56:5691
18. Yatagai H, Yamamoto Y, Maruyama K (1980) J Am Chem Soc 102:4548
19. For the synthesis of 1-alkenyl-9-BBN, see Brown HC, Scouten CG, Liotta R (1979) J Am Chem Soc 101:96; for 2-alkenyl-9-BBN, see Kramer GW, Brown HC (1977) J Organomet Chem 132:9
20. Verdone JA, Mangravite JA, Scarpa NM, Kuivila HG (1975) J Am Chem Soc 97:843
21. (a) Kramer GW, Brown HC (1977) J Organomet Chem 132:9; (b) Brown HC, Liotta R, Kramer GW (1979) J Am Chem Soc 101:2966

22. Wang KK, Gu YG, Liu C (1990) J Am Chem Soc 112:4424
23. Soderquist JA, Colberg JC, Valle LD (1989) J Am Chem Soc 111:4873
24. Still WC (1977) J Am Chem Soc 99:4836
25. Sharma S, Olhlschlager AC (1988) Tetrahedron Lett 29:261
26. Brown HC, Blue CD, Nelson DJ, Bhat NG (1989) J Org Chem 54:6064
27. (a) Linstrumelle G, Krieger JK, Whitesides GM (1976) Org; Synth; 55:103; (b) Zweifel G, Lewis W, On HP (1979) J Am Chem Soc 101:5101

Section 24.2

1. Singaam B, Rangaishenvi MV, Brown HC, Goralski CT, Hasha DL (1991) J Org Chem 56:1543
2. Brown HC, Ravindran N (1976) J Am Chem Soc 98:1785
3. (a) Yamamoto Y, Yatagai H, Moritani I, (1975) J Am Chem Soc 97:5606; (b) Yamamoto Y, Yatagai H, Sonoda A, Murahashi S-I, Moritani I (1977) ibid. 99:5652
4. Campbell JB Jr, Brown HC (1980) J Org Chem 45:549
5. Brown HC, Campbell JB Jr (1980) J Org Chem 45:550
6. Ciganek E (1984) Org React (N.Y.) 32:1 and references cited therein
7. Henrick CA (1977) Tetrahedron 33:1845
8. Rossi R (1977) Synthesis 817
9. Tsai DJS, Matteson DS (1981) Tetrahedron Lett 22:2751
10. Yamamoto Y, Saito Y, Maruyama KJ (1985) Organomet Chem 292:311
11. Liu C, Wang KK, (1986) J Org Chem 51:4733
12. General references for allylborane and allylboronate additions to carbonyl compounds: (a) Hoffmann RW (1982) Angew Chem Int Ed Engl 21:555; (b) Matteson DS (1986) Synthesis 973; (c) Yamamoto Y (1987) Acc Chem Res 20:243; (d) Pelter A, Smith K, Brown HC (1988) Borane reagents. Academic, New York; (e) Bubnov YN, Zheludeva VI, Ignatenko AV (1989) J Organomet Chem 359:151 and references therein
13. Related diene synthesis using silane-substituted allylic organometallics: (a) Chan T-H, Li J-S (1982) J Chem Soc Chem Commun 969; (b) Sato F, Suzuki Y, Sato M (1982) Tetrahedron Lett 23:4589; (c) Reetz MT, Wenderoth B (1982) ibid. 23:5259; (d) Murai A, Abiko A, Shimada N, Masamune T (1984) ibid. 25:4951; (e) Ikeda Y, Ukai J, Ikeda N, Yamamoto H (1987) Tetrahedron 43:731 and references therein
14. (a) Hudrlik PF, Peterson D (1975) J Am Chem Soc 97:1464; (b) Chan TH (1977) Acc Chem Res 10:442; (c) Ager DJ (1984) Synthesis 384; (d) Colvin EW (1988) Silicon reagents in organic synthesis. Academic, New York
15. Montury M, Psaume B, Gore J (1980) Tetrahedron Lett 21:163
16. (a) Westmijze H, Vermeer P (1979) Synthesis 390; (b) Wang KK, Nikam SS, Ho CD (1983) J Org Chem 48:5376; (c) Wang KK, Liu C (1985) J Org Chem 50:2578; (d) Danheiser RL, Carini DJ, Fink DM, Basak A (1983) Tetrahedron 39:935; (e) Danheiser RL, Tsai Y-M, Fink DM, (1987) Org Synth 66:1
17. Kramer GW, Brown HC, (1977) J Organomet Chem 132:9; (b) Brown HC, Liotta R, Kramer GW (1979) J Am Chem Soc 101:2966
18. Liu C, Wang KK (1986) J Org Chem 51:4733
19. (a) Wang KK, Liu C, Gu YG, Burnett FN (1991) J Org Chem 56:1914; (b) Gu YG, Wang KK (1991) Tetrahedron Lett 32:3029

20. (a) Kitching W, Olszowy HA, Drew GM, Adcock W (1982) J Org Chem 47:5153; (b) Hwu JR, Wang N (1989) Chem; Rev; 89:1599
21. Brown HC, Zweifel G (1959) J Am Chem Soc 81:247
22. (a) Johnson F (1968) Chem Rev 68:375; (b) Hoffmann RW (1989) Chem Rev 89:1841
23. Yamamoto Y, Saito Y, Maruyama K (1985) J Organomet Chem 292:311
24. Soderquist JA, Santiago B (1989) Tetrahedron Lett 30:5693
25. (a) Cope AC, Nealy DL, Scheiner P, Wood G (1965) J Am Chem Soc 87:3130; (b) Allan RD, Cordiner BG, Wells RJ (1968) Tetrahedron Lett 6055
26. Bosone E, Farina P, Guazzi G (1983) Synthesis 942
27. Colberg JC, Rane A, Baquer J, Soderquist JA (1993) J Am Chem Soc 115:6065
28. (a) Scott WJ, Stille JK (1986)J Am Chem Soc 108:3033 and references cited therein; (b) Satoh M, Miyaura N, Suzuki A (1986) Chem Lett 1329; (c) Miyaura N, Yamada K, Suginome H, Suzuki A (1985) J Am Chem Soc 107:972; (d) Negishi E-I, Luo F-T (1983) J Org Chem 48:1560; (e) Molander GA, Zinke PW (1986) Organometallics 5:2161; (f) Bjorkling F, Norin T, Unelius CR, Miller RB (1987) J Org Chem 52:292 and references cited therein; (g) Jabri N, Alexakis A, Normant JF (1983) Bull Soc Chim Fr II:321, 332; (h) Tsai DJS, Matteson DS (1981) Tetrahedron Lett 22:2751; (i) Yamamoto Y, Saito Y, Maruyama K (1985) J Organomet Chem 292:311; (j) Ikeda Y, Ukai J, Ikeda N, Yamamoto H, (1987) Tetrahedron 43:723, 731 and references cited therein
29. (a) Block E, Aslam M, Eswarakrishnan V, Gebreyes K, Hutchinson J, Iyer R, Laffitte JA, Wall A (1986) J Am Chem Soc 108:4568; (b) Djahanbini D, Cazes B, Gore J (1984) Tetrahedron 40:3645; (c) Djahanbini D, Cazes B, Gore J (1984) ibid. 41:867; (d) Trost BM, Fortunak JM (1980) J Am Chem Soc 102:2841; (e) Cuvigny T, Fabre JL, Herve du Penhoat C, Julia M (1983) Tetrahedron Lett 24:4319
30. Wang KK, Gu YG, Liu C (1990) J Am Chem Soc 112:4424
31. Zweifel G, Backlund SJ (1978) J Organomet Chem 156:159
32. Ditrich K, Bube T, Stürmer R, Hoffmann RW (1986) Angew Chem Int Ed Engl 25:1028
33. Satrsangi PD, Wang KK (1992) Tetrahedron Lett 33:5025
34. Matuszak CA, Matuszak AJ (1976) J Chem Educ 53:280
35. (a) Review: Petrzilka M, Grayson JI (1981) Synthesis 753; (b) Gundermann KD, Holtmann P (1966) Angew Chem Int Ed Engl 5:668; (c) Cohen T, Mura AJ Jr, Shull DW, Fogel ER, Ruffner RJ, Falck JR (1976) J Org Chem 41:3218; (d) Hopkins PB, Fuchs PL (1978) ibid. 43:1208; (e) Trost BM, Vladuchick WC, Bridges AJ (1980) J Am Chem Soc 102:3554 and earlier work cited therein; (f) Trost BM, Lavoie AC (1983) ibid. 105:5057; (g) Bridges AJ, Fischer JW (1983) Tetrahedron Lett 24:445, 447; (h) Liotta CL, Verbicky JW Jr (1985) ibid. 26:1395; (i) Proteau PJ, Hopkins PB (1985) J Org Chem 50:141; (j) Chan TH, Prasad CVC (1987) ibid. 52:110 and earlier references therein; (k) Chou T, Lee S-J, Peng M-L, Sun D-J, Chou S-SP (1988) ibid. 53:3027; (l) Chou S-SP, Tsai C-Y (1988) ibid. 53:5305; (m) McDougal PG, Oh Y-I, VanDerveer D (1989) ibid. 54:91; (n) Evans DA, Bryan CA, Sims CL (1972) J Am Chem Soc 94:2891; (o) Cohen T, Kosarych A, (1982) J Org Chem 47:4005; (p) Kozikowski AP, Huie EM (1982) J Am Chem Soc 104:2923; (q) Overman LE, Petty CB, Ban T, Huang GT (1983) ibid. 105:6335; (r) Ihara M, Kawaguchi A, Chihiro M, Fukumoto K, Kametani T (1986) J Chem Soc Chem Commun 671
36. Pearson WH, Lin K-C, Poon Y-F (1989) J Org Chem 54:5814

37. (a) Pourcelot G, Cadiot P (1966) Bull Soc Chim Fr 208:3016; Pourcelot G, Cadiot P, Willemant ACR (1961) Acad Sci 252:1630; (b) Brandsma L, Jonker C, Berg MH (1965) Recl Trav Chim Pays Bas 84:560; (c) Bridges AJ, Thomas RD (1983) J Chem Soc Chem Common 485
38. Sharma S, Oehlschlager AC (1988) Tetrahedron Lett 29:261
39. (a) Semmelhack MF, Fewkes EJ (1987) Tetrahedron Lett, 28:1497; (b) Katzellenbogen JA, Lenox RS (1973) J Org Chem, 38:326; (c) Riley RG, Silverstein RM, Katzellenbogen JA, Lenox RS (1974) ibid. 39:1957; (d) Hosomi A, Iguchi H, Endo M, Sakurai H (1979) Chem Lett 977; (e) Hosomi A, Araki Y, Sakurai H (1983) J Org Chem 48:3122; (f) Bubnov N, E'Tinger MY (1985) Tetrahedron Lett 26:2797
40. Brown HC, Randad RS (1990) Tetrahedron 46:4463
41. Fujita K, Schlösser M (1982) Helv Chim Acta 65:1258
42. Klusener PAA, Hommes HH, Verkuijsse HD, Brandsma L (1985) J Chem Soc Chem Commun 1677
43. Brown HC, Jadhav PK (1984) J Org Chem 49:4089

Section 24.3

1. Brown HC, Molander GA (1981) J Org Chem 46:645
2. Wang KK, Wang Z, Gu YG (1993) Tetrahedron Lett 34:8391
3. Hudrlik PF, Peterson D (1975) J Am Chem Soc 97:1464
4. (a) Furuta K, Ishiguro M, Haruta R, Ikeda N, Yamamoto H (1984) Bull Chem Soc Jpn 57:2768; (b) Yamakado Y, Ishiguro M, Ikeda N, Yamamoto H, (1981) J Am Chem Soc 103:5568
5. Brown HC, Sinclair JA (1977) J Organomet Chem 131:163

Section 24.4

1. Wang KK, Wang Z, Gu YG (1993) Tetrahedron Lett 34:8391
2. Hudrlik PF, Peterson D (1975) J Am Chem Soc 97:1464
3. Jones ERH, Skattebol L, Whiting MC (1958) J Chem Soc 1054

Section 24.5

1. Bailey WJ, Baylouny RA (1962) J Org Chem 27:3476
2. (a) Hasselmann D, Loosen K (1978) Angew Chem Int Ed Engl 17:606; (b) Kopecky KR, Lau M-P (1978) J Org Chem 43:525; (c) Graham WD, Green JG, Pryor WA (1979) ibid. 44:907; (d) Brinker UH, Fleischhauer I (1979) Angew Chem Int Ed Engl 18:396
3. (a) Murata I, Nakazawa T, Kato M, Tatsuoka T, Sugihara, Y (1975) Tetrahedron Lett 1647; (b) Scott LT, Brunsvold WR (1979) J Org Chem 44:641; (c) Sugimoto A, Yoneda S (1982) J Chem Soc Chem Commun 376
4. Andemichael YW, Wang KK (1992) J Org Chem 57:796
5. (a) Bross H, Schneider R, Hopf H (1979) Tetrahedron Lett 21, 29; (b) Balme G, Malacria M, Gore J (1979) 7; (c) Balme G, Malacria M, Gore J (1981) Chem Res Synop 244
6. Clinet JC, Linstrumelle G (1977) Nouv J Chem 1:373
7. Okamura WH, Peter R, Reischl W (1985) J Am Chem Soc 107:1034 and references cited therein

Section 24.6

1. (a) Myers AG, Kuo EY, Finney NS (1989) J Am Chem Soc 111:8057; (b) Myers AG, Dragovich PS (1989) ibid. 111:9130; (c) Nagata, R, Yamanaka H, Okazaki E, Saito I (1989) Tetrahedron Lett 30:4995; (d) Nagata R, Yamanaka H, Murahashi E, Saito I (1990) ibid. 31:2907; (e) Nicolaou KC, Maligres P, Shin J, de Leon E, Rideout D (1990) J Am Chem Soc 112:7825
2. (a) Lee MD, Dunne TS, Siegel MM, Chang CC, Morton GO, Borders DB (1987) J Am Chem Soc 109 3464; (b) Lee MD, Dunne TS, Chang CC, Ellestad GA, Siegel MM, Morton GO, McGahren WJ, Borders DB (1987) ibid. 109:3466; (c) Lee MD, Manning JK, Williams DR, Kuck NA, Testa RT, Borders DB (1989) J Antibiot 42:1070
3. (a) Golik J, Clardy J, Dubay G, Groenewold G, Kawaguchi H, Konishi M, Krishnan B, Ohkuma H, Saitoh K-I, Doyle TW (1987) J Am Chem Soc 109:3461; (b) Golik J, Dubay G, Groenewold G, Kawaguchi H, Konishi M, Krishnan B, Ohkuma H, Saitoh K-I, Doyle TW (1987) ibid. 109:3462
4. (a) Myers AG, Proteau PJ, Handel TM (1988) J Am Chem Soc 110:7212 and references cited therein; (b) Wender PA, McKinney JA, Mukai C (1988) ibid. 112:5369
5. (a) Konishi M, Ohkuma H, Tsuno T, Oki T, VanDuyne GD, Clardy J (1990) J Am Chem Soc 112:3715; (b) Konishi M, Ohkuma H, Matsumoto K, Tsuno T, Kamei H, Miyaki T, Oki T, Kawaguchi H, VanDuyne GD, Clardy J (1989) J; Antibiot; 42:1449; (c) Sugiura Y, Shiraki T, Konishi M, Oki T (1990) Proc Natl Acad Sci USA 87:3831; (d) Porco JA Jr, Schoenen FJ, Stout TJ, Clardy J, Schreiber SL (1990) J Am Chem Soc 112:7410; (e) Nicolaou KC, Smith AL, Wendeborn SV, Hwang C-K, (1991) ibid. 113:3106
6. (a) Westmijze H, Vermeer P (1979) Synthesis 390; (b) Danheiser RL, Carini DJ, Fink DM, Basak A (1983) Tetrahedron 39:935; (c) Danheiser RL, Tsai Y-M, Fink DM (1987) Org; Synth; 66:1; (d) Wang KK, Liu C, Gu YG, Burnett FN, Sattsangi PD (1991) J Org Chem 56:1914; (e) Wang KK, Gu YG, Liu C, (1990) J Am Chem Soc 112:4424
7. Clinet J-C, Linstrumelle G (1981) Synthesis 875
8. Brown HC, Sinclair JA (1977) J Organomet Chem 131:163
9. Andemichael YW, Gu YG, Wang KK (1992) J Org Chem 57:794
10. (a) Clinet J-C, Linstrumelle G (1977) Nouv; J; Chem; 1:373; (b) Sueiras J, Okamura WH, (1980) J Am Chem Soc 102:6255
11. (a) Krishnamurthy S, Brown HC (1975) J Org Chem 40:1864; (b) Gu YG, Wang KK (1991) Tetrahedron Lett 32:3029
12. Hudrlik PF, Peterson D (1975) J Am Chem Soc 97:1464
13. Wang KK, Wang Z, Sattsangi PD (1996) J Org Chem 61:1516
14. Andemichael YW, Huang Y, Wang KK (1993) J Org Chem 58:1651

Section 24.7

1. Wang KK, Liu B, Petersen JL (1996) J Am Chem Soc 118:6860
2. (a) Furuta K, Ishiguro M, Haruta R, Ikeda N, Yamamoto H, (1984) Bull Chem Soc Jpn 57:2768; (b) Yamakado Y, Ishiguro M, Ikeda N, Yamamoto H, (1981) J Am Chem Soc 103:5568
3. Kramer GW, Brown HC, (1974) J Organomet Chem 73:1
4. Brown HC, Sinclair JA (1977) J Organomet Chem 131:163

5. (a) Arnold Z, Holy A (1961) Collect Czech Chem Commun 26:3059; (b) Robertson IR, Sharp JT (1984) Tetrahedron 40:3095
6. (a) Saito I, Yamaguchi K, Nagata R, Murahashi E (1990) Tetrahedron Lett 31:7469; (b) Wang KK, Liu B, Lu Y-D, (1995) Tetrahedron Lett 36:3785
7. Wang KK, Wang Z, Gu YG (1993) Tetrahedron Lett 34:8391
8. Hudrlik PF, Peterson D (1975) J Am Chem Soc 97:1464
9. (a) Stiles M, Burckhardt U, Haag A (1962) J Org Chem 27:4715; (b) Blomquist AT, Bottomley CG (1965) J Am Chem Soc 87:86
10. Wang KK, Shi C, Petersen JL (1998) J Org Chem 63:4413
11. Derguini F, Balogh-Nair V, Nakanishi K (1979) Tetrahedron Lett 20:4899

25 Reduction

25.1 With 9-BBN·THF

9-BBN reduces aldehydes and ketones rapidly and cleanly to the corresponding alcohols [1]. The reaction is carried out by the dropwise addition of an essentially stoichiometric quantity of 9-BBN solution in THF to the aldehyde or ketone in THF solution at 25 °C. After completion of the reaction, the product carbinol is isolated by either of the following procedures. After the reaction is over, the reaction mixture is treated with alkaline hydrogen peroxide to oxidize the 9-BBN moiety, and the alcohol is separated from the *cis*-1,5-cyclooctanediol by distillation. Alternatively and more conveniently, THF is removed and *n*-pentane is added to the reaction mixture. The addition of 1 mol of ethanolamine precipitates the 9-BBN moiety as the adduct, displacing the carbinol to the pentane layer quantitatively. After removal of pentane, the desired alcohol is obtained in quantitative yields by distillation (Scheme 25.1). The process is an excellent neutral workup procedure for compounds having acid-and base- sensitive groups. In addition to pentane, ether and benzene also work quite satisfactorily.

Scheme 25.1

Both aldehydes and ketones consume 1 equiv of 9-BBN (3–5% excess) for complete reduction.

Cyclic and bicyclic ketones such as cyclohexanone, 2-methylcyclohexanone, 4-*tert*-butylcyclohexanone, and *nor*camphor undergo complete reduction in 30–60 min. Hindered ketones, like diisopropyl ketone and camphor, need 6–12 h for complete reduction at 25 °C. In refluxing THF, the complete reduction of these ketones is achieved in 1 h. A highly hindered ketone, such as 2,2,4,4-tetramethyl-3-pentanone, is inert and remains ineffective even in refluxing THF after 24 h.

The reduction of monocyclic ketones with disiamylborane and di-3-pinanylborane takes place from the less hindered side to yield predominantly the less stable of the two possible isomers. In contrast, 9-BBN exerts little influence on the direction taken by the reduction. With 3-methyl- and 4-*tert*-butylcyclohexanones, the products are predominantly the more stable of the two possible isomers. Reduction of bicyclic ketones such as *nor*camphor and camphor proceeds with preferential attack of the 9-BBN from the less hindered side, yielding the less stable of the two possible isomers (Eq. 25.1).

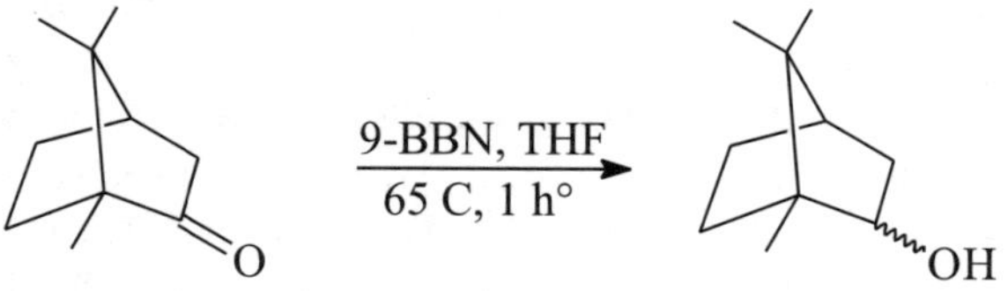

100% (81% isolated)
75% *exo*

(25.1)

The stereochemical outcome after the reduction are summarized in Table 25.1 [1].

Table 25.1 Stereochemistry of the reduction of representative cyclic and bicyclic ketones with 9-BBN in THF at 25 °C [1]

Ketone	Total yield (%)	Percentage *cis*:*trans*
2-Methylcyclohexanone	100	40:60
3-Methylcyclohexanone	98	88:12
4-*tert*-Butylcyclohexanone	99	8:92
		endo:*exo*
Norcamphor	100	91:9
Camphor	100	25:75

Anthraquinone reacts with 2 equiv of 9-BBN to give cleanly 9,10-dihydro-9,10-anthracenediol (Eq. 25.2). However, *p*-benzoquinone gives unsatisfactory results.

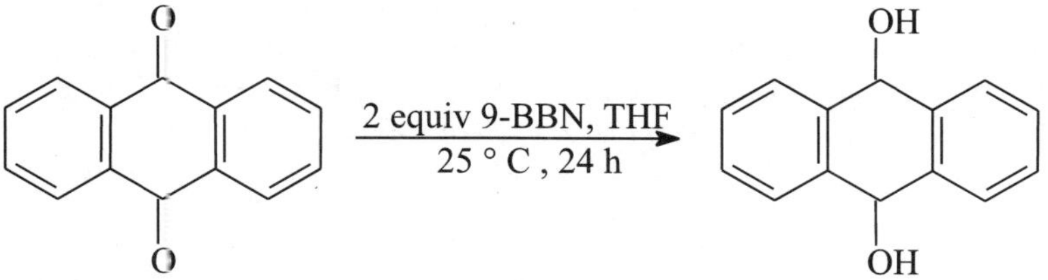

79% isolated (25.2)

Carboxylic acids liberate hydrogen rapidly and quantitatively (<5min) and further reduction is very slow. However, complete reduction of *n*-hexanoic acid and *n*-octanoic acid have been achieved at 65 °C in <24 h (Eq. 25.3). The same reaction with benzoic acids is very slow.

$$CH_3(CH_2)_6COOH \xrightarrow[65\ ^\circ C,\ 18\ h]{4\ equiv\ 9\text{-}BBN,\ THF} CH_3(CH_2)_6CH_2OH \quad (25.3)$$

92% (83% isolated)

Acetic anhydride consumes two hydrides very rapidly for reduction, with slow reduction thereafter. Both succinic anhydride and phthalic anhydride react only at a moderate rate.

Acid chlorides react sluggishly with borane and thexylborane and are inert with disiamylborane. However, hexanoylchloride (Eq. 25.4) and benzoylchloride undergo rapid and quantitative reduction with 2 equiv of 9-BBN to the corresponding alcohols.

$$CH_3(CH_2)_4COCl \xrightarrow[25\ ^\circ C,\ 12\ h]{2\ equiv\ 9\text{-}BBN,\ THF} CH_3(CH_2)_4CH_2OH \quad (25.4)$$

81% isolated

Ethylhexanoate, methylhexanoate, and γ-butyrolactone (Eq. 25.5) are reduced in refluxing THF with 2 equiv of 9-BBN.

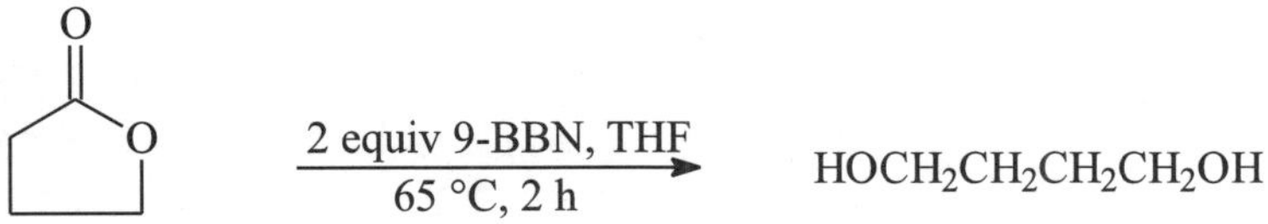

98% (25.5)

n-Octylbromide and *p*-bromotoluene are completely inert toward 9-BBN. Although the reaction of epoxides with 9-BBN is quite sluggish and requires 3–8 days for completion, the introduction of a catalytic (7.5 mol%) of lithium borohydride converts 1,2-butylene oxide to 98% of 2-butanol (Eq. 25.6).

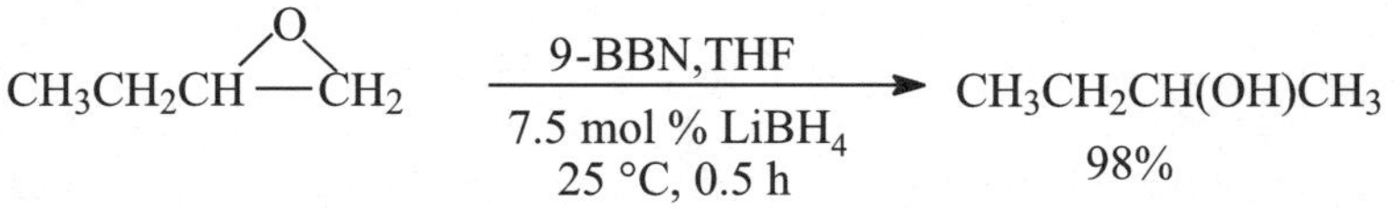

(25.6)

Primary amides liberate only one of the two possible hydrogens, and further reduction is very slow. Tertiary amides are rapidly reduced to give alcohols as the major product (Eq. 25.7). Nitriles and nitrobenzenes are reduced slowly. 1-Nitropropane is inert. Azobenzene is inert while azoxybenzene is slowly reduced to the azobenzene stage.

$$C_6H_5CONMe_2 \xrightarrow[25\,°C,\ 4\,h]{9\text{-BBN, THF}} \underset{80\%}{C_6H_5CH_2OH} + \underset{20\%}{C_6H_5CO_2NMe_2} \qquad (25.7)$$

The reduction of tertiary amides with borane–THF proceeds to give amines and with disiamylborane affords aldehydes. Consequently, it is possible to get the desired product using the appropriate reagent (Chart 25.1).

Cyclohexanone oxime undergoes slow reduction to *N*-cyclohexylhydroxylamine. Phenylisocyanate undergoes rapid reduction to the amine stage, and further reduction is very slow.

Pyridine and pyridine *N*-oxide undergo slow reduction with 9-BBN. The behavior of four hydroborating agents toward pyridine *N*-oxide is illustrated in Chart 25.2 [1].

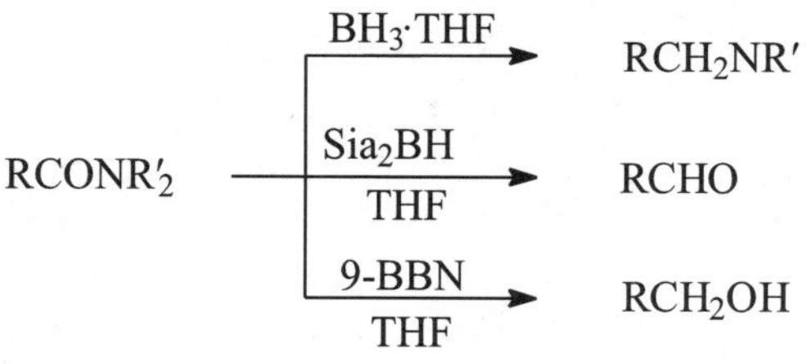

Chart 25.1

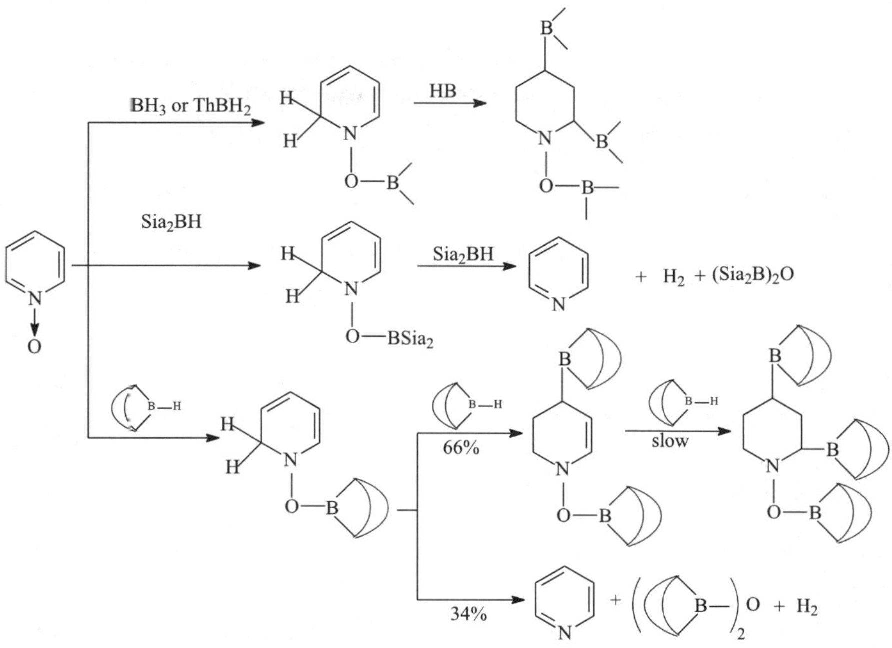

Chart 25.2

Among the sulfur compounds, only dimethylsulfoxide is reduced at a moderate rate whereas disulfide, sulfide, sulfone, tosylate, and sulfonic acid are inert to 9-BBN under the conditions. These observations are similar to those realized with borane, thexylborane, and disiamylborane.

The results are summarized in Table 25.2 [1].

The competition experiments [1] of a mixture of hexanal and 2-heptanone reveal the preferential reduction of aldehyde of over 85% ($k_{hexanal}/k_{hepatanone} = 27$) Cyclohexanone undergoes reduction at a faster rate than does cyclopentanone by a factor of 3.2. Cyclopentene is hydroborated with total exclusion of an ester, an acid, and an epoxide. Acid chlorides are reduced, quantitatively, without any significant attack on the esters (≤2%).

The reactivity of various functional groups toward 9-BBN thus is classified into five broad categories as: (1) very rapid-reduction aldehyde and ketone; (2) rapid reduction–reaction olefin, quinone, tertiary amide, acid anhydride, acid chloride, and lactone; (3) slow-reduction ester, epoxide, and oxime; (4) very slow-reduction carboxylic acid, sulfoxide, and azoxy; and (5) inert (no reaction) nitro (both aliphatic and aromatic), azo, sulfide, disulfide, sulfone, sulfonic acid, tosylate, and halogen (alkyl and aryl).

The relative rates of reduction obtained by competition experiments (Table 25.3) [2] reveal that many aldehydes and ketones have widely different reactivity toward $(9\text{-BBN})_2$, though these show the same first-order rate constant (*vide supra*).

The relative behavior of 9-BBN, which is an acidic reducing agent, has been compared with extensively studied, nonacidic sodium borohydride [3] (Chart

Table 25.2 Products of reduction of selected organic compounds containing representative functional groups with 9-borabicyclo[3.3.1]nonane in tetrahydrofuran [1]

Compound	Time (h)	Temp (°C)	9-BBN: compound	Products	Yield (%)
Hexanal	0.5	25	1	1-Hexanol	100
Cyclohexanone	1	25	1	Cyclohexanol	100
2-Methylcyclo-hexanone	1	25	1	2-Methylcyclo-hexanol	100
2,2,4,4-Tetramethyl-3-pentanone	24	65	4	2,2,4,4-Tetramethyl-3-pentanol	<1
				2,2,4,4-Tetramethyl-3-pentanone	96
Camphor	12	25	1	Borneols	100
	1	65	1	Borneols	100
Cinnamaldehyde	2	0	1	Cinnamyl alcohol	98
Anthraquinone	24	25	2	9,10-Dihydro-9,10-anthracenediol	79
Hexanoic acid	18	65	4	1-Hexanol	92
Octanoic acid	18	65	4	1-Octanol	92
Benzoyl chloride	6	25	4	Benzyl alcohol	90
Hexanoyl chloride	18	25	2	1-Hexanol	92
Ethyl hexanoate	24	25	4	1-Hexanol	75
	4	65	2	1-Hexanol	100
Methyl heptanoate	4	65	2	1-Heptanol	99
γ-Butyrolactone	2	65	2	1,4-Butanediol	98
Cyclohexene oxide	24	25	4	Cyclohexanol	30
1,2-Butylene oxide	0.5	25	4	2-Butanol	98
				1-Butanol	2
n-Octylbromide	24	25	4	*n*-Octane	0
				n-Octyl bromide	100
p-Bromotoluene	24	25	4	Toluene	0
				p-Bromotoluene	96
N,*N*-Dimethyl Benzamide	3	25	4	Benzyl alcohol	80
				N,*N*-dimethyl-benzyl amine	20
Nitrobenzene	24.0	25	4	Nitrobenzene	90
Di-*n*-butyl disulfide	24.0	25	4	Di-*n*-butyl disulfide	98

[a] In the presence of 7.5 mol % of lithium borohydride.

25.3) [2]. The data show that reduction of carbonyls with 9-BBN is less susceptible to steric effects than is sodium borohydride, in spite of the bulky nature of the former reagent. For example, cyclooctanone is reduced by 9-BBN by a factor of 50 slower than cyclohexanone. Whereas the factor is as much as 2,050 for reduction with sodium borohydride [3c].

	$(9\text{-BBN})_2$	$NaBH_4$
Cyclopentanone	26	4.35
Cyclohexanone	100	100
Cycloheptanone	11	0.64
Cyclooctanone	2	0.049

Chart 25.3

Table 25.3 Relative reactivities for the reduction of representative aldehydes and ketones with $(9\text{-BBN})_2$ in THF at 25 °C [2]

Compound	Relative rate	
	1-Hexene = 100	Hexanal = 100
p-Methoxybenzaldehyde	1,154	142
p-Tolualdehyde	911	112
Hexanal	813	100
Benzaldehyde	672	87
p-Chlorobenzaldehyde	494	64
2,2-Dimethylpropanal	479	62
Cyclohexanone	197	25.5
Cyclopentanone	51	6.61
p-Methoxyacetophenone	32.1	4.15
Cycloheptanone	22	2.85
Norbornanone	19.6	2.54
p-Methylacetophenone	16.8	2.18
Acetone	13.8	1.78
2-Butanone	11.3	1.46
Acetophenone	9.8	1.27
2-Heptanone	9	1.17
3-Methyl-2-butanone	8.9	1.15
Benzophenone	7.2	0.93
p-Chloroacetophenone	7	0.9
3,3-Dimethyl-2-butanone	6.9	0.89
Cyclooctanone	3.9	0.5
Camphor	3.6	0.46
3-Pentanone	3	0.39
2,4-Dimethyl-3-pentanone	0.94	0.122

This huge difference in reactivity is certainly not owing to electronic factors only, since it cannot be large in these two ketones. Moreover, camphor is reduced more slowly, by a factor of 6, over that of *nor*bornanone by 9-BBN, whereas the factor is more than 900 in case of sodium borohydride reduction (Chart 25.4).

	$(9\text{-BBN})_2$	**$NaBH_4$**
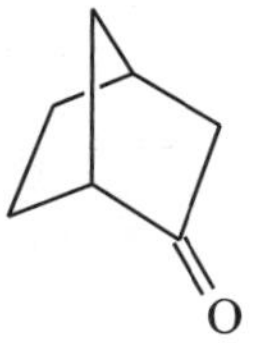	1	1
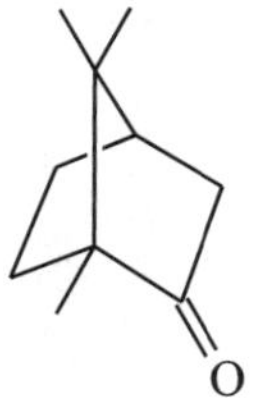	0.18	0.011

Chart 25.4

The lower susceptibility of 9-BBN to steric effects is explained that the boron of the 9-BBN monomer gets coordinated to carbonyl oxygen while the bulky cyclooctyl moiety keeps away from the alkyl groups of the substrate. The acyclic ketones behavior toward the rate of reduction is also similar (*vide supra*). For example, the introduction of two methyl at the same α position of acetone (3-methyl-2-butanone) leads only to a moderate decrease, whereas introduction of methyl groups on both the α positions (3-pentanone) brings about a considerable decrease in the rate of reduction.

	Relative rate
Acetone	1.00
3-Methyl-2-butanone	0.67
3-Pentanone	0.20
2,4-Dimethyl-3-pentanone	0.069

Chart 25.5

Since the change in electronic factor is comparable in both the cases, it thus demonstrates that the 9-BBN monomer approaches the carbonyl group from the less hindered face. This fact is further confirmed in the reduction of methyl *tert*-butylketone, which is reduced by a factor of 2 slower than is acetone. On the other hand, 2,4-dimethyl-3-pentanone is reduced by a factor of 15 slower than is acetone (Chart 25.5).

The study of electronic factor reveals that electron-releasing substituent in *p*-substituted benzaldehyde and acetophenone increase the rate of reduction. The electron-withdrawing substituents decrease the rate. This again provides the evidence that carbonyl oxygen gets complexed with the boron of 9-BBN monomer. The opposite electronic effects are observed with sodium borohydride reduction [4].

The steric and electronic effects thus suggest acidic 9-BBN as reductant gets coordinated with the carbonyl oxygen, and such an interaction makes the boron atom electron rich. Consequently, this facilitates the transfer of hydrogen with the bonded electron pair, as shown in Chart 25.6.

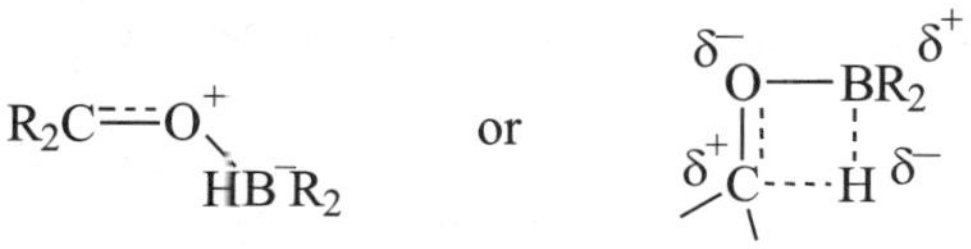

Chart 25.6

It has been established [5] by competition experiments involving 9-BBN that the carbonyl of aldehyde or ketone is reduced rapidly and cleanly to the corresponding alcohol, much faster than the hydroboration of olefins ($k_{cyclohexanone}/k_{cyclopentene} = 37$). This is complementary to the behavior of borane, which shows greater reactivity toward alkenes. This characteristic of 9-BBN has been utilized for the selective 1,2-reduction of α,β-unsaturated aldehydes and ketones to the corresponding allylic alcohols (Chart 25.7) [5].

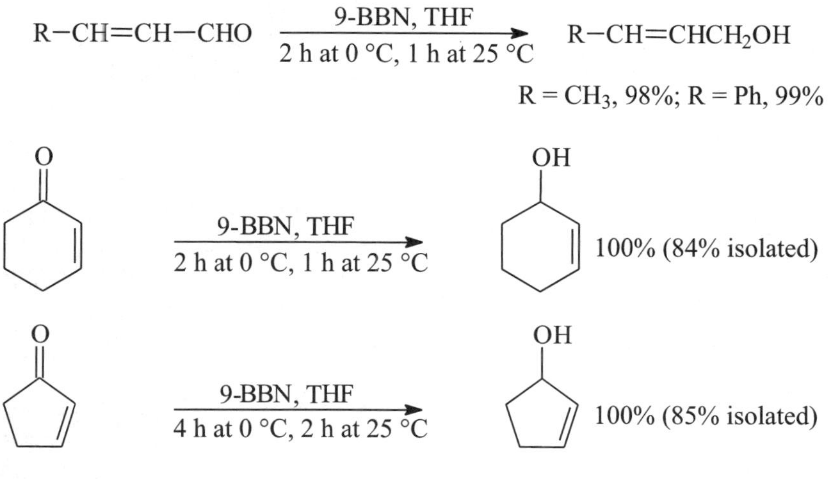

Chart 25.7

The product alcohol is isolated either by oxidation or alternatively and conveniently by the addition of ethanolamine (Eq. 25.8).

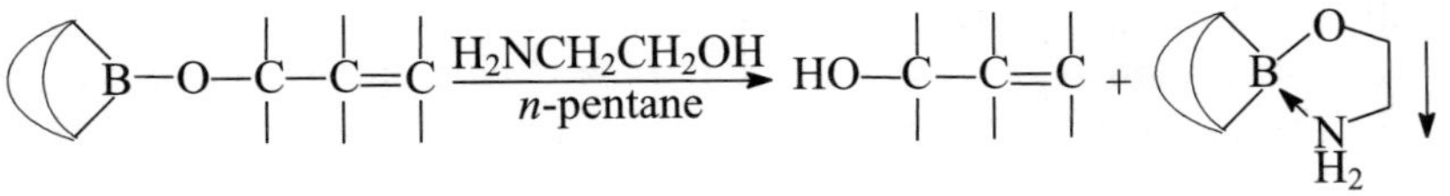

(25.8)

The results of reduction of 2-cyclopentenone with various reducing agent (Table 25.4) [5] clearly reveals the superiority of 9-BBN over these reagents.

Table 25.4 Reduction of 2-cyclopentenone with various reducing agents [5]

Reagent	Product composition (%)			
	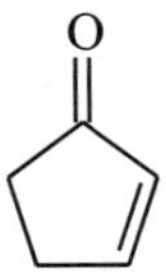	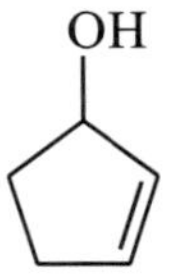	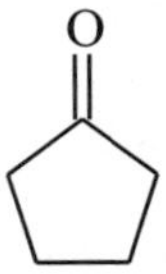	
$LiAlH_4$, THF, 0 °C[a]	0	14	2.5	83.5
$LiAlH(OCH_3)_3$, THF, 0 °C[a]	0	90.5	0	9.5
LiAlH(O-*tert*-Bu)$_3$, THF, 0 °C[a]	0	0	11.2	88.2
$NaBH_4$, EtOH, 78 °C[a]	0	0	0	100
AlH_3, THF, 0 °C[a]	0	90	6.1	3.9
i-Bu$_2$AlH, C_6H_6, 0 °C[b]	0.5	99	0	0.5
9-BBN, THF, 0 °C	0	100	0	0

[a] Data from Brown HC, Hess HM (1969) J Org Chem 34:2206.
[b] Data from Wilson KE, Seidner RT, Masamune S (1970) J Chem Soc Chem Commun 213.

The earlier studies and the competition experiments [5b] with equivalent molar, enal, or enone: 9-BBN (3–5% excess) reveal that enals or enones are selectively reduced to allylic alcohols (Eq. 25.9) in the presence of other organic functional groups such as nitro, halogen, epoxide, carboxylic acid, amide, ester, nitrile, sulfide, disulfide, sulfoxide, sulfone, tosylate, azo, etc.

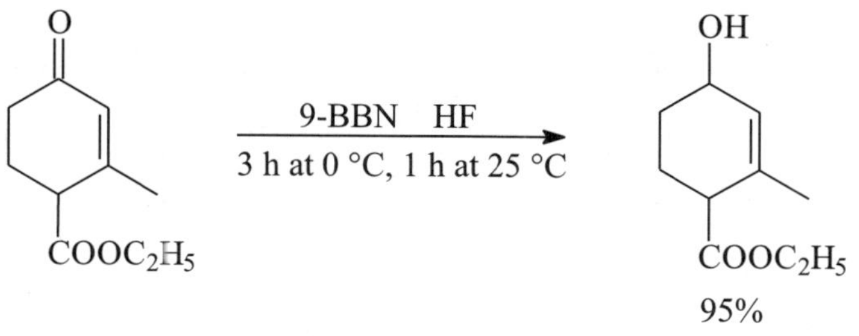

(25.9)

The selective 1,2-reductions of some α,β-unsaturated aldehydes and ketones is summarized in Table 25.5 [5b].

Phenalene attracts considerable attention from chemists because of its ability to generate an anion, neutral radical, and cation, all of which are aromatic and stable in solution [6]. This molecule is prepared by the reduction of phenalenone. Phenalenone, synthesized by the method of Fieser and Hershberg [7], when stirred overnight with excess of 9-BBN in THF at room temperature affords [8] phenalene (Eq. 25.10) in net yields of ~90%. Phenalene is sensitive to silica gel, so Florisil, a much more acidic absorbent than silica gel, has been used to get the satisfactory purification.

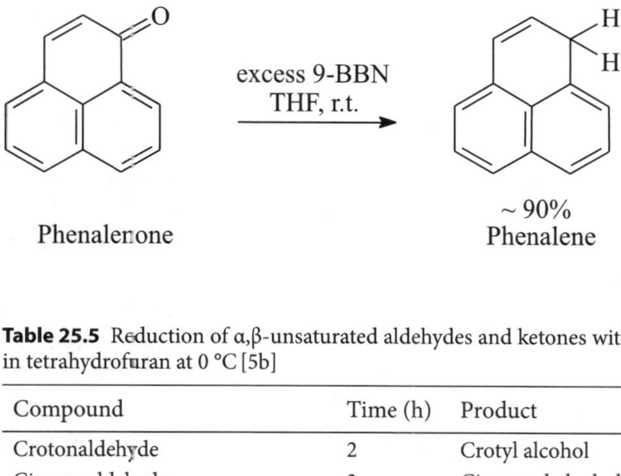

(25.10)

Table 25.5 Reduction of α,β-unsaturated aldehydes and ketones with 9-borabicyclo[3.3.1]nonane in tetrahydrofuran at 0 °C [5b]

Compound	Time (h)	Product	Yield (%)
Crotonaldehyde	2	Crotyl alcohol	98
Cinnamaldehyde	2	Cinnamyl alcohol	99
3-Penten-2-one	3	3-Penten-2-ol	75
1-Acetylcyclohexene	2	1-Cyclohexene-1-ethanol	90
2-Cyclohexenone	2	2-Cyclohexenol	100
2-Cyclopentenone	4	2-Cyclopentenol	99
3-Methyl-2-cyclopentenone	2	3-Methyl-2-cyclopentenol	76
o-Nitrocinnamaldehyde	2	*o*-Nitrocinnamyl alcohol	76
4-Carbethoxy-3-methyl-2-cyclohexenone	4	4-Carbethoxy-3-methyl-2-cyclo-hexenol	95

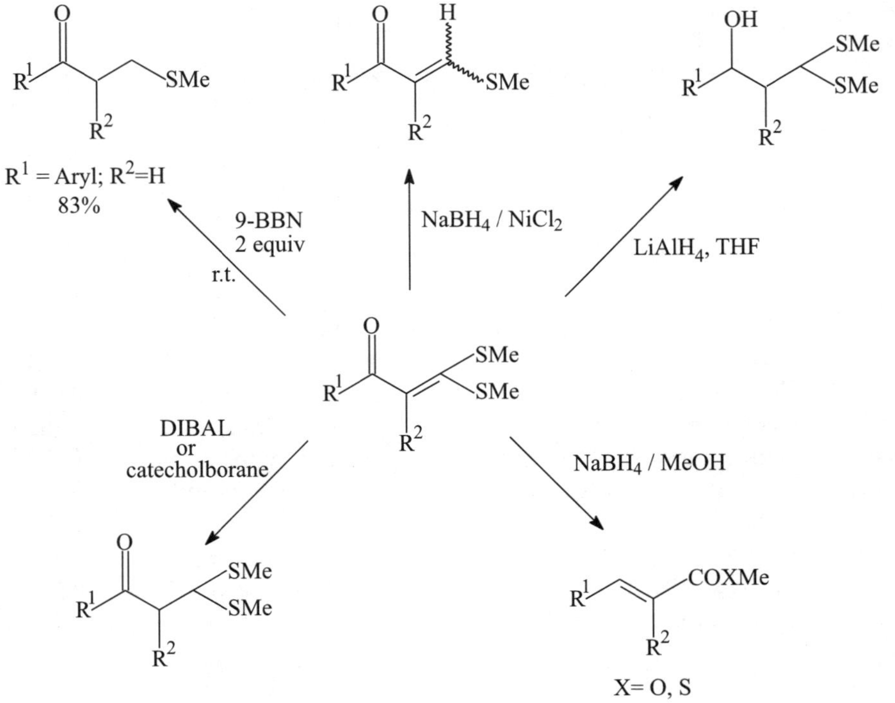

Chart 25.8

The α-oxoketone dithioacetal reduction is reported [9] to follow different pathways, depending on the nature of reducing agents and reaction condition (Chart 25.8). The electrophilic reducing agent, 9-BBN effects the conjugate reduction to afford the β-methylthioketone [10].

9-BBN exhibits good selectivity toward carboxylic group of the amino acids in proteins. Consequently, it reduces the γ-carboxyl group of the glutamate and unhindered C-terminal carboxylate group with only marginal reduction of the β-carboxyl group of aspartate in proteins, lysozyme, and myoglobin [11]. Unlike BH_3-THF [12], 9-BBN does not cleave the disulfide bonds that are responsible for the tertiary structure of protein.

25.2 With 9-BBN·Py

9-Borabicyclo[3.3.1]nonane-pyridine (9-BBN·Py), prepared [13] readily from 9-BBN dimer and pyridine in pentane solution (Eq. 25.11), has been found to cleanly reduce the aldehyde group in the presence [14] of the ketone (Table 25.6) [13].

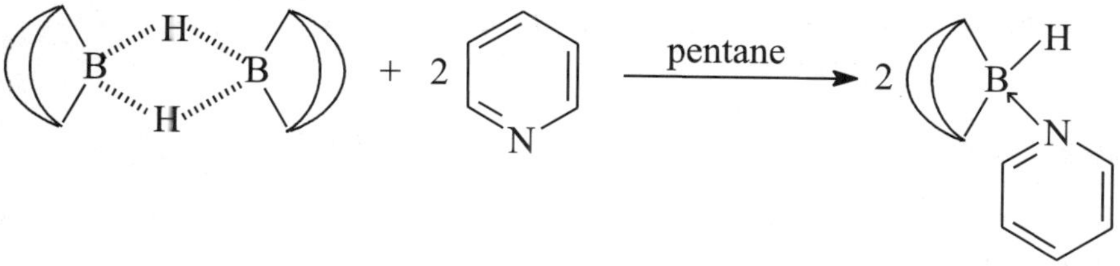

(25.11)

Table 25.6 Selective reduction of aldehydes with respect to ketone with 9-BBN·Py in Et_2O at 25 °C [13]

Compounds	Product	Mol %
Cyclohexanone	Cyclohexanone	98.5
+	Cyclohexanol	1.5
Benzaldehyde	Benzaldehyde	6.0
	Benzylalcohol	93
Acetophenone	Acetophenone	96
+	1-Phenylethanol	2
Benzaldehyde	Benzaldehyde	4
	Benzylalcohol	94
3-Pentanone	3-Pentanone	96
+	3-Pentanol	2.5
Octanal	Octanal	4.5
	1-Octanol	94.5

The functional groups, such as ester, lactone, *N,N*-dialkylamide, nitrile, alkylhalide, benzylic halide, epoxide, alkene, alkyne, and nitroalkane are inert. Alcohols, water, and carboxylic acid produce only hydrogen. No further reaction occurs in *B*-acyloxy-9-BBN·Py with excess of the reagent 9-BBN·Py. Acid chlorides and anhydrides are, however, reduced rapidly. Thus, with exception of these groups, the selective reduction of aldehyde, in the presence of nearly all other functional groups, can be achieved (Eq. 25.12).

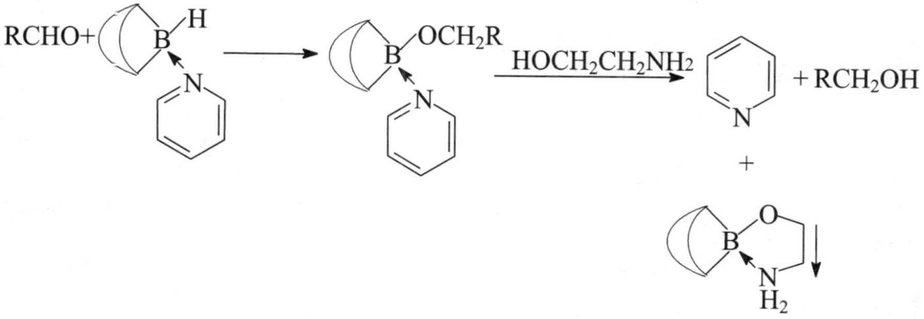

(25.12)

The times required for the reduction of aldehydes and ketones are summarized in Table 25.7 [13].

Table 25.7 Reduction of aldehydes and ketones by 9-BBN·Py in THF solutions at 25 °C [13]

Compounds	Time (h)	Percentage reduced
Benzaldehyde	1	95
	1.5	100
Cinnamaldehyde	1	86
	1.5	98
Cyclohexylcarboxaldehyde	1	91
	3	100
Octanal	1	91
	2	99
Hexanal	1	94
	2	100
Propanal	1	85
	1.5	97
Cyclohexanone	1	0
	1.5	3
2-Methylcyclohexanone	2	0
2-Hexanone	1	5
	12	43
Dicyclopropyl ketone	1.5	8
3-Pentanone	1	5
	7	10
Acetophenone	1	3
	9	5
Phenylacetone	1	6
	8	46

25.3 With Li 9-BBNH

It is reported that the mild reducing character of lithium borohydrides [15] is altered by the introduction of alkyl groups. The alkyl groups enhance the hydride donor activity tremendously. Lithium trialkylborohydride [16] is a remarkably powerful and selective reducing agent. In addition, various alkali metal alkylborohydrides have been prepared [17]. Thus, a "basic" dialkylborohydride lithium 9-boratabicyclo[3.3.1] nonane (Li 9-BBNH) [18] in THF is prepared by refluxing at 65 °C 1 equiv of 9-BBN with excess of finely divided lithium hydride (0.5 equiv excess) in THF for 24h, followed by stirring for another 24 h at room temperature. The reagent has been examined (in THF, room temperature) for its approximate rates and stoichiometry, in order to explore its reducing characteristics [18]. The rate of hydrogen evolution with alcohols decreases in the following order primary >> secondary > tertiary. Phenol, 2,6-di-*tert*-butylphenol and thiols liberate hydrogen instantly and quantitatively. 2,6-Di-*tert*-butylphenol is known to be inert to 9-BBN [5], the parent compound of Li 9-BBNH.

n-Hexylamine is inert to Li 9-BBNH.

Aldehydes and ketones of wide variety such as caproaldehyde, benzaldehyde, 2-heptanone, and norcamphor utilize 1 equiv of hydride in THF and are reduced rapidly and quantitatively to the corresponding alcohols at 15±2 °C. The hindered ketone, 2,2,4,4-tetramethyl-3-pentanone, which is inert to 9-BBN [5], however, is rapidly reduced with Li9-BBNH. Reduction of camphor gives 91% isoborneol and 9% borneol (Eq. 25.13).

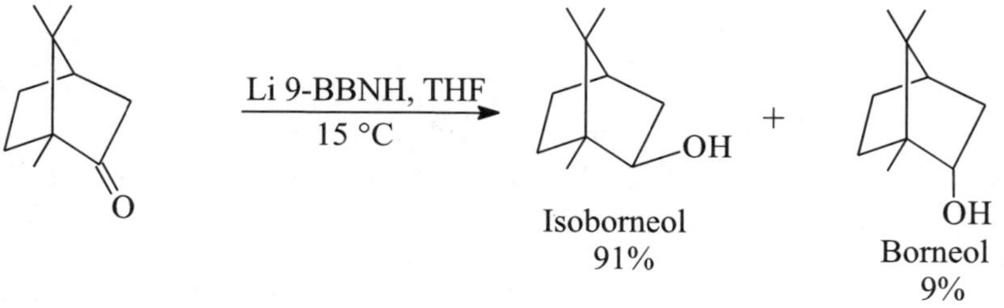

(25.13)

The stereochemical outcomes of Li9-BBNH reduction are summarized in Table 25.8 [18].

Cinnamaldehyde is rapidly reduced to the cinnamyl alcohol (97%), without attack on the double bond.

The reduction of carboxylic acid is very sluggish. Anhydrides—acetic anhydride, succinic anhydride, phthalic anhydride—consume 2 equiv of hydride and

Table 25.8 Stereochemistry of reduction of representative cyclic and bicyclic ketones with Li 9-BBNH in THF at 15±2 °C [18]

Ketone	Time (h)	Total yield (%)	Percentage *cis:trans*
2-Methylcyclohexanone	1	94	63:37
4-*tert*-Butylcyclohexanone	1	100	24:76
			endo:exo
Norcamphor	1	100	94:6
Camphor	1	100	9:91

afford equimolar mixture of corresponding carboxylic acids and alcohols (Eq. 25.14). Acid chlorides—caproylchloride and benzoyl chloride—too consume 2 equiv of the hydride to give the corresponding alcohol in quantitative yield (Eq. 25.15).

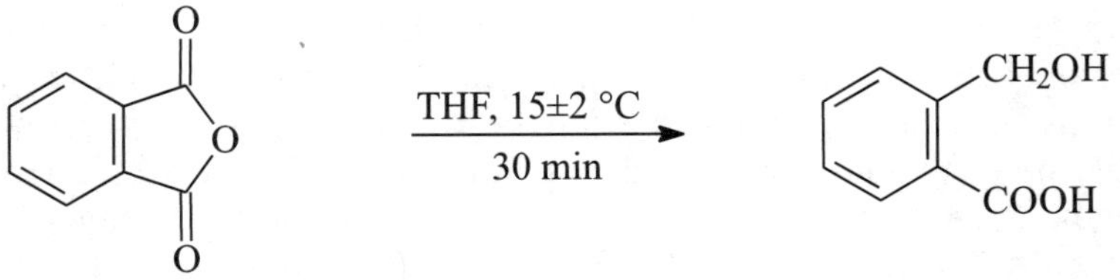

(25.14)

$$C_6H_5COCl \xrightarrow[30\ \text{min}]{\text{THF, } 15\pm2\ °\text{C}} C_6H_5CH_2OH \qquad (25.15)$$

Esters such as ethylcaproate, ethylbenzoate, phenylbenzoate and lactones, γ-butyrolactone, and phthalide utilize 2 equiv of hydrides. Esters are reduced to the corresponding alcohols, while lactones afford the diols in almost quantitative yield.

Esters are selectively reduced in the presence of the carboxylic acid (Chart 25.9) [18].

$$\underset{5\ \text{mmol}}{PhCOOEt} + \underset{5\ \text{mmol}}{n\text{-}C_5H_{11}COOH} + \underset{10\ \text{mmol}}{\text{Li 9-BBNH}} \xrightarrow[15\ °\text{C, } 0.5\ \text{h}]{\text{THF}} \underset{5\ \text{mmol}}{PhCH_2OH} + \underset{\text{Not detected}}{n\text{-}C_6H_{13}OH}$$

$$\underset{5\ \text{mmol}}{n\text{-}C_7H_{15}COOEt} + \underset{5\ \text{mmol}}{n\text{-}C_5H_{11}COOH} + \underset{10\ \text{mmol}}{\text{Li 9-BBNH}} \xrightarrow[15\ °\text{C, } 0.5\ \text{h}]{\text{THF}} \underset{5\ \text{mmol}}{n\text{-}C_8H_{17}OH} + \underset{\text{Not detected}}{n\text{-}C_6H_{13}OH}$$

Chart 25.9

Alkylbromides react at a moderate rate to Li 9-BBNH. Thus, in the presence of *n*-octylbromide, ethylacetate has been selectively reduced (Eq. 25.16).

$$\underset{5\ \text{mmol}}{CH_3COOEt} + \underset{5\ \text{mmol}}{n\text{-}C_8H_{17}Br} + \underset{5\ \text{mmol}}{\text{Li 9-BBNH}} \xrightarrow[25\ °\text{C, } 15\ \text{min}]{\text{THF}} \underset{9.21\ \text{mmol}}{2\ EtOH} + \underset{4.9\ \text{mmol}}{n\text{-}C_8H_{17}Br} \qquad (25.16)$$

Isopropenylacetate utilizes 3 equiv of the hydride, resulting the reduction of the acetate group to ethanol and isopropenyl group to 2-propanol (Eq. 25.17).

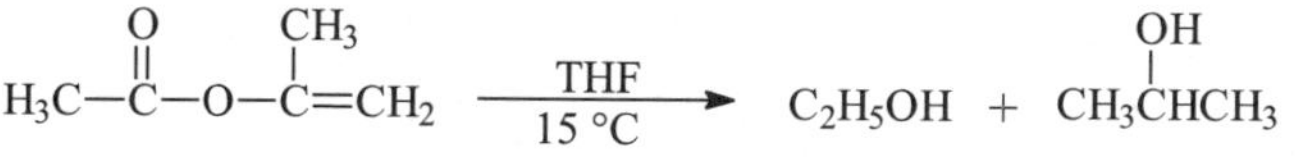
(25.17)

The reactivity order of epoxides is in accordance with their steric requirements. 1,2-Butylene oxide, styrene oxide, cyclohexene oxide, and 1-methyl-1,2-cyclohexene oxide consume 1 equiv of hydride within 0.25, 1.0, 1.5, and 24 h, respectively. The ring opening of the epoxide occurs from the less hindered side and affords the Markovnikov alcohol, in quantitative yield (Chart 25.10) [18].

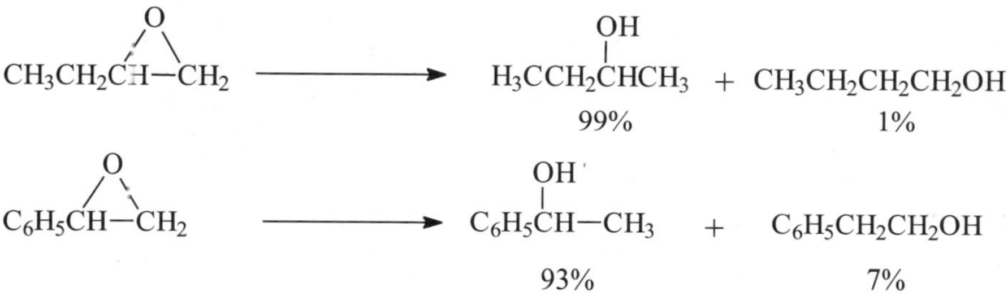

Chart 25.10

Acetals, ketals, and ortho esters are inert toward Li 9-BBNH.

Primary amides undergo reaction very slowly. Tertiary amides consume 2 equiv of hydride slowly to produce the corresponding amines (Eq. 25.18).

$$\underset{\text{2 equiv}}{n\text{-}C_5H_{11}CON(CH_3)_2} + \underset{\text{2 equiv}}{\text{Li 9-BBNH}} \xrightarrow[\text{24 h}]{\text{THF, 25 °C}} \underset{94\%}{n\text{-}C_6H_{13}NMe_2} \quad (25.18)$$

It is significant to note that $LiEt_3BH$ [19] or 9-BBN [5] reduces tertiary amides to the corresponding alcohols, whereas Li9-BBNH reduces *N,N*-dimethylamides to the corresponding tertiary amines.

Benzonitrile consumes 2 equiv of hydride to give benzylamine within 12 h, whereas an aliphatic nitrile, capronitrile, reacts sluggishly.

Nitropropane involves 1 equiv of hydrogen and forms white precipitates with 0.6 equiv of hydride consumption. The reaction probably involves the active α-hydrogen of nitropropane. Nitrobenzene undergoes slow reduction (48 h) utilizing 2.5 equiv of hydride, 1 equiv for hydrogen evolution, and 1.5 equiv for reduction.

Azobenzene is inert, whereas azoxybenzene is reduced very sluggishly. However, both are reduced very rapidly by lithium triethylborohydride [19]. No reduction is observed of cyclohexanone oxime, which means it provides another method for protection of the carbonyl group toward Li 9-BBNH [18].

Phenyl isocyanate consumes 1 equiv of the hydride and rapidly provides formanilide, in quantitative yield (Eq. 25.19) [18].

$$C_6H_5N{=}C{=}O \xrightarrow[\text{1 h}]{\text{THF, 15±2 °C}} HCONHC_6H_5 \quad (25.19)$$

Pyridine reduction is very slow, whereas pyridine *N*-oxide undergoes rapid reduction and utilizes 3 equiv of hydride.

Both, aliphatic and aromatic disulfides such as diphenyl disulfide and di-*n*-butyldisulfide undergo rapid reduction to the thiol stage, each consuming 2 equiv of the hydride, 1 equiv for hydrogen evolution and 1 equiv for reduction. However, methylphenyl sulfide is inert toward Li 9-BBNH. Sulfoxides, sulfones, and sulfonic acids are inert to this reagent.

A secondary tosylate, cyclohexyltosylate, is inert, whereas *n*-octyltosylate, the primary one, consumes 1 equiv of hydride within 3 h to afford *n*-octane. This selective conversion is useful for synthetic purposes.

n-Alkylchlorides are inert to Li9-BBNH, whereas *n*-alkylbromides react sluggishly, and *n*-alkyliodides react rapidly. *n*-Hexyliodide and *n*-octylbromide each consumes 1 equiv of hydride within 1 and 6.0 h, respectively, whereas *n*-octylchloride is inert, showing the order of the reaction as R-I > R-Br >> R-Cl [18]. The rate of reduction in case of bromides is primary alkylbromide >> secondary alkylbromide. A secondary alkylbromide is almost inert to Li 9-BBNH. Therefore, it is possible to reduce alkyliodide or bromide selectively in the presence of alkylchloride. And it is also possible to reduce a primary alkylhalide in the presence of the secondary.

In summary, the reactivity of various functional groups toward Li 9-BBNH is classified into four broad categories [18]: (1) rapid- or fast-reduction aldehyde, ketone, ester, lactone, acylchloride, acid anhydride, epoxide, disulfide, *n*-alkyliodide, and tosylate; (2) slow-reduction tertiary amide, alkylbromide, and aromatic nitrile; (3) sluggish-reduction carboxylic acid, aliphatic nitrile, primary amide, nitro and azoxy compounds, and secondary alkylbromide and tosylate; (4) inert olefin, oxime, alkylchloride, sulfoxide, azo-compound, sulfide, sulfone, and sulfonic acid.

25.4 With *B*-Siamyl-9-BBN

B-Siamyl-9-BBN, [*B*-(3-methyl-2-butyl)-9-BBN] [20], is prepared by the hydroboration of 2-methyl-2-butene with 9-BBN. The reagent reduces a variety of functionalized aldehydes to the corresponding alcohols (Table 25.9; Chart 25.11) [21].

Table 25.9 Reduction of all aldehydes to alcohols with *B*-siamyl-9-BBN in THF at reflux temperature [21]

Product	GC yield (%)
$CH_3(CH_2)_4CH_2OH$	103
$(CH_3)_3CCH_2OH$	101
$(CH_3)_2C{=}CH(CH_2)_2C(CH_3){=}CHCH_2OH$	100
$C_6H_5CH{=}CHCH_2OH$	82
$C_6H_5CH_2OH$	97
p-$ClC_6H_4CH_2OH$	92
p-$CH_3OC_6H_4CH_2OH$	97
p-$(CH_3)_2NC_6H_4CH_2OH$	92
p-$O_2NC_6H_4CH_2OH$	76

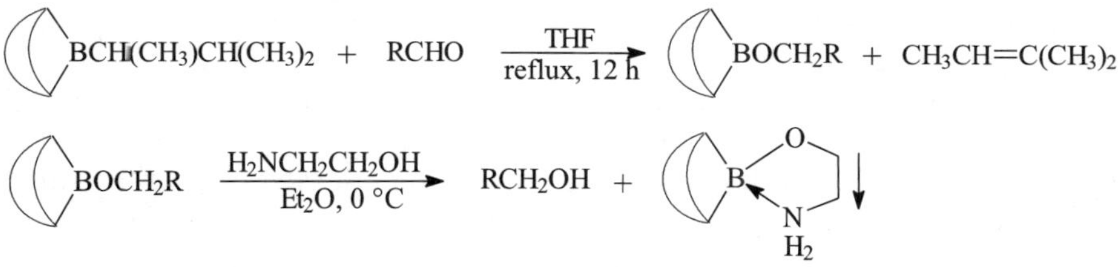

Chart 25.11

Further, it is reported [21] the ability to selectively reduce the aldehydes even in the presence of unhindered ketones. Competition experiment of benzaldehyde and acetophenone with 1 equiv of *B*-siamyl-9-BBN results in a >95% reduction of the aldehyde in 2 h, with no detectable reduction of ketone. The ketones are reduced at least 100–200 times slower are than the aldehydes [21].

25.5 With Lithium 9,9-Di-*n*-Butyl-9-Borabicyclo[3.3.1]nonanate

It has been found that lithium-9-methyl-9-*n*-butyl and lithium-9,9-di-*n*-butyl-9-borabicyclo[3.3.1]nonanates are capable of selective removal of halogen atom from *tert*-alkyl, benzyl, and allylic halides to afford the corresponding hydrocarbons in excellent yields, without concomitant attack on the secondary, primary, and aryl halide (Chart 25.12) [22].

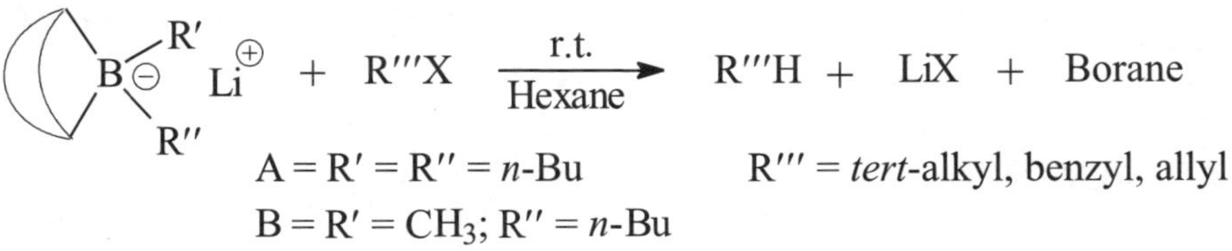

Chart 25.12

The, lithium di-*n*-butyl ate complex [A] of 9-BBN is found [23] to be better than the mixed ate complex [B]; the white gel of di-*n*-butyl of the ate complex is prepared [23] by treatment of *B*-*n*-butyl-9-BBN with 1 equiv of *n*-butyllithium in pentane or hexane. There is a significant effect of the solvent in the reaction. In benzene or hexane the reduction of benzylchloride is completed within 1.5 h, whereas in THF 85% of benzyl chloride is recovered [23].

The reaction is carried out by adding alkyl halide to the white gel of lithium 9,9-di-*n*-butyl-9-borabicyclo[3.3.1]nonanate suspended in hexane, and the reaction is quenched with alkaline hydrogen peroxide. The results are summarized in Table 25.10 [23]. Tertiary alkylhalides are smoothly converted to the corresponding alkanes, whereas primary and secondary halides are inert under the reaction conditions.

The reactivities of organotin hydrides [24] and chromium (II) complex [25] toward alkyl halides are also in the order of tertiary > secondary > primary alkylhalides. However, this trend is much stronger with the ate complex of 9-BBN than that with these reagents. Consequently, the high selectivity, gentleness, and convenience exhibited by the reagent has the practical synthetic application (Eq. 25.20) Benzylic halides are also reduced easily (entries 11–13, Table 25.10), whereas aryl and vinylhalides are inert (entries 19, 20). Benzylic geminal dihalides are reduced stepwise (entries 16, 17). The reaction of 1-phenylallylchloride [26] which also contains 33% of cinnamyl chloride with an equiv of ate complex of 9-BBN affords mixture of allylbenzene (36%) and β-methylstyrene (66%).

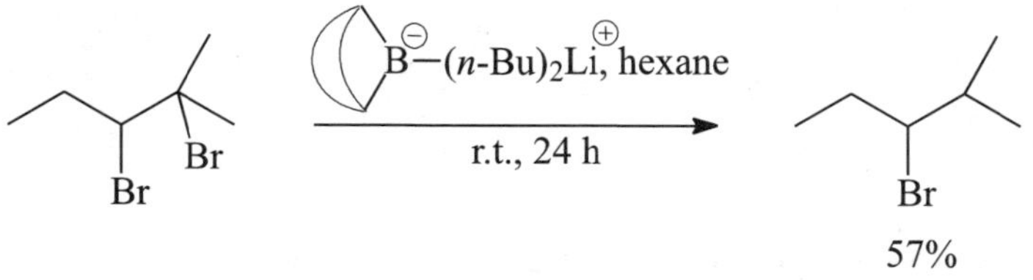

(25.20)

The hydride source in the reduction is the bridged α-carbon as reduction of benzylchloride with lithium di-*n*-butyl-9-BBN ate complex gives exclusively bicyclo[3.3.0]oct-1-yldibutylborane [27] (oxidation of this gives *cis*-bicyclo[3.3.0]octan-1-ol, 79%) and *n*-butanol (200%) in addition to toluene. The

Table 25.10 Reduction of various halides with lithium 9,9-di-*n*-butyl-9-borabicyclo[3.3.1]nonanate in hexane at 20 °C [23]

Entry	Substrate	Time (h)	Product	Yield (%)
1	*n*-$C_4H_9C(Br)(CH_3)(C_2H_5)$	3	*n*-$C_4H_9CH(CH_3)(C_2H_5)$	98
2	*n*-$C_4H_9CCl(CH_3)(C_2H_5)$	3	*n*-$C_4H_9CH(CH_3)(C_2H_5)$	94
3		3	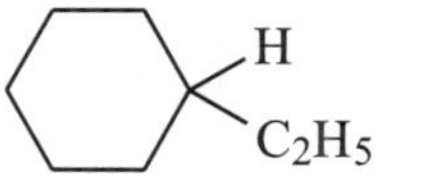	90
4	1-Bromoadamantane	4	Adamantane	100
5	$C_{10}H_{21}Br$	16	$C_{10}H_{22}$	1
6	$C_{10}H_{21}Br$	6 (at 60 °C)	$C_{10}H_{22}$	8
7	$C_9H_{19}CHBrCH_3$	16	$C_{10}H_{22}$	3
8	$C_9H_{19}CHBrCH_3$	6 (at 60 °C)	$C_{10}H_{22}$	39
9	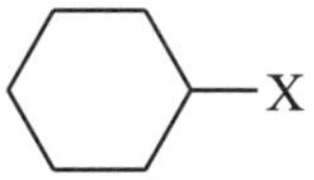 (X = Cl, Br, I)	18	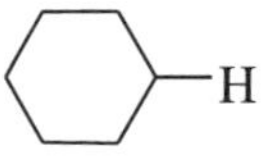	0
10	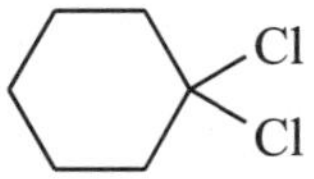	16	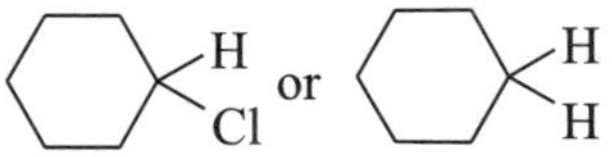	0
11	$C_6H_5CH_2Cl$	0.5	$C_6H_5CH_3$	100
12	$C_6H_5CHClC_6H_5$	18	$C_6H_5CH_2\ C_6H_5$	
13	$C_6H_5CHBrCH_3$	18	$C_6H_5C_2H_5$	81
14	$C_6H_5CH_2CH_2Br$	18	$C_6H_5C_2H_5$	2
15	$C_6H_5CHBrCH_2Br$	18	$C_6H_5CH_2CH_2Br$	60
16	$C_6H_5CHCl_2$	18	$C_6H_5CH_2Cl$	49
17	$C_6H_5CHCl_2$ [a]	18	$C_6H_5CH_3$	50
18	$C_6H_5CH{=}CHCH_2Br$	18	$C_6H_5CH{=}CHCH_3$	90[b]
19	C_6H_5Cl	18	C_6H_6	0
20	$C_6H_5CH{=}CHBr$	16	$C_6H_5CH{=}CH_2$	0

[a] Two equiv of reagent are used.
[b] Alkylbenzene (10%) is produced.

possibility that the reducing species is R_3BHLi, a known and efficient reagent for dehalogenation [28] and is often derived from the reaction of a trialkylborane to a bulky alkyllithium [29], is ruled out because of the absence of IR absorption of the B–H bond of such a borohydride (at 4–6 μ) [30].

The reactivity order of alkylhalides that is tertiary > secondary > primary suggests the reaction does not proceed *via* S_N^2 mechanism but *via* radical or carbocation like mechanism. To discriminate between the two mechanisms: 1,1-dimethyl-5-hexenyl chloride is reacted with tributyltin hydride (which involves radicals) and with the lithium di-*n*-butyl-9-BBN ate complex. It is well known that 5-hexenyl radicals undergo cyclization to give cyclopentylmethyl radicals [31], whereas hexenyl cation cyclizes to cyclohexyl cation [32]. As reduction with ate complex (Chart 25.13) affords none of the cyclopentyl derivatives, thus rules out the radical mechanism.

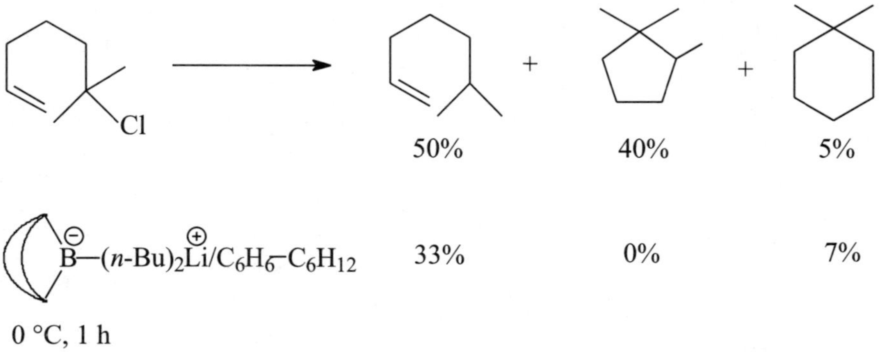

Chart 25.13

Further, the reduction of 1,7,7-trimethylbicyclo[2.2.1]hept-2-ylchloride (2-chlorocamphane) with the lithium di-*n*-butyl-9-BBN ate complex proceeds with skeleton rearrangement to afford 2,2,3-trimethylbicyclo[2.2.1]heptane, isocamphane. In contrast, reaction of 2-chlorocamphane with tributyltinhydride affords exclusively 1,7,7-trimethylbicyclo[2.2.1]heptane, bornane (Scheme 25.2). Consequently, the rearrangements of alkyl halides with the ate complex of 9-BBN are assumed to proceed through the intermediacy of a carbocation.

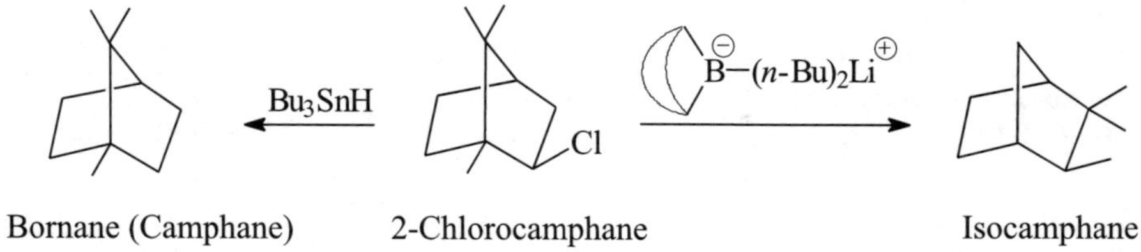

Scheme 25.2

The reduction of *cis*-and *trans*-4-*t*-butyl-1-methylcyclohexyl chloride with lithium di-*n*-butyl-9-BBN in hexane gives 4-*t*-butyl-1-methylcyclohexane, with partial inversion of configuration in cyclohexane, whereas in benzene both *cis* and *trans* isomers afford thermodynamically stable *trans*-4-*t*-butyl-1-methylcyclohexane, predominantly (Scheme 25.3) [23].

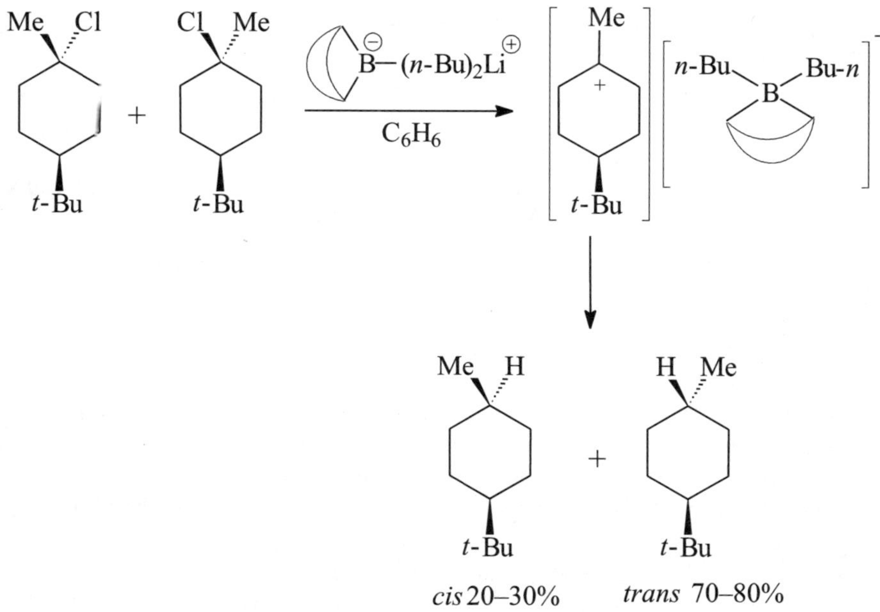

Scheme 25.3

The predominant formation of thermodynamically stable *trans* isomer from *cis*-and *trans*-chloro compounds in benzene (homogeneous condition) is explained [23] due to the trap of the dissociated carbocation intermediate, as shown in Scheme 25.3. The partial inversion of configuration under heterogeneous conditions in hexane (ate complex is insoluble in hexane) is attributed to the nucleophilic attack of the second molecule of the ate complex to the carbocation, which still has a very weak interaction with the leaving 9-BBN moiety.

The lithium di-*n*-butyl ate complex of 9-BBN also exhibits high chemoselectivity, where aldehyde is preferentially reduced in the presence of a ketone. The reagent also discriminates between the regioisomers of ketone, as shown in Chart 25.14 [33].

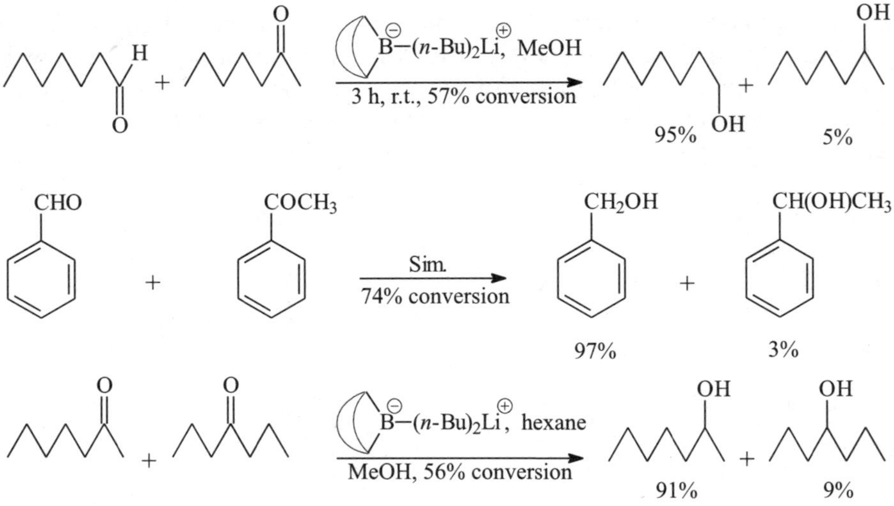

Chart 25.14

Cyclohexenone is reduced to cyclohexanone, whereas excess reagent leads to the formation of cyclohexanol. Cyclohexanone is preferentially reduced in the presence of benzyl chloride. Under the reaction conditions, methyl benzoate and benzylcyanide do not undergo reduction. Consequently, the lithium di-*n*-butyl ate complex of 9-BBN permits [33] the chemoselective reduction of aldehydes in the presence of ketones, esters, and nitriles.

A further study of the rate of reduction of various ketones reveals that the reactivity varies with electronic and positional environments of carbonyl groups (Table 25.11) [33].

Table 25.11 Relative reactivity of ketones toward lithium-di-*n*-butyl ate complex of 9-BBN [33]

Ketone	R[a]	Ketone	R[a]
Cyclohexanone	1	Phenylethyl ketone	0.22
Cycloheptanone	0.5	Phenyl-*n*-propyl ketone	0.06
Cyclooctanone	0.4	Phenylisopropyl ketone	0.03
p-Chloroacetophenone	1	2-Heptanone	0.78
Acetophenone	0.51	3-Heptanone	0.43
p-Methoxyacetophenone	0.18	4-Heptanone	0.03

The equimolar mixture (2-mmol scale) of the ate complex and substrate in hexane are stirred for 5 h, and then the reaction mixture is quenched.
[a] R = yield of the alcohol based on ketone. R = 1 indicates formation of alcohol in 100% yield.

A noteworthy discrimination is between phenylethyl ketone and phenyl *n*-propyl ketone and between 3-heptanone and 4-heptanone. The presence of methanol enhances regioselectivity, although it decreases conversion. Translation of these intermolecular results of ketones to an intramolecular diketones are demonstrated in Chart 25.15 [33].

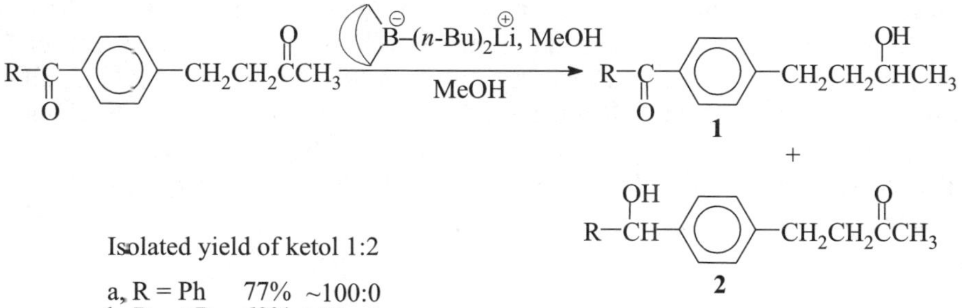

Isolated yield of ketol 1:2

a,	R = Ph	77%	~100:0
b.	R = *i*-Pr	69%	>99:<1
c.	R = *n*-Pr	59%	97:3
d.	R = Et	57%	90:10
e.	R = Me	60%	83:17

Chart 25.15

The lithium-di-*n*-butyl-9-BBN ate complex exhibits a "reversed" regioselectivity in the reduction of epoxides [34]. For example, aromatic epoxides are reduced from the most hindered site, whereas aliphatic epoxides are attacked at the least hindered position (Chart 25.16) [34].

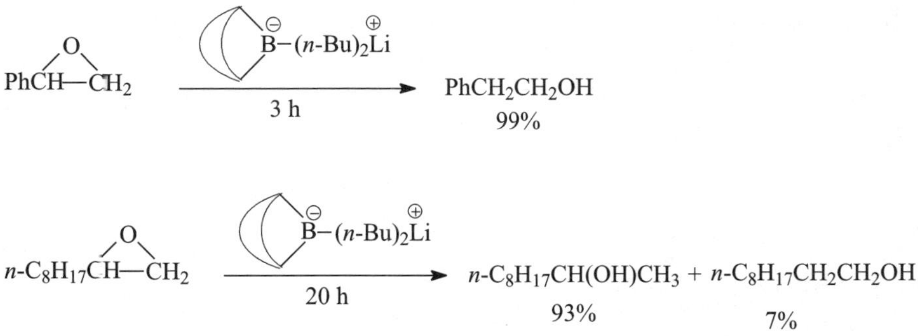

Chart 25.16

These results contrast with the reduction of $LiAlH_4$ or $LiEt_3BH$, which show Markovnikov ring opening or the mixed hydrides, such as $LiAlH_4{\cdot}AlCl_3$ or $BH_3{\cdot}BF_3$, which show *anti*-Markovnikov ring coupling [35–37].

25.6 With 9-BBN Derivatives (as Catalysts)

For reduction, lithium borohydride [38] has an advantage over sodium borohydride, as it is readily soluble in simple ether solvents. Lithium borohydride reduces selectively the esters in the presence of other reducible groups. The reduction of esters proceeds smoothly but relatively slowly. However, it is found [39] that in the presence of 10 mol% of lithium-9-boratabicyclo[3.3.1]nonane, Li 9-BBNH, or lithium triethylborohydride, $LiEt_3BH$ strongly catalyzes the reduction of esters by lithium borohydride in diethyl ether. Catalytic effects have also been observed with *B*-OMe-9-BBN, $LiB(OMe)_2$-9-BBN, and $LiEt_3BOMe$ (Table 25.12) [39].

Table 25.12 Rate of reduction of ethyl caproate by $LiBH_4$ in the presence of various catalysts in ether at 25 °C [39a]

	Percentage reaction					
Catalyst	0.5 h	1 h	2 h	4 h	8 h	24 h
No catalyst	17	28	41	65	100	100
$LiEt_3BH$	80	100	100			
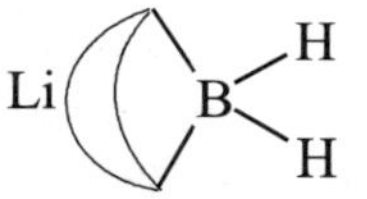	100	100				
$LiEt_3BOMe$	83	98	100			
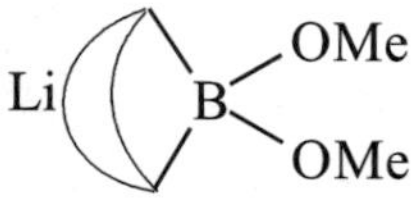	100	100				
BF_3·OEt_2	21	35	50	73	100	100
BH_3·THF	10	14	18	26	53	62
n-Bu_3B	22	98	100			
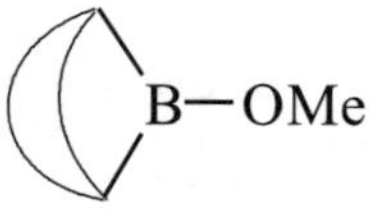	100	103				
n-$OctB(OMe)_2$	92	100				
$(MeO)_3B$	52	100				
$(PhO)_3B$	14	30	45	68	98	
(*n*-DodO)$_3$B[b]		26	46		100	

[a] [Ester] = 1.0M, [$LiBH_4$] = 1.0 M, [catalyst] = 0.1 M.
[b] *n-Dod* *n*-dodecyl.

Further, it is found that *B*-OMe-9-BBN is for more powerful than any other species examined (Table 25.13). Another advantage of this reagent is its easy removal from the product. *B*-Methoxy-9-BBN forms the hydroxy derivative, which is soluble in sodium hydroxide as the ate complex. Consequently, it provides a practical method for the reduction of esters in the presence of reducible groups such as chloro and nitro (Table 25.13) [39].

Table 25.13 Reduction of esters by lithium borohydride in refluxing ether in the presence of *B*-methoxy-9-BBN or methyl borate [39]

Ester	Catalyst	Product	Reduction time (h)	Yield (%)
Ethyl caproate	$(MeO)_3B$	1-Hexanol	1.0	82
Methyl stearate	*B*-MeO-9-BBN	1-Octadecanol	0.5	97
Ethyl cyclohexane-carboxylate	$(MeO)_3B$	(Hydroxymethyl)-cyclohexane	1.0	80
Ethyl pivalate	$(MeO)_3B$	2,2-Dimethylpropanol	1.0	81
Ethyl-1-adamantanecarboxylate	$(MeO)_3B$	1-Adamantanecarbinol	1.0	90
Ethyl benzoate	*B*-MeO-9-BBN	Benzyl alcohol	2.0	81
Ethyl-4-chlorobenzoate	*B*-MeO-9-BBN	4-Chlorobenzyl alcohol	1.0	90
Ethyl-3-chloro propionate	$(MeO)_3B$	3-Chloro-1-propanol	0.5	84
Ethyl-4-nitrobenzoate	*B*-MeO-9-BBN	4-Nitrobenzyl alcohol	0.5	91

Ester = 20 mmol (2 M), $LiBH_4$ = 11 mmol (1.1 M), catalyst = 2 mmol (0.2 M).

References

1. Brown HC, Krishnamurthy S, Yoon NM (1976) J Org Chem 41:1778
2. Brown HC, Wang KK, Chandrasekharan J (1983) J Am Chem Soc 105:2340
3. (a) Brown HC, Wheeler OH, Ichikawa K (1957) Tetrahedron 1:214; (b) Brown HC, Ichikawa K (1957) ibid. 1:221; (c) Brown HC, Ichikawa K, (1962) J Am Chem Soc 84:373; (d) Brown HC, Muzzio J (1966) ibid. 88:2811
4. Bowden K, Hardy M (1966) Tetrahedron 22:1169
5. Krishnamurthy S, Brown HC (1975) J Org Chem 40:1864; (b) Krishnamurthy S, Brown HC (1977) ibid. 42:1197
6. Reid DH (1965) Q Rev Chem Soc 19:274
7. Fieser LF, Hershberg EB (1938) J Am Chem Soc 60:1658
8. Boudjouk P, Johnson PD (1978) J Org Chem 43:3979
9. Rao CS, Chakrasali RT, Ila H, Junjappa H (1990) Tetrahedron 46:2195
10. Gammill RB, Sobieray DM, Gold PM (1981) J Org Chem 46:3555
11. Atassi MZ, Rosenthal AF, Vargas L (1973) Biochim Biophys Acta 303:379
12. (a) Rosenthal AF, Atassi MZ (1967) Biochim Biophys Acta 147:410; (b) Atassi MZ, Rosenthal AF (1969) Biochem J 111:593

13. Brown HC, Kulkarni SU (1977) J Org Chem 42:4169
14. The other reagents used for chemoselective reduction of aldehydes in the presence of ketones are (a) *n*-$Bu_4BH(OAc)_3$: Nutaitis CF, Gribble GW (1983) Tetrahedron Lett 24:4287; (b) *n*-Bu_4NBH_3CN: Hutchins RO, Kandasamy D (1973) J Am Chem Soc 95:6131; (c) LiAlH (O-*t*-Bu)$_3$: Sell CS (1975) Aust J Chem 28:1383; (d) $KBH(OAc)_3$: Tolstikov GA, Odinokov VN, Galeeva RI, Bakeeva RS, Akhunova VR (1979) Tetrahedron Lett 50:4851; (e) $NaBH_4$/RSH: Maki Y, Kikuchi K, Sugiyama H, Seto S (1977) ibid. 263; (f) $(CH_3)_2CHOH/Al_2O_3$: Posner GH, Runquist AW, Chapdelaine MJ (1977) J Org Chem 42:1202; (g) BH_3/LiCl: Yoon NM, Cha JS (1978) J Korean Chem Soc 22:259; (h) *n*-Bu_3SnH/silica gel: Fung NYM, de Mayo P, Schauble JH, Weedon AC (1978) J Org Chem 43:3977; (i) $NaBH(OAc)_3$: Gribble GW, Ferguson DC (1975) J Chem Soc Chem Commun 535; (j) $LiBH_4$/molecular sieve: Risbood PA, Ruthven DM (1979) J Org Chem 44:396; (k) BH_3/RNH_2: Andrews GC (1980) Tetrahedron Lett 21:697; (l) Et_4NBH_4: Sorrell TN, Pearlman PS (1980) ibid. 21:3963; (m) $(Ph_3P_2)CuBH_4$: Fleet GWJ, Harding PJC, (1981) ibid. 22:675; (n) $EtCH(OMgBr)_2$: Babler JH, Invergo BJ (1981) ibid. 22:621; (o) $NaBH(OAr)_3$: Yamaguchi S, Kabuto K, Yasuhara F (1981) Chem Lett 461; (p) NaOCHO: Babler JN, Sarussi SJ (1981) J Org Chem 46:3367; (q) $Li(Et_3CO)_3AlH$: Krishnamurthy S (1981) ibid. 46:4628; (r) Gribble GW (1996) In: Abdel-Magid AF (ed) Reduction in organic synthesis: recent advances and practical applications. ACS symposium series 641. American Chemical Society, Washington, D.C. pp 167–200
15. Yoon NM, Cha JS (1977) J Korean Chem Soc 21:108
16. (a) Krishnamurthy S (1974) Aldrichimica Acta 7:55; (b) Krishnamurthy S, Brown HC (1976) J Am Chem Soc 98:3383; (c) Krishnamurthy S, Brown HC, (1976) J Org Chem 41:3064; (d) Brown HC, Kim SC (1977) Synthesis 635
17. (a) Brown HC, Singaram B, Mathew CP (1981) J Org Chem 46:2712, 4541; (b) Brown HC, Krishnamurthy S, Hubbard JL (1978) J Am Chem Soc 100:3343; (c) Brown HC, Krishnamurthy S, Hubbard JL (1979) J Organomet Chem 166:273
18. Brown HC, Mathew CP, Pyun C, Son JC, Yoon NM (1984) J Org Chem 49:3091
19. Brown HC, Kim SC, Krishnamurthy S (1980) J Org Chem 45:1
20. Midland MM, Tramontano A, Zderic SA (1977) J Organomet Chem 134:C17
21. Midland MM, Tramontano A (1978) J Org Chem 43:1470
22. Yamamoto Y, Toi H, Murahashi S-I, Moritani I (1975) J Am Chem Soc 97:2558
23. Toi H, Yamamoto Y, Sonoda A, Murahashi S-I, (1981) Tetrahedron 37:2261
24. Doyle MP, McOsker CC, West CT (1976) J Org Chem 41:1393
25. (a) Erickson RE, Holmquist RK (1969) Tetrahedron Lett 4209; (b) Hanson JR, Premucic E (1968) Angew Chem, Int. Ed. 7:247
26. Caserio FF, Dennis GE, Dewoffe RH, Young WG (1955) J Am Chem Soc 77:4182
27. Kramer GW, Brown HC (1976) J Am Chem Soc 98:1964
28. Brown HC, Krishnamurthy S (1973) J Am Chem Soc 95:1669
29. (a) Corey EJ, Albonico SM, Koelliker U, Schaaf TK, Varma RK (1971) J Am Chem Soc 93:1491; (b) Negishi E, Chiu K, Yoshida T (1975) J Org Chem 40:1676; (c) Reetz MT, Stephane W (1977) Angew Chem, Int. Ed. 16:44
30. (a) Negishi E, Katz J-J, Brown HC (1972) J Am Chem Soc 94:4026; (b) Kramer GW, Brown HC (1975) J Organomet Chem 90:C1
31. (a) Juria M (1972) Acc Chem Res 4:386; (b) Walling C, Cioffari A (1972) J Am Chem Soc 94:6059

32. Bartlett PD, Closson WD, Cogdell TJ (1965) J Am Chem Soc 87:1308
33. Yamamoto Y, Toi H, Sonoda A, Murahashi S-I, (1976) J Am Chem Soc 98:1965
34. (a) Yamamoto Y, Toi H, Sonoda A, Murahashi S-I, (1976) J Chem Soc Chem Commun 672; (b) Gramer GW, Brown HC (1977) Heterocycles 7:487
35. Eliel EL, Rerick MN (1960) J Am Chem Soc 82:1362
36. Krishnamurthy S, Schubert RM, Brown HC (1973) J Am Chem Soc 95:8486
37. Brown HC, Yoon NM (1968) J Chem Soc Chem Commun 1550
38. Brown HC, Choi YM, Narasimhan S (1981) Inorg Chem 20:4454
39. Brown HC, Narasimhan S (1982) J Org Chem 47:1604

26 Asymmetric Reduction

The synthesis of optically active compounds is challenging problem before organic chemists. The main interest lies in the biologically active and pharmaceutical compounds. A large number of drugs, agrochemicals, food additives, and flavoring agents are being prepared by total synthesis. The racemic compounds thus formed in the course of reaction sequence are normally resolved at the end of the synthesis. This procedure thus leads to the wastage of half of the synthetic product, which is discarded. Moreover, resolution is usually a tedious and laborious process.

Asymmetry in organic molecules [1] and concept of asymmetric synthesis [2] are known for nearly a century. The tragedy of thalidomide babies is one of the necessities to get enantiomerically pure pharmaceuticals, emphasizing the desirability of achieving asymmetric synthesis [3].

In an asymmetric reaction [4, 5] the reactants combine to form diastereomeric states, and one of the two reactants must have chiral center to induce asymmetry at the reaction site. Generally, asymmetry is created when the sp^2 carbon is converted to sp^3 carbon, and one such functionality in organic compounds involves the conversion of prochiral >C=O to asymmetric >CH–OH.

In the last three decades, the asymmetric reductions of carbonyl compounds have been actively investigated by organic chemists [6].

The chosen asymmetric reducing agent is required to mimic the action of enzymes. Consequently, the reagent must be catalytic, selective, and yield products of high enantiomeric purity. In addition, the desired reagents should be inexpensive, easy to handle, tolerant of "typical" organic solvents and functional groups, effective on a wide range of substrates, allow for the easy isolation of products, and be available in both enantiomeric forms.

Trialkylboranes, however, do not contain an active hydrogen on boron, yet are made to react (e.g., with carbonyl moiety) under forcing conditions of heat or by careful selection of organoborane structure. The reaction proceeds through a bimolecular six-membered state in which the β-hydrogen of the organoborane is transferred to the carbonyl carbon (Eq. 26.1) [7]. Organoboranes are tolerant of many functional groups and thus can be used in the presence of wide variety of system [8].

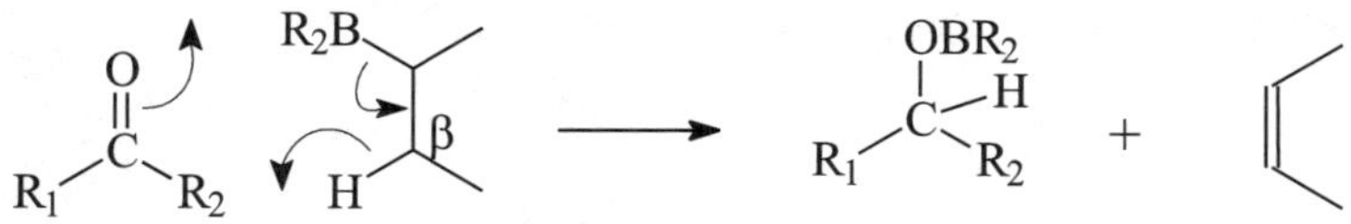 (26.1)

In addition to asymmetric hydroboration of C=C, the hydroborating reagents also exhibit remarkable asymmetric induction in the reduction of the carbonyl moiety. Chiral organoboranes derived *via* hydroboration with 9-BBN are proven to be highly useful for the asymmetric reduction. The characteristics of asymmetric induction in the reduction of prochicral carbonyl compounds of individual chiral organoboranes are discussed in this chapter.

26.1 Alpine-Borane

26.1.1 Reduction of Aldehydes

Optically active primary-1-deuterio alcohols are important class of compounds, as these have been extensively studied for mechanistic studies of chemical and biochemical reactions [1]. However, their preparations are tedious or inefficient [2]. The preparation of these alcohols generally involves the fermenting yeast reduction of the corresponding deuterated aldehydes. Although high optical purity is achieved, the process is tedious and not amenable to large-scale preparation.

Midland has found that certain *B*-alkyl-9-BBN compounds reduce benzaldehyde at a much faster rate than the corresponding trialkylboranes, under mild conditions [4]. Midland has further demonstrated [5] that chiral organoborane, *B*-3-pinanyl-9-BBN, prepared from (+)-α-pinene (92%, *ee*) and 9-BBN, reduces benzaldehyde-α-d [6] with high enantiomeric yield, approaching those obtained from enzyme. The process is extremely simple and can be performed on a large scale. The reaction is completed within 2 h at room temperature, with slight excess of *B*-3-pinanyl-9-BBN. The excess organoborane is destroyed with volatile aldehydes. The generated pinene is removed under vacuum and can be recycled. 9-BBN is precipitated by the addition of 1 mol of ethanolamine [7, 12]. The process affords (*S*)-(+)-benzyl-α-d-alcohol when (+)-α-pinene is used (Chart 26.1).

The systematic studies conducted by Midland and coworkers [8] have shown that among the various *B*-alkyl-9-BBN obtained from (+)-α-pinene, (–)-β-pinene, (–)- camphene, and (+) 3-carene (Chart 26.2), *B*-3-pinanyl-9-BBN (Alpine-Borane) shows remarkable enantioselectivity as illustrated in Table 26.1 [8].

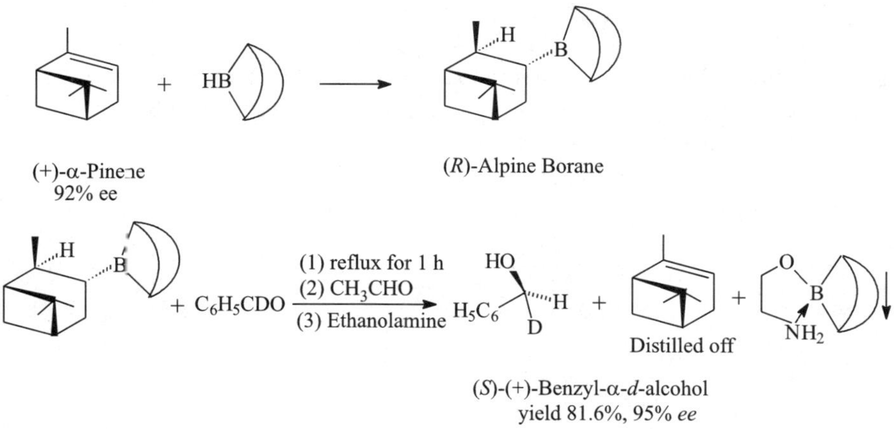

Chart 26.1

Table 26.1 Reduction of benzaldehyde 1-d with chiral *B*-alkyl-9-BBN reagents [8]

Reagent from	ee (%)	Configuration of benzyl-(α)-d-alcohol
(+)-α-Pinene	90	*S*-(+)
(−)-β-Pinene	47	*S*-(+)
(−)-Camphene	75	*R*-(−)
(+)-3-Carene	61	*S*-(+)

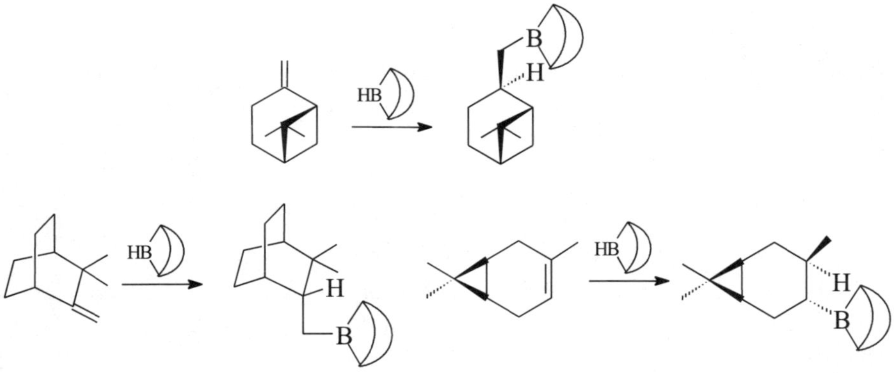

Chart 26.2

The, *B*-3-pinanyl-9-borabicyclo[3.3.1]nonane or Midland's reagent or Alpine-Borane is commercially available [9].

As indicated in Table 26.1, (*S*)-(+)-benzyl-α-d-alcohol (Eq. 26.2) of 90% *ee* is obtained by using 92% enantiomerically pure α-pinene; the results thus represent an essentially complete asymmetric reduction. Such an exceptionally high degree of asymmetric induction in the reduction of deuterated aldehyde thus far is unknown in nonenzymatic system.

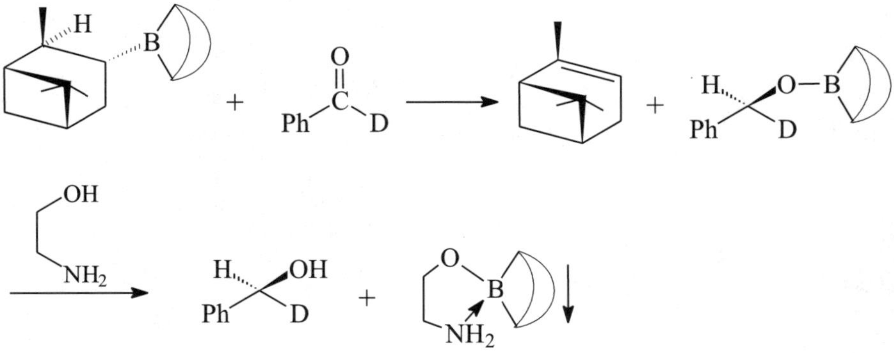

(26.2)

Now both (+) and (−)-α-pinene of optical purities approaching 100% *ee* are readily available [10].

It is observed that tertiary β-hydrogen (relative to boron) preferentially reacts in the presence of secondary or primary β-hydrogen [4]. Consequently, the hydrogen added via the hydroboration process is the reducing hydrogen. The chiral organoborane obtained by the deuterioboration of (+)-α-pinene with 9-BBN-9-D, quantitatively, transfers deuterium to benzaldehyde (Eq. 26.3).

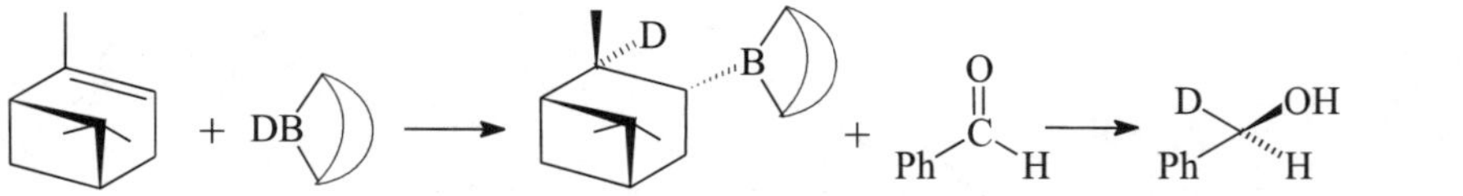

(*R*)-(−)Benzyl-α-*d*-alcohol

(26.3)

Therefore, this avoids the necessity of prior preparation [6] of 1-deuterated aldehydes. The absolute configuration of the major products are consistently (*R*)- when deuterated Alpine-Borane (from (+)-α-pinene) is used for the reduction. The product of (*S*)-configuration is readily obtained by using deuterated *B*-3-pinanyl-9-BBN prepared from (−)-α-pinene and 9-BBN-9-D. On the other hand, reduction of deuterated aldehyde with Alpine-Borane from (+)-α-pinene affords (*S*)-alcohols and (*R*)-alcohols are obtained when (−)-α-pinene is used (Chart 26.3).

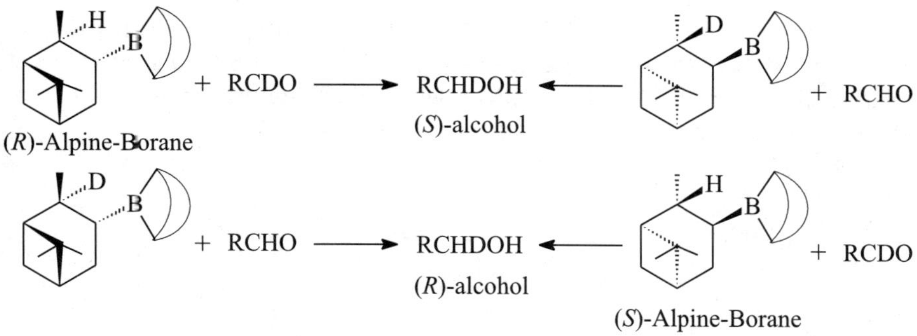

Chart 26.3

The results of reduction of aldehydes with Alpine-Borane are summarized in Table 26.2 [8].

Table 26.2 Reduction of aldehydes with deuterated *B*-3-pinanyl-9-BBN [8]

Aldehydes	Alcohols	*ee* (%)[a]	*ee* (%)[b]	*ee* (%)[c]
$CH_3CH_2CH_2CHO$	$CH_3(CH_2)_2CHDOH$	83	101	101
$CH_3(CH_2)_4CHO$	$CH_3(CH_2)_4CHDOH$	64	83	89
$(CH_3)_3CCHO$	$(CH_3)_3CCHDOH$	70	91	98
$C_6H_5CH{=}CHCHO$	$C_6H_5CH{=}CHDOH$	60	78	84
$(CH_3)_2C{=}CH(CH_2)_2C(CH_3){=}CHCHO$		58	75	81
C_6H_5CHO	C_6H_5CHDOH	70	91	98
p-ClC_6H_4CHO	*p*-ClC_6H_4CHDOH	87	94	101
p-$O_2NC_6H_4CHO$	*p*-$O_2NC_6H_4CHDOH$	86	93	100
p-$CH_3C_6H_4CHO$	*p*-$CH_3C_6H_4CHDOH$	76	83	89
p-$CH_3OC_6H_4CHO$	*p*-$CH_3OC_6H_4CHDOH$	70	76	82
p-$(CH_3)_2NC_5H_4CHO$	*p*-$(CH_3)_2NC_6H_4CHDOH$	61	66	71

[a] The percent *ee* is measured by carbinyl proton area, using $Eu(hfc)_3$. The undeuterated product contributes to both the peaks and that results to low values.
[b] The percent *ee* values are corrected because different samples of 9-BBN-9-d were used for deuteration. The value of first entry is corrected for 90% deuteration, next five for 87% deuteration and remaining values for 96% deuteration.
[c] The corrected values as (+)-α-pinene used was 93% *ee* (100% *ee* for the first entry)

The results in Table 26.2 show that steric effects have little influence on the reduction. The electron-withdrawing groups in benzaldehyde increase the enantiomeric purities, which surprisingly are accompanied by an enhancement in the rate of reduction. Normally, the faster the reaction, the less selective it should be. The loss of selectivity is attributed to a competitive dehydroboration–reduction mechanism that leads to inactive product (Scheme 26.1) [11, 12]. Therefore, in the faster reduction dehydroboration becomes remote, and the product is obtained from chiral reagent (Scheme 26.1).

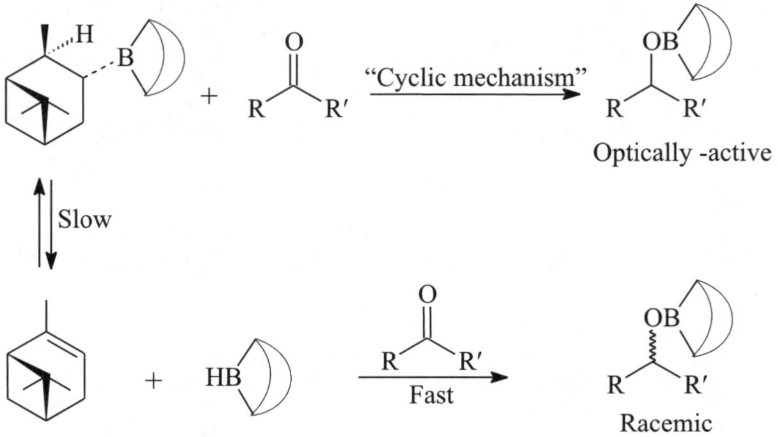

Scheme 26.1

Consequently, a deuterated *B*-3-pinanyl-9-BBN finds applications in the reduction of variety of aromatic, aliphatic, and α,β-unsaturated aldehydes to the corresponding chiral primary 1-deuterio alcohols either with (*R*)- or (*S*)-configuration.

The following model (Fig. 26.1) explains the delivery of deuterium for chiral reduction of aldehydes from Alpine-Borane prepared from (+)-α-pinene.

The reagent is also used to prepare chiral tritium labeled alcohol [13].

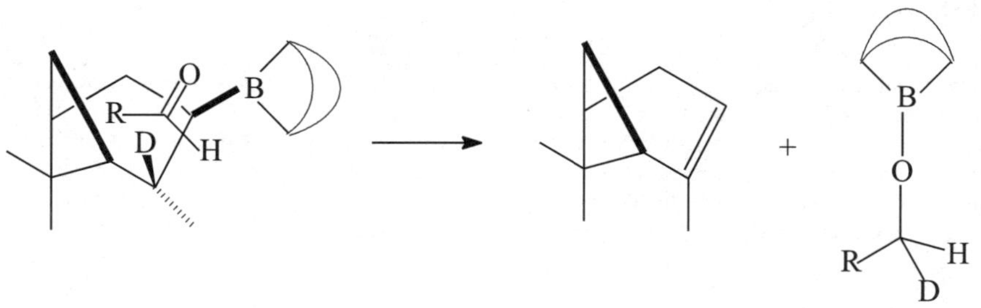

Fig. 26.1 Deuteride delivery by Alpine-Borane-A model for chiral reduction of aldehyde

26.1.2
Reduction of Ketones

The availability of both enantiomeric pinenes and the ability to recycle the pinene ligands make Alpine-Boranes attractive reagents for asymmetric reduction. The reagents are the most stereoselective reducing agents [1–4]. Consequently, the ketones of moderate steric bulk [5] and those containing electron-withdrawing groups [1], α-haloketones, and α-ketoesters [3, 4a, b, 5] are reduced to the corresponding alcohols in high asymmetric induction. The cyanoketones give the amino alcohols in high *ee* [4c]. However, the reduction of typical ketones is less straightforward; alkyl ketones and arylalkylketones are reduced very slowly and often several days or even weeks are required for complete reaction. In general, the longer the reaction time, the lower the enantiomeric purity.

26.1.2.1
Reduction with Alpine-Borane, under Pressure

Reduction of Prochiral Ketones

Similar to the reduction of aldehydes, reduction of ketones with Alpine-Borane also involves two competing reaction pathways, a bimolecular β-hydride elimination process (cyclic mechanism) affording optically active product [6], and a dehydroboration–reduction sequence yielding racemic product [2] (Scheme 26.1).

It is found that the undesired dissociation process is suppressed by carrying out the reaction at elevated pressure (up to 6.08×10^8 Pa) along with the enhancement of the rate of desired reaction [7]. In selected cases enantiomeric efficiencies approaching 100% are achieved. Acetophenone is reduced in 3 days with 100% *ee* at 2.03×10^8 Pa (Table 26.3) [8] and in less than 24 h at 6.08×10^8 Pa.

Table 26.3 Asymmetric reduction of prochiral ketones with Alpine-Borane at 6.08×10^8 Pa [8]

Ketone	Pressure (Pa)	Reaction time (days)	Percentage *ee*[a]	Percentage reduction
Acetophenone	1.01×10^5	8	87	80
	6.08×10^8	2.5	92 (100)	78
2-Octanone	1.01×10^5	7	57	–
	6.08×10^8	3.5	63 (68)	90
3-Methyl-2-butanone	1.01×10^5	29	52	50
	6.08×10^8	10	79 (86)	71

[a] 92% *ee* α-pinene is used. The values in *parentheses* are corrected for the pinene purity.

The results of the reduction of ketones at 6.08×10^{8} Pa at 25 °C are shown in Table 26.4 [8].

Table 26.4 Asymmetric reduction of prochiral ketones with Alpine-Borane at 6.08×10^{8} Pa [8]

Ketone	Reaction time	Isolated yield (%)	Percentage *ee*[a] (6.08×10^{8} Pa)	Percentage *ee* (1.01×10^{5} Pa)	Configu-ration
	1 day	63	58 (63)	57	*S*
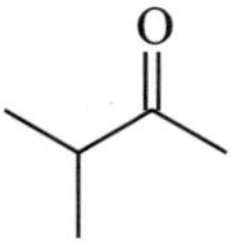	1 day	47	83 (90)	67	*S*
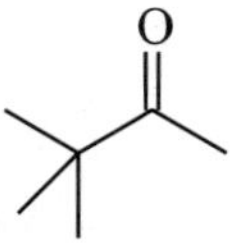	9 days	No reaction			
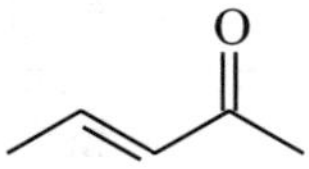	11 h	67	45 (49)	–	*S*
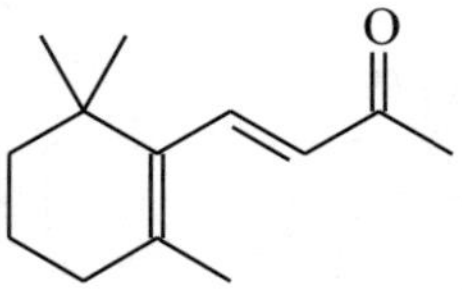	18 h	69	45 (49)	38	*S*
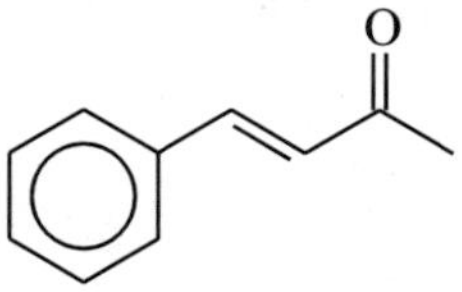	23 h	90	65 (71)	58	*S*
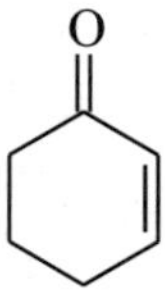	1 day		38 (41)	34	*S*
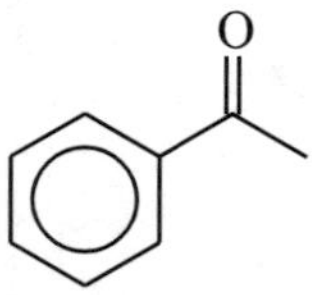	1 day	80	92 (100)	87	*S*

[a] 92% *ee* pinene is used. The values in *parentheses* are corrected for the pinene purity.

Table 26.4 *(continued)* Asymmetric reduction of prochiral ketones with Alpine-Borane at 6.08×10^8 Pa [8]

Ketone	Reaction time	Isolated yield (%)	Percentage *ee*[a] (6.08×10^8 Pa)	Percentage *ee* (1.01×10^5 Pa)	Configuration
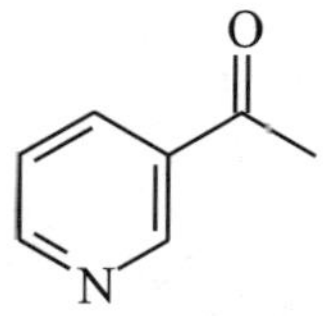	1.5 days	67	92 (100)	90	*S*
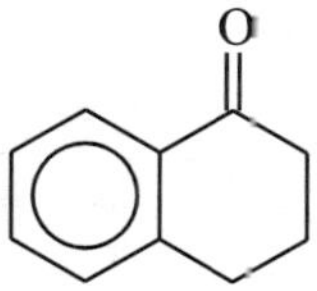	3 days	43	82 (89)	52	*S*
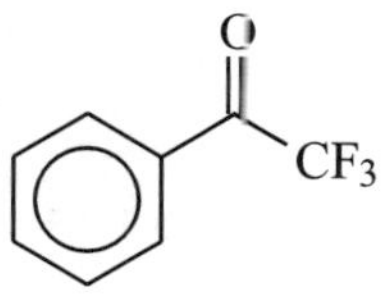	3 days	46	50 (54)	18	*R*
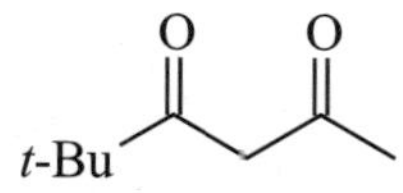	1.5 days	60	57 (62)	52	*R*
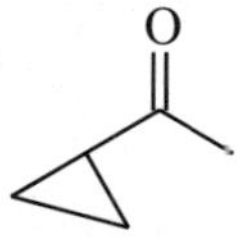	5.5 days	65	69 (75)	–	*S*

[a] 92% *ee* pinene is used. The values in *parentheses* are corrected for the pinene purity.

The asymmetric induction decreases at elevated temperature at 6.08×10^8 Pa (Table 26.5) [8].

The results of Table 26.4 along with other data [8] describe that the relative steric requirements of the groups on ketone are categorized as: very small (C≡CH, C≡N, H, D), small (CH_3, $COOCH_3$), medium (*n*-alkyl, *trans*-RHC=CH), medium large (CF_3, *i*-Pr), large (Ar), and too large (*t*-butyl). Effective asymmetric reductions are achieved when groups from nonadjacent categories are attached to the carbonyl moiety [8].

The ketone approaches the borane, such that boat-like transition state is formed (Fig. 26.2). Consequently, the absolute configuration of the product alcohol can be predicted from the model [8].

Table 26.5 Asymmetric reduction of prochiral ketones at elevated temperatures and pressure [8]

Ketone	Pressure (Pa)	Reaction time	Temp (°C)	Percentage *ee*[a]	Percentage reduction
Acetophenone	1.01×10^5	8 days	25	87	80
	6.08×10^8	2.5 days	25	92	78
	6.08×10^8	13 h	40	88	88
	6.08×10^8	14 h	66	71	100[b]
2-Octanone	1.01×10^5	7 days	25	57	–
	6.08×10^8	3.5 days	25	63	90
	6.08×10^8	16 h	55	46	100[b]
α-Tetralone	1.01×10^5	29 days	25	52	50
	6.08×10^8	10 days	25	79	71
	6.08×10^8	16 h	57	38	70

[a] 92% *ee* pinene is used. In all cases the *S* alcohol is obtained.
[b] No ketone is detected.

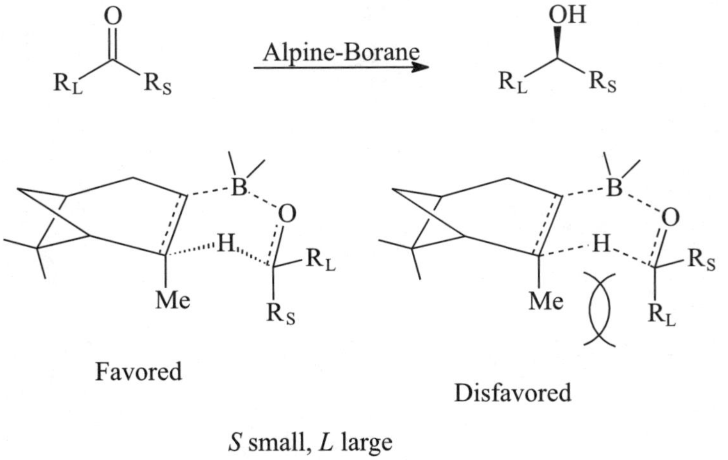

Fig. 26.2 Cyclic transition states for asymmetric reduction of ketone

Reduction of Ketones with Chiral Centers

The reduction of ketones containing chiral center is studied using (*R*)-Alpine-Borane obtained [8] from (+)-α-pinene. The chiral center plays significant role in determining the selectivity and rate of reduction. *d*-Carvone gets reduced at 6.08×10^8 Pa in 90% yield in 3 days and provides two epimeric alcohols in a 4.6:1 ratio, predominantly equatorial alcohol (Chart 26.4). At 6.08×10^8 Pa *l*-carvone shows no reaction with (*R*)-Alpine-Borane, even after 5 days. Models indicate that for *l*-carvone to approach (*R*)-Alpine-Borane, either the sterically demanding vinylmethyl group or isopropylene group interacts with the 2-methyl group of the pinene or bicyclononane moiety of 9-BBN.

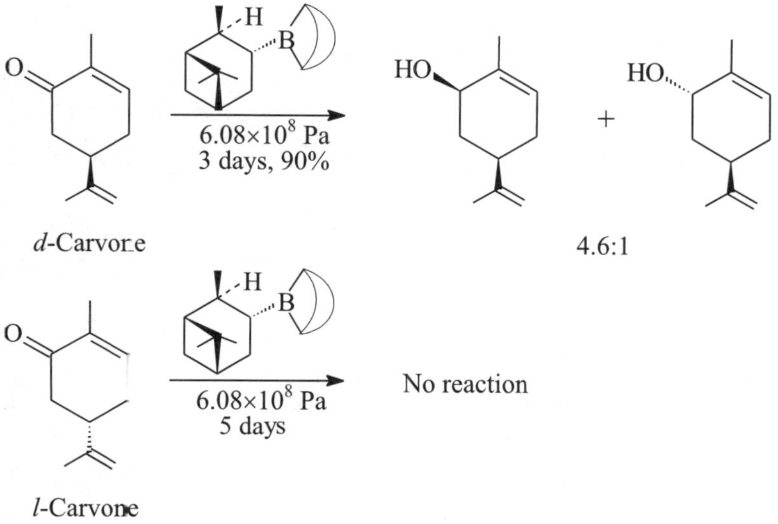

Chart 26.4

Methyl-substituted cyclohexanones show no discrimination between enantiomers. The steric interactions are complicated by changes in the conformations of cyclohexyl rings and by the position of the substituents. The results of reduction of chiral ketone are summarized in Table 26.6 [8].

26.1.2.2
Reduction with Alpine-Borane as Neat or in Excess

26.1.2.3
Reduction of Prochiral Ketones

Brown and Pai have reported [3] that the dissociation problem is also overcome by carrying out the reaction with the Alpine-Borane either as neat reagent or concentrated solutions (≥2 M). The reduction of a wide range of prochiral carbonyl compounds with good to excellent asymmetric induction has been achieved. The reduction is performed at 25 °C by using either a 100 or 40% excess (for reactive substrate) of the reagent, synthesized by the hydroboration of commercial available 92% *ee* (+)-α-pinene with 9-BBN. On completion of the

Table 26.6 Reduction of chiral ketones with Alpine-Borane [8]

Ketone structure	Alcohol reduction products[a] A	Alcohol reduction products[a] B	Percentage *ee* (A/B)	Ratio of A:B
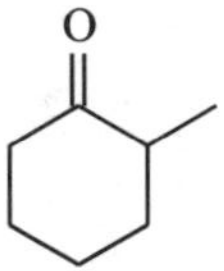	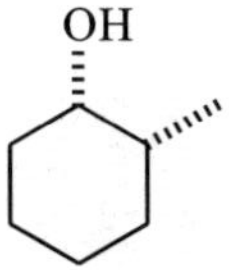	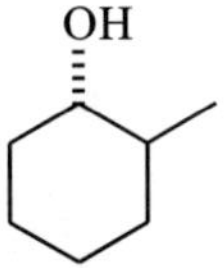	63/68	1:1
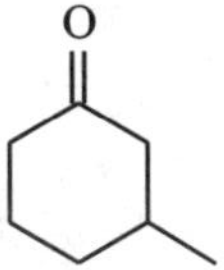	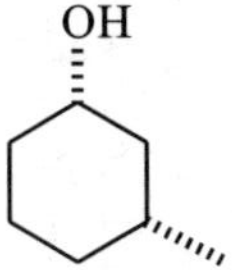	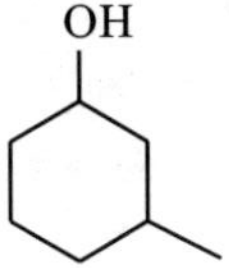	0/0	1:1
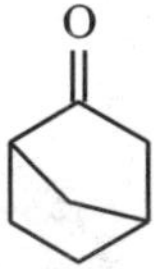	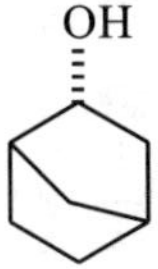	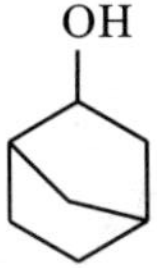	0/0	1:1
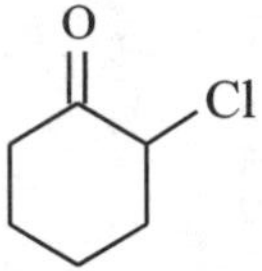	Decomposition			
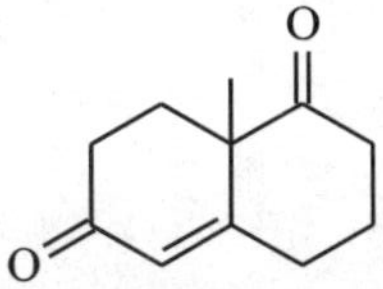	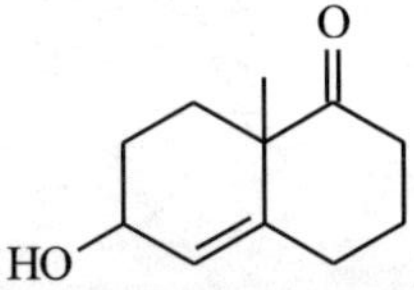	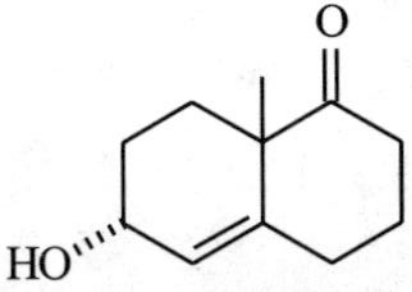	0/0	1:1

Table 26.6 *(continued)* Reduction of chiral ketones with Alpine-Borane [8]

Ketone structure	Alcohol reduction products[a] A	B	Percentage *ee* (A/B)	Ratio of A:B
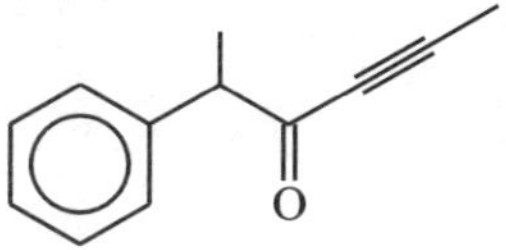	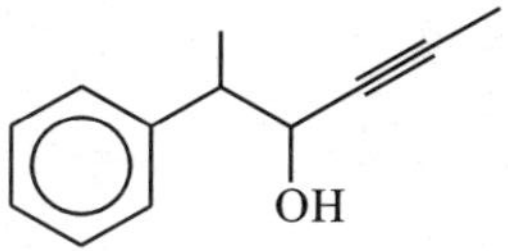	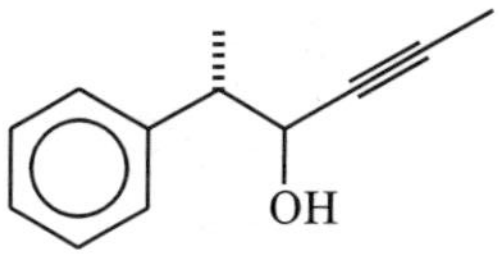	92/92	2.4:1
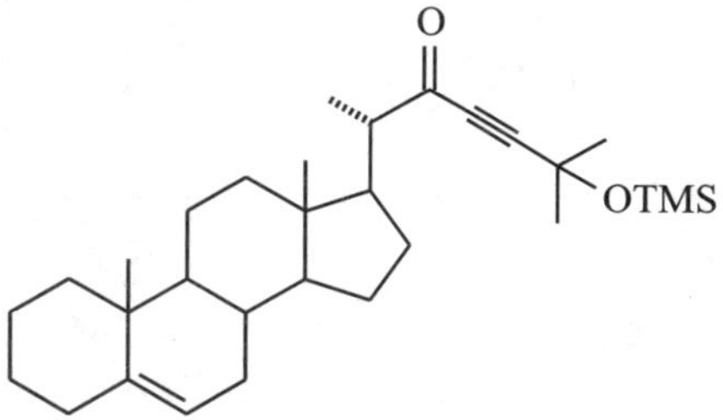	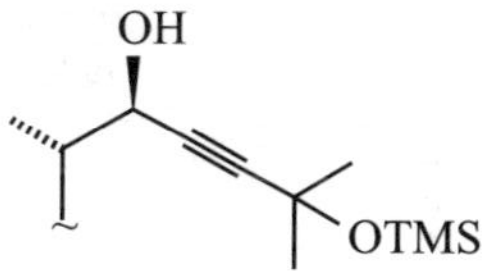	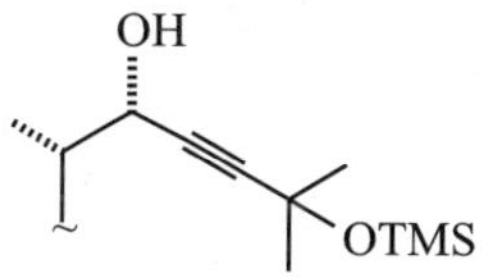	[b]	125:1
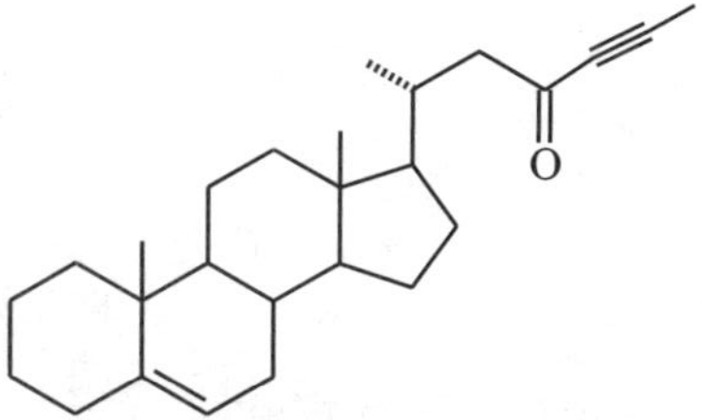	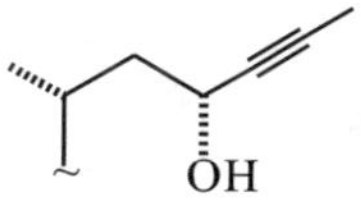	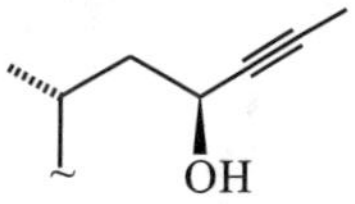	[b]	24:1

[a] Excess neat Alpine-Borane is used. Ketones 1–5 are reduced with 2 equiv, and ketones 6–8 are reduced with 1.5 equiv.
[b] Only one enantiomer of the ketone is used.

reaction, the excess reagent is destroyed with volatile acetaldehyde or propionaldehyde, and liberated α-pinene is pumped off under vacuum. The free alcohol is liberated from its borinate ester by exchange with ethanolamine [9], where 9-BBN-ethanolamine adduct precipitates out (Chart 26.5).

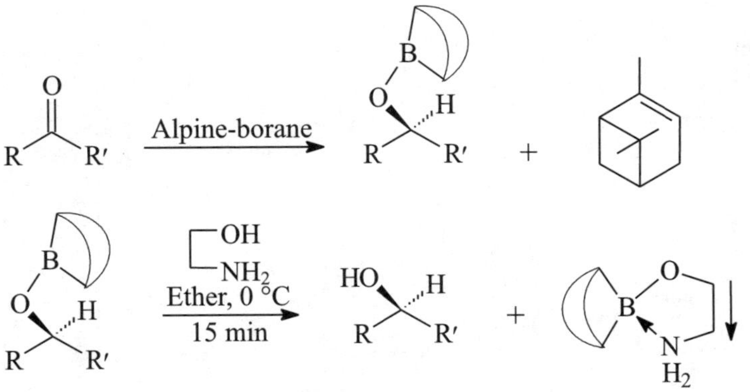

Chart 26.5

The reduction utilizing 100% excess of the neat reagents is slow [5], and most ketones require 7–14 days for complete reaction. The optical induction increases with steric inequality of the two groups attached to the carbonyl functionality (Table 26.7) [3, 5].

Table 26.7 Reduction of prochiral simple ketones with neat Alpine-Borane (from 92% *ee* (+)-α-pinene) at 25 °C [3, 5]

Ketone	Reagent (molar equiv)	Reaction time (days)	Yield (%)	Percentage *ee* Observed	Percentage *ee* Correction	Absolute configu-ration
2-Butanone	2	10	90	40	43	*S*
3-Methyl-2-butanone	2	14	78	57	62	*S*
3,3-Dimethyl-2-butanone	2	40	40	0.6	0.7	*S*
2-Octanone	2	07	65	44	48	*S*
Acetophenone	1	14	68	80	87	*S*
Acetophenone	2	07	68	78	85	*S*
Acetophenone	2	3–4 at 45 °C	85	63	68	*S*

26.1.2.4 Reduction of α-Haloketones

The electron-withdrawing substituents on the carbonyl compounds are known to increase the rate of reduction [1, 2]. Higher reduction rate sometimes also increases the optical induction, owing to the reaction operating via cyclic mechanism rather than dissociation mechanism.

The α-haloalkylaromatic ketones (entries 1–7, Table 26.8) react with 100% excess of neat Alpine-Borane and afford the corresponding halohydrins in excellent enantiomeric purities with high chemical yields. The α-chloroacetophenone as compared with α-iodoacetophenone has poor solubility in the neat reagent and takes 6–8 days for completion. The model proposed by Midland [1b] correctly predicts the (*R*)-configuration. The (*R*)-configuration is a consequence of CIP convention rather than a change in the mode of attack.

With the iodo compound (entry 3, Table 26.8), reduction is accompanied by deiodination. Reduction of α,*p*-dibromoacetophenone and α-bromo-*p*-cyanoacetophenone is carried out in highly concentrated (~5 M), partially heterogeneous solution in THF, as these compounds are insoluble in neat reagent. The α,α,α-trifluoroacetophenone is very sluggish to react and takes 45 days to achieve ~90% reaction, with only 32% optical induction. The sluggish reduction is attributed to the steric bulk of trifluoromethyl or to its powerful electron-withdrawing capacity. The former might make it difficult to approach the reagent, while the latter might prevent coordination.

The chiral bromohydrins are valuable intermediates for further elaboration, for example, in the synthesis of the alkaloid ubine [10].

Table 26.8 Reduction of prochiral α-halo ketones with neat Alpine-Borane (from 92% *ee* (+)-α-pinene) at 25 °C using 100% excess reagent [3, 4a]

Entry	Ketone	Time (days)	Yield (%)	Percentage *ee* Observed	Percentage *ee* Corrected	Absolute configuration
1	α-Chloroacetophenone	6–8	91	88.5	96.2	*R*
2	α-Bromoacetophenone	4	95	86	93	*R*
3	α-Iodoacetophenone	2	60	86	93	*R*
4	α,*p*-Dibromoacetophenone[a]	3	95	88	96	*R*
5	α-Bromo-*p*-cyano-acetophenone[a]	2–3	60	88	96	*R*
6	α-Bromo-2′-acetonaphthone[a]	3–4	90	83	90	*R*
7	α,α,α-Trifluoroacetophenone	45	57	32	35	*R*
8	1-Bromo-3-methyl-2-butanone	14	60	61	66	*R*
9	3-Bromo-3-methyl-2-butanone	No reaction				

[a]Reaction is carried out in ~5 M THF solution.

The aliphatic α-haloketones do not give good results. For example, 3-bromo-3-methyl-2-butanone fails to react, and 1-bromo-3-methyl-2-butanone yields the corresponding bromohydrin in 61% *ee*.

26.1.2.5
Reduction of Ketoesters

The reaction is extended [3] to the reduction of α-ketoesters, another class of ketones, possessing powerful electron-withdrawing ester group. The reduction of α-ketoesters with 40% excess of neat Alpine-Borane proceeds rapidly at 25 °C (Eq. 26.4; Table 26.9) [3, 4b].

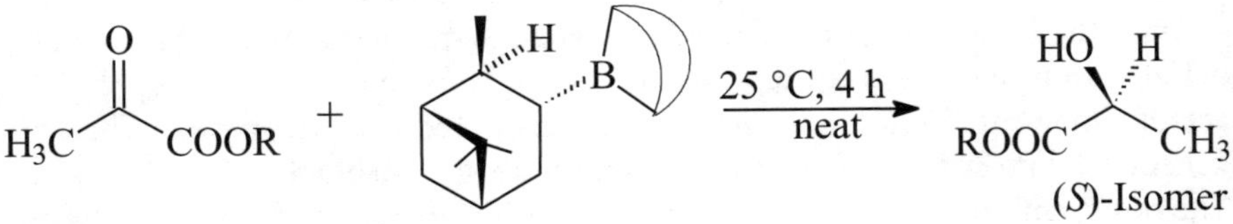

(26.4)

Except for *t*-butyl ester, the increase in the steric bulk has no significant affect (entries 1, 2, 4, Table 26.9). However, increase in the steric bulk of alkyl chain (Table 26.9) results in increase in asymmetric induction. On the other hand, branching of the alkyl chain leads to a drastic decrease in the reaction rate as well as in asymmetric induction (entry 11, Table 26.9). Separating the isopropyl group from the reaction center by a methylene group leads to both increase in the reaction rate and increase in asymmetric induction (entry 12, Table 26.9) [3].

The corresponding aromatic ketoesters such as methyl, isopropyl, and *tert*-butylbenzoylformates are reduced to the corresponding mandelic esters (entries 14–16, Table 26.9) with excellent asymmetric induction, possessing (*R*)-configurations (Eq. 26.5) [4b]. The formation of product with *(R)*-configuration is in accordance with Midland's model, which also suggests that phenyl group is bulkier than the ester moiety.

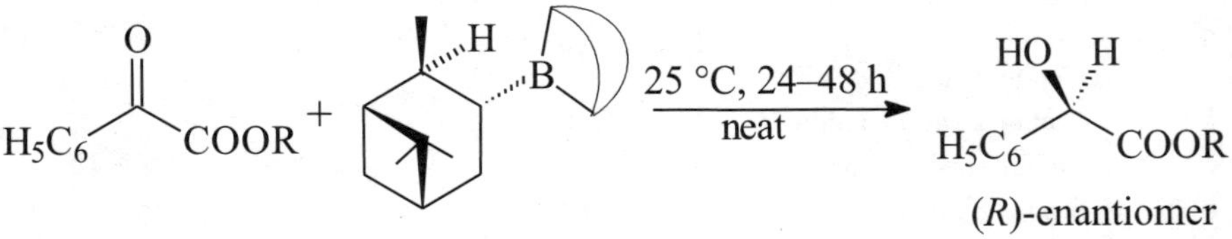

(26.5)

Table 26.9 Reduction of α-ketoesters with 40% excess of neat Alpine-Borane (from 92% *ee* (+)-α-Pinene) [3, 4b]

Entry	α-Ketoester	Reaction conditions		Yield (%)	Percentage *ee*		Absolute configu-ration
		Temperature (°C)	Time (h)		Observed	Corrected	
1	Methylpyruvate	25	4	64	79	86	*S*
2	Ethylpyruvate	25	4	81	76	83	*S*
3	Ethylpyruvate	0	24	81	82	89	*S*
4	Isopropyl-pyruvate	25	4	72	72	78	*S*
5	*tert*-Butyl-pyruvate	25	4	98	85	92	*S*
6	*tert*-Butyl-pyruvate	0	24	98	92	100	*S*
7	*tert*-Butyl-2-oxobutyrate	0	24	71	92	100	*S*
8	Methyl-2-oxopentanoate	25	10–20	80	88	96	*S*
9	Ethyl-2-oxopentanoate	25	4	77	88	96	*S*
10	*tert*-Butyl-2-oxopentanoate	0	24	79	92	100	*S*
11	Methyl-3-methyl-2-oxobutanoate	25	8 days	70	10	11	*R*
12	Ethyl-4-methyl-2-oxopentanoate	25	20	50	75	82	*S*
13	*tert*-Butyl-4-methyl-2-oxopentanoate	0	24	72	92	100	*S*
14	Methyl-benzoylformate	25	24	95	83	90	*R*
15	Isopropyl-benzoylformate	25	48	91	88	96	*R*
16	*tert*-Butyl-benzoylformate	25	48	89	92	100	*R*

The unexpected high asymmetric induction realized with *tert*-butyl esters in both aliphatic and aromatic systems may be due to increased +I effect of the *tert*-butyl group.

The substrate with two electron-withdrawing substituents, such as ethyl α-bromopyruvate, is made to react with a slightly less than stoichiometric amount of Alpine-Borane. The reaction is very fast and is completed in less than half an hour. The product alcohol, however, has a disappointing low enantiomeric purity of only 7.6% *ee* (Eq. 26.6).

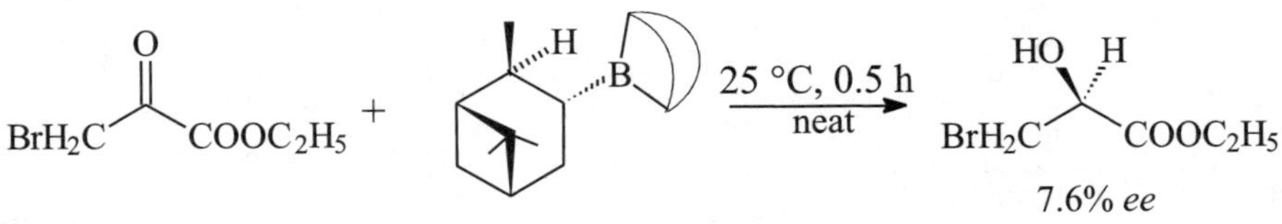

(26.6)

The low asymmetric induction is because of the comparable size of both bromomethyl and ethyl carboxyl groups, and the reagent has no preferred mode of approach.

The other substrates with different electron-withdrawing substituents such as α-hydroxyacetone, α-acetoxyacetophenone, and phenyl glyoxal monohydrate are completely inert to Alpine-Borane. Unlike α-ketoesters, the β-ketoesters react very slowly, e.g., reduction of ethylacetoacetate with 100% excess of neat Alpine-Borane takes 3–5 days for completion, with 50.6% asymmetric induction.

26.1.2.6
Reduction of α,β-Unsaturated Ketones

Among the α,β-unsaturated ketones, only the *trans*-4-phenyl-3-buten-2-one affords the excellent induction (Eq. 26.7). The product alcohols have the (*S*)- configuration (Table 26.10) [3, 5].

Table 26.10 Reduction of prochiral α,β-unsaturated ketones with neat Alpine-Borane (from 92% *ee* (+)-α-pinene) at 25 °C [3]

Ketone	Reaction condition			Percentage *ee*		
	Reagent (molar equiv)	Time (days)	Yield (%)	Observed	Corrected	Absolute configuration
3-Buten-2-one	1.4	5	30	60	65	*S*
1-Acetyl-1-cyclohexene	1.4	12	90	59	64	*S*
3-Methyl-2-cyclohexenone	1.4	12	70	18	19.6	*S*
trans-4-Phenyl-3-buten-2-one	1.4	10	80	89	97	*S*

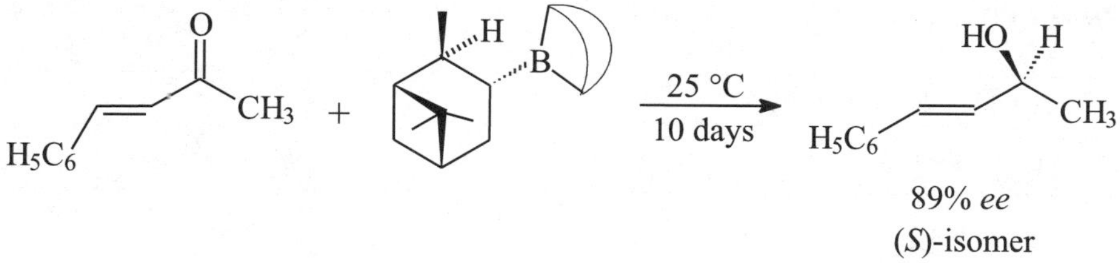

(26.7)

26.1.2.7
Reduction of Propargyl Ketones

It is significant to mention that most of the studies for asymmetric reduction are conducted on acetophenone as a substrate. This is because most of the reducing agents uniformly fail for nonaromatic ketones. To circumvent this problem several research groups have investigated the reduction of propargyl ketones [11], as the resulting propargylic alcohols are very useful in organic synthesis [12].

Midland and coworkers [13, 14] achieved the reduction of sterically less congested propargyl ketones with Alpine-Borane. The reduction is accomplished using 2 equiv of 0.5-M solutions of Alpine-Borane (Table 26.11) [14]. Terminal acetylenic ketones and acetylenic ketoesters are completely reduced after 8 h at room temperature. Internal acetylenic ketones require 1–4 days at room temperature for complete reductions. The optically active chromanyl substrates (entries 7, 8) yield diastereomeric alcohols with (*R*,*R*):(*R*,*S*) ratios of 85:15 for internal and 91:9 for the terminal acetylenes with the Alpine-Borane derived from (+)-α-pinene of 100% *ee*. The reagent obtained from pure (−)-α-pinene affords 18:82 ratio of the two diastereomeric internal propargylic alcohols.

Table 26.11 Reduction of alkynyl ketones with 100 excess, 0.5 M Alpine-Borane (from 92% *ee* (+)-α-pinene) in THF [14]

	Ketone RCOC≡CR′			
	R	R′	Yield	*ee*[a]
1	C_6H_5	n-C_4H_9	72	89
2	CH_3	C_6H_5	98	72 (78)
3	n-C_3H_7	n-C_6H_{13}	68	77[b]
4	n-C_5H_{11}	n-C_4H_9	66	78 (85)
5	n-C_5H_{11}	H	70	92[b]
6	$CH(CH_3)_2$	H	78	91 (99)

[a] The numbers in *parentheses* are corrected for 92% *ee* α-pinene.
[b] 100% optically pure (+)-α-pinene is used.

Table 26.11 *(Continued)* Reduction of alkynyl ketones with 100 excess, 0.5 M Alpine-Borane (from 92% *ee* (+)-α-pinene) in THF [14]

	Ketone RCOC≡CR′			
	R	R′	Yield	*ee*[a]
7	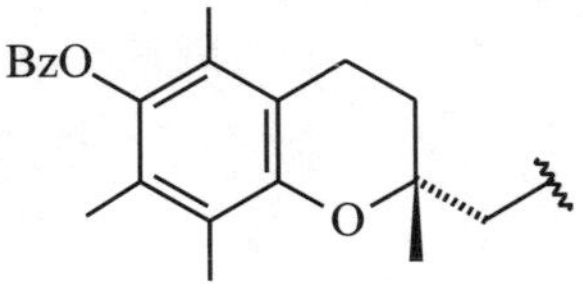	CH_3	77	85:15
8	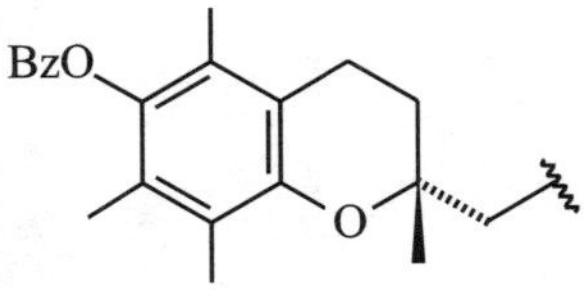	H	75	91:9
9	$CH_3O_2CCH_2CH_2$	*n*-C_8H_{17}	75	90 (98)
10	CH_3	$CO_2C_2H_5$	59	71 (77)
11	CH_3CH_2	$CO_2C_2H_5$	58	88 (96)
12	*n*-C_5H_{11}	$CO_2C_2H_5$	72	85 (92)
13	C_6H_5	$CO_2C_2H_5$	64	92 (100)
14	*Z*-$C_5H_{11}CH{=}CHCH_2$	$CO_2C_2H_5$	73	90
15	*Z*-$C_8H_{17}CH{=}CH$	$CO_2C_2H_5$	62	98
16	$C(CH_3)_3$	CH_3	0	–
17	CH_3	$C(CH_3)_3$	62	73[b]
18	$CH_3OOCCH_2CH_2$	$SiMe_3$	–	89

[a] The numbers in *parentheses* are corrected for 92% *ee* α-pinene.
[b] 100% optically pure (+)-α-pinene is used.

The mildness and selectivity of the reagent are illustrated in the reduction of vinylacetylenic ketone. This ketone is extremely sensitive to acid- or base-catalyzed isomerization of the *cis* double bond. (*R*)-Alpine-Borane (from 100% *ee* (+)-α-pinene) provides the (*S*)-enantiomer of alcohol in >98% *ee*, while the reagent from (–)-α-pinene (90% *ee*) gives 87% *ee* of (*R*)-alcohol (Eq. 26.8) [14].

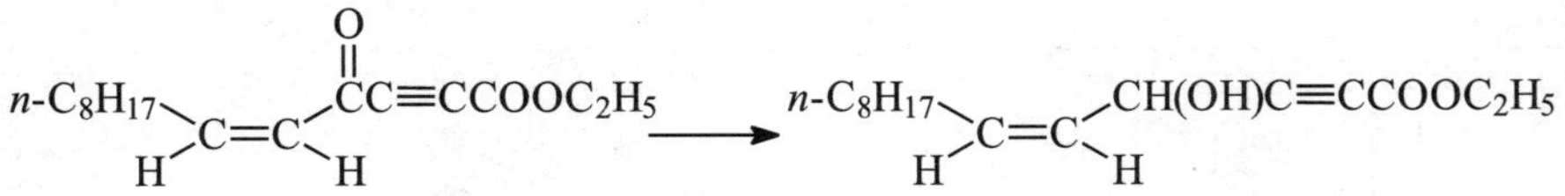

(26.8)

The base-sensitive propargyl alcohol is separated by precipitating 9-BBN as its ethanolamine adduct. In many cases the product propargyl alcohol is entrapped in the precipitate. This problem is overcome by oxidative workup, and propargylic alcohol is separated from *cis*-1,5-cyclooctanediol by distillation or precipitation of the diol from hexane or ether.

Midland and coworkers [1c] have reported that reduction of alkynyl ketones affords excellent chemical and optical yields that approached 100% in many cases. In general, under Midland's reaction condition, for example, 4-phenyl-3-butyn-2-one takes 48 h for complete reduction at 25 °C with 100% excess of Alpine-Borane. On the other hand Brown has reported [3] the reduction of ketones in 8–12 h, using a 40% excess of the neat reagent (Scheme 26.2; Table 26.12), and products show a substantially higher optical rotation as compared to reported by Midland [1c].

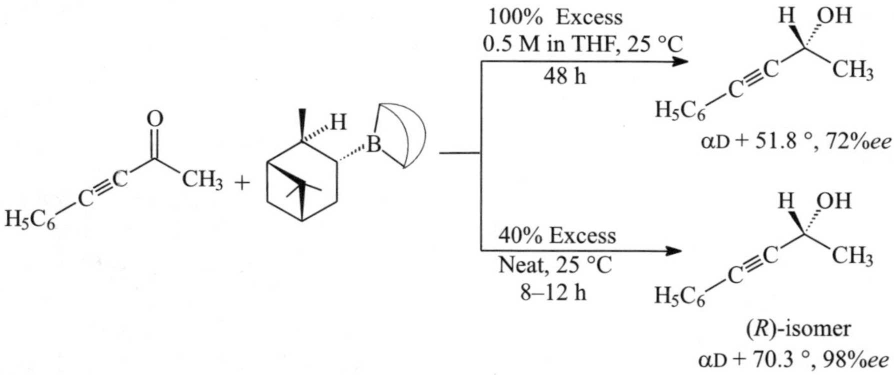

Scheme 26.2

For low-molecular-weight acetylenic ketones such as 3-butyn-2-one, it is advantageous to use a slight excess of the ketone rather than the reducing reagent. Though the reaction is slowed down, the asymmetric induction is not affected. The advantage of excess ketone lies in the avoidance of destroying the excess reagent using acetaldehyde, which produces ethanol as the side product; since the two alcohols have fairly close boiling points, the separation of the product needs careful distillation.

The reduction products have the same absolute configuration. Thus, the acetylenic moiety exerts the same influence as hydrogen in aldehydes, i.e., it behaves as if it were smaller than the alkyl group [1c].

The results are summarized in Table 26.12 [3].

Table 26.12 Reduction of prochiral α,β-acetylenic ketones with 40% excess of neat Alpine-Borane (from 92% (+)-α-pinene) at 25 °C [3]

Ketone	Time (h)	Yield (%)	Percentage *ee*	
			Observed	Corrected
3-Butyn-2-one	4	80	73	79
4-Methyl-1-pentyn-3-one	4	87	91	99
4-Phenyl-3-butyn-2-one	8–12	95	98	106.5

The ketone with *t*-butyl group fails to undergo reduction. However, these ketones are reduced when subjected to high pressure [14] (Eq. 26.9).

$$\mathrm{Me_3C\overset{O}{\overset{\|}{C}}C{\equiv}CSiMe_3} \xrightarrow[\substack{2.5\text{ days,}\\ 6.08\times10^8\text{ Pa}\\ 71\%}]{\text{Alpine-Borane}} \mathrm{Me_3C\overset{OH}{\overset{|}{C}}HC{\equiv}CSiMe_3} \quad (100\%\ ee) \tag{26.9}$$

Alpine-Borane is an extremely chemoselective; benzoyl chloride, γ-valerolactone, ethylpropiolate and phthalic anhydride are not reduced by the neat reagent [14] over a period of several days. Aldehydes are reduced up to 10^3 times faster than ketones [15].

The chemoselectivity shown by Alpine-Borane allows one to achieve selective reduction of propargyl ketone in the presence of methyl ketone [14] (Eq. 26.10).

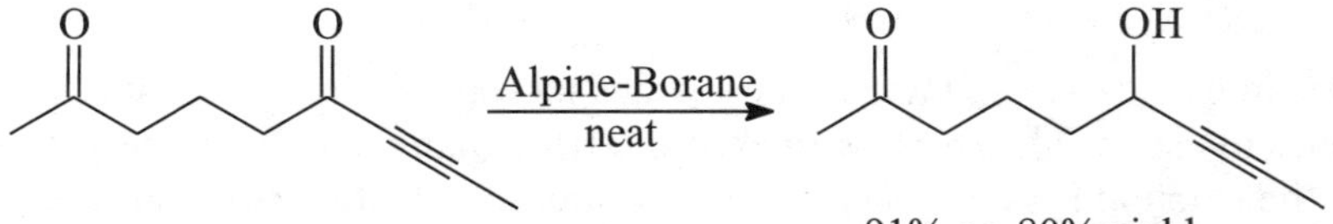

91% *ee*, 80% yield

(26.10)

The selectivity of the reagent is based purely on steric grounds rather than on an electronic effects, though electronic effects do change the rate of reduction.

The stereochemical outcome of the reduction is predicted by the following model [14] for Alpine-Borane prepared from (+)-α-pinene. The hydride transfer occurs from the boat-like transition state in which the acetylene occupies the axial position (Fig. 26.3).

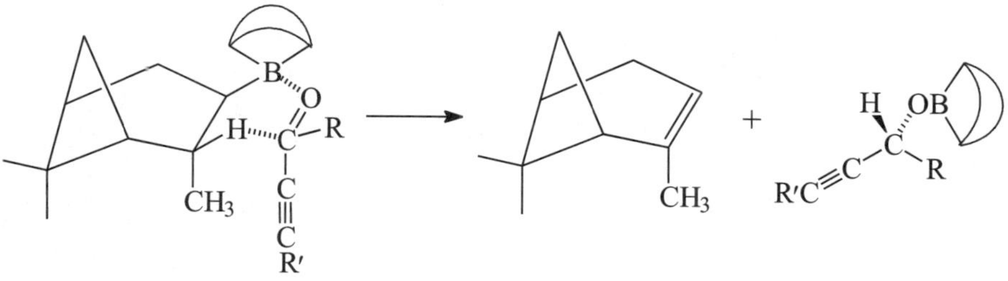

Fig. 26.3 Midland's model for a reduction of prochiral α, β-acetylenic ketones

The acetylene handle can be moved through a series of CH_2 units, using potassium-3-aminopropylamide (KAPA) without affecting the chiral center [16] (Eq. 26.11).

(26.11)

Reduction of a carbon–carbon triple bond of acetylene produces the optically active *cis* or *trans* allylic alcohols. The functionality and chirality of the alcohol center may then be transferred by sigmatropic rearrangement [17] or by S_N2 displacement [18] to remote carbon centers (Chart 26.6) [14].

26.1.2.8
Reduction of Propargylic Ketones with α-Chiral Centers

An efficient stereoselective reduction of 22-keto-23-acetylenic steroid to *anti*-Cram product 22-(*R*)-hydroxy-23-acetylenic steroid and Cram product 22-(*S*)- hydroxy-23-acetylenic steroid has been achieved using (*R*)-Alpine-Borane [(+)-α-pinene, 92% *ee*] (125:1, *R*:*S*) and L-selectride (lithium tri-*sec*-butylborohydride) (1:11, *R*:*S*), respectively [19] (Chart 26.7). (*S*)-Alpine-Borane (2 M in THF) prepared from (–)-α-pinene (92% *ee*) provides unexpectedly low 1:2.7, *R*:*S* ratio due to the influence of the α-chiral center at C-20 of the steroid, and also the reduction is much slower than with (*R*)-Alpine-Borane.

Using neat (*S*)-Alpine-Borane the reaction is 67% complete in 92 h, and 22-(*S*)- is obtained in a 7:1 ratio. It is thus apparent that for (*R*)-Alpine-Borane the double asymmetric inductions are working together, whereas in (*S*)-Alpine-Borane they are working in opposite direction. A similar change in selectivities is observed in the asymmetric reduction of these acetylenic ketones using (+)- and (–)-*N*-methylephedrine/LAH [18]. The faster rate of reduction with (*R*)-

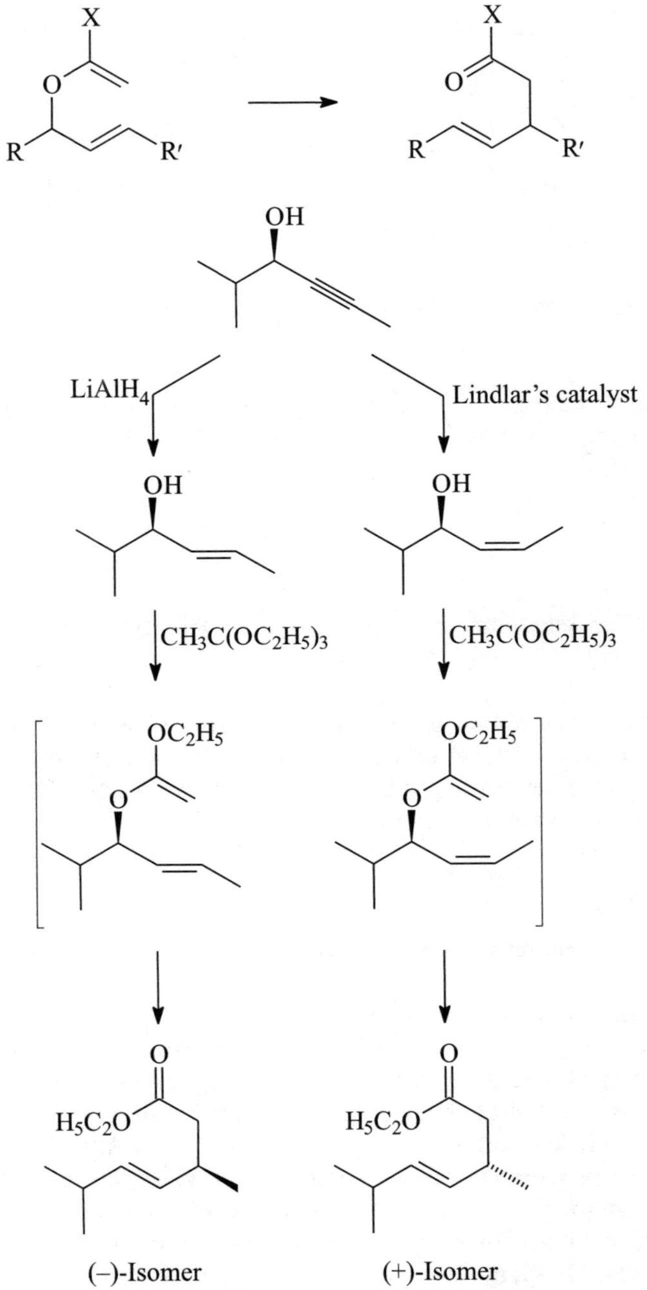
X
O
R
R′
X
O
R
R′
OH
LiAlH4
Lindlar's catalyst
OH
OH
CH3C(OC2H5)3
CH3C(OC2H5)3
OC2H5
O
OC2H5
O
O
H5C2O
O
H5C2O
(–)-Isomer
(+)-Isomer

Chart 26.6

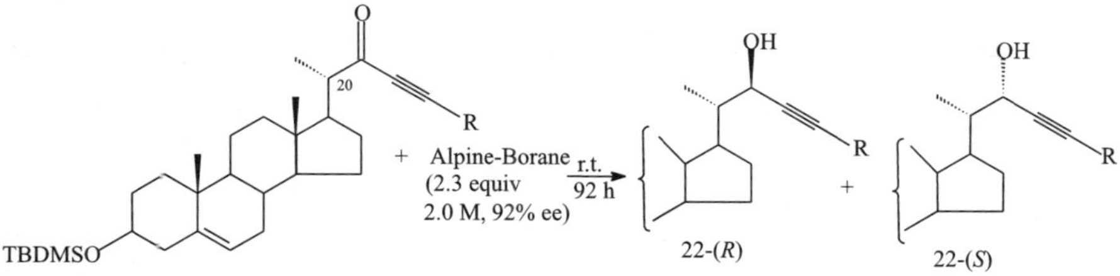

R-Alpine-Borane	22-(*R*):22-(*S*)	Isolated yield
R = CH_3	125:1	95%
R = $C(CH_3)_2OTHF$	125:1	96%
R = $C(CH_3)_2OTBDMS$	125:1	96%
L-Selectride		
R = $C(CH_3)_2OTBDMS$	1:11	100%

Chart 26.7

Alpine-Borane reveals that there is a synergistic effect of the two chiral centers on the rate of reduction. This suggests that it should be possible to selectively reduce the (*S*)-C-20 epimer in the presence of (*R*)-C-20 epimer with (*R*)-Alpine-Borane.

Consequently, α-chiral site reinforces or diminish enantioselectivity as well as influence the rate of asymmetric reduction of α-chiral alkynyl ketones with Alpine-Borane. This type of influence of an asymmetric induction has not been observed in the reduction of β-chiral alkynyl ketones [14, 20].

26.1.2.9 Reduction of Acylcyanide

The reduction of acylcyanide using neat (*R*)-Alpine-Borane affords the corresponding (*R*)-β-amino alcohols [4c]. The reduction of acylcyanide and subsequent workup is not a straightforward process. The reaction of benzoylcyanide with neat Alpine-Borane (1.5 equiv) is complete within 2 h. The cyanohydrin–9-BBN adduct builds up to maximum, and then decreases with the appearance of a 9-BBN–benzyl alcohol adduct. Apparently, the 9-BBN–cyanohydrin adduct undergoes an elimination reaction to give benzaldehyde, which then undergoes reduction. The results indicate that the desired bimolecular reduction process can compete with the elimination reaction.

The clean reduction, however, is achieved [4c] using 15 mmol of neat solution of (*R*)-Alpine-Borane and acylcyanide (10 mmol). The progress of the reaction is monitored by ^{1}H NMR (appearance of α-pinene signal at 5.2 ppm). The reaction mixture is then treated with methanol containing cobaltous chloride [21], followed by the addition of sodium borohydride (Eq. 26.12) [4c].

$$\underset{10\ \text{mmol}}{H_5C_6-\overset{\overset{\displaystyle O}{\|}}{C}-CN} + \underset{15\ \text{mmol}}{(R)\text{-Alpine-Borane}} \xrightarrow[\text{(2) 3M, HCl}]{\text{(1) }CoCl_2,\ NaBH_4} \underset{\substack{(-)\text{-}(R)\text{-2-Amino-1-}\\ \text{phenylethanol}}}{C_6H_5CHOHCH_2NH_2} \qquad (26.12)$$

The results are summarized in Table 26.13.

Table 26.13 Yields and ee of substituted 2-ethanolamine from acylcyanides using neat Alpine-Borane (from 92% *ee* (+)-α-pinene) [4c]

RCHOH CH_2NH_2	Percentage yield	Percentage *ee*[a]	Configuration
C_6H_5	77	90 (98)	*R*
o-ClC_6H_4	74	86 (94)	*R*
p-$CH_3C_6H_4$	78	87 (95)	*R*
m-$CH_3C_6H_4$	84	85 (92)	*R*
n-C_6H_{13}	86	77 (84)	*R*

[a] The numbers in *parentheses* are corrected for 92% *ee* α-pinene. Sodium borohydride–cobaltous chloride reduction is performed as soon as the reduction of carbonyl is complete. Warming of the reaction mixture or longer times leads to the formation of undesired product.

26.2 NB-Enantrane

The synthesis of (*S*)-enantiomer of propargyl alcohol is achieved by Alpine-Borane obtained from (–)-α-pinene. However, (–)-α-pinene is an expensive reagent. Consequently, structurally related nopol, 6,6-dimethylbicyclo[3.1.1]hept-2-en-2-ethanol, which is a low-cost and commercial alternative to (–)-α-pinene is used. Nopol, after conversion to benzyl ether is hydroborated [1, 2] with 9-BBN on reflux in THF overnight to give the borane NB-Enantrane [2] (Scheme 26.3).

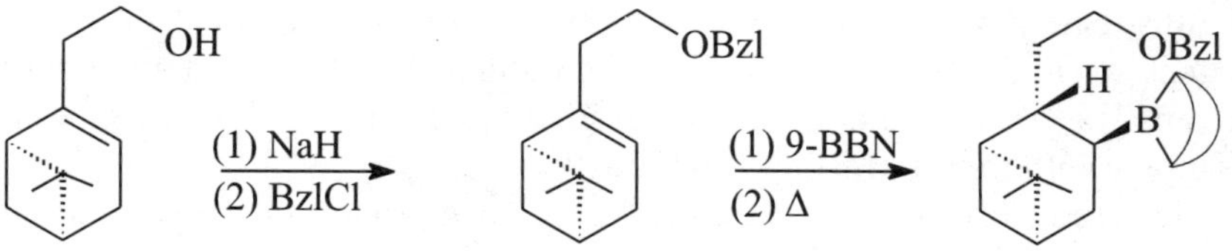

NB-Enantrane

Scheme 26.3

The reduction of α,β-acetylenic ketones is accomplished in 24–48 h at room temperature by using twofold excess of NB-Enantrane and running the reaction without solvent. Both chemical and enantiomeric yields are high (Table 26.14) and provide (*S*)-propargyl alcohols. The reduction fits the steric model proposed for Alpine-Borane reduction [3]. Nopol benzyl ether liberated after the reduction may be easily isolated during purification of the product and recycled.

Table 26.14 Reduction of α,β-acetylenic ketones with NB-Enantrane [1]

RCOC≡CR′		Isolated yield %	Percentage *ee*	Configuration
R	R′			
n-C_5H_{11}	H	74	95	*S*
n-C_5H_{11}	CH_3	79	91	*S*
n-C_5H_{11}	$SiMe_3$	81[a]	96	*S*
n-C_5H_{11}	$CO_2C_2H_5$	74	91	*S*
C_2H_5	C_2H_5	77	94	*S*
Cyclohexyl	n-C_5H_{11}	84	96	*S*
CH_3	C_6H_5	87	86	*S*

[a] During oxidative workup, the trimethylsilyl group is removed.

26.3 Eapine-Borane and Prapine-Borane

Brown and coworkers [1] have found that NB-Enantrane is effective only for the reduction of α,β-acetylenic ketones. The reduction of other ketones is too slow to be of any practical use. The retarded rate is attributed to the steric bulk at the 2 position since no internal coordination has been detected by ^{11}B NMR (δ 86 ppm) [2]. On the other hand, Alpine-Borane has proven to be versatile reagent for the asymmetric reduction of variety of ketones. Consequently, two reagents *B*-(*iso*-2-ethylapopinocampheyl)-9-borabicyclo[3.3.1]nonane (Eapine-Borane) and *B*-(*iso*-2-*n*-propylapopinocampheyl)-9-BBN (Prapine-Borane) having increasing steric requirement at the 2 position, are prepared by the hydroboration [3] of 2-ethyl- and 2-*n*-propylapopinene.

(–)-2-Ethylapopinene is prepared from the commercially available, inexpensive homoallylic alcohol, (–)-nopol $[\alpha]_D^{23}$ –37 °C (neat), as shown (Eq. 26.13) [4].

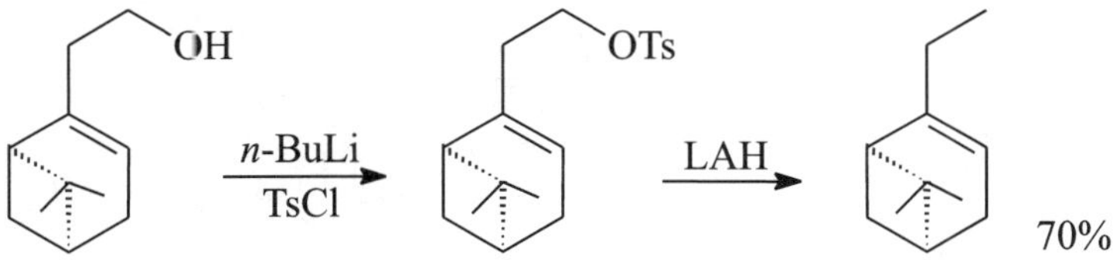

(–)-Nopol (–)-2-Ethylapopinene

(26.13)

(+)-Nopol is not commercially available. Thus, (+)-2-ethylapopinene is prepared from (+)-α-pinene by the following sequence (Eq. 26.14) in high optical purity [4].

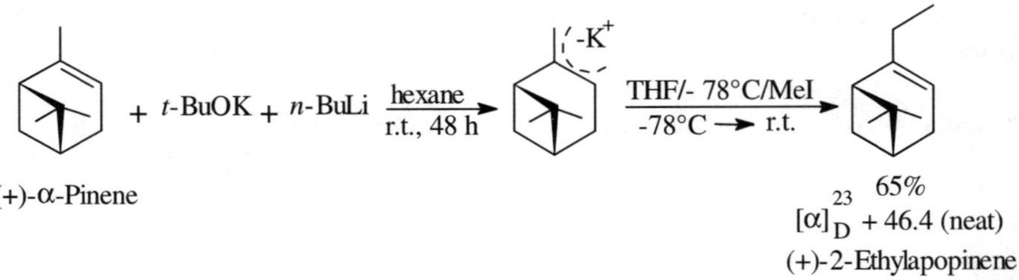

(26.14)

The reagent Eapine-Borane, prepared (Eq. 26.15) from (–)-2-ethylapopinene by hydroboration with 9-BBN have been studied for its reaction with representative ketones and is found to react at faster rate than NB-Enantrane, but at slightly slower rate than Alpine-Borane (Table 26.15) [1].

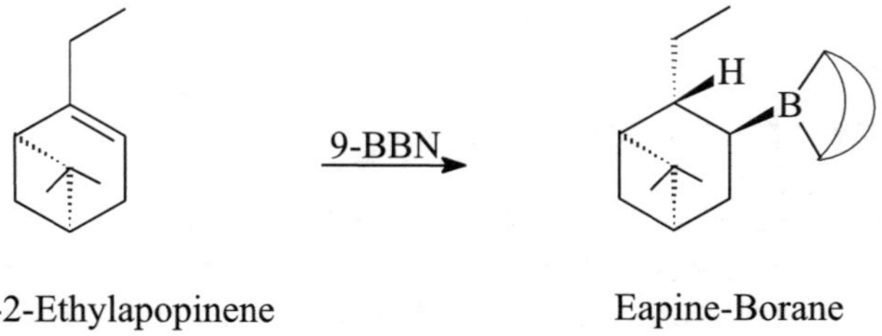

(26.15)

Table 26.15 Reaction of Eapine-Borane with representative ketones at 25 °C [1]

Ketone	Reagent equiv	Reaction time (days)	Percentage *ee*	Percentage *ee* with Alpine-Borane
3-Methyl-2-butanone	2	20	38	62
2,2-Dimethylcyclopentanone	2	7	3	20
Acetophenone	2	20	78	85
3-Acetylpyridine	3	15	96	93
2-Chloroacetophenone	2	15	72	96
Methylbenzoylformate	1.4	3	90	90
trans-4-Phenyl-3-buten-2-one	1.4	15	32	56
2-Cyclohexenone	2	15	36	30
4-Phenyl-3-butyn-2-one	1.4	16 h	89	82

The high asymmetric induction achieved with Eapine-Borane for the α,β-acetylenic ketones and α-ketoesters led an examination of the reduction of a series of α,β-acetylenic ketones (Eq. 26.16) and α-ketoesters (Eq. 26.17), and the comparative reduction data of Eapine-Borane with Alpine-Borane are summarized in Tables 26.16 and 26.17. It should be mentioned that Eapine-Borane offers no advantage for the reduction of aromatic α-ketoesters. Thus, Eapine-Borane is an efficient reagent for the chiral reduction of α,β-acetylenic ketones (Table 26.16) [1] and of alkyl α-ketoesters (Table 26.17) [1], of appreciable steric difference between the two groups on both sides of the carbonyl group.

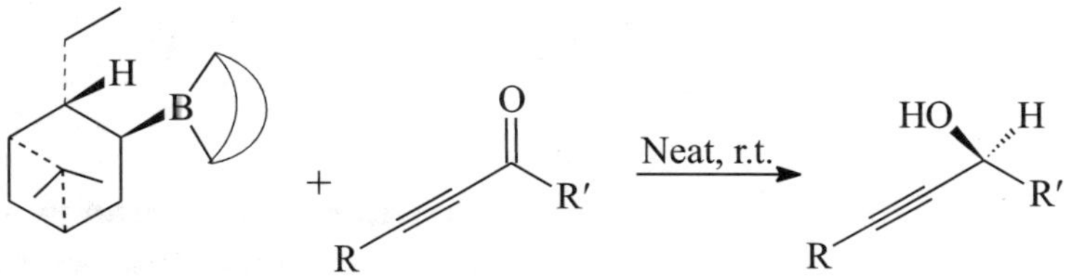

1.4 equiv

(26.16)

Table 26.16 Reaction of Eapine-Borane with representative α,β-acetylenic ketones at 25 °C [1]

Ketone	Reagent equiv	Reaction time (h)	Percentage *ee* (configuration)	Percentage *ee* with Alpine-Borane
3-Butyn-2-one	1.4	12	82 (*S*)	77
1-Octyn-3-one	1.4	48	96 (*S*)	88
3-Hexyn-2-one	1.4	36	88 (*S*)	80
3-Nonyn-2-one	1.4	72	88 (*S*)	82
5-Methyl-3-hexyn-2-one	1.4	72	88 (*S*)	88
4-Phenyl-3-butyn-2-one	1.4	16	89 (*S*)	82

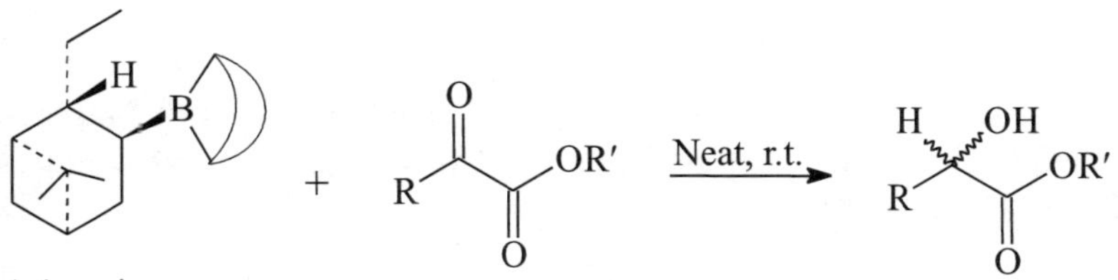

1.4 equiv

(26.17)

Table 26.17 Reaction of Eapine-Borane with α-ketoesters at 25 °C [1]

Ketone	R	R′	Reagent equiv	Reaction time	Percentage *ee* (configuration)	Percentage *ee* with Alpine-Borane (configuration)
Methylpyruvate	Me	Me	1.4	4 h	97 (*R*)	92 (*S*)
Ethylpyruvate	Me	Et	1.4	4 h	95 (*R*)	91 (*S*)
Ethyl 4-methyl-2-oxovalerate	*t*-Bu	Et	1.4	7 days	82 (*R*)	51 (*S*)
Methyl-benzoylformate	Ph	Me	1.4	3 days	90 (*S*)	90 (*R*)
Ethylbenzoylformate	Ph	Et	1.4	3 days	89 (*S*)	93 (*R*)

Prapine-Borane is prepared neat from (–)-2-*n*-propylapopinene by hydroboration with solid 9-BBN, using the same procedure as used for the preparation of Alpine-Borane and Eapine-Borane. (–)-2-*n*-Propylapopinene is prepared in 84% yield from nopyltosylate by treatment with dimethylcuparate, prepared in situ [5] from methyllithium and cuprous iodide (Scheme 26.4).

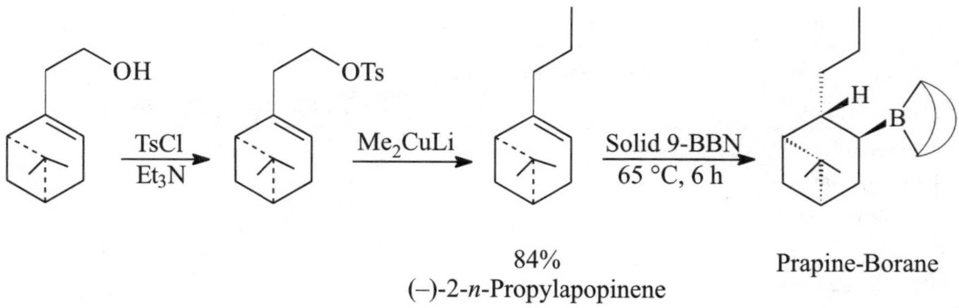

Scheme 26.4

The rate of reduction of representative α,β-acetylenic ketones with Prapine-Borane is slightly slower than Eapine-Borane but with improved chiral induction (Eq. 26.18; Table 26.18) [1].

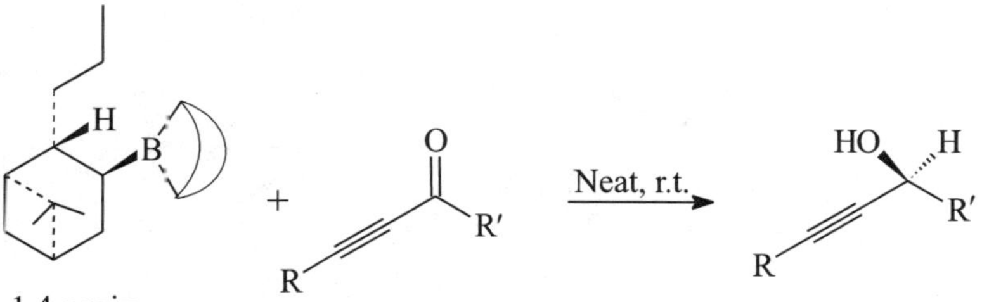

1.4 equiv

(26.18)

Table 26.18 Reaction of Prapine-Borane with representative α,β-acetylenic ketones at 25 °C [1]

Ketone	Reagent equiv	Reaction time (h)	Percentage *ee* With Prapine-Borane	With Eapine-Borane	With Alpine-Borane
3-Butyn-2-one	1.4	12	82 (*S*)	82 (*S*)	77
1-Octyn-3-one	1.4	48	99 (*S*)	96 (*S*)	88
3-Hexyn-2-one	1.4	36	88 (*S*)	88 (*S*)	80
3-Nonyn-2-one	1.4	72	91 (*S*)	88 (*S*)	82
5-Methyl-3-hexyn-2-one	1.4	72	88 (*S*)	88 (*S*)	88
4-Phenyl-3-butyn-2-one	1.4	16	96 (*S*)	89 (*S*)	82

(+)-2-*n*-Propylapopinene is prepared, in 78% yield, via Schlösser metalation [2] of (+)-α-pinene [3], followed by treatment with ethyliodide (Scheme 26.5).

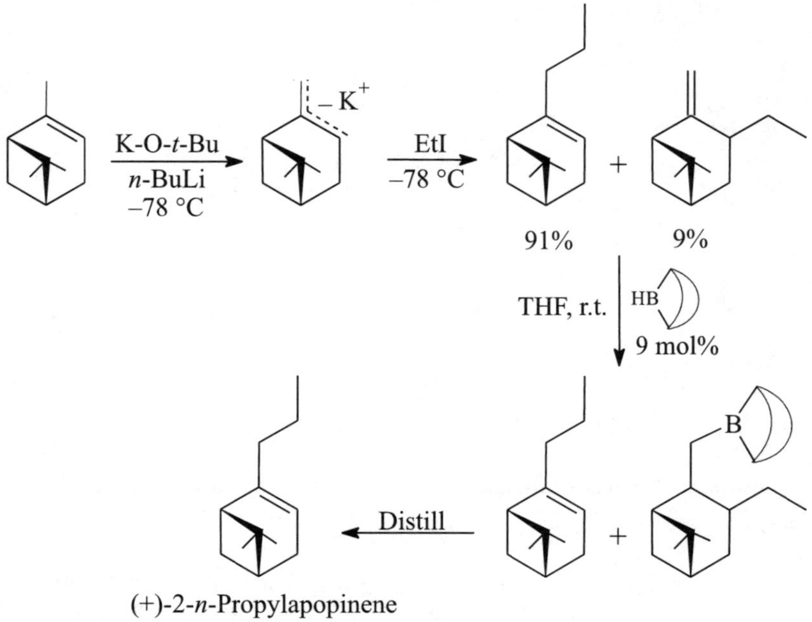

Scheme 26.5

Prapine-Borane obtained from (+)-2-*n*-propylapopinene reacts with 4-phenyl-3-butyn-2-one and provides the (*R*)-propargylic alcohol in 96% *ee* (Eq. 26.19). Thus, both enantiomers of propargylic alcohols are synthesized in very high *ee*.

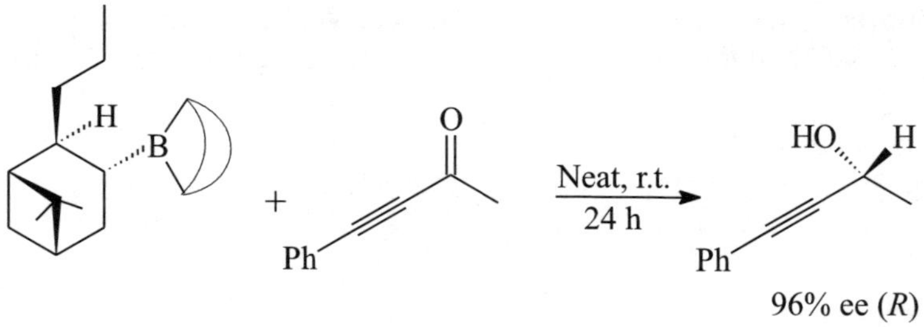

(26.19)

However, unlike Eapine-Borane and Alpine-Borane; Prapine-Borane fails to provide improved asymmetric induction in the reduction of α-keto esters (Table 26.19) [1].

Table 26.19 Reaction of Prapine-Borane with α-keto esters at 25 °C [1]

Ketone	Reagent equiv	Reaction time	Percentage ee (configuration)	Percentage ee with Eapine-Borane (configuration)
Ethylpyruvate	1.4	4 h	89 (*R*)	95 (*R*)
Methylbenzoylformate	1.4	3 days	86 (*S*)	90 (*S*)

26.4 NB-Enantride

The lithium *B*-(isopinocampheyl)-9-borabicyclo[3.3.1]nonane hydride (Alpine-Hydride), a highly hindered trialkylborohydride containing an asymmetric group is obtained by treatment of Alpine-Borane with *t*-BuLi at –78 °C (Eq. 26.20) [1]. The ketones are reduced rapidly and quantitatively to the corresponding alcohols [1] at –78 °C

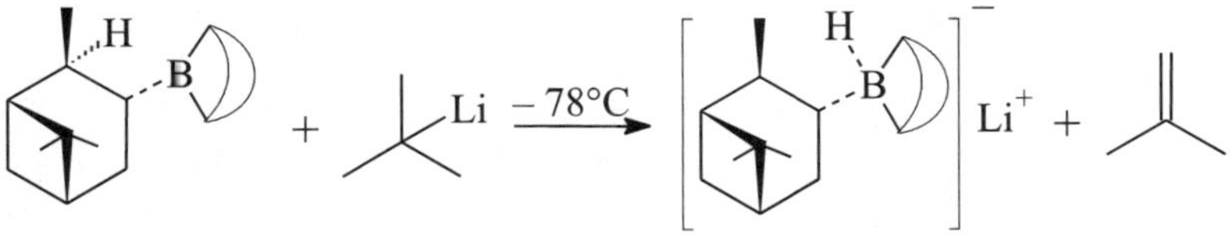

Alpine-Hydride

(26.20)

with Alpine-Hydride and affords (*R*)-alcohols (3–37% *ee*) as the major enantiomers when the reagent is prepared from (+)-α-pinene. The low asymmetric induction is attributed to the transfer of the hydride from the reagent, which takes place one atom away from the chiral center. Midland and Kazubsky [2] have modified the reagent, preparing it by treatment of NB-Enantrane with *t*-BuLi (Eq. 26.21) [2, 3].

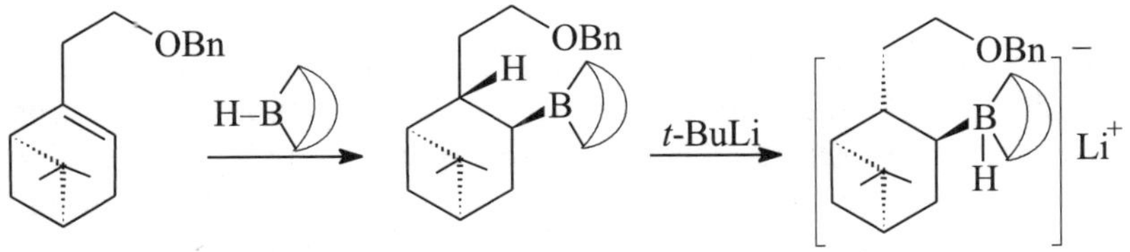

NB-Enantrane NB-Enantride

(26.21)

It is found that incorporation of an oxygen into the chiral ligand provides a fixed coordination site for lithium and hence to a more rigid and thus more sterically demanding transition state. The NB-Enantride reduces acetophenone to (*S*)-phenylethanol (Eq. 26.22), whereas the Alpine-Borane prepared from (−)-α-pinene gives the (*R*)-enantiomer [4].

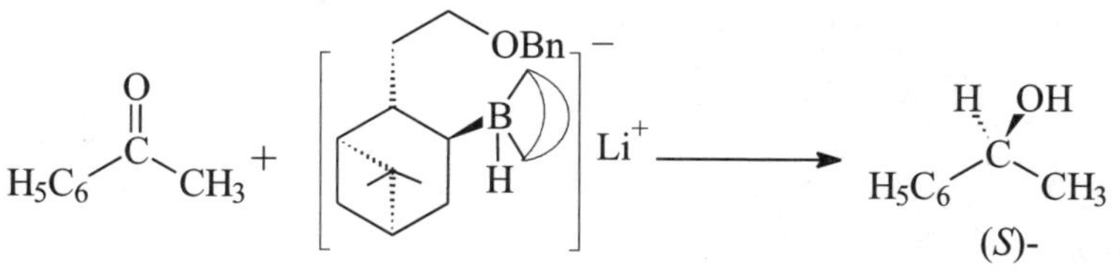

(26.22)

The results of asymmetric reduction with NB-Enantride are summarized in table 26.20 [2].

Table 26.20 Asymmetric reduction of ketones with NB-Enantride [2]

Ketone	Percentage (configuration)	Ketone	Percentage (configuration)
Acetophenone	70 (*S*)	3,3-Dimethyl-2-butanone	2 (*S*)
α,α,α-Trifluoro-acetophenone	50 (*R*)	4-Methyl-2-pentanone	30 (*S*)
Butyrophenone	54 (*S*)	3-Methyl-2-butanone	68 (*S*)
β-Ionone	20 (*S*)	2-Butanone	76 (*S*)
4-Heptyn-3-one	30 (*R*)	2-Octanone	79 (*S*)
4-Phenyl-3-butyn-2-one	10 (*S*)		

The reagent NB-Enantride is highly efficient in case of reduction of straight-chain aliphatic ketones. The reduction of 2-octanone with 79% *ee* is the highest achieved by any chemical reducing agent, thus far.

The effective asymmetric reduction of NB-Enantride for aliphatic ketones is highly significant, as most asymmetric reducing agents are effective only for aromatic ketones [2]. A trend of decreasing asymmetric induction is observed with increasing size of the alkyl groups. The reduction products are isolated in 70–80% yields.

Midland and coworkers [5] have prepared several analogs of lithium NB-Enantride in order to probe the structural features, which contribute to make this reagent effective. The oxygen of ether plays a role in obtaining high selectivity, since changing to nitrogen or sulfur results in drop in selectivity. However, replacing $OCH_2C_6H_5$ group with methyl, i.e., the reagent prepared from 2-ethylapopinene [6] shows the same selectivity as observed for NB-Enantride. This suggests that the selectivity is due to steric effect rather than a complexation of lithium to the ether oxygen.

Substitution at the 11 position (Fig. 26.4) also results in drastic lower enantioselectivity. It has been found that lithium cation is vital to high selectivity. Changing the ion to any other species leads to a reagent with poorer selectivity and much slower reducing power. The absolute configuration of the product may be predicted using the mnemonic model (Fig. 26.4).

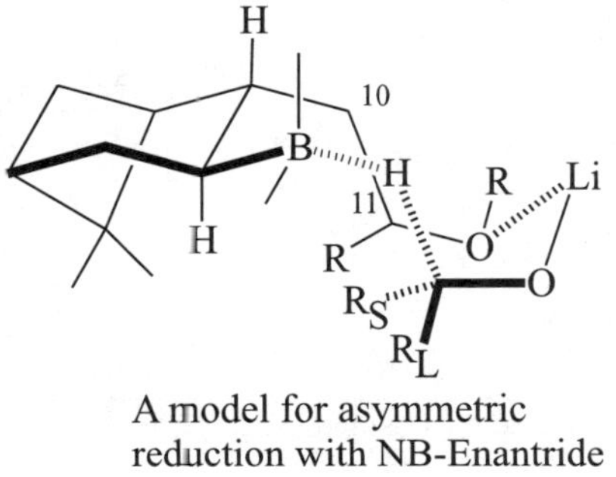

Fig. 26.4 A model for asymmetric reduction with NB-Enantride

26.5 Eapine-Hydride

Brown and coworkers [1] have reported that lithium *B-iso*-2-ethylapopinocampheyl-9-borabicyclo[3.3.1]nonyl hydride (Eapine-Hydride), prepared by hydroboration of 2-ethylapopinene with 9-BBN followed by treatment with *tert*-butyllithium, is as effective as NB-Enantride for the chiral reduction of prochiral ketones (Scheme 26.6; Table 26.21) [1].

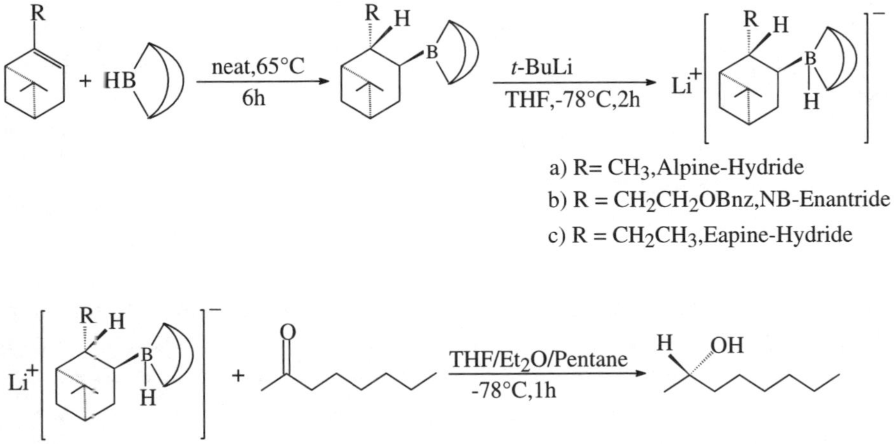

Scheme 26.6

Table 26.21 Reduction of representative ketones with chiral trialkylborohydride [1]

Class of ketone[a]	Ketone	% ee				
		Alpine-Hydride	NB-Enantride		Eapine-Hydride	
		−78 °C	−78 °C	−100 °C	−78 °C	−100 °C
I	2-Octanone	33	62	79	70	77
I	3-Methyl-2-butanone*	36	–	68	–	77
I	Acetylycyclohexane	27	65	80	70	80
II	2,2-Dimethyl-cyclopentanone*	01	0.5	–	07	–
III	Acetophenone*	20	63	70	56	61
V	2-Chloroacetophenone*	04	41	–	48	–
X	4-Phenyl-3-butyn-2-one*	05	–	10	05	–

Those marked with asterisks are the selected representatives of the class indicated.
[a] Brown HC, Park WS, Cho BT, Ramachandran PV (1987) J Org Chem 52:5406.

26.6 K-Glucoride

The synthesis of potassium 9-alkoxy-9-boratabicyclo[3.3.1]nonanes (K9-OR-9-BBNH) [1, 2] involves the reaction of 9-BBN with alcohols (Eq. 26.23), followed

(26.23)

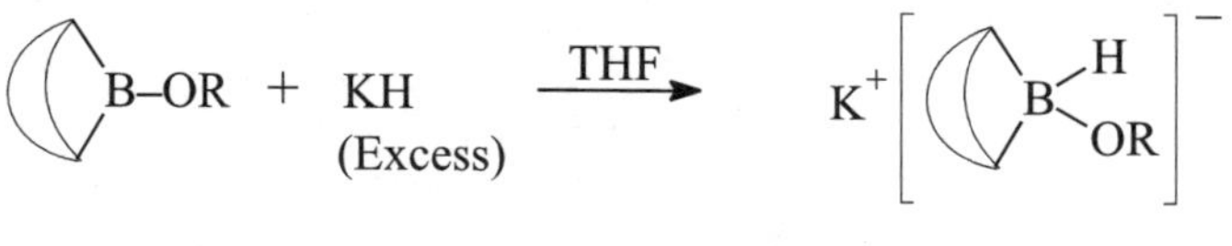

(26.24)

by the conversions of resulting borinic esters into the corresponding potassium dialkylmonoalkoxy borohydrides by treatment with excess potassium hydride (Eq. 26.24). Consequently, Brown and coworkers [3] have synthesized a series of chiral-9-alkoxy-9-borabicyclo[3.3.1]nonane derivatives, by the reaction of 9-BBN with several readily available chiral alcohols, such as (–)-*iso*-pinocampheol (**1**), (+)- menthol **2**), (–)-4-isocaranol (**3**), (+)-*trans*-2-methylcyclopentanol (**4**), and (–)-1,2,5,6-di-*O*-*iso*-propylidene-α-D-glucofuranose (**6**) and also a chiral borinic ester possessing a cyclic chiral dialkylboryl moiety, (+)-2-(cyclohexyloxy)-4,8-dimethyl-2-borabicyclo[3.3.1]nonane (**5a**) (Chart 26.8).

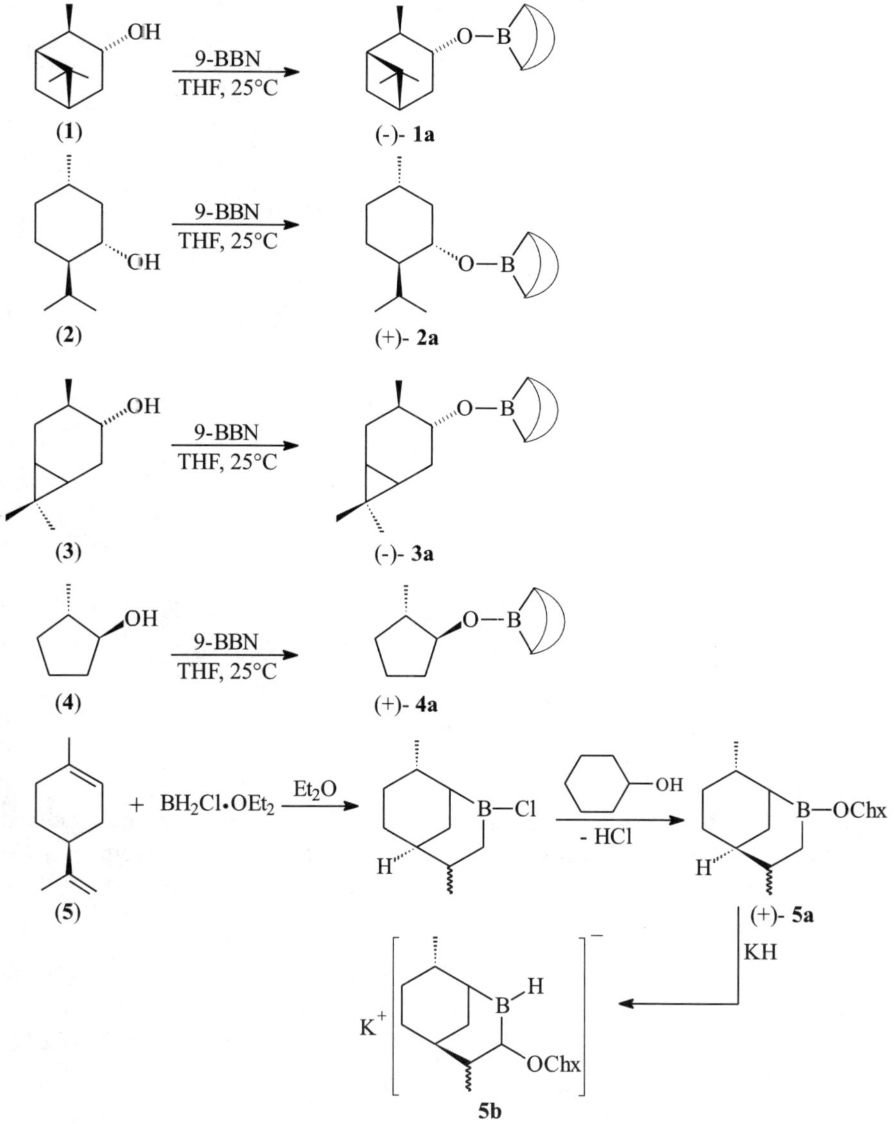

Chart 26.8

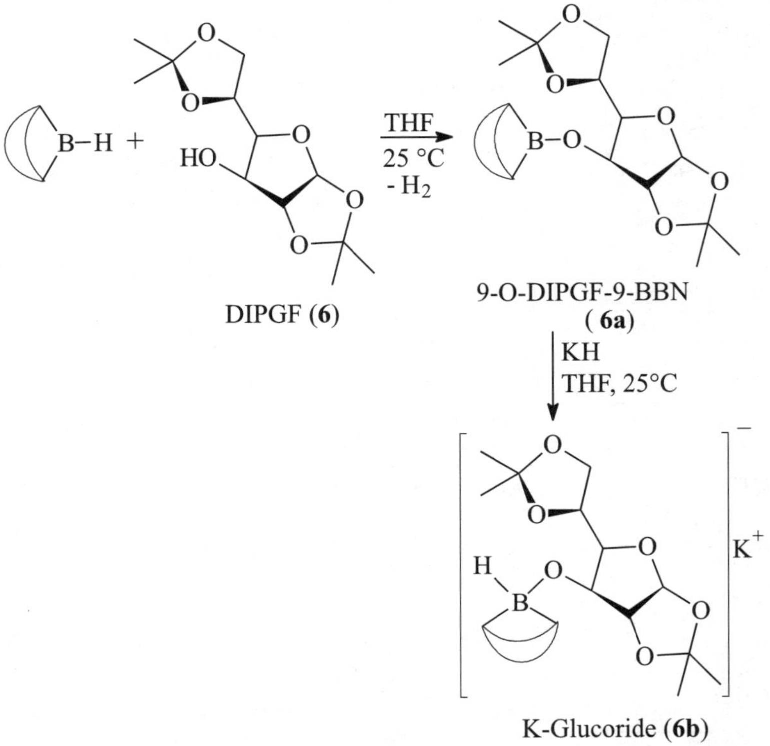

Chart 26.8 *(continued)*

These chiral borinic esters are readily converted into the corresponding chiral dialkylmonoalkoxyborohydrides by treatment with excess potassium hydride (Eq. 26.25) in THF at 25 °C with one exception of (+)-9-(menthyloxy)-9-BBN (9-*O*-Men-9-BBN),

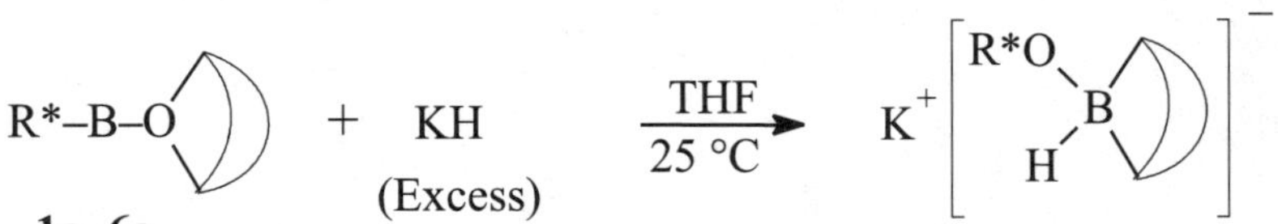

(26.25)

1a–6a

1b–6b

R* = Chiral alkyl

which requires 15 days at 65 °C (refluxing THF). These chiral dialkylmono alkoxyborohydrides are stable at 25 °C for several months. These reagents are tested against acetophenone and 2-methyl-2-butanone for chiral reductions (Table 26.22) [3].

Table 26.22 Asymmetric reduction of acetophenone and 3-methyl-2-butanone with chiral dialkylmonoalkoxyborohydride [3]

Chiral dialkylmono-alkoxy borohydrides	Acetophonone			3-Methyl-2-butanone		
	Time (h)	Yield (%)	Percentage *ee*	Time (h)	Yield (%)	Percentage *ee*
6b	24	98	78 (*R*)	6	96	39 (*R*)
1b	24	95	47 (*S*)	6	98	61 (*S*)
2b	24	90	12 (*S*)	6	98	40 (*R*)
3b	24	97	34 (*R*)	6	93	28 (*S*)
4b	24	98	26 (*R*)	6	91	37 (*R*)
5b	24	96	3 (*R*)	6	98	14 (*R*)

Among these asymmetric reducing agents, potassium 9-*O*-(1,2:5,6-di-*O*-isopropylidene-α-D-glucofuranosyl-9-BBNH (K-9-*O*-DIPGF-9-BBNH; K-glucoride; 6B) prepared [4] from 1,2:5,6-di-*O*-isopropylidene-α-D-glucofuranose (DIPGF) and 9-BBN followed by reaction with excess potassium hydride (Eq. 26.25) and K 9-*O*-IPC-9-BBNH (**1b**) provide the highest optical yield 78 and 61% *ee*, respectively (Table 26.22).

In addition, it is reported that sodium borohydride when modified with DIPGF in the presence of isobutyric acid gives 83% *ee*, in the reduction of propiophenone [5]. Consequently, K-glucoride is studied for asymmetric reduction and has found to provide very high optical yields (90–100% *ee*) in the reduction of α-keto esters and pivalophenone [4, 6]. The utility of K-glucoride is extended for stereoselective reduction of cyclic and bicyclic ketones (Table 26.23) [7], and it exhibits 94–99.7% stereoselectivities favoring the thermodynamically less stable alcohol. The high isomeric purity (99.7%) in case of 2-*tert*-butylcyclohexanone is due to large steric requirements [2] of glucoride moiety (Eq. 26.26).

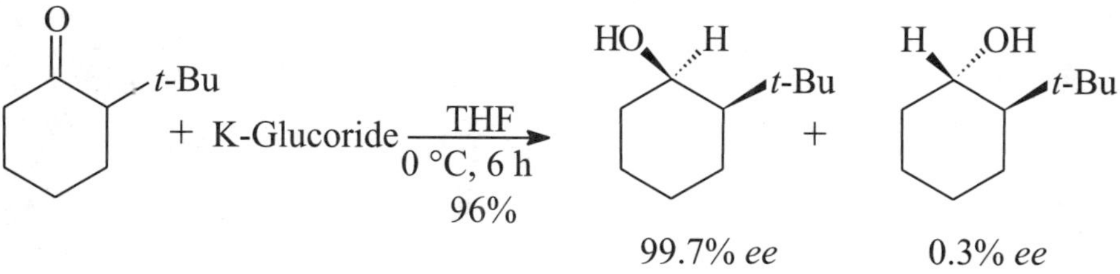

(26.26)

Table 26.23 Stereoselective reduction of representative cyclic and bicyclic ketones with K-glucoride in THF [7]

Ketone	Temp (°C)	Time (h)	Yield (%)	Less stable alcohol (%)
2-Methylcyclohexanone	0	3	98	97
	−78	6	97	98
2-Phenylcyclohexanone	0	3	97	95
	−78	6	96	96
2-*tert*-Butylcyclohexanone	0	6	96	99.7
4-*tert*-Butylcyclohexanone	0	3	98	92
	−78	6	98	94
Norcamphor	0	3	97	94
	−78	6	94	96
Camphor	0	24	90	96

The effect of reaction temperature on asymmetric reduction taking a propiophenone as a model reveals (Table 26.24) [7] that the maximum 92% *ee* is achieved at −78 °C, and at −100 °C the reaction is too slow to obtain the product alcohol.

Table 26.24 Effect of reaction temperature on the reduction of propiophenone in THF [7]

Temp (°C)	Time (h)	Yield (%)	Percentage *ee*	Configuration
0	6	97	76	*R*
−25	10	98	80	*R*
−78	16	93	92	*R*

The reduction of sterically hindered aliphatic ketones with K-glucoride (Table 26.25) [7] also proceeds with high asymmetric induction, and all alcohols are enriched in the (*R*)-enantiomers.

Table 26.25 Asymmetric reduction of representative aliphatic ketones with K-glucoride in THF [7]

Ketone	Temp (°C)	Time (h)	Yield (%)	Percentage *ee*	Absolute configuration
2-Butanone	−78	6	99	3	*R*
3-Methyl-2-butanone	−78	6	98	39	*R*
3,3-Dimethyl-2-butanone	−78	16	95	70	*R*
2-Octanone	−78	6	97	27	*R*
Cyclohexylmethyl ketone	−78	10	95	23	*R*
2,2-Dimethylcyclopentanone	−78	48	88	84	*R*
Spiro[4.4]nonan-1-one	−78	96	75	82	*R*
2,2-Dimethylcyclohexanone	−50	48	92	64	*R*

Similarly, asymmetric reduction of alkylaromatic ketones gives high optical yields (Table 26.26) [7]. It is noteworthy that the optical yields obtained by K-glucoride in the reduction of hindered ketones such as pivalophenone is 97–100% *ee* and 87% *ee* for isobutyrophenone are considerably higher than the values (44 and 71% *ee*, respectively) obtained with highly promising Binal-H reagent [8] (Eq. 26.27).

Table 26.26 Asymmetric reduction of representative alkyl aromatic ketones with K-glucoride in THF at –78 °C [7]

Ketone	Time (h)	Yield (%)	Percentage ee	Absolute configuration
Acetophenone	16	95	78	*R*
Propiophenone	16	93	92	*R*
Butyrophenone	20	95	87	*R*
Isobutyrophenone	36	96	87	*R*
Valerophenone	20	94	85.4	*R*
Pivalophenone	40	93	100	*R*
2′-Methylacetophenone	24	93	91	*R*

(26.27)

In the asymmetric reduction of α-ketoesters, the asymmetric induction depends on the steric inequality of two moieties attached to the carbonyl group [9]. The reduction of hindered α-ketoesters yields the corresponding α-hydroxyesters with optical purities approaching 100% *ee* (Eq. 26.28). This contrasts strongly with the 11% *ee* realized in the reduction of **7** (R= *i*-Pr) with neat Alpine-Borane [9b]. Alpine-Borane (from (+)-pinene) yields (*S*)-alcohols in the reduction of both alkylaromatic ketones and α-ketoesters. On the other hand K-glucoride yields (*R*)-alcohols with aralkylketones and (*S*)-alcohols with α-keto esters (Table 26.27) [7].

Table 26.27 Asymmetric reduction of representative α-ketoesters with K-glucoride in THF at −78 °C [7]

Ketone	Time (h)	Yield (%)	α-Hydroxyester Percentage ee	α-Hydroxyester Absolute configuration
Methyl pyruvate	6	80	86	*S*
Ethyl pyruvate	6	75	85	*S*
Isopropyl pyruvate	8	78	87	*S*
tert-Butyl pyruvate	8	72	79	*S*
Ethyl 2-oxobutanoate	6	80	90	*S*
Ethyl 2-oxopentanoate	6	81	111	*S*
Methyl 3-methyl-2-oxobutanoate	8	83	113	*S*
Ethyl 3-methyl-2-oxobutanoate	8	85	99	*S*
Methyl 3,3-dimethyl-2-oxobutanoate	10	85	113	*S*
Ethyl 3,3-dimethyl-2-oxobutanoate	10	87	98	*S*
Ethyl 4-methyl-2-oxopentanoate	6	83	93	*S*
Methyl benzoylformate	10	85	89	*S*
Ethyl benzoylformate	10	80	93	*S*
Isopropyl benzoylformate	10	83	92	*S*
Ethyl α-oxo-1-naphthaleneacetate	10	78	96	*S*

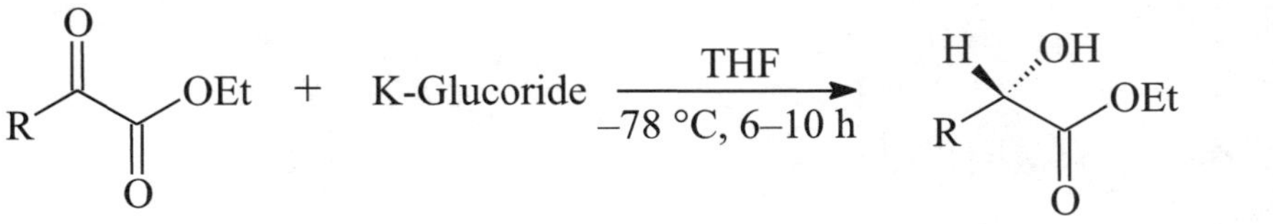

(7)

86–98% *ee* (*S*)-isomer

(26.28)

The asymmetric reduction of some other ketones with K-glucoride is summarized in Table 26.28 [7].

26.7 K-Xylide

K-Glucoride is inefficient for the reduction of prochiral unhindered aliphatic ketones. Hence, Cho and Chun [1] have synthesized various chiral borohydrides using monosaccharide units: potassium 9-*O*-(1,2:5,6-di-*O*-isopropylidene-α-D-allofuranosyl)-9-boratabicyclo[3.3.1]nonane from 1,2:5,6-di-*O*-isopropylidene-α-D-allofuranose (DIPAF) a C-3 epimer of DIPGF; potassium 9-*O*-(1,2:5,6-di-*O*-cyclohexylidene-α-D-glucofuranosyl)-9-boratabicyclo[3.3.1]nonane utilizing

Table 26.28 Asymmetric reduction of some other ketones with K-glucoride in THF [7]

Ketone	Temp (°C)	Time (h)	Yield (%)	Percentage ee	Absolute configuration
3,3-Diethylpentanone	0	60	88	25	–
1,1,1-Triphenylacetone	25	48	87	7	*S*
2,2,2-Triethylacetophenone	0	60	75	34	–
2,2,2-Triphenylacetophenone	25	48	85	4	*R*
2′,4′,6′-Trimethylacetophenone	–25	60	50	35	*R*
4-Chlorobenzophenone	–78	48	88	11.5	
Ethyl 2,2-dimethylacetoacetate	–78	8	80	43	
2-Acetylfuran	–78	12	98	103	*S*
2-Acetylthiophene	–78	12	97	42	*R*
3-Acetylpyridine	–78	12	97	70	*R*
2-Chloroacetophenone	–78	12	82	77	*S*
2,2,2-Trifluoroacetophenone	–78	12	95	48	*S*
trans-4-Phenyl-3-buten-2-one	–78	10	92	60	*R*
4-Phenyl-3-butyn-2-one	–78	10	95	58	–

1,2:5,6-di-*O*-cyclohexylidene-α-D-glucofuranose (DCHGF), bearing bulkier groups at the 4 position of the furanose ring as compared to that of DIPGF; and potassium 9-*O*-(1,2-*O*-isopropylidene-5-deoxy-α-D-xylofuranosyl)-9-boratabicyclo[3.3.1]nonane from 1,2-isopropylidene-5-deoxy-α-D-xylofuranose (IPDXF) derived from α-D-xylose, possessing a small methyl substituent at the 4 position of the furanose moiety (Chart 26.9).

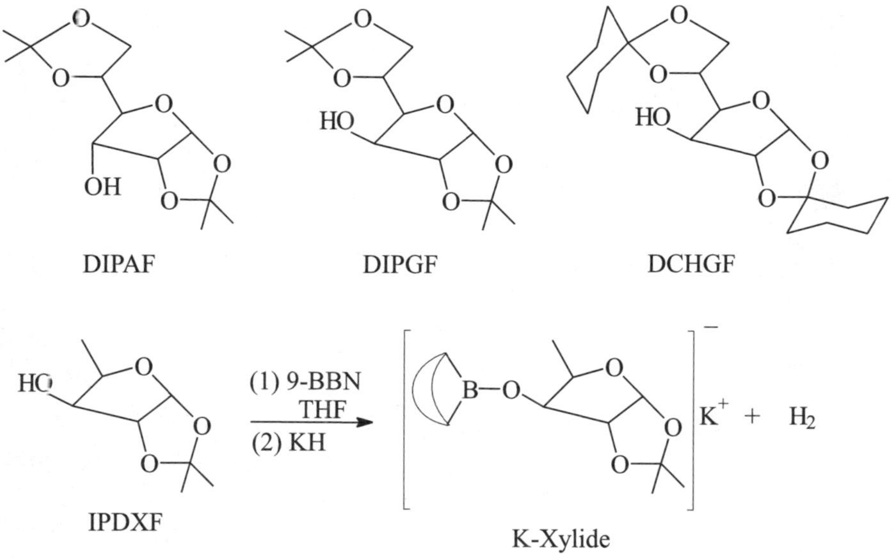

Chart 26.9

Borohydride of DIPAF provides low optical induction, 57% *ee* for acetophenone and 3% for 3-methyl-2-butanone and product alcohol with opposite configurations (*S*-isomers) as compared to (*R*)-isomers obtained by K-glucoride. Borohydride of DCHGF gives almost the same optical induction as obtained by K-glucoride [2] against the representative classes of ketones [1, 3]. These results reveal that steric effects at C-4 position of furanose ring moiety in K-glucoride and borohydride of DCHGF do not play a significant role in optical induction. On the other hand, borohydride of IPDXF possessing a small substituent as methyl group at the 4 position of furanose ring moiety shows high optical induction for asymmetric reduction of some prochiral ketones. This stable chiral borohydride, potassium 9-*O*-(1,2-*O*-isopropylidene-5-deoxy-α-D-xylofuranosyl)-9-boratabicyclo[3.3.1]nonane is synthesized from 9-BBN and a monosaccharide, IPDXF. IPDXF is prepared by reductive cleavage of 1,2-isopropylidene-3,5-anhydro-α-D-xylofuranose obtained from α-D-xylose with lithium aluminum hydride [4]. The reaction of 9-BBN and IPDXF yields the corresponding borinic ester, which on treatment with excess potassium hydride gives the desired potassium 9-*O*-(1,2-*O*-isoporpylidene-5-deoxy-α-D-xylofuranosyl)-9-boratabicyclo[3.3.1]nonane, named for general use as K-xylide.

The reagent is highly effective for optical induction in the reduction of hindered prochiral aliphatic ketones (Eqs. 26.29, 26.30; Table 26.29) [1].

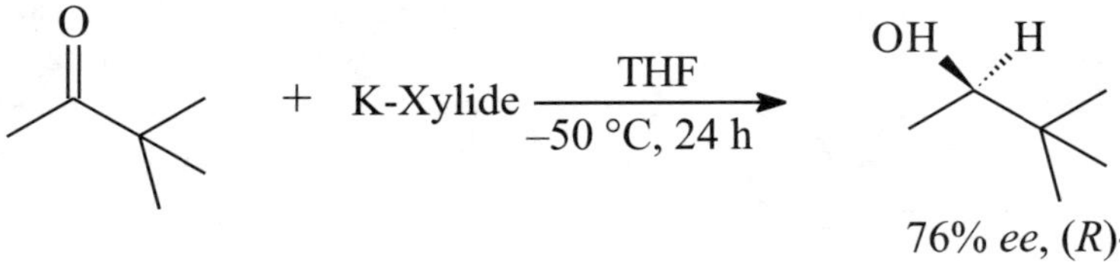

(26.29)

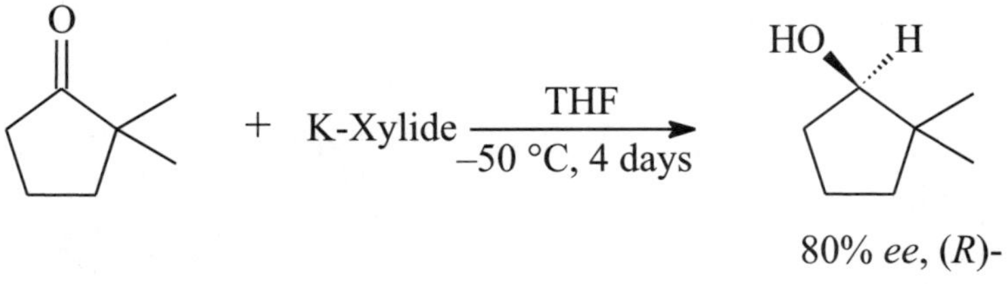

(26.30)

The reagent K-xylide also reduces alkyl aromatic ketones smoothly and in high yields. (Eqs. 26.31, 26.32; Table 26.30) [1].

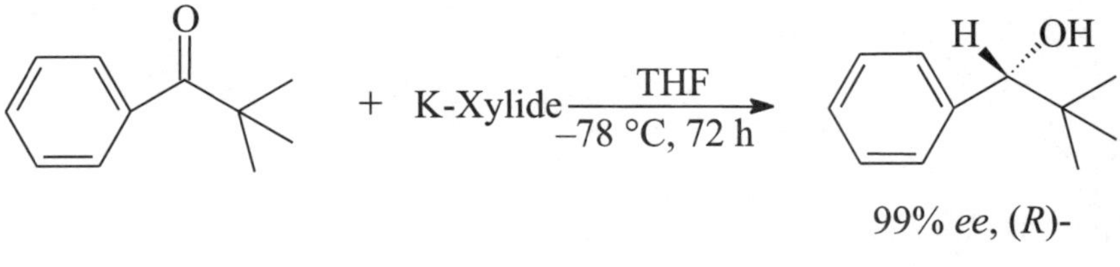

(26.31)

Table 26.29 Asymmetric reduction of representative aliphatic ketones with K-xylide [1]

Ketones	K-Xylide				Percentage *ee*	
	Time (h)	Yield (%)	Percentage *ee*	Configu-ration	K-Glucoride	Boro-hydride of DCHGF
2-Heptanone	20	98	36	*R*	26	38
3-Methyl-2-butanone	20	97	46	*R*	39	58
4-Methyl-2-pentanone	36	98	65	*R*	–	51
3,3-Dimethyl-2-butanone	24	98	76	*R*	70	58
2,2-Dimethyl-3-pentanone	4 days	95	27	*R*	–	–
2,2-Dimethyl cyclopentanone	4 days	92	80	*R*	84	80

Table 26.30 Asymmetric reduction of representative alkyl aromatic (RCOAr) ketones with K-xylide in THF at –78 °C [1]

	K-Xylide				Percentage *ee*	
Ketones	Time (h)	Yield (%)	Percentage *ee*	Configu-ration	K-Glucoride	Boro-hydride of DCHGF
Acetophenone	24	96	70	*R*	78	70
Propiophenone	30	97	86	*R*	92	
Butyrophenone	36	98	82	*R*	87	
Isobutyrophenone	48	98	89	*R*	87	
Pivalophenone	72	95	99	*R*	100	
2′-Methylacetophenone	24	95	93	*R*	91	

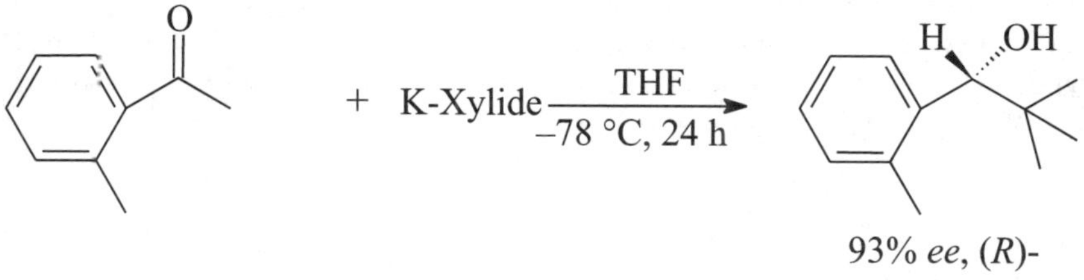

(26.32)

The reduction of alkylaromatic ketone again reveals that steric effect at position 4 of the furanose ring moiety in K-xylide or K-glucoride is not a significant factor for the asymmetric induction.

The asymmetric reduction of functionalized ketones with K-xylide also affords comparable results with that of K-glucoride or borohydride of DCHGF (Eq. 26.33; Table 26.31) [1].

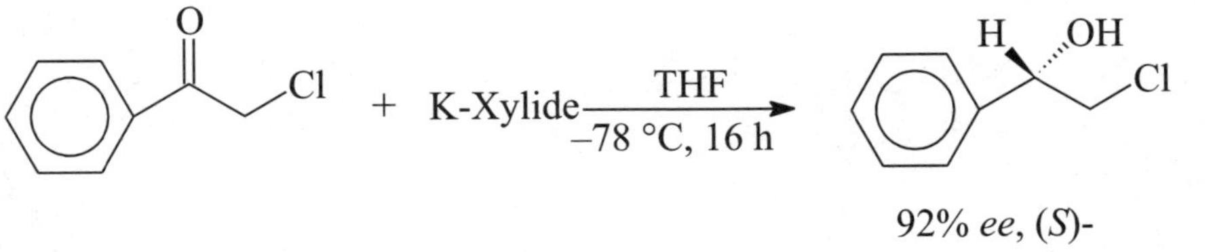

(26.33)

Table 26.31 Asymmetric reduction of representative functionalized ketones with K-Xylide in THF at –78 °C [1]

Ketones	K-Xylide				Percentage *ee*	
	Time (h)	Yield (%)	Percentage *ee*	Configu-ration	K-Glucoride	Boro-hydride of DCHGF
2-Chloroacetophenone	16	99	92	*S*	77	78
Methylbenzoylformate	50	96	60	*S*	92	92
3-Acetylpyridine	36	98	62	*R*	70	55
4-Phenyl-3-butyn-2-one	22	98	52	*R*	61	62

The opposite configuration notation of the product alcohols are due to the CIP sequence rules. The directions of the asymmetric induction are consistent to afford (*R*)- isomers by *si* facial addition of hydride for both aliphatic and alkylaromatic ketones. However, for α-haloketone and α-ketoester, *re* facial addition of hydride provides (*S*)-isomeric alcohols. The direct comparison of K-xylide asymmetric reductions reveals that the reagent resembles very closely to that of K-glucoride.

26.8 K9-*O*Thx-9-BBNH

Potassium 9-(2,3-dimethyl-2-butoxy)-9-boratabicyclo[3.3.1]nonane (K9-*O*Thx-9-BBNH) prepared [1] from 9-BBN and 2,3-dimethylbutan-2-ol (Scheme 26.7) is very stable, as no disproportionation is observed over more than one year when its solution in THF is stored under nitrogen.

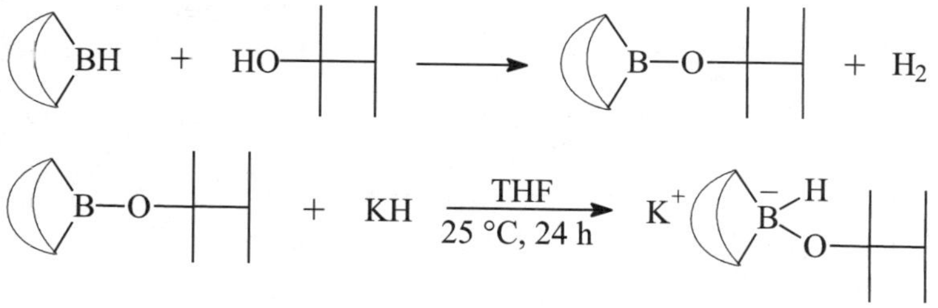

Scheme 26.7

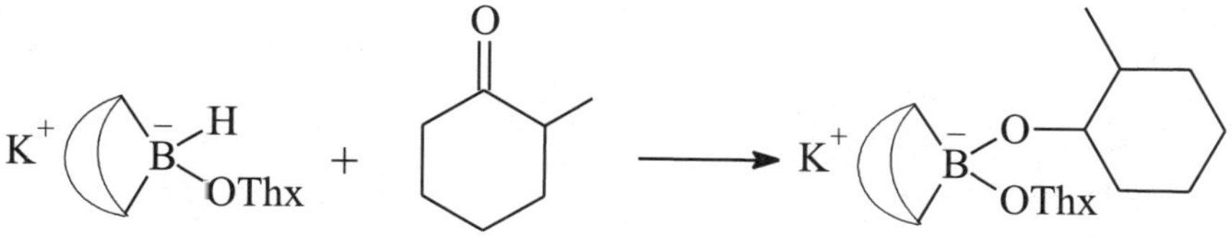

98.5 *cis*

(26.34)

The reagent achieves high stereoselective reduction (Eq. 26.34) of cyclic ketones.

The recovery of the product alcohol is very simple as complex K 9-*O*Thx-9-BBNOR is readily converted into the hydroxy ate complex by treatment with a slight excess over theoretical amount of water (Eq. 26.35).

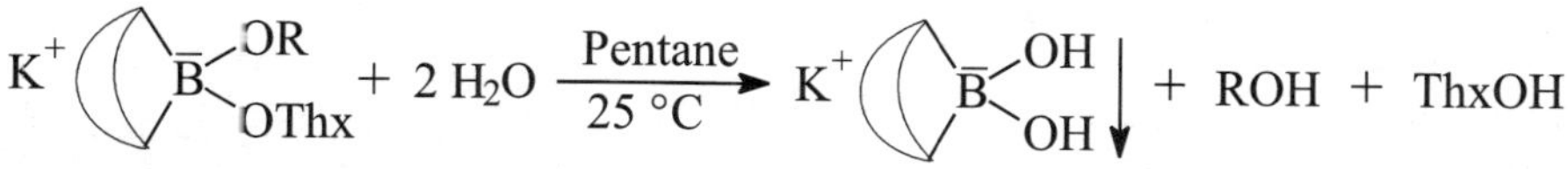

(26.35)

The comparison data of K9-*O*Thx-9-BBNH [1], Li-*s*-Bu_3BH [2], and $LiSia_3BH$ [1] are presented in Table 26.32 [1].

26.9 Lithium Di-*n*-Butyl Ate Complex of 9-BBN

The lithium di-*n*-butyl ate complex of 9-BBN, prepared [1] from *B*-*n*-butyl-9-BBN and *n*-butyllithium exhibits high stereo-, chemo-, and regioselectivities in the reduction of cyclohexanones. Consequently, both *cis*- and *trans*-4-methylcyclohexanols are independently obtained from 4-methylcyclohexanone by a mere change in the additives (Chart 26.10; Table 26.33) [2].

Table 26.32 The comparison data of reduction of cyclic ketones with K9-*O*Thx-9BBNH, Li-*s*-Bu_3BH, and $LiSia_3BH$ in THF at 0 °C [1]

	Ratio of less stable isomer (%)		
Ketone	K9-*O*Thx-9-BBNH[a,b]	Li-*s*-Bu_3BH	$LiSia_3BH$
Cyclohexanones			
2-methyl-	98.5	99.3	99.4
3-methyl-	90	85	98
4-methyl-	85.5	80.5	93
4-*tert*-butyl-	87	87.5	96.5
3,3,5-trimethyl-	>99.9	99.8	99
Norcamphor	95	99.6	99
Camphor	97.5	99.6	>99.9

[a] A 2:1 ratio for reagent:ketone is utilized.
[b] The yields of alcohols are quantitative.

Table 26.33 Stereoselective reduction of representative cyclic ketones with lithium di-*n*-butyl ate complex of 9-BBN [2]

Ketone	Entry	Additive	Ketone 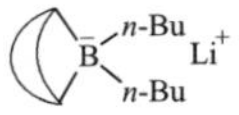 additive	Time (h)	Conversion (%)	Major isomer (%)
2-Methylcyclohexanone	1	None	1:1	0.5	–	–
	2	None	1:1	18	35	*cis* (92)
	3	MeOLi	1:1:1	3	59	*trans* (89)
3-Methylcyclohexanone	4	None	1:1	0.5	62	*trans* (91)
	5	MeOH	1:2:1	3	100	*trans* (86)
4-Methylcyclohexanone	6	None	1:1	0.5	63	*cis* (77)
	7	None	1:1	18	77	*cis* (48)
	8	None	2:1	0.5	54	*cis* (47)
	9	MeOH	1:2:1	3	91	*cis* (84)
	10	MeOLi	1:1:1	5	94	*trans* (90)
	11	MeOLi	1:1:1	0.5	96	*trans* (78)
	12	LiBr	1:1:1	0.5	83	*trans* (79)
4-*tert*-Butylcyclohexanone	13	MeOH	1:2:1	1.5	100	*cis* (88)
	14	MeOLi	1:1:1	3	89	*trans* (92)

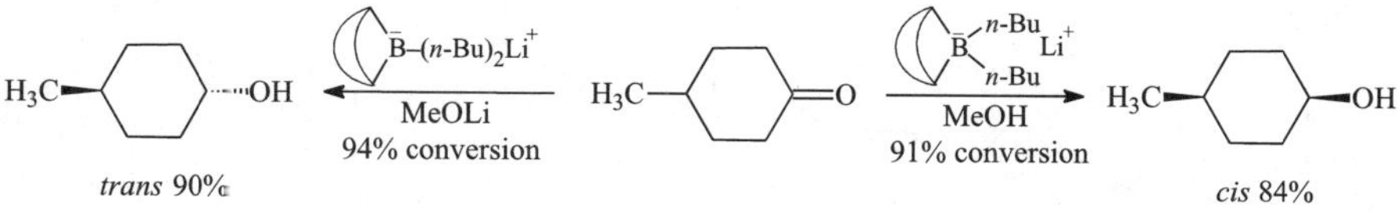

Chart 26.10

As is apparent (Table 26.33), the reduction of methylcyclohexanones (entries 1, 2, 4, 6, and 7) requires longer reaction times to achieve satisfactory conversions. This results in the equilibration of epimeric alcohols, leading to higher formation of more stable alcohols (entries 6 and 7). The isomerization is because of the intermolecular hydride transfer, which is similar to Meerwein–Pondorf–Verley reduction of ketones with aluminum alkoxides (Scheme 26.8).

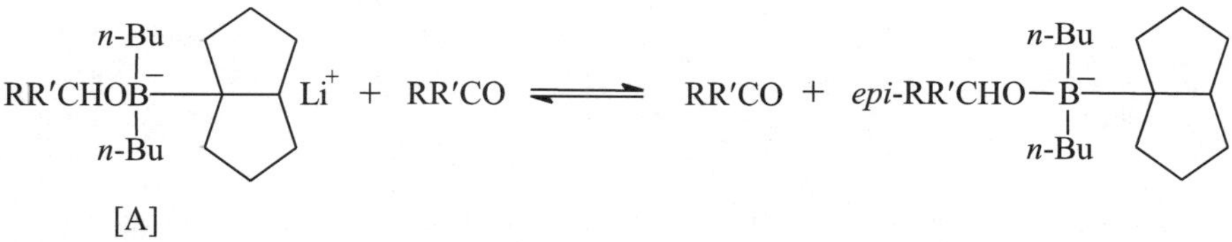

Scheme 26.8

It is possible to suppress the equilibration by addition of methanol, which assists in the decomposition of [A] and prevents hydride transfer. In contrast, MeOLi or LiBr accelerates the isomerization to afford the more stable alcohols.

26.10 Li 9-BBNH

The lithium 9-boratabicyclo[3.3.1]nonane (Li 9-BBNH) is prepared [1] by refluxing 1 equiv of 9-BBN with excess of finely divided lithium hydride. The reagent reduces [2] cyclic ketones, and the hydride adds from the less hindered side. For example, camphor on reduction with Li 9-BBNH affords the less stable *exo* isomer isoborneol in 91% yield (Eq. 26.36).

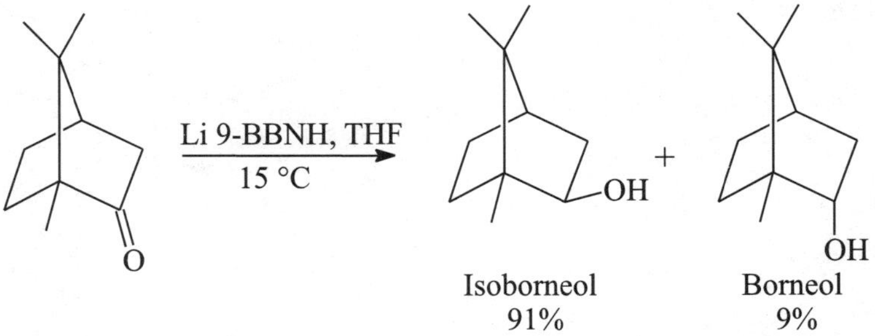

(26.36)

The results are summarized in Table 26.34 [2].

Table 26.34 Reduction of cyclic and bicyclic ketones with Li 9-BBNH in THF at 15±2 °C [2]

Ketone	Time (h)	Total yield (%)	Isomeric ratio
2-Methylcyclohexanone	1	94	*cis:trans*, 63:37
4-*tert*-Butylcyclohexanone	1	100	*cis:trans*, 24:76
Norcamphor	1	100	*endo:exo*, 94:6
Camphor	1	100	*exo:endo*, 91:9

The results reveal that Li 9-BBNH attacks more from the less hindered side of the carbonyl group than 9-BBN [3] (*vide supra*).

26.11 Comparative Data of Asymmetric Reducing Agents

Brown and coworkers [1] have critically examined the data of 20 asymmetric reducing agents *viz.* *B*-Ipc-9-BBN (Alpine-Borane), THF [2]; *B*-Ipc-9-BBN, neat [3]; *B*-Ipc-9-BBN-6kbr [4]; NB-Enantrane [5]; Ipc_2BCl [6]; BH_3-AMDPB (2:1) [7]; (*R*,*R*)-or (*S*,*S*)-2,5-dimethylborolane [8]; NB-Enantride [9]; $LiBH_4$-DBC-*t*-BuOH [10]; $NaBH_4$-IBA-DIPGF [11]; K-Glucoride [12]; $LiAlH_4$-Darvon Alc [13]; $LiAlH_4$-MEP-ArOH [14]; $LiAlH_4$-diamine [15]; $LiAlH_4$-aminobutanol [16]; Binal-H [17]; $LiAlH_4$-DBP-EtOH [18]; $LiAlH_4$-MEP-NEA [19]; $LiAlH_4$-MEP-EAP [20]; and TBADH enzyme [21]. The assembled available data of these 20 reagents are divided for the asymmetric reduction on nine distinct classes of ketones. The comparison is made by selecting one representative ketone for each class. Except for acetophenone, the corresponding data are not available for most of the representative ketones. Consequently, Brown and coworkers [1] have conducted the asymmetric reduction of nine representative ketones with six of the more promising reagents, namely Alpine-Borane, neat (A); Ipc_2BCl (B); BH_3-AMDPB (2:1) (C); NB-Enantride (D); K-glucoride (E); and Binal-H (F). The results are presented in Table 26.35 [1].

Table 26.35 Asymmetric reduction of representative ketones [1]

Class	Ketone	Percentage *ee* of alcohol products					
		A	B	C	D	E	F
Acyclic	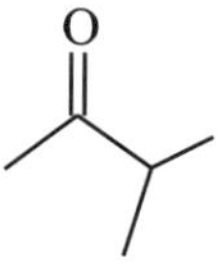	62	32	60	68	36	78
Cyclic	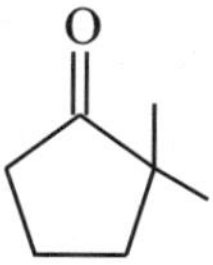	19.6	98	96	1	84	11
Aralkyl	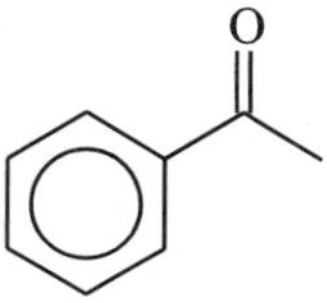	87	98	94	70	78	95
Heterocyclic	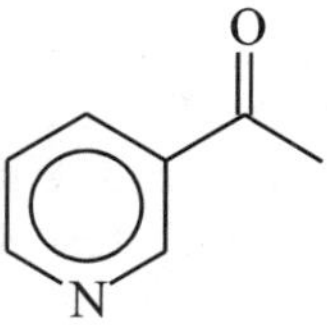	93	92	73	8	70	–
α-Halo	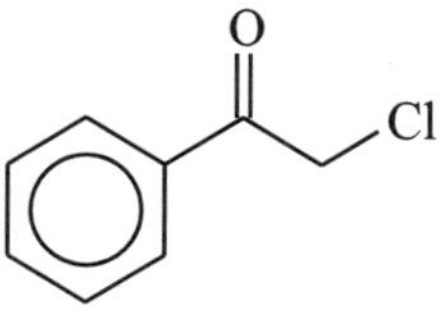	96	95	96	41	77	95
α-Ketoester	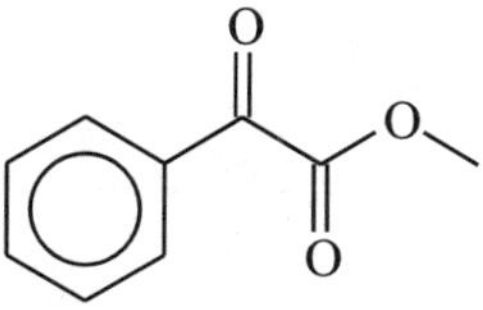	90	70	25	33	92	24
Acyclic conjugated ketone	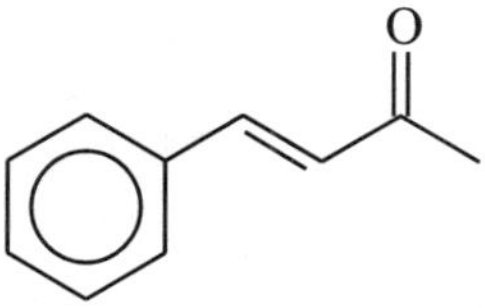	56	12	6	13	60	70

Table 26.35 *(continued)* Asymmetric reduction of representative ketones [1]

		Percentage *ee* of alcohol products					
Class	Ketone	A	B	C	D	E	F
Cyclic conjugated ketone	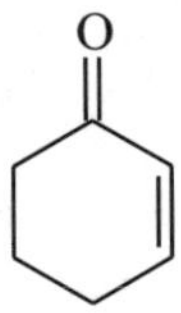	30.4	36	35	–	–	–
Conjugated ynone	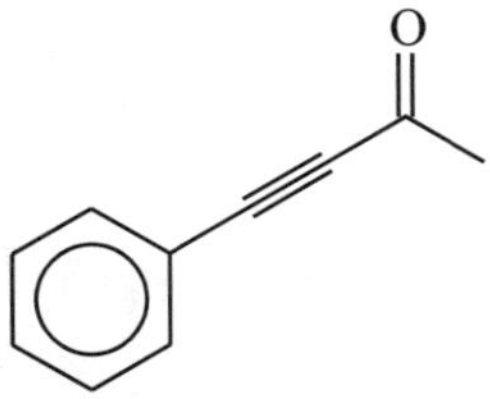	83	21	7	10	61	89

The optical yields of the alcoholic products are analyzed either by measuring the rotation of the isolated material (where the rotation for 100% material appeared well established) or by capillary GC analysis of the corresponding (*R*)-(+)-α-methoxy-α-(trifluoromethyl) phenylacetic acid (MTPA) [22], or, also commonly the (–)-menthylchloroformate (MCF) derivatives [23].

On the basis of data of Table 26.35, and the available data from the literature, the selected preferred asymmetric reducing agents for 10 structurally different classes of ketones are suggested (Table 26.36) for each class of compounds.

Table 26.36 Preferred asymmetric reducing agents [1]

Classes of ketones		Preferred reagents (in order of effectiveness)
I. Acyclic	a. Unhindered	2,5-Dimethylborolane *B*-Ipc-9-BBN, 6 kbar TBADH enzyme NB-Enantride
	b. Hindered	2,5-dimethylborolane Ipc_2BCl BH_3-AMDPB (2:1)
II. Cyclic		Ipc_2BCl BH_3-AMDPB (2:1) K-Glucoride

Table 26.36 *(continued)* Preferred asymmetric reducing agents [1]

Classes of ketones		Preferred reagents (in order of effectiveness)
III. Aralkyl	a. Unhindered	*B*-Ipc-9-BBN, 6 kbar Ipc_2BCl $LiAlH_4$-DBP-EtOH Binal-H $LiAlH_4$-diamine BH_3-AMDPB (2:1)
	b. Hindered	K-Glucoride $LiAlH_4$-diamine Ipc_2BCl
IV. Heterocyclic		*B*-Ipc-9-BBN, 6 kbar *B*-Ipc-9-BBN, neat Ipc_2BCl
V. α-Halo		*B*-Ipc-9-BBN, neat BH_3-AMDPB (2:1) Ipc_2BCl Binal-H
VI. α-keto esters		K-Glucoride *B*-Ipc-9-BBN, neat
VII. β-Ketoesters		$LiBH_4$-DBC-*t*-BuOH Ipc_2BCl
VIII. Acyclic conjugated enones		Binal-H $LiAlH_4$-MEP-NEA $LiBH_4$-DBC-*t*-BuOH $LiAlH_4$-aminobutanol
IX. Cyclic conjugated enones		$LiAlH_4$-aminobutanol $LiAlH_4$-MEP-EAP
X. Conjugated ynones		Binal-H NB-Enantrane $LiAlH_4$-MEP-ArOH *B*-Ipc-9-BBN, THF

References

1. (a) Le Bel JA (1874) Bull Soc Chim Fr 22:337, 11; Le Bel JA (1894) Bull Soc Chim Fr 22:295; (b) van't Hoff JH (1975) La Chimie dans L'Espace. Bazendijk, Rotterdan, The Netherlands, p 13–14; Bull Soc Chim Fr 23:295. English translations of both Le Bel and van't Hoff's work are available: Richardson GM (ed) (1901) The foundation of stereochemistry. American Book, New York; Benfey OT (ed) (1963) Classics in theory of chemical combinations. In: Classics of science, vol. 1. Dover, New York, reprinted (1981) by Krieger, Malabar, Florida
2. Fischer E (1984) Chem Ber 27:3231

3. Thalidomide is an excellent example of drug whose *S*-isomer possesses teratogenic activity, while the *R*-isomer is potent mutagen. Blaschke G, Kraft HP, Fickentscher K, Koehler F (1979) Arzneim Forsch 29:1640; Blaschke G, Kraft HP, Markgraf H (1980) Chem Ber 113,:2318
4. (a) Morrison JD, Mosher HS (1971) Asymmetric organic reactions. Prentice Hall, New York; (b) Valentine D Jr, Scott JW (1978) Synthesis 329; (c) ApSimon JW, Seguin RP (1979) Tetrahedron 35:2797; (d) Kagan HB, Fiaud JC (1978) Top Stereochem 10:175; (e) Brown HC, Jadhav PK, Mandal AK (1981) Tetrahedron 37:3547; (f) Corey EJ, Guzman-Perez A (1998) Angew Chem Int Ed 37:37
5. Brown HC, Pai GG (1985) J Org Chem 50:1384
6. Mosher HS, Morrison JD (1983) Science 221:1013. For an extensive review of the asymmetric synthesis, see Morrison, JD (ed) Asymmetric synthesis, vols. 1–4. Academic, New York; (a) Imai T, Tamura T, Yamamuro A, Sato T, Wollmann TA, Kennedy RM, Masamune S (1986) J Am Chem Soc 108:7402; (b) Corey EJ, Bakshi RK, Shibata S (1987) ibid. 109:5551; (c) Corey EJ, Bakshi RK, Shibata S, Chen C-P, Singh, V K (1987) ibid. 109:7925. (d) Itsuno S, Nakano M, Miyazaki K, Masuda H, Ito K, Hirao A, Nakahama S (1985) J Chem Soc Perkin Trans I 2039; (e) Itsuno S, Nakano M, Ito K, Hirao A, Owa M, Kanda N, Nadahama S (1985) ibid. 2615. (f) Brown HC, Chandrasekharan J, Ramachandran PV (1988) J Am Chem Soc 110:1539. (g) Brown HC, Cho BT, Park WS (1988) J Org Chem 53:1231 and references within these papers; (h) For a comparison of relative effectiveness of various reducing agents, see Brown HC, Park WS, Cho BT, Ramachandran PV (1987) J Org Chem 52:5406; (i) Mikhailov BM, Bubnov YN, Kiselev VG (1966) J Gen Chem USSR 36:65; (j) Corey EJ, Helal CJ (1998) Angew Chem Int Ed 37:1986
7. Mikhailov B, Bobnov YN, Kiselev VG (1966) J Gen USSR 36:65; Midland MM, Zderic SA (1982) J Am Chem Soc 104:525
8. Brown HC, Subba Rao BC (1956) J Am Chem Soc 78:2582. For reviews of the uses of organoboranes in synthesis, see (b) Brown HC (1962) Hydroboration. Benjamin, New York; (c) Brown HC (1972) Boranes in organic chemistry. Cornell University, Ithaca, New York. (d) Brown HC, Kramer GW, Levy AB, Midland MM (1975) Organic synthesis via boranes. Wiley InterScience, New York

Section 26.1.1

1. Arigoni D, Eliel EL (1969) In: Eliel EL, Allinger NL (eds) Topics in stereochemistry, vol 4. Wiley InterScience, New York, pp 127–244; Verbit L (1970) In: Streitweiser R, Taft RW (eds) Progress in physical organic chemistry, vol. 7. Wiley InterScience, New York, p 51–127; Lau KSY, Wong PK, Stille JK (1976) J Am Chem Soc 98 (1976) 5832; Caspi E, Eck CR (1977) J Org Chem 42:767 references cited therein
2. (a) Althouse VE, Feigl DM, Sanderson WA, Mosher HS (1966) J Am Chem Soc 88:3595; (b) Varma KR, Caspi E (1968) Tetrahedron 24:6365; (c) Wolfe S, Rauk A (1966) Can J Chem 44:2591; (d) Reich CJ, Sullivan GR, Mosher HS (1973) Tetrahedron Lett 1505; (e) Streitwieser A Jr, Wolfe JR Jr, Shaeffer WD (1959) Tetrahedron 6:338; (f) Gerlach H (1966) Helv Chim Acta 49:2481;
3. Midland MM, Tramontano A, Zderic SA (1978) J Organomet Chem 156:203
4. Midland MM, Tramontano A, Zderic SA (1977) J Organomet Chem 134:C17
5. Midland MM, Tramontano A, Zderic SA (1977) J Am Chem Soc 99:5211

6. Burgstahler AW, Walker DE Jr, Kuebrich JP, Schowen RL (1972) J Org Chem 37:1272. Also see Chancellor T, Quill M, Bergbreiter DE, Newcolmb MV (1978) ibid. 43:1245; Seebeck D, Erickson BW, Singh L (1966) ibid. 31:4303
7. Krishnamurty S, Brown HC (1977) J Org Chem 42:1197
8. Midland MM, Greer S, Tramontano A, Zderic SA (1979) J Am Chem Soc 101:2352
9. *B*-3-Pinanyl-9-borabicyclo[3.3.1]nonane is commercially available from Aldrich Chemical Co. under the trademark Alpine-Borane.
10. (a) Brown HC, Yoon MM (1977) Isr J Chem 15:12.; (b) Brown HC, Jadhav PK, Desai MC (1982) J Org Chem 47:4583
11. Midland MM, Tramontano A (1978) J Org Chem 43:1470
12. Midland MM, Petre JE, Zderic SA (1979) J Organomet Chem 182:653
13. Parry RJ, Trainor DA (1978) J Am Chem Soc 100:5243

Section 26.1.2

1. (a) Midland MM, Tramontano A, Zderic SA (1977) J Am Chem Soc 99:5211; (b) Midland MM, Greer S, Tramontano A, Zderic SA (1979) ibid. 101:2352; (c) Midland MM, McDowell DD, Hatch RL, Tramontano A (1980) ibid. 102:867; (d) Midland MM, Tramontano A, Kazubski A, Graham RS, Tsai DJS, Cardin DB (1984) Tetrahedron 40:1371
2. (a) Midland MM, Petre JE, Zderic SA, Kazubski A (1982) J Am Chem Soc 104:528; (b) Midland MM, Tramontano A, Zderic SA (1977) J Organomet Chem C17:134; Midland MM, Tramontano A, Zderic SA (1978) ibid. 156:203
3. Brown HC, Pai GG (1985) J Org Chem 50:1384
4. (a) Brown HC, Pai GG (1983) J Org Chem 48:1784; (b) Brown HC, Pai GG, Jadhav PK (1984) J Am Chem Soc 106:1531; (c) Midland MM, Lee PE (1985) J Org Chem 50:3237
5. Brown HC, Pai GG (1982) J Org Chem 47:1606
6. Midland MM, Zderic SA (1982) J Am Chem Soc 104:525
7. Midland MM, McLoughlin JI, (1984) J Org Chem 49:1316
8. Midland MM, McLoughlin JI, Gabriel J (1989) J Org Chem 54:159
9. Krishnamurthy, S, Brown HC (1977) J Org Chem 42:1197
10. Ranieri RL, McLaughlin JL (1977) Lloydia 40:173
11. (a) Cohen N, Lopresti RJ, Neuekom C, Saucy G (1980) J Org Chem 45:582; (b) Brinkmeyer RS, Kapoor VM (1977) J Am Chem Soc 99:8339; (c) Vigneron JP, Bloy V (1979) Tetrahedron Lett 2683; (d) Mukaiyama T, Suzuki K, Soai K, Sato T (1979) Chem Lett 447; (e) Noyori R, Nishizawa M, Yamada M (1981) Tetrahedron Lett 22:247
12. (a) Overman LE, Bell KL (1981) J Am Chem Soc 103:1851; (b) Pirkle WH, Boeder CW (1978) J Org Chem 43:2091; (c) Fried J, Sih JC, Lin CH, Dalven P, Cooper GF (1972) J Am Chem Soc 94:4343; (d) Fried J, Lin CH (1973) J Med Chem 16:429; (e) Partridge JJ, Chadha, NK, Uskokovic MR (1973) J Am Chem Soc 95:7171; (f) Fried J, Sih JC (1973) Tetrahedron Lett 3899; (g) Johnson WS, Frei B, Gopalan AS (1981) J Org Chem 46:1512; (h) Chan K, Specian AC Jr, Saucy G (1978) ibid. 43:3435; (i) Stork G, Poinier JM (1983) J Am Chem Soc 105:1073
13. Midland MM, McDowell DC, Hatch RL, Tramontano A (1980) J Am Chem Soc 102:867
14. (a) Midland MM, Tramontano A, Kuzubski A, Graham RS, Tsai DJS, Cardin DB (1984) Tetrahedron 40:1371
15. Midland MM, Tramontano A (1978) J Org Chem 43:1470

16. Midland MM, Halterman RL, Brown CA, Yamaichi A (1981) Tetrahedron Lett 22:4171
17. (a) Cohen N, Lopresti RJ, Neukom C, Saucy G (1980) J Org Chem 45:582, and refs therein; (b) Johnson WS, Werthemann L, Bartlett WR, Brocksom TJ, Li TT, Faulkner DJ, Peterson MR (1970) J Am Chem Soc 92:741; (c) Hirano Y, Djerassi C (1982) J Org Chem 47:2420
18. Sardina FJ, Mourino A, Castedo L (1983) Tetrahedron Lett 24:4477
19. Midland MM, Kwon YC (1984) Tetrahedron Lett 25:5981
20. Midland MM, McDowell DC, Hatch RL, Tramontano A (1980) J Am Chem Soc 102:867
21. Satoh T, Suzuki S, Suzuki Y, Miyaji Y, Imai Z (1969) Tetrahedron Lett 4555

Section 26.2

1. (a) Midland MM, Kazubski A (1982) J Org Chem 47:2814; (b) Midland MM, Tramontano A, Kazubski A, Graham RS, Tsai DJS, Cardin DB (1984) Tetrahedron 40:1371
2. NB-Enantrane is a trademark of Aldrich Chemical Co. Nopol is also available from Aldrich Chemical Co.
3. Midland MM, McDowell DC, Hatch RL, Tramontano A (1980) J Am Chem Soc 102:867

Section 26.3

1. Brown HC, Ramachandran PV, Weissman SA, Swaminathan S (1990) J Org Chem 55:6328
2. Brown HC, Ramachandran PV (1989) J Org Chem 54:4504
3. Brown HC, Pai GG (1985) J Org Chem 50:1384
4. Brown HC, Randad RS, Bhat KS, Zaidlewicz M, Weissman SA, Jadhav PK, Perumal PT (1988) J Org Chem 53:5513
5. Johnson CR, Dutra CR (1973) J Am Chem Soc 95:7777

Section 26.4

1. Krishnamurthy S, Vogel F, Brown HC (1977) J Org Chem 42:2534
2. Midland MM, Kazubski A (1982) ibid. 47:2495
3. The reagent is available from Aldrich under the trademark NB-Enantride
4. Brown HC, Pai GG (1982) J Org Chem 47:1606
5. Midland MM, Kazubski A, Woodling RE (1991) J Org Chem 56:1068
6. Brown HC, Randad RS, Bhat KS, Zaidlewicz M, Weissman SA, Jadhav PK, Perumal PT (1988) J Org Chem 53:5513

Section 26.5

1. Ramachandran PV, Brown HC, Swaminathan S (1990) Tetrahedron Asym 1:433

Section 26.6

1. Brown HC, Cha JS, Nazer B (1984) J Org Chem 49:2073
2. Brown HC, Cha JS, Nazer B, Brown CA (1985) J Org Chem 50:549
3. (a) Brown HC, Park WS, Cho BT (1986) J Org Chem 51:3278; (b) Brown HC, Park WS, Cho BT (1987) Bull Korean Chem Soc 8:2, 211; (c) Brown HC, Park WS, Cho BT (1987) ibid. 8:276
4. Brown HC, Park WS, Cho BT (1986) J Org Chem 51:1934
5. Hirao A, Itsuno S, Owa M, Nagami S, Mochizuki H, Zoorov HHA, Niakahama S, Yamazaki M (1981) J Chem Soc Perkin Trans I 900

6. Brown HC, Cho BT, Park WS (1986) J Org Chem 51:3396
7. Brown HC, Cho BT, Park WS (1988) J Org Chem 53:1231
8. Noyori R, Tomino I, Tanimoto, Y, Nishizawa M (1984) J Am Chem Soc 106:6709; Noyori R, Tomino I, Yamada M, Nishizawa M (1984) ibid. 106:6717
9. Brown HC, Pai GG, Jadhav PK (1984) J Am Chem Soc 106:1531; (b) Brown HC, Pai GG (1985) J Org Chem 50:1384; (c) Midland MM, Zderic SA (1982) J Am Chem Soc 104:525

Section 26.7

1. Cho BT, Chun YS (1992) Tetrahedron Asym 3:73
2. Cho BT (1990) J Korean Chem Soc 34:313
3. For ketones see Brown HC, Park WS, Cho BT, Ramachandran PV (1987) J Org Chem 52:5406
4. Shah RH, Schaeffer HJ, Murray DH (1965) J Pharm Sci 54:15

Section 26.8

1. Brown HC, Cha JS, Nazer B (1984) J Org Chem 49:2073
2. Brown HC, Krishnamurthy S (1972) J Am Chem Soc 94:7159

Section 26.9

1. Yamamoto Y, Toi H, Sonoda A, Murahashi S-I (1976) J Am Chem Soc 98:1965
2. (a) Yamamoto Y, Toi H, Murahashi S-I, Moritani I (1975) J Am Chem Soc 97:2558; (b) Damico R (1964) J Org Chem 29:1971

Section 26.10

1. Brown HC, Singaram B, Mathew CP (1981) J Org Chem 46:2712
2. Brown HC, Mathew CP, Pyun C, Son JC, Yoon NM (1984) J Org Chem 49:3091
3. Brown HC, Krishnamurthy S, Yoon NM (1975) J Org Chem 40:1864

Section 26.11

1. Brown HC, Park WS, Cho BT, Ramachandran PV (1987) J Org Chem 52:5406
2. Reagent A, *B*-Ipc-9-BBN, THF. This reagent is available from Aldrich Chemical Co. as Alpine-Borane: (a) Midland MM, Greer, S, Tramontano A, Zderic SA (1979) J Am Chem Soc 101:2352; (b) Midland MM, Tramontano A, Kazubski A, Graham RS, Tsai DJS, Cardin DB (1984) Tetrahedron 40:1371; (c) Tramontano A (1980) Ph.D. Thesis, University of California, Riverside
3. Reagent B, *B*-Ipc-9-BBN, neat: (a) Brown HC, Pai GG (1982) J Org Chem 47:1606; (b) Brown HC, Pai GG (1983) ibid. 48:1784; (c) Brown HC, Pai GG, Jadhav PK (1984) J Am Chem Soc 106:1531: (d) Brown HC, Pai GG (1985) J Org Chem 50:1384
4. Reagent C, *B*-Ipc-9-BBN, 6 kbar: Midland MM, McLoughlin JI (1984) J Org Chem 49:1316
5. Reagent D, NB-Enantrane is a registered trademark of Aldrich Chemical Co. Midland MM, Kazubski A (1982) J Org Chem 47:2814
6. Reagent E, Ipc_2BCl, is available from Aldrich Chemical Co. (a) Chandrasekharan J, Ramchandran PV, Brown HC (1985) J Org Chem 50:5446; (b) Brown HC, Chandrasekharan J, Ramachandran PV (1986) ibid. 51:3394

7. Reagent F, BH_3-AMDPB (2:1), AMDPB=(*S*)-(–)-2-amino-1,1-diphenyl butan-1-ol: (a) Hirao A, Itsuno S, Nakahama S, Yamazaki N (1981) J Chem Soc Chem Commun 315; (b) Itsuno S, Ito K, Hirao A, Nakahama S (1983) ibid. 469; (c) Itsuno S, Hirao A, Nakahama S, Yamazaki N (1983) J Chem Soc Perkin Trans I 1673; (d) Itsuno S, Ito K, Hirao A, Nakahama S (1984) J Org Chem 49:555; (e) Itsuno S, Nakano M, Miyazaki K, Masuda H, Ito K, Hirao A, Nakahama S (1985) J Chem Soc Perkin Trans I 2039
8. Reagent G, (*R,R*)-2-5-Dimethylborolane: Imai T, Tamura T, Yamamuro A, Sato T, Wollman TA, Kennedy RM, Masamune S (1986) J Am Chem Soc 108:7402
9. Reagent H, NB-Enantride, is a registered trademark of Aldrich Chemical Co. Midland MM, Kazubski A (1982) J Org Chem 47:2495
10. Reagent I, $LiBH_4$-DBC-*t*-BuOH: (a) Soai K, Oyamada H, Yamanoi T (1984) J Chem Soc Chem Commun (1984) 413; (b) Soai K, Yamanoi T, Hikima H, Oyamada H (1985) ibid. 138
11. Reagent J, $NaBH_4$-IBA-DIPGF, IBA = isobutyric acid, DIPGF = 1, 2:5, 6-di-*O*-isopropylidene-α-D-glucofuranose: (a) Hirao A, Itsuno S, Owa M, Nagami S, Mochizuki H, Zoorov HHA, Niakahama S, Yamazaki N (1981) J Chem Soc Perkin Trans I 900; (b) Morrison JD, Grandbois ER, Howard SI (1980) J Org Chem 45:4229
12. Reagent K, K-glucoride, potassium 9-*O*-(1,2,5,6-di-*O*-isopropylidene-α-D-glucofuranosyl)-9-boratabicyclo[3.3.1]nonane: (a) Brown HC, Park WS, Cho BT (1986) J Org Chem 51:1934; (b) Brown HC, Park WS, Cho BT (1986) ibid. 3278; (c) Brown HC, Cho BT, Park WS (1986) ibid. 3396
13. Reagent L, $LiAlH_4$-Darvon Alc, Darvon Alc=[2*S*,3*R*]-(+)-4-(dimethyl amino)-3-methyl-1,2-diphenyl-2-butanol: (a) Yamaguchi S, Mosher H, Pohland A (1972) J Am Chem Soc 94:9254; (b) Yamaguchi S, Mosher H (1973) J Org Chem 38:1870; (c) Brinkmeyer RS, Kapoor VM (1977) J Am Chem Soc 99:8339; (d) Cohen N, Lopresti RJ, Neukom C, Saucy G (1980) J Org Chem 45:582
14. Reagent M, $LiAlH_4$-MEP-ArOH, MEP=*N*-methylephedrine, ArOH-3,5-dimethylphenol: Vigneron JP, Jacquet I (1976) Tetrahedron 32:939; Vigneron JP, Bloy V (1979) Tetrahedron Lett 2683
15. Reagent N, $LiAlH_4$-diamine, diamine=(*S*)-2-(2,6-xylidinomehtyl) pyrrolidine: Asami M, Mukaiyama T (1979) Heterocycles 12:499
16. Reagent O, $LiAlH_4$-aminobutanol, aminobutanol=(*S*)-4-anilino-3-(methylamino)-1-butanol: (a) Sato T, Goto Y, Fujisawa T (1982) Tetrahedron Lett 23:4111; (b) Sato T, Gotoh Y, Wakabayashi Y, Fujisawa T (1983) ibid. 24:4123
17. Reagent P, Binal-H: (a) Noyori R, Tomino R, Tomino I, Tanimoto Y (1979) J Am Chem Soc 101:3129; (b) Nishizawa M, Yamada M, Noyori R (1981) Tetrahedron Lett 22:247; (c) Noyori R, Tomino I, Tanimoto Y, Nishizawa M (1984) J Am Chem Soc 106:6709; (d) Noyori R, Tomino I, Yamada M, Nishizawa M (1984) ibid. 6717
18. Reagent Q, $LiAlH_4$-DBP-EtOH, DBP=(*S*)-(–)-10,10′-dihydroxy-9,9′-biphenanthryl: Yamamoto K, Fukushima H, Nakazaki M (1984) J Chem Soc Chem Commun 1490
19. Reagent R, $LiAlH_4$-MEP-NEA, MEP=*N*-methylephedrine, NEA=*N*-ethylaniline: Terashima S, Tanno N, Koga K (1980) J Chem Soc Chem Commun 1026
20. Reagent S, $LiAlH_4$-MEP-EAP, MEP=*N*-methylephedrine, EAP=2-(ethylamino)pyridine: Kawasaki M, Suzuki Y, Terashima S (1984) Chem Lett 239
21. Reagent T, TBADH, *Thermoanaerobium brockii* alcohol dehydrogenase: Keinam E, Hafeli EK, Seth KK, Lamed R (1986) J Am Chem Soc 108:162

22. (a) Dale JA, Dull DL, Mosher HS (1969) J Org Chem 34:2543; (b) Dale JA, Mosher HS (1973) J Am Chem Soc 95:512
23. (a) Westley JW, Halpern B (1968) J Org Chem 33:3978; (b) Pirkle WH, Hanske JR (1977) ibid. 42:2436

27 Cleavage of Ethers

The usefulness of a mild and selective ether cleavage reagents in the synthesis of polyfunctional natural products and biologically active molecules is well documented. Boron halides are extensively used for this purpose [1–4], and borontribromide is favored because of its effectiveness under mild conditions [1, 4]. However, the presence of three reactive sites in borontribromide generally induces difficulties with polyfunctional molecules. To overcome this difficulty, it is found [5] that methylene chloride solution of *B*-Br-9-BBN readily cleaves, in excellent yield, a variety of ethers. The reagent is selective when two different alkyl groups are present in an ether. These results are summarized in Table 27.1 [5].

Table 27.1 Cleavage of ethers with *B*-Br-9-BBN [5]

Ether	Bromide formed	Yield (%)
Di-*n*-butylether	*n*-Butylbromide	100
Isobutyl-*n*-butylether	*n*-Butylbromide	97
	Isobutylbromide	<3
sec-Butyl-*n*-butylether	*sec*-Butylbromide	100
t-Butyl-*n*-butylether	*t*-Butylbromide	100

The clear cleavage pattern emerges: the tertiary alkyl group is converted more readily into the corresponding bromide than the secondary one, and the latter more readily than a primary group. Consequently, *t*-butyl-*n*-butyl ether gives exclusively *t*-butylbromide, whereas *sec*-butyl-*n*-butyl ether leads only to *sec*-butylbromide. Isobutyl-*n*-butyl ether yields predominantly *n*-butylbromide.

The reaction of *B*-Br-9-BBN with ether involves the inversion of configuration at the C–O bond. This is illustrated [5] in the cleavage of (*R*)-2-octylmethyl ether, $[\alpha]_D^{25}$ −8.14, (76.5% optical purity) to yield (*S*)-2-octylbromide, $[\alpha]_D^{25}$ +24.54 (55.6% optical purity) (Eq. 27.1).

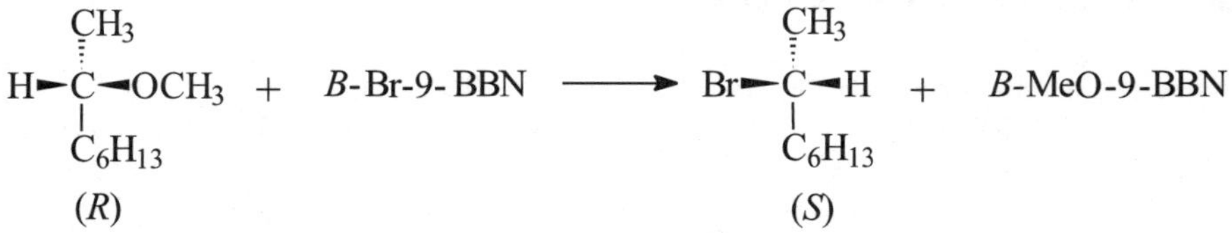

(27.1)

The cleavage of some aromatic ether are sketched in Chart 27.1.

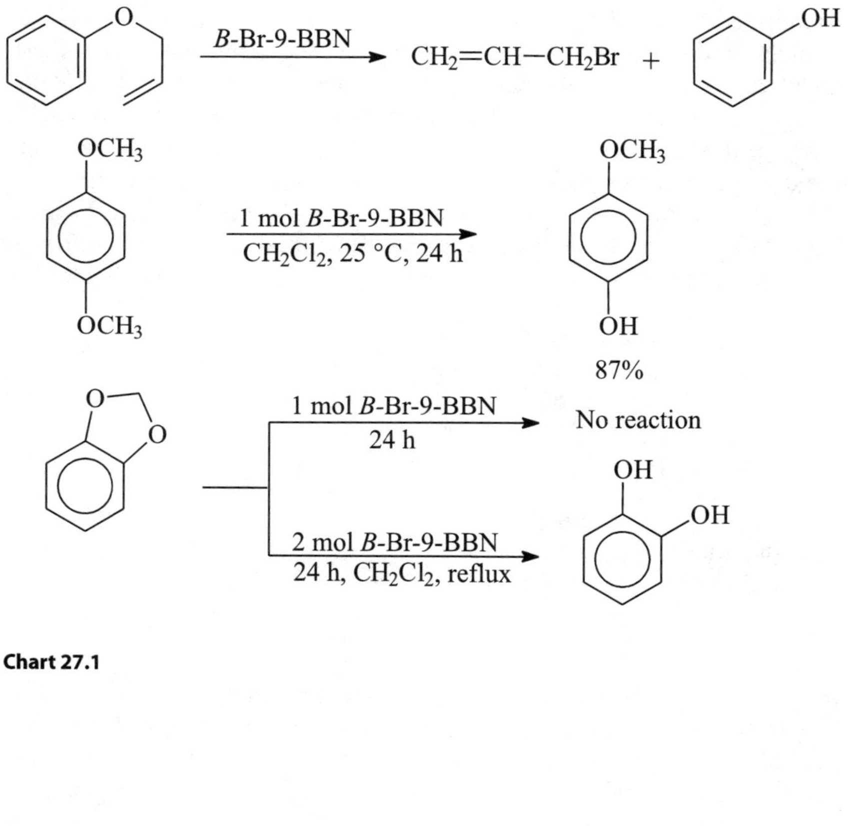

Chart 27.1

Relative rates of cleavage of some ether are presented in Table 27.2 [5].

Table 27.2 Relative rates of cleavage of ethers at room temperature [5]

Ether	Relative Rate
Di-*n*-butyl ether	1.0
Isobutyl-*n*-butyl ether	0.66
sec-Butyl-*n*-butyl ether	Too fast
Phenylmethyl ether (anisole)	0.14
Phenylethyl ether (phenelote)	0.09
Hydroquinone dimethyl ether	1.0
Phenylallyl ether	>0.76

References

1. (a) Taub T, Girotra NN, Hoffsommer RD, Kuo CH, Slates HL, Weber S, Wendler NL (1968) Tetrahedron 24:2443; (b) Vlattas I, Harrison IT, Tökes L, Fried JH, Cross AD (1968) J Org Chem 33:4176; (c) Grieco PA, Nishizawa M, Burke SD, Marinovic N (1976) J Am Chem Soc 98:1612; (d) Grieco PA, Nishizawa M, Oguri T, Burke SD, Marinovic N (1977) ibid. 99:5773
2. Burwell RL Jr (1954) Chem Rev 54:615
3. Gerrard W, Lappert MF (1958) Chem Rev 58:1081
4. (a) Benton FL, Dillon TE (1942) J Am Chem Soc 64:1128; (b) McOmie JFW, Watts ML (1963) Chem Ind 1658; (c) McOmie JFW, Watts ML, West DE (1968) Tetrahedron 24:2289
5. (a) Bhatt MV (1978) J Organomet Chem 156:221; (b) Bhatt MV, Kulkarni SU (1983) Synthesis 249

28 *trans*-Metalation

The synthesis of stereodefined substituted alkene is one of the major challenges in organic synthesis, as many naturally occurring compounds such as pheromones and hormones contain mono-, di-, or trisubstituted double bonds. In many cases a small amount of the wrong isomer acts an inhibitor of their biological activity. Thus, a highly stereospecific synthesis of such compounds is required. The organometallic compounds are such class of compounds that leads to the synthesis of desired molecules.

Among the metalation reactions, allylic metalation occupies an especially significant position due to its considerable synthetic utility [1]. Due to "self-consumption" of α-halogen–substituted allylic metals, this area received negligible attention [2]. A notable achievement in this area is the metalation of allyl chloride, using LDA [3]. In spite of the considerable effects to react allylic [4] and nonallylic [5] carbenoids with boron substrates, the problem of selectivity (α- versus γ-product) remains unsolved. Brown and Rangaishenvi [6] have shown that (α-chloroallyl)lithium, generated *in situ* from allylhalide and LDA, undergoes facile *trans* metalation with boron substrate to afford allylboron (Eq. 28.1), and its utility in organic synthesis has been amply demonstrated (Eq. 28.2) [7].

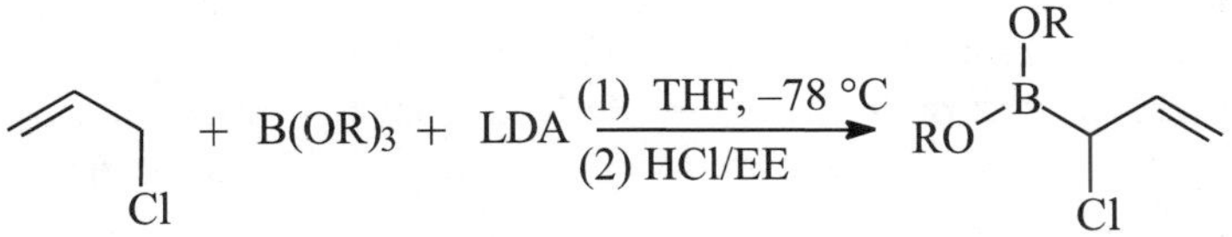

(28.1)

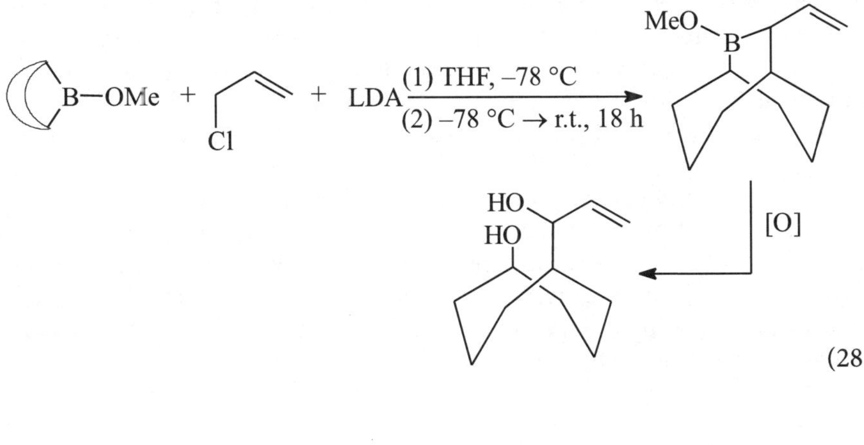

(28.2)

Brown and Jayaraman [8] have explored the commercially available 9-methoxy-9-BBN to capture the carbenoid, (α-chloroallyl)lithium generated *in situ*. Among the various bases tested for the purpose, lithium dicyclohexylamide ($LiNChx_2$) is the most effective to afford the desired intermediate (**3a**, Chart 28.1). The dicyclohexylamine liberated is easily removed by precipitation with addition of MeI or $BF_3{\cdot}OEt_2$. The use of superbase [9], a mixture of (KO-*t*-Bu and *n*-BuLi), however, results in *trans* esterification between the *B*-OMe and *O*-*t*-Bu groups and yield a mixture of **3a** and **3c**.

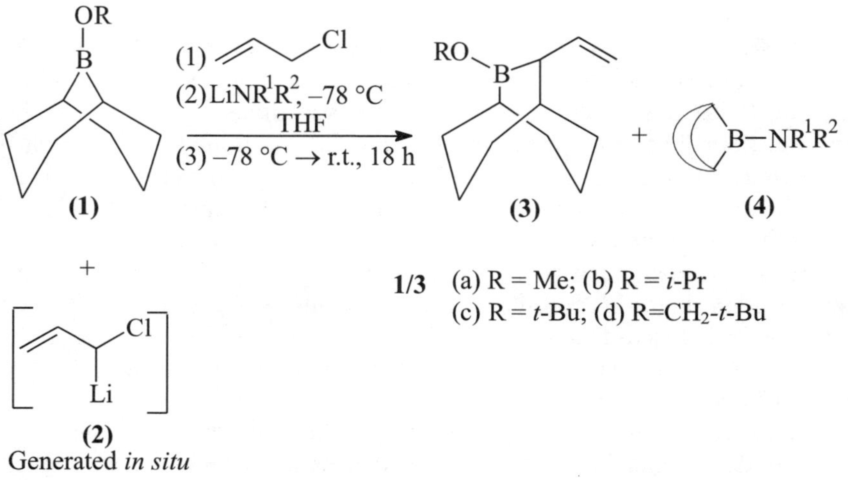

Chart 28.1

The other bases such as, $LiNMe_2$, $LiNEt_2$, and $LiNPh_2$ afford major products, **4**, in which the OMe group has been nucleophilically displayed by the NR_2 group. On the contrary, lithium amides bases derived from the hindered secondary amines are highly proton selective to give the desired allylic borinate (**3a**) in high yield (Table 28.1) [8].

The effectiveness of these bases is correlated with the pK_a of respective amines [10]. Lithium amides derived from secondary amines whose pK_a values are in the vicinity of 30 (e.g., HN-*t*-$BuSiMe_3$ and $HN(SiMe_3)_2$) are not suitable for efficient metalation. However, amines like HN-*i*-Pr_2, $HNChx_2$, and HN-*i*-PrChx, with pK_a values around 35 bring efficient metalation. Moreover, the studies have revealed [8] that bulkiness of OR, as in **1b–d**, gives lower yield of desired allylic borinate [11]. Consequently, combination of 9-methoxy-9-BBN and $LiNChx_2$ are found to afford the desired allylic borinate (**3a**) intermediate in excellent yields (Table 28.1) [8]. The **3a** intermediate serves as a versatile intermediate in the synthesis of cyclooctane compounds (Chart 28.2) and thus solves the long-standing problem of high-degree strain arising from the powerful transannular interactions present in such systems [12].

Table 28.1 In situ metalation of allylichloride with LiNR[1]R[2] with capture by 9-MeO-9-BBN [8]

Base used	Product(s) (ratio)	Yield (%)
$LiNMe_2$	**3a**:**4** (5:95)	91
$LiNEt_2$	**3a**:**4** (10:90)	93
Lithium pyrrolidide	**4**	90
LiN-*i*-Pr_2	**3a**	76
LiN-*s*-Bu_2	**3a**	80
LiN-*i*-PrChx	**3a**	85
$LiNChx_2$	**3a**	90
LiTMP[a]	**3a**	75
LiN-*t*-$BuSiMe_3$	**3a**:**4** (40:60)	90
$LiNPh_2$	**4**	92

[a] *TMP* 2,2,6,6-tetramethylpiperidide.

The mechanism of formation of **5** and **8** are shown in Chart 28.3.

The synthetic utility of allylic borinate gets further impetus when a second homologation provides dienyl borinate (**11**), which is formed by the migration of allyl group (bond a) in preference to the secondary alkyl group (bond b). It is observed that in such reactions, the allyl group undergoes migration in preference to the secondary alkyl group. The compound **11** on oxidation provides **8**, a substrate useful for oxy-Cope rearrangement. In conclusion, this process forms a simple methodology for the synthesis of eight-membered natural products, like ophiobolane, taxane, fusicoccin, etc.

The hydroboration–metalation methodology is another method of prime importance in organic synthesis. A convenient one-pot synthesis of (*Z*)-2-alkenylsilanes and tin (Scheme 28.1), and their synthetic applications are reported [13].

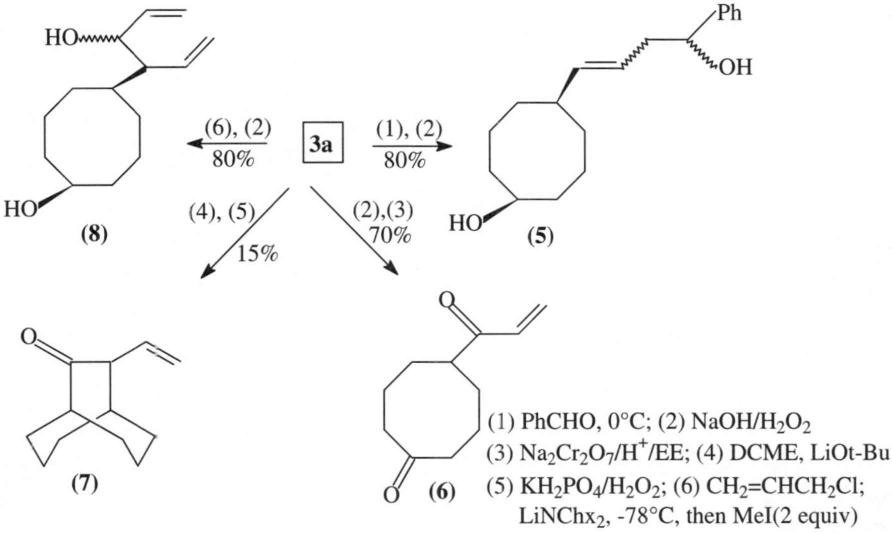

Chart 28.2

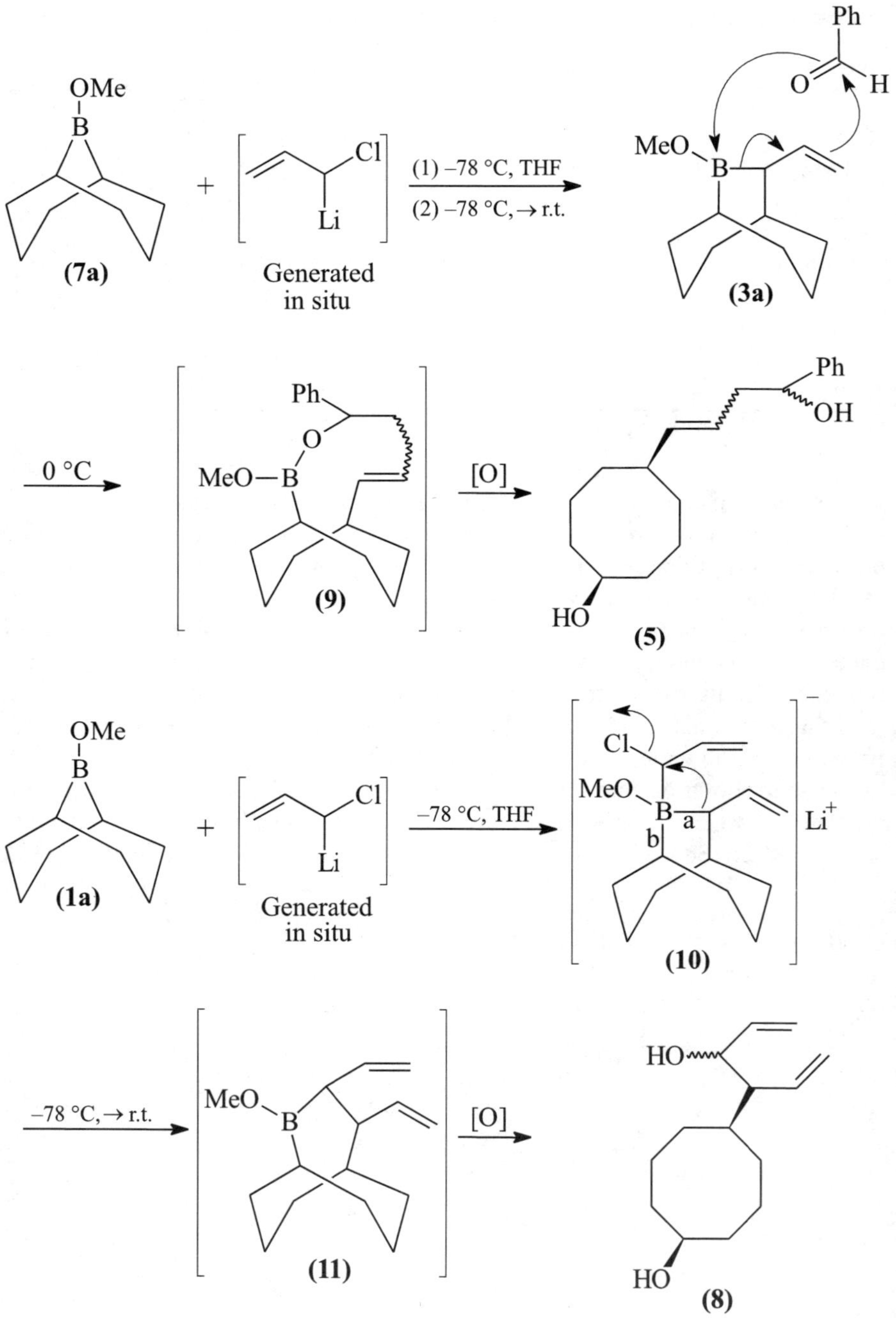
OMe
B
(7a)
Cl
Li
Generated in situ
(1) –78 °C, THF
(2) –78 °C, → r.t.
Ph
O
H
MeO
B
(3a)
0 °C
Ph
O
MeO—B
(9)
[O]
Ph
OH
HO
(5)
OMe
B
(1a)
Cl
Li
Generated in situ
–78 °C, THF
Cl
MeO
B
a
b
Li+
(10)
–78 °C, → r.t.
MeO
B
(11)
[O]
HO
HO
(8)

Chart 28.3

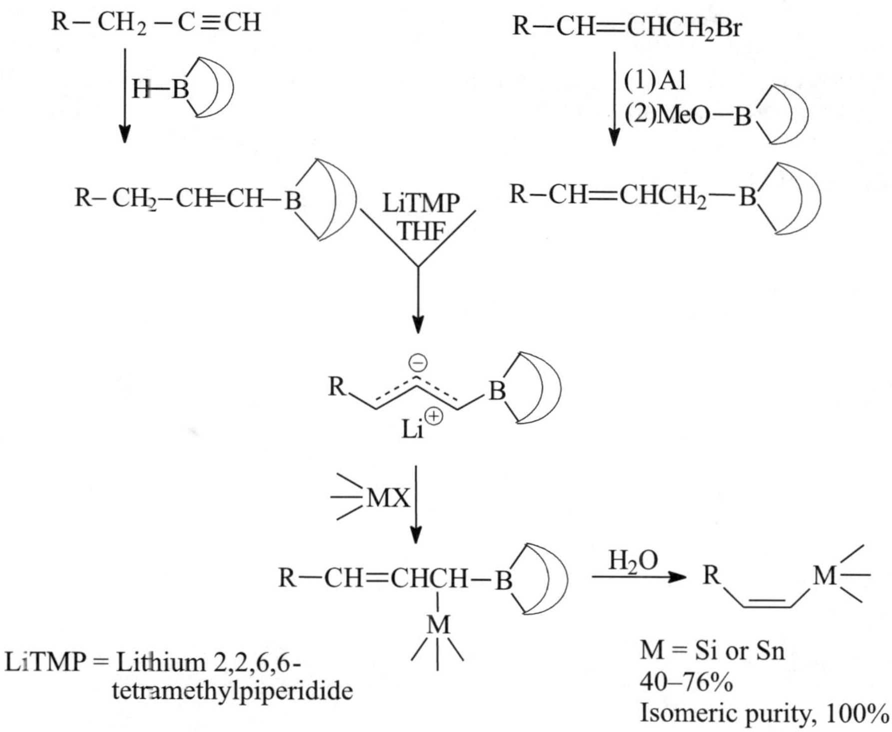

Scheme 28.1

The *trans*-metalation process has been employed for the synthesis of 9-vinyl-9-BBN, by the reaction of *B*-Br-9-BBN either with vinyltributyltin or tetravinyltin (Eq. 28.3) [14]. Singleton and his group have extensively used 9-vinyl-9-BBN as a dienophile for Diels-Alder reactions (*vide supra*) and have described it as omniphilic dienophile [15] because of its high reactivity.

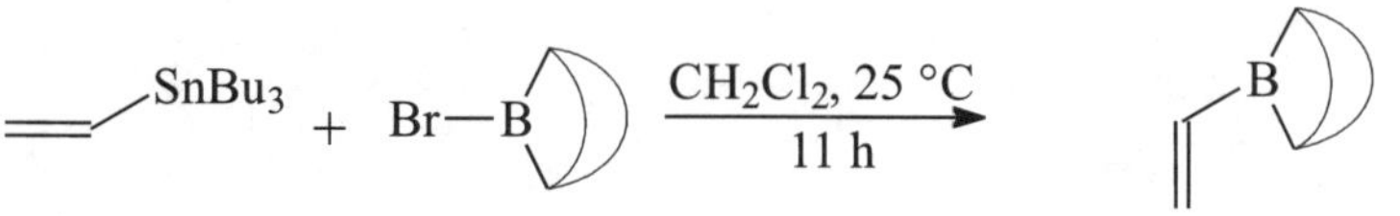

(28.3)

Tin–boron *trans*-metalation has also been used to synthesize *B*-2-propenyl-9-BBN (Eq. 28.4) [16], 9-*cis*-propenyl-9-BBN (Eq. 28.5) [16], 9-*cis*-2-chlorovinyl-9-BBN (Eq. 28.6) [17], *B*-cyclopentenyl- and *B*-cyclohexenyl-9-BBN (Eq. 28.7) [18], *B*-1′,7′,9′-decatrienyl-9-BBN (Eq. 28.8) [19], *trans*-1,2-bis(9-borabicyclo[3.3.1]nonylethylene (Eq. 28.9) [20], and bis(9-borabicyclo[3.3.1]nonylacetylene (Eq. 28.10) [21].

(28.4)

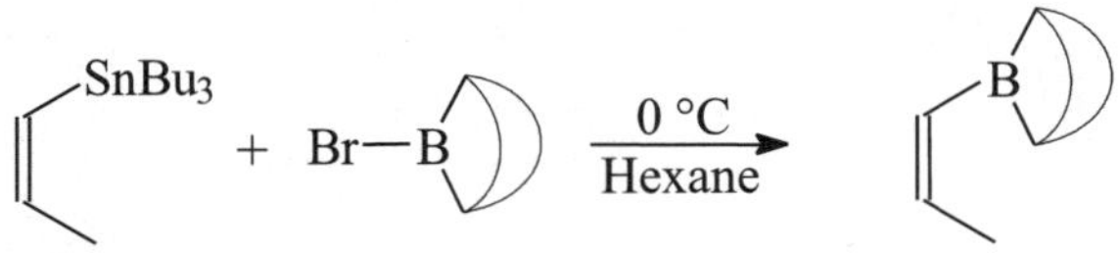

(28.5)

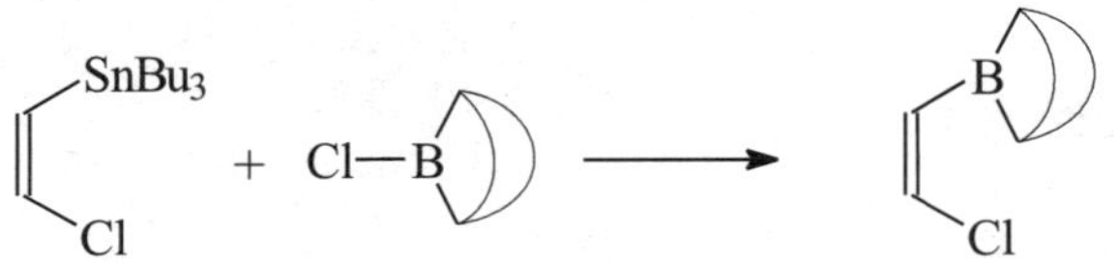

(28.6)

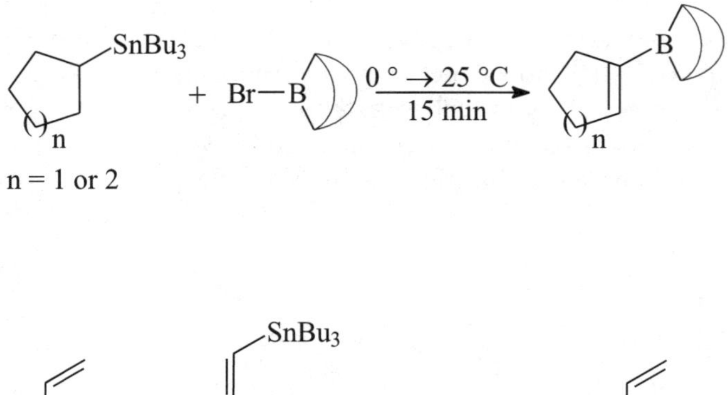

n = 1 or 2

(28.7)

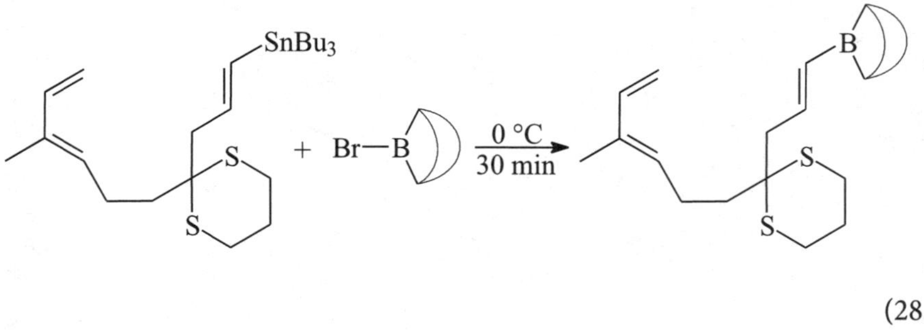

(28.8)

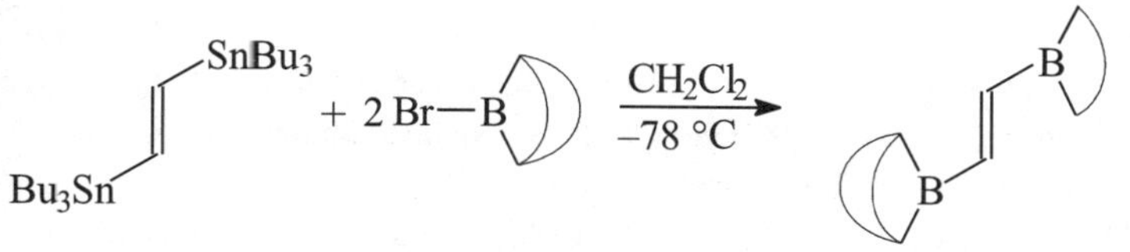

(28.9)

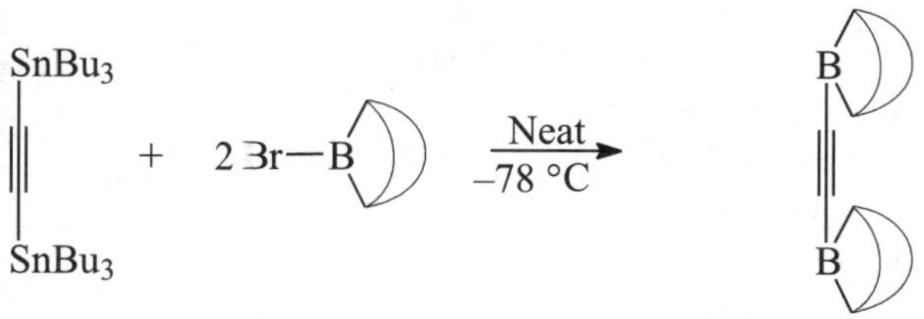

(28.10)

Lithium–boron *trans* metalation also occurs and is employed to prepare air-sensitive crystalline solid, [(trimethylsilyl)ethynyl]-9-BBN·THF (Eq. 28.11) [22]. This borane is, however, indefinitely stable when stored in a freezer under an inert atmosphere.

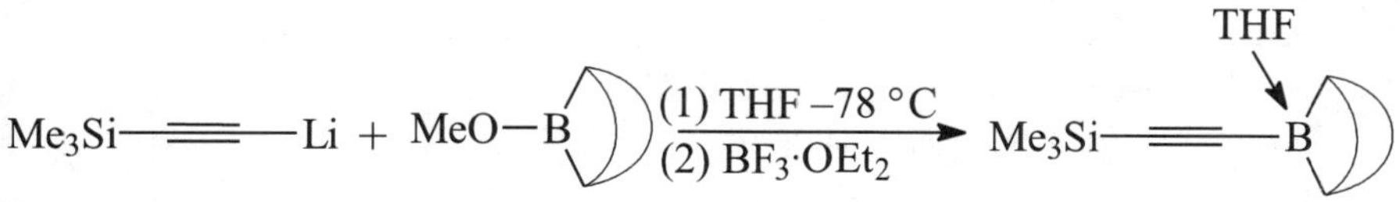

(28.11)

Tin–boron *trans*-metalation has been used to prepare *in situ* (for Diels–Alder reaction), this air sensitive crystalline solid, [(trimethylsilyl)ethynyl]-9-BBN (Eq. 28.12) [22].

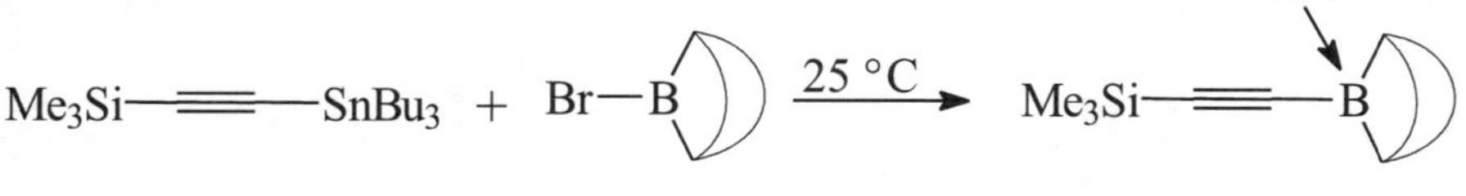

(28.12)

References

1. Biellmann JF, Ducep JB (1989) Org React 27:1344
2. (a) Wenkert E, Bakuzis P, Dynak JN, Swindell CS (1979) Synth Commun 9:11; (b) Olofson RA, Dougherty CM (1973) J Am Chem Soc 95:582
3. (a) Macdonald TL, Narayanan BA, O'Dell DE (1981) J Org Chem 46:1504; (b) Julia M, Verpeaux J-N, Zahneisen T (1990) Synlett 769; (c) Mallaiah K, Satyanarayana J, Ila H, Junjappa H (1993) Tetrahedron Lett 34:3145
4. Hara S, Imai S, Hara T, Suzuki A (1982) Synth Commun 12:813
5. Larson GL, Arguelles R, Rosario O, Sandoval S (1980) J Organomet Chem 198:15
6. Brown HC, Rangaishenvi MV (1990) Tetrahedron Lett 31:7113, 7115
7. Brown HC, Rangaishenvi MV, Jayaraman S (1992) Organometallics 11:1948
8. Brown HC, Jayaraman S (1993) J Org Chem 58:6791
9. Schlösser M (1998) Pure Appl Chem 60:1627
10. Fraser RR, Mansour TS (1984) J Org Chem 49:3442
11. Brown HC, Jayaraman S (1993) Tetrahedron Lett 34:3997
12. Petasis NA, Patane MA (1992) Tetrahedron 48:5757
13. Yatagai H, Yamamoto, Y, Maruyama K (1980) J Am Chem Soc 102:4548
14. Singleton DA, Martinez JP (1990) J Am Chem Soc 112:7423; (b) Singleton DA, Martinez JP (1991) Chemtracts Organic Chemistry 4:41; (c) Singleton DA, Martinez JP, Ndip GM (1992) J Org Chem 57:5768
15. Singleton DA, Martinez JP, Watson JV (1992) Tetrahedron Lett 33:1017
16. Singleton DA, Kim, K, Martinez JP (1993) Tetrahedron Lett 34:3071
17. Singleton DA, Leung S-W, Martinez JP, Lee Y-K (1997) Tetrahedron Lett 38:3163
18. Lee Y-K, Singleton DA (1997) J Org Chem 62:2255
19. Singleton DA, Lee Y-K (1995) Tetrahedron Lett 36:3473
20. Singleton DA, Redman AM (1994) Tetrahedron Lett 35:509
21. Singleton DA, Schulmeier BE, Hang C, Thomas AA, Leung S-W, Merrigan SR (2001) Tetrahedron 57:5149
22. Singleton DA, Leung S-W (1992) J Org Chem 57:4796

29 Separation of Isomers

Diels-Alder reaction of cyclopentadiene and butadiene affords a mixture of *exo*-5-vinyl-2-norbornene (**1a**) and *endo*-5-vinyl-2-norbornene (**1b**) [1]. Preparative GC separation [1] of these isomers encounters difficulties in obtaining the individual isomer in pure form and in large quantities. An alternative approach of separation via thermal isomerization [2], in which **1b** gets transferred to 4,7,3a,7a-tetrahydro-1*H*-indene, whereas **1a** remains unchanged, is also not successful. This is because it is difficult to prevent **1a** being contaminated by unreacted **1b**. As no other method is available for their separation, Inoue has reported [3] that hydroboration of **1** with 9-BBN, followed by oxidation with alkaline hydrogen peroxide results in the formation of alcohols **2a** and **2b**. The iodoether cyclization only of the *endo* isomer takes place. The sequence of approach is delineated in Scheme 29.1.

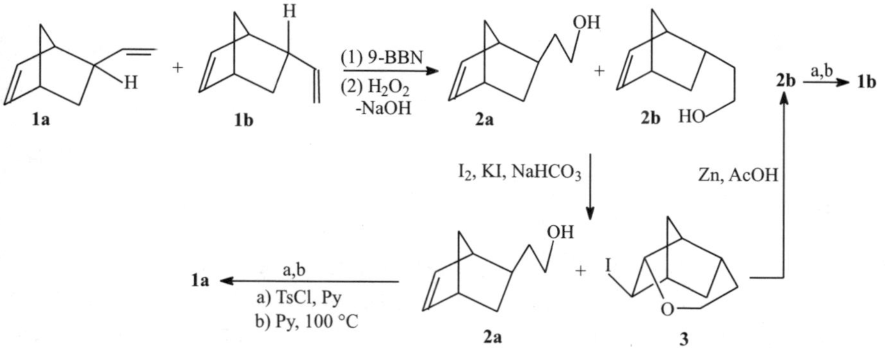

Scheme 29.1

After cyclization of **2b** to **3**, **2a** is separated easily from **3**, and **1a** is regenerated from **2a**, following the treatment with tosylchloride in pyridine, followed by elimination in pyridine at 100 °C. The *endo* isomer **1b** is generated by conversion of **3** to **2b**, using zinc in acetic acid, and is then followed by tosylation and elimination (Scheme 29.1) [1].

For the preparation of chiral reducing agent Prapine-Borane, the required (+)-2-*n*-propylapopinene, is prepared from (+)-α-pinene via Schlösser metalation, followed by treatment with ethyliodide. The undesired olefin is removed by its conversion to 9-BBN derivative and the desired (+)-2-*n*-propylapopinene is distilled off (Scheme 29.2) [4] (*vide supra*).

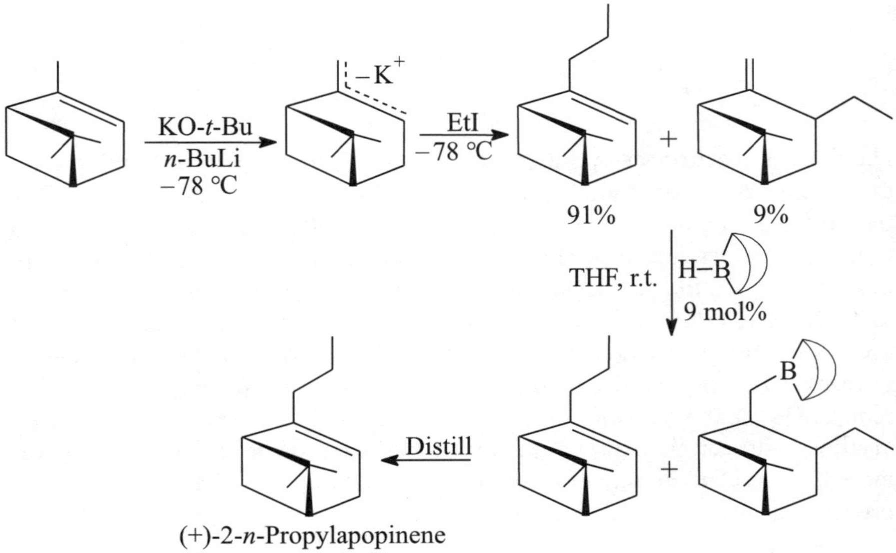

Scheme 29.2

References

1. Maeda T, Muranaka M, Hamanaka S, Ogawa M (1974) Nippon Kagaku Kaishi 1587
2. (a) Bobyleva AA, Belikova NA, Kalinichenko AN, Baryshnikov AT, Dubitskaya NF, Pekhk TI, Lippmaa ET, Plate AF (1980) Zh Org Khim 16:1645; (b) Just G, Linder U, Pritzkow W, Roellig M (1975) J Prakt Chem 317:979
3. Inoue Y (1987) Bull Chem Soc Jpn 60:1954
4. Brown HC, Ramachandran PV, Weissman SA, Swaminathan S (1990) J Org Chem 55:6328

30 Diels-Alder Reaction

The chemists have long believed that the trivalent boron atom with its empty *p* orbital should behave in a fashion to that of the traditional organic electron-withdrawing group [1]. This idea is attractive when applied to the activation of Diels-Alder dienophiles, because the intermediate boron compounds can be transformed to an array of different functional moieties, none of which can usually be obtained by a direct Diels-Alder reaction. Matteson [2] and Evans [3] have reported the reactions of vinylboronic esters with dienes under strenuous reaction conditions but with low yields of the products. Similarly, vinyldichloroborane gives low yield [4] on reaction with cyclopentadiene.

In 1990, Singleton and Martinez [5] discovered that the dialkylboron group is dramatically more activating than are esters of boronic esters, i.e., esters in normal Diels-Alder reactions, and that vinylboranes are exceptionally reactive and are regioselective and stereoselective dienophiles. The discovery is based on the utilization of diarylboron groups in anion chemistry [6, 7]. It contrasts reaction of butadiene or isoprene with methyl acrylate [8], nitroethylene [9], phenylvinylsulfone [10a], dibutylvinylboronate, or even vinylboronic esters substituted with a second electron-withdrawing group [10b], all require prolonged heating (100–150 °C). However, the Diels-Alder reactions of vinyl-9-BBN with butadiene and isoprene are both >80% in a day at 25 °C. Vinyl-9-BBN is 200 times more reactive than methylacrylate is toward butadiene [5], but less reactive than Lewis acid–complexed dienophiles [8] and α,β-unsaturated Fisher carbene complexes [11]. However, among the simple neutral substituents, the activating effect of dialkylboron group is approached only by the acyl and sulfonyl halides [12].

Vinyl-9-BBN is easily prepared (Eq. 30.1) by 1.0-M solution of *B*-bromo-9-BBN in methylene chloride with 1 equiv of vinyltributyltin, followed by distillation of reaction mixture [b.p. 28–30 °C (0.25 mm)] [5]. Vinyl-9-BBN can be stored for weeks in a freezer under an inert atmosphere and is extremely pyrophoric.

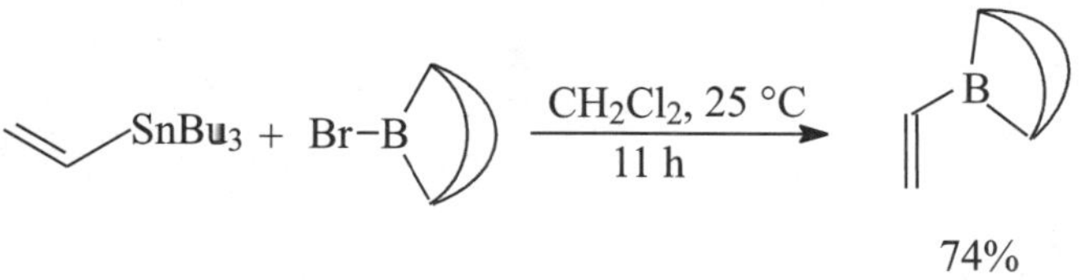

(30.1)

However, Singleton has pointed out [13] that low yields of the product of Diels-Alder reaction and sometimes formation of complex mixtures are associated with previously synthesized vinyl-9-BBN. The reaction gives 3-cyclohexen-1-ols, which are now essentially as available as 3-cyclohexencarboxylates from Diels-Alder reaction. Consequently, Singleton has described a better synthetic method [13] to prepare vinyl-9-BBN for one-pot Diels-Alder synthesis of 3-cyclohexen-1-ols. A mixture of 1.5 mmol of commercial 1.0 M *B*-Br-9-BBN in CH_2Cl_2 and 1.9 mmol of vinyltributyltin is stirred at 25 °C for 5 min and 1.0 mmol of 2-*tert*-butylbutadiene is added. The mixture after stirring for 20 h at 20 °C is oxidized to afford 82% of 4-*tert*-butyl-3-cyclohexen-1-ol, with no detectable formation (>98:2) of the other isomer. It is important to note that excess of vinyltributyltin is used to remove adventitious protic acid, and that a greater than usual amount of H_2O_2/NaOH is used to allow for reaction with the byproduct bromotributyltin. Tetravinyltin may be used in place of vinyltributyltin and is fairly economical source of vinyl groups for these reactions.

Vinyl-9-BBN undergoes Diels-Alder reactions with dienes as summarized in Table 30.1 [5].

These reactions display an outstanding regioselectivity e.g., vinyl-9-BBN forms the normal "*para* isomer" with isoprene, the selectivity comparable to that of methylacrylate–$AlCl_3$ (95:5) [14] and is greater than the reaction of methylacrylate (70:30) or dibutylvinylboronate (75:25) [3]. In contrast, with *trans*-piperylene there is regiospecific formations of normally disfavored [15] "*meta* isomers". Molecular mechanics calculations reveal that both *endo* and *exo* transition states leading to the normal "*ortho*" isomers are sterically hindered by the bulky bicyclo[3.3.1]nonane group [5, 16]. The steric effect is also responsible for the regiospecific formation of 2-substituted octahydronaphthalenes from 1-vinylcyclohexene. The *endo* regioselectivity in the reactions [5] of piperylene and 1-vinylcyclohexene with vinyl-9-BBN is also much greater than observed with common neutral dienophiles [17].

Table 30.1 Rate constants, products and yields for Diels-Alder reactions with vinyl-9-BBN [5]

Diene	Rate constant ($M^{-1} s^{-1}$) or reaction conditions	Yield (%)	Product(s) and ratio
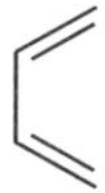	$3.2 \pm 0.4 \times 10^{-6}$ (25 °C)	82	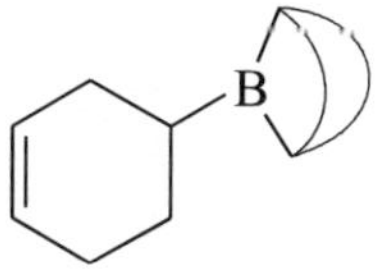
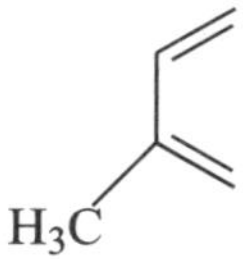	$3.0 \pm 0.6 \times 10^{-6}$ (25 °C)	79	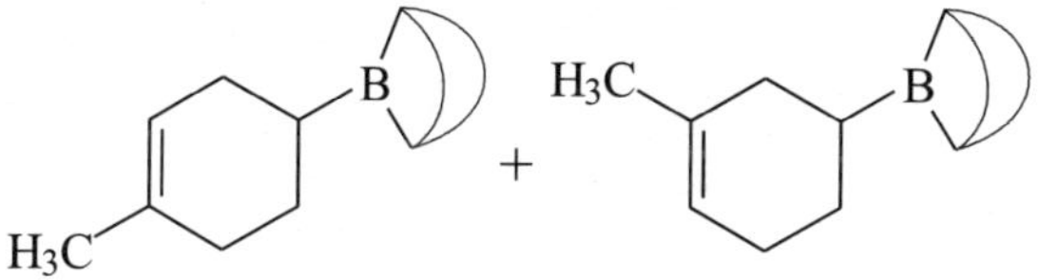 93:7
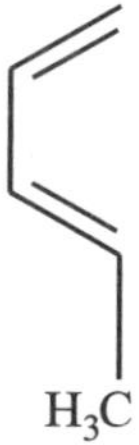	$4.6 \pm 0.5 \times 10^{-7}$ (25 °C)	71	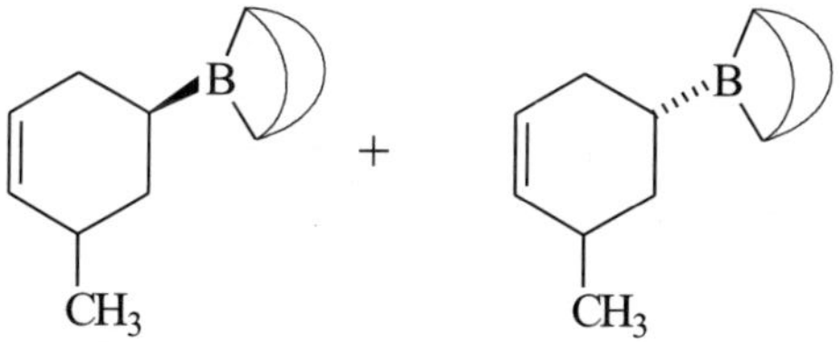 92:8

Table 30.1 *(continued)* Rate constants, products and yields for Diels-Alder reactions with vinyl-9-BBN [5]

Diene	Rate constant ($M^{-1} s^{-1}$) or reaction conditions	Yield (%)	Product(s) and ratio
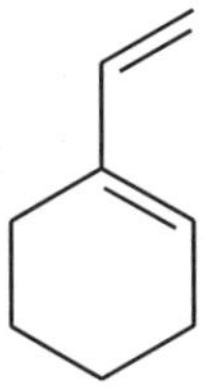	48 h, 55 °C	72	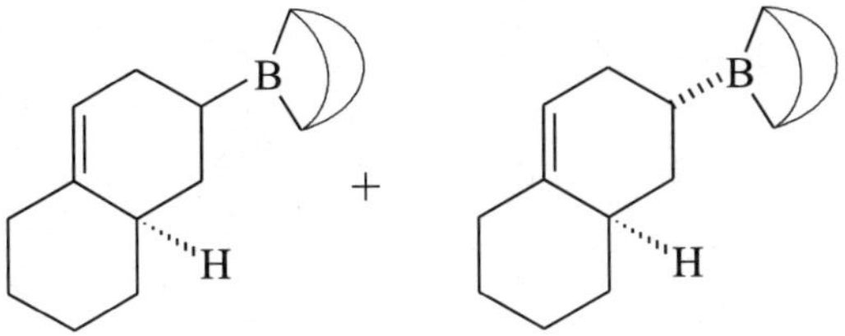 96:4
	3.5 h, 25 °C	86	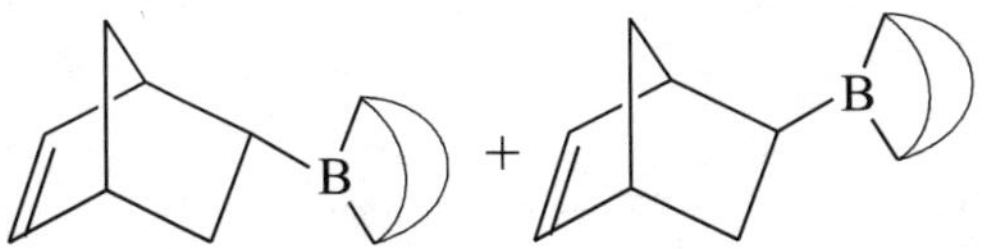 2:1

The regioselectivity is >98:2, >95:5, and 90:10 *para*:*meta* with 2-*t*-butyl butadiene, 2-phenylbutadiene, and 2-chlorobutadiene, respectively, (Chart 30.1) [18]. The models suggest that there is steric interaction between the bulky bicyclo[3.3.1]nonane group and substituent in the 2 position of the diene in the *endo*–transition state, leading to the *meta* isomer, with exception of 2-ethoxybutadiene. Singleton and coworkers [19] have conducted the relative regioselective studies of vinyl-3,6-dimethylborepane, trivinylborane, and vinyl-9-BBN and have found that vinyl-9-BBN is the most regioselective dienophile [19], in keeping with principally the steric control of regioselectivity (Table 30.2) in Diels-Alder reactions of vinyl boranes.

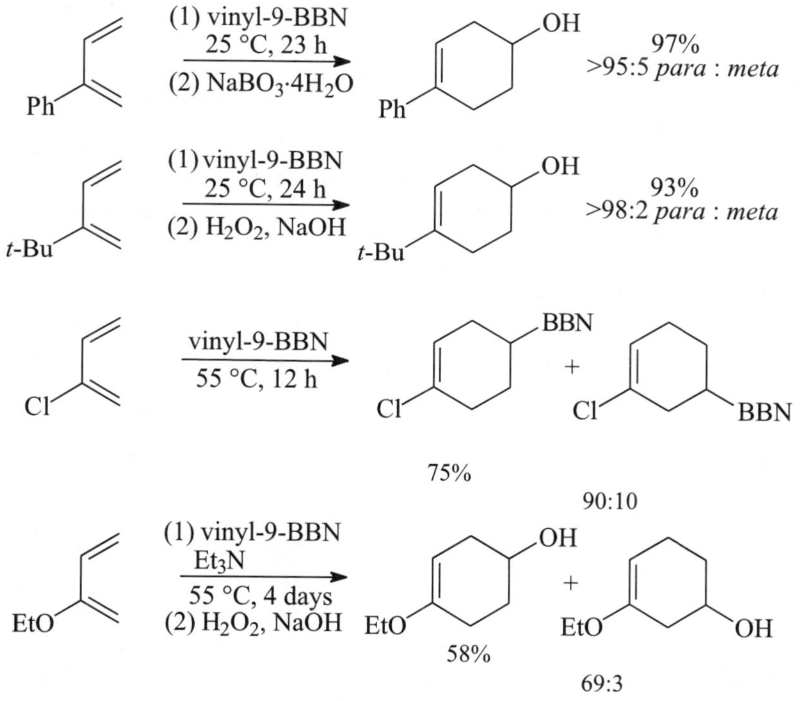

Chart 30.1

Table 30.2 Regioselectivity of Diels-Alder reactions with vinylboranes [19]

Diene	Vinyl-9-BBN	Vinyldimethyl-borane	Vinyl-DMB	Trivinylborane
Myrcene	96:4[a]	61:39	70:30	62:38
2-Phenylbutadiene	>98:2[a]	84:16	84:16	90:10
2-*t*-Butylbutadiene	>98:2[a]	90:10	90:10	>98:2
Piperylene	>98:2[b]	50:50	65:35	49:51

[a] The ratios of "*para*":"*meta*" (1,4:1,3) isomers.
[b] The ratio of "*meta*":"*ortho*" (1,5:1,2) isomers.

Singleton and coworkers [18] have conducted a detailed study on the reactivity of vinyl-9-BBN, with wide range of dienes including hexachlorocyclopentadiene and methyl-2,4-pentadienoate. It is found that the rate of Diels-Alder reactions with vinyl-9-BBN is uniquely insensitive to diene substituent effects; and high reactivity, regioselectivity, and *endo* stereoselectivity are observed in both electron-rich and electron-poor dienes. The ambiphilic reagents favoring reactions with both electron-rich and electron-poor species are known for many reactions, including a report on an ambiphilic diene for Diels-Alder reactions [13]. The high and relatively invariant reactivity of vinylboranes with electron-rich and electron-poor and inactivated dienes led Singleton to describe vinylboranes as *omniphilic* dienophiles [18].

The relative reactivity of vinyl-9-BBN toward various dienes is summarized in Table 30.3 [18].

Table 30.3 Diene structure versus reactivity for vinyl-9-BBN and maleic anhydride [18]

Diene	$k \times 10^{-6}$ (25 °C, $M^{-1} s^{-1}$) Vinyl-9-BBN	k_{rel} (25 °C) Vinyl-9-BBN	k_{rel} (25 °C) Maleic anhydride	k_{rel} (20 °C) Tetracyano-ethylene
2,3-Dichlorobuta-diene	1.1	0.34	0.0047	–
2-Chlorobutadiene	4.3	1.3	0.1 (5.3×10^{-6} $M^{-1} s^{-1}$)	0.002 (10×10^{-6} $M^{-1} s^{-1}$)
Butadiene	3.2	1	1	1
Isoprene	3.0	0.94	3	45
2-(m)-Ethoxybutadiene	0.53	0.17	10	1,750
2-Phenylbutadiene	230	72	8.8	191
2-*t*-Butylbutadiene	40	13	28	–
Cyclopentadiene	65	22	1,160	2.1×10^{6}

In the one-pot process of Diels-Alder reaction, the *in situ* preparation of vinyl-9-BBN, from vinyltributyltin or tetravinyltin and *B*-Br-9-BBN has been utilized for the preparation of various cyclohexenols, and this is compared with other vinyldialkylboranes for yield and regiospecificity (Table 30.4) [13].

Diels-Alder reactions of *cis*-2-chlorovinyltributylstannate are greatly accelerated by the presence of catalytic amount of 9-Cl-9-BBN [20]. The catalytic cycle shown in Scheme 30.1 in which 9-Cl-9-BBN acts not as a Lewis acid but as a source of the dienophile's activating group. Boron–tin *trans* metalation [21] between *cis*-2-chlorovinyltributylstannate and 9-Cl-9-BBN yields activated dienophile 9-*cis*-2-chlorovinyl-9-BBN. This activated dienophile reacts with the diene to give the cyclic product after the elimination of 9-Cl-9-BBN, and the catalytic cycle continues [20].

Table 30.4 Reaction conditions, products, and yields for Diels-Alder reactions of vinylboranes formed *in situ* [13]

Diene	Vinylborane source[a]	Temperature (°C)/time (h)	Yield (%) (ratio)	Product(s)
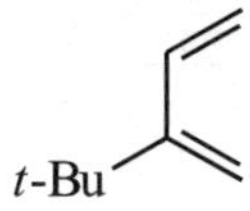	A B	25/20 55/16	82 (>98:2) 75 (90:10)	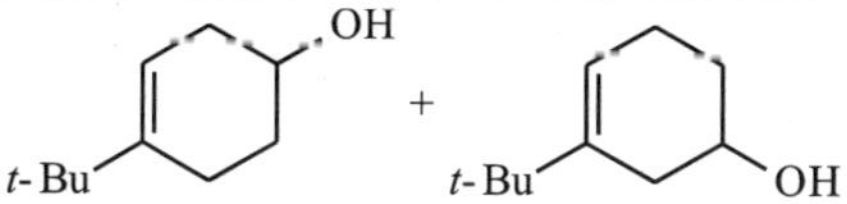
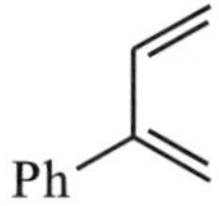	A B C	25/16 25/22 55/28	86 (98:2) 84 (84:16) 87 (91:9)	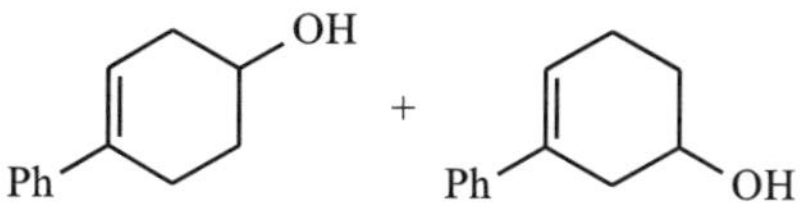
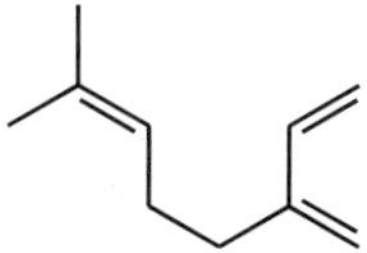	A D	55/36 65/12	81 (96:4) 92 (61:39)	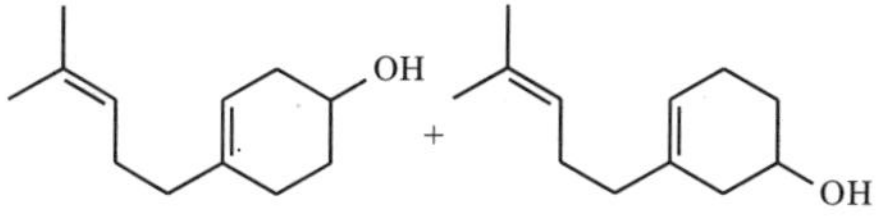
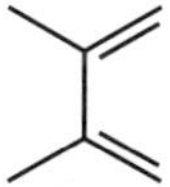	B	100/7	78	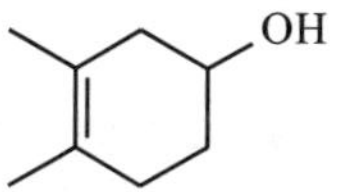
	B	25/22	81 (84:16)	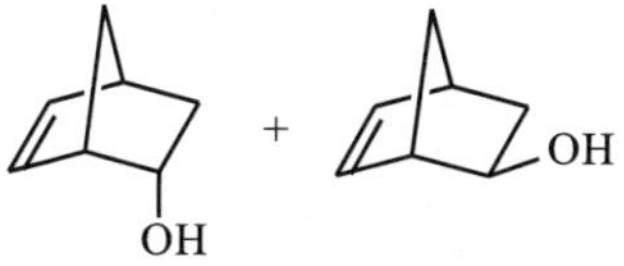

[a] Vinylborane source A: Br-9-BBN, CH_2=$CHSnBu_3$; B: $BrBMe_2$, CH_2=$CHSnBu_3$; C: (+)-camphene, H_2BBr, CH_2=$CHSnBu_3$; D: Br-9-BBN, $(CH_2{=}CH)_4Sn$.

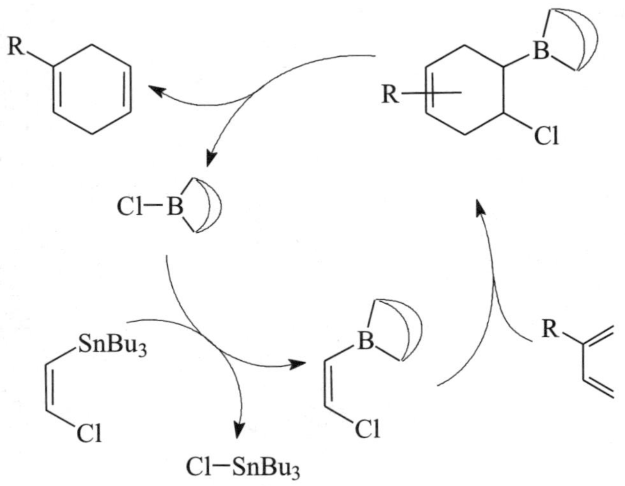

Scheme 30.1

Singleton and Redman [22] have found that *trans*-1,2-bis(catecholboryl)-ethylene is an air-stable crystalline solid, whereas *trans*-1,2-bis(9-BBN)ethylene (Chart 30.2) is less stable and is maintained in solution only; as its acetonitrile complex, both reagents function as *trans*-dihydroxyethylene equivalents for Diels-Alder reactions. Cyclohexenediols obtained from reactions of catecholboryl derivatives are summarized in Table 30.5 [22].

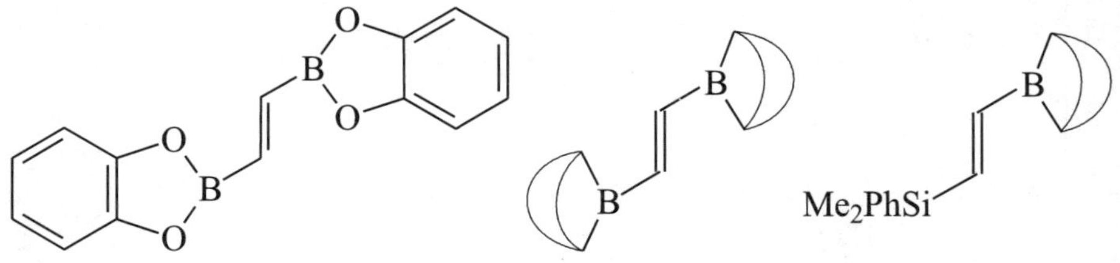

Chart 30.2

The reactivity and selectivity of *trans*-2-phenyldimethylsilylvinyl-9-BBN is almost comparable with 2-trimethylsilylvinyl-9-BBN [23]. The results are summarized in Scheme 30.2 [22]. The use of this reagent as a dihydroxyethylene equivalent is indirect, and it has the advantage as the latent hydroxyl groups are automatically differentiated, and this differentiation may be used for further stereochemical elaboration (Chart 30.3). Another advantage of using this re-

Table 30.5 Cyclohexenediols from reactions of *trans*-1,2-bis(catecholboryl) ethylene [22]

Diene	Reaction conditions and yield	Product	Diene	Reaction conditions and yield	Product(s)
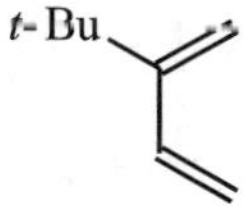	100 °C, 1h 70%	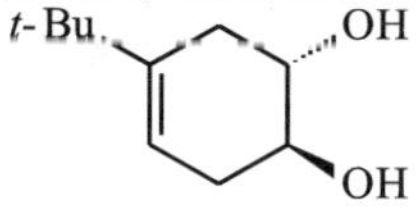		Et_3N 100 °C, 1h 71%	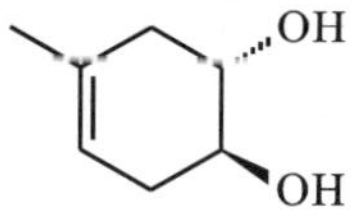
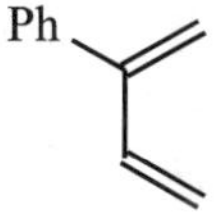	Xylene, Et_3N 100 °C, 24h 69%	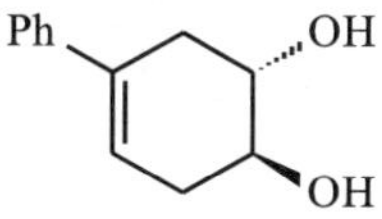	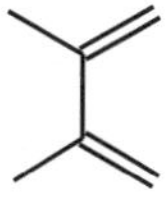	100 °C, 1h 68%	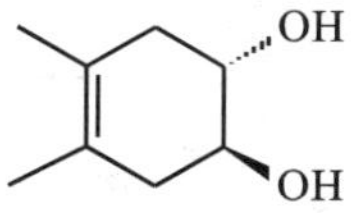
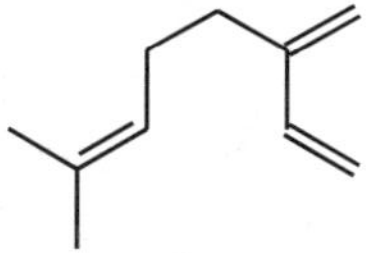	Et_3N 100 °C, 5 h 88%	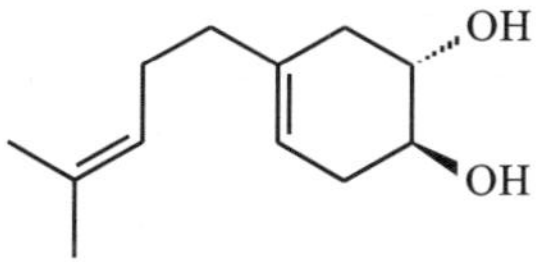		Et_3N 100 °C, 4 h 50% 2:1 mixture of isomers	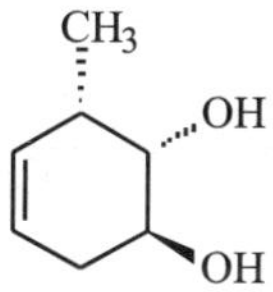
	Et_3N 100 °C, 4h 82%	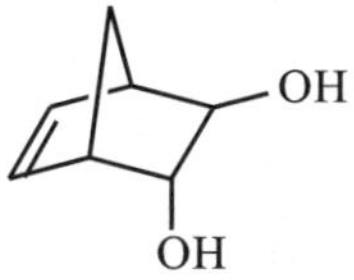			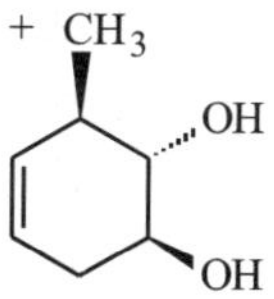

agent is that alternative borane conversions of the initial Diels-Alder adduct are possible, which makes it more versatile than catecholboryl derivative. It is to be noted that basic hydrogen peroxide oxidation affords poor yields, while oxidation with $NaBO_3$ and trimethylamine oxide [24] are inconsistent.

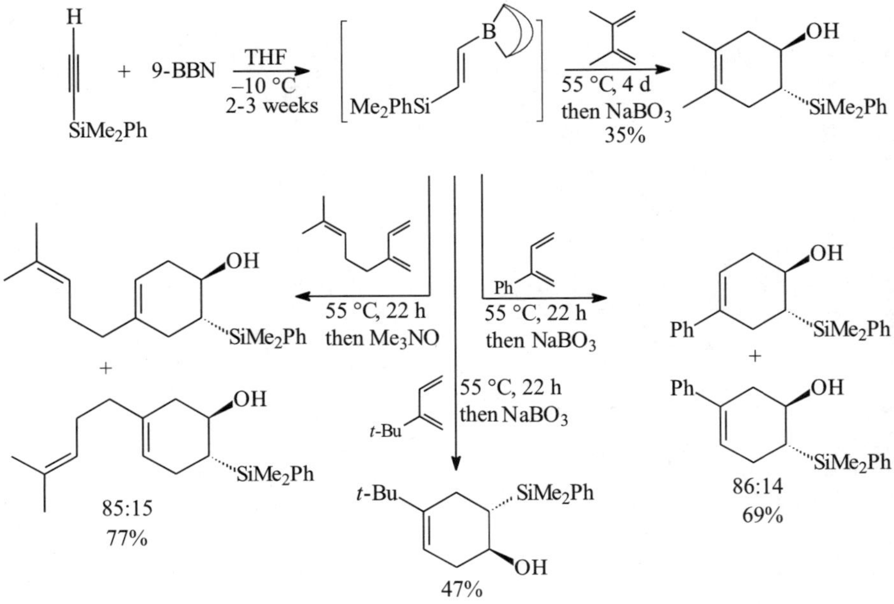

Scheme 30.2

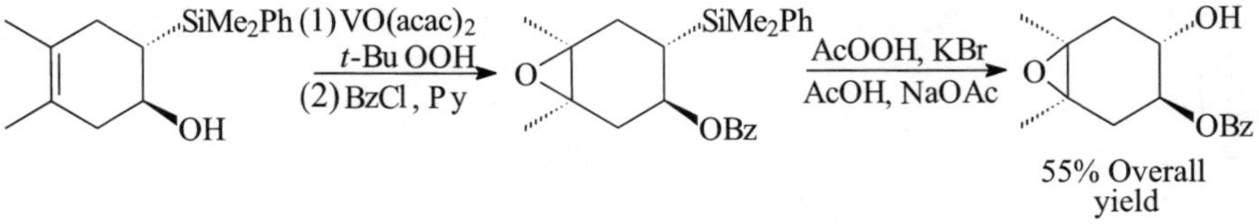

Chart 30.3

As compared with the intermolecular Diels-Alder reaction, the applicability of intramolecular Diels-Alder reactions [25] is more limited. The synthesis of requisite trienes for intramolecular Diels-Alder reactions is itself a significant task, and the incorporation of a ketene equivalent, for example, adds to the burden. Singleton and Lee [26] have reported the first examples of intramolecular Diels-Alder reaction on vinylboranes, which provide a two-step, very highly stereoselective decalin synthesis. The precursor to vinylborane is obtained by the Gassman one-pot triene synthesis [27] (Scheme 30.3).

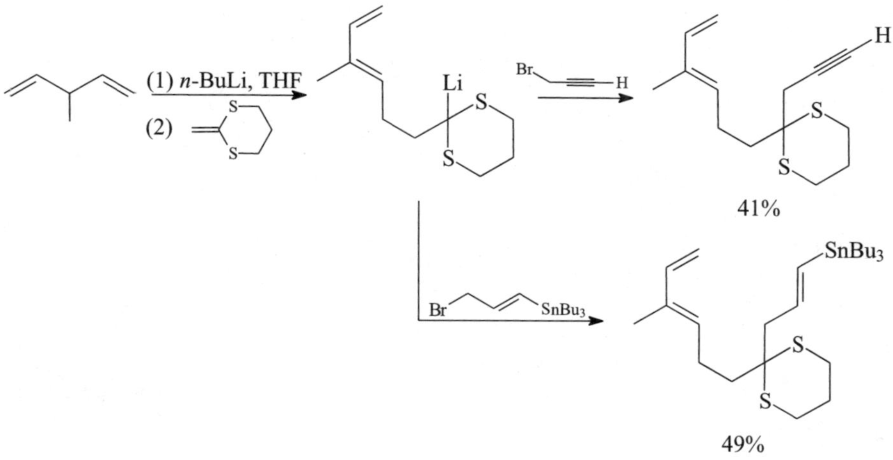

Scheme 30.3

1,7,9-Decatrienylboranes generated *in situ*, from either hydroboration of acetylene derivative with dicyclohexylborane or boron–tin exchange using 9-bromo-9-BBN with tributylstannate derivative, followed by heating to 75 °C in a sealed tube and oxidation, affording a single isomer (Chart 30.4) [26]. The reactions proceed with high stereoselective intramolecular fashion to afford the "*endo*" Diels-Alder product.

It is significant to mention that cycloalkenylboranes (Chart 30.5) for the synthesis of bicyclics are generated [28] *in situ*. The, boron–tin exchange reaction takes place between cyclopentenyl- and cyclohexenyltributylstannates [29] with BBr_3, BCl_3 or 9-bromo-9-BBN [13]. The cycloalkenyl-9-BBN does not undergo Diels-Alder reactions with dienes, but cycloalkenylbromoboranes are uniquely reactive with simple dienes at 25 °C and exhibit exceptional selectivity [28].

In Diels-Alder reactions, acetylenes reactivity is very low. Consequently, many ingenious dienophiles acetylene equivalents have been developed [30–32]. However, several of these synthons are so unreactive that they are useless with all but the most reactive dienes, and an addition number would generally require prohibitively vigorous reaction conditions with complex acyclic dienes. Among the reactive acetylene equivalents; 1,2-bis(phenylsulfonyl)ethylenes and related compounds are in common use [33], but the final elimination step is often inefficient. 2-Phenylsulfonylnitroethylene is highly reactive and gentle. The gentle conditions (Bu_3SnH) for the final elimination [31] make it a valuable equivalent of acetylene, but it has seen little use, possibly owing to its more difficult availability. Though a number of efficient intramolecular transition-metal

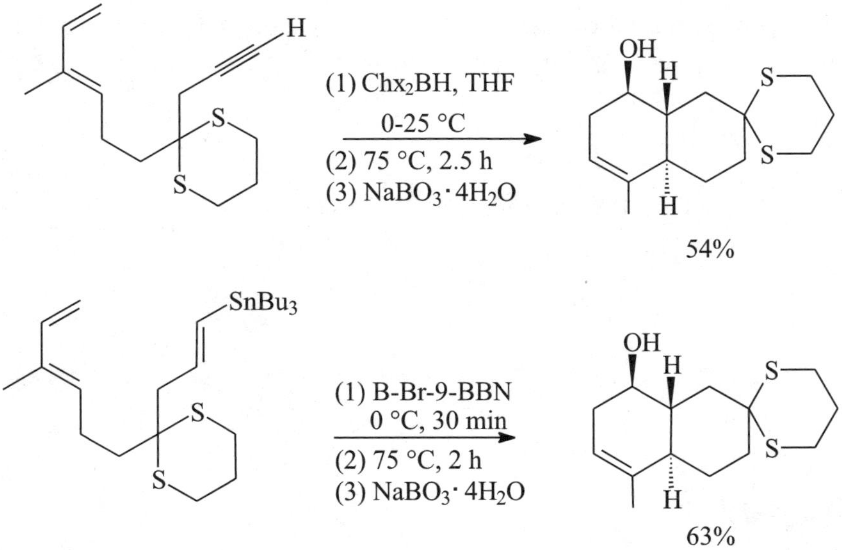

Chart 30.4

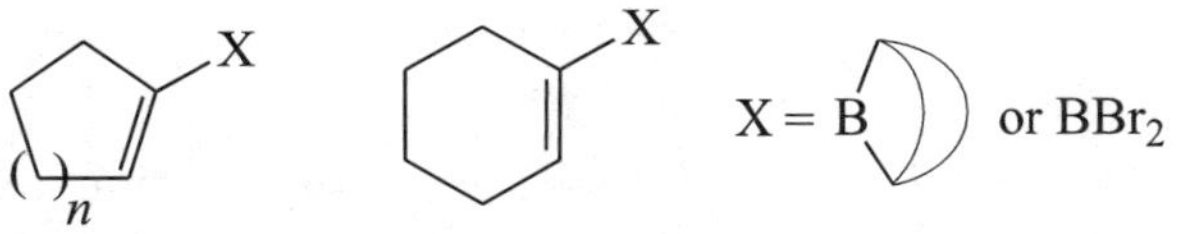

Chart 30.5

catalyzed [4+2]cyclo additions of alkynes are known [34], a lack of generality requires ready availability of acetylene equivalents. Singleton and Martinez [23] have reported the synthesis of 2-trimethylsilylvinylboranes and its utility as synthetic equivalents of 2-trimethylsilylvinyl alcohol and acetylene.

The hydroboration of trimethylsilylacetylene with 9-BBN is reported [35] to afford only 9% yield of 2-trimethylsilylvinyl-9-BBN, in the absence of solvent. However, Singleton and Martinez [23] have successfully achieved the synthesis of 2-trimethylsilylvinyl-9-BBN (Eq. 30.2) in 63% yield by reacting 9-BBN with 1.5 equiv of trimethylsilylacetylene in dilute (≈0.1 M) solution in THF for 4 days at 5 °C, followed by distillation of the reaction mixture [b.p. 60 °C (0.005 mmol)]. 2-Trimethylsilylvinyl-9-BBN is highly air sensitive and pyrophoric but is considerably more stable than unsubstituted vinylborane and can be stored at room temperature for weeks without noticeable decomposition.

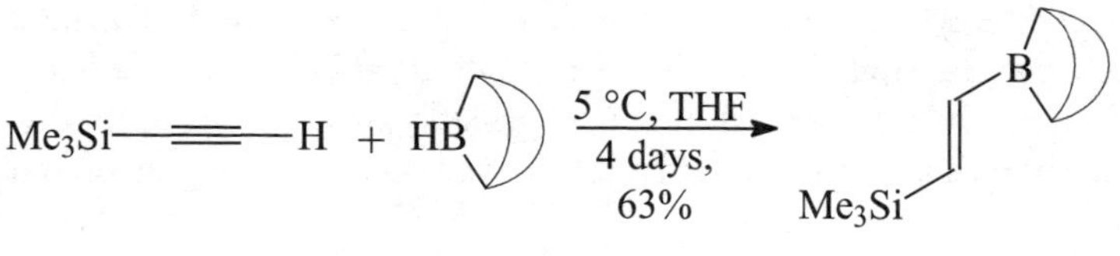

(30.2)

It reacts with *trans*-piperylene with high regioselectivity and *endo* stereoselectivity; no *exo* product is observed (Table 30.6) [23]. However, the major isomer is reversed from that observed with vinyl-9-BBN, which has afforded only *meta* substituted products [5].

The purified alcohols are converted to the corresponding 1,4-cyclohexadienes, quantitatively and cleanly on warming with 1 mol% of *p*-toluenesulfonic acid.

The boron adduct, obtained from 2-*tert*-butylbutadiene and 2-trimethylsilylvinyl-9-BBN, on treatment with 6 equiv of acetaldehyde affords 1-*tert*-butyl-5-trimethylsilyl-1,3-cyclohexadiene (Scheme 30.4). This diene in the presence of oxygen is converted to aromatic derivative, a versatile substrate for electrophilic substitutions.

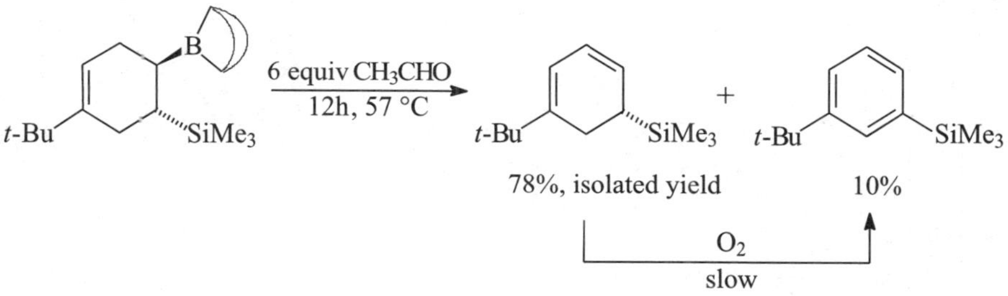

Scheme 30.4

Thus, convenient synthesis of trimethylsilylvinyl-9-BBN, its high reactivity, and the simple and efficient transformations of the adducts to 1,3-cyclohexadienes makes this an attractive acetylene equivalent in Diels-Alder reactions.

Diels-Alder reactions of simple 2-alkyl- or 2-alkoxybutadienes with activated dienophiles lead to preferential formation of the 1,4-disubstituted ("*para*") product (Scheme 30.5) over the 1,3-disubstituted ("*meta*") product. Although the regiochemistry of more complex reactions is not always so predictable, the electronic preference for the "*para*" product in simple reactions has been invariable.

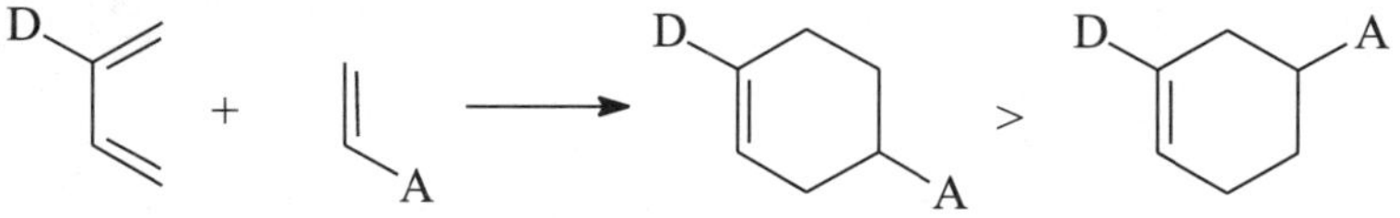

D = R, OR; A = CHO, C(O)R, CO_2R, SO_2R, CN, BR_2, $P(O)(OEt)_2$, NO_2 etc

Scheme 30.5

Table 30.6 Reaction conditions, products and yields for Diels-Alder reactions with 2-trimethylsilylvinyl-9-BBN [23]

Diene	Reaction conditions	Product(s) and ratio	Yield (%)
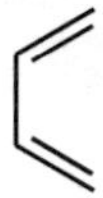	22 h, 50 °C	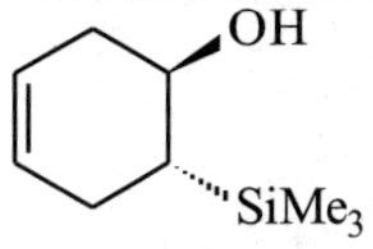	85
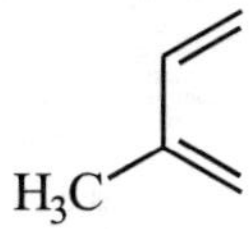	24 h, 45 °C	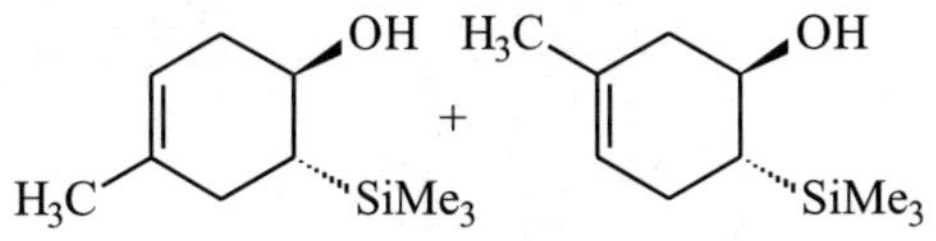 85:15	73
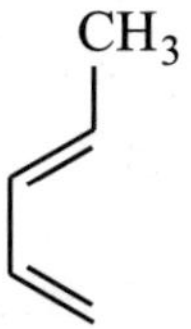	144 h, 85 °C	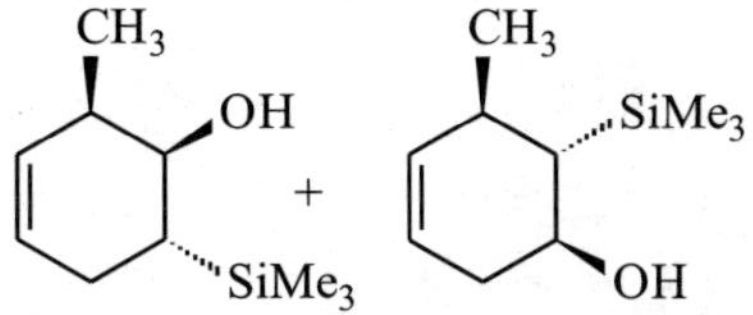 88:12	63

Table 30.6 *(continued)* Reaction conditions, products and yields for Diels-Alder reactions with 2-trimethylsilylvinyl-9-BBN [23]

Diene	Reaction conditions	Product(s) and ratio	Yield (%)
	20 h, 25 °C	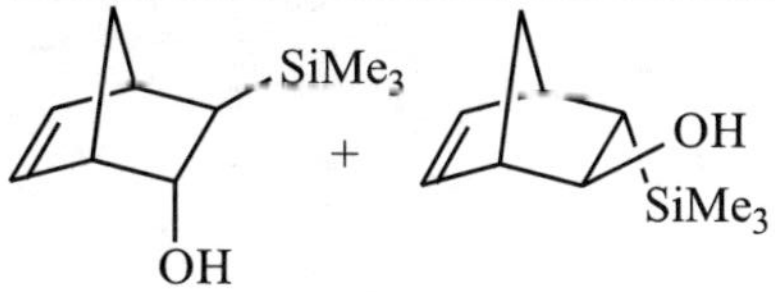 3:1	87
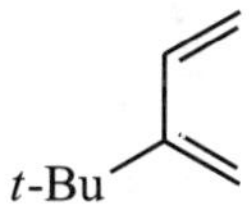	11 h, 25 °C	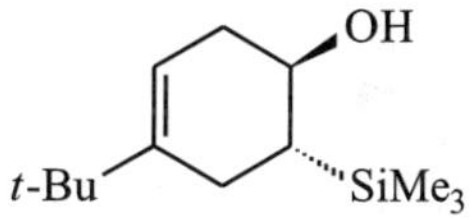 >98%	87

An unprecedented electronic preference for the "*meta*" product in Diels-Alder reaction of [(trimethylsilyl)ethynyl]-9-BBN has been reported [36]. [(Trimethylsilyl) ethynyl]-9-BBN·THF is prepared (Eq. 30.3) [36] in 97% yield by the reaction of (trimethylsilyl) ethynyllithium and *B*-methoxy-9-BBN. The reagent [(trimethylsilyl) ethynyl]-9-BBN·THF is crystalline but is air sensitive. However, the reagent is indefinitely stable in a freezer under an inert atmosphere.

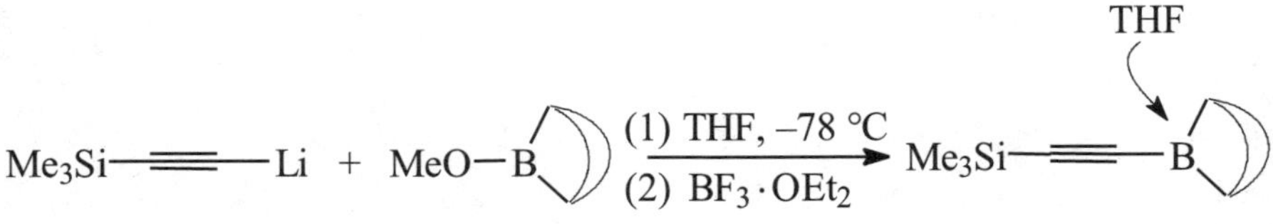

(30.3)

The difficulties associated with the isolation and manipulation of this air-sensitive reagent are avoided by its *in situ* formation (Eq. 30.4) [36] from [(trimethylsilyl) ethynyl]tributyltin [37] and 9-bromo-9-BBN. The reaction is instantaneous and quantitative at 25 °C.

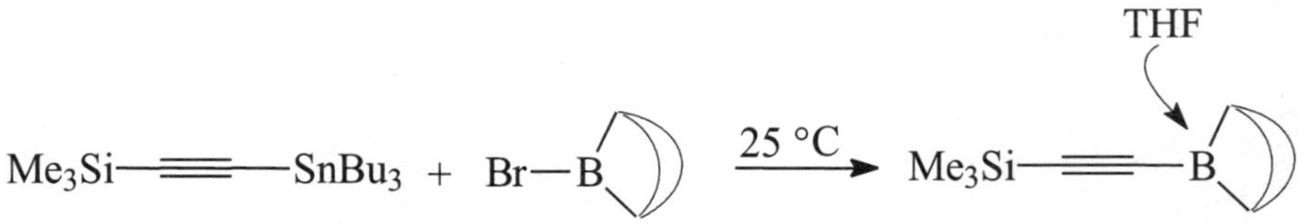

(30.4)

The Diels-Alder reactions of the reagent with dienes are summarized in Scheme 30.6 [36].

The novel regioselectivity [38] of these reactions has been found to be consistent with an *ab initio* prediction of advanced bonding to boron in a [4 atom + 3 atom] transition state [36].

Bis(9-borabicyclo[3.3.1]nonyl)acetylene is rapidly prepared [39] *in situ*, utilizing neat bis(tributylstannyl)acetylene and 2 equiv of 9-bromo-9-BBN at –78 °C. The reagent is an effective dienophile and undergoes reactions with simple acyclic dienes at 110 °C. The 1,4-cyclohexadienes are obtained in good yields after debromination of the Diels-Alder adducts with acetic acid (Chart 30.6) [39].

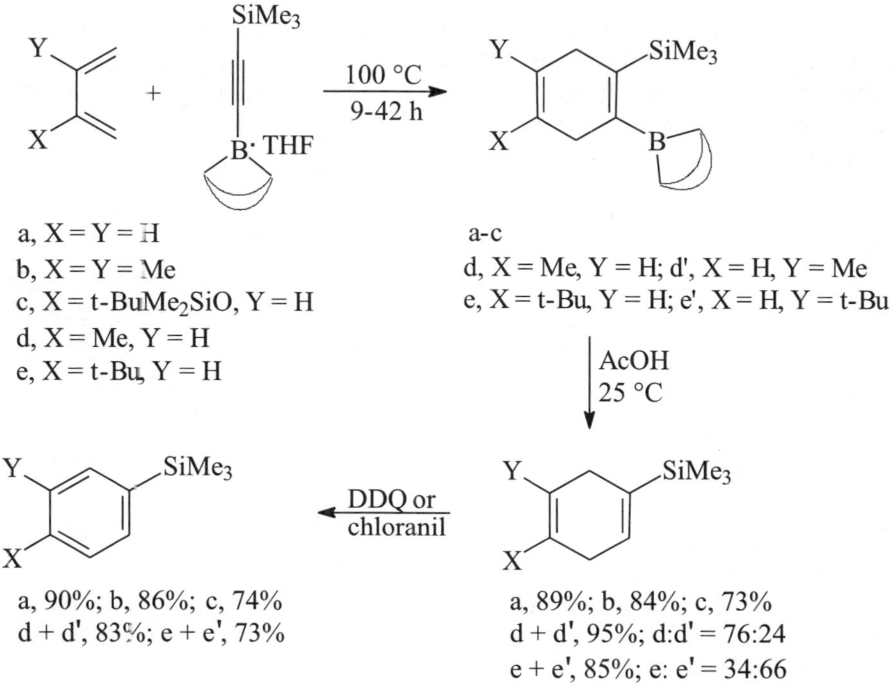

Scheme 30.6

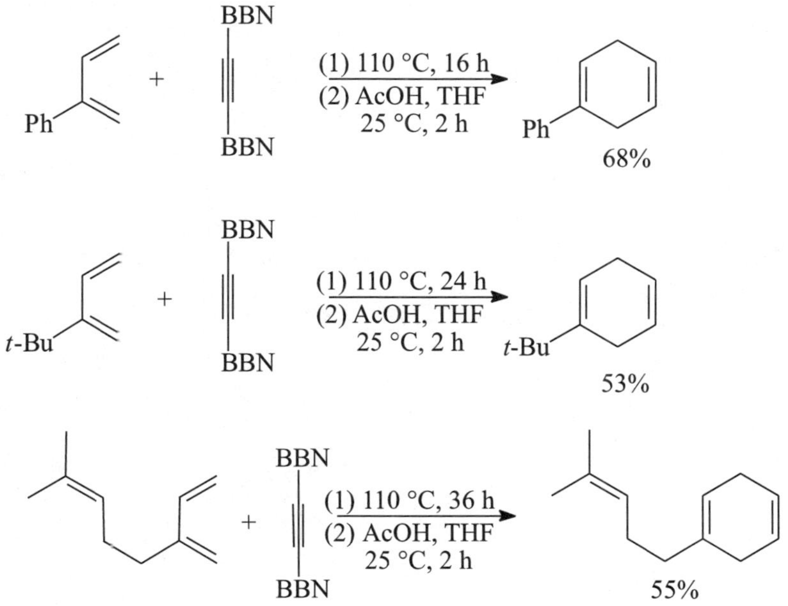

Chart 30.6

Singleton and coworkers [40] have studied the regioselective behaviors of six 2-alkenyl- and *cis*-1-alkenylboranes toward dienophiles for Diels-Alder reaction. The boranes are generated *in situ* from the reaction of BBr_3, BCl_3, 9-BBN or *B*-chloro-3,6-dimethylborepane [41] with 2-propenyltributyltin or *cis*-1-propenyltributyltin.

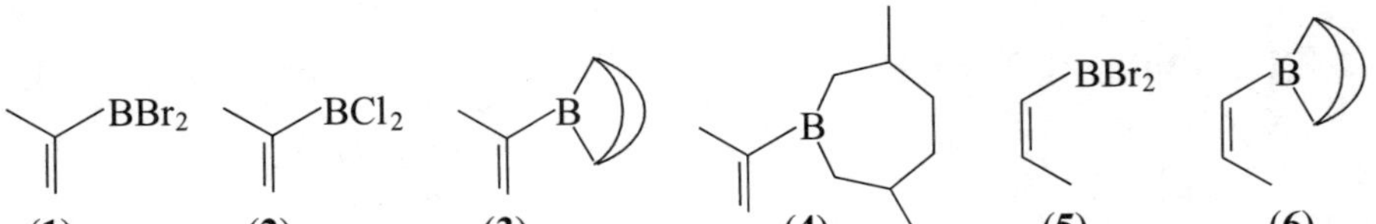

(1) **(2)** **(3)** **(4)** **(5)** **(6)**

The reactivity and selectivity of the alkenylboranes are summarized in Table 30.7 [40].

The relative reactivity of 2-propenylborane is **1** > **2** > **3** ≈ **4**; **1** reacts rapidly with simple dienes at room temperature, whereas **3** and **4** require longer reaction times at 55 °C.

The data in Table 30.7 [40] reveal that the alkenyldihalo- and alkenyldialkylboranes consistently favor regiochemically reversed products. The alkenyldihaloboranes affords the unusual "*meta*" (1,3) as the major products, whereas the alkenyldialkylboranes affords the normal "*para*"(1,4) as the major products. The regioselectivities of **3** and **5** are consistently outstanding (90:10 in the worst case scenario).

Table 30.7 Reaction conditions, products and yields for Diels-Alder reactions of alkenylboranes [40]

Diene	Alkenylborane[a]	Reaction conditions	Yield (%) (ratio)	Products(s)
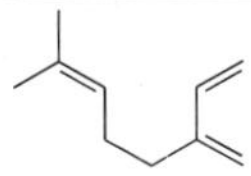	**1** **2** **3**	25 °C, 3 h 25 °C, 22 h 55 °C, 30 h	77 (100:0) 74 (80:20) 68 (10:90)	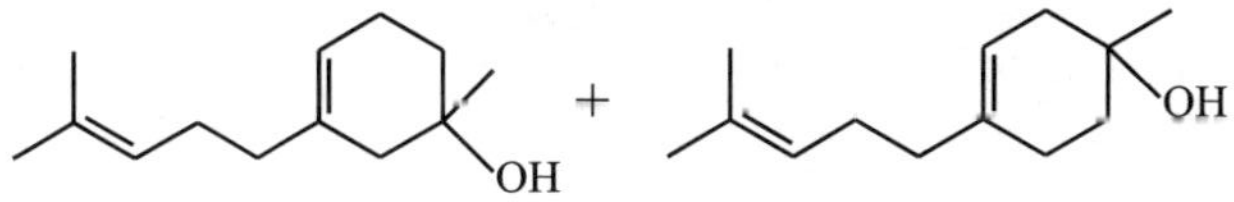
	5 **6**	25 °C, 24 h 90 °C, 4 days	75 (100:0) 55 (31:69)	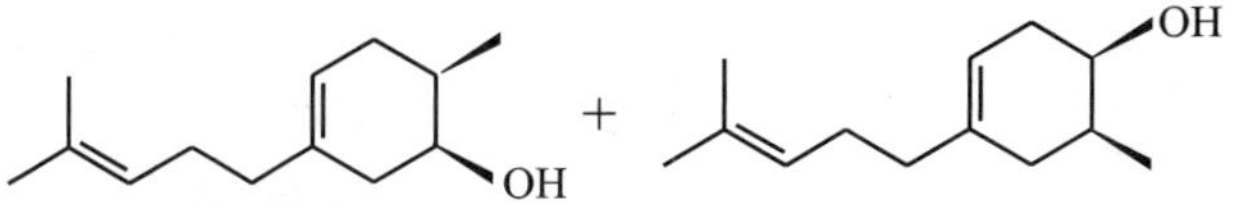
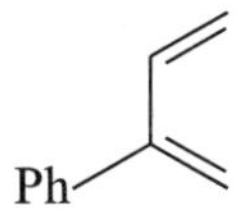	**1**	25 °C, 2 h	90 (87:13)	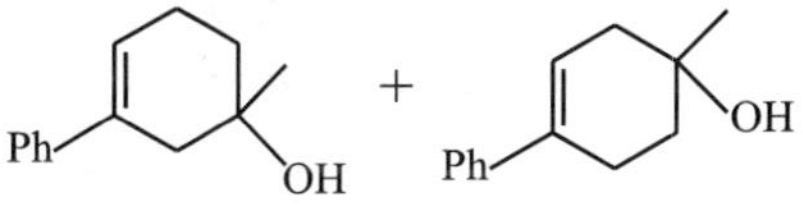
	5	25 °C, 24 h	90 (87:13)	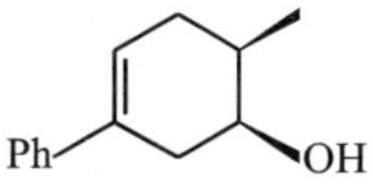
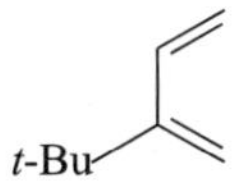	**1** **3** **4**	25 °C, 4.5 h 25 °C, 12 h 55 °C, 5 days	77 (68:32) 72 (0:100) 72 (14:86)	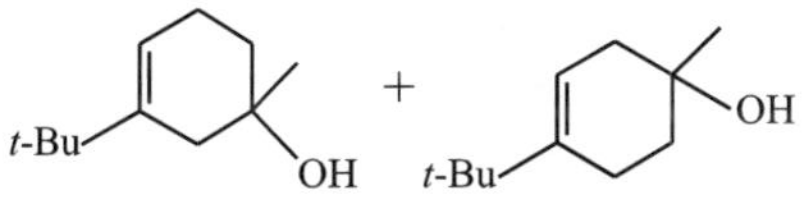
	5 **6**	25 °C, 24 h 45 °C, 3 days	74 (90:10) 55[b] (24:76)	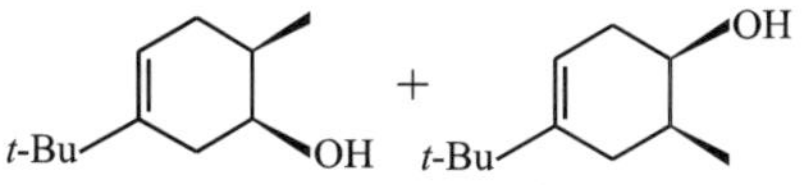

[a] A 1.3-fold excess of alkenylborane is used to remove adventitious protic acid.
[b] This reaction is not taken to completion.

Table 30.7 *(continued)* Reaction conditions, products and yields for Diels-Alder reactions of alkenylboranes [40]

Diene	Alkenylborane[a]	Reaction conditions	Yield (%) (ratio)	Products(s)
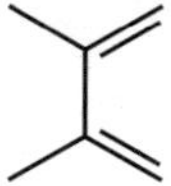	**1**	25 °C, 4 h	75	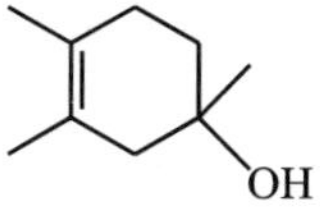
	5	25 °C, 24 h	80	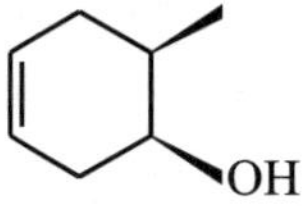
	**3**	55 °C, 2 days	70 (93:7)	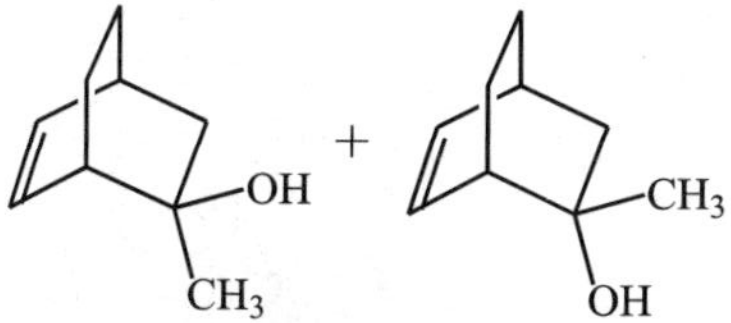

[a] A 1.3-fold excess of alkenylborane is used to remove adventitious protic acid.
[b] This reaction is not taken to completion.

References

1. Matteson DS (1960) J Am Chem Soc 82:4228
2. Matteson DS, Waldbillig JO (1963) J Org Chem 28:366
3. Evans DA, Scott WL, Truesdale LK (1972) Tetrahedron Lett 121; Evans DA, Golob AM, Mandel NS, Mandel GS (1978) J Am Chem Soc 100:8170
4. Coindard G, Braun J (1972) Bull Chim Soc Fr 817
5. (a) Singleton DA, Martinez JP (1990) J Am Chem Soc 112:7423; (b) Singleton DA, Martinez JP (1991) Chemtracts Organic Chemistry 4:41
6. (a) Pelter A, Buss D, Colclough E (1987) J Chem Soc Chem Commun 297; (b) Pelter A, Singaram B, Williams L, Wilson JW (1983) Tetrahedron Lett 24:623
7. Cooke MP Jr, Widener RK (1987) J Am Chem Soc 109:931
8. Inukai T, Kojima T (1967) J Org Chem 32:872
9. Kaplan RB, Shechter H (1961) J Org Chem 26:982
10. (a) Carr RVC, Paquette LA (1980) J Am Chem Soc 102:853; (b) Martinez-Fresneda P, Vaultier M (1989) Tetrahedron Lett 30:2929
11. Wulff WD, Yang DC (1983) J Am Chem Soc 105:6726
12. Snyder HR, Anderson HV, Hallada DP (1951) J Am Chem Soc 73:3258
13. Singleton DA, Martinez JP, Ndip GM (1992) J Org Chem 57:5768
14. Inukai T Kojima T (1966) J Org Chem 31:1121
15. Inukai T, Kojima T (1967) J Org Chem 32:869
16. Brown FK, Houk KN (1985) J Am Chem Soc 107:1971
17. Guner OF, Ottenbrite RM, Shillady DD, Alston PV (1988) J Org Chem 53:5348
18. Singleton DA, Martinez JP, Watson JV (1992) Tetrahedron Lett 33:1017
19. Singleton DA, Martinez JP, Watson JV, Ndip GM (1992) Tetrahedron 48:5831
20. Singleton DA, Leung S-W, Martinez JP, Lee Y-K (1997) Tetrahedron Lett 38:3163
21. (a) Good CD, Ritter DM (1962) J Am Chem Soc 84:1162; (b) Brinckman FE, Stone FGA (1960) J Am Chem Soc 82:6218; (c) Singleton DA, Martinez JP, Ndip GM (1992) J Org Chem 57:5768
22. Singleton DA, Redman AM (1994) Tetrahedron Lett 35:509
23. Singleton DA, Martinez JP (1991) Tetrahedron Lett 32:7365
24. Kabalka GW, Hedgecock HC (1975) J Org Chem 40:1776
25. For a leading review, see Roush WR (1991) Chapter 4.4. In: Trost B, Fleming I (eds) Comprehensive organic synthesis, vol. 5. Pergamon, New York
26. Singleton DA, Lee Y-K, (1995) Tetrahedron Lett 36:3473
27. Gassman, PG, Singleton, DA (1986) J Org Chem 51:3075
28. Lee Y-K, Singleton DA (1997) J Org Chem 62:2255
29. Behling JR, Babiak KA, Ng JS, Campbell AL, Moretti R, Koerner M, Lipshutz BH (1988) J Am Chem Soc 110:2641
30. (a) Kumar PR (1989) J Chem Soc Chem Commun 509; (b) Paquette LA, Moerck RE, Harirchian B, Magnus PD (1978) J Am Chem Soc 100:1597; (c) Davis AP, Whitham GH (1980) J Chem Soc Chem Commun 639; (d) De Lucchi O, Lucchini V, Pasquato L, Modena G (1984) J Org Chem 49:596; (e) Nakayama J, Nakamura Y, Hoshino M (1985) Heterocycles 23:1119; (f) For a review see De Lucchi O, Modena G (1984) Tetrahedron 40:2585
31. Ono N, Kamimura A, Kaji A (1986) Tetrahedron Lett 27:1595; Ono N, Kamimura A, Kaji A (1988) J Org Chem 53:251

32. Hermeling D, Schafer HJ (1984) Angew Chem Int Ed Engl 23:233
33. De Lucchi O, Fabbri D, Cossu S, Valle G (1991) J Org Chem 56:1888
34. (a) Jolly RS, Luedtke G, Sheehan D, Livinghouse T (1990) J Am Chem Soc 112:4965; (b) Wender PA, Jenkins TE (1989) J Am Chem Soc 111:6432
35. Soderquist JA, Colberg JC, DelValle L (1989) J Am Chem Soc 111:4873
36. Singleton DA, Leung S-W, (1992) J Org Chem 57:4796
37. Logue MW, Teng K (1982) J Org Chem 47:2549
38. A trimethylsilyl group has little effect on the regioselectivity of Diels-Alder reaction. For example see Hayama T, Tomoda S, Takeuchi Y, Nomura T (1984) J Org Chem 49:3235; Kloek JA (1981) J Org Chem 46:1951; Hermeling D, Schäfer HJ (1984) Angew Chem Int Ed Engl 23:233
39. Singleton DA, Schulmeier BE, Hang C, Thomas AA, Leung S-W, Merrigan SR (2001) Tetrahedron 57:5149
40. Singleton DA, Kim K, Martinez JP (1993) Tetrahedron Lett 34:3071
41. Brown HC, Zaidlewicz M (1976) J Am Chem Soc 98:4917

31 Suzuki Reaction

The Suzuki reaction is palladium-catalyzed coupling of organic halides or triflates with organoboranes under basic conditions. The organic halides and triflates react in the order of I > Br > OTf >> Cl. The reaction provides a highly versatile procedure for the construction of new carbon–carbon bond that tolerates many functional groups [1].

The cross-coupling reaction of *B*-R-9-BBN occurs readily with 1-haloalkenes or haloarenes in the presence of catalytic amount of $PdCl_2$(dppf) and bases such as sodium hydroxide, potassium carbonate, and potassium phosphate and affords the corresponding alkenes or arenes. The reaction does not take place in the absence of base due to poor nucleophilic character of R of *B*-R-9-BBN, and the reaction also fails with weak base such as sodium acetate.

A series of reactions are performed with iodobenzene and *B*-octyl-9-BBN to establish the reaction conditions (Eq. 31.1; Table 31.1) [2].

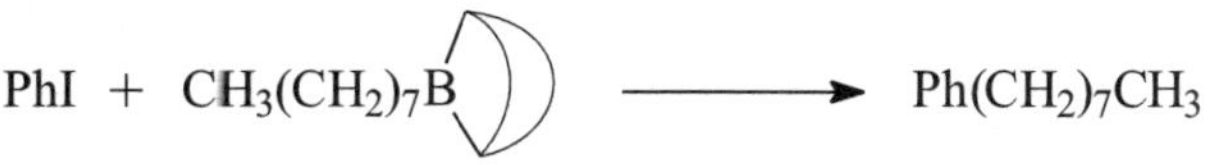

(31.1)

Table 31.1 Reaction conditions: cross-coupling of *B*-octyl-9-BBN with iodobenzene [2]

Catalyst	Base (equiv)	Solvent	Temp (°C)	Yield (%)
$PdCl_2$(dppf)	NaOH (3)	$THF–H_2O$ (5:1)	65	99
$PdCl_2$(dppf)	TlOH (1.5)	$THF–H_2O$ (5:1)	20	79
$PdCl_2$(dppf)	NaOMe (1.5)	THF	65	98
$PdCl_2$(dppf)	NaOMe (1.5)	THF–MeOH (5:1)	65	18
$PdCl_2$(dppf)	K_2CO_3 (2)	DMF	50	98
$PdCl_2$(dppf)	K_3PO_4	DMF	50	94
$Pd(PPh_3)_4$	NaOH (3)	$THF–H_2O$ (5:1)	65	84
$Pd(PPh_3)_4$	NaOH (3)	$Benzene–H_2O$	80	97

A comparison of various *B*-octylboron compounds in the reaction with iodobenzene demonstrates that hydroborating reagents such as 9-BBN, Sia_2BH, Chx_2BH and borane are effective in the Suzuki reaction (Table 31.2) [2].

Table 31.2 Coupling reaction of iodobenzene with various alkylboron compounds [2]

Borane	Yield (%)	Borane	Yield (%)
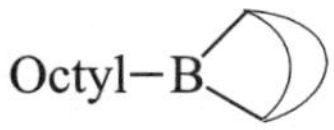	99	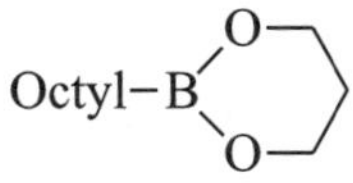	1
Octyl-B(Sia)$_2$	82	(2-Butyl)$_3$B	40
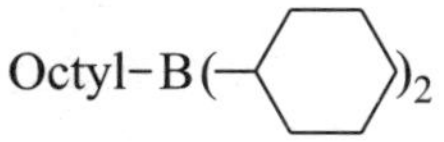	93	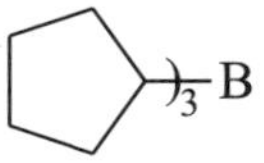	65
(Octyl)$_3$B	98	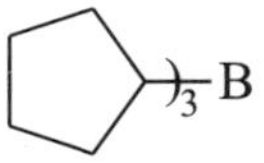	55

In the absence of base-sensitive groups in both alkylboranes and organic halides, $PdCl_2$(dppf) and sodium hydroxide in THF–H_2O works nicely (Table 31.3, procedure A). For functionalized alkylboranes and halides, powdered sodium methoxide suspended in THF (procedure B) accelerates the reaction. However procedure B is less effective for vinylic halides. A promising coupling is achieved employing either potassium carbonate (procedure C) or potassium phosphate (procedure D), both suspended in DMF at 50 °C. The results also reveal that coupling reaction is applicable to iodides and bromides of vinyl and aryl halides (Chart 31.1) [2].

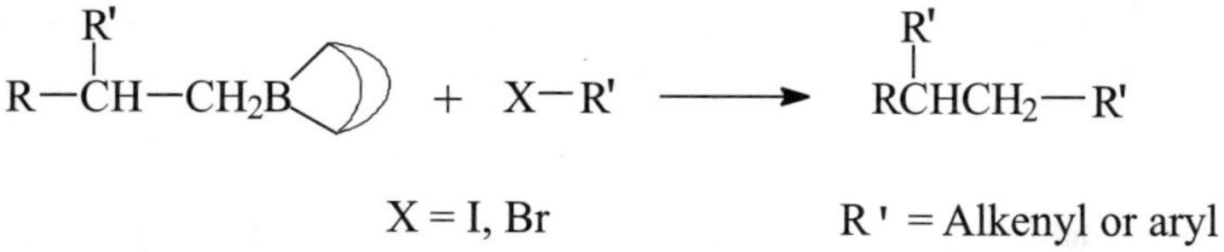

Chart 31.1

The results are summarized in Tables 31.3 and 31.4 [2]. A wide variety of functional groups such as cyano, ester, and carbonyl, etc., are tolerated under these conditions. It is significant to mention that none of the coupling products arising from cyclooctyl group is detected in the crude reaction mixture.

The reaction is very useful in the stereodefined synthesis of 1,5-alkadienes. Consequently, the selective hydroboration of a terminal double bond of triene, followed by coupling with ethyl-(*E*)-2-bromocrotonoate in DMF using $PdCl_2$(dppf) and K_3PO_4 affords the corresponding trienic ester in 60% yield. Similarly, coupling with (*Z*)-3-bromo-2-butanol tetrahydropyranyl ether affords a 67% yield of farnesol tetrahydropyranyl ether (Chart 31.2) [2].

Table 31.3 Cross-coupling reaction of 9-alkyl-9-BBN with haloarenes [2]

Entry	Haloarene	Alkene	Product	Percentage yield		
				Procedure		
				A	B	C
1	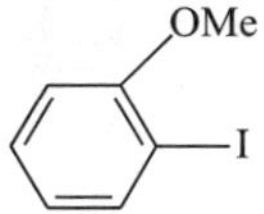	1-Octene	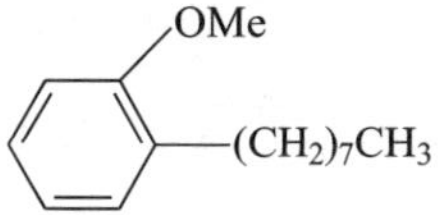	90	71	
2	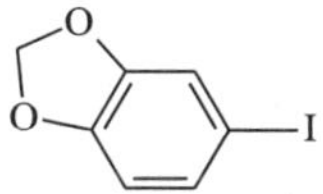	1-Octene	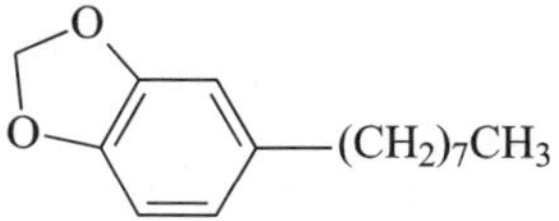		78	
3	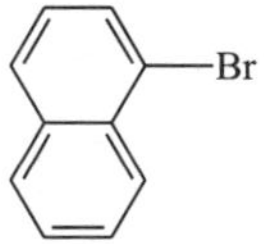	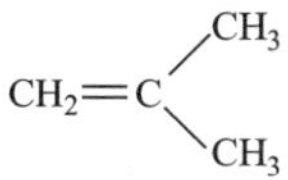	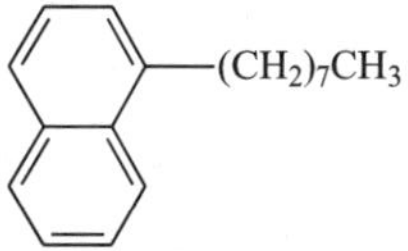	88	71	
4	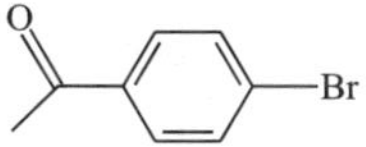	CH_2=CH$(CH_2)_8CO_2$Me	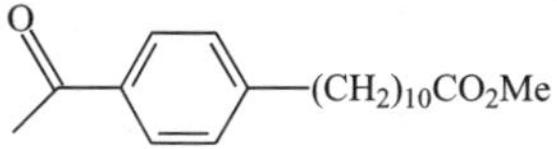			88
5	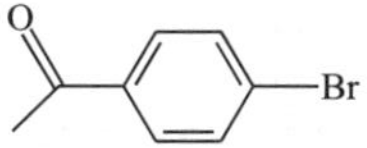	CH_2=CH$(CH_2)_8$CN	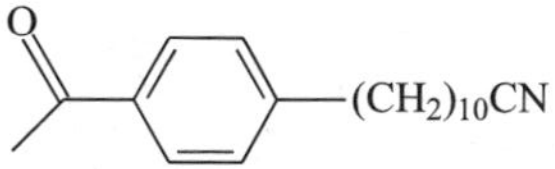			98

The coupling reaction with haloarene (1 equiv) conducted under the following conditions. Procedure A: $PdCl_2$(dppf) (3 mol%) and NaOH (3 equiv) in THF at 65 °C. Procedure B: $PdCl_2$(dppf) (3 mol%) and NaOH (1.5 equiv) in THF at 65 °C. Procedure C: $PdCl_2$(dppf) (3 mol%) and K_2CO_3 (2 equiv) in DMF–THF at 50 °C.

[a] A 30% yield of 7-(4-acetylphenyl)-2-methylheptane-1,2-diol is also obtained.

Table 31.3 *(continued)* Cross-coupling reaction of 9-alkyl-9-BBN with haloarenes [2]

				Percentage yield		
				Procedure		
Entry	Haloarene	Alkene	Product	A	B	C
6	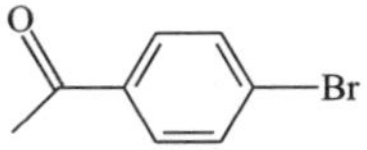	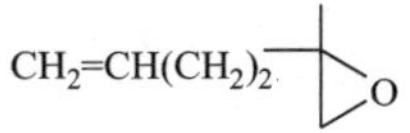	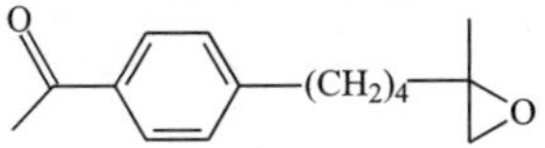			52[a]
7	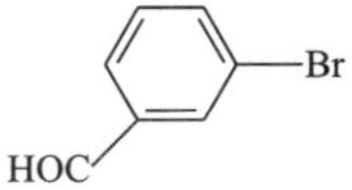	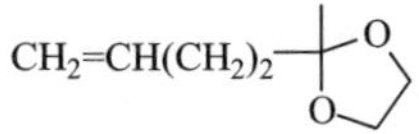	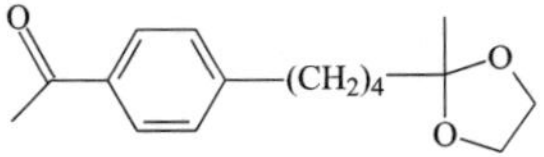			77
8	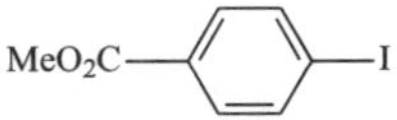	CH_2=CH$(CH_2)_8CO_2$Me	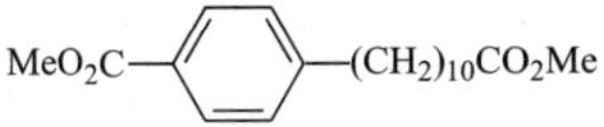		82	84
9	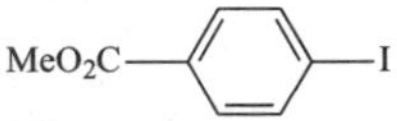		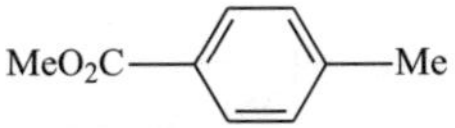		77	49

The coupling reaction with haloarene (1 equiv) conducted under the following conditions. Procedure A: $PdCl_2$(dppf) (3 mol%) and NaOH (3 equiv) in THF at 65 °C. Procedure B: $PdCl_2$(dppf) (3 mol%) and NaOH (1.5 equiv) in THF at 65 °C. Procedure C: $PdCl_2$(dppf) (3 mol%) and K_2CO_3 (2 equiv) in DMF–THF at 50 °C.

[a] A 30% yield of 7-(4-acetylphenyl)-2-methylhextane-1,2-diol is also obtained.

Table 31.4 Cross-coupling reaction of 9-alkyl-9-BBN with halo-1-alkenes [2]

Entry	Haloalkene	Alkene	Product	Percentage yield Procedure A	C	D
1	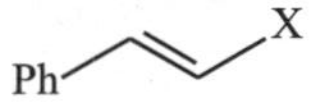	1-octene	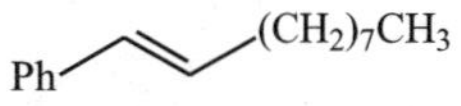	85 (X = Br)[a], 70 (X = I)[a]	82[b]	
2	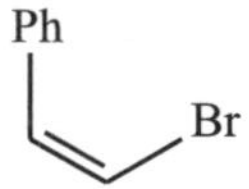	1-octene	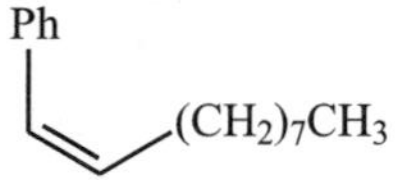	90[b]		
3	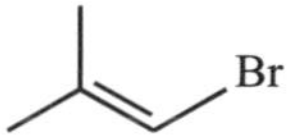	1-octene	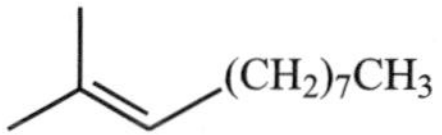	94		
4	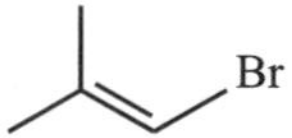	$CH_2{=}CH(CH_2)_8CO_2Me$	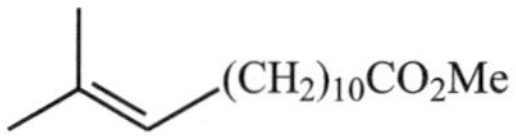		69	92
5	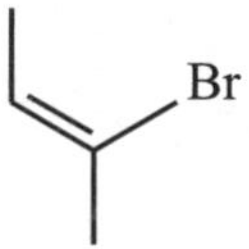	1-octene	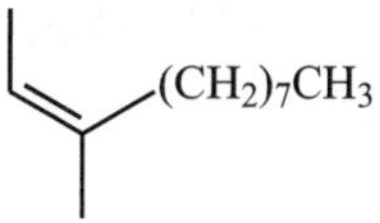	98		
6	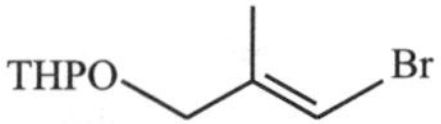	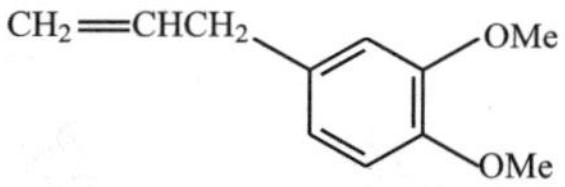	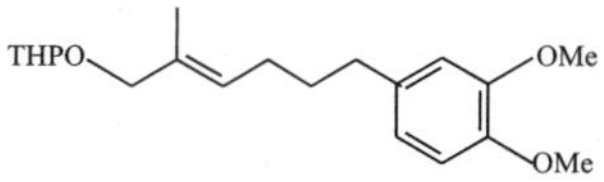	80		

Procedure A: $PdCl_2$(dppf)(3 mol%) and NaOH (3 equiv) in THF at 65 °C. Procedure C: $PdCl_2$(dppf) (3 mol%) and K_2CO_3 (2 equiv) in DMF–THF at 50 °C. Procedure D: $PdCl_2$(dppf) (3 mol%) and K_2CO_3 (1 equiv) in DMF–THF at 50 °C.

[a] β-Bromostyrene (*E*/*Z* = 1/99) gives decenylbenzene (*E*/*Z* = 3/97).

[b] β-Halostyrene (*E*/*Z* = 99/1) gives decenylbenzene (*E*/*Z* = 99/1).

Table 31.4 *(continued)* Cross-coupling reaction of 9-alkyl-9-BBN with halo-1-alkenes [2]

Entry	Haloalkene	Alkene	Product	Percentage yield Procedure A	C	D
7	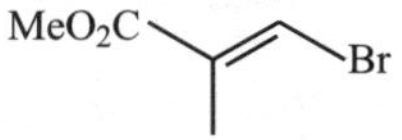	CH_2=CH$(CH_2)_8$CN	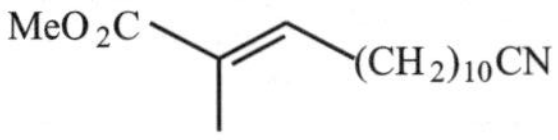		80	
8	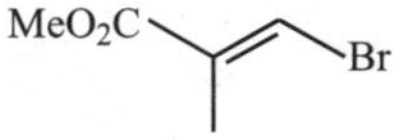	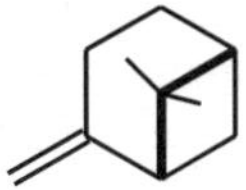	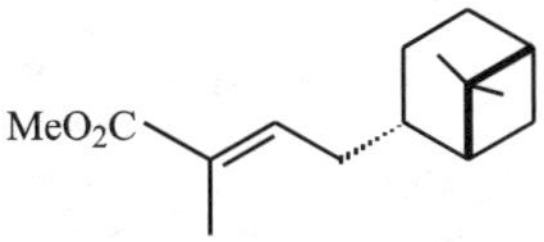		73	79
9	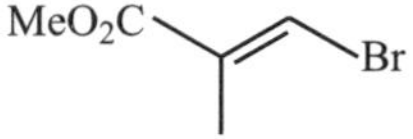	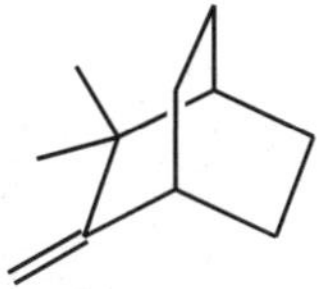	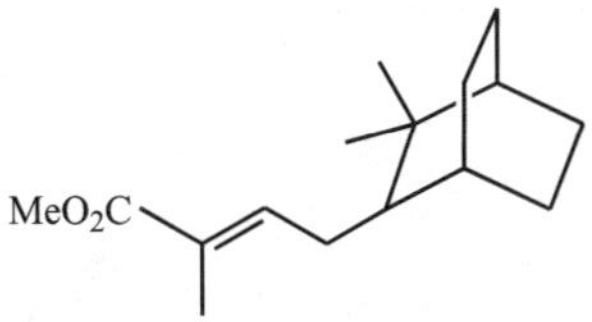		85	
10	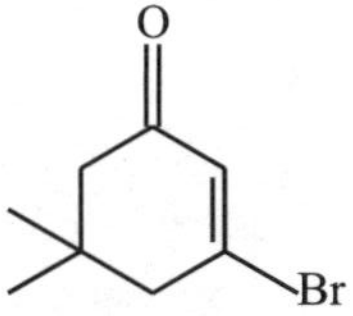	CH_2=CH$(CH_2)_8$CN	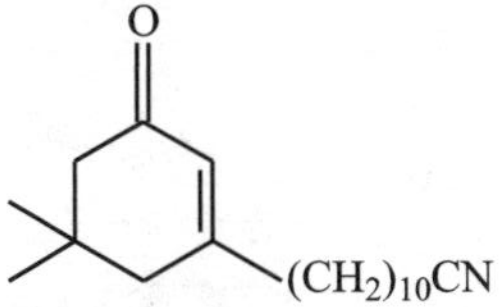		81	72

Procedure A: $PdCl_2$(dppf)(3 mol%) and NaOH (3 equiv) in THF at 65 °C. Procedure C: $PdCl_2$(dppf) (3 mol%) and K_2CO_3 (2 equiv) in DMF–THF at 50 °C. Procedure D: $PdCl_2$(dppf) (3 mol%) and K_2CO_3 (1 equiv) in DMF–THF at 50 °C.

[a] β-Bromostyrene (*E*/*Z* = 1/99) gives decenylbenzene (*E*/*Z* = 3/97).

[b] β-Halostyrene (*E*/*Z* = 99/1) gives decenylbenzene (*E*/*Z* = 99/1).

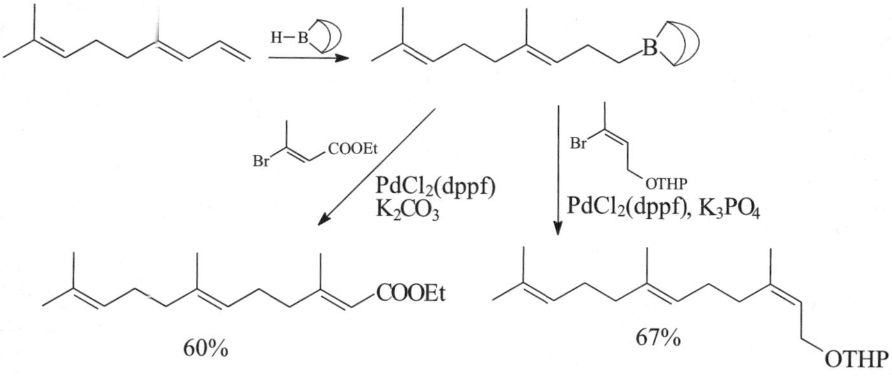

Chart 31.2

It has been reported that the hydroboration of 20(21)-methylene steroid, prepared from pregrenolone acetate by Wittig reaction, with 9-BBN produces predominantly (20*R*)-21-boryl steroid [3]. This boryl steroid on cross-coupling with ethyl(*E*)-β-bromomethacrylate affords the corresponding steroid with extended side chain and having natural configurations at C-20 (Eq. 31.2).

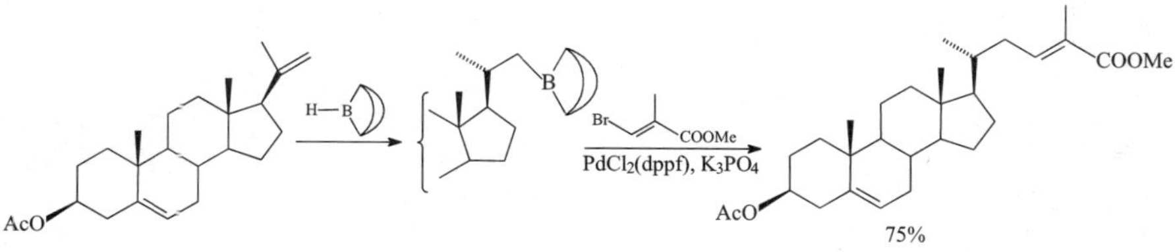

(31.2)

A combination of hydroboration and intramolecular cross-coupling provides a convenient method for the synthesis of five- and six-membered cycloalkenes and benzo-fused cycloalkanes and pyran derivatives (Scheme 31.1) [2].

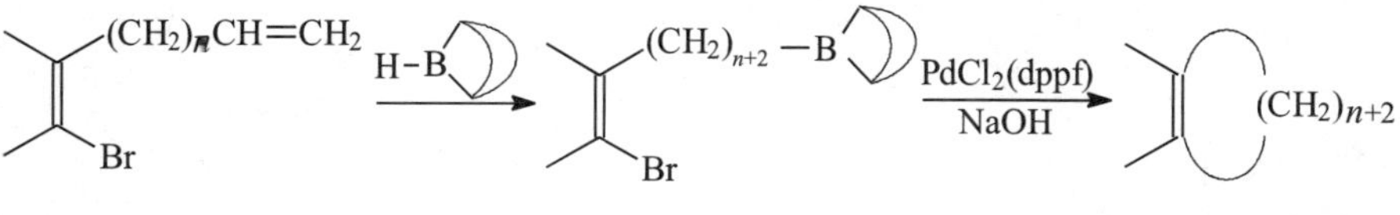

Scheme 31.1

The results are summarized in Table 31.5 [2].

Table 31.5 Cyclization of bromoalkenes *via* hydroboration and intramolecular cross-coupling reaction [2]

Haloalkene	Product	Yield (%)
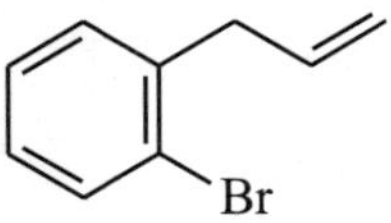	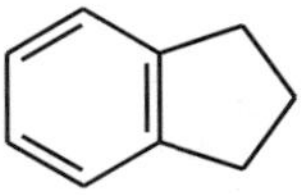	86
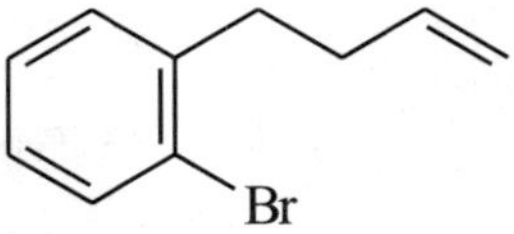	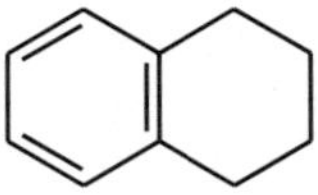	66
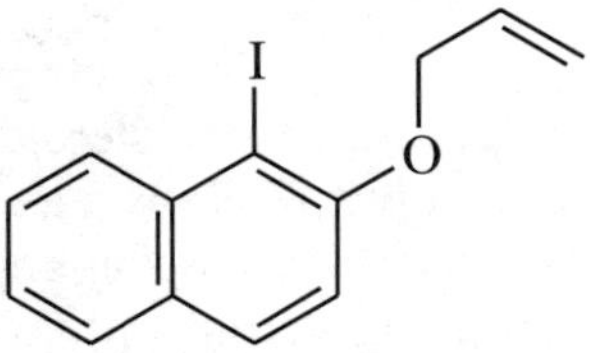	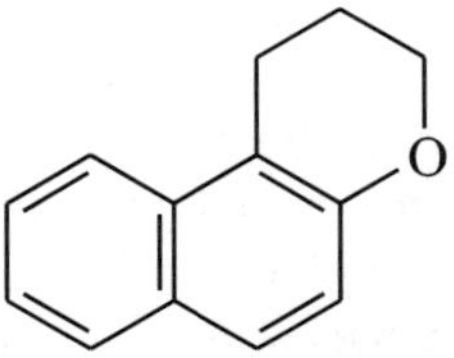	70
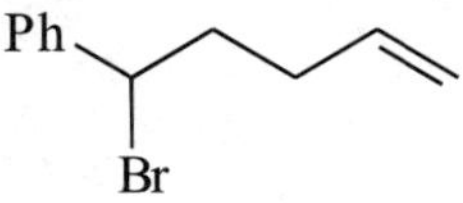	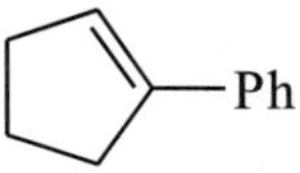	67
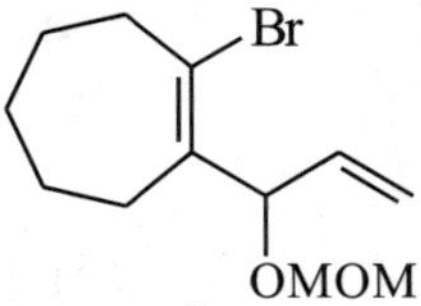	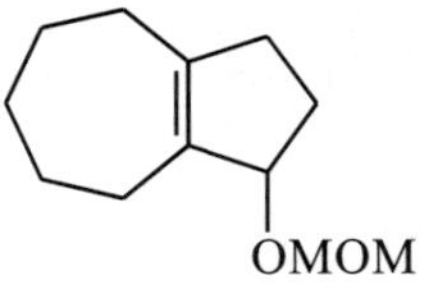	68
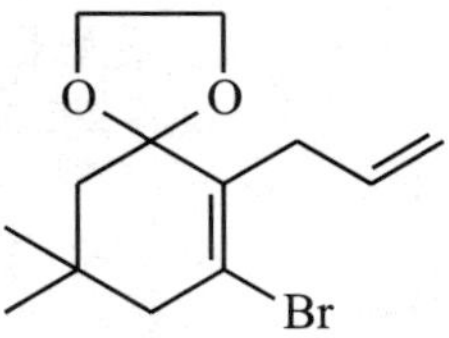	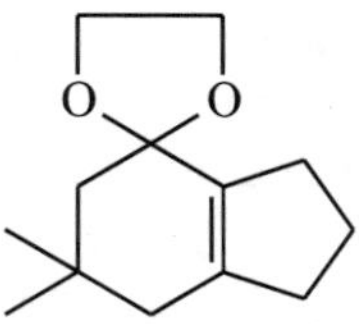	84
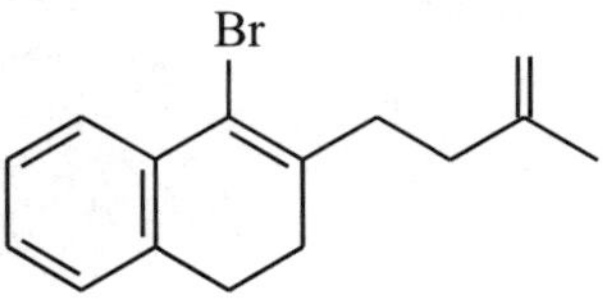	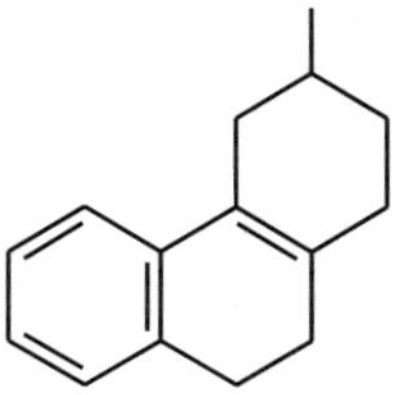	77

The stereodefined exocyclic olefins (Eq. 31.3) are readily synthesized by hydroboration of dienic bromide, followed by intramolecular coupling.

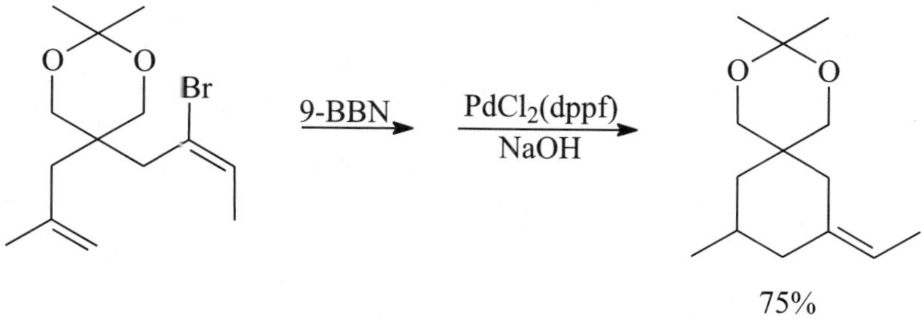

(31.3)

The reaction is extended to the synthesis of arylacetylenes and stereodefined enynes. The reactions sequence involves the preparation of thermally stable lithium complexes from *B*-OMe-9-BBN and alkyllithium in THF at –78 °C. These complexes undergo Pd-catalyzed Suzuki coupling to both aromatic and olefinic substrates to produce a variety of alkynyl derivatives and enynes, with complete retention of the double bond geometry (Eq. 31.4) [4a].

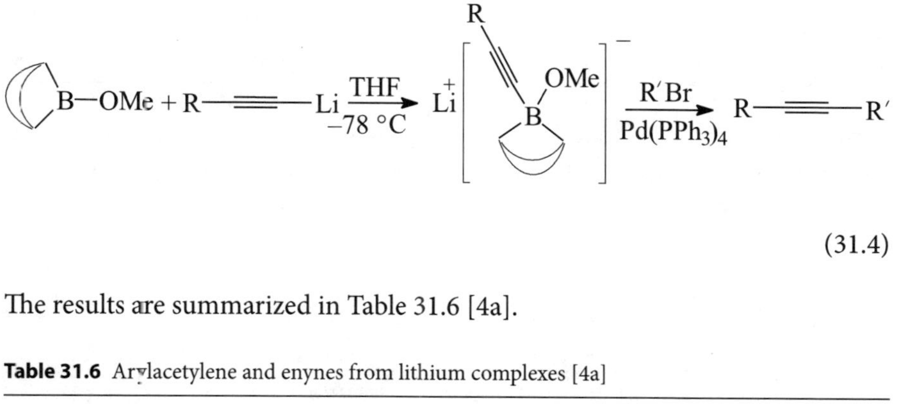

(31.4)

The results are summarized in Table 31.6 [4a].

Table 31.6 Arylacetylene and enynes from lithium complexes [4a]

R	R′	Yield (%)
n-Bu	C_6H_5	92
$SiMe_3$	C_6H_5	64
Ph	C_6H_5	94
n-Bu	*p*-$MeOC_6H_4$	68
$SiMe_3$	CH_2=CC_6H_5	88
t-Bu	*cis*-CH=CH-*t*-Bu	56
$SiMe_3$	*trans*-CH=CH-*n*-Bu	55

Soderquist and coworkers [4] have prepared 9-oxa derivatives of *B*-OMe-9-BBN and submitted its lithium complex to Suzuki cross-coupling to a variety of aryl and vinylbromides. The yields of both arylacetylenes and stereodefined enynes (Eq. 31.5) match or exceed those obtained in the 9-BBN [4a] process (Table 31.7) [4b].

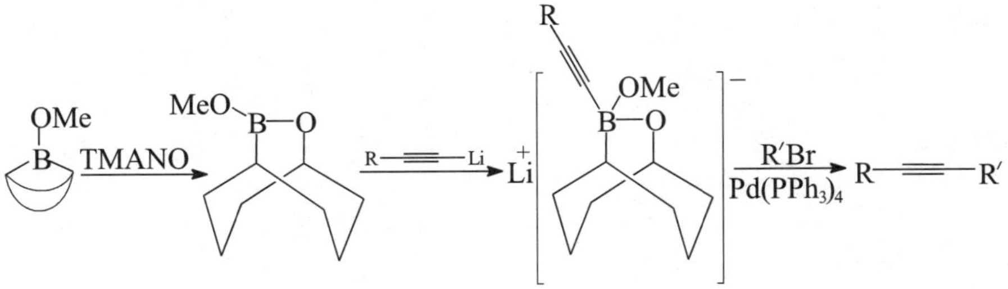

(31.5)

Table 31.7 Suzuki coupling with alkynylborate complexes [4b]

R	R′	Yield %
Me	C_6H_5	88
$SiMe_3$	*cis*-CH=CH-*n*-Bu	83
$SiMe_3$	*trans*-CH=CH-*n*-Bu	59
$SiMe_3$	*p*-$CH_3OC_6H_4$	62

The Markovnikov vinylborinates prepared from 9-oxa-10-borabicyclo[3.3.2]decanes undergo facile Suzuki coupling to alkylbromide and provide unsymmetrical 1,1-disubstituted alkenes (Table 31.8; Eq. 31.6) [5].

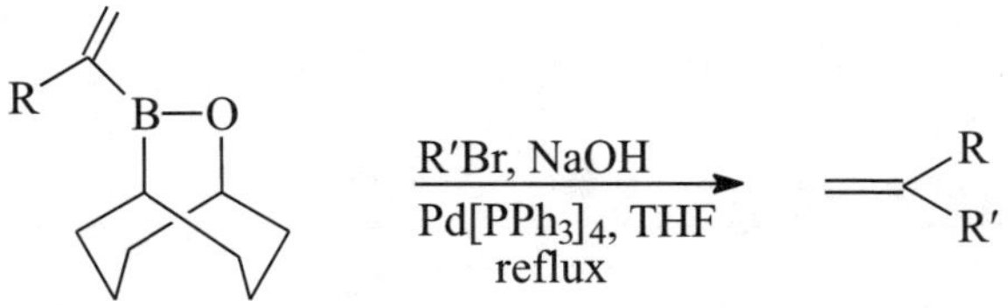

(31.6)

Table 31.8 Dienes and styrenes through Suzuki coupling [5]

R	R′	Yield %
t-Bu	CH_2=CPh	80
t-Bu	*cis*-CH=CHBu	59
n-Hx	*trans*-CH=CHMe	63
$Me_3SiCH_2CH_2$	*p*-Anisyl	68

The significance of Soderquist's 9-oxa-10-borabicyclo[3.3.2]decyl (OBBD) derivatives obtained by selective oxidation of 9-BBN derivatives with TMANO is demonstrated in the synthesis of (*S*)-(–)-frontalin [6] (Scheme 31.2), the aggregation pheromone of the pine beetle (*Dendroctonus frontalis*) by Suzuki coupling [7].

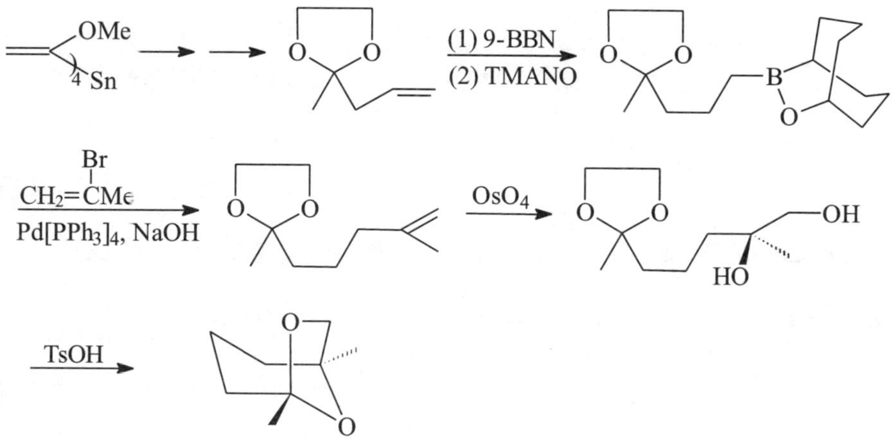

Scheme 31.2

Both carbon- and heterocyclic six-membered ring systems with an exocyclic carbon–carbon double bond have been synthesized (Table 31.9, 25–76%) from α,ω-dienes through double Suzuki cross-coupling of their dihydroboration adducts with either aromatic or olefinic vinylidene dibromide in a one-pot Pd-catalyzed sequence (Scheme 31.3) [8].

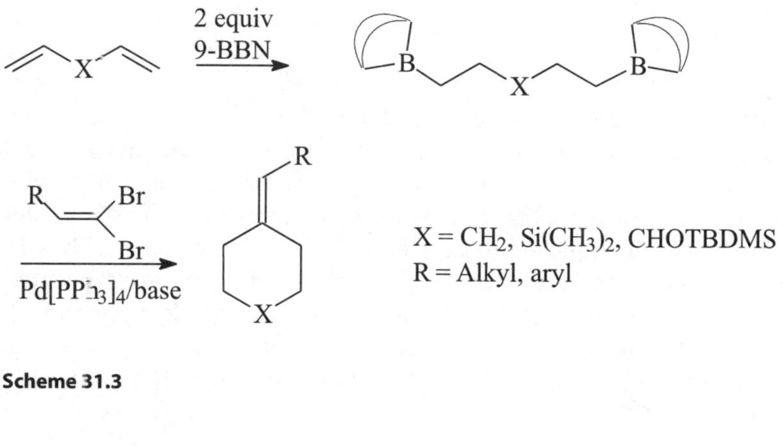

Scheme 31.3

gem-Dibromovinylidene intermediates are prepared either from 1-bromoalkynes *via* modified Zweifel method [9] or from aldehydes *via* the Fuchs-Corey method [10] (Eq. 31.7, R = Ar, 50–60%).

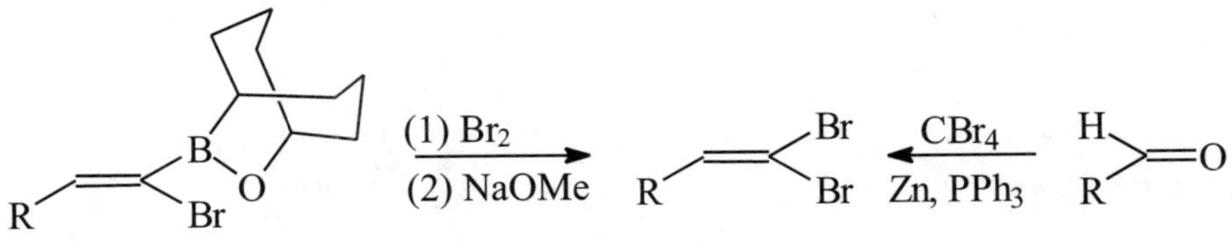

(31.7)

The results of double Suzuki coupling are summarized in Table 31.9 [8].

Table 31.9 Synthesis of six-membered ring via double Suzuki coupling [8]

X	R	Yield (%)	Time (h)
CH_2	Ph	57	4
CH_2	*p*-Biphenyl	42	2
CH_2	*p*-Tolyl	40	8
CH_2	*p*-Anisyl	76	8
CH_2	*n*-Bu	44	8
$Si(CH_3)_2$	Ph	48	1.25
$Si(CH_3)_2$	Biphenyl	60	2
$Si(CH_3)_2$	2-Furyl	36	8
$Si(CH_3)_2$	*p*-Tolyl	25	17
$Si(CH_3)_2$	*p*-Anisyl	60	1.25
$Si(CH_3)_2$	*n*-Bu	48	2
CHOTBDMS	Ph	38	3
CHOTBDMS	Me	64	3

Soderquist [11] has reported the synthesis of both *erythro* and *threo* isomers of *B*-(3,3-dimethyl-1,2-dideuterio-1-butyl)-9-BBN from 3,3-dimethyl-1-butyne through a hydroboration–deuteronolysis–hydroboration sequence employing first 9-BBN-H and then 9-BBN-D, or in reverse order, respectively. Whitesides' protocol [12] has been employed to determine the stereochemistry of B → Pd alkyl group transfer in the Suzuki coupling of both *erythro* and *threo* derivatives of 9-BBN to PhBr. It is found that transfer occurs with complete retention of configuration with respect to carbon (Scheme 31.4). The retention process suggests a four-centered hydroxy μ_2-bridged transition state model (Fig. 31.1).

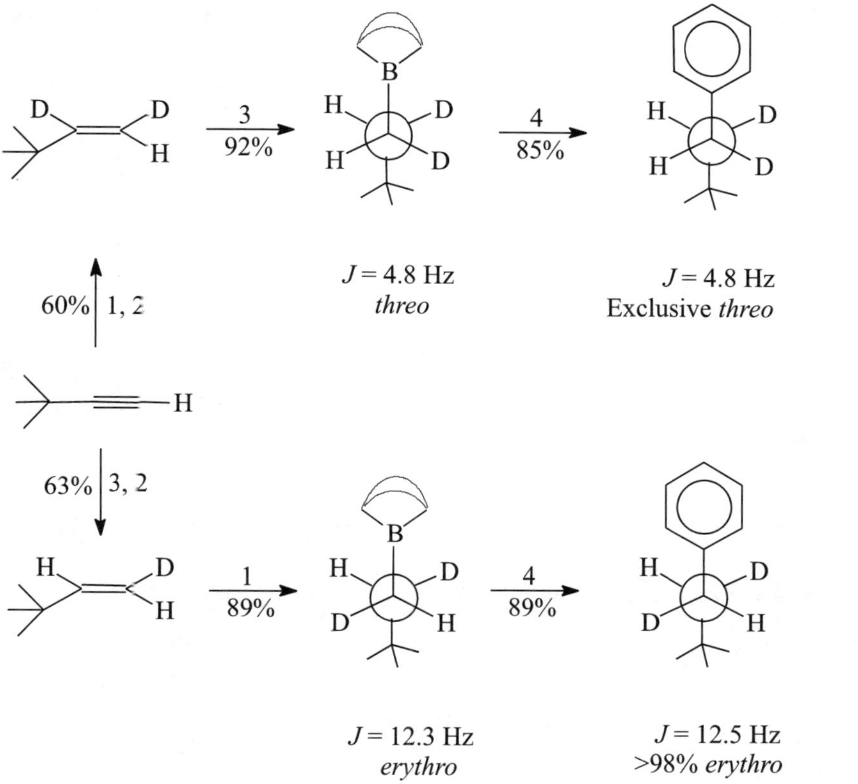

(1) $(9\text{-BBN-D})_2$, (2) CD_3COOD, (3) $(9\text{-BBN-H})_2$, (4) $PhBr/NaOH/Pd(PPh_3)_4$

Scheme 31.4

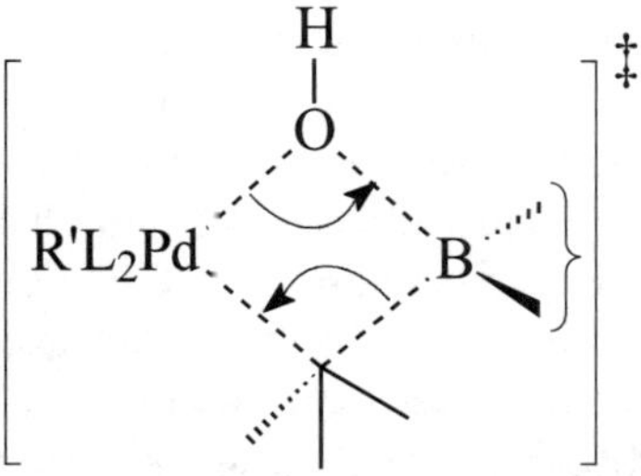

Fig. 31.1 Four- centered transition state – a model for alkyl group transfer *via* an Se[2] (coord)[1b] process

Soderquist [11] has further conducted the competitive rate studies (Chart 31.3) of *B*-R-9-BBN and corresponding 9-oxa-10-borabicyclo[3.3.2]decanes (OBBD).

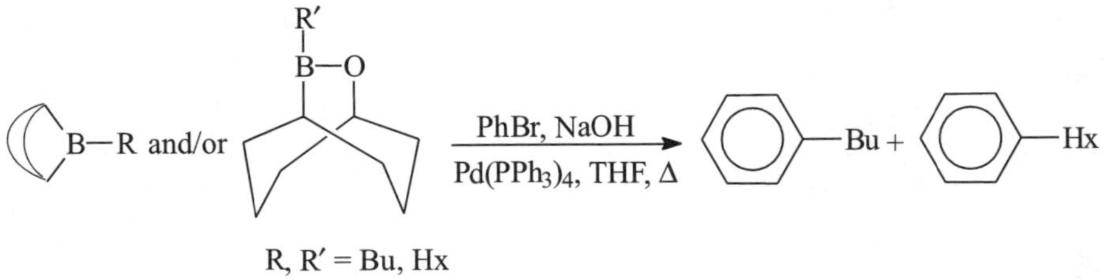

Chart 31.3

The results of these competitive studies are summarized in Table 31.10 [11].

Table 31.10 Competitive coupling for *B*-R-9-BBN and R′-OBBD derivatives [11]

Entry	*B*-R-9-BBN	R′-OBBD derivatives	Time (h)	Yield (%)	
	R	R′		PhBu	PhHx
1	Bu, Hx	–	1	40	48
2	Hx	Bu	12	0	86
3	Bu	Hx	12	84	0
4		Bu, Hx	12	44	49

The data reveal that as expected, little difference is observed in Bu versus Hx coupling (Table 31.10, entries 1 and 4). However, when *B*-R-9-BBN/R′ of OBBD pairs are compared (Table 31.10, entries 2 and 3), only the 9-BBN derivatives undergo coupling. This remarkable and previously unknown feature of Suzuki coupling, clearly, has potential synthetic importance for selective coupling.

Vinylsilanes undergo numerous and often unique stereoselective transformations, which make them important synthons in chemical syntheses [14]. Vinylsilanes can be efficiently prepared through a tin-mediated version of the Crandall reductive alkylation of epoxides [15]. However, for 1,3-alkadienylsilanes and related systems, this approach is neither promising nor practical. In Stille coupling [16] both stereochemical drift and competitive homocoupling have been observed, which lead to low yields. These problems as well as the formation of regioisomeric products are also observed in the Heck coupling of vinyltrimethylsilane to vinylic substrates [17]. Suzuki coupling of silylated vinylboranes [18], however, overcomes these difficulties through the use of vinyl-9-BBN derivatives such as *trans*-2-trimethylsilylvinyl-9-BBN [19]. *trans*-2-Trimethylsilyl vinyl-9-BBN are efficiently prepared (82%) through a new dehydroborylation process [20]. Consequently, Soderquist has reported [21] that aryl, vinyl, and alkynyl bromides undergo efficient Pd-catalyzed Suzuki coupling with *trans*-1-(9-borabicyclo[3.3.1]non-9-yl)-2-(trimethylsilyl)ethene under basic conditions to give the corresponding *trans*- styryl-, dienyl-, and enynylsilanes (Eq. 31.8; Table 31.11) [21].

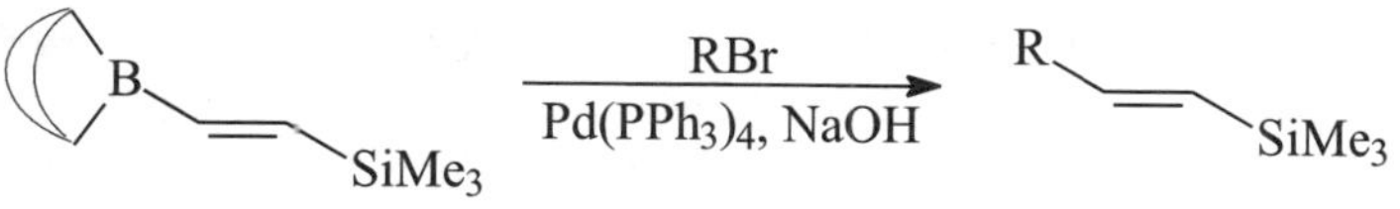

(31.8)

Table 31.11 *trans*-1-Silylstyrenes, dienes, and enynes [21]

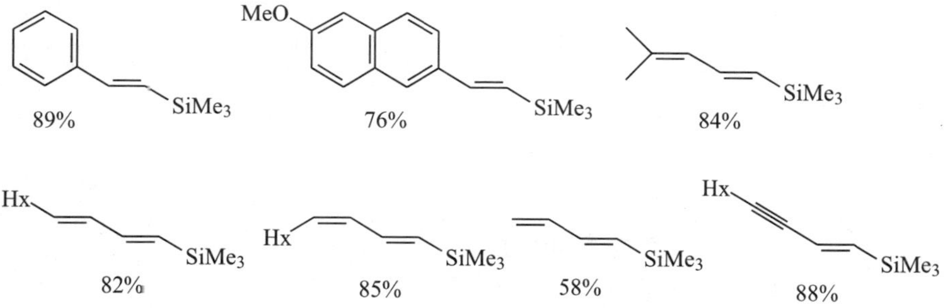

Several unusual silametacyclophanes, macrocyclic cage compounds, have been prepared by multiple Pd(0)-catalyzed Suzuki coupling reactions of 9-BBN adducts of allylsilane and bromobenzene. For example, reaction of 9-BBN adduct of methyltriallylsilane and 1,3,5-tribromobenzene leads to 4-methyl-4-sila[$3^{4,10}$][7] metacyclophane, 4-(2-propenyl)-4-methyl-4-sila[7]metacyclophane, which results from an intramolecular coupling of two legs of the silane and β-elimination of the third leg and 4-(3-phenylpropyl)-4-methyl-4-sila[7]metacyclophane (Eq. 31.9) [22]. The coupling reaction is carried out using Pd(dppf)Cl_2, a Pd(0) catalyst that has been reported to minimize the β-elimination [23]. The reaction in addition to formation of metacyclophanes, produces a large amount of polymers.

Coupling of bis-9-BBN adduct of allyldimethylsilane and 1,3-dibromobenzene in presence of NaOH yields the corresponding silacyclophane in 32% (Eq. 31.10). It should be noted that neither 1,2- nor 1,4-dibromobenzene affords the corresponding silacyclophanes under similar conditions. In addition, no reaction takes place between allylmethylsilane and bromobenzene, presumably because of Pd(0) insertion into the Si-H bond [24].

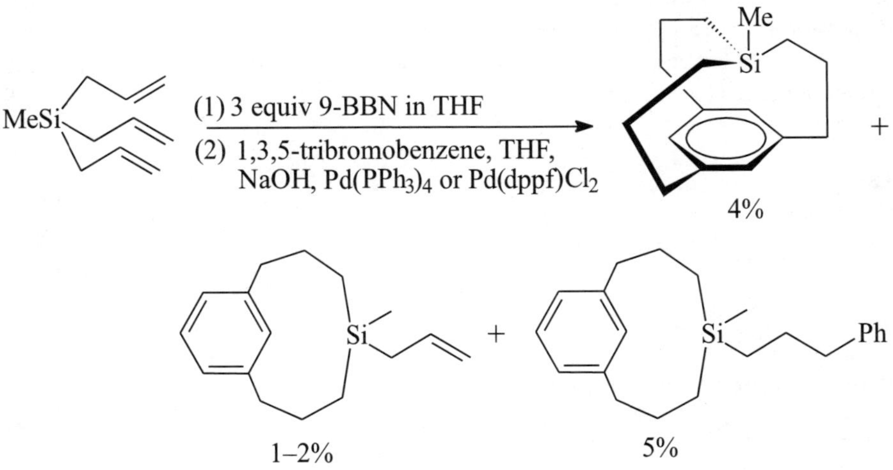

(31.9)

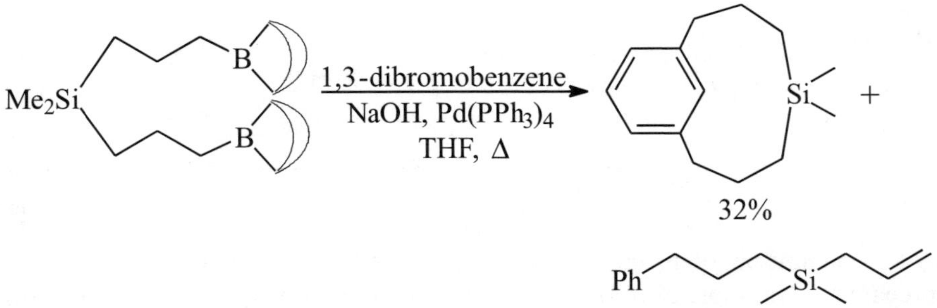

(31.10)

The Suzuki reaction is a powerful carbon–carbon bond-forming reaction method for the rapid introduction of diverse substituent onto an aromatic ring. Substituted arylacetic acids represent an important class of cyclooxygenase inhibitors. These inhibitors that are built on the phenylacetic acid core, typically incorporating three elements of variability: an α-alkyl group, R′; alkyl, aryl or heteroaryl substitution on the phenyl ring, R^2; and acid or amide functionality (Scheme 31.5). To synthesize this variety of molecules requires successful construction of a combinatorial library of a class of compounds and that depends on the availability of general and high yielding strategies, the general one being the solid support [25]. The linkage to solid support must be compatible with the basic enolate alkylation and Suzuki reaction conditions, yet at the end of the synthesis it must be labile for nuecleophilic cleavage of the final product from the solid support. The seldom-used acylsulfonamide linker, developed by Kenner for peptide synthesis, fulfills these requirements [26]. Under basic conditions, the acylsulfonamide ($pK_a \sim 2.5$) is deprotonated, preventing nucleophilic cleavage; however, once solid-support synthesis is complete, treatment with diazomethane leads to the formation of an N-methylated derivative, which is activated for nucleophilic displacement.

The sulfonamide-derivatized support is readily prepared by treating aminomethylated resin with 4-carboxybenzenesulfonamide, *N,N*-diisopropylcarbodiimide, and 1-hydroxybenzotriazole. Treatment of the sulfonamide resin with pentafluorophenyl active ester of 4-bromophenylacetic acid and 4-(dimethylamino) pyridine then gives the acylsulfonamide. Subsequent treatment of acylsulfonamide with excess LDA (15 equiv) in THF at 0 °C results in rapid deprotonation to give the trianion.

The trianion on treatment with activated alkyl halides such as methyl iodide or benzyl bromide or highly inactivated alkylhalides such as isopropyl iodide results in rapid and complete alkylation—incorporation of an α-alkyl group R^1. Subsequently, the Suzuki coupling is performed using $Pd(PPh_3)_4$ as the catalyst, 2 M aqueous Na_2CO_3 as the base, and THF as the solvent at reflux for 24–40 h. Good conversion is observed both for alkyl-9-BBN derivatives and for arylboronic acids that are electron poor or electron rich as well as *ortho* substituted; the process is incorporation of R^2 group. It is important to note that no protodehalogenation takes place in any of this coupling reaction.

The final step in the synthesis is nucleophile-mediated cleavage of the desired product from the solid support. Acylsulfonamide activation is accomplished by first rinsing the acylsulfonamide derivative with 5% trifluoroacetic acid in THF to ensure complete protonation, followed by treatment with CH_2N_2 in Et_2O. Nucleophilic cleavage of the product from support is then achieved by treatment with either hydroxide at room temperature or an amine in THF or dioxane at elevated temperature (incorporation of X group) (Scheme 31.5) [27].

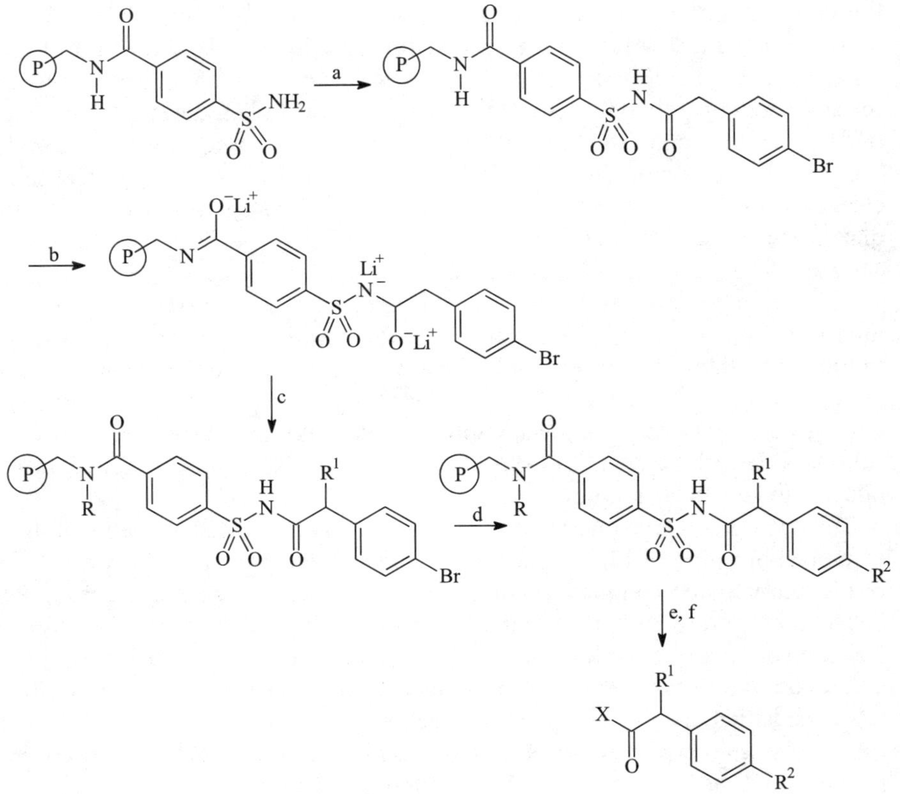

(a) Pentafluorophenyl-4-bromophenylacetate, 4-(dimethylamino) pyridine; (b) LDA, THF, 0 °C; (c) alkyl halide, 0 °C; (d) alkyl-9-BBN or arylboronic acid, $Pd(PPh_3)_4$, Na_2CO_3, THF, 65 °C; (e) CH_2N_2; (f) HO^- or amine.

Scheme 31.5

The results are summarized in Table 31.12 [27].

Table 31.12 Substituted arylacetic acid and amide derivatives [27]

Entry	R^1	Derivative R^2	Nucleophile	Yield (%)
a	H	$(Me)_2CHCH_2$	H_2O	100
b	Me	$(Me)_2CHCH_2$	H_2O	96
c	Me	$(Me)_2CHCH_2$	$BnNH_2$	96
d	Bn	$(Me)_2CHCH_2$	$BnNH_2$	98
e	Et	$(Me)_2CHCH_2$	$BnNH_2$	92
f	*i*-Pr	$(Me)_2CHCH_2$	$BnNH_2$	91
g	Me	$(Me)_2CHCH_2$	Piperidine	96
h	Me	$(Me)_2CHCH_2$	Aniline	0
i	H	Ph	H_2O	93
j	Me	Ph	$BnNH_2$	95
k	Me	4-F_3CPh	$BnNH_2$	87
l	Me	4-MeOPh	$BnNH_2$	88
m	Me	2,4-Cl_2Ph	$BnNH_2$	88

As shown in Table 31.12, hydroxide cleanly provides the carboxylic acids ibufenac (entry a), ibuprofen (entry b), and felbinac (entry i), whereas nucleophilic amines such as benzylamine and piperidine provide high yields of the amide derivatives.

The utilization of solid-support in Suzuki coupling is further exemplified [28] in the synthesis of amino acid esters (Scheme 31.6).

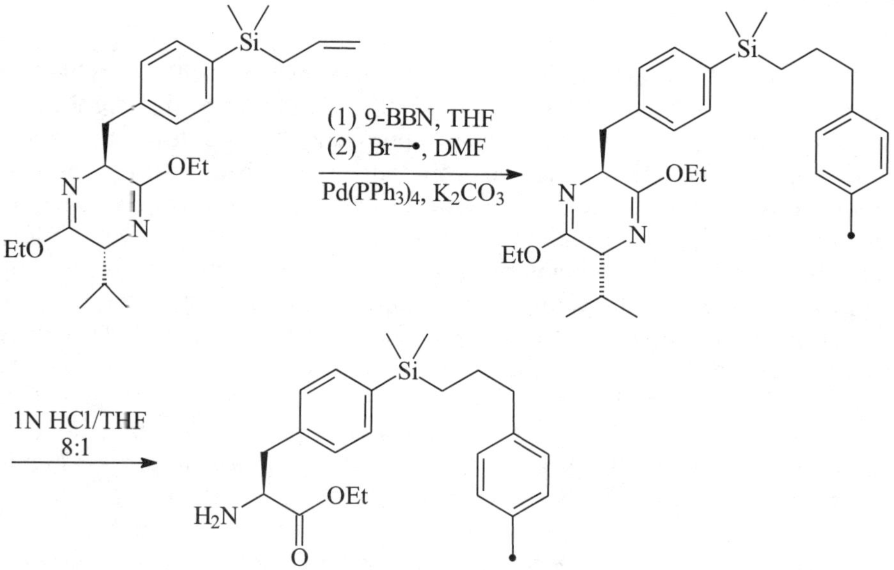

Scheme 31.6

Hydroboration of allylsilane with 9-BBN forms the corresponding borane, which is coupled [28] to bromostyrene resin with $Pd(PPh_3)_4$ and K_2CO_3 in DMF to give bislactin. Acid hydrolysis of the bislactin affords chiral amino ester. It is noted that Boc-protected amino acids can be generated first, followed by hydroboration and Suzuki coupling of the allylsilane.

Another solid-support Suzuki coupling is the hydroboration of vinyl polystyrene with 9-BBN in THF to provide [29] polymer supported borane. This borane is coupled to give different aryl, vinyl, or alkyl iodides. $Pd(OAc)_2$ is employed as the catalyst for aryl iodides and $PdCl_2$(dppf) in other cases to yield the coupled adduct in 55–85% yield (Scheme 31.7) on the basis of the mass of the product cleaved or by comparison with IR integration to standards.

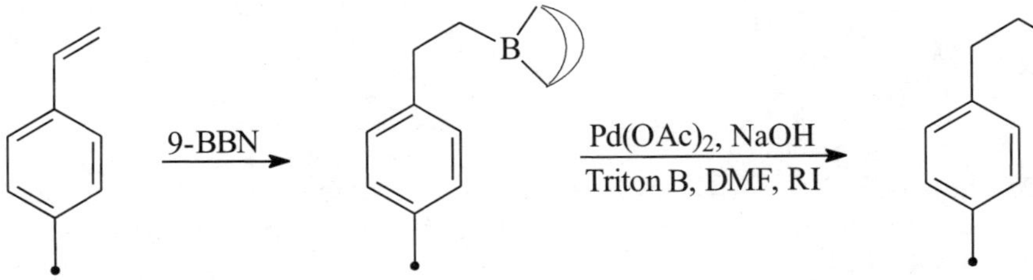

Scheme 31.7

General and efficient syntheses of prostaglandins have been the subject of much effort over the past four decades [30]. In addition to the lengthy Corey's synthesis [31], two other popular approaches are the three-component coupling process [32] and the two- component (conjugate addition) process [30]. However, these processes are associated with many limitations, e.g., one-pot, three-component coupling faces the problems of enolate equilibrium and β-alkoxide elimination and in the two-component approach, the limiting approach is the availability of the enantiopure α-alkylcyclopentenones [33]. These problems are avoided by the use of Suzuki reaction as the organoborane reagents are easily prepared (*in situ*) and in most cases display little reactivity with other functionalities. In addition, Pd(0)-catalyzed coupling occurs under mildly basic conditions and is tolerant of a wide range of functionality (ketone, aldehyde, ester, nitrile, alcohol, etc.), making the overall process highly adept in the synthesis of delicate compounds such as prostaglandins (Schemes 31.8, 31.9). Consequently, PGE_1, precursor **2** is synthesized from the readily available [34] cyclopentenone by employing Suzuki coupling reaction (Scheme 31.9) [35].

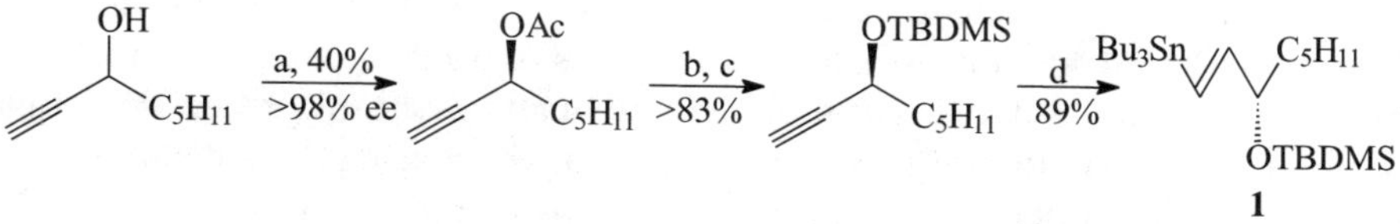

(a) SP-435, isopropenyl acetate, 25 °C; (b) NaCN, MeOH; (c) TBDMSCl, imidazole, DMF; (d) Bu_3SnH, AIBN, 110 °C

Scheme 31.8

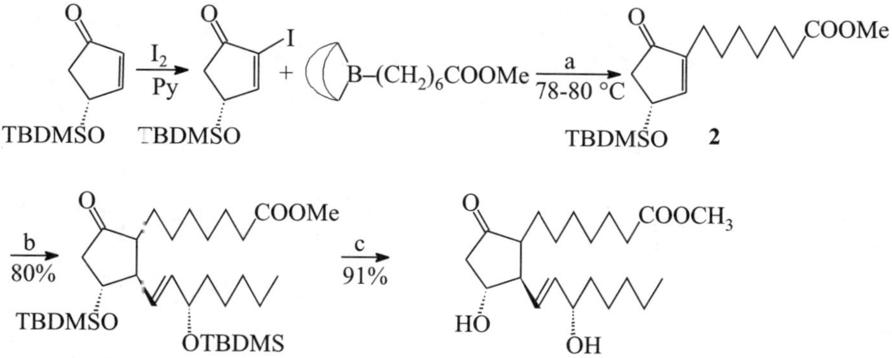

(a) 1.5 equiv of borane, $PdCl_2$(dppf), 5 mol%, Ph_3As 10 mol% CS_2CO_3 (1.8 equiv), DMF/THF/H_2O, 25 °C; (b) (i) vinyl cuprate derived from stannate 1, THF, –78 °C, (ii) saturated aq. NH_4Cl; (c) HF, pyridine, CH_3CN

Scheme 31.9

It is pertinent to mention that presence of water is absolutely necessary, the factor not normally observed with the Suzuki reaction; the other protic solvents like methanol prove unsatisfactory. The PGE_1 methylester is obtained in good overall yield (54%).

L-Tryptophan is an essential amino acid that provides the biosynthetic precursor for many naturally occurring indole alkaloids [36]. The ability to synthesize enantiomerically pure tryptophan analogs substituted in the benzenoid ring constitutes a useful approach toward the total synthesis of chiral indole alkaloids of natural and nonnatural origin. Consequently, it provides a general entry into the nonnatural amino acid derivatives of potential pharmacologic value [37]. Hino has demonstrated [38] the powerful utility of cyclic tryptophan tautomers as intermediates for direct functionalization of chiral tryptophans. Cyclic tautomers allow standard electrophilic substitution at 5 position of the benzenoid ring due to deactivation of the reactive C-2 and C-3 positions. Consequently, analog **3** is used for the synthesis of 5-nitro [39], 5-bromo [40], 5-chloro [39], and 5-hydroxytryptophan [40] analogs. The cyclic tautomer **3** is used for the preparation of L-tryptophan derivatives bearing 5-alkyl and 5-aryl substituents. Treatment of **3** with 4 equiv of iodine monochloride affords the 5-iodo species **4** in 81% yield. Cross-coupling of **4** with arylboronic acids or *B*-alkyl-9-BBN derivatives, catalyzed by 3 mol% [bis(1,1′-diphenylphosphino) ferrocene]palladium (II) chloride give the 5-aryl- and 5-alkyl cyclic tryptophan tautomers **5a–f** in 59–79% yields (Scheme 31.10; Table 31.13) [41]. The cyclic tautomers are easily decyclized and deprotected to the corresponding 5-aryl- and 5-alkyl-L-tryptophan [41].

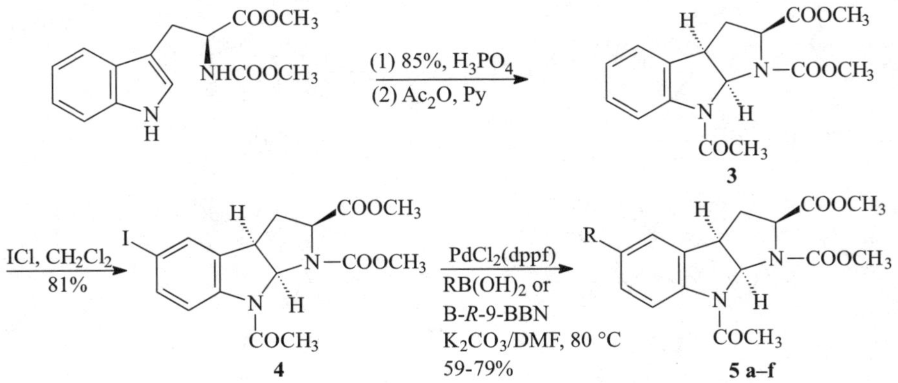

Scheme 31.10

Table 31.13 Preparation of cyclic tryptophan analogs **5** [41]

Entry	*B*-R-9-BBN or $RB(OH)_2$	R	Yield (%)
5a	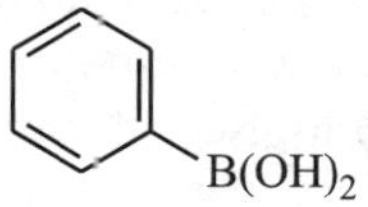	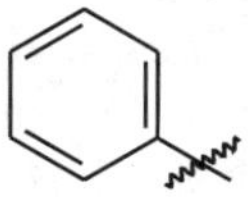	64
5b	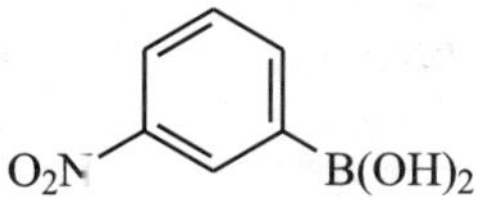	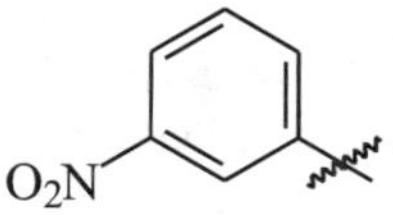	64
5c	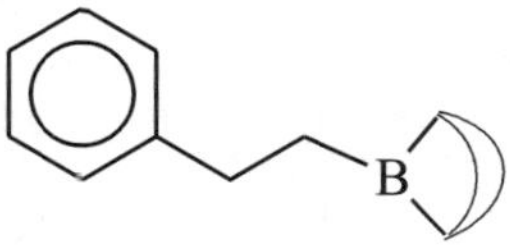	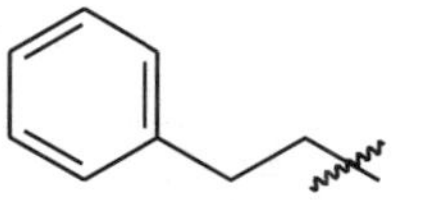	79
5d	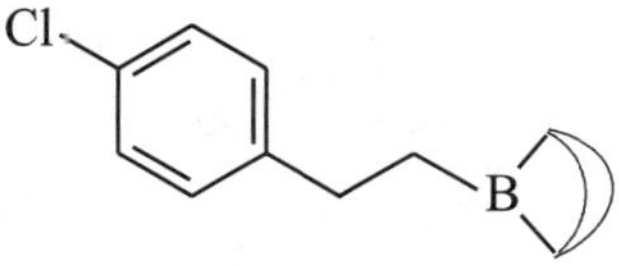	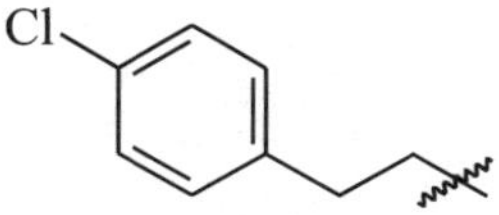	74
5e	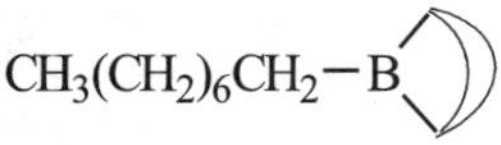	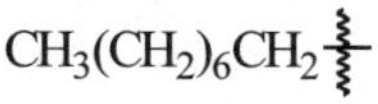	59
5f	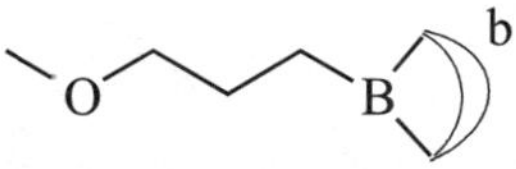 b		59

Prepared as described from Brown HC, Lynch GJ (1981) J Org Chem 46:531.

The epothilones are a class of cytotoxic macrolides that have emerged as highly promising new anticancer agents. Although structurally dissimilar to paclitaxel (Taxol), the epothilones, apparently, function through an analogous mechanism involving inhibition of cellular division by stabilization of microtubule assemblies, thereby leading to cell death. The naturally occurring compound, epothilone B (**6**, Epo B) is the most potent member of this family. Recently, the Bristol-Myers Squibb Company (BMS) has advanced the 15-desoxy-15-aza analog of epothilone B (**7**, Aza-Epo B) as a clinical candidate for cancer therapy [42, 43]. Aza-Epo B, has been synthesized [44] by Suzuki coupling and subsequent macrolactamization as outlined in Scheme 31.11.

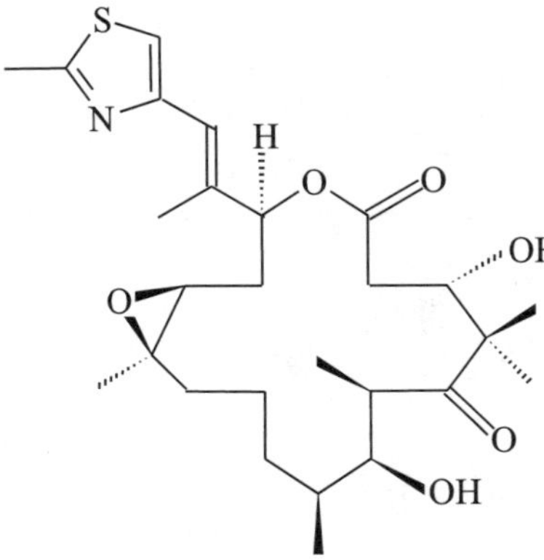

Epothilone B (EpoB)
6

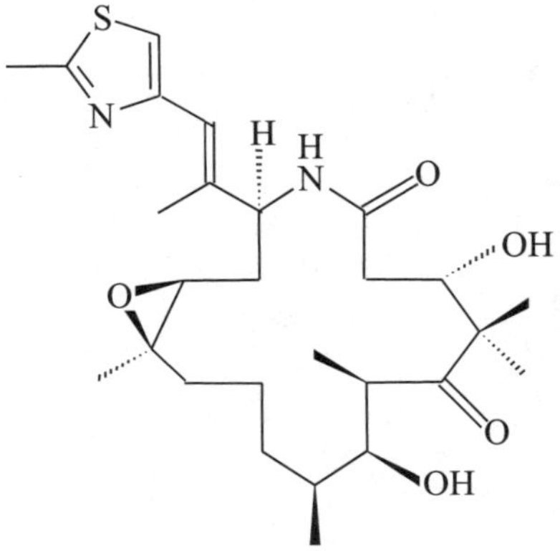

Aza-epothilone B (Aza-EpoB)
7

Polyhydroxylated piperidines ("azasugars" or "alkaloid sugars mimics") are a novel class of glycomimetics compounds. They exhibit potent biological activity [45] such as inhibition of oligosaccharide-processing enzymes called glycosidases and glycotransferases. These enzymes target glycosidic linkage of oligosaccharides and glycopeptides by stabilizing an intermediate oxonium ion, thus facilitating the lysis and modification of the anomeric center [46]. Azasugars are thought to be good inhibitors due to their ability to mimic the transition state oxonium ion as a result of the heterocyclic nitrogen being protonated at physiological pH [46d].

Johnson and coworkers [47] have synthesized (1→6)-, (1→4)-, and (1→1)-linked aza-*C*-disaccharides; D-azaMan-β-(1→6)-D-Man(**8**), D-azaMan-β-(1→6)-D-Glc(**9**), D-azaMan-β-(1→4)-D-Talo(**10**) and D-azaMan-β-(1→1)-D-β-Glc(**11**). The polyhydroxylated piperidine ring is constructed employing vinylbromide, (3*R*,4*R*,5*S*,6*S*)-3-[(*N*-benzyloxycarbonyl)amino]-1-bromo-4-[(*tert*-butyldimethylsilyl)oxy]-5,6-(*O*-isopropylidenedioxy)cyclohex-1-ene(**12**), as a common intermediate for these sugars. Vinylbromide (**12**) required for Suzuki coupling is synthesized *de novo* from 3-bromocyclohex-3,5-dien-1,2-diol as outlined in Scheme 31.12 [47].

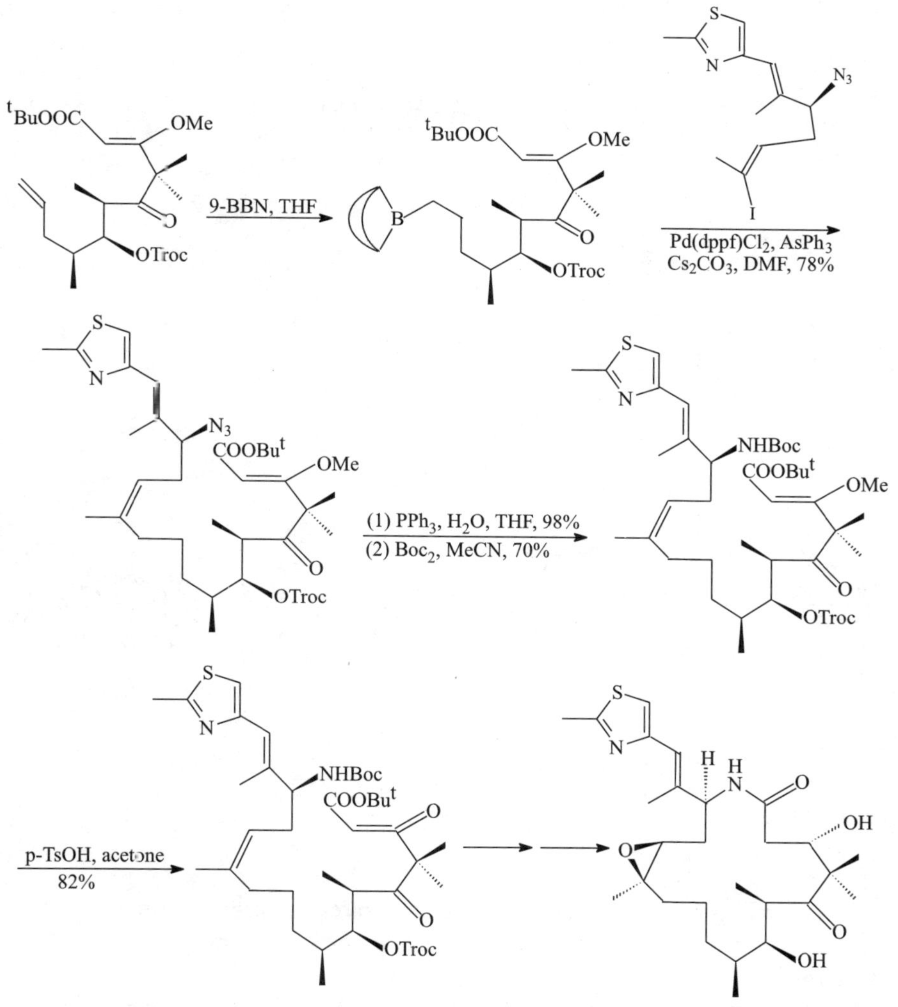

7

Scheme 31.11

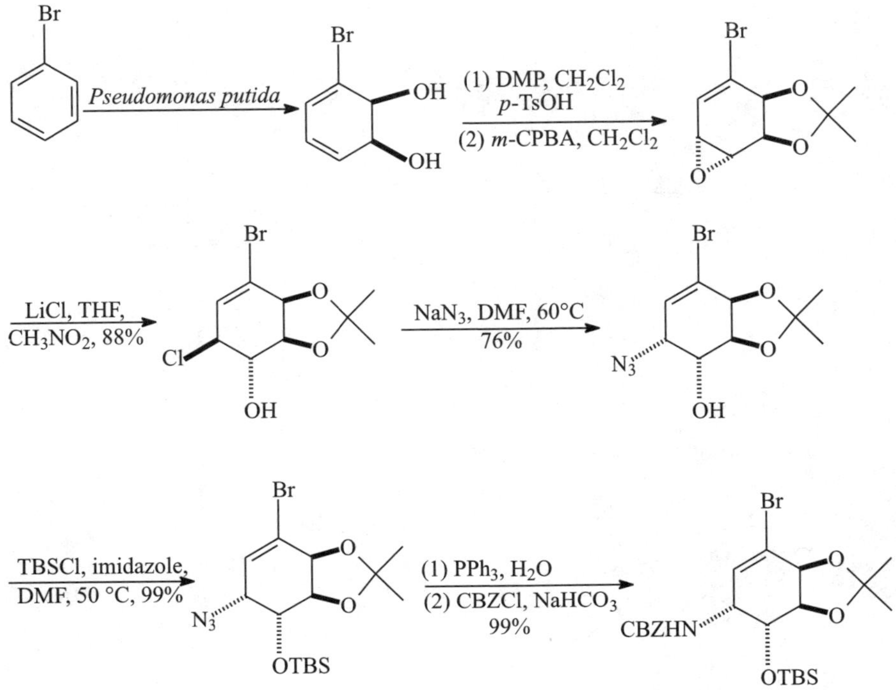

(**12**)

Scheme 31.12

The coupling partners, olefinated carbohydrates required for hydroboration are prepared *via* methylenation of the aldehyde or ketone moiety of the monosaccharides.

In the synthesis of methyl-2,3-*O*-isopropylidene-4-*O*-(methoxymethyl)-6-methylene-α-D-mannopyranoside (**13**), the hydroxyl groups of methyl-α-D-mannopyranosides are selectively protected: 6-hydroxyl group as silyl ether; 2,3-hydroxyl groups as acetonides; and 4-hydroxyl group as methoxymethylether. The silyl group at the C-6 position is then selectively cleaved to get primary alcohol. Oxidation of the alcohol to aldehyde followed by Wittig methylenation yields the desired olefin (**13**, Scheme 31.13).

In the synthesis of methyl-2,3,4-tri-*O*-(methoxymethyl)-6-methylene-α-D-glucopyranoside (**14**), the 6-hydroxyl group of methyl-α-D-glucopyranoside is protected as its silylether, while the remaining three hydroxyl groups by their methoxymethylethers. The silyl group is removed by tetrabutylammonium fluoride (TBAF) to produce primary alcohol. The alcohol is oxidized using modified Swern oxidation, followed by Wittig methylenation to produce the olefin (**14**, Scheme 31.14).

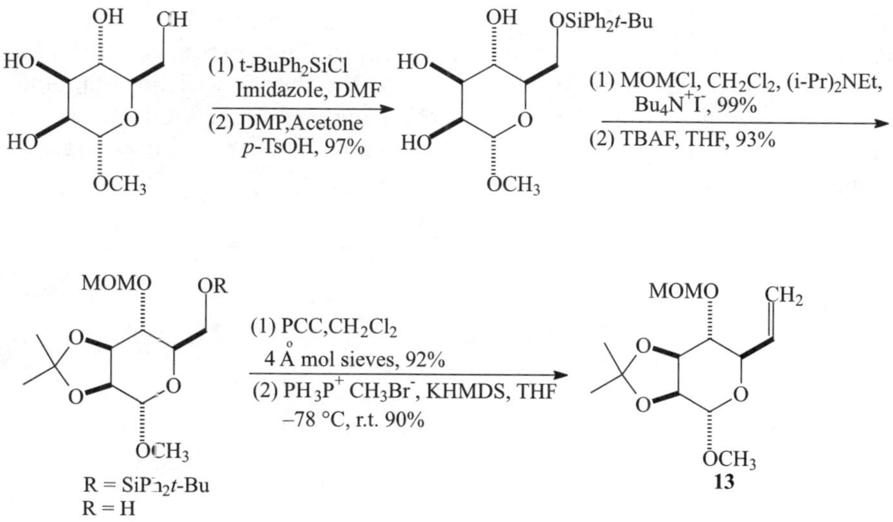

13

Scheme 31.13

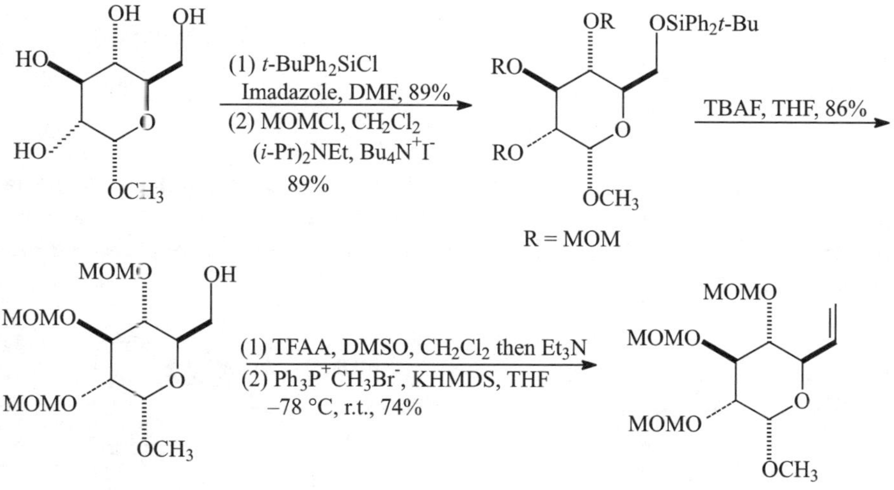

14

Scheme 31.14

Methyl-6-*O*-(*tert*-butyldiphenylsilyl)-2,3-*O*-isopropylidene-4-deoxy-4-methylene-α-D-*lyxo*-hexopyranoside (**15**) is prepared from methyl-α-D-mannopyranoside whose 6-hydroxyl group is protected as *tert*-butyldiphenylsilylether and 2,3-hydroxyl group as 2,3-acetonide. The oxidation of the C-4–OH group gives the corresponding ketone. The Wittig reaction affords the poor yield of the olefin due to epimerization and formation of elimination products, resulting from the basic nature of the Wittig conditions. The alternative Tebbe titanium–aluminum also gives only the modest yield. However, dimethyltitanocene [48] generated *in situ* affords the desired olefin derivative in 81% yield, without any epimerization or elimination products (Scheme 31.15).

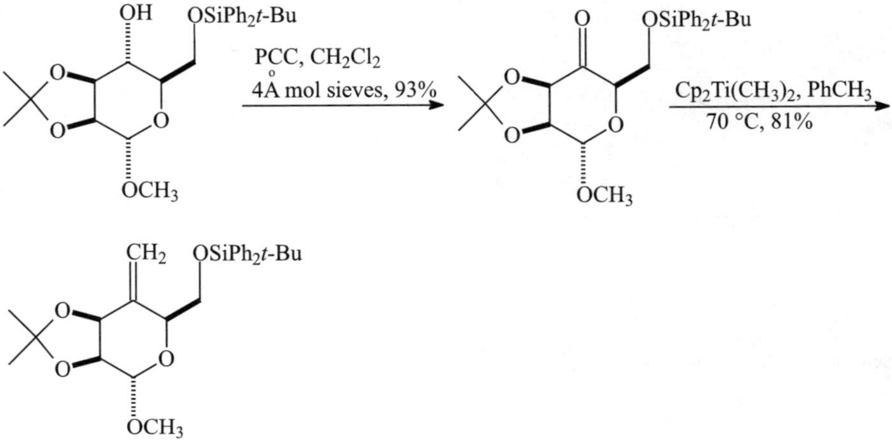

15

Scheme 31.15

2,6-Anhydro-3,4,5,7-tetra-*O*-(methoxymethyl)-D-gluco-hept-1-enitol (**16**) is prepared from tetra(methoxymethylether) of δ-D-gluconolactone. The carbonyl function of lactone is olefinated with dimethyltitanocene to provide enol ether (Scheme 31.16).

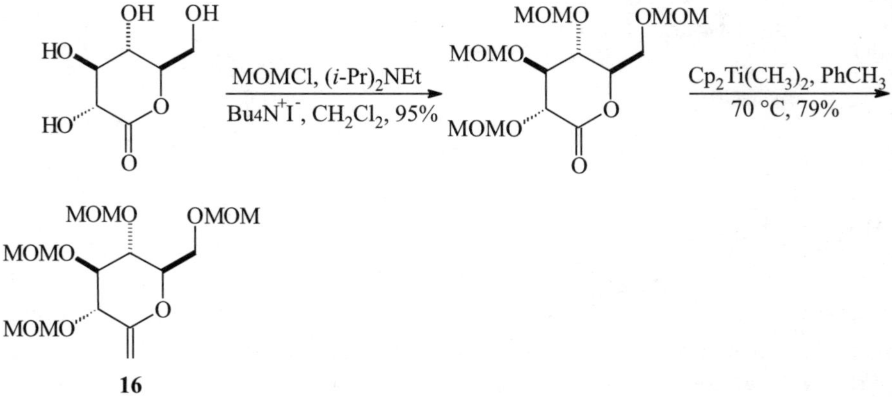

16

Scheme 31.16

In the synthesis of aza-*C*-disaccharide-D-azaMan-β-(1→6)-D-Man, hydroboration of methyl-2,3-*O*-isopropylidene-4-*O*-(methoxymethyl)-6-methylene-α-D-mannopyranoside (**13**) with 9-BBN in THF produces the corresponding borane (**13a**). The Suzuki coupling of borane and vinylbromide is achieved, using $PdCl_2$(dppf), 3 M aqueous K_3PO_4, and DMF. The resulting olefin **17** is transformed into protected disaccharide mimic **20** in three steps requiring only one chromatographic separation. Ozonolysis followed by DMS workup affords ketoaldehyde, which is in equilibrium with its closed form hemiacetal **18**. The chemoselective reduction of aldehyde moiety produces ketoalcohol **19**, which is converted to protected azasugar **20** *via* an intramolecular reductive amination. Acid preprotonation (6 N HCl, CH_3OH) produces aza-*C*-disaccharide-D-aza-Man-β-(1→6)-D-Man as its HCl salt (Scheme 31.17) [47].

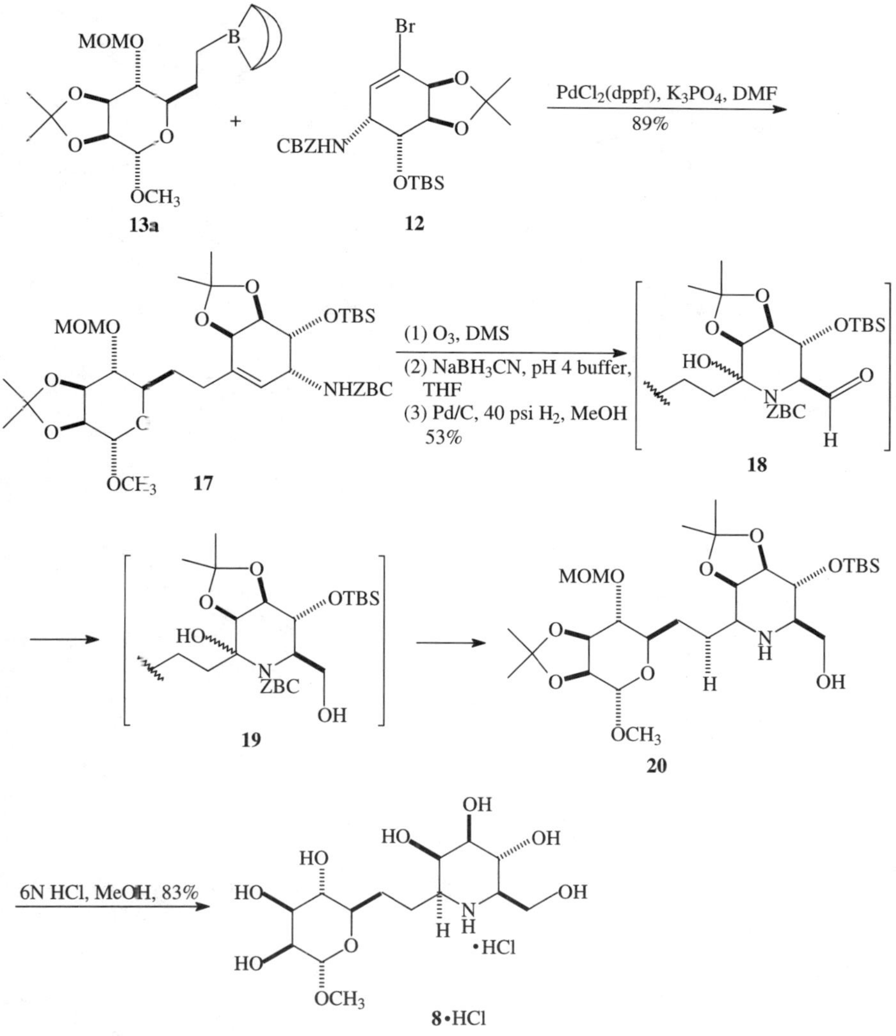

Scheme 31.17

Similarly, synthesis of D-azaMan-β-(1→6)-D-Glc (**9**, Scheme 31.18), D-aza-Man-β-(1→4)-D-Talo (**10**, Scheme 31.19), D-azaMan-β-(1→1)-D-*β*-Glc (**11**, Scheme 31.20) are achieved [47] as their HCl salts.

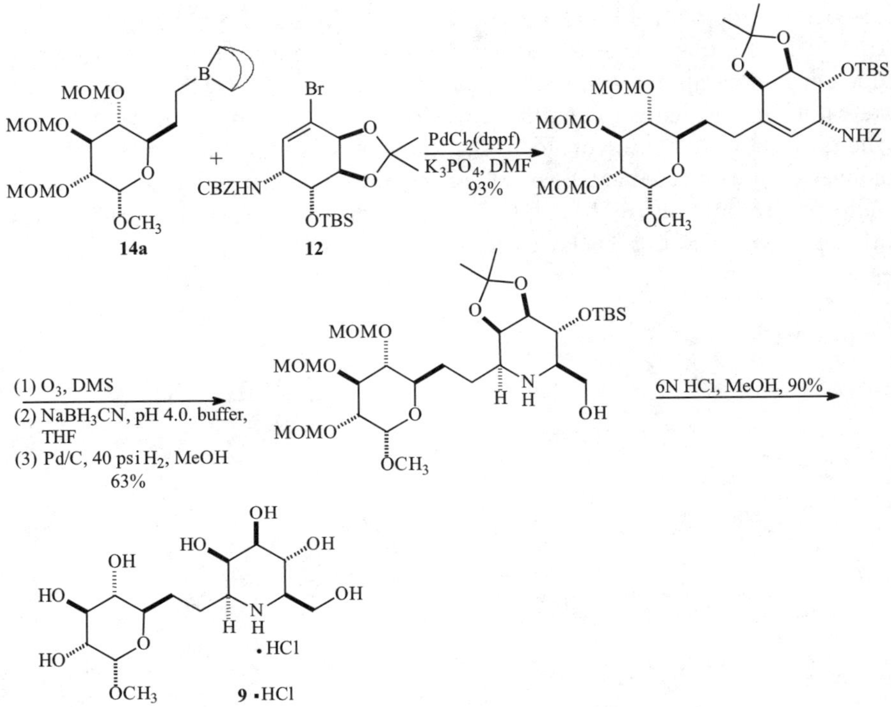

Scheme 31.18

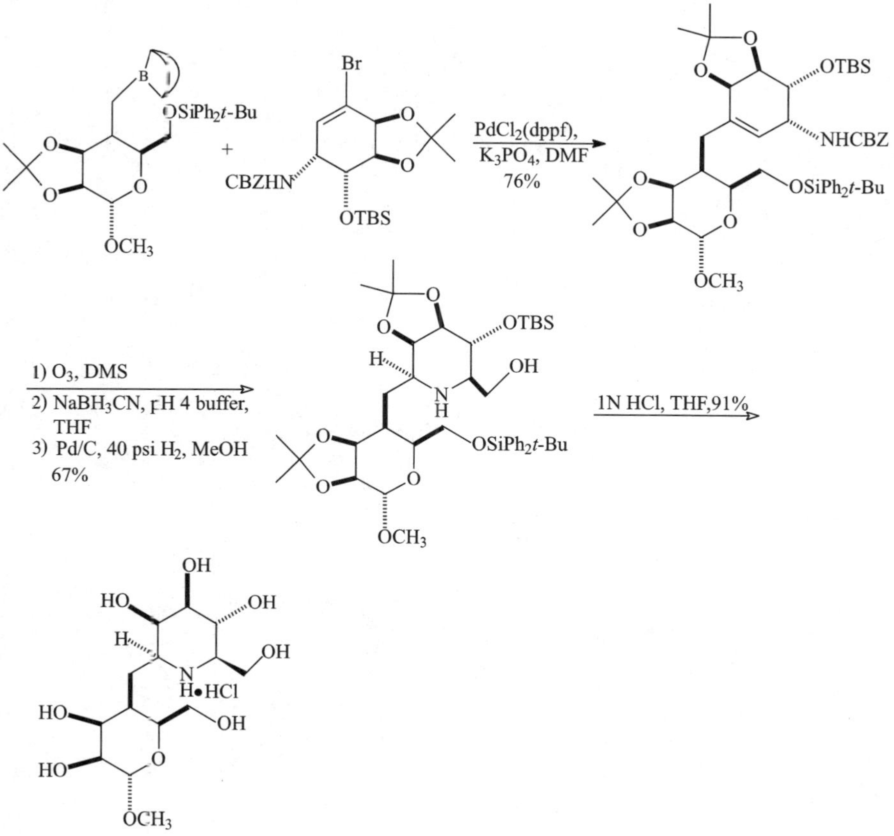

10.HCl

Scheme 31.19

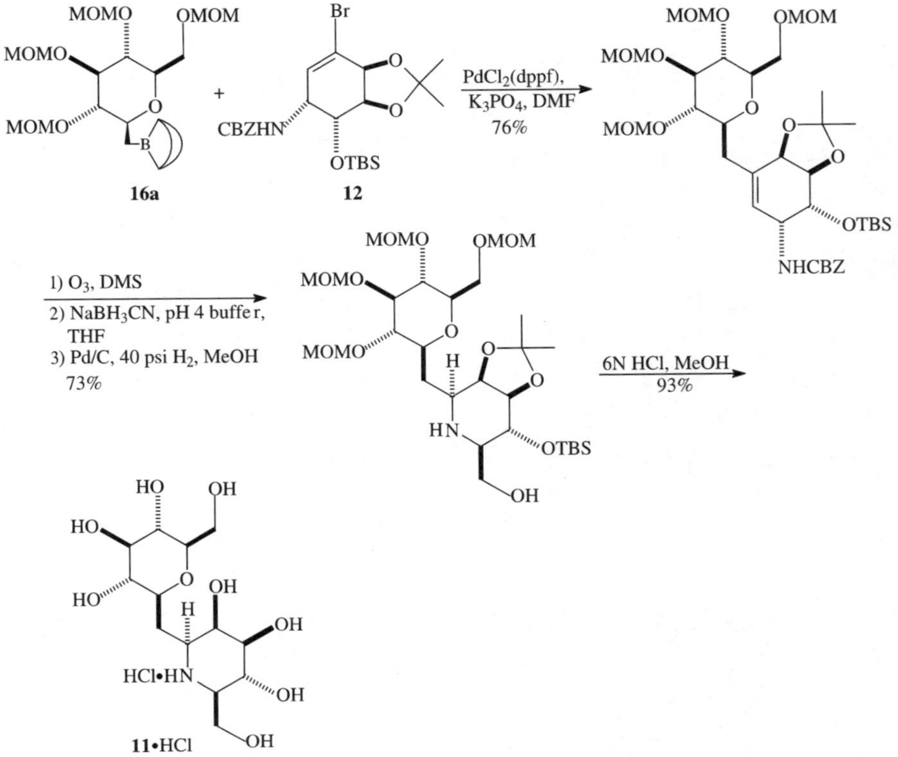

Scheme 31.20

31.1 Mechanism of the Suzuki Catalytic Cycle

The Suzuki catalytic cycle consists of a series of sequential reactions whose rates vary with the individual reacting species [49, 50]. In addition, the type of organoborane used must also be considered for successful coupling of even simple alkylboranes. Either, trialkylboranes (BR_3) or borinate esters (R_2BOR') are efficient coupling partners under the standard basic (NaOH) conditions [1, 2]. However, more oxygenated derivatives (i.e., boronates [$RB(OR')_2$] require thallium bases [51, 52]. As the successful coupling of alkylboranes is sensitive to specific conditions and substrates employed, Soderquist reexamined [11] this process under the standard conditions (i.e., $Pd(PPh_3)_4$, NaOH) to determine (1) the stereochemistry of the alkyl group transfer [53], (2) the effect of alkyl versus alkoxy boron litigation on the rate coupling, (3) the actual role of added base, and (4) the rate-limiting step in the catalytic process. Soderquist has reported [11] that the catalytic cycle proposed earlier [1], as illustrated in Fig. 31.2, only

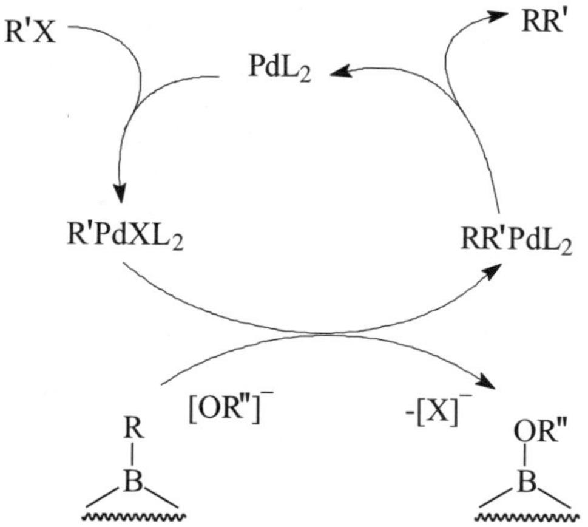

Fig. 31.2 The Suzuki catalytic cycle

superficially represents which otherwise is highly complex process particularly when the base is included.

In the coupling process, the Lewis acidity of the boron plays an important role with *B*-R-9-BBN (**21**) forming $[HO(R)\text{-9-BBN}]^{-1}$ (**22**) complex with the added base, in marked contrast to their *B*-alkyl-9-oxa-10-borabicyclo[3.3.2]decane counterparts (R-*B*-OBBD) (**23**), which do not. This behavior parallels their coupling rates with the exclusive reaction of **21** over **23** in competitive experiments (Table 31.10) [11].

The new features which emerge when base is added are illustrated in Fig. 31.3.

The results reveal [11] that the formation of $BrPdPh(PPh_3)_2$ (**24**) at 27 °C is rate-limiting for 9-R-9-BBN derivatives (**21**). These organoboranes are primarily present in their hydroxyborate complexes (**22**). The reaction between **22** and **24** is rapid, probably displacing bromide and forming a hydroxo μ_2-bridged intermediate $PhL_2Pd \leftarrow (OH)$*B*-R-9-BBN (**25a**) intermediate, which facilitates the alkyl B → Pd transmetalation with retention of configuration through a four-centered transition states (Fig. 31.1). This intermediate is expected to possess lower energy than that of related species derived from oxygenated organoboranes (**23**), owing to the greater Lewis acidity of *B*-R-9-BBN (**21**). The resulting $PhPdR(PPh_3)_2$ rapidly gives PhR and regenerates $Pd(PPh_3)_2$. The boron byproduct (*B*-OH-9-BBN) effectively competes with **21** for base, thus the optimal **21**/OH^- stoichiometry is 1:2 (Eq. 31.11), which ensures that **22** will be present to continue the cycle once $BrPdPh(PPh_3)_2$ (**24**) is regenerated.

The alternative pathway, involving the key hydroxo-μ2-bridged species (**23b**), dominates for the less acidic OBBD derivative **23**. Here, it is the formation of $HOPdPh(PPh_3)_2$ from **24** that is rate limiting (Fig. 31.3) [11].

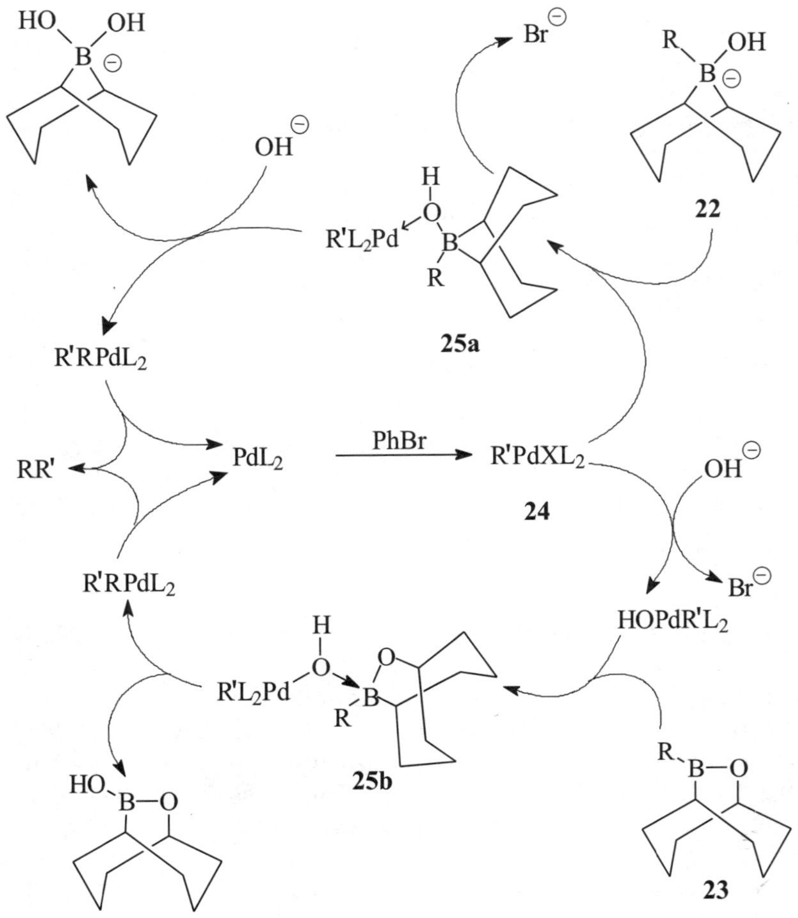

Fig. 31.3 Modified Suzuki catalytic cycle, illustrating the roles played by added base

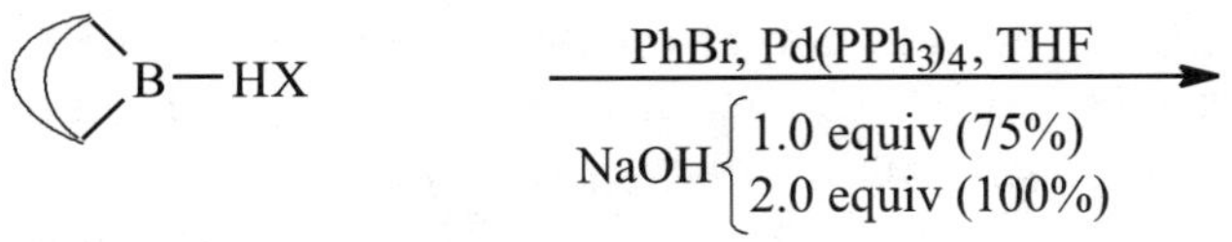

(31.11)

References

1. (a) Suzuki A, Brown HC (2002) Organic synthesis via boranes, vol. 3, Suzuki coupling. Aldrich Chemical, Milwaukee, Wisconsin; (b) Miyaura N, Suzuki A (1995) Chem Rev 95:2457
2. (a) Miyaura N, Ishiyama T, Sasaki H, Ishikawa M, Satoh M, Suzuki A (1989) J Am Chem Soc 111:314; (b) Miyaura N, Ishiyama T, Ishikawa M, Suzuki A (1986) Tetrahedron Lett 27:6369
3. (a) Bottin J, Fetizon M (1971) J Chem Soc Chem Commun 1087; (b) Midland MM, Kwon YC (1981) J Org Chem 46:229
4. (a) Soderquist JA, Matos K, Rane A, Ramos J (1995) Tetrahedron Lett 36:2401; (b) Soderquist JA, Rane AM, Matos K, Ramos (1995) J Tetrahedron Lett 36:6847
5. Soderquist JA, Ramos J, Matos K (1997) Tetrahedron Lett 38:6639
6. Mori K (1989) Tetrahedron 45:3233 and references cited therein
7. Santiago B, Soderquist JA (1992) J Org Chem 57:5844
8. Soderquist JA, Leon G, Colberg JC, Martinez I (1995) Tetrahedron Lett 36:3119
9. Fisher RP, On HP, Snow JT, Zweifel G (1982) Synthesis 127
10. Fuchs PL, Corey EJ (1972) Tetrahedron Lett 36:3769
11. Matos K, Soderquist JA (1998) J Org Chem 63:461
12. Whitesides GM, Bock PL, Boschetto DJ, Rasmussen JR, Demers JM (1974) J Am Chem Soc 96:2814; (b) see also Kabalka GW, Newton RJ Jr, Jacobus J (1979) J Org Chem 44:4185
13. For selective couplings employing the electrophile partner, see (a) Roush WR, Moriarty KJ, Brown BB (1990) Tetrahedron Lett 31:6509; (b) Rivera I, Colberg JC, Soderquist JA (1992) Tetrahedron Lett 33:6915
14. Fleming I, Dunoguẽs J, Smithers R (1989) In: Kende AS (ed) Organic reactions, vol. 37. Wiley, New York, pp 57–576
15. (a) Soderquist JA, Lopez, C (1991) Tetrahedron Lett 32:6305; (b) Crandall JK, Lin L-HC (1967) J Am Chem Soc 67:4526, 4527; (c) For a detailed discussion, see Crandall JK, Apparu M (1983) Org React 29:345
16. Stille JK, Groh BL (1987) J Am Chem Soc 109:813
17. Karabelas K, Hallberg A (1988) J Org Chem 53:4909 and references cited therein
18. (a) Soderquist JA, Colberg JC (1989) Synlett 25; (b) Soderquist JA, Leõn-Colõn G (1991) Tetrahedron Lett 32:43
19. (a) Soderquist JA, Colberg JC, Del Valle L (1989) J Am Chem Soc 111:4873; (b) Singleton DA, Martinez JP (1991) Tetrahedron Lett 32:7365
20. Colberg JC, Rane A, Vaquer J, Soderquist JA (1993) J Am Chem Soc 115:6065
21. Soderquist JA, Colberg JC (1994) Tetrahedron Lett 35:27
22. Kwochka WR, Damrauer R, Schmidt MW, Gordon MS (1994) Organometallics 13:3728
23. Hayashi T, Konishi M, Kobori Y, Kumada M, Higuchi T, Hirotsu K (1984) J Am Chem Soc 106:158
24. Ojima I (1989) Chapter 25. In: Patai S, Rappoport Z (eds) The chemistry of functional groups: the chemistry of organic silicon compounds. Wiley, New York
25. Leznoff CC (1978) Acc Chem Res 11:327
26. Kenner GW, McDermott JR, Sheppard RC (1971) J Chem Soc Chem Commun 636
27. Backes BJ, Ellman JA (1994) J Am Chem Soc 116:11171
28. Lee Y, Silverman RB (1999) J Am Chem Soc 121:8407
29. Vanier C, Wagner A, Mioskowski C (1999) Tetrahedron Lett 40:4335

30. (a) Bindra JS, Bindra R (1977) Prostaglandin synthesis. Academic, New York; (b) Collins PW, Djuric SW (1993) Chem Rev 93:1533
31. Corey EJ, Weinshenker NM, Schaaf TK, Huber W (1969) J Am Chem Soc 91:5675
32. (a) Noyori, R, Suzuki M (1984) Angew Chem Int Ed Engl 23:847; (b) Noyori R, Suzuki M, Yangisawa A (1988) J Am Chem Soc 110:4718; (c) Noyori R, Suzuki M (1990) Chemtracts Org Chem 173; (d) Johnson CR, Penning TD (1988) J Am Chem Soc 110:4726
33. (a) Babiak KA, Ng JS, Dygos JH, Weyker CL, Wang Y, Wong C-H (1990) J Org Chem 55:3377; (b) Sato F, Okamoto S, Kobayashi Y, Kato H, Hori K, Takahashi T, Tsuji J (1988) J Org Chem 53:5590
34. Levin JI (1989) Tetrahedron Lett 30:13
35. Johnson CR, Braun MP (1993) J Am Chem Soc 115:11014
36. Dalton DR (1979) The alkaloids. Dekker, New York, pp 415–628
37. Zembower DE, Gilbert JA, Ames MM (1993) J Med Chem 36:305
38. Hino T, Taniguchi M (1978) J Am Chem Soc 100:5564
39. Taniguchi M, Gonsho A, Nakagawa M, Hino T (1983) Chem Pharm Bull 31:1856
40. Taniguchi M, Anjiki T, Nakagawa M, Hino T Chem Pharm Bull 32:2544
41. Zembower DE, Ames MM (1994) Synthesis 1433
42. Vite GD, Borzilleri RM, Kim S-H, Johnson JA (1999) Patent WO99/102514
43. Borzilleri RM, Zheng X, Schmidt R, Johnson J, Kim S-H, DiMarco JD, Fairchild CR, Gougoutas JZ, Lee FYF, Long BH, Vite GD (2000) J Am Chem Soc 122:8890
44. Stachel SJ, Lee CB, Spassova M, Chappell MD, Bornmann WG, Danishefsky SJ, Chou T-C, Guan Y (2001) J Org Chem 66:4369
45. (a) Fleet GWJ, Karpus A, Dwek RA, Fellows LE, Tyms AS, Petursson S, Namgoong SK, Ramsden NG, Smith PW, Son JC, Wilson FX, Witty DR, Jacob GS, Rademacher TW (1988) FEBS Lett 237:128; (b) Winchester B, Fleet GWJ (1992) Glycobiology 2:199; (c) Elbein AD (1987) Ann Rev Biochem 56:497
46. (a) Jensen JL, Tsuang SC, Uslan AH (1986) J Org Chem 51:816; (b) Liu PS (1987) J Org Chem 52:4717; (c) Cogoli A, Semenza G (1975) J Bol Chem 250:7802; (d) Sinnot ML (1990) Chem Rev 90:1171
47. Johns BA, Pan YT, Elbein AD, Johnson CR (1997) J Am Chem Soc 119:4856
48. (a) Petasis NA, Bzowej EI (1990) J Am Chem Soc 112:6392; (b) Csuk R, Glänzer BI (1991) Tetrahedron 47:1655
49. Smith GB, Dezeny GC, Hughes DL, King AO, Verhoeven TR (1994) J Org Chem 59:8151
50. Rivera I, Soderquist JA (1991) Tetrahedron Lett 32:2311
51. (a) Sato M, Miyaura N, Suzuki A (1989) Chem Lett 1405; (b) Wright SW, Hageman DL, McClure LD (1994) J Org Chem 59:6095
52. Armstrong RW, Beau J-M, Cheon SH, Christ WJ, Fujioka H, Ham W-H, Hawkins LD, Jin H, Kang SH, Kishi Y, Martinelli MJ, McWhorter WW Jr, Mizuno M, Nakata M, Stutz AE, Talamas FX, Taniguchi M, Tino JA, Ueda K, Uenishi J-I, White JB, Yonaga M (1989) J Am Chem Soc 111:7225
53. Ridgway BH, Woerpel KA (1998) J Org Chem 63:458

32 Miscellaneous Reactions

The hydroboration-oxidation studies of 1,4-epoxy-1,4-dihydronaphthalene (**1**) with various hydroborating agents such as borane-methyl sulfide (BMS), dicyclohexylborane, disiamylborane, and 9-BBN have yielded interesting results [1]. The reaction of **1** with dicyclohexylborane or disiamylborane affords the *exo* alcohol **2** without epoxy ring opening (Eq. 32.1). On the other hand, 9-BBN hydroboration of **1** (1:1 mole ratio) in THF at 25 °C, followed by oxidation affords a homoallylic alcohol **3** after opening of the epoxy ring (Eq. 32.2). BMS, however, affords a mixture of both the alcohols.

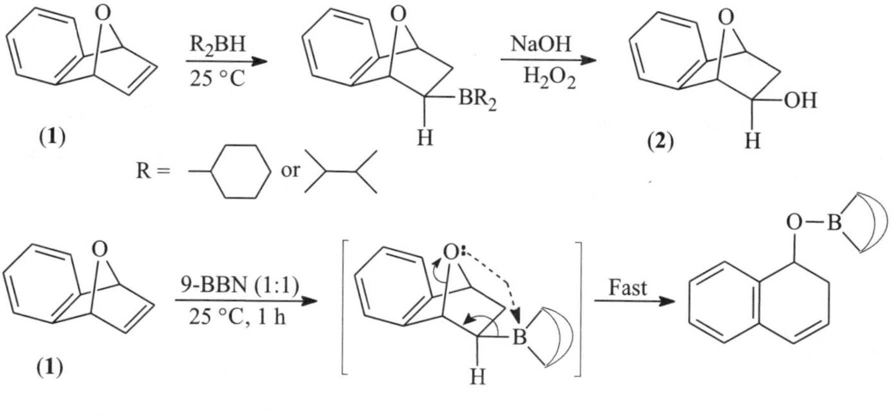

(32.1)

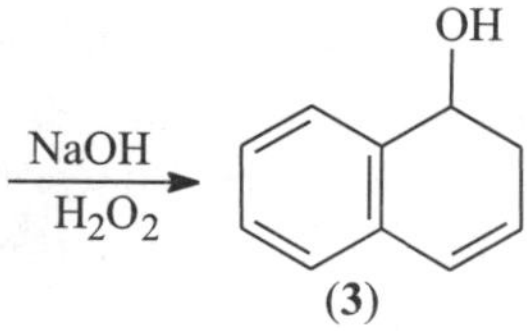

(32.2)

These studies, therefore, have considerable importance on the mechanistic pathways. It is known that organoboranes having electronegative substituents at the β position readily undergo elimination [2–8].

In case where the substituent is a strong leaving group (X = Cl, Br, I, or OTs), the reaction involves a *trans*-elimination, catalyzed even by a weak base THF (Eq. 32.3). However, when the substituent is not a strong leaving group but possesses good donor properties (X = OR, OAc), the reaction takes place through *cis*-elimination (Eq. 32.4).

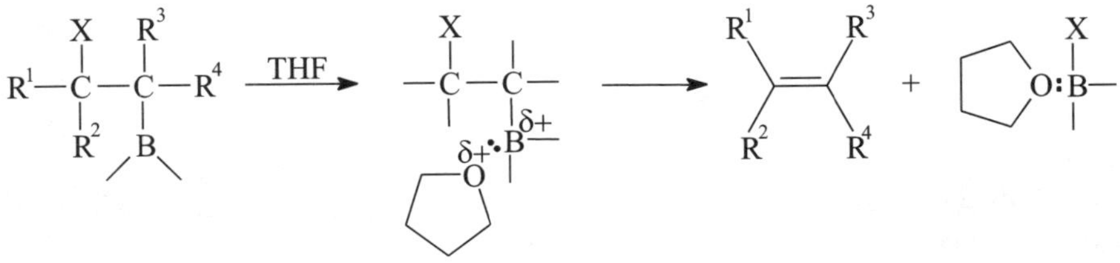

(32.3)

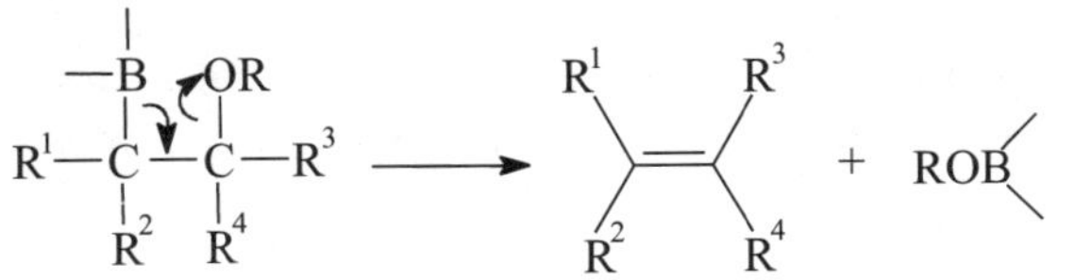

(32.4)

As the reaction of **1** is not catalyzed by THF, thus, rejecting the *trans* pathway for elimination process. The reaction, indeed, involves the *cis* pathway and relatively exposed boron atom of 9-BBN readily facilitates elimination (Eq. 32.5).

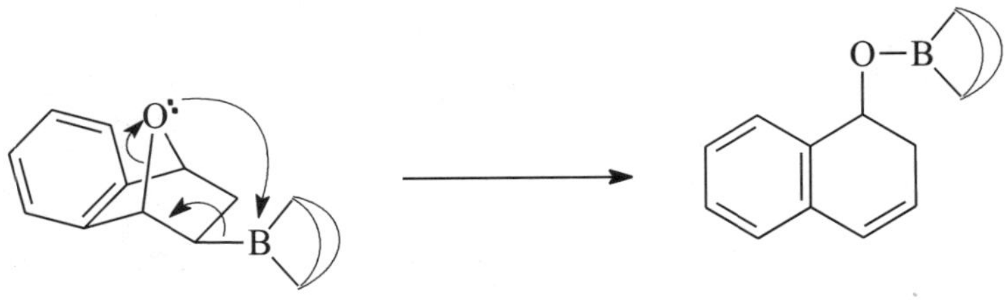

(32.5)

On the other hand, in the derivatives of dicyclohexylboranes and disiamylboranes, the boron atom, that is surrounded by bulky substituents, finds it difficult to achieve coordination with the 7-oxa substituent. These derivatives on oxidation afford the epoxy-hydroxy compound **2**. Consequently, the preparation of the ring opened compound **3** [1] *via* hydroboration–oxidation is superior to the literature method [9].

The reaction of 9-BBN with alkynyl trimethylstannate **4** yields a new alkene (**5**, R = Bu-*t*), and a new allene (**6**, R = Me_3Si or Me_3Sn) (Scheme 32.1) [10], and none of the product is consistent with the expected *cis*-addition of the B–H bond to the C≡C bond in **4**, nor it is in agreement with the mechanism involving hydrostannation of alkynes [11, 12].

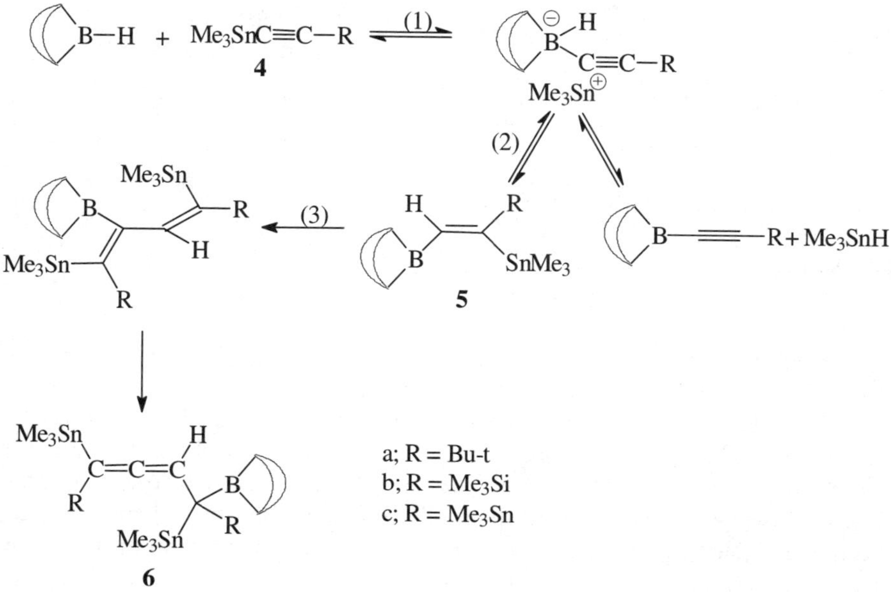

a; R = Bu-t
b; R = Me_3Si
c; R = Me_3Sn

(1) THF, 20-25°C

(2) 20-25°C, 0.2 h (R = Me_3Sn); 60°C, 0.2 h (R = Me_3Si); 60°C, 12 h (R = t-Bu)

(3) Me_3Sn-C≡C-R; 20-25°C, 0.2 h (R = Me_3Sn); 60°C, 0.2 h (R = Me_3Si); no reaction for R = t-Bu.

Scheme 32.1

Brown and coworkers [13] have conducted exhaustive studies of the hydroboration of α,ω-dienes: 1,3-butadiene, 1,4-pentadiene, 1,5-hexadiene, 1,7-heptadiene, 1,8-octadiene, 1,8-nonadiene, 1,9-decadiene, 1,11-dodecadiene, and 1,13-tetradecadiene with 2 molar equiv of 9-BBN, followed by the redistribution of the resulting dumbbell-shaped trialkylboranes with 1 molar equiv of borane–dimethylsulfide complex (BMS).

The hydroboration of 1,3-butadiene is carried out by adding the diene to the slurry of 9-BBN (2 molar equiv) in THF at –78 °C. The reaction mixture is slowly warmed to 25 °C, and BMS is added and redistribution is carried out at 25 °C for 14–18 h. The initially formed unstable [14] boracyclane, borolane, undergoes a rapid ring opening to form 1,6-diboracyclodecane (Scheme 32.2) [13].

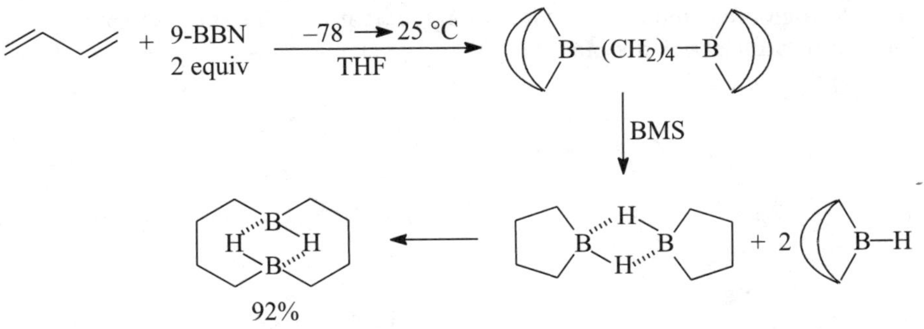

Scheme 32.2

The hydroboration of 1,5-hexadiene with 9-BBN/BH_3·THF [15] or 9-BBN/$B(OCH_3)_3$ [15] affords the borepane ring system in quantitative yields. The 9-BBN/BH_3·THF [15] gives a mixture of 9-BBN and borepane in solution, and fractional distillation separation is unsuccessful. The 9-BBN/$B(OCH_3)_3$ affords *B*-methoxyborepane and *B*-OCH_3-9-BBN, which are separated by fractional distillation. The hydroboration of 1,5-pentadiene with 9-BBN and redistribution of dumbbell-shaped dibora adduct with BMS leads to the formation of borepane and regeneration of 9-BBN (Scheme 32.3). However, none of the attempted methods, such as fractional crystallization from different solvents, selective complexation with amines or distillation effect a complete separation of the two boracyclanes [13].

Scheme 32.3

The hydroboration of 1,6-heptadiene with 9-BBN and redistribution of the dumbbell-shaped dibora adduct with BMS results in the regeneration of the 9-BBN, but no cyclization product is identified. Instead, a polymer separates out as a white globular species. Te depolymerization at 175–200 °C (0.01 mm) yields 95% of boracyclanes. The oxidation of boracyclanes gives 1,7-heptanediol and 1,5-heptanediol in ratio of 3:1 and minor isomers <2 % (Scheme 32.4) [13].

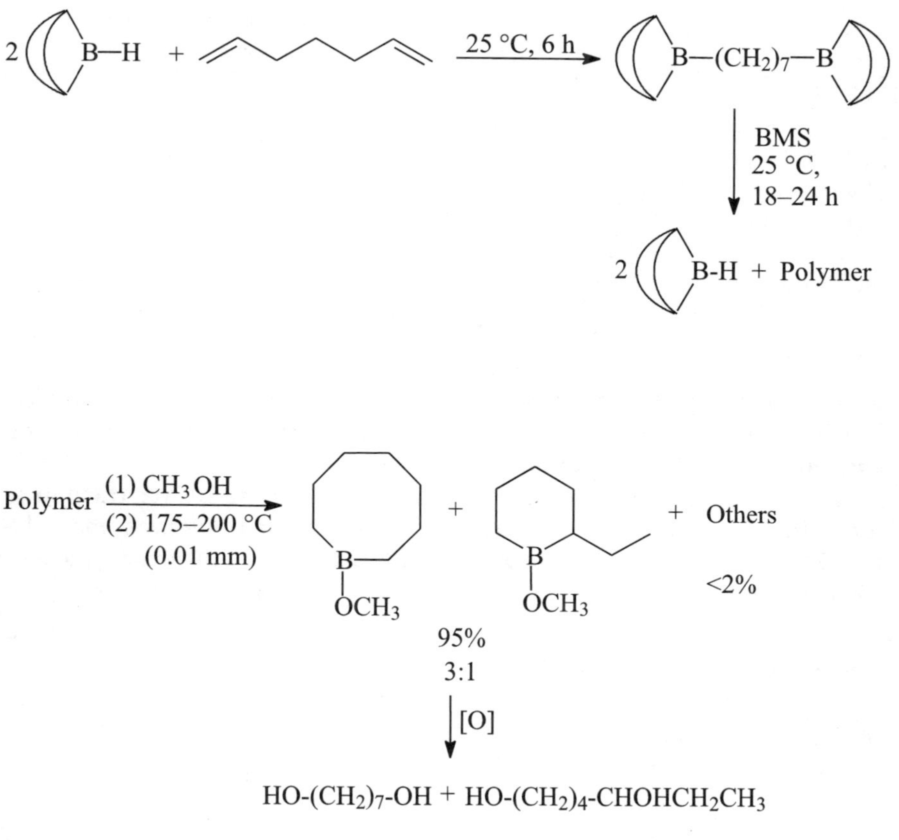

Scheme 32.4

The formation of 1,5-heptanediol and 2-ethylcyclohexanone *via* DCME reaction proves that the depolymerization is accompanied by isomerization.

The hydroboration of each of 1,7-octadiene, 1,8-nonadiene, 1,9-decadiene, 1,11-dodecadiene, and 1,13-tetradecadiene with 2 equiv of 9-BBN affords a linear polymer in each case. Methanolysis, removal of *B*-OCH_3-9-BBN and depolymerization at 175–200 °C under reduced pressure, affords the boracyclanes. But unlike the 1,6-heptadiene experiment, in all cases except 1,13- tetradecadiene, the isomerized product is the major portion of the distillate. (Table 32.1) [13].

Table 32.1 Hydroboration–redistribution–depolymerization of representative diene with 9-BBN-BMS [13]

Diene	Depolymerization temperature (°C)	Total yield of boracyclanes (%)	Percentage composition[a]: Parent 2-alkyl: Other Boracyclane-derivatives isomers		
1,5-Hexadiene	–	100	100	–	–
1,6-Heptadiene	175–180	95	75	25	<2
1,7-Octadiene	185–190	93	11	67	22
1,8-Nonadiene	190–200	98	15	71	14
1,9-Decadiene	175–180	88	12	65	23
1,11-Dodecadiene	180–185	91	0	65	35
1,13-tetradecadiene	175–195	95	38	44	18

[a] Determined by oxidation and GC analysis of the diol.

Both the parent *B*-methoxyboracyclanes and the corresponding isomerized product *B*-methoxy-2-*n*-alkylborinane are identified by their conversion to cyclic ketones *via* DCME procedure [16] and oxidation.

The oxidation affords terminal diols and 1,5-diols. The DCME reaction affords the mixture of ketones in each case, and only the isomerized, major ketone 2-alkylcyclohexanone is isolated, identified, and determined for percentage yield (Scheme 32.5) [13].

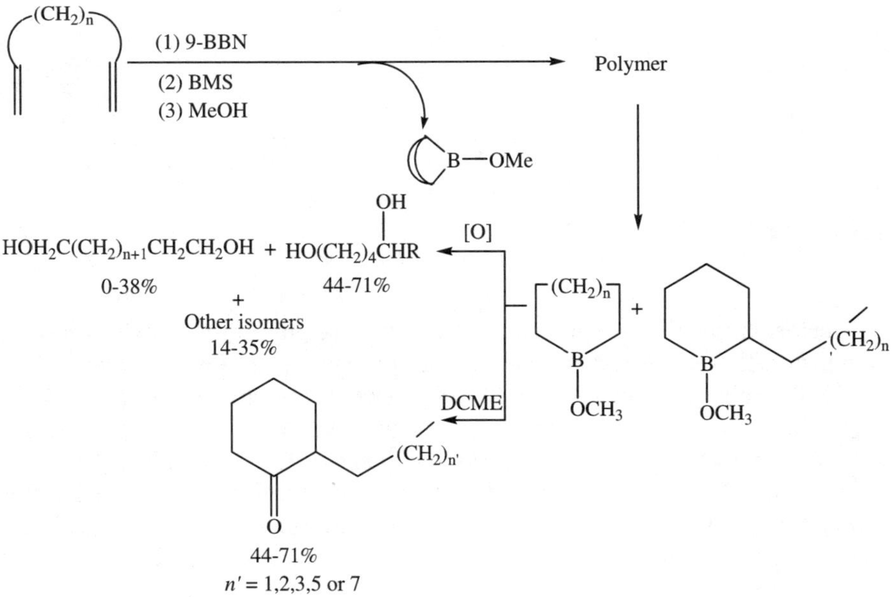

Scheme 32.5

9-(Phenylseleno)-9-borabicyclononane prepared by the reaction of 9-BBN with benzeneselenol (Eq. 32.6) undergoes a novel conjugate addition to a variety of α, β-unsaturated ketones to afford β-selenoboron enolates, which are converted to unsaturated ketols (Eq. 32.7) on treatment with aldehydes and subsequent oxidative elimination [17].

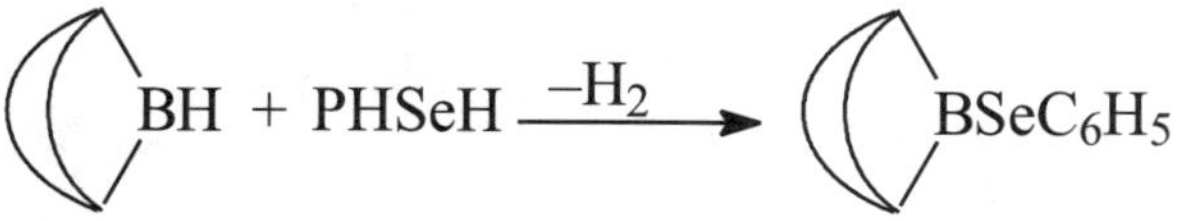

(32.6)

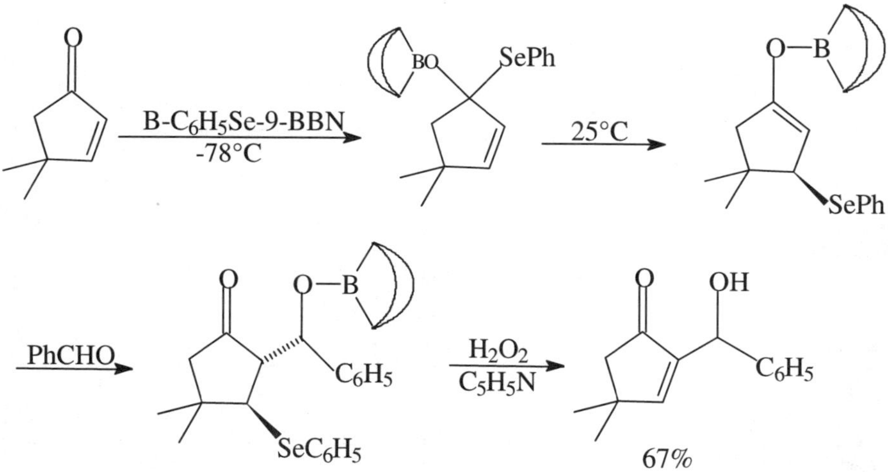

(32.7)

Regio- and stereospecific transannulation of the parallel C=C and N=N has been realized to afford the corresponding adducts in good yields (Chart 32.1) [18].

Nucleophilic attack of anion of N-unsubstituted imidazolines **9a–c** or N-substituted imidazolines (**10**) on THF or 2-methyl tetrahydrofuran, (used as solvent) after activation with 9-BBN triflate affords alkylated product with moderate to good stereoselectivities [19] (Scheme 32.7; Table 32.3).

The N-unsubstituted **9a-c** and **10b** and N-substituted **10a–d** 2-alkyl imidazolines are prepared as described in Scheme 32.6 (Table 32.2) [19].

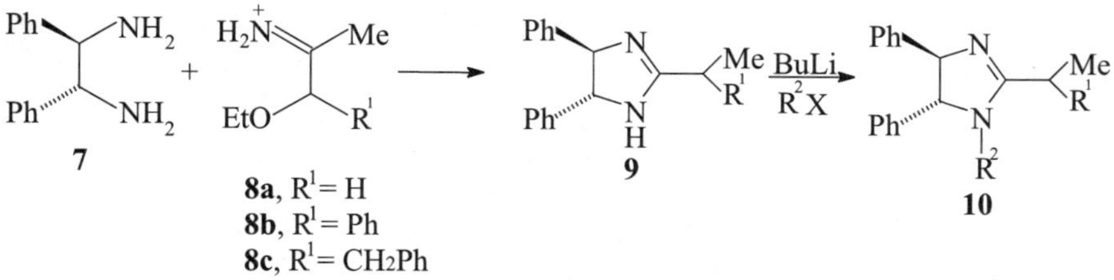

Scheme 32.6

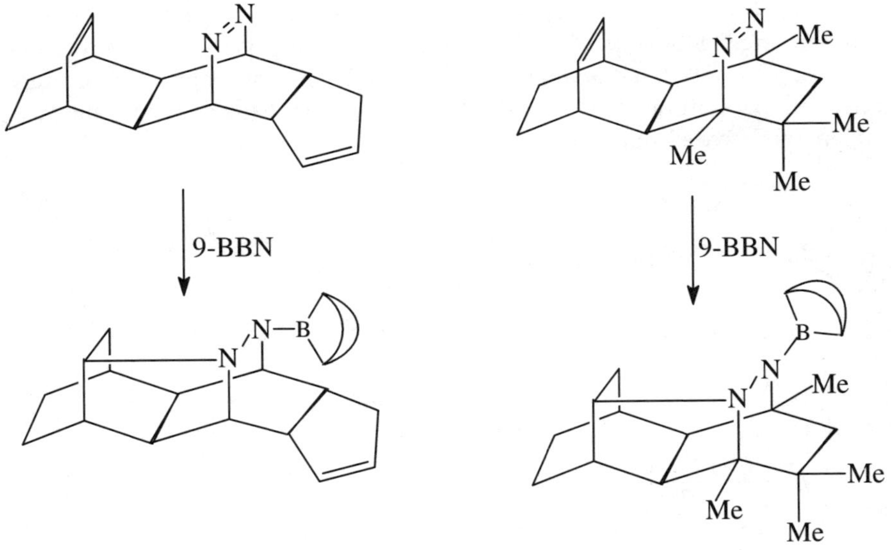

Chart 32.1

Table 32.2 Preparation and N-alkylation of imidazolines [19]

8, R^1	**9**, yield %	R^2X	**10**, yield %
a, H	**a**, 81	MeI	**a**, 81
a, H	–		**b**, 90
a, H	–	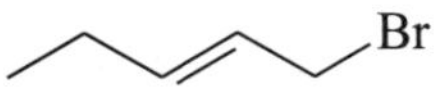	**c**, 88
a, H	–	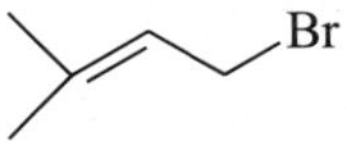	**d**, (86
b, Ph	**b**, 78	–	–
c, CH_2Ph	**c**, 79	–	–

The cleavage of 2-methyltetrahydrofuran (Scheme 32.7; entry 3, Table 32.3) is regioselective [19].

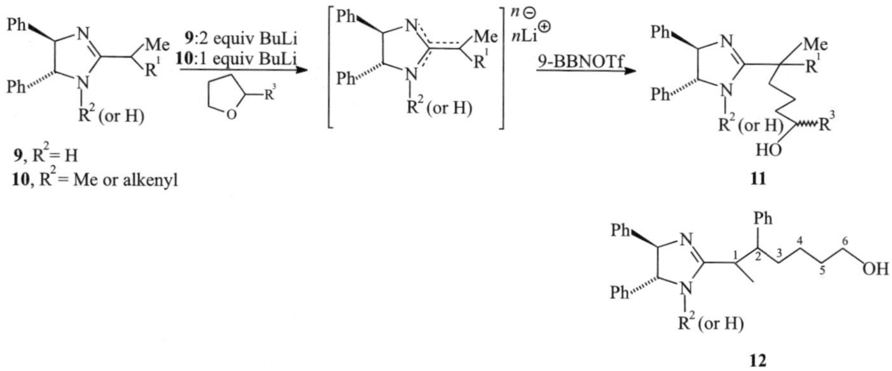

Scheme 32.7

Table 32.3 Alkylation of imidazolines **9** and **10** in the presence of 9-BBNOTf [19]

Entry	Starting material	THF derivative R^3	Temp for 9-BBNOTf[a] introduction (°C)	Product yield (%)	Percentage de
1	**9a**	H	0	**11a** (89)	0
2	**9a**	H	–78	**11a** (87)	>95
3	**9a**	Me	–78	**11b** (87)	>95[b]
4	**9b**	H	–78	**11c** (67)	>95
5	**9c**	H	–78	**12** (81)	>95[c]
6	**10a**	H	–78	**11d** (22)	0
7	**10a**	H	–78	**11d** (30)	67
8	**10a**	H	–78	**11d** (79)	0
9	**10b**	H	–78	**11e** (48)	64
10	**10d**	H	–78	**11f** (36)	40

[a] After addition of 9-BBNOTf, the reaction is generally complete within 5 min.
[b] This product is a 50:50 mixture of diastereomers at C-5.
[c] Mixture of diastereomers at C-1.

One-pot intramolecular reaction between trialkyl-(1-methyl-2-indolyl)borate and allylpalladium intermediate has been reported [20]. This simultaneous 1,2-alkylmigration from boron to carbon provides a novel pathway to afford [*b*]-annelated indole derivatives in 10–56% yields (Scheme 32.8). Trialkylindolylborate **14** is generated *in situ* by the hydroboration of diene **13** with 9-BBN in THF followed by treatment with 2-lithio-1-methylindole. The THF solution of borate **14** containing 10 mol% of $Pd(Ph_3P)_4$ was heated under reflux for 1–3 h to afford [*b*]-annelated indole **15** [20].

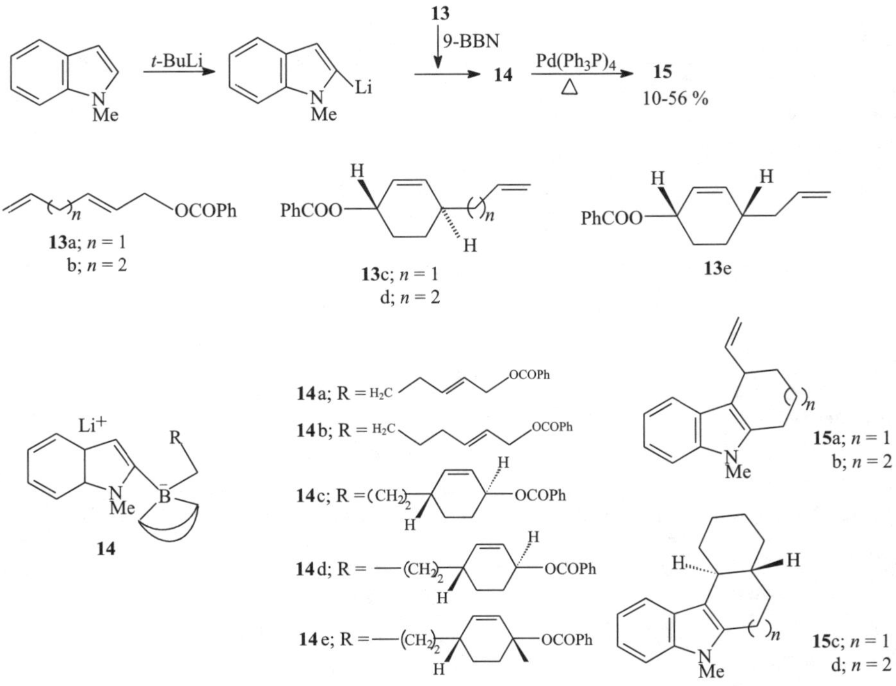

Scheme 32.8

The pathway leading to indole **15d** involves initial generation of allylpalladium complex **16a** and a subsequent intramolecular nucleophilic attack of indole on the allylpalladium intermediate from the opposite side of the palladium [21], with simultaneous 1,2-alkyl migration from boron to carbon [22] (Scheme 32.9).

There are no precedents of such use of the 1,2-alkyl migration, a very common sequence in organoborane chemistry [23], for intramolecular cyclization. Ishikura and Terashima have further developed [24] a novel one-pot procedure for [*a*]-annelated indoles *via* cyclic trialkyl(indol-2-yl)borate as a key intermediate (Scheme 32.10).

The results for the construction of [*a*]-annelated indoles are summarized in Table 32.4 [24].

Treatment of alkylborane **23**, obtained *via* hydroboration of **20** with 9-BBN, with NaOMe, prior to lithiation, was essential as it serves as a tentative boron-protecting group via methoxyborate **24**. Then series of sequences as delineated in Scheme 32.11 give cyclic indolylborate (**21**). In absence of NaOMe, alcohol **25** is obtained solely after an oxidation in 75% yield from indole.

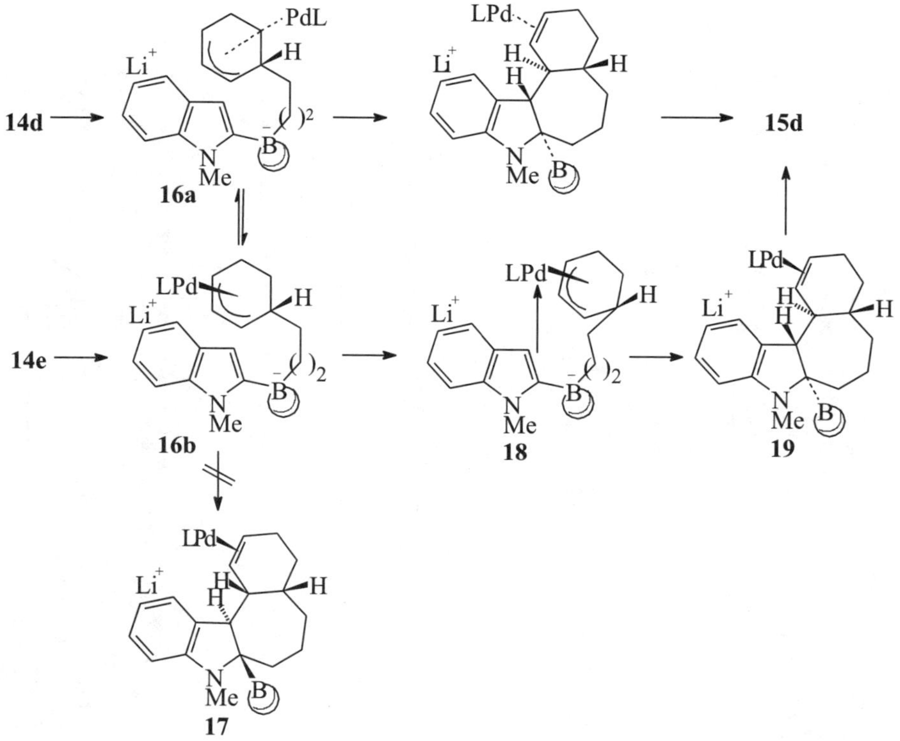

Scheme 32.9

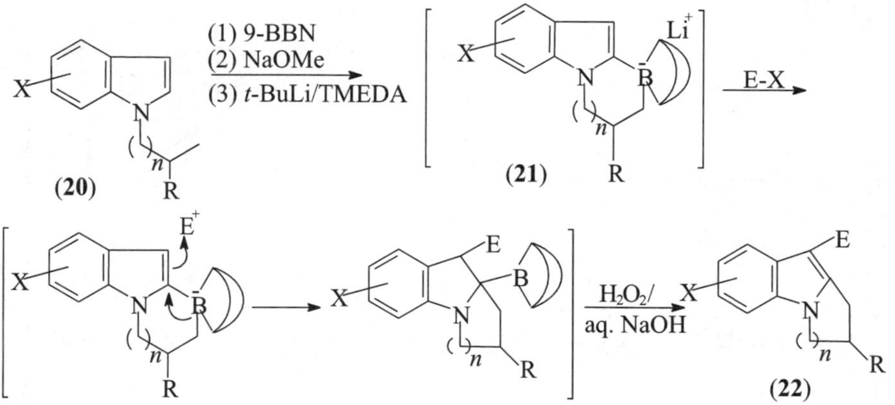

Scheme 32.10

Table 32.4 Formation of [*a*]-annelated indole **22** from indole **20** [24]

n	X	E-X	Isolated yield (%) of **22**
1	H	H_2O	38 (E = H, R = H)
1	H	H_2O	62 (E = H, R = H)
1	H	H_2O	60 (E = H, R = Me)
2	H	H_2O	60 (E = H, R = H)
3	H	H_2O	40 (E = H, R = H)
1	5-Me	H_2O	60 (E = H, R = H)
1	5-OMe	H_2O	60 (E = H, R = H)
1	5-NO_2	H_2O	20 (E = H, R = H)
1	7-Me	H_2O	30 (E = H, R = H)
1	H	MeI	60 (E = Me, R = H)
1	H	ICH_2CN	25 (E = CH_2CN, R = H)
1	H	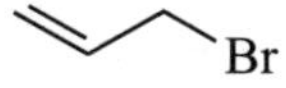	58 (E = $CH_2CH{=}CH_2$, R = H)
1	H	 OAc/$PdCl_2(Ph_3P)_2$	56 (E = $CH_2CH{=}CH_2$, R = H)
1	H	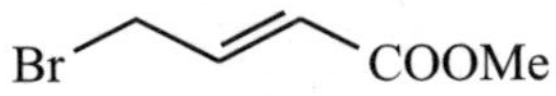	54 (E = CH_2-CHOH-CH_2COOMe, R = H)
1	H	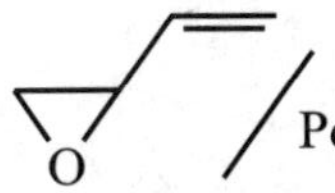 /$PdCl_2(Ph_3P)_2$	58 (E = CH_2-CH=CH-CH_2OH, R = H)
1	H	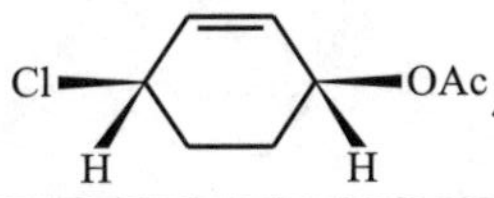 /$PdCl_2(Ph_3P)_2$	52 (E = 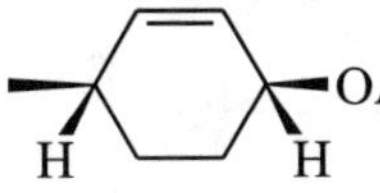, R = H)

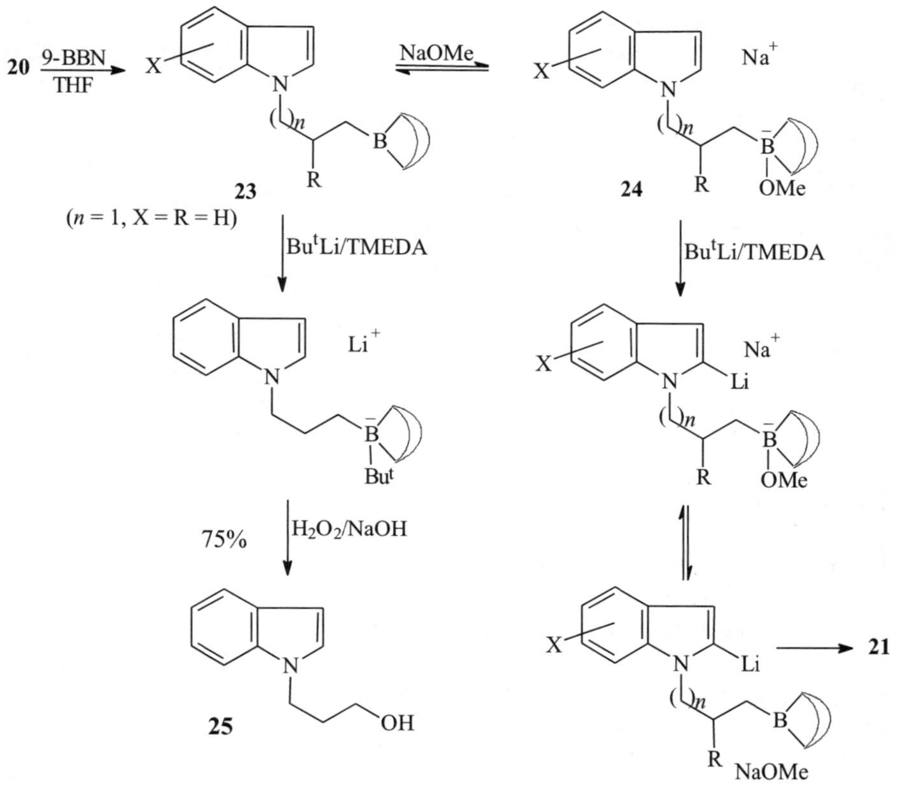

Scheme 32.11

The selective deprotonation of methyl group of 9-methyl-9-BBN by lithium 2,2,6,6-tetramethylpiperidide and the reaction of the resulting carbanion with an electrophile gives chain extended or other products (Scheme 32.12) [25].

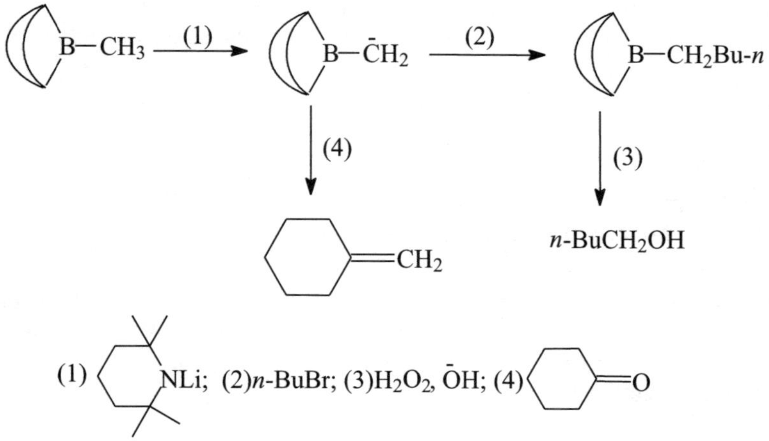

Scheme 32.12

The 9-ClCH=CHCH$_2$-9-BBN undergoes condensation with unsaturated aldehydes to afford the corresponding unsaturated-*trans*-chlorohydrin (Eq. 32.8) [26].

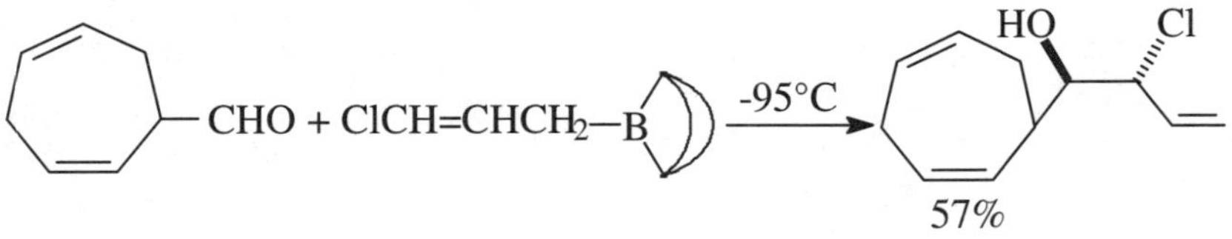

(32.8)

References

1. Brown HC, Vara Prasad JVN (1985) J Org Chem 50:3002
2. Brown HC, Cope OJ (1964) J Am Chem Soc 86:1801
3. Pasto DJ, Synder R (1966) J Org Chem 31:2773, 2777
4. Pasto DJ, Hichman J (1968) J Am Chem Soc 90:4445
5. Mikhailov BM, Schegoleva TA (1959) Izv Akad Nauk SSSR Ser Khim 546
6. Pasto DJ, Cumbo CC (1964) J Am Chem Soc 86:4343
7. Cristol SJ, Parungo FP, Plorde DE (1965) J Am Chem Soc 87:2870
8. Brown HC, Knights EF (1968) J Am Chem Soc 90:4440
9. Jeffrey AM, Jerina DM (1972) J Am Chem Soc 94:4048
10. Wrackmeyer B, Bihlmeyer C (1981) J Chem Soc Chem Commun 1093
11. Neumann WP (1967) Die Organische Chemie des Zinns. Enke, Stuttgart
12. Poller RC (1970) The chemistry of organ tin compounds. Logos, London
13. Brown HC, Pai GG, Naik RG (1984) J Org Chem 49:1072
14. Brown HC, Negishi E (1979) J Am Chem Soc 93:6682
15. Bruke PL, Negishi E, Brown HC (1973) J Am Chem Soc 95:3654
16. (a) Brown HC, Carlson BA (1973) J Am Chem Soc 95:6876; (b) Carlson BA, Brown HC (1974) J Organomet Chem 67:C39
17. Leonard WR, Livinghouse T (1985) J Org Chem 50:730
18. Brand U, Huenig S, Peters K, von Schnering HG Chem Ber 124:1187
19. Dalko PI, Langlois Y (1992) Tetrahedron Lett 33:5213
20. Ishikura M, Terashima M, Okamura K, Date T (1991) J Chem Soc Chem Commun 1219
21. Tsuji J (1986) Tetrahedron Lett 42:4361
22. Levy AB (1979) Tetrahedron Lett 4021
23. (a) Brown HC, Basavaiah D, Singh SM, Bhat NG (1988) J Org Chem 53:246: (b) Hongxun D, Weike Z, Junchai B (1987) Tetrahedron Lett 28:2599; (c) Pelter A, Smith K, Brown HC (1988) Borane reagents. Academic, London
24. Ishikura M, Terashima M (1992) Tetrahedron Lett 33:6849
25. Rathke MW, Kow R (1972) J Am Chem Soc 94 (1972) 6854
26. Hertweck C, Boland W (1997) Tetrahedron 53 (1997) 14651

Subject Index

A

$A(^1,^2)$ strain 25
$A(^1,^3)$ strain 24
acetic anhydride, reduction 399
– with 9-BBN·THF 399
acid chlorides, reduction 399
– with 9-BBN·THF 399
acyclic ketone 404
alcohols 52, 194
– asymmetric synthesis 194
aldehydes 41, 42, 199, 200, 215, 216, 217, 218
– conversion from B-acyloxy-9-borabicyclo[3.3.1]nonane with Li-9-BBNH; LTDEA 215, 216, 217
– rate constants 41
– relative reactivity 42
– stereoregulated reaction of α-silyl or α-stannyl substituted crotyl-9-BBN 199, 200
aldehydes, ketones 184
– with propargylic boranes 184
aldehydes, ketones reduction 410
– with 9-BBN·Py THF 410
aldehydes reduction 415, 428, 431
– with alpine-borane 428
– with B-siamyl-9-BBN 415
– with deuterated alpine-borane 431
1,5-alkadienes 524
– synthesis, stereodefined 524
alkenes 22, 28, 29, 30
– hyperconjugative effects 28
– inductive effects 28
– rate constants, hydroboration 22
– reaction rate 28
– reactivities; $(9\text{-BBN})_2$, Sia_2BH, $HBBr_2{\cdot}SMe_2$ 29
– reactivity order 29
– steric effects 28
alkenylboranes 518, 519, 520
– reactivity and selectivity toward Diels-Alder reaction 518, 519, 520
alkylation 221, 327, 328, 329
– 9-BBN, organolithium reagents 327
– B-chloro-9-BBN, organolithium reagents 328
– B-OMe-9-BBN, organolithium reagents 329
– B-R-9-BBN 254
– α-alkyl-α-chloroacetonitriles 256
– α-bromoketones 221
alkylborohydrides 325
alkyl halides 416
alkynes 29, 31, 39, 151, 154, 512
– [4+2]cyclo addition 512
– conversion, to alcohols 151
– conversion to allylic alcohols 154
– halogen effect 39
– mono-, dihydroboration 31
– rate constants 29
– reactivity; $(9\text{-BBN})_2$, Sia_2BH, $HBBr_2{\cdot}SMe_2$ 29
– reactivity order 29
alkynylborinates 317
allenes 135, 162, 561
– hydroboration–derivatization–oxidation 135, 162

- synthesis, boron and tin substituted 561
allenyl 9-BBN 177
- conversion, kinetic to thermodynamic product 177
allenylboration 167, 168, 169, 266
- aldehydes, ketone 167
- imines 266
allyboranes 358
- geometry of the double bond 354, 358
allylamines 79, 80, 81
- anti selectivity 79, 82
- electronic factors, catalyzed hydroboration 81
- syn selectivity, catalyzed hydroboration 80
allylation 195, 196
- diastereoselective 195, 196
allylboranes 153
allylboration 155, 156, 157, 158, 159, 160, 161
- acid anhydride 159
- acid chloride 158
- aldehydes, ketones 155
- esters 160
- N,N-amides 160
allyl halide 34, 37
- electronic effect 34
- inductive, hyperconjugative 37
allylic acetates 82
allylic amines 79
- catalyzed hydroboration 79
allylic borinate 492, 493
- synthesis of cyclooctane compounds 492, 493, 494
allylic chloride 35
- chlorine influence 35
allylic metalation 491
allylic trifluoroacetates 82
allylsilane 542
- conversion, to bislactin 542
allylsilanes 77
- hydroboration, regiochemistry 77
- stereoselective hydroboration 77
Alpine-Borane 448
- chemoselective reduction of propargyl ketone in presence of methyl ketone 448
Alpine-hydride 459, 461
- synthesis 459, 461
amides 157
amines 49, 50, 51, 263, 265
- anti-Cram product 263
- complexation, solvent effect 50
- complex formation 49
- Cram-erythro 265, 266
- Cram-threo 265, 266
- Cram product 263
- kinetic behavior 51
2,6-anhydro-3,4,5,7-tetra-O-(methoxymethyl)-D-gluco-hept-1-enitol 550
annelated indoles 567, 568, 570
- synthesis 567, 568, 570
aromatic nitriles 219
- selective reduction 219
arylacetic acid 538
- as cyclooxygenase inhibitors 538
arylacetylene 531
α-arylation 223, 254
- bromoacetone 223
- with B-R-9-BBN 253, 254
asymmetric induction 261, 262, 264, 265, 430, 442, 451, 458
asymmetric reducing agent 478
- preferred class of ketones 478
asymmetric reduction 435, 467
- Binal-H reagent 467
- of ketones, steric group category 435
aza-C-disaccharide-D-azaMan-β-(1→6)-d-Man 551
azasugars 546
- glycomimetics compounds 546

B

B-(3,3-dimethyl-1,2-dideuterio-1-butyl)-9-BBN 534
B-(Z)-1-alkenyl-9-BBN 335

B-1-alkynyl-9-BBN 163, 165
B-1-alkynyl-9-BBN·THF derivatives 334
– preparation 334
B-1-alkynyl–9-BBN·THF 333
B-1-allenyl-9-BBN 170, 172
– chemoselectivity; aldehydes, ketones 170
– reaction with esters 170
– regioselectivity, ketones 172
B-1-hexynyl-9-BBN 165, 230
– chemoselectivity; aldehydes, ketones 165
– conjugate addition 230
B-1′,7′,9′-decatrienyl-9-BBN 496
– synthesis 496
B-2-propenyl-9-BBN 496
– synthesis 496
B-3-methyl-2-butyl-9-BBN 44
B-3-pinanyl-9-BBN 44
– asymmetric reducing 44
B-alkyl-9-BBN 45, 428
– rate of reduction 45
B-alkynyl-9-BBN 164
B-allenyl-9-BBN 167, 169, 171
– chemoselectivity, aldehydes, ketones 171
– reaction with carbonyl compounds 169
B-allyl-9-BBN 156
– acetone, allylboration 156
9-BBN 7, 11, 12, 13, 60, 61, 62, 63, 64, 65, 66, 67, 68, 268, 401
– amines interaction 12, 13
– basic solvents 7
– complexation 7
– cyclopropyl derivatives, hydroboration 66
– functional groups reactivity 401
– hydroboration; rate constants, of cis–trans alkenes 63
– hydroboration rate, 1-alkene 61
– hydroboration rate, internal olefins 61
– hydroboration rate; cis-alkenes, trans-alkenes 62
– olefins reactivity 67, 68
– pyridine complex 11
– reaction of 3-substituted azanorbornenes 268
– regiospecificity 60
– styrene, site of hydroboration 65
– styrene derivatives, hydroboration 65
– thermal stability 10
– with PhS(Rt)CuM 332
– α-methyl substituent, hydroboration rate 64
$(9\text{-BBN})_2$ 5, 7, 9, 19, 20, 43, 50, 53, 54, 55
– ^{1}H NMR 5
– alkenes, alkynes, aldehydes, ketones, alcohols, amines; kinetics 50
– coordination of oxygen; phenol, alcohol 54
– dissociation 19, 20
– mass spectrum 5
– melting point 5
– molar solubility 7
– protonolysis, benzylic alcohols 55
– protonolysis rate; phenol, alcohol 54
– protonolysis rates, phenols 55
– protonolysis rates; alcohols, phenols 53
– reaction of alkenes, alkynes, aldehydes, ketones, alcohols, amines; mechanistic pathways 50
– recrystallization 9
– relative reduction 43
$(9\text{-BBN-D})_2$ 14
– synthesis 14
B-benzyl-9-BBN 330
B-Br-9-BBN 488
– cleavage of ether 488
B-cis-2-chlorovinyl-9-BBN 496
B-cis-propenyl-9-BBN 496
– synthesis 496

B-Cl-9-BBN 506
- catalytic cycle, Diels Alder 506
B-cyclohexenyl-9-BBN 496
- synthesis 496
B-cyclopentenyl-9-BBN 496
- synthesis 496
B-ethynyl-9-BBN 165
B-hexyl-9-BBN 283
B-isopropyl-9-BBN 330
B-methoxy-9-BBN 283
B-methyl-9-BBN 571
- deprotonation of methyl group 571
B-OMe-9-BBN 196, 207, 328
- as an additive 196
- bromination 207
B-p-methoxyphenyl-9-BBN 330
B-R-9-BBN 330, 331, 332, 345, 346
- boiling points 330
- mixed (phenylthio)alkylcuprates with 9-BBN 331, 332
- synthesis 327
- to exocyclic olefins 345, 346
- to trisubstituted alkenes from vinyliodides 345
B-R*-9-BBN 202, 245
- alkylation, ethyl bromoacetate 245
- optically active 202, 245
B-sec-alkyl-9-BBN 272
- free radical substitution of the α-hydrogen by bromine 272
B-substituted-9-BBN 13
- monoxidation 13
B-tert-butyl-9-BBN 330
B-[(trimethylsilyl)ethynyl]-9-BBN·THF 497, 516
- synthesis 497, 516
B-X-9-BBN 274, 310, 311
- enolate geometry, effect of leaving group 310, 311
B-X-9-BBN·SMe_2 274
- synthesis 274
base-induced elimination 287
benzaldehydes 45, 46, 47
- boat conformation, reaction of B-alkyl-9-BBN 47
- reduction with B-3-pinanyl-9-BBN, rate constants 46
- reduction with B-n-octyl-9-BBN, rate constants 45
BH_3 59, 60, 146, 147
- cycloalkenes, hydroboration 146, 147
- hydroboration 59
- regiospecificity 60
bicyclo[3.3.1]nonan-9-one 224
bis(9-borabicyclo[3.3.1]nonyl)acetylene 516
- in Diels-Alder reaction 516
- preparation 516
bis(9-borabicyclo[3.3.1]nonylacetylene 496
- synthesis 497
bis-9-BBN adduct 537
- for Suzuki coupling 537
borane-1,4-thioxane (BOT) 8, 9
- 1,5-Cyclooctadiene 8
- ^{11}B NMR 8
- hydroboration 8
- neat reaction of 1,5-cyclooctadiene 9
- synthesis 8
borane-dimethylsulfide complex (BMS) 9, 234, 292, 561, 562
- redistribution of boron of 9-BBN 561, 562
borepane 562
- synthesis, separation from 9-BBN 562
borinane 292, 293, 294
- separation from 9-BBN 293, 294
borneol 411
boroenolate 206
- aldol reaction, aldehydes 206
(β-Borylvinyl)silane 118
bromoboration reaction 275
3-bromocyclohex-3,5-dien-1,2-diol 546
Bu_3SnH 511

C

camphor-borneol, reduction 398
- with 9-BBN·THF 398
camphor-isoborneol, reduction 411
- with Li9-BBNH·THF 411
carbamates 82
(carbethoxymethyl)dimethylsulfonium bromide 246
- to esters, with B-alkenyl-9-BBN 246, 247, 248
carbonylation 213, 214, 215
- B-alkyl-9-BBN 213, 214, 215
- lithium-tri-t-butoxyaluminohydride 214
carboxylic acids, reduction 218, 399
- with 9-BBN·THF 399
charge polarization 122
chemoselectivity 419
chiral allylic alcohols 78, 82
- catalyzed hydroboration 82
- diastereoselective hydroboration 78
- uncatalyzed hydroboration 82
chiral allylic amines 78
- diastereoselective hydroboration 78
1-chloroalkenes 40
- regiochemistry of hydroboration 40
1-chloroalkynes 40, 123
- regiochemistry of hydroboration 40
(α-chloroallyl)lithium 491
- trans metalation with boron 491
2-chlorocamphane to isocamphane 418
- lithium di-n-butyl-9-BBN ate complex 418
chromium (II) complex 416
Chx_2BH 145, 178
- allenes, hydroboration 178
- cis-4-methyl-2-pentene, hydroboration 145
Chx_2BX 310, 311
- enol geometry, effect of leaving group 310, 311
cis-3-hexene 21
- rate constants, hydroboration 21
cis- and trans-1,3-pentadienes 128
cis- and trans-alkenes 23
- rate constants, hydroboration 23
- relative reactivity 23
cis-bicyclo[3.3.0]-oct-1-yldialkylboranes 298, 299
- conversion to 1-substituted derivatives 299, 300, 301
cis-elimination 560
- β-elimination of weak leaving group 560
cis-vinyl-9-BBN 154
citronellal 352
comparative reduction 462, 474
- Alpine-Hydride, NB-Enantride, Eapine-Hydride 462
- with K9-OThx-9BBNH, Li-s-Bu_3BH, $LiSia_3BH$ 474
comparative reduction of acetylenic ketones 457, 458
- Alpine-Borane, Eapine-Borane, Prapine-Borane 457, 458
comparative reduction of cyclic ketones 403
- $(9\text{-BBN})_2$ vs $NaBH_4$ 403
competitive coupling 536
- B-R-9-BBN, R′-OBBD derivatives 536
conversions 338
- carbonyl compounds; alkenes 338
Cram (Felkin) addition 263
Crandall reductive alkylation 536
- epoxides 536
cross-coupling 226, 523
- B-octyl-9-BBN 523
- palladium-catalyzed carbonylative 226
cross-coupling reaction 225, 523
- B-R-9-BBN 523
crotyl organometallic compounds 264
cyclic allenes 136
- conversion to allylderivatives via 9-BBN 136

- vinylboranes via Sia_2BH 136
cyclic hydroboration 291
cyclization of bromoalkenes 530
- via hydroboration 530
cycloheptene 110
- vs tetrahydrooxepin 110
cyclohex-3-en-1-ols 162, 502
cyclohexene 18, 21, 22, 110, 111
- rate constants 18
- rate constants, hydroboration 21, 22
- vs dihydropyran, hydroboration 110, 111
cyclopentene 18, 21, 109, 110, 111
- rate constants 18
- rate constants, hydroboration 21
- vs dihydrofurans, hydroboration 109
- vs dihydrothiophene, hydroboration 111
1-cyclopentyl-2-decanone 225

D

D-azaMan-β-(1→1)-D-β-Glc 546, 552
D-azaMan-β-(1→4)-D-Talo 546, 552
D-azaMan-β-(1→6)-D-Glc 546, 552
D-azaMan-β-(1→6)-D-Man 546
DCHGF 469, 470
DCME reaction 563, 564
1,7,9-deca-trienylborane 511
- synthesis 511
dehydroboration–reduction mechanism 432
dehydroborylation 113
dehydroborylation, transition state 115
deiodination 441
depolymerization 564
deuterated dienyne-allene 380
deuteride delivery 432
dialkylation 256
- dichloroacetonitrile 256
- α-alkyl-α-chloroacetonitriles 256
dialkylborohydrides 325
- IR and ^{11}B NMR 325
1,3-diaxial interaction 195
diazabicyclooctane (DABCO) 293
1,1-diboraadduct 153
- trans-vinyl-9-BBN, dehydroboration 153
1,6-diboracyclodecane 561
diborane 3, 8, 17, 60
1,2-diboryl, conversion to 1,1-diboryl 119
Diels-Alder dienophiles 501
Diels-Alder reaction 499, 510
- intermolecular, intramolecular 510
dienes 126, 353, 561
- aldehyde enamines 353
- monohydroboration 126
dienyl borinate 493
- synthesis 493
- synthesis of cyclooctane products 493, 494
dihydrofuran 101, 102, 103, 109
- hydroboration 101, 102, 103
- mesomeric effect 109
- relative reaction 109
- vs cyclopentenes, hydroboration 109
dihydropyran 104, 110
- hydroboration 104
- vs cyclohexene, hydroboration 110
dihydrothiophene 103, 111
- hydroboration 103
- vs cyclopentene, hydroboration 111
dihydrothiopyran 111
- vs cyclohexene, hydroboration 111
2,5-dimethyl-2,4-hexadiene 128
DIPGF 468
disiamylborane (Sia_2BH) 4, 60, 61, 62, 64, 65, 69, 70
1,1-disubstituted alkenes 532
- synthesis 532
double Suzuki coupling 534

E

Eapine-Borane 454, 455, 456
– asymmetric reduction of ketones 454
– reduction of α,β-acetylenic ketones 455
– reduction of α-ketoesters 456
– synthesis 454
Eapine-hydride 461
– synthesis 461
effectiveness of LiNR[1]R[2]-9-methoxy-9-BBN 492
– lithium-boron trans-metalation 492, 493, 494
empirical rules 68
– alkyl group effect, hydroboration 68
enamines 97, 193, 339, 340, 341, 342, 343, 344
– conversion, to amino alcohols 193
– hydroboration 339
– hydroboration, directive effects 97
– regiochemistry, hydroboration 97
– to (E)-alkenes, with BMS 339, 340, 341, 342
– to (Z)-alkenes, with 9-BBN 339, 340, 341, 342
– to alkenes 338
– to cycloalkenes 343
– to heterocyclic alkenes 344
enolborinates 310
– steric control 310
γ,δ-enones 227
– synthesis 227
enyne 137
– hydroboration of C=C, 9-BBN 137
– hydroboration of C≡C, $HBBr_2 \cdot SMe_2$ 137
enynes 531
epothilone B 546
epothilones 546
– anticancer agents 546
epoxide reduction 413, 421
– lithium-di-n-butyl-9-BBN ate complex 421
– with Li-9-BBNH·THF 413
1,4-epoxy-1,4-dihydronaphthalene 559
– epoxy ring opening 559
– hydroboration-oxidation 559
esters reduction in presence of carboxylic acid 412
– with Li 9-BBNH·THF 412
ether 487, 489
– cleavage pattern 487
– rates of cleavage 489
ethyl caproate 422
exo-heterocyclic, olefins 533
– Suzuki reaction 533
exocyclic olefin
– 345
(E)-β,γ-unsaturated amides 259
– synthesis 259

F

farnesol tetrahydropyranyl ether 524
Fisher carbene complexes 501
Fuchs-Corey method 534

G

Galvinoxyl 283
Gassman synthesis 511
– trienes 511
gem-Dibromovinylidene 534
glycomimetics compounds 546

H

halo-1-alkenes 352, 527, 528
– Suzuki coupling 527, 528
haloalkene 32, 34
– rate constants 32
– reactivity 34
haloalkynes 38, 39, 120, 121, 122, 123, 124, 125, 352
– ^{13}C NMR 124
– charge polarization, rate of hydroboration 122
– electronic, steric effects 123
– halogen effects, hydroboration 122
– inductive effect 125
– mesomeric effect 125

- rate constants 38, 39
- relative reactivity 120, 121
haloarenes 525, 526
- Suzuki coupling 525, 526
haloboration-conjugate addition 280
$HBCl_2$ 146
- cycloalkenes, hydroboration 146
Heck coupling 536
heterocyclic olefins 190, 192
- vs carbocyclic olefins, hydroboration 190, 192
1-hexyne 117
- vs 1-(trimethylsilyl)-1-hexyne 117
homoallylic alcohol 156
homologated ketones 220
- synthesized 220
hydroboration 27, 69, 70, 71, 72, 73, 74, 76, 83, 119, 126, 134, 206, 564
- 1-alkylcycloalkenes 71
- 3-alkylcycloalkenes 72
- alkyl group effect 27
- alkynylsilanes 119
- allyl halides 73
- allylic alkenes, functionalized 73
- catecholborane-rhodium complex 83
- cycloalkenes 70
- cyclopentenes 70
- haloalkenes, rate order 74
- hyperconjugative effects 69
- internal allenes 134
- phenyl 1-alkenyl ketones 206
- relative reactivities, cis–trans-haloalkenes 76
- steric effects 69
hydroboration–dehydroboration–hydroboration 173
hydroboration–deuteronolysis–hydroboration 534
hydroboration–halogenation reaction 273
hydroboration–oxidation 191, 238
- of allylamines 238
5-hydroxytryptophan 544
hydrozirconation 303
hyperconjugation 26, 27

I

Imidazolines 565, 566, 567
- inductive effect 27
- preparation, N-alkylation 566
IPDXF 469, 470
IR 20, 27
- quantitative spectrometry 20, 27
isoborneol
- 411
isocamphane
- 418

K

K-glucoride 466, 467
- asymmetric reduction of isobutyrophenone 467
- asymmetric reduction of pivalophenone 467
- reduction of alkylaromatic ketones 467
- reduction of cyclic, bicyclic ketones 466
K-xylide 470, 471
- reduction of alkylaromatic ketones 471
- synthesis 470
K 9-sec-Am-9-BBNH 218
KAPA 152
- long-chain alcohols, synthesis 152
- to terminal alkyne from internal alkyne 449
ketone reduction 397, 440, 462, 476
- with 9-BBN·THF 397
- with chiral trialkylborohydride 462
- with Li 9-BBNH 476
- with neat Alpine-Borane 440
ketones 41, 42, 43, 132, 308, 309, 433, 565
- allylboration 132
- asymmetric reduction 433
- from β-selenoboron enolates 565
- rate constants 41
- reduction; effect of methyl substitution 43
- relative reactivity 42

– stereoselective enolization 308, 309
ketone stereoselective reduction 474
– with lithium di-n-butyl ate complex of 9-BBN 474
kinetic of protonolysis 51
kinetic reduction 41
kinetics 52
– methanol, ethanol, isopropyl alcohol 51
– protonolysis of $(9\text{-BBN})_2$, alcohols 52
– protonolysis of $(9\text{-BBN})_2$, phenol 52
kinetics of alkynes 27
kinetics of complex formation 48
kinetics of haloalkenes 31
kinetics of haloalkynes 38

L

L-Tryptophan 543
Li 9-BBNH 216, 414
– functional group reactivity category 414
lithiation 568
lithium-tri-t-butoxyaluminohydride 214
lithium dialkyl ate complexes of 9-BBN 297
– oxidation 297
lithium trialkylborohydride 410
lithium trimethoxyaluminohydride 214
LTDEA 217
– preparation 217

M

Markovnikov vinylborinates 532
Meerwein–Pondorf–Verley reduction 475
(–)-menthylchloroformate (MCF) 478
3-methyl-1,2-butadiene 133
methyl-2,3,4-tri-O-(methoxymethyl)-6-methylene-α-D-glucopyranoside 548
methyl-2,3-O-isopropylidene-4-O-(methoxymethyl)-6-methylene-α-D-mannopyranoside 548
4-methyl-4-sila[3^4,^{10}][7] metacyclophane, 4-(2-propenyl)-4-methyl-4-sila[7]metacyclophane 537
methyl-6-O-(tert-butyldiphenylsilyl)-2,3-O-isopropylidene-4-deoxy-4-methylene-α-D-lyxo-hexopyranoside 549
methylene cyclohexane 571
– synthesis 571
methyl hyperconjugation 24, 25
methyl ketones 222, 223, 224
methylmethylene-d_2-cyclopentane 345
methyl vinyl ketone (MVK) 229
– synthesis 229
Midland's reagent 429
model for chiral reduction of aldehyde 432
monohydroboration 126, 127, 128, 129, 133, 175
– 1,3-hexadiene 129
– 1,3-pentadiene 128, 129
– 3-methyl-1,2-butadiene 133
– allenes 175
– conjugated vs nonconjugated dienes 127
– dienes 126

N

N,N-dimethylamides 413
(3R, 4R, 5S, 6S)-3-[(N-benzyloxycarbonyl)amino]-1-bromo-4-[(tert-butyldimethylsilyl)oxy]-5,6-(O-isopropylidenedioxy) cyclohex-1-ene 546
N-(benzyloxycarbonyl)-1,2,3,6-tetrahydropyridine 107
– hydroboration 107
N-(benzyloxycarbonyl)-3-pyrroline 103
– hydroboration 103
N-alkylation in presence of 9-BBNOTf
– 567

N-protecting 238
N-substituted-3-pyrrolines 108
- hydroboration 108
N-Tosyl-5-aminocyclooctanol 193
$NaBH_4$ 43
- relative reduction 43
NB-Enantrane 453, 459
- reduction of α,β-acetylenic ketones 453
- synthesis 459
NB-Enantride 459, 461
- synthesis 459, 461
nitriles 219
- with K-9-sec-amyl-9-BBNH conversion, to aldehydes 219
norcamphor
- 411

O

olefinated carbohydrates 548
- Wittig methylenation, synthesis of 548, 549
olefinic homopropargylic alcohols 169
- via B-allenyl-9-BBN 169
olefins 19, 143, 144, 145, 215, 242
- conversion, to aldehydes 215
- conversion, to alcohols 143
- hydroboration-carbonylation 215
- hydroboration-oxidation 144, 145
- rate constants 19
omniphilic dienophile 495
one-pot procedure 223
one-pot synthesis 493
organoborane-carbonyl oxygen complex 47
organoboranes 559, 560
- β-elimination of electronegative substituents 559, 560
organotin hydrides reactivity 416
9-oxa-10-borabicyclo[3.3.2]decyl (OBBD) derivatives 533
- in Suzuki coupling 533
9-oxa-B-OMe-9-BBN 532
- for Suzuki coupling 532

P

palladium-catalyzed coupling 523
$Pd(PPh_3)_3$-catalyzed cross-coupling 181
pentene 110
phenalene 407
phenalenone 407
- conversion to phenalene 407
phenols 52
phenyl conjugation 24, 25
phenylpropene 24, 67
- hydroboration 24
- methyl hyperconjugation 67
- phenyl conjugation 67
4-(3-phenylpropyl)-4-methyl-4-sila[7] metacyclophane 537
9-(phenylseleno)-9-borabicyclononane 565
α-pinene 428, 430, 437
- via Schlösser metalation 500
piperylene 502
pivalaldehyde 164
pivalophenone 465
pivalophenone, isobutyrophenone reduction 467
- with Binal-H reagent 467
polyhydroxylated piperidines 546
potassium-3-aminopropylamide (KAPA) 151, 449
- internal C≡C to terminal C≡C 151
potassium 2,6-di-tert-butylphenoxide 256
- dialkylation, alkylation 256
Prapine-Borane 456, 500
- synthesis 456
Prapine-Borane reaction 459
- α-keto esters 459
preferential reduction 401
- categories of functional groups 401, 402
- of acid chlorides in presence of esters 401
- of hexanal in presence of 2-heptanone 401
pregrenolone acetate 529

- synthesis, steroidal side chain 529
propargylic-allenic rearrangement 183
propargylic alcohol 163
propargylic boranes 184, 185, 186
- chemoselectivity; aldehydes, ketones 185
- chemoselectivity; ketones, esters 186
- relative reactivity; aldehydes, ketones 184
2-propenylborane 518
- relative reactivity 518
propenylfuran 84, 87
- hydroboration 84
propenylpyridine-BF_3 95
- hydroborating agents 95
2-propenylthiophene 87
- hydroboration 87
propionaldehyde 164
prostaglandin 542
- synthesis 542

Q

Quinuclidine (QN) 12

R

R-(+)-α-methoxy-α-(trifluoro-methyl) phenylacetic acid (MTPA) 478
R-(−)-Benzyl-α-d-alcohol 430
redistribution 564
reduction 44, 403, 404, 406, 409, 412, 413, 420, 422, 423
- B-OMe-9-BBN as catalyst 422
- electronic effects 44
- Li-9-BBNH as catalyst 422
- of 2-cyclopentenone with various reducing agents 406
- of 2-heptanone in presence of 4-heptanone 420
- of acid chlorides to alcohol 412
- of aldehydes in presence of ketone 409
- of aldehydes in presence of ketones, esters, nitriles 420
- of cyclohexanone in presence of benzyl chloride 420
- of enals or enones in presence of other organic functional groups 406
- of esters in presence of chloro-, nitrogroups 423
- of ethylacetate in presence of n-octylbromide 412
- of tertiary amides to amines 413
- relative rate of ketones 404
- relative reactivity of aldehydes, ketones; $(9\text{-BBN})_2$ 403
- relative reactivity of ketones 420
- with 9-BBN, $NaBH_4$ reactivity difference; camphor, norbornanone 404
1,2-reduction 405
reduction of esters by lithium borohydride 423
relative rate of ketone 403
relative reactivity; aldehydes, ketones 184
- propargylic boranes 184

S

Schwartz's reagent, $H(Cl)ZrCp_2$ 303
Schwartz's reagent, hydrozirconation 303
selective coupling 536
B-selenoboronenolates 565
- synthesis 565
(S)-(−)-frontalin 533
- synthesis 533
Sia_2BH 60, 61, 62, 65, 145, 147
- cis-4-methyl-2-pentene, hydroboration 145
- cycloalkenes, hydroboration 147
- hydroboration rate, 1-alkene 61
- hydroboration rate, internal olefins 61
- hydroboration rate; cis-alkenes, trans-alkenes 62
- regiospecificity 60
- styrene, hydroboration 65

– α-methyl substituent, hydroboration rate 64
sigmatropic rearrangement 449
– transfer of chirality 449
– transfer of functionality 449
silametacyclophanes 537
silylalcohols 173
– hydroboration–oxidation 173
α-silyl- or -stannyl-substituted crotyl-9-BBN 198
six-membered ring 534
– via Suzuki coupling 534
stereoselective hydroboration 148
stereoselective synthesis 356, 357
– 1,3-butadienes 356, 357
– 2-[(trimethylsilyl) methyl]-1,3-butadienes 356
Stille coupling 536
substituted enamines 96
– hydroboration 96
substituted β-amino 238
sulfonamide-derivatized 539
superbase 492
– mixture of (KO-t-Bu and n-BuLi) 492
Suzuki coupling 525, 526, 527, 528, 532, 534, 537
– 1,3-dibromobenzene, reaction with bis-9-BBN adduct 537
– of 9-alkyl-9-BBN with halo-1-alkenes 527, 528
– of 9-alkyl-9-BBN with haloarenes 525, 526
– synthesis, trans-1-silylenynes 537
– synthesis, trans-1-silylstyrenes 537
– synthesis of dienes 532
– synthesis of six-membered ring 534
– synthesis of styrenes 532
– transition state model 534, 535
Suzuki cross-coupling 529
– intramolecular 529
Suzuki reaction 523, 529, 531, 533, 539, 540, 541, 542, 543, 544, 545
– arylacetylenes; synthesis 531, 532
– cycloalkenes, synthesis via hydroboration-intramolecular coupling 529, 530
– enynes, stereodefined synthesis 531, 532
– exo-heterocyclics olefins synthesis 533
– in presence of water 543
– reactivity order of organic halides, triflates 523
– solid-support 539, 540, 541, 542
– stereodefined exo cyclic olefins, synthesis via hydroboration; intramolecular coupling 531
– tryptophan analogs 544, 545
synthesis of 1,3-dienes 354
synthesis via B-1-Alkynyl-9-BBN 163

T

terminal allenylsilanes 177
– hydroboration 177
1-tert-butyl-5-trimethylsilyl-1,3-cyclohexadiene 513
tertiary amides 400, 413, 414
– to alcohol; 9-BBN·THF 400
– to alcohol with $LiEt_3BH$ 413
– to aldehydes; Sia_2BH 400
– to amines; BM_3·THF 400
– to tertiary amines with Li-9-BBNH 414
tetrabutylammonium fluoride (TBAF) 548
tetrahydrooxepin 110
– vs cycloheptene, hydroboration 110
tetramethylethylenediamine (TMEDA) 293, 330
thermal arrangement 119
THF 17, 560
– catalytic effect 17
– trans-elimination of β-substituents, Cl, Br, I 560
1,2 transposition of oxygen functionality 152

trans,trans-1,9-di-9-BBN-1,8-nonadiene 114
trans,trans-dideuterated diene 360
trans-1,2-bis(9-BBN)ethylene 508
- trans-dihydroxyethylene equivalents for Diels-Alder reaction 508
trans-1,2-bis(9-borabicyclo[3.3.1]nony lethylene 496
- synthesis 497
trans-1,2-bis(catecholboryl)ethylene 508, 509
- conversion to, cyclohexenediols 509
trans-dihydroxyethylene equivalents for Diels-Alder reaction 508
trans-1-deuterio-1-alkene 338
trans-2-phenyldimethylsilylvinyl-9-BBN 508
- reactivity and selectivity toward Diels-Alder reaction 508
trans-2-trimethylsilyl-vinylborane 176
- synthesis 176
trans-2-trimethylsilylvinyl-9-BBN 537
- preparation 537
trans-3-TMS allylic alcohols 176
trans-3-trimethylsilyl(TMS) allylic alcohols 176
trans-chlorohydrin 572
- synthesis 572
trans-disubstituted olefin 153
trans-elimination 560
- weak base catalyzed β-elimination of strong leaving group 560
trans-styryl-, dienyl-, enynylsilanes 537
trans-vinyl-9-BBN 153
trans-vinylboranes 113
- dehydroborylation of diboryl adduct 113
transannulation 565
- of C=C, N=N with 9-BBN 565
transition state 64
transition state model 534, 535
1,3,5-tribromobenzene 537
trimethylamine-N-oxide (TMANO) 14, 223, 313
- oxidation of B-substituted 9-BBN 14
trimethylsilylacetylene 512
- hydroboration 512
1-(trimethylsilyl)-1-hexyne 117
- vs 1-hexyne, hydroboration 117
trimethylsilylvinyl-9-BBN 508, 512, 513, 514, 515
- 1,4-disubstituted products, Diels Alder 513
- acetylene equivalent in Diels-Alder reaction 513
- Diels-Alder reaction 514, 515
- synthesis 512
tryptophan analogs 544, 545
- via Suzuki reaction 544, 545

U

α,β-unsaturated aldehydes, ketones reduction 407
- with 9-BBN·THF 407

V

vinyl-9-BBN 495, 501, 502, 503, 504, 505, 506
- "meta isomer" with trans-piperylene, Diels-Alder 502
- "para isomer" with isoprene, Diels-Alder 502
- Diels-Alder reaction 502
- Diels-Alder reaction, rate constants 503, 504
- Diels-Alder reaction; regioselectivity, stereoselectivity 503, 504
- in situ preparation from tetravinyltin 506
- in situ preparation from vinyltributyltin 506
- preparation 501
- regioselectivity 502

– regioselectivity of Diels-Alder reaction 505
– relative reactivity towards dienes 506
– stereoselectivity of Diels-Alder reaction 506
– synthesis from B-Br-9-BBN and tetravinyltin 495
– synthesis from B-Br-9-BBN and vinyltributyltin 495
vinylborane 40
– site of hydroboration 40
vinylboranes 501, 506, 536
– ambiphilic, omniphilic dienophiles 506
– regioselective, stereoselective dienophiles 501
– regioselective dienophiles 501
– via Crandall reductive alkylation of epoxide 536
vinyl chloride 36, 37
– hydroboration 36
– inductive, conjugation 37
2-vinylfuran 84, 85
– hydroboration 84, 85
vinyl halide 40
– site of hydroboration 40
vinylpyridine-BF_3 86, 91, 92, 93, 94
– hydroborating agents 92, 93, 94
– hydroboration 91
vinylpyridines 84, 88, 89, 90
– hydroborating agents 88, 89, 90
– hydroboration 84
vinylsilanes 173, 536
2-vinylthiophene 84, 85
– hydroboration 84, 85
vinyltributyltin 501

W

Wacker-type oxidation 197
Wittig methylenation 548
– synthesis, olefinated carbohydrites 548, 549
Wittig reaction 529, 550

Z

Zweifel method 534